Transportation Decision Making

Transportation Decision Making

Principles of Project Evaluation and Programming

Kumares C. Sinha
Samuel Labi

BICENTENNIAL
1807
WILEY
2007
BICENTENNIAL

John Wiley & Sons, Inc.

Library of Congress Cataloging-in-Publication Data:

Sinha, Kumares C. (Kumares Chandra)
  Transportation decision making : principles of project evaluation and
programming / Kumares Sinha, Samuel Labi.
       p. cm.
  Includes index.
  ISBN: 978-0-471-74732-1 (cloth)
  1. Transportation engineering.   I. Labi, Samuel, 1962- II. Title.
  TA1145.S487 2007
  625.068′4—dc22
                                              2006022829

Printed in the United States of America

10 9 8 7 6 5

To our past, present, and future students, from whom we are continually learning.

CONTENTS

PREFACE

The core of transportation decision making is the evaluation of transportation projects and programs in the context of available funding. For this reason, the principles and procedures of project evaluation and programming are of interest to transportation engineers and planners, policymakers and legislators, transportation agency administrators, facility managers and service providers, environmental groups, and the general public. This is a critical issue for governments everywhere. Each year, several trillions of dollars are invested worldwide in transportation facilities with a view to enhancing transportation system mobility, security, and safety, and to spurring economic development while minimizing environmental and other adverse impacts. In most countries, the sheer size of existing transportation assets and investment levels, coupled with the multiplicity of transportation system impacts and stakeholders, necessitates a comprehensive, yet integrated and consistent approach to evaluating such impacts.

The authors' intention is to fill the need for a single source of information to cover all key areas of transportation system evaluation. This was done partly by synthesizing information available in various evaluation reports, primers, syntheses, and manuals to form a single comprehensive text that provides a holistic approach to decision making in transportation project development and programming. Recognizing the evolution of trends in transportation-related areas and new research findings, this book seeks to provide transportation academics and practitioners with a solid set of methodologies for evaluating transportation alternatives on the basis of a comprehensive range of impact types.

This text is the outcome of more than 50 combined years of close contact with the subject through teaching, research, training of agency personnel, and consulting. The first author began his career in the late 1960s just as the issues of environment, sustainability, and quality of life were emerging in the national consciousness. The second author, a 21st-century professional, encounters the field in its current mature form.

It is hoped that this book will be useful for college instructors and students in the areas of transportation engineering, planning, management, policy analysis, and related courses. In addition, consultants and other private organizations involved in transportation project development and evaluation, public entities such as state and local (city and county) departments of transportation, regional planning agencies, and metropolitan planning organizations will find the book useful. Furthermore, the text serves as a helpful reference guide that can be employed by domestic and international development agencies in assessing and evaluating transportation projects.

The first didactic strategy adopted is one of subject modularity combined with logical transition as the user navigates through the entire book. Chapters 1 to 4 present introductory material to transportation systems evaluation: namely, the chain of the decision-making process at a typical agency, performance measures for evaluation, travel demand estimation, and costing for transportation projects. Chapters 5 to 8 are devoted to the tangible impacts of transportation—travel time, safety, and vehicle operating cost, and how these priceable impacts are combined for use in economic efficiency evaluation. In Chapters 9 to 17 we discuss the developmental and environmental impacts of transportation and therefore address issues such as business attractions, air quality, noise, ecology, water resources aesthetics, energy, land use, and social aspects. To demonstrate how these performance measures all fit together, we present, in Chapter 18, approaches for multicriteria evaluation in making decisions. In Chapter 19 we discuss how agencies can manage their information properly for the enhancement of decision making, and in Chapter 20 discuss the techniques for programming transportation investments in the long term with a view to achieving systemwide goals.

The second didactic strategy is the use of a common presentation format for every chapter. Each chapter begins with definitions, descriptions, and discussions of basic concepts. The authors then explain how such constructs are used to develop a step-by-step methodology for assessing and interpreting the impacts, and present available tools and software for evaluation. Finally, background legislation and recommended mitigation measures are discussed. At the end of each chapter is a list of references to available information in print or electronic media. Recognizing that only a limited amount of impact assessment information can be included, a set of useful resources is also provided. These include URL addresses to several Internet resources, including international and domestic agencies, transportation centers, and research institutions. Also, the authors maintain a Web page for this book at http://bridge.ecn.purdue.edu/~srg/book/. Periodically, updated or additional material, data for analysis, work examples, and so on, are added to the site for the benefit of the reader.

A book of this scope could not be undertaken without the generous assistance of knowledgeable friends and colleagues who graciously contributed their time in reviewing various chapters and making helpful suggestions. We extend our gratitude to Edward A. Beimborn of University of Wisconsin-Milwaukee; Arun Chatterjee of University of Tennessee; Louis F. Cohn of University of Louisville; Asif Faiz of the World Bank; Tien Fang Fwa of National University of Singapore; David J. Forkenbrock of University of Iowa; Robert Gorman of Federal Highway Administration; Chris T. Hendrickson of Carnegie Mellon University; Joseph E. Hummer of North Carolina State University; David Jukins of Capital District Transportation Committee, Albany, New York; Matthew Karlaftis of National Technical University–Athens; Patrick McCarthy of Georgia Institute of Technology; Lance A. Neumann of Cambridge Systematics; Mitsuru Saito of Brigham Young University; Edward C. Sullivan of California Polytechnic State University; John Weaver of Indiana Department of Transportation; James Bethel, Jon. D. Fricker, Rao S. Govindraju, Robert Jacko, Fred L. Mannering, Thomas Morin, Loring Nies, and Andrew P. Tarko of Purdue University. We gratefully acknowledge the indispensable help of the following students in preparing various portions of this text: Abhishek Bhargava, Adeline Akyeampong, Konstantina Gkritza, Siew Hwee Kong, Issa Mahmodi, Jung Eun Oh, and Vandana Patidar. The assistance of the following students is also deeply appreciated: Panagiotis Anastasopoulos, Muhammad Bilal, Phyllis Chen, Michael Inerowitz, Muhammad Irfan, Menna Noureldin, Adadewa Okutu, Lakwinder Singh, and Ahmad Soliman. We are also grateful to Karen Hatke, Dorothy Miller, and Rita Adom for their immense assistance. The preparation of the book was partially supported by the endowment from Edgar B. and Hedwig M. Olson. The book could not be completed without gentle pressure from James Harper and Bob Hilbert of John Wiley & Sons. Finally, we are thankful to our families for their patience and support.

Kumares C. Sinha
Samuel Labi
West Lafayette, Indiana

Transportation
Decision Making

CHAPTER 1

Introductory Concepts in Transportation Decision Making

The beginning is the most important part of the work.
—Plato (427–347 B.C.)

INTRODUCTION

The transportation system in many countries often constitutes the largest public-sector investment. The economic vitality and global competitiveness of a region or country are influenced by the quantity and quality of its transportation infrastructure because such facilities provide mobility and accessibility for people, goods, and services, and thereby play an important role in the economic production process. The new millennium is characterized by continued growth in commercial and personal travel demand, and transportation agencies and providers strive to keep their assets in acceptable condition so as to offer desirable levels of service in the most cost-effective manner and within available resources. Consistent with such efforts is the need for best-practices evaluation and monitoring of the expected impacts of alternative investment decisions, policies, and other stimuli on the operations of existing or planned transportation systems and their environments. Such impacts may involve economics (such as quantified benefits and costs); economic development (such as job increases); environmental or ecological impacts (such as air, water, or noise pollution, community effects, and land-use shifts); and technical impacts (such as changes in facility condition, vulnerability and longevity, network mobility and accessibility, and facility and user safety and security). Methodologies for assessing such impacts generally depend on the types of impacts under investigation, the scope, and the project type and size; and a variety of disciplines typically are involved, including operations research, engineering, environmental science, and

economics. It is important to view the evaluation of transportation projects and programs from a broad perspective, at both the project and network levels, that generally comprises overall system planning, project development, multiyear programming, budgeting, and financing. Furthermore, due cognizance should be taken of emerging or continuing trends in the transportation sector, as such trends often necessitate review of the traditional portfolio of impact types and scopes. In this chapter, we discuss the various phases involved in a typical transportation development process, and the importance of evaluation particularly at project development and programming phases.

1.1 OVERALL TRANSPORTATION PROGRAM DEVELOPMENT

In its most complex form, the development of a transportation program may involve an entire network of various facility types spanning multiple modes. In its simplest form, it may comprise a single project at a specified location. Regardless of its scope, the entire sequence of transportation development generally comprises the phases of network-level planning, development of individual projects, programming, budgeting, and financing (Figure 1.1). This sequence may have variations, depending on the existing practices of the implementing country, state, or agency.

1.1.1 Network-Level Planning

Network-level planning involves an estimation of travel demand for a general network-wide system on the basis of past trends and major shifts in the socioeconomic environment. In the United States, the transportation planning process comprises metropolitan and state-level planning, each of which is required to have short- and long-term transportation improvement programs (TIP). Various aspects of network-level systems planning include environmental inventories as well as inputs from the management systems for pavements, bridges, public transportation, intermodal facilities, safety, and congestion. These management systems help identify the candidate projects for improvements in facility condition, safety enhancement, and congestion mitigation. Transportation plans include long-range capital (e.g., new construction, added lanes) plans and a set of strategies for preservation and effective operations of all facilities on the network. A transportation plan is typically accompanied by a financial plan that not only involves the cash flows associated with needed physical improvements but also validates the feasibility of the transportation plan. Certain large MPOs are also required to develop a strategy for long-range congestion mitigation

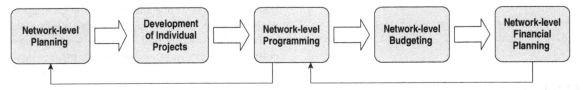

Figure 1.1 Phases of overall transportation development process.

and air quality management. Network-level systems planning is a continuous process that consists of:

- Inventory of current transportation facilities and use (travel)
- Analysis and forecast of population, employment, land use, travel data, and facility needs
- Establishing and evaluating alternatives for future facility physical components or policies

The evaluation step of network-level planning includes an assessment of conformity of the developed plan with other existing transportation improvement plans of the agency. At this phase, the major players are the federal, state, and regional agencies, as well as local governments and citizen groups. Special-interest groups also become involved through townhall meetings, public hearings, and other forums. Network-level planning yields a collection of selected projects that takes due cognizance of network-level needs. Relevant issues to be considered include the expected impacts of the network-level plan on existing land-use patterns, cooperation between various agencies, and a clear definition of the need for the proposed system. Legislation that needs to be considered at this step is related to issues such as air quality and energy conservation.

1.1.2 Project Development

This process is applied to each candidate project identified in the network, identification being through the long-range plan or through the various management systems. For each candidate, project development involves design, construction, management, operation, and postimplementation evaluation. At certain agencies, project development includes, as a first step, a project-level plan that is essentially a review of an existing overall transportation system plan for a region or network that includes the project corridor or area. In Section 1.2, we discuss the transportation project development phase in the context of an overall transportation program development process.

1.1.3 Programming

Programming involves the formulation of a schedule that specifies what activity to carry out and when. This is typically accomplished using tools such as ranking, prioritization, and optimization; the goal typically is to select the project types, locations, and timings such that some network-level utility is maximized within a given budget. Such utility, in the context of safety management, for example, could be a systemwide reduction in travel fatal crashes per dollar of safety investment. In the context of congestion management, the utility could be a systemwide reduction in travel delay per dollar of congestion mitigation investment; and in the context of bridge or pavement management, the utility could be a systemwide increase in facility condition, security, or longevity per dollar of facility preservation investment.

1.1.4 Budgeting

Although budgeting and programming are intertwined, programming yields a mix of projects to be undertaken during a given period, typically one to four years. Thus, setting the investment needs and budgeting involves a reconciliation of what work is needed and what resources will be available.

1.1.5 Financial Planning

An increasingly important aspect of transportation program development is financial planning. A financial plan or program is a specification of cash flows into (and in some cases, out of) a transportation facility over its entire period of implementation and operation, or part thereof. This step follows logically from the development of a program budget.

1.2   THE PROCESS OF TRANSPORTATION PROJECT DEVELOPMENT

A transportation *project development process* (PDP) can be defined as the sequence of activities related to the planning, design, construction, management, operation, and evaluation of a single transportation facility (Mickelson, 1998). PDP is a project-level endeavor that takes its input from an overall network-level transportation plan.

The process for developing transportation projects varies from agency to agency, due to differences in

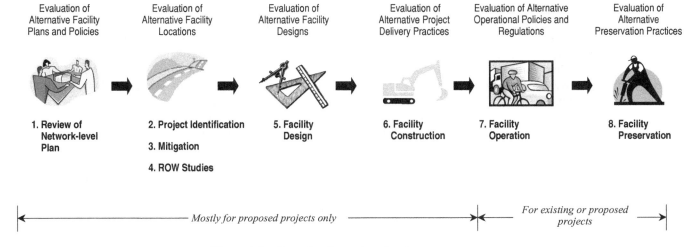

Figure 1.2   Steps for the project development process.

local requirements and conditions. The project development process is complex and resource intensive because it involves consideration of sensitive social, economic, environmental, cultural, and public policy issues. However, the overall PDP effort can be greatly facilitated by adopting good practices. The PDP often involves all levels of government: national, state (or provincial), and local. As illustrated in Figure 1.2, a PDP comprises several steps: a review of the network-level plan, particularly how it relates to the project in question, location planning and site selection, engineering design, construction, operations, and preservation. The tools for transportation systems evaluation are applicable at each of the PDP steps which are discussed briefly below.

1.2.1 PDP Steps

(*a*) *Review of the Overall Network-Level Plan with Focus on the Project Area*   Overall network-level planning can be considered implicitly as an initial step of the PDP and is a continuous process. Even when a project involves only a single mode, its planning must be carried out in a multimodal context. Multimodal transportation planning defines transportation demand and supply problems for an integrated network that comprises all available modes, selecting alternative actions to mitigate any problems identified, evaluating such actions on the basis of their costs and effectiveness, and selecting the action that best satisfies technical, economic, and environmental considerations and meets community goals.

(*b*) *Project Identification and Scoping*   This phase involves an individual portion (corridor, link, or node)

of a network-level plan, includes location planning, and typically takes three to five years, depending on the project complexity. In general, the following steps are involved:

1. Evaluation of existing modal facilities and further study of the need and purpose of the proposed improvement
2. Collection and analysis of social, economic, and environmental data
3. Definition of alternative project corridors, links, or nodes
4. Informal public meetings
5. Draft environmental impact report
6. Location public hearings
7. Final report and environmental impact statement approval
8. Location approval

The project identification step includes the most sensitive aspects of a PDP. The heightened emphasis on the social, economic, and environmental impacts necessitates a comprehensive and objective approach to the collection and analysis of data relating to such impacts. Federal laws and regulations that need to be considered at this step concern ecology, natural resource (i.e., land, water, energy, etc.) conservation, air pollution, historic facility preservation, archeological resources, civil rights, property relocation and acquisition, and other factors. As a result, the influence of special-interest groups such as the Sierra Club, the Environmental Defense Fund, and the Center for Law in the Public Interest could be most visible at the project development step. Although the involvement of special-interest groups typically leads to increased project development time and cost, particularly for controversial projects, it should nevertheless be carried out. Another federal requirement at this step is the major investment study (MIS) for the proposed project corridor or surrounding subarea, especially when the project has a high cost estimate or is expected to have significant adverse impacts. The United States Department of Transportation (USDOT) (1994) provides details on the various issues that should be addressed by an MIS. Coordination among various state and local agencies is critical at the project identification and scoping phase.

(*c*) *Mitigation*   This involves refinement of the project development plans and is carried out after approval of the location design. Such refinement is often necessary to reduce adverse impacts that are identified through public involvement and other means.

(*d*) *Right-of-Way Issues*   Activities at this stage include land surveys, development of right-of-way plans, acquisition, compensation, or relocation of affected property. The Uniform Relocation Assistance and Real Property Acquisition Policies Act (1970) establishes procedures that must be followed when there is a need to acquire property falling within the right-of-way of federal funded highway or transit projects. The legislation seeks to ensure equitable and fair compensation to affected persons.

(*e*) *Facility Design (Including Preparation of Contract Documents)*   This step involves preparation of detailed construction plans and drawings, technical and general specifications, and a schedule of quantities. In many cases, the design step of (PDP) also includes an invitation to bid, bid evaluation and selection, and preparation of contract award documents. This step may take two to five years, depending on project type and size, and typically includes:

- *Engineering design*
  - Engineering design studies and review
  - Public hearings on design
  - Final design
  - Approval of final design
  - Development or refinement of detailed plans and specifications
  - Project cost estimation
- *Contract administration*
  - Preparation of contract documents and invitation to bid
  - Evaluation of submitted bids and selection of best bidder
  - Contract award

At this step, the federal laws that need to be considered concern ecology, resource conservation (e.g., land, water, energy), and air pollution. Other federal requirements that need to be considered at the design step relate to design standards, policies, and specifications. For highway projects for instance, the Federal Highway Administration has established design standards, policies, and specifications based on American Association of State Highway and Transportation Officials (AASHTO) work to address highway-related issues such as pavement and geometrics (AASHTO, 1993; 2004), asset management and preservation (AASHTO, 2003), and traffic monitoring (AASHTO, 1990), among others. At the design step, relevant areas for evaluation include alternatives on material type (such as asphalt vs. concrete) and identification of optimal facility preservation practices over the life of a facility. Life-cycle costing, which can help identify optimal designs, is now a standard feature during the evaluation of design alternatives. Also, the need to consider the tort consequences of transportation design and operations is deservedly gaining increased attention (Cooley, 1996). Evaluation concerns in contract administration include alternative contractual practices (such as warranties vs. traditional contracts).

(*f*) *Facility Construction*   Sometimes referred to as *project implementation*, the actual construction of a project may take two to five years. Depending on the type of contract, the transportation agency shoulders varying degrees of supervisory and quality control responsibilities. An example of the evaluation of transportation construction alternatives can involve the estimation of the costs and benefits of total highway closure vis-à-vis partial closure during facility reconstruction (Nam et al., 1999).

(*g*) *Facility Operation*   The use of a facility is associated with a significant number of impacts on the ecology, agency resources, noise, and air pollution, among others. Evaluation of alternative operational policies can be used to identify best practices that would yield minimal cost and maximum benefits in terms of environmental degradation, mobility, safety, accessibility, and agency resources. Examples include studies that have evaluated the overall impacts of stimuli such as changes in rail operating policies, post-9/11 changes in air and transit security measures, changes in highway speed limits, implementation of truck-only highway lanes, and so on.

(*h*) *Facility Preservation*   After a project is constructed, it needs continuous rehabilitation and maintenance. Life-cycle cost analysis may be used to determine the most cost-effective schedule of rehabilitation and preventive maintenance treatments over the project's remaining life (Colucci-Rios and Sinha, 1985; Markow and Balta, 1985; Murakami and Turnquist, 1985; Tsunokawa and Schofer, 1994; Li and Madanat, 2002; Lamptey, 2004), or to determine the optimal funding levels to be set aside for preventive maintenance activities within periods of rehabilitation (Labi and Sinha, 2005). Also, it may be required to identify, at a given time, the most cost-effective practices, including treatment type (such as microsurfacing vs. thin asphaltic overlay), material type (such as bitumen vs. crumb rubber for crack sealing), work source (such as contractual vs. in-house), or work procedure.

1.2.2 Federal Legislation That Affects Transportation Decision Making

Figure 1.3 presents a time line of the historical developments in federal legislation related to transportation planning and programming. In what constituted the first formal recognition of the need to consider the consequences of transportation development and public input in transportation decisions, the Federal Highway Act of 1962 established the continuous, comprehensive, and cooperative (3C) planning process for metropolitan areas.

Prior to the 1960s, probably the only federal legislative actions that affected PDPs (informal at the time) were the Rivers and Harbor Act of 1899 and the Fish and Wildlife Coordination Act of 1934. The 1960s were characterized by increased concern for the environmental impacts of human activity. In that decade, the PDP became formalized and the Land and Water Conservation Act and Wilderness Act were passed. The Historic Preservation Act in 1966 mandated that transportation agencies evaluate the impacts of their decisions on historic

resources, publicly owned recreational facilities, wildlife refuges, and sites of historic importance. In 1969, the National Environmental Policy Act (NEPA) was passed and has since had a profound impact on transportation decision making. NEPA established a national environmental policy geared at promoting environmentally sound and sustainable transportation decisions. The type and scale of environmental studies required for each project depend on the certainty and expected degree of impact. For projects expected to have a significant impact, an environmental impact statement (EIS) is required. Categorical exclusion (CE) reports are prepared for projects that do not have any significant impact on the human and natural environment. Environmental assessment (EA) and finding of no significant impact (FONSI) reports or statements are prepared for projects where the scale of environmental impact is uncertain. If an EA suggests that there could be a significant impact, an EIS is prepared. Otherwise, a FONSI is prepared as a separate document.

In the 1970s, important legislation that affected the PDP included the Clean Air Act of 1970, the Endangered Species Act of 1973, and the Resource Conservation and Recovery Act of 1976. Also, in the wake of the 1973 oil crisis, energy conservation became a major criterion in the evaluation of transportation decisions. The 1970 Uniform Relocation Assistance and Real Property Acquisition Policies Act made available a set of procedures for compensation to owners of properties physically affected by transportation projects and required that for any new transportation facility in a metropolitan area, several alternatives involving TSM strategies be developed.

PDP-related legislation in the following decade seemed to focus primarily on funding issues but contained clauses that reinforced the importance of evaluating

Up to 1969	1970s	1980s	After 1990
Rivers and Harbors Act, 1899 Fish and Wildlife Coordination Act Federal-Aid Highway Act of 1950 Federal Aid Highway Act of 1956 Federal Aid Highway Act of 1962 Urban Mass Transportation Act National Historic Preservation Act National Environmental Policy Act Land and Water Conservation Act Wilderness Act Civil Rights Act	Uniform Relocation Assistance and Real Property Acquisition Policies Act Environmental Quality Improvement Act Clean Air Act Federal Water Pollution Control Act/Clean Water Act Resource Conservation and Recovery Act Wild and Scenic River Act Marine Protection Research and Sanctuaries Act Coastal Zone Management Act Endangered Species Act Archeological Resources Protection Act	Coastal Barrier Resources Act Comprehensive Environmental Response, Compensation and Liability Act Farmland Protection Policy Act Safe Drinking Water Act Surface Transportation and Uniform Relocation Assistance Act of 1987	Intermodal Surface Transportation Efficiency Act Americans with Disabilities Act National Highway Systems Act Transportation Equity Act of the 21st Century Safe, Accountable, Flexible, Efficient Transportation Equity Act – A Legacy for Users

Figure 1.3 Historical developments in federal legislation related to the transportation project development process.

the environmental impact of transportation decisions. Furthermore, the 1980s legislation appeared to give much importance to accessibility criteria, as a dominant share of the funding went to complete metropolitan connections to the interstate highway system.

The Intermodal Surface Transportation Efficiency Act (ISTEA) of 1991 had a significant impact on transportation decision making. It provided a foundation for subsequent establishment of the national highway system (NHS) and spawned several programs that emphasized aesthetics, mobility, and air quality impacts, such as the Scenic Byways Program and the Congestion Mitigation and Air Quality Program. Also, ISTEA brought about changes in the processes of planning, programming, coordination, and public involvement. It mandated the inclusion of management system outputs (this requirement was subsequently removed but is nevertheless being pursued by individual states) and made the entire PDP process more flexible and open to innovation (Mickelson, 1998). From the perspective of accessibility impacts, the NHS Act of 1995 helped to enhance linkages between intermodal facilities. Other legislation in the 1990s, such as the Americans with Disabilities Act, focused on the socioeconomic impacts of transportation projects or renewed the emphasis on water and air quality impacts.

The Transportation Efficiency Act of the 21st Century (TEA-21), passed in 2001, required state highway agencies to streamline the environmental clearance process in order to expedite project development. A key difference of TEA-21 from its predecessors is its consolidation of 16 previous planning factors into seven broad imperatives for inclusion in the planning process:

1. Support the economic vitality of the metropolitan area, especially by enabling global competitiveness, productivity, and efficiency
2. Increase the safety and security of the transportation system for motorized and nonmotorized users
3. Increase the accessibility and mobility options available to people and for freight
4. Protect and enhance the environment, promote energy conservation, and improve the quality of life
5. Enhance the integration and connectivity of the transportation system, across and between modes, for people and freight
6. Promote efficient system management and operation
7. Emphasize the preservation of the existing transportation system

The Safe, Accountable, Flexible, Efficient Transportation Equity Act—A Legacy for Users (SAFETEA-LU) of 2005 includes provisions for environmental stewardship and incorporates changes aimed at improving and streamlining the environmental clearance process for highway transit and multimodal transportation projects. Also included is an appropriate mechanism for integrating air quality and transportation planning requirements to facilitate the transportation project development process. In addition, SAFETEA-LU establishes highway safety improvement as a core program tied to strategic safety planning and performance. Fundamental in SAFETEA-LU are provisions aimed at reducing congestion, which will in turn save time and fuel, decrease vehicle emissions, lower transportation costs, allow more predictable and consistent travel times, and provide safer highways.

1.3 IMPACTS OF TRANSPORTATION SYSTEM STIMULI

1.3.1 Types of Transportation Stimuli

Synonymous with the words *change* and *intervention*, a *stimulus* may be defined as "an agent that directly influences the operation of a system or part thereof" and may be due to deliberate physical or policy intervention by an agency or to the external environment (Figure 1.4). External stimuli may be natural or human-made. Natural stimuli include severe weather events and earthquakes; human-made stimuli include facility overloads, interventions (facility repair by the owner or agency), and disruptions (terrorist attacks). Also, in the context of transportation decision making, stimuli may be categorized as *physical stimuli* (change in the physical structure) and *regulatory stimuli* (institutional policy or regulation of transportation infrastructure use). An example of a change in physical structure is the construction of a new road or the

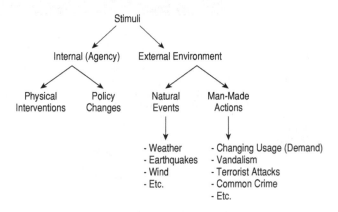

Figure 1.4 Classification of transportation stimuli.

addition of new lanes. Examples of institutional policy and regulation are speed limit and seat belt laws, respectively.

1.3.2 Impact Categories and Types

Identification of the various types and levels of impacts arising from a stimulus is a key aspect of transportation system evaluation and decision making. Given the multiplicity of stakeholders in transportation decision making, it is vital that all possible impact types be duly considered. Therefore, the various categories and types of impacts expected to occur in response to transportation system changes need to be identified prior to detailed analyses of the impacts. For example, the construction of a new transit line may affect (1) travelers (by decreasing their travel time), (2) the transit agency (by introducing a need for the agency to maintain the system after it has been constructed), (3) persons living near the transit line (by creating a noise pollution source), and (4) travelers on the network (by offering them new travel choices, and possibly changing their origin–destination patterns). In Table 1.1 and the sections that follow, we present briefly various categories and types of impacts of transportation system stimuli.

(*a*) *Technical Impacts*   These impacts typically constitute the primary motive for undertaking improvements in a transportation system. The secondary (but no less important) impacts are the consequences or side effects of the stimulus. Technical impacts are described below.

Facility Condition: An improvement in the condition of a facility leads to a host of impacts, such as increased service life, reduction in vehicle operating costs, and decreased vulnerability to natural or human-made threats. There are established standards of facility characteristics and conditions that must be met, failing which a facility owner may suffer increased operational or safety liability risks.

Vehicle Operating Costs: In the course of using transportation facilities, vehicles consume fuel, lubricants, and other fluids; "soft" replacements such as wiper blades and tires; "hard" replacements such as alternators and batteries; and experience general vehicle depreciation due to accumulated weather and usage effects. VOCs are categorized as *running costs* (whose values are typically a function of vehicle speed) and *nonrunning costs* (whose values are largely independent of speed). In a network-level estimation of VOCs, it is important to recognize that networks having only new and small vehicles (on one extreme) would incur far lower average vehicle operating costs than would a network having only old and large trucks (on the other extreme). As such, the changing

Table 1.1   Impact Categories and Types

Categories of Impact	Impact Types
"Technical"	Facility condition
	Travel time
	Vehicle operating cost
	Accessibility, mobility, and congestion
	Safety
	Intermodal movement efficiency
	Land-use patterns (including urbanization)
	Risk and vulnerability
Environmental	Air quality
	Water resources
	Noise
	Wetlands and ecology
	Aesthetics
Economic efficiency	Initial costs
	Life-cycle costs and benefits
	Benefit–cost ratio
	Net present value
Economic development	Employment
	Number of business establishments
	Gross domestic product
	Regional economy
	International trade
Legal	Tort liability exposure
Sociocultural	Quality of life

composition of the network-level vehicle fleets, as well as the relationship between running cost and age (for each vehicle class), are important (Heggie, 1972). The changing fleet composition is best tracked using cohort analysis (Mannering and Sinha, 1980).

Travel-Time Impacts: For a given project, the travel-time impact is the product of the reduction in travel time (in vehicle-hours) and the value of travel time per unit vehicle and per unit hour. If vehicle occupancies are known, the analysis can be done in terms of persons rather than vehicles.

Accessibility, Mobility, and Congestion: For already developed transportation networks, a desired impact of system improvements [e.g., lane additions, high-occupancy-vehicle (HOV) and bus rapid transit (BRT) facilities, intelligent transportation system (ITS) implementation, ramp metering, signal timing revisions] may be the mitigation of traffic congestion. On the other hand, in rural areas of developing countries, system improvement may be expected to provide accessibility to markets, health centers, agricultural extension facilities, and so on. In both cases, system improvements can lead to enhanced mobility of people, goods, and services.

Safety: Increased transportation system safety is typically due to diverse safety enhancement efforts including physical changes to a system and institutional changes such as educating the facility users and enforcing the operating laws and regulations. Safety enhancement may be due to direct implementation of such changes to address safety concerns (e.g., guardrail construction) or may be a secondary benefit of a larger project scope (e.g., pavement resurfacing, which enhances safety by improving skid resistance in addition to its primary objective of increasing pavement strength and service life).

Intermodalism: Physical or institutional changes in a transportation system can have profound effects on the efficiency or effectiveness of the overall intermodal transportation network in a region. For example, provision of additional links for a mode, or imposing or relaxing restrictions on the types and quantities of loads, can profoundly change the overall economics of freight delivery.

Land-Use Patterns: It is well known that changes in a transportation system cause shifts in land-use patterns, and vice versa. For example, highway construction and transit line extensions have been linked to changes in the extent and distribution of residential, commercial, and industrial developments.

Risk and Vulnerability: Recent world events have led to increased awareness of the need to assess the risk and vulnerability of existing transportation facilities or changes thereto. Thus, there are increasing calls to evaluate the impacts of system improvement (or deterioration) based not only on traditional impact criteria but also on the vulnerability of the facility to failure in the event of human-made or natural disasters.

(b) Environmental Impacts
Air Quality: Transportation-related legislation passed over the past three decades has consistently emphasized the need to consider air quality as a criterion in the evaluation of transportation systems.

Water Resources: Construction and operations of a transportation system can cause a significant reduction in both the quantity and quality of water resources, and it is often necessary to evaluate the extent of this impact prior

and subsequent to project implementation. Construction or expansion of airport runway and highway pavements and other surface transportation facilities lead to reduction in the permeable land cover, reduced percolation of surface water, and consequent reduced recharge of underground aquifers. Surface runoff from such facilities often results in increased soil erosion, flooding, and degraded water quality.

Noise: The noise associated with transportation system construction and operation has been linked to health problems, especially in urban areas, and often merits analysis at the stages of preimplementation (i.e., the planning stage) and postimplementation evaluation and monitoring.

Ecology: The construction and operation of transportation facilities may lead to the destruction of flora and fauna and their habitat, such as wetlands. For a comprehensive evaluation of ecological impacts, a basic knowledge of ecological science, at a minimum, is needed.

Aesthetics: Transportation projects typically have a profound visual impact on the surrounding built or natural environment. Such impacts may be in the form of a good or bad blend with the surrounding environment, or obscuring an aesthetically pleasant natural or human-made feature.

(c) Project Economic Efficiency Impacts
Initial Cost: The cost of designing, constructing, preserving, and operating a transportation facility is an important "impact" of the facility. Of these, the construction cost is typically dominant, particularly for a new project. The definition of construction and preservation costs can be expanded to include the cost of associated activities, such as administrative work, work-zone traffic control, work-zone impacts to facility users (such as safety and delay), and diversions.

Life-Cycle Costs and Benefits: The life-cycle approach involves the use of economic analysis methods to account for different cost and benefit streams over time. The life-cycle approach makes it possible to consider the fact that an alternative with high initial cost may have a lower overall life-cycle cost. TEA-21 required the consideration of LCCA procedures in the evaluation of NHS projects (FHWA, 1998).

(d) Economic Development Impacts Economic development benefits of transportation projects are increasingly being recognized as a criterion for consideration in the evaluation of such projects. The impacts of transportation

facilities in a regional economy may be viewed by examining their specific roles at each stage of the economic production process.

(*e*) *Legal Impacts* The operation of transportation facilities is associated with certain risk of harm to operators, users, and nonusers. With the removal of sovereign immunity in most states, agencies are now generally liable to lawsuits arising from death, injury, or property damage resulting from negligent design, construction, or maintenance of their transportation facilities. The growing problem of transportation tort liability costs is considered even more critical at the present time, due to increasing demand and higher user expectations vis-à-vis severe resource constraints. It is therefore useful to evaluate the impact of a change in a transportation system (project or policy) on the exposure of an agency to possible tort.

1.3.3 Dimensions of the Evaluation

It is important to identify the dimensions of the evaluation, as doing so would help guide the scope of the study and to identify the appropriate performance measures to be considered in the evaluation. The categories of the dimensions are presented in Table 1.2. The possible levels of each dimension are also shown.

(*a*) *Entities Affected* In carrying out project evaluation for purposes of decision making, it is essential to consider not only the types of impacts but also the various entities that are affected, as discussed below.

Users: User impacts include the ways in which persons using a transportation system (vehicle operators and passengers) are directly affected by a change in the system. User impacts typically include vehicle operating costs, and travel time, and safety.

Nonusers (Community): Consideration of the effect of transportation systems on nonusers is necessary to ensure equity of system benefits and costs to the society at large. These impacts often include noise and air pollution, other environmental degradation, dislocation of farms, homes, and businesses, land-use shifts, and social and cultural impacts.

Facility Operator: Operators of transportation systems, such as shippers, truckers, highway agencies, and air, rail, water, and land carriers, may be affected by physical changes (e.g., improvements) and institutional changes (e.g., deregulation, speed limits) in a transportation system. This typically occurs through increased or decreased resources for operations (and in the case of rail operators, for facility preservation).

Agency: The impacts on a transportation agency are typically long term in nature and are related to the costs of subsequent agency activities. For example, system improvements may lead to lower costs of maintenance and tort liability in the long run.

Government: These impacts concern the change in the nature or level of the functioning of the city, county, state, or national government due to a change in a transportation system. For example, a new type of infrastructure, policy, or regulation for the system may lead to the establishment of a new position, office, or department to implement or monitor implementation of the change.

(*b*) *Geographical Scope* A well-designed study area is critical in transportation evaluation studies because the outcome of the analysis may very well be influenced by the geographical scope of the impacts. As shown in Figure 1.5, spatial scopes for the analysis may range from point or segmentwide (local), to facility- or corridorwide, to areawide (city, county, district, state, etc.) As the geographical scope of an evaluation widens, the impact of the transportation project not only diminishes but also becomes more difficult to measure, due to the extenuating effects of other factors. Specific geographical scopes are typically associated with specific impact types and affected entities. For example, in the context of air pollution, carbon monoxide concentration is a local problem, whereas hydrocarbons are a regional problem, and the emission of greenhouse gases is a global problem. Also, each geographical impact may be short, medium, or long term, in duration but wider geographical scopes are typically more associated with longer terms, as impacts often take time to spread or be felt over a wider area.

Table 1.2   Evaluation Scopes of Impacts

Dimension (Scope)	Levels
Entities affected	Users
	Nonusers (Community)
	Agency
	Facility Operator
	Government
Geographical scope of impacts	Project
	Corridor
	Regional
	National and global
Temporal scope of impacts	Short term
	Medium term
	Long term

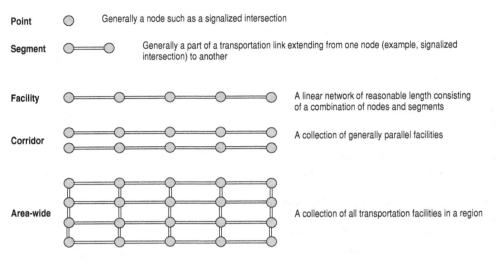

Figure 1.5   Spatial scopes of transportation systems evaluation.

Table 1.3   Suggested Relationships between Project Impact Categories and Dimensions

Impact Category	Temporal Scope of the Evaluation	Parties That Are Directly Concerned or Affected[a]			
		Users	Nonusers (Community)	Transportation Agency or Operator	Governmental
Technical (system preservation	Short term	P	—	P	—
and operational effectiveness)	Medium term	P	—	P	—
	Long term	P	—	P	—
Environmental	Short term	—	P, C	—	C, R
	Medium term	—	C, R	—	C, R
	Long term	—	C, R	—	C, R
Economic efficiency	Short term	P	—	P	—
	Medium term	P	—	P	—
	Long term	P	—	P	—
Economic development	Short term	—	—	—	C
	Medium term	—	C, R, N	—	C, R
	Long term	—	C, R, N	—	C, R, N
Safety and security	Short term	P	P, C	P, C	C
	Medium term	P	P, C	P, C	C, R
	Long term	P	P, C	P, C	C, R
Quality of life and sociocultural	Short term	—	P	—	—
	Medium term	—	P, C	—	—
	Long term	—	P, C, R	—	—

[a]P, project; C, corridor; R, regional; N, national or global.

(*c*) *Temporal Scope*   A transportation system stimulus may have impacts that last only a relatively short time (e.g., dust pollution during facility construction) or may endure for many decades after implementation (e.g., economic development). Obviously, the temporal scope of the evaluation will depend on the type of impact under investigation and is also sometimes influenced by (or related to) the geographical scope of the evaluation and the entity affected. Temporal distribution of impacts can also be classified by the occurrence in relation to the time of the stimulus: during-implementation impacts vs. postimplementation impacts. For example, construction dust and topsoil disturbance constitute during-implementation impacts, whereas traffic noise during highway operation is a postimplementation impact. For the purpose of grouping impacts from a temporal perspective, the categories used are short, medium, or long term.

Table 1.3 presents the relationships among the various impact categories, temporal scopes of evaluation, and parties most affected by (or concerned with) the impact.

1.4 OTHER WAYS OF CATEGORIZING TRANSPORTATION SYSTEM IMPACTS

Depending on the viewpoint of the decision maker, there are several alternative or additional ways of categorizing the impacts of transportation stimuli (Manheim, 1979; Meyer and Miller, 2001) as discussed below.

(*a*) *Direct vs. Indirect Impacts*   Direct benefits and costs are those related directly to the goals and objectives of the transportation stimulus and affect the road users and agency directly, whereas indirect impacts are generally by-products of the action and are experienced by society as a whole. For example, a major objective of speed-limit increases may be to enhance mobility (a direct impact), but may result in indirect impacts such as increased fuel use or increased frequency or severity of crashes.

(*b*) *Tangible vs. Intangible Impacts*   Unlike intangible benefits and costs, tangible benefits and costs can be measured in monetary terms. Examples of tangible impacts are construction cost and increase in business sales due to an improved economy. Examples of intangible impacts are increased security (due, for example, to transit video surveillance) or the aesthetic appeal of a rehabilitated urban highway. The intangibility of certain impacts precludes an evaluation of all impacts on the basis of a single criterion such as economic efficiency. Therefore, in evaluating a system that produces both tangible and intangible impacts, the techniques of scaling the multiple criteria are useful. An alternative way is to monetize intangible performance measures using the concept of willingness to pay: for example, how much people would pay to see a specific improvement in the aesthetic appeal of a bridge in their community, and then use economic efficiency to assess and evaluate all impacts.

(*c*) *Real vs. Pecuniary Impacts*   In assessing the impacts of transportation systems, it is important to distinguish between real costs or benefits [i.e., some utility that is completely lost to (or gained from) the world] and pecuniary costs or benefits (i.e., some utility that is related

only to the movement of money around the economy). Real costs represent a subtraction from community welfare. An example is the cost of fatal crashes on the streets of a city. Pecuniary costs are costs borne by people or communities that are exactly matched by pecuniary benefits received elsewhere, so that although there is a redistribution of welfare, there is no change in total community welfare. The same definitions apply in the case of real and pecuniary benefits. An example is the increase in business relocations to a city due to improved transportation infrastructure. This would be at the expense of competing cities (located in the region) from which the businesses are expected to relocate; thus, there is no net welfare gain for the region. Failure to distinguish between real and pecuniary costs can lead to double counting of costs. It has been recommended that strictly pecuniary effects could be excluded from the evaluation. However, such effects could be included in the evaluation if the analyst seeks to investigate the redistributional impacts of the transportation system among population subgroups or among cities in a region.

(*d*) *Internal vs. External Impacts*   For jurisdictional and administrative reasons, it may be worthwhile to consider whether system impacts are internal or external to the study area or analysis period defined at the initial stages of the evaluation procedure. Often, the benefits or costs of transportation system actions are felt beyond the study region or analysis period. For example, enhancement in air quality due to transportation improvements in a region may benefit another region located downwind. Also, the economic impacts of transportation system improvement may start to be realized only after the analysis period has expired.

(*e*) *Cumulative vs. Incremental Impacts*   Cumulative costs or benefits are the overall costs and benefits from a preidentified initial time frame and include the impacts of the transportation stimuli. On the other hand, incremental costs and benefits are those impacts associated only with the transportation stimuli and are determined as the total impact after application of the stimuli less the the existing costs and benefits before application.

(*f*) *Other Categorizations*   Heggie (1972) grouped transportation impacts from the perspectives of consumption of scarce resources, creation of additional consumption, and generation of non-monetary costs and benefits. Also, Manheim (1979) categorized transportation system impacts in two different ways: the party affected and the resource type consumed in constructing, preserving, and operating a transportation system.

Changes in a transportation system may result in desired outcomes with regard to some impact types and undesired outcomes with regard to others. For example, a new road in a town may yield improved travel time and accessibility but may have adverse impacts on pedestrian safety or the ecology. Stakeholders often have conflicting perceptions of the benefits of a transportation system change. As such, it is important to develop a methodology that incorporates all the various impacts, including social and cultural issues, to arrive at a single, balanced, impartial, and final decision. Unfortunately, in real practice, final decisions are sometimes made without regard to (or giving only minimal consideration to) the foregoing impacts.

In some countries or regions where transportation projects are sponsored by multilateral lending agencies, it may be required to measure the impacts, mostly in terms of economic benefits in which case economic efficiency impacts assume a dominant role in the evaluation process. In such cases, impact types such as vehicle operating costs, initial (construction) and preservation costs, and increased farm productivity are often given the highest priority in the evaluation process.

1.5   ROLE OF EVALUATION IN PDP AND BASIC ELEMENTS OF EVALUATION

1.5.1   Role of Evaluation in PDP

As seen in Figure 1.1, each step of the transportation project development process requires evaluation of alternative actions so that the best decision can be made to address the requirements of that step. The most visible (and probably the best known) traditional step that involves explicit evaluation of alternatives is the network-level or systems planning step, where projects are identified. The next common steps are those for site selection and facility design. With regard to impact type, the most common evaluation criterion that has traditionally been used for all steps is economic efficiency. Depending on the scale of a project, other criteria including environmental, economic development, and socioculture are also considered. In recent times, there are increasing calls to include system effectiveness and equity evaluation criteria such as system vulnerability and social justice. Evaluation of public projects therefore needs to give due cognizance to such concerns.

At any step of the PDP, any evaluation process should seek not only to identify the most optimal course of action, but also to investigate what-if scenarios because transportation systems are often characterized by significant risk and uncertainty. The sensitivity analysis should be for various levels of factors, such as system use (e.g., traffic volumes) and economic climate (e.g., interest rates), and should help reveal trade-offs between

competing objectives. Given the importance of public participation in the decision-making process and the multiplicity of stakeholders, another important role of evaluation is consensus building. Performance measures for decision making are typically derived from conflicting interests and considerations. Evaluation can therefore generate an impartial solution that yields the highest "benefits" while incurring the least possible "cost" to all parties affected.

1.5.2   Reasons for Evaluation

Evaluation studies are typically needed for at least one of the following reasons (USDOT, 1994; Forkenbrock and Weisbrod, 2001)

1. *Assessment of proposed investments.* For decision-support purposes, an agency may seek to determine the impacts of several alternative project attributes (such as operating policies, designs, or locations). Methods used to determine these impacts range from questionnaire surveys to comprehensive analytical or simulation models. The output of such studies is typically a prediction of the outcomes expected relative to base-case scenarios.

2. *Special transportation development programs.* In some cases, the evaluation seeks to measure the effectiveness of a specific stimulus on a specific aspect of the transportation system, such as the impact of seat belt use on teen fatalities.

3. *Fulfillment of regulatory mandate.* Impact assessments are often required to ensure compliance with government regulations and policies.

4. *Postimplementation evaluation.* It is useful to assess the actual impacts that are measured after project implementation and to evaluate such findings vis-à-vis the levels predicted at the pre-implementation phase as well as base-year levels. Unfortunately, few agencies typically invest time and resources in such efforts.

5. *Public education.* In controversial project cases or for the purposes of public relations, a transportation agency may carry out the evaluation with the objective of increasing general public awareness of the expected benefits to the citizenry.

1.5.3   Measures of a Project's Worth

The choice of any particular evaluation parameter depends on the decision maker, the type of problem, and the available alternative actions that can be undertaken. In the course of evaluation, the relative and absolute assessment of the worth of a particular course of action is debated in relation to the existing situation or other alternatives. Two questions are raised:

1. How should worth be measured?
2. What unit of measure should be used?

The worth of a project differs for different stakeholders. For a given project, therefore, there are several (sometimes conflicting) measures of worth that often have different units. Some measures may not be easily quantifiable on a numerical scale. The challenge here is to bring to a common and commensurate scale the various aspects of worth and identifying the trade-off relationships that exist between them. For example, a proposed transit line may enhance accessibility but may involve destruction of some natural habitats. The question then would be how much ecological damage can be tolerated to gain a certain level of accessibility.

In public project decision making, it is sought to select the *best possible* alternative—one that can be considered a good (rather than optimal) choice, which means that it may not be possible to arrive at a true optimal solution because all conflicting interests may not be fully satisfied. However, the achieved solution can be a consensus solution that represents a good balance of all possible concerns known at the time of decision making.

After all possible courses of action have been screened for their appropriateness, adequacy, and feasibility for implementation, the resulting feasible courses of action are defined as *alternatives*. The alternatives are evaluated on the basis of the three E's or 3E triangle: efficiency, effectiveness, and equity (Figure 1.6). These may be considered the *overall goals* of evaluation.

1. *Efficiency:* indicates the relative monetary value of the return from a project with respect to the investment required. By evaluating efficiency, the analyst seeks to ascertain if the transportation project is yielding its money's worth. Therefore, efficiency involves economic analysis and accompanying concepts of life-cycle agency and user costing. However, the range of performance measures to be considered is much wider than that implied by efficiency considerations alone, as it includes nonmonetary or nonquantifiable performance measures.

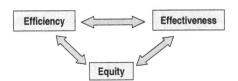

Figure 1.6 Basis of evaluation: the 3E triangle of overall goals.

2. *Effectiveness:* represents the degree to which an alternative is expected to accomplish a given set of tasks: in other words, just how well it attains the specified objectives. A clear understanding of the goals and objectives of the project is important to analyze its effectiveness. Effectiveness can include both monetary and nonmonetary or nonquantifiable benefits and costs, such as social well-being and aesthetic appeal.
3. *Equity:* can be measured in terms of both social and geographical equity in the distribution of both costs and benefits related to an alternative. Although equity can be incorporated within the effectiveness consideration, it may also be evaluated separately. Equity issues include whether low-income or minority populations bear a disproportionate share of the adverse impacts or whether they receive a proportionate share of the benefits of a transportation system change. Federal legislation such as the Civil Rights Act of 1964 and the Americans with Disabilities Act as well as the environmental justice requirements have led to the increased importance of equity considerations.

1.6 PROCEDURE FOR TRANSPORTATION SYSTEM EVALUATION

Most transportation agencies have established procedures that they follow in evaluating alternative policies or physical improvements to their assets. At some agencies, such procedures are not documented and thus vary from one decision maker to another. Formally documented evaluation procedures enable rational, consistent, and defensible decision making. Figure 1.7 presents the general steps that could be used to carry out the evaluation of alternative transportation system actions.

Step 1: Identify the Evaluation Subject   The subject of evaluation depends on which step of the transportation project development process is involved. At the project identification step, an action can be new construction or modification of an existing asset. If the step under investigation is construction, the subject of evaluation could be an innovative system of construction delivery. Also, if the investigation pertains to the operations step, the evaluation subject could be a change in service attributes or operations policy, such as changes in the operation of a BRT system or a change in truck weight restrictions.

Step 2: Identify the Concerns of the Decision Makers and Other Stakeholders   The next step is to identify stakeholders or affected parties, which could include the

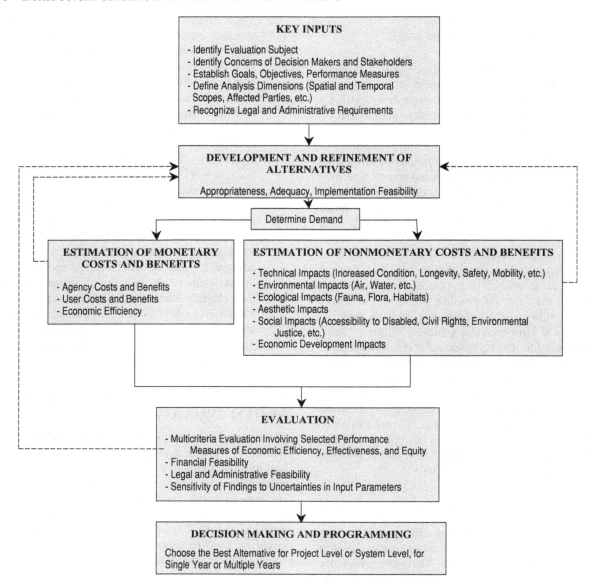

Figure 1.7 Procedural framework for transportation systems evaluation.

transportation agency that is responsible for the upkeep of the facility; the users of the facility, who reap direct benefits; and nonusers or the society as a whole. This step of the evaluation is important because it serves as a prelude to the development of the appropriate performance measures and dimensions of the evaluation (temporal and spatial scopes, etc.).

Strong local opposition can severely impede the possibility or progress of a transportation project or policy implementation. As such, early public involvement in the PDP is vital. Public involvement helps to identify stakeholders, affected parties, and interest groups (and their concerns); helps to identify the impacts that may have been overlooked by the planners; and also helps to determine the information needed to measure and mitigate the impacts expected. For major capital improvements, the involvement of the public is particularly necessary because such projects tend to have severe adverse impacts on the community and the environment. Before soliciting the input of the public, the decision makers need to decide on the timing, type, and level of public participation, to maximize the effectiveness of that effort.

Public participation can yield favorable results when the transportation agency interacts with the public in a way that demonstrates sincerity that the input from the public is valued and would be given due consideration. Public participation can also be used as a didactic instrument: to educate the public about favorable but not-so-obvious impacts of the proposed development. Public participation also affords the decision maker knowledge of public perceptions regarding the trade-offs among the various performance measures, including mobility and accessibility, air quality, and the economy. In soliciting public participation, the agency should remind the public that the best solution may not satisfy all interest groups and that a healthy compromise may be needed. The elements of effective public involvement include (NHI, 1995):

- Offering each interest group, a level of involvement and a type of interaction consistent with its requirements. Levels of interaction should range from Web site comments to detailed work sessions with the appropriate staff.
- Establishment of a proactive rather than a reactive program to inform the public and interest groups through the use of town hall meetings, print media, the Internet, and other mechanisms.
- Soliciting advice from representatives of citizens' associations and interest groups.

Step 3: Identify the Goals and Objectives of Transportation Improvement After the concerns of decision makers and other stakeholders have been identified, the objectives and goals of the evaluation process should be established to form the basis for the development of performance measures (measures of effectiveness). Goals are set to cover not only agency objectives, but also the perspectives of users, nonusers, and the government. The goals of the affected community provide an indication of the relative importance of various nonuser impacts and how the locality might react to such impacts. The most common community goals are mobility, safety, accessibility, and security. Other impacts of interest to the community are more long term in nature: environmental improvements, economic development impacts, downtown revitalization, and arresting urban sprawl. The early definition of goals helps not only in reaching an early consensus and compromise among conflicting interests, but also in identifying specific issues about the consensus reached. At this stage of the evaluation process, any documented material on regional or metropolitan goals and objectives should be collected and reviewed for consideration. Such efforts should include solicitation of information and perspectives from all stakeholders.

Step 4: Establish the Performance Measures for Assessing Objectives After identifying the goals and objectives for the proposed transportation stimuli, the performance measures or measures of effectiveness (MOEs) for each impact type can then be established. Examples of MOEs for multiple criteria that may be associated with a transportation action include the number of fatal crashes reduced (safety impacts), the number of jobs created (economic development impacts), the extent of natural habitat area damaged (ecological impacts), and the benefit–cost ratio (economic efficiency impacts).

Step 5: Establish the Dimensions for Analysis (Evaluation Scopes) The analyst should establish the boundaries of regions affected in the analysis: project, corridor, subarea, systemwide, regional, national, or even international. For any given impact type and temporal scope, different spatial scopes may have different approaches to the evaluation as well as different MOEs. The importance of certain impact types may differ from one spatial scope to another and even across temporal scopes.

Step 6: Recognize the Legal and Administrative Requirements Legal and administrative requirements typically encountered in a PDP include local ordinances, state statutes, and federal program requirements concerning the environment, safety, equity, and access. As discussed in Section 1.2.2, several laws and regulations have been passed over the last few decades in a bid to ensure efficiency in the decision-making process; to protect the environment, ecology, historical treasures, scenic beauty and so on; and to ensure equity. Also, the process of transportation system investment involves a multiplicity of administrative issues that need to be addressed. As such, the evolution of a transportation system stimulus from the conceptual stage through implementation involves a sequence that consists of formal notifications and requests; submission of engineering, economic, environmental, and other studies; approvals of requests and studies; and other administrative processes.

Identification and documentation of requisite legal and administrative processes is important because it helps the decision maker to define the various duties to be carried out by the transportation agency and the expected duties of other parties responsible for approving, reviewing, or commenting on the actions of the agency. Also, legal requirements need to be identified because they affect the establishment of performance measures and constraints and therefore have a great potential to influence the narrowing down of possible actions to selected alternatives and may even influence the choice of the best alternative.

Step 7: Identify Possible Courses of Action and Develop Feasible Alternatives All possible courses of action should be identified and should then be screened for their appropriateness, adequacy, and feasibility for implementation. The resulting feasible actions are defined as *alternatives*. Criteria that may be used to screen the possible courses of action are as follows:

- *Appropriateness*. Does the course of action address the specific goals or objectives sought by the decision maker? Does the alternative respond directly to other secondary considerations, such as community goals and needs?
- *Adequacy (of each alternative)*. Does the course of action address the intended goals and objectives adequately? In other words, is the performance offered by the alternative within the standards for the performance measures?
- *Implementation feasibility*. Is it physically feasible to implement the alternative? Is there enough right-of-way? Is there sufficient technological know-how? Is the cost of implementing the alternative within the means of the agency?

On the basis of the above criteria step 7 should be carried out to ensure due responsiveness to existing goals and needs, generation of a suitable number of alternatives, and a transparent sequence of development the alternatives.

(a) *Responsiveness of Alternatives to Local Goals and Needs* For a corridor project that goes through several communities, the local goals and needs in each community should be considered along with corridorwide objectives. Traditionally, alternatives have been developed by considering single physical facilities and operating strategies. At the current time, however, multimodal approaches are increasingly being used in the development of alternatives. As such, any alternative is not considered as an independent entity but as a part of a larger network of multimodal facilities. Such an approach encourages consideration of a possible mix of modes, physical facilities (e.g., access policies and location), and operating strategies.

(b) *Optimal Range of Alternatives* How many alternatives should be established? The least number of alternatives is two: One is to carry out a proposed activity, and the other is to do nothing. Inclusion of the do-nothing or "no-build" alternative in the list of alternatives is required by NEPA, while at least one alternative involving transportation system management is required by major investment studies (MIS) procedures. The number of alternatives should be large enough to enable identification of trade-offs across the various performance goals and objectives. Alternatives that involve transportation demand management and pricing are not formally required

by legislation but offer a low-cost benchmark and should be considered as much as possible (NHI, 1995). A fallback alternative should be provided where feasibility of the "best" alternative becomes questionable for any reason. The number of alternatives should not be too many but rather, should be manageable.

(c) *Open and Documented Development of Alternatives* The development of alternatives typically involves three steps:

1. *Conceptual development:* where details are sketchy but enough is known to state the intention to carry out the transportation development
2. *Detailed development:* where enough detail is developed to support analyses
3. *Final development:* involves a systematic process of evaluating and modifying the detailed alternatives (decisions are documented at this stage)

The need for inclusion and transparency in the development of alternatives cannot be overemphasized. Each of the steps mentioned above should be carried out collaboratively with the parties affected, with stakeholders, and with interest groups, and the results of each step should be open to full review and participation by the general public. Each alternative will need to be defined by its associated levels of performance measures. Performance measures may include general location, operating policies, institutional setting, and financial strategy.

Developed alternatives may differ by transportation mode, location (in terms of siting, routing, or alignment), facility or service type, area served by facility or service, effectiveness expected from the alternative stimulus (i.e., change in the performance associated with the stimulus), overall operating policies, institutional setting, and financial strategy. In a few cases, alternatives may also differ by analysis periods, a situation that should be avoided, especially where it is difficult to annualize effectively the impacts of the various alternatives.

No method exists that would, at all times, assure identification of all alternatives because the conception of alternatives is a product of endeavor that is only too human. As such, group thinking and brainstorming involving persons from diverse backgrounds and disciplines are helpful. In developing alternatives it should be realized that some alternatives are physical whereas others are policy-oriented, and some involve little capital outlay whereas others involve a large investment. Some alternatives pertain to transportation supply, whereas others pertain to transportation demand. Also, some alternatives primarily involve the physical highway facility, whereas others primarily involve the vehicle operator (driver), the vehicle, or the driving environment. Some alternatives involve little or no cost to the agency but a high cost to society or road users, whereas others may involve high cost to the agency and little or no cost to the users and or nonusers. It is important to consider all feasible alternatives, and a thorough discussion of the merits and demerits of each alternative would help justify the choice of the best alternative.

Step 8: Estimate the Agency and User Costs After alternatives have been developed, their costs should be estimated. Initial costs are still used by most agencies to make implementation decisions. However, the costs could be estimated over the life cycle of the facility (or service life of the stimulus under investigation) and therefore economic analysis principles should be used. Only those costs that differ by alternative should be used. Agency costs include construction costs, preservation (rehabilitation and maintenance) costs, and operating costs. User costs comprise work-zone costs (such as queuing delay) and costs associated with normal facility operation (such as vehicle operating costs). In Chapters 3 and 4, we discuss how agency and user costs can be developed for purposes of systems evaluation.

Step 9: Estimate Other Benefits and Costs The nonmonetary impacts due to each alternative should then be estimated. Impact types to be considered should be consistent with the established performance measures, objectives, and goals, as discussed in Section 1.3 and Chapter 2. Such impacts, whose estimations are required by law, include air quality, water resources, historic preservation, and others. Chapters 3 to 17 provide detailed procedures for the estimation of such impacts.

Step 10: Compare the Alternatives The evaluation of alternatives is simply an assessment of their respective costs, benefits, and cost-effectiveness used to make a selection of the best alternative. All performance measures (measures of effectiveness) may be successive hierarchical categories of system objectives, system goals, and overall system goals (economic efficiency, effectiveness, and equity). Furthermore, each alternative should be evaluated on the basis of its financial, legal, and administrative feasibility. Finally, economic and technical inputs (such as interest rates and the costs and effectiveness of interventions) are not constant over time, but rather, are subject to marked variations in response to foreseen and unforeseen conditions The evaluation process therefore should include a sensitivity analysis (what-if scenarios) for deterministic problems and a probabilistic analysis that incorporates the probability distributions of various input parameters, whereby an alternative that may seem optimal for a current or given set of conditions might be found to be far from optimal under a different (but not unlikely) set of conditions.

Another important consideration in evaluation is the role of system demand. There are several temporal physical and operational attributes of transportation systems that influence (and may be affected by) a proposed change to the system, such as facility condition, use, and so on. Usage forecasts are seen as particularly important because they have a profound influence on performance measures such as economic efficiency impacts, air quality, and energy consumption. It is therefore important that the evaluation process be accompanied by reliable predictions of travel demand changes in response to the transportation system stimulus. Again, as an input parameter, future demand is not known with certainty and could be subjected to some probabilistic analysis.

Faced with the costs and benefits (expressed in terms of the performance measures) that are associated with each of several alternatives, on what basis should a decision maker choose the best alternative? In other words, what *comparison criteria* should be used? Obviously, the choice of criterion or criteria for evaluation depends on the nature of the performance measures that are being considered. In helping a decision maker compare that which is sacrificed (cost) to that which is gained (benefit or effectiveness), an evaluation compares the input costs to the outcomes, whether or not such outcomes are priced. The outcomes of each strategy could be a reduction in subsequent facility preservation or operating costs, community benefits, economic or financial returns, public satisfaction, or progress toward stated objectives. From an economist's viewpoint, an evaluation could be carried out in three ways:

- The maximum benefits for a given level of investment (the maximum benefit approach)
- The least cost for effective treatment of problems (least cost approach)
- The maximum cost-effectiveness (a function that maximizes benefits and minimizes costs)

(a) *Benefits-Only Comparison Criteria* This approach is often used for evaluation of capital investment projects that typically involve a single large investment that is associated with significant elements of uncertainty and where the alternatives have equal costs. Furthermore, this approach is appropriate for such projects, where it is difficult to identify cost related performance measures or to provide a scale for such measures, due in part to the complex nature of such projects and their relatively long duration and spillover effects.

(b) *Costs-Only Comparison Criteria* In cases where benefits are expected to be similar across alternatives or where it is difficult to measures the benefits, the evaluation criteria are comprised of costs only. For many years, this has been the practice at many agencies, where decision makers compared alternatives solely on the basis of initial or life-cycle agency costs.

(c) *Comparison Criteria Involving Both Costs and Benefits* To arrive at a fair comparison of alternatives, both benefits and costs should be considered. If all the performance measures chosen can be expressed adequately in monetary terms, economic efficiency criteria such as benefit–cost ratio or net present value can be used. An alternative is deemed superior if its cost-effectiveness value exceeds that of other alternatives. If the performance measures consist of monetary and nonmonetary measures, an approach is to carry out multicriteria decision making where both monetary and nonmonetary criteria are expressed weighted, scaled, and amalgamated to derive a single objective function. The alternative action that yields the most favorable value of the objective function is deemed the best option.

1.6.1 Good Practices in Evaluation

A multitude of literature provides indications of good practices that could be followed in the evaluation of alternative transportation systems for the purpose of decision making, and include the following:

1. *Focus on the problem at hand.* The types of impacts (performance measures) to be considered and the dimensions of the impacts (temporal, geographical scope, and entities affected) should be pertinent to the problem under investigation.
2. *Relationship between the consequences of the alternatives and the established goals and objectives.* It should be possible to relate the performance levels associated with the alternatives to minimum levels of performance measures.
3. *Comprehensive list of appropriate criteria.* There is a need to consider a wide range of performance criteria (impact types) so that all stakeholders (decision makers, interest groups, affected parties, etc.) are duly represented. The desired characteristics of criteria used for decision making should be adequate to indicate the degree to which the overall set of goals is met. The list should be operational (must be useful and meaningful to understand the implications of the alternatives and to make the problem more tractable), nonredundant (should be defined to avoid double counting of consequences), and minimal (the number of criteria should not be so large as to obfuscate the evaluation and decision process).

4. *Clear definition of evaluation criterion or objective function.* Due to the multiplicity of stakeholders and the diversity of their interests, it is important to incorporate all key performance measures so that the evaluation results may be acceptable to all major parties concerned. Also, because there may be differences in the units of performance measures, they should be brought to dimensionless and commensurate values before they are incorporated into the objective function.
5. *Clear definition of constraints.* The performance measures that are used to build the objective function also individually present constraints within which the decision must be made. Such constraints arise largely from legal or administrative requirements and technical considerations and are often due to the influence of the stakeholders. For example, it may be required that an increase in emissions due to the system improvement should not exceed a certain maximum, or that the average condition of a physical facility or network of facilities should exceed some minimum.
6. *Ability to carry out trade-offs between performance measures.* For example, how much vulnerability can be reduced by a given budget, or how much would it cost to ensure a given level of risk?
7. *Ability to carry out sensitivity analysis with respect to key evaluation input variables.* The sensitivity of findings to uncertainties and value-based assumptions, and the adequacies of alternatives and impacts involved, will need to be considered.
8. *Clear presentation of evaluation process and results.* In decision making for public projects, several performance criteria involve subjective judgment, such as quality of life and convenience. As such, evaluation documents tend to contain lengthy and detailed statements of the influence of each impact type on each alternative, requiring decision makers to read and digest a large amount of information. The documented result of a comprehensive evaluation should therefore be presented in a very pleasing and easy-to-read manner. Key findings should be highlighted, and the reader should be able to navigate through the evaluation report with as much ease as possible.

SUMMARY

Traditionally, the evaluation of transportation systems has aimed at analyzing the economic efficiency of alternative proposed engineering plans and/or designs by comparing the monetary benefits and costs. Where the analysis involves cash flow over time, the economic principles of discounting and compounding are used to convert cash streams into time-independent values. This approach makes it possible to include only those evaluation criteria that could be expressed in monetary terms. Thus, vital nonmonetary criteria such as environmental impacts, economic development impacts, and sociocultural impacts are excluded in engineering economic analysis. At agencies, where there is no requirement to consider non-monetary criteria, this traditional practice continues at the present time. The evaluation of transportation systems, however, is currently evolving from the traditional approach and is increasingly being adapted to include nonmonetary criteria. In this chapter, we presented a framework for comprehensive evaluation of transportation alternatives and outlined key inputs to evaluation, important relationships between evaluation and other planning activities, and the basic components of evaluation itself. As shown, the estimation of impacts depends on a clear definition of the characteristics of modal alternatives and the local context in terms of the goals, the concerns of stakeholders, and the legal and other administrative requirements. Public involvement is desirable in all phases, particularly when developing key inputs, designing and refining alternatives, and evaluation. Decision making involves choices about the combinations of alternatives to pursue, including anticipated levels of performance measures, collateral mitigation measures, funding, and other relevant issues.

EXERCISES

1.1. Identify the various types of impacts of transportation system changes, and give one example of each.

1.2. Describe the role of evaluation in the transportation development process.

1.3. What are the elements of the 3E triangle, and what do they represent?

1.4. List some of the common measures of effectiveness for assessing community objectives of transportation projects.

1.5. List the phases of a typical evaluation work plan.

1.6. What are the basic principles for developing transportation alternatives?

REFERENCES

AASHTO (1990). *AASHTO Guidelines for Traffic Data Programs*, American Association of State Highway and Transportation Officials, Washington, DC.

———— (1993). Life cycle cost analysis for pavements Chap. 3 in *Guide for Design of Pavement Structures*, Part. 1, American Association of State Highway and Transportation Officials, Washington, DC.

———— (2003). *AASHTO Guide for Asset Management*, American Association of State Highway and Transportation Officials, Washington, DC.

———— (2004). A Policy on Geometric Design of Highways and Streets, 5th Edition, American Association of State Highway and Transportation Officials, Washington, DC.

Colucci-Rios, B., Sinha, K. C. (1985). *Optimal Pavement Management Approach Using Roughness Measurements*, Transp. Res. Rec. 1048, Transportation Research Board, National Research Council, Washington, DC, pp. 14–23.

Cooley, A. G. (1996). Risk management principles of transportation facility design engineering, *J. Transp. Eng.*, Vol. 122, No. 3.

FHWA (1998). *Life-Cycle Cost Analysis in Pavement Design: In Search of Better Investment Decisions*, Pavement Div. Interim Tech. Bull., Federal Highway Administration, U.S. Department of Transportation, Washington, DC.

Forkenbrock, D. J., Weisbrod, G. E. (2001). *Guidebook for Assessing the Social and Economic Effects of Transportation Projects*. NCHRP Rep. 456, Transportation Research Board, National Research Council, Washington, DC.

Heggie, I. (1972). *Transport Engineering Economics*, McGraw-Hill, New York.

Labi, S., Sinha, K. C. (2005). Life-cycle cost effectiveness of flexible pavement preventive maintenance, *J. Transp. Eng.*, Vol. 131, No. 10, pp. 114–117.

Lamptey, G. (2004). Optimal scheduling of pavement preventive maintenance using life cycle cost analysis, M.S. thesis, Purdue University, West Lafayette, IN.

Li, Y., Madanat, S. (2002). A steady-state solution for the optimal pavement resurfacing problem, *Transp. Res.*, Pt. A36, pp. 525–535.

Manheim, M. (1979). *Fundamentals of Transportation Systems Analysis*, Vol. 1, *Basic Concepts*, MIT Press, Cambridge, MA.

Mannering, F. L., Sinha, K. C. (1980). *Methodology for Evaluating the Impacts of Energy, National Economy, and Public Policies on Highway Financing and Performance*, Transp. Res. Rec. 742, Transportation Research Board, National Research Council, Washington, DC, pp. 20–27.

Markow, M., Balta, W. (1985). *Optimal Rehabilitation Frequencies for Highway Pavements*, Transp. Res. Rec. 1035, Transportation Research Board, National Research Council, Washington, DC, pp. 31–43.

Meyer, M. D., Miller, E. J. (2001). *Urban Transportation Planning—A Decision-Oriented Approach*, McGraw-Hill, Boston, MA.

Mickelson, R. P. (1998). *Transportation Development Process*, NCHRP Synth. Hwy. Pract. 267, National Academy Press, Washington, DC.

Murakami, K., Turnquist, M. A. (1985). A dynamic model for scheduling maintenance of transportation facilities, presented at the Transportation Research Board 64th Annual Meeting, Washington, DC.

Nam, D., Lee, J., Dunston, P., Mannering, F. (1999). *Analysis of the Impacts of Freeway Reconstruction Closures in Urban Areas*, Transp. Res. Rec. 1654, Transportation

Research Board, National Research Council, Washington, DC, pp. 161–170.

NHI (1995). *Estimating the Impacts of Transportation Alternatives*, FHWA-HI-94-053, National Highway Institute, Federal Highway Administration, U.S. Department of Transportation, Washington, DC.

Tsunokawa, K., Schofer, J. L. (1994). *Trend Curve Optimal Control Model for Highway Pavement Maintenance: Case Study and Evaluation*, Transp. Res. Rec. 28A, Transportation Research Board, National Research Council, Washington, DC. pp. 151–166.

USDOT (1994). *Major Investment Studies: Questions and Answers*, Memorandum, U.S. Department of Transportation, Washington, DC.

CHAPTER 2

Performance Measures in Transportation Evaluation

Give no decision till both sides thou'st heard.
—*Phocylides, sixth century* B.C.

INTRODUCTION

Performance may be defined as the execution of a required function. Performance measures represent, in quantitative or qualitative terms, the extent to which a specific function is executed. As such, transportation performance measures reflect the satisfaction of the transportation service user as well as the concerns of the system owner or operator and other stakeholders.

Performance measures are needed at various stages of the transportation program or project development process for the purposes of decision making and at various hierarchical levels of transportation management and administration. At one extreme (top level), performance measures are used for assessing systemwide plans and programs; at the other extreme (bottom level), they are used to select desirable solutions for a specific localized problem.

The establishment of performance measures has been fostered by various legislative impetuses, particularly the 1991 Intermodal Surface Transportation Efficiency Act (ISTEA). The need for meaningful performance measurement in government has also been advocated by several professional organizations over the past decades. These include the 1989 Governmental Accounting Standards Board (GASB) resolution, which encouraged state and local governments to develop indicators in four categories: input, output, outcome and service quality, and efficiency (GASB, 1989).

2.1 TRANSPORTATION SYSTEM GOALS, OBJECTIVES, AND PERFORMANCE MEASURES

The development of performance measures derives from a hierarchy of desired system outcomes. This hierarchy starts with the broad overall goals of efficiency, effectiveness, and equity; under these broad goals are the goals of system preservation, economic development, environmental quality protection, and so on; and under each goal is a set of objectives, and for each objective, performance measures are established (Figure 2.1).

Identification of goals and objectives is a key prerequisite to the establishment of performance measures and therefore influences the evaluation and decision outcome. Diversity in system goals and objectives is desirable because it reflects different expectations (held by various stakeholders) of what the transportation system should be achieving. Goals and objectives are typically developed through extensive examination of top-level agency requirements, by soliciting the perspectives of the users and other stakeholders and by outreach to the general public. Definitions of the various levels of the hierarchy are provided as follows:

- An *overall goal* is a broad description of what the transportation action is generally meant to achieve. As mentioned in Chapter 1, there are three overall goals: efficiency (is the output worth the input?), effectiveness (is the action producing the desired outcomes?),

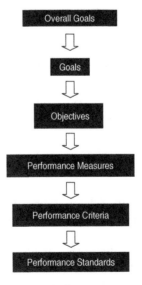

Figure 2.1 Hierarchy of desired outcomes for transportation system projects and programs.

and equity (are diverse segments of the population receiving a fair share of the action's benefits?).

- A *goal* is a desired end state toward which effort is directed, and is derived from the overall goals. From the perspective of effectiveness, for example, goals may involve the physical condition, operational characteristics, or external effects of the transportation system. Goals associated with physical condition include system preservation; goals associated with system operations include mobility, accessibility, and safety; and goals associated with external impacts include environmental conservation and economic development.

- An *objective* is a specific statement that evolves from a goal and is geared toward achieving that goal. For example, if a goal is to enhance regional air transportation mobility, a corresponding objective could be to reduce air travel time.

- A *performance measure* is an objective that is stated in measurable terms. Synonyms include *performance indicator, performance attribute,* or *service attribute.* For the goal of air transportation mobility enhancement and the objective of reducing air travel time, for example, a performance measure could be the air traveler delay.

- A *performance criterion* is a specific definition attached to a performance measure. For example, a criterion could be to minimize average transfer time for air travelers over the regional network or airports over a given period.

- A *performance standard* is a fixed value of a performance criterion that clearly delineates a desired state from an undesired state. For example, the average passenger transfer time should not exceed 90 minutes. Synonyms include *threshold, trigger,* or *minimum level of service.* A performance standard therefore specifically defines the least desired level of the performance criterion.

At many transportation agencies, performance measures for improvement projects are generally derived from the agency's overall goals or objectives. For instance, at Delaware's state transportation agency, performance measures are tied to the agency's goals, strategies, policies, and long-range transportation plans in a tiered fashion (Abbot et al., 1998). Literature on performance measures (Cambridge Systematics, 2000; Shaw, 2003) provides typical groups or categories of goals and objectives that have been identified by transportation agencies for performance-based management. These include system condition and performance, operational efficiency, accessibility, mobility,

economic development, quality of life, safety, and environmental and resource conservation. Examples of typical goals and objectives are shown in Table 2.1.

2.2 PERFORMANCE MEASURES AT THE NETWORK AND PROJECT LEVELS

The application of performance measures to transportation systems evaluation can occur at two levels:

1. *Network level* or *system level.* At this level, evaluation is used in programming and priority setting (determining the optimal use of limited funds for the entire network of transportation facilities), estimating funding levels needed to achieve specified systemwide targets (such as average facility condition or average user delay), and estimating the systemwide performance impacts of alternative funding levels, investment strategies, or policies.

2. *Project level* or *facility level.* Here, the intent is to select an optimum policy, physical design, or preservation strategy for a specific transportation facility, much as a pavement section, bridge, or transit terminal, at a given time or over the facility life cycle. Project-level evaluation is typically more comprehensive, deals with technical variables and design issues, and requires more detailed information than at the network level.

Performance measures used at the network level are typically used in a context that differs from those at the project level. For air transportation, for example, a project-level goal may be to assess the change in average plane delay in response to a specific project such as expansion of runway capacity; while at the network level, the goal may be to assess the average plane delay (averaged across an entire network of airports) in response to changes in nationwide transportation security policies. However, it must be noted that network- and project-level evaluation are often interdependent: Depending on its internal practices, an agency may carry out evaluation using a top-down approach (from network level to project level) or a bottom-up approach (from project level to network level). In the top-down approach, for example, performance targets can be established for the entire network, and then using project-level performance measures, specific projects can be identified to achieve network-level performance targets. In the bottom-up approach, project-level performance measures are first used to estimate the impacts of alternative actions (and their respective timings) at each facility, and then the corresponding impact of each set of actions at the network level is determined. It must be recognized that the optimal decisions

Table 2.1   Typical Goals, Objectives, Performance Measures, and Performance Criteria

Overall Goals	Goals	Objectives	Performance Measures	Performance Criteria
Efficiency	Improve system financial performance	Enhance economic attractiveness of the system Enhance economic viability (financial feasibility) of the system	Reduce initial or life cycle costs for agency or users or both Maximize benefit cost ratio or net present value Maximize economic efficiency Enhance financial feasibility of project construction and preservation	Initial cost Life cycle agency cost Life cycle user cost User costs at workzones Benefit cost ratio or net present value Cost per new person-trip per mile Feasibility of funding project construction (yes/no) Feasibility of project life-cycle preservation (yes/no)
Effectiveness	Improve system physical condition Improve system operational performance	Maintain condition of physical transportation infrastructure at a certain minimum level Improve technical feasibility (operational effectiveness) so that transportation system provides desired service that maximizes mobility, accessibility, and intermodalism	Improve construction techniques and materials to minimize construction delays and improve service life of transportation improvements *Mobility*: decrease congestion and delay at arterials, freeways, and intersections *Accessibility*: improve transit frequency and reduce waiting times and walking distance *Intermodal connectivity*	Average facility condition index (either for each facility or average for all facilities in network) Average or total delay Average traffic speed or density Average travel time Transit frequency Average delay time in intermodal transfers
	Safety of system users and nonusers	Enhance safe use of the transportation system for the benefit of road users (drivers and pedestrians) and nonusers Minimize the incidence of tort liability associated with use of the transportation system	Reduce the frequency and/or rates of fatalities, injuries, and property damage associated with use of the transportation system Reduce the frequency and payment amounts associated with tort liability	Fatal crashes per 100 million vehicle-miles traveled Number of injury or property-damage crash rates Annual safety-related tort payments (amounts and frequency)
	Economic development and land-use impacts of the system	Improve transportation services to enhance economic competitiveness of a region, thus attracting new businesses or retaining existing businesses Promote land-use patterns that foster progressive community development	Increase employment Increase business output and productivity Increase the number of businesses Change in land-use patterns (toward a prespecified desired land-use mix)	Number of jobs created Increase in gross regional product Increase in business sales Changes in land-use ratios (residential, industrial, commercial, and agricultural)

(continued overleaf)

Table 2.1 (*continued*)

Overall Goals	Goals	Objectives	Performance Measures	Performance Criteria
	Environmental quality and resource conservation	Minimize adverse environmental impacts or enhance environmental quality, including ecology, water quality and quantity, air pollution, noise, and privacy Reduce energy use or enhance energy efficiency Minimize damage to cultural heritage, such as historical sites and archeological treasures	Reduce air and noise pollution Reduce environmental degradation Improve aesthetics and general environmental quality Avoid damage to sites of cultural interest	Tons of carbon monoxide emitted per year Average energy consumed per vehicle per mile per year Percentage of green space, open space, and parkland Intrusion of cultural treasures sites
Equity	Improve quality of life	Enhance general quality of life and community well-being Promote social equity Promote environmental justice	Enhance community cohesion Enhance accessibility to social services Provide transportation opportunities for handicapped and other socially disadvantaged groups Increase recreational opportunities	Number of displaced persons, farms, businesses, and homes Benefits per income group

Source: Adapted from Cambridge Systematics (2000).

for project-level evaluation may not necessarily translate to optimal decisions at the network level.

2.3   PROPERTIES OF A GOOD PERFORMANCE MEASURE

Generally, a suitable performance measure should have the following properties (Turner et al., 1996; Cambridge Systematics, 2000):

- *Appropriateness*. The performance measure should be an adequate reflection of at least one goal or objective of the transportation system action. It should be applicable to an individual mode or a combination of modes. The appropriateness of a performance measure helps guarantee its relevance because its reporting would provide the needed information to decision makers.
- *Measurability*. It should be possible (and easy) to measure the performance measure in an objective manner and to generate the performance measure levels with available analytical tools and resources. Measurement results should be within an acceptable degree of accuracy and reliability.
- *Dimensionality*. The performance measure should be able to capture the required level of each dimension

associated with the evaluation problem. For example, it should be of the appropriate spatial and temporal scales associated with the transportation action and should address the perspectives of the parties affected. The performance measure should be comparable across time periods or geographic regions.
- *Realistic*. It should be possible to collect, generate or extract reliable data relating to the performance measure without excessive effort, cost, or time.
- *Defensible*. The performance measure should be clear and concise so that the manner of assessing and interpreting its levels can be communicated effectively within a circle of decision makers and to the stakeholders and general public. This is often possible when the performance measure is clear and simple in its definition and method of computation.
- *Forecastable*. For planning purposes, it should be possible to determine the levels of the performance measure reliably at a future time using existing forecasting tools.

It is important that the list of selected measures be comprehensive, yet manageable, to facilitate a meaningful analysis. Transportation agencies that seek to select performance measures are concerned particularly with the practicality of performance measures in terms of their usefulness, data availability and forecasting ability, flexibility across modes, data precision, dimensions, and other attributes. Poister (1997) and Shaw (2003) provided examples of performance indicators that have been used in past evaluation of highway projects, while Cambridge Systematics (2000) presented a perspective of how performance measures could be formulated and used in project evaluation.

2.4 DIMENSIONS OF PERFORMANCE MEASURES

Performance measures can be viewed from the perspective of several dimensions, such as the goals or objectives, transportation mode, facility type, temporal scope, spatial scope, and so on. For example, performance measures may be classified by their applicability to multimodal vs. single-mode evaluations or to freight vs. passenger transportation. Also, performance measures may differ by facility type. For example, the impact of transit guideway projects are measured using specific performance measures that differ from those used for transit terminals, even though the overall goals may be the same. Also, performance measures that are used when evaluation is being carried out over a short time frame may differ from those that are used for a long time frame. For example, performance jump (immediate improvement in facility performance) could be used for the short-term evaluation of physical, policy, or operational interventions; while deterioration rate reduction or extension in facility life may be used to measure the effectiveness of interventions over relatively longer evaluation periods. With regard to spatial scope, the measures of performance for a given impact type may differ, depending on whether the analysis is being carried out at project level, statewide network level, or even regional level. A case in point is air pollution impacts: pollutant types and parameters used to evaluate local pollution differ from those used to evaluate regional pollution. Performance measures may also be categorized by the planning and programming jurisdiction to which they are most relevant, and by the perspective of user, agency, or operator. A classification of possible dimensions of performance measures is shown as Table 2.2.

2.5 PERFORMANCE MEASURES ASSOCIATED WITH EACH DIMENSION

For the transportation program or project under evaluation, the analyst should identify the appropriate dimensions for the evaluation, and should then establish the relevant performance measures associated with each dimension. A discussion of performance measures based on various dimensions is presented below.

2.5.1 Overall Goals

Efficiency-related performance measures involve an assessment of how much return can be achieved for a given input. Examples include the savings in travel costs per dollar of investment, benefit−cost ratio, and net present value. Performance measures for the overall goal of effectiveness are used to assess the degree to which operational goals are being attained. Equity-related performance measures help assess the extent to which specific benefits and/or costs (monetary or nonmonetary) are being shared across

Table 2.2 Dimensions of Performance Measures

Dimension	Example
Overall goals	Economic efficiency, effectiveness, and equity
Objectives	Preservation of system condition, operational efficiency, economic development, quality of life, safety, and environment
Sector concerns	Private (profit) and public (service)
Flow entity	Freight and passenger
Modal scope	Multimodal and single mode
Specific mode	Highway, urban transit, railway, waterway, and pipeline intermodal
Entity and stakeholder affected	Agency, user, or nonuser
Spatial scope	Urban, rural, citywide vs. intercity
Level of agency responsibility	State, district, local
Time frame	Long and short terms
Level of refinement	Primary and secondary indicators
Intended use	Policy, programming, implementation, postimplementation review
Level of use of information	Management and operational levels

various particular demographic or geographic groups of the affected population or region and help to ensure that no group suffers a disproportionate level of hardship due to the transportation project. Examples of equity-based performance measures include those that can be related to environmental justice or how well the expected adverse community impacts can be mitigated.

2.5.2 System Objectives

Most transportation agencies have established a portfolio of performance measures for their agency goals (which generally include objectives involving system preservation, agency cost, operational efficiency, mobility, safety, and environmental preservation). Network-level performance measures that are based on overall system goals and objectives are presented in Table 2.3.

(*a*) *Preservation of the System Physical Condition   System preservation* refers to the set of activities geared toward ensuring a minimum level of physical condition of transportation facility or rolling stock and is generally considered to be a vital aspect of transportation management. For an assessment of the extent to which this goal is being achieved, the following general performance measures can be used:

- Percentage of system units or segments that have been maintained at or a certain minimum or target level of condition or that are operating above a certain specified level of service threshold
- Average level of service, physical condition, or structural or functional sufficiency of the system

General Appendix 2 presents specific examples of these performance measures.

Data on system physical condition and operation, which can be used to derive levels of established performance measures, are generally available at most transportation agencies.

(*b*) *System Operational Performance*   This includes *operational effectiveness* (the degree to which the transportation system provides a desired service that maximizes mobility, accessibility, and intermodalism; and *operational efficiency* (the extent to which the resources are used to produce a given level of transportation output). The public sector is typically interested in operational effectiveness, whereas the private sector (comprising shippers and carriers and other businesses whose operations are heavily linked to the transportation system) is interested in operational efficiency, particularly from a monetary standpoint. Operational efficiency could be viewed in the flow entity dimension; as such, its performance

measures may be grouped into those applicable to passenger or freight movement, or both.

Accessibility: An important function of any transportation system is to provide for people accessibility to residences; places for employment, recreation, shopping, and so on; and for goods and services, accessibility to points of production and distribution. Any performance measure for accessibility should reflect the ease with which passengers and goods reach their destinations. Performance measures for accessibility as illustrated in General Appendix 2, include:

- The ability of a facility to handle specific types of passengers or freight
- The capacity of specific intermodal facilities for freight and passengers
- The ease of access to the transportation system
- The ease of connecting at transfer facilities
- The percentage of the population or freight-generating businesses located within a certain distance or travel time from a specific transportation facility

Mobility: Performance measures associated with mobility may apply to passenger or freight transportation. As illustrated in General Appendix 2, these may include:

- The travel time, level of service, travel speed, delay, congestion
- The average speed vs. peak-hour speed
- The transfer time at intermodal transfer terminals, hours of delay
- The percentage of a facility that is not heavily congested during peak hours

Data on travel time and congestion-related measures are typically estimated with existing analytical or simulation models, while mode shares and levels of service (intermodal connecting times) can be ascertained using surveys of individual facility users or businesses.

(*c*) *System Financial Performance*   Transportation systems aim to enhance accessibility and mobility at a reasonable cost to both agencies and users. Benefits could be expressed in terms of the reduction in agency or user costs or both, relative to a base case (which is typically the do-nothing scenario). Performance measures for system financial performance may include:

- The initial cost per unit dimension of transportation facility
- The preservation cost per unit dimension of transportation system

Table 2.3 Examples of Network-Level Performance Measures Based on Highway System Goals and Objectives

Objective	Facility or Category	Performance Measures
System preservation	Pavement	Percentage of highway miles built to target design
		Average roughness or overall pavement index value for state highways, by functional class
		Percentage of highways rated good to excellent
		Percentage of roads with score of 80 or higher on overall highway maintenance rating scale
		Percentage of total lane miles rated fair or better
		Miles of highway that need to be reconstructed or rehabilitated
	Bridge	Percentage of highway bridges rated good or better
		Percentage of highway mainline bridges rated poor
		Number of bridges that need to be reconstructed or rehabilitated
Operational efficiency	Construction, maintenance, and operation	Cost per lane-mile of highway constructed, by functional class and material type
		Cost per unit of highway maintenance work completed; labor cost per unit completed
	Cost-effectiveness	Cost per percentage point increase in lane-miles rated fair or better on pavement condition
		Cost per crash avoided by safety projects
Accessibility	Roadway	Percentage of population residing within 10 minutes or 5 miles of public roads
		Percentage of bridges with weight restrictions
		Miles of bicycle-compatible highways rated good or fair
Mobility	Travel speed	Average speed vs. peak-hour speed
	Delay, congestion	Hours of delay
		Percentage of limited-access highways in urban areas not heavily congested during peak hours
	Amount of travel	Vehicle-miles of travel (VMT) on highways
		Percentage of VMT at specific road classes
		Percentage passenger-miles traveled (PMT) in private vehicles and public transit buses at specific road classes
Economic development	Support of economy by transportation	Percentage of wholesale and retail sales occurring in significant economic centers served by unrestricted market artery routes
Quality of life	Accessibility, mobility	Percentage of motorists satisfied with travel times for work and other trips
Safety	Number of vehicle collisions	Vehicular crashes per 100 million VMT
		Fatality or injury rates per 100 million VMT
		Crashes involving injuries per 1000 residents
		Crashes involving pedestrians or bicyclists
		Number of pedestrians killed on highways
	Facility condition–related	Percent change in miles in high-accident locations
		Percent crash reduction due to highway construction or reconstruction projects
		Reduction in highway crash due to safety improvement projects
		Number of railroad-crossing accidents
		Percentage of motorists satisfied with snow and ice removal or roadside appearance
		Risk (vulnerability) and consequence of facility element failure
	Construction-related	Number of crashes in highway work zones
Resource and environment	Fuel use	Highway VMT per gallon of fuel

Source: Adapted from Poister (1997).

- The total life-cycle agency costs
- The user cost per unit dimension or per unit use (travel volume) of transportation system
- The total life-cycle user costs and benefits

To enable equitable comparison across time, these performance measures are expressed in constant rather than current dollars after duly correcting for inflationary effects. Furthermore, in assessing system financial performance, some analysts may combine agency costs with user costs to obtain an overall picture of the monetary costs.

(d) System Safety and Security

Safety of System Use: Transportation system safety includes the safety of those using the system (vehicle operators and passengers), those affected by the use of the system (pedestrians), and those involved in the system preservation and operations (field personnel of the agency or its contractors). Performance measures for transportation safety can be measured in terms of frequencies or rates (per mile, per annual average daily traffic, or per vehicle-mile traveled) of all crashes or various categories of crashes (fatal, injury, or property damage).

For highway, rail, water, or air transportation, performance measures for safety include the number of crashes or rate of crashes (per facility dimension, use, or usage dimension such as VMT); for all crash severity types or patterns, or for each crash severity type or pattern; and for vehicles or pedestrians or both. Additional performance measures for transit safety can include crime and vandalism rates.

Defining performance measures for safety helps agencies to determine the effectiveness of safety related projects: for example, crash reduction due to shoulder or lane widening.

Security from Extraordinary Events: At many agencies, facility vulnerability is increasingly assuming a key role as a performance measure for evaluating projects aimed at enhancing facility resilience to (or recovery from) human-made or natural disasters and for purposes of emergency evacuation planning. A suitable performance measure is the vulnerability rating, which is based on the likelihood and consequence of a harmful event.

1. The *likelihood* is based on external factors such as the population and the visibility or national importance of the transportation system (for human-made attacks) and water flow rate or seismic histories (for natural disasters such as flood or earthquake failures, respectively).
2. The *consequence* of failure is evaluated on the basis of the exposure of the facility: for example, the level of usage. It indicates the degree of catastrophe that would result in the event of failure of the transportation facility.

For example, a facility may have a low likelihood of failure but a high consequence of failure (such as a new heavily traveled and well-built city bridge) or a high likelihood of failure but low consequence of failure (such as a lightly used and weak county bridge in a flood- or earthquake-prone area). As illustrated in Figure 2.2, both the event likelihood and its consequence are used to establish the value of the vulnerability rating performance measure. Threat types include human-made attacks, earthquakes, flooding, system fatigue, and major collisions.

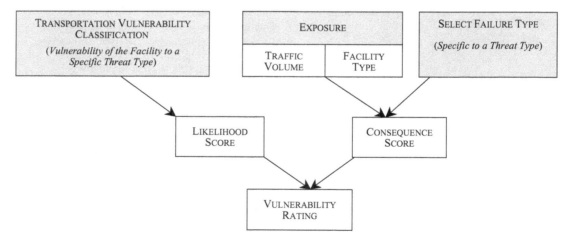

Figure 2.2   Generalized procedure for developing vulnerability ratings. (Adapted from New York State DOT, 1996–2002.)

(*e*) *Economic Development and Land Use*   Most transportation improvements are geared toward enhancing operational effectiveness, but the end goal may be the provision of a top-class transportation infrastructure for the region so as to retain existing businesses or to attract new ones. As illustrated in General Appendix 2, performance measures associated with economic development may include:

- Number of businesses
- Business sales
- Employment (number of jobs)
- Per capita income
- Acreage and proportions of commercial, residential, and agricultural land areas

(*f*) *Environmental Quality and Resource Conservation* Most transportation actions affect the environment and require the consumption of natural resources. Performance measures for environmental impacts are typically expressed in terms of the amount of environmental damage (e.g., pollutant emissions, noise, water quality, habitat degradation). Performance measures for environmental quality and resources conservation may include:

- Acreage of wetlands affected
- Pollutant emissions and concentrations,
- Noise and vibration levels
- Energy consumption

(*g*) *Quality of Life*   Transportation facilities are expected to contribute to the overall quality of life of residents in a region. Quality of life typically captures attributes such as overall well-being, community spirit, social equity, privacy, aesthetics, and concern for the disadvantaged. General Appendix 2 presents a set of performance measures related to the quality of life in a community.

2.5.3 Sector Concerns and Interests

In the private sector, *profit* is the primary measure of performance. For example, the operators of a toll facility may be interested primarily in whether the revenue collected provides sufficient return after deducting the costs of operation, maintenance, and debt service. Also, transportation providers, shippers, truckers, and others in the transportation industry ensure that they are providing their transportation services at a reasonable profit. For the public sector, the primary motive is *service* to the general public, which is typically measured on the basis of operational effectiveness (i.e., mobility, accessibility, safety, and so on.). For publicly subsidized transit

services, the performance measures may also include such items as the deficit per passenger serviced, the operating ratio, and the revenue per vehicle-mile or vehicle hour.

2.5.4 Flow Entity (Passenger and Freight)

From the perspective of passengers, measures that can be used to assess the performance of a transportation project or policy may include the delay per passenger, out-of-pocket costs, and travel-time reliability. For freight operations, facility performance measures may include loading time and inventory time and cost (which depend on inventory size and type), and travel-time reliability. General Appendix 2 presents performance measures that could be used to evaluate system improvements from the perspective of freight and passenger operational efficiency.

2.5.5 Type of Transportation Mode

Although the general objectives (and associated performance measures) of delay reduction, safety enhancement, system preservation, and other dimensions appear to be consistent across the various modes of transportation, there are specific performance measures that may be unique to each mode.

(*a*) *Highway*   For highway systems, typical performance measures include the percentage of the highway network that experiences congestion, the percentage of time that a given highway corridor suffers from congestion, and the incident frequency or severity for the network or at a highway segment or intersection. For a given mode, performance measures may vary by the component system type. For example, traffic density is used to evaluate basic freeway sections, weaving areas, ramp junctions, and multilane highways; while delay is often used to evaluate two-lane highways, intersections, and interchanges, and speed is used for freeway facilities and arterials (Shaw, 2003). In Europe, the OECD (2001b) established a set of performance indicators for the road sector.

General Appendix 2 presents examples of performance measures that could be used to assess the extent to which highway systems help achieve the goals and objectives of operational efficiency, accessibility, mobility and economic development, quality of life, and safety and the environment.

Also, examples of performance measures for specific highway management systems (highway, bridge, congestion, and safety) are provided in General Appendix 2.

(*b*) *Rail and Urban Transit*   For rail transportation in North America, the values of the following performance measures for each regional rail freight carrier are published on a weekly basis: the total cars on line, average train speed, average terminal dwell time, and bill of lading timeliness. For passenger rail transportation, performance measures include on-time arrivals (the number and percentage of on-time rail services that exit or arrive at their destinations within an agreed threshold) and total trip delay (resulting from rail vehicle breakdown, or loading and unloading passengers at terminals). Delay can be expressed in several ways: for example, total delay, delay per vehicle, delay per delayed vehicles, delay per passenger, delay per day, delay per mile, delay per passenger per day, or delay per passenger per mile per day. Other performance measures for rail transportation are the frequency and rate of major incidents, complaints, and trip cancellations. Other rail performance measures can also relate to revenue, cost, or productivity, such as the revenue, cost, or output per resource input (e.g., employee, person-hour, railcar, time).

Performance measurement for urban rail and bus systems has become fairly standardized, due in part to long-standing reporting requirements for transit operators receiving financial assistance from Federal Transit Administration (FTA). Examples and details of performance measures for urban transit are available in the literature (Sinha and Jukins, 1978; Fielding, 1987). A summary of these measures is presented in Table 2.4.

(c) Air   For air transportation, arrival delays are monitored and published routinely for each airline. For airport facilities, typical performance measures can be categorized as described below.

Operational Adequacy: An important item for airport operation is the gate delay, which can be represented by the demand–capacity ratio. Federal Aviation Administration (FAA) guidelines specify the demand–capacity ratio thresholds at which an airport should begin planning to resolve capacity constraints or to implement these plans. At the network level, performance measures related to air transportation capacity include the percentage of system airports that operate at or above a specified level of their annual operational capacity, the percentage of a region (by area, population, or number of business centers) that is within a specified distance or travel time from the nearest system airport, and the percentage of system airports with adequate automobile parking facilities.

Physical Adequacy: Performance measures in this respect include whether the runway and taxiway separations of an airport meet the current FAA guidelines, whether an airport has runway safety areas on its primary runway that meet established standards, whether an airport meets pavement condition standards on its primary runways, whether an airport has shared airspace resulting in operating restrictions, and whether an airport has any obstruction that may affect its operations. At the network level, performance measures involve the percentage of system airports that have the foregoing characteristics.

Environmental and Land-Use Compatibility: It is essential that the operation of airports does not result in environmental degradation or pose a nuisance to abutting land uses. From this perspective, performance measures include the following: whether an airport has worked with surrounding municipalities to adopt height zoning based on federal guidelines, whether an airport is recognized in local comprehensive plans and/or regional vision statements for a community, whether an airport has a noise management plan, and whether the airport complies with state or federal guidelines regarding "airport influence maps" and public disclosure.

Financial Performance: Measures used to evaluate the financial performance of an airport may include the operating ratio, the level of subsidy, and the amount of revenue generated in relation to the number of passengers served. At the network level, performance measures involve the percentage of system airports that have the foregoing characteristics.

Accessibility: Accessibility standards are set for different types of aircraft and aviation facilities. Intermodal links are important for air transportation of goods, and access to the region's airports via alternative transportation modes is important for passengers. Performance measures to assess the ability of an airport to provide adequate ground and air access include the extent to which a region, its population, and its major business centers are within a 30-minute drive time of the airport; whether an airport is served by public transportation; and whether an airport has intermodal transfer capabilities. At the network level, performance measures could involve the percentage of system airports that satisfy the characteristics discussed above.

2.5.6 Number of Transportation Modes Involved

A performance measure may be associated with only a single mode or with two or more modes. For example, the delay encountered in freight transfer from rail to truck transportation is a multimodal performance measure, whereas the delay encountered from one rail terminal to another is a single-mode performance measure. General Appendix 2 presents possible performance measures that could be used to evaluate the effectiveness

Table 2.4   Summary of Transit Performance Measures

Goal Category	Category	Performance Measure
System preservation	Transit vehicle	Miles between road calls for transit vehicles
		Age distribution of vehicles
		Capacity or remaining useful life index
Operational efficiency	Financial	Fare recovery rate of urban transit system
		Cost per passenger-mile of travel (PMT) in urban areas
		Cost per VMT in urban areas
		Cost per revenue-mile in urban areas
		Cost per PMT in rural areas
		Cost per VMT in rural areas
		Cost per revenue-mile in rural areas
		Total transit operating expenditure per transit-mile
		Grant dollars per transit trip
	Ridership	Transit ridership per capita
		Transit ridership-to-capacity ratio
		Transit ridership per VMT
		Transit ridership per route-mile
		Transit ridership per revenue-mile
		Transit peak load factor
		PMT on intercity rail and bus service
	Operational	Number of peak-period vehicles
		Revenue vehicle hours per transit employee
		Average wait time to board transit
		Ratio of number of transit incidents to investment in transit security
Accessibility	Access to and amount of transit	Percentage of population with access to (or within a specified distance from) transit service
		Percentage of urban and rural areas with direct access to bus service
		Percent of workforce that can reach work site in transit within a specified time period
		Access time to passenger facility
	Service characteristics	Route-miles (or seat-miles or passenger-miles) of transit service
		Frequency of transit service
		Route spacing
		Percentage of total transit trip time spent out of vehicle
	Facility characteristics	Transfer distance at passenger facility
		Availability of intermodal ticketing and luggage transfer
		Existence of information services and ticketing
	Parking, pickup/delivery	Volume–capacity ratio of parking spaces during daily peak hours for bus or other passenger terminal lots
		Parking spaces per passenger
		Parking spaces available loading and unloading by autos
		Number of pickup and discharge areas for passengers
Mobility	Transit	On-time performance of transit
		Frequency of transit service
		Average wait time to board transit
		Number of public transportation trips

(*continued overleaf*)

Table 2.4   (*continued*)

Goal Category	Category	Performance Measure
Economic development	Transit	Passengers per capita within urban service area
		Number of commuters using transit park-and-ride facilities
		Number of demand–response trip requests
		Percentage of transit demand–response trip requests met
		Economic indicator for people movement
		Percentage of region's unemployed or poor who cite transportation access as a principal barrier to seeking employment
		Percentage of wholesale and retail sales in the significant economic centers served by market routes
Quality of life	Transit accessibility, mobility	Customer satisfaction with commute time
		Customer perception of quality of transit service
Safety	Transit	Transit collisions (injures or fatalities) per PMT
		Transit collisions (injures or fatalities) per VMT
		Number of intercity bus collisions
		Crimes per 1000 passengers
		Ratio of number of transit collisions to investment in transit security
Environmental and resource conservation	Air pollution	Tons of pollutants generated
		Air quality rating
		Number of days for which air pollution is in an unhealthful range
		Customer perception of satisfaction with air quality
	Fuel use	Fuel consumption per VMT

Source: Adapted from Sinha and Jukins (1978); Poister (1997); Cambridge Systematics (2000).

of improvements at intermodal facilities. For intermodal connections (also called terminals), including rail–road crossings, rail depots (rail–highway), harbors and water ports (water–rail and water–highway; Figure 2.3), and airports (air–rail and air–highway), performance measures include:

- The percentage of time that congestion is experienced
- The incident frequency or severity
- The average time delay in passengers or freight
- The reliability of time taken for intermodal transfers

2.5.7 Entity or Stakeholder Affected

The perspectives of various affected entities and stakeholders often differ significantly. For example, an agency may be interested primarily in facility preservation and financial solvency, whereas users may be more focused on travel time and accessibility. Adjacent businesses and residents may be more concerned with physical and operational impact such as relocation collisions from vehicles, pollution, and accessibility to raw materials, labor, and product distribution points. Environmental groups

Figure 2.3   Multimodal performance measures at intermodal terminals include average delay of freight transfer. (Courtesy of Kevin Walsh, Creative Commons Attribution 2.0.)

typically focus on damage to the ecology, wetlands, and water resources. Furthermore, specific advocacy groups may be particularly interested in safety or accessibility for disadvantaged users, for example. For a transportation project or action to be implemented successfully, it is important to consider the perspectives of all affected stakeholders as part of the evaluation process.

2.5.8 Spatial Scope

As explained earlier in Section 2.3, certain performance measures are more appropriate for network-level evaluation, whereas others are more appropriate for project-level evaluation. Even within these levels, performance measures have to be appropriate for specific spatial scopes, such as statewide, countrywide, citywide, areawide, or corridorwide, or for a specific segment or intersection of a specific mode or terminal (for multimodal systems).

2.5.9 Level of Agency Responsibility

For a given set of other dimensions, performance measures may differ by the level of agency responsibility; state and local agencies may have different measures, as they typically have different perspectives regarding the intended benefits of transportation system actions. For example, the local economic development effect of a corridor improvement may not be an added benefit at the state level because the gain expected may simply be a shift from one local area to another.

2.5.10 Time Frame and Level of Refinement

There can be some performance measures that relate to immediate consequences (primary impacts) of the transportation action, whereas others are impacts that occur in the wake of the primary impacts: that is, secondary impacts. For example, construction of a new bypass may result in immediate impacts, such as a reduction in travel time, whereas secondary impacts, such as increased business productivity due to the travel-time reduction, will take some time to be noticed.

2.6 LINKING AGENCY GOALS TO PERFORMANCE MEASURES: STATE OF PRACTICE

There is widespread explicit or implicit use of the performance measures concept at transportation agencies all over the world. The current generation of performance measures is outcome oriented, tied to strategic objectives, and is focused on quality and customer service. For example, in the state of Delaware, the highway agency's performance measures are connected to the agency's goals, strategies, policies, and long-range transportation plans (Abbott et al., 1998). Also, the state transportation

agency of Minnesota uses a performance measures pyramid that has a top layer comprising policy-based system-level performance measures reflecting outcome targets over a 20-year period; a second layer comprising performance measures specific to districts and transportation modes with long-term impacts; a third layer of performance measures specific to business plans, with a planning horizon of approximately two years; and a fourth layer of performance measures for systems operations that are associated with work plans with a planning horizon of one year or less. The fourth layer contains measures for project-level evaluation. The state transportation agency of California (Caltrans) uses a similar pyramid that consists of three tiers of performance measures for the purpose of monitoring the progress of its strategic plan. The apex of the Caltrans pyramid consists of a set of performance measures that are derived from the agency's strategic goals. The second tier is comprised of performance measures geared toward evaluating products and services provided to customers in terms of quality, efficiency, and customer satisfaction. The third tier consists of performance measures for process and output quantities.

The OECD (2001a) discussed the institutional aspects of intermodal freight transportation, thus laying the groundwork for possible development of measures for assessing the performance of intermodal transportation facilities. Pickrell and Neumann (2000) presented various ways to link performance measures with decision making. Baird and Stammer (2000) developed a model that incorporated an agency's mission, vision, goals, stakeholder perspectives, and system preservation and outcomes. Kassof (2001) reinforced the need to amalgamate the several performance measures and stressed the importance of "omnidirectional alignment" of performance management systems (i.e., vertical alignment of goals, strategies, policies, programs, projects, and measures) so as to span the organizational hierarchy and horizontal alignment to span geographical units (such as districts or functional divisions). Poister (2004) emphasized the importance of performance measures and identified how they can be used in strategic planning at the executive level of an agency. TransTech Management, Inc. (2003) identified modal performance measures that help provide transportation agencies and transportation project managers with the information they need to support transportation-project planning, design, and implementation.

2.7 BENEFITS OF USING PERFORMANCE MEASURES

The establishment of clear performance measures helps agencies to assess the degree to which a program, project, or policy will be or has been successful in achieving its intended goals and objectives in terms of improved system benefits. In effect, performance measures help transportation agencies monitor facility performance, identify and undertake requisite remedial measures, and plan for future investments. By adopting performance measures for transportation project and program evaluation, an agency can reap the following benefits:

1. *Clarity and transparency of decisions.* When the performance measures are objective and unbiased, transportation actions can be evaluated and selected in a rational and unbiased manner, thereby enhancing agency accountability.

2. *Attainment of policy goals.* The use of performance measures provides a basis upon which attainment of agency goals and objectives can be assessed, and provides a link between the ultimate outcomes of policy decisions and the more immediate actions of the agency. For example, the average waiting time for water vessel unloading for a given year can be compared with established thresholds so that any necessary improvements can be identified and implemented.

3. *Internal and external agency communications.* The use of performance measures provides a rational and objective language that can be understandable by various stakeholders and can be used to describe the level of progress being made toward the established goals and objectives (Pickrell and Neumann, 2000). For example, the average air traveler delay is a performance measure that is readily understood by the aviation operator, facility owner, air travelers, and the general public.

4. *Monitoring and improvement of agency business processes.* Performance measures can be used to evaluate the degree to which established strategic or tactical targets (yardsticks or benchmarks) have been achieved (Shaw, 2003). As such, they are useful for decision making regarding continuation of specific operational strategies. Performance measures therefore help not only to define or redefine goals and objectives, but also assist in network performance reviews for program development and for the facility planning stages of the project development process.

SUMMARY

Performance measures are needed at various stages of the transportation development process for the purpose of evaluating the various possible courses of action at each stage and also at various hierarchical levels of transportation management and administration and consequently, for decision making. Performance measures also assess the degree to which the investment program selected has been successful in achieving agency goals and objectives

in terms of improved system benefits. Performance measures therefore enable agencies to monitor facility performance, identify and undertake requisite remedial measures, and plan for future investments. They also assist in ensuring internal agency clarity, communications and transparency, internal agency efficiency and effectiveness, and monitoring and improvement of agency business processes. Performance measures therefore not only aid in defining or redefining goals and objectives but are also helpful during the system of facility planning stages of the transportation development process. The identification of goals and objectives is a key prerequisite to the establishment of performance measures and therefore influences the evaluation and decision outcome. Selection of appropriate performance measures depends on the type of transportation facility, the stage of the transportation development process at which evaluation is being carried out, whether the transportation stimulus under investigation is a policy or a physical intervention, whether the evaluation is preimplementation or postimplementation, and whether it is a network-level problem or a project-level problem. A suitable performance measure should be appropriate, measurable, realistic, defensible, and forecastable and should address all dimensionality aspects of the evaluation. It is important that the final set of measures selected be comprehensive, yet manageable, to facilitate meaningful analysis. The current generation of performance measures at most agencies are derived from agency goals that are outcome oriented, tied to strategic objectives, and focused on quality and customer service.

EXERCISES

2.1. For a proposed rail transit system to connect suburbs to downtown, list the possible goals, objectives, performance measures, and performance criteria.

2.2. What are the attributes of (**a**) an individual performance measure for purposes of systems evaluation, and (**b**) a set of performance measures?

2.3. You have been asked to evaluate the performance of a new air terminal that was constructed five years ago. What performance measures would you consider in such an evaluation? Defend your choice of performance measures.

2.4. It is proposed to widen an existing arterial street to make way for an HOV facility. List appropriate performance measures from the point of view of (**a**) the owner (local highway agency), (**b**) facility users, and (**c**) nonusers who are affected by the system.

2.5. An increase in air travel has made it necessary to expand the regional airport in the city of Townsville. You are asked to evaluate the proposed expansion project on behalf of the city. What types of performance measures would you select?

2.6. Consider a transportation company that provides bus transit service to the elderly and handicapped in a rural county in a contract with the county government. Develop a set of performance measures from the perspectives of the transportation company, the county government, and the service users.

REFERENCES

Abbott, E. E., Cantalupo, J., Dixon, L. B. (1998). *Performance Measures: Linking Outputs and Outcomes to Achieve Goals*, Transp. Res. Rec. 1617, Transportation Research Board, National Research Council, Washington, DC, pp. 90–95.

Baird, M. E., Stammer, R. E., Jr. (2000). *Conceptual Model to Support Systematic Use of Performance Measures in State Transportation Agencies*, Transp. Res. Rec. 1706, Transportation Research Board, National Research Council, Washington, DC, pp. 64–72.

Cambridge Systematics. (2000). *A Guidebook for Performance-Based Transportation Planning*, NCHRP Rep. 446, National Academy Press, Washington, DC.

Fielding, G. (1987). *Managing Public Transit Strategically*, Jossey-Bass, San Francisco, CA.

GASB. (1989), *Resolution on Service Efforts and Accomplishments Reporting*, Governmental Accounting Standards Board Norwalk, CT.

Kassoff, H. (2001). Implementing performance measurement in transportation agencies, in *Performance Measures to Improve Transportation Systems and Agency Operations*, Transportation Research Board, National Research. Council, Washington, DC.

New York State DOT (1996–2002). *Vulnerability Manuals for Bridge Safety Assurance Program*, State Department of Transportation, New York.

OECD (2001a). *Intermodal Freight Transport: Institutional Aspects*, Organization for Economic Cooperation and Development, Paris.

_____ (2001b). *Performance Indicators for the Road Sector*, Organization for Economic Cooperation and Development, Paris.

Pickrell, S., Neumann, L. (2000). Linking performance measures with decision-making, *Proc. 79th Annual Meeting of the Transportation Research Board*, National Research Council, Washington, DC.

Poister, T. H. (1997). *Performance Measurement in State Department of Transportation*, NCHRP Synth. Hwy. Pract. 238 National Academy Press, Washington, DC.

_____ (2004). *Strategic Planning and Decision Making in State Departments of Transportation*, NCHRP Synth. Hwy. Pract. 326, National Academy Press, Washington, DC.

Shaw, T. (2003). *Performance Measures of Operational Effectiveness for Highway Segments and Systems*, NCHRP Synth. Hwy. Pract. 311, National Academy Press, Washington, DC.

Sinha, K. C., Jukins, D. (1978). *Definition and Measurement of Urban Transit Productivity*, Tech. Rep. prepared for the Urban Mass Transportation Administration of the U.S. Department of Transportation by the School of Civil Engineering, Purdue University, West Lafayette, IN.

TransTech Management, Inc. (2003). *Strategic Performance Measures for State Departments of Transportation: A Handbook for CEOs and Executives*, Transportation Research Board, National Research Council, Washington, DC.

Turner, S. M., Best, M. E., Shrank, D. L. (1996). *Measures of Effectiveness for Major Investment Studies*, Tech. Rep. SWUTC/96/467106-1, Southwest Region University Transportation Center, Texas Transportation Institute, College Station, TX.

Figure 3.1 Relationships between demand, supply, and traffic volume.

CHAPTER 3

Estimating Transportation Demand

To go beyond is as wrong as to fall short.
—*Confucius (551–479 B.C.)*

INTRODUCTION

Transportation demand estimation is a key aspect of any transportation evaluation process because it provides a basis for predicting the needs for transportation in terms of passenger, freight, or vehicle volumes expected for a facility. Such forecasts are vital in evaluating alternative actions at every stage of the transportation development process. The decision to proceed with a project is often dictated by the levels of usage predicted for the proposed facility. Then at the facility design stage, the sizing of a proposed transportation facility and the scope of the proposed operational policies are influenced by the expected levels of demand. Furthermore, decision making to select and implement system policies is influenced by the expected levels of trip making. For example, the user benefits and costs, cash flow patterns, economic efficiency, effectiveness, and equity are all influenced by the volume of traffic using the facility. Finally, knowledge of the expected levels of demand in each future year is also useful for developing agency cost streams for preserving facilities whose deterioration and performance are influenced by the level of use.

As shown in Figure 3.1, the demand for transportation, which is derived from socioeconomic activities (e.g., commercial, industrial, educational, medical, and agricultural entities), is ultimately manifested in the form of traffic volume on the facility, such as the number of passengers and the freight tonnage. It is often appropriate to establish different levels of travel demand that correspond to different levels of supply attributes (cost, time, and so on).

The volumes of traffic observed or predicted at a system is therefore the interaction of travel demand and system supply.

It is thus important to be able to predict levels of transportation demand and system supply (performance) at any time in a facility's life or changes in these attributes in response to changes in socioeconomic characteristics, system price, system technology, and so on. Classical topics in transportation economics, such as demand modeling, supply functions, market equilibrium, price elasticity, production costs, and pricing, are therefore important in the evaluation of transportation system impacts.

In this chapter, we first discuss some basic concepts in economic demand theory and present methods for estimating aggregate project-level transportation demand. We then discuss the related topics of transportation supply and elasticity and explain how these concepts can help in estimating transportation demand or changes in demand.

3.1 TRANSPORTATION DEMAND

The demand for transportation is the number of trips that individuals or firms are prepared to make under a given set of conditions (i.e., trip time, cost, security, comfort, safety, etc.). The demand for transportation is often described as a *derived* demand because trips are typically undertaken not for the sake of simply traveling around but because of an expected activity at the end of a journey, such as work, shopping, returning home, or picking up or delivering goods. In this section we discuss methods for estimating travel demand.

3.1.1 Basic Concepts in Transportation Demand Estimation

The demand for any specific transportation facility or service depends on the characteristics of the activity system and the transportation system. An *activity system* is defined as the totality of social, economic, political, and other transactions taking place over space and time in a particular region (Manheim, 1979). Changes in an activity system may be represented by economic or population growth, relocation of commercial, industrial, or other organizational entities into (or out of) an area, or increased (or decreased) scale of operations by entities already existing in an area. A *transportation system* is a collection of physical facilities, operational components, and institutional policies that enable travel between various points in a transportation network. The physical and operational components of a transportation system include the guideway, vehicle, transfer facilities, and facility management systems, while institutional components include pricing policies. The characteristics of a transportation network that are relevant to travel choice (and hence demand estimation) are termed *service attributes* and include travel time, travel cost (out-of-pocket expenses), safety and security, and comfort and convenience. *Demand functions* or *demand models* quantify the willingness of trip makers to "purchase" (i.e., undertake) a trip at various "prices" (i.e., levels of service attributes associated with the trip) under prevailing socioeconomic conditions. In its simplest formulation, a demand function is a two-dimensional model such as the classic demand–price curve. In a more complex formulation, demand is a multidimensional function of several explanatory variables (often including price) that represent the service attributes and trip-maker characteristics. These include a class of demand functions that estimate the expected total demand given the total trip-maker population and the probability that an individual (or group or individuals) will choose a particular transportation mode over another.

Figure 3.2(*a*) illustrates a simple aggregate function for transportation demand between a given origin and a destination at a specific time of day, for transit for a specific trip purpose (work trips), and for only one service attribute: trip price. The figure shows, for various trip prices, the associated levels of trip-making demand, and therefore provides an indication of the number of transit work trips that people are willing to undertake at various levels of the transit service attribute (in this case, trip fare).

Where the demand model predicts the shares of a travel alternative (such as mode, route, and so on), the service attributes that are specific to the alternative are termed *alternative-specific attributes*. These often include travel time, comfort, convenience, User attributes (income levels, household size, etc.), which describe socioeconomic characteristics and therefore do not vary by mode, are termed *generic attributes*. A demand function that estimates demand on the basis of more than one service attribute belongs to the class of *multiattribute demand functions*, and can be represented by a graph showing the relationship between demand and any single service attribute at constant levels of other service attributes. In

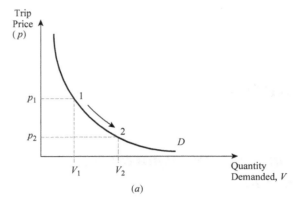

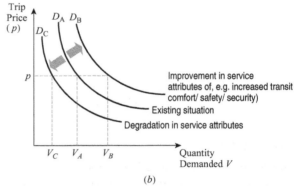

Figure 3.2   (*a*) Demand curve; (*b*) shifts in the curve.

such simplified cases, a change in the demand may be reflected as one of the following two situations:

1. A change in quantity demanded for a transportation service due to a change in the attribute selected (e.g., increase in trip fare). Demand changes in such cases are represented by an upward or downward *slide* along the demand curve (illustrated as $1 \rightarrow 2$ in Figure 3.2*a*). Demand curves of this nature apply primarily to competitive market conditions where travel demand is adequately responsive to changes in service attributes.
2. A *shift* in the single-attribute demand curve at a given level of the trip attribute in question (such as trip price) due to a change in the other trip or service attributes (such as trip time, comfort, accessibility, and security) of the transportation product or its rivals (Figure 3.2*b*).

For example, improvement of a transit system to reach more areas and to enhance passenger security and comfort would lead to an increase in transit demand even if the fare is kept the same—the single-attribute demand curve shifts to the right ($D_A \rightarrow D_B$). The same result would be obtained if there is decreased attractiveness of a rival good such as auto travel through for instance increased parking fees and tolls. By a similar reasoning, a reduction in the quality of transit service attributes, an increase in the attractiveness of auto travel, or a decrease in area employment would lead to reduced demand for transit travel even if transit fares remain the same ($D_A \rightarrow D_C$). In Section 3.1.2 we discuss factors that typically cause such shifts in the transportation demand curve.

In the example above, the single attribute (the variable on the ordinate axis of the demand function) is the trip price or fare. In a bid to simplify a multiattribute demand function, the ordinate could be expressed as a single composite cost variable that is an agglomeration of other *trip or service attributes*, such as trip fare, time, discomfort, safety and security, out-of-pocket expenses, and other "sacrifices" that each traveler incurs in making a trip. Therefore, various costs incurred by the trip maker can collectively represent the *user cost* that will be incurred by the trip maker.

3.1.2 Causes of Shifts in the Transportation Demand Curve

As explained in Section 3.1.1, there could be a change in the demand for a transportation facility even when its price remains the same, and this is reflected as a shift in the demand curve for that transportation facility. Factors that cause such demand shifts discussed below.

- *Sudden change in customer preference* (season, life style, etc.). For example, more people seem to ride transit in the winter season.
- *Change in the level of the attribute of interest (e.g., price increase) of related goods.* For complementary products, a decrease in the price of a product increases the demand for the other product, shifting the latter's demand curve to the right (e.g., parking spaces, automobile use). For rival products, an increase in price of a product increases the demand for its rival product, shifting the latter's demand curve to the right (e.g., transit and auto).
- *Change in regional income.* An increase in income shifts the demand curve for *normal* goods to the right. A normal good is one whose demand increases as a person's income increases.
- *Change in the number of potential consumers.* An increase in population or market size shifts the demand curve to the right.
- *Expectations of an impending change in the level of the attribute of interest.* For example, a news report predicting higher prices in the future can cause a shift in the demand curve at the current price as customers

purchase increased quantities in anticipation of the price change.

3.1.3 Categorization of Demand Estimation Models

Demand models or functions can be either aggregate or disaggregate. *Aggregate demand functions* directly estimate the demand of a group of trip makers (such as a group of individuals, households, firms, or residents in a region or in a given class) in response to future changes in conditions. Alternatively, the decision processes of individual travelers or shippers can be modeled directly using *disaggregate demand functions*, and then summed up for all travelers and shippers to obtain the aggregate predicted demand. The disaggregate approach is based directly on the assumption that the trip makers seek to maximize their utility. It is also possible to develop demand models for a specific trip type and route and to estimate the probability that an individual or firm will undertake the trip given their characteristics and the attributes of the various modes of the transportation system. For the purpose of sketch planning, the aggregate approach, which estimates overall demand directly, is generally more appropriate than the disaggregate approach and has been used widely in past practice to estimate the predicted demand for transportation facilities.

Demand models may also be categorized by their stochastic nature. *Deterministic demand models* assume that the analyst has perfect information in order to predict travel demand, while *stochastic demand models* account for such lack of perfect information by introducing a random or probabilistic element into the demand model. This typically involves adding a random error variable in the demand model and implies that the utility assigned by the traveler to each travel alternative (and consequently, the precise choice of the traveler as to whether or not to travel) is unknown. Where data are available, it may be more appropriate to use stochastic demand models, particularly (1) when there exist some service attributes that are important to travelers but whose utilities are typically not explicitly represented in the demand modeling process, such as transportation security and safety, and convenience; (2) when travelers are not aware of all alternatives that are available to them or may not have correct or updated information on the levels of attributes of the alternatives; and (3) when a traveler's behavior is influenced by factors that change with time, such as weather.

3.1.4 Aggregate Methods for Project-Level Transportation Demand Estimation

Transportation improvements are typically carried out for a specific facility in a network, such as links (e.g., highway segments, rail corridors, air travel corridors) or nodes (e.g., airports, water ports, bus terminals, transit stations). Analysts may seek to estimate aggregate transportation demand at a link between two nodes (population or activity centers ranging from small areas that differ by land use to large cities), for a segmental facility within the link, or for a nodal facility. There are two general ways of doing this: The first involves the use of network methods that simultaneously estimate the demand for all links in the parent regional or urban network of that facility type on the basis of the trip productions and attractions and trip distributions at various points in the network. This approach yields demand models with predictive capabilities that account for any changes that may occur at other facilities in the network and affect demand at the facility in question. The second approach considers only the data for a link or nodal facility and yields total demand for the facility only. A discussion of each approach is presented here.

(a) Demand Estimation Based on the Attributes of the Entire Parent Network   The four-step transportation planning model (TPM), shown in Figure 3.3, is currently the cost widely used model for estimating the link-by-link for an urban or regional network demand. Besides its applicability to entire networks rather than just a single origin–destination route, the attractiveness of the TPM framework lies in its ability to estimate not only overall demand but also demand with respect to trip type, mode, and route. In recent years, this framework has been extended to statewide transportation planning involving passengers and freight. The TPM estimates expected demand on the basis of the attributes of the activity system (such as employment and population) that generates such demand and the characteristics of the transportation system (that serves this demand). The end product of TPM is the demand on each link in a network at "equilibrium" conditions.

Step A1: Establish the Market Segmentation   This step provides a basis for carrying out demand estimation separately for different attributes, such as flow units (passenger vs. freight) and commodity types. Other segmentation criteria (e.g., trip purpose, or mode) could be considered at this stage or may be accounted for in subsequent steps of the framework. It is essential to design a market segmentation process so as to enable the analyst to predict the new demand patterns reliably and ultimately to capture the expected effects of the new system or policy.

Step A2: Establish Traffic Analysis Zones   Trip makers are typically classified by certain characteristics. Urban travelers, for example, can be classified by income, automobile availability, household size, and trip purpose, and most commonly, geographical location. The common procedure involves dividing the study area into traffic analysis zones and then characterizing each zone by each attribute of the entities that demand transportation.

Step A3: Estimate the Number of Generated Trips   This step estimates the total passenger or freight transportation demand for all modes and routes into and out of each zone. This process is carried out on the basis of trip productions and trip attractions. For passenger transportation, variables in trip production equations typically include residential and household characteristics, while variables in trip attraction equations typically include employment types and levels, and floor space by business type (e.g., educational, commercial, or industrial). Analysts may determine the expected number of trips to be generated using information available in ITE's *Trip Generation Handbook* (ITE, 2003). This publication presents average rates and regression equations for each land-use category, such as ports and terminals, industrial area, residential area, institutions, medical facilities, offices, lodging, retail, services, and recreational facilities. For freight, trip generation rates developed by Cambridge Systematics (1996) may be used.

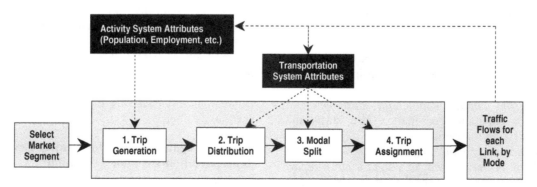

Figure 3.3   Four-step transportation planning model.

Step A4: Estimate Trip Distribution   This step identifies specific origins and destinations of trips generated. Trips can be distributed using any of several methods. However, the most common method is the gravity model, which estimates trip making between two points directly as a function of the trip generation potential of any two points and indirectly as some measure of trip-making impedance (such as distance or travel time) between the two points. Such impedance, referred to as the *friction factor*, should be calibrated for the area of interest, time of day, and so on. The number of trips between any pair of zones i and j is given by

$$T_{ij} = P_i \frac{A_j F_{ij} K_{ij}}{\sum_{j=1}^{n} A_j F_{ij}} \qquad (3.1)$$

where P_i are the trip productions from zone i, A_j the trip attractions to zone j, K_{ij} is an adjustment factor for trip interchanges between zone i and j, and F_{ij} is the friction factor, a measure of travel impedance between i and j given by $F_{ij} = t_{ij}^{-\alpha}$, where t_{ij} is the travel time between i and j and α is a coefficient.

Step A5: Determine the Modal Split   These models predict the shares of overall demand taken by each available mode and may be carried out before or after the trip distribution step. The most common modal split models are of the logit or probit forms.

Step A6: Assign the Traffic   For each bundle of demand associated with an origin–destination pair and mode, this step predicts, the route to be undertaken by that bundle. Traffic assignment can be carried out either on the basis of various techniques associated with user or system equilibrium.

Example 3.1   A transportation improvement program is planned in a metropolitan area for implementation in year 2020. Figure E3.1 shows the main corridors in the area. You are asked to estimate the passenger travel demand along the corridors. Instead of a simple trend analysis or two-point gravity model, it is preferred to use a network demand model and to incorporate supply characteristics. Three neighborhoods or population centers (1, 2, and 3) are considered for the network. The tables below provide the following information: zone-to-zone person-trips for the base year, zone-to-zone travel times and costs (for auto and transit, at the base and horizon years); and utility functions for auto and transit, zonal socioeconomic characteristics, and trip generation models. The trips shown in all tables are person-trips in hundreds.

1. *Base year* (2000) Table E3.1.1 shows the base year zone-to-zone person-trips, travel times, and friction factors.

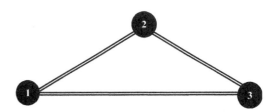

Figure E3.1   A simple network example.

Table E3.1.1   Base-Year Zone-to-Zone Person Trips, Travel Time, and Friction Factors[a]

From Zone:	To Zone: 1	2	3	Total Trip Productions
1	TT = 1 NT = 40 FF = 0.753	TT = 9 NT = 110 FF = 1.597	TT = 4 NT = 150 FF = 0.753	300
2	TT = 11 NT = 50 FF = 0.987	TT = 2 NT = 20 FF = 0.753	TT = 17 NT = 30 FF = 0.765	100
3	TT = 6 NT = 110 FF = 1.597	TT = 12 NT = 30 FF = 0.765	TT = 3 NT = 10 FF = 0.753	150
Total trip attractions	199	161	190	

[a]TT, travel time in minutes; NT, number of trips; FF, friction factor ($\alpha = 2$).

2. *Horizon year* (2020) Provision of the transit service Trip generation models (from the trip generation phase):

Productions: $P_i = -10 + 2.0X_1 + 1.0X_2$,   where X_1 = number of cars and X_2 = number of households.

Attractions: $A_j = -30 + 1.4X_3 + 0.04X_4$,   where X_3 = employment and X_4 = area of commercial area in hectares.

Table E3.1.2 shows the socioeconomic characteristics of each zone in terms of the number of cars, number of households, employment, and area of commercial activity; the travel time and friction factors between zone centroids for the year 2020 are shown in Table E3.1.3, and the zone-to-zone travel times and costs for auto and transit are given in Table E3.1.4.

Table E3.1.2   Zonal Socioeconomic Characteristics (Projected) in Horizon Year

Zone	Cars, X_1	Households, X_2	Employment, X_3	Commercial Area, X_4
1	280	200	420	4100
2	220	150	560	800
3	190	110	220	600

The utility functions for auto and transit, which are used in the mode choice models, are as follows:

Auto: $U_{\text{auto}} = 2.50 - 0.5CT_A - 0.010TT_A$
Transit: $U_{\text{transit}} = -0.4CT_T - 0.012TT_T$

Table E3.1.3   Horizon Year Zone-to-Zone Person Trips, Travel Time, and Friction Factors

From Zone:	To Zone: 1	2	3
1	TT = 2 FF = 0.753	TT = 12 FF = 0.987	TT = 7 FF = 1.597
2	TT = 13 FF = 0.987	TT = 3 FF = 0.753	TT = 19 FF = 0.765
3	TT = 9 FF = 1.597	TT = 16 FF = 0.765	TT = 4 FF = 0.753

where CT_A and TT_A are the cost and travel time for auto travel, respectively, and CT_T and TT_T are the cost and travel time for transit travel, respectively, where TC = travel costs in dollars and TT = travel time in minutes.

SOLUTION

1. *Trip generation.* The projected trip productions P_i and attractions A_j for each zone for the year 2020 are shown in Table E3.1.5.

    Total number of trips produced

    = total number of trips attracted

    = 1810 (trip balancing)

2. *Trip distribution.* Calculate the zone-to-zone trips for the base year 2000 with the use of the gravity model. (Assume that $K_{ij} = 1.0$ for all zones and use zonal trip productions and attractions, and friction factors from Table E3.1.1.)

Table E3.1.4   Horizon Year Zone-to-Zone Travel Time and Cost for Auto and Transit[a]

From Zone:	To Zone: 1 Auto	Transit	2 Auto	Transit	3 Auto	Transit
1	TT = 3 CT = \$0.5	TT = 5 CT = \$1.0	TT = 12 CT = \$1.0	TT = 5 CT = \$1.5	TT = 7 CT = \$1.4	TT = 12 CT = \$2.0
2	TT = 13 CT = \$1.2	TT = 15 CT = \$1.8	TT = 3 CT = \$0.8	TT = 6 CT = \$1.2	TT = 19 CT = \$1.2	TT = 26 CT = \$1.9
3	TT = 9 CT = \$1.7	TT = 20 CT = \$2.0	TT = 16 CT = \$1.5	TT = 21 CT = \$2.0	TT = 4 CT = \$0.7	TT = 8 CT = \$1.1

[a]TT, travel time in minutes; CT, travel cost in dollars.

Table E3.1.5   Trip Productions and Attractions for Year 2020

	Zone		
	1	2	3
Trip productions, P_i	750	580	480
Trip attractions, A_j	722	786	302

Table E3.1.7   Adjustment Factors (K_{ij})

	To Zone:		
From Zone:	1	2	3
1	0.47	0.99	1.45
2	1.27	1.06	0.72
3	1.47	0.98	0.23

Table E3.1.6   Calculated Trip Table (2000) Using the Gravity Model [equation (3.1)]

	To Zone:			
From Zone:	1	2	3	P_i
1	85	111	104	300
2	39	19	42	100
3	75	31	44	150
A_j	199	161	190	550

Table E3.1.8   Calculated Trip Table (2020) Using the Gravity Model [equation (3.1)]

	To Zone:			
From Zone:	1	2	3	P_i
1	105	396	249	750
2	288	247	45	580
3	329	143	9	480
A_j	722	786	303	1810

Table E3.1.6 shows the trip interchanges calculated between the various zones after row and column factoring. The adjustment factors K_{ij} are calculated as follows:

$$K_{ij} = \frac{T_{ij}(\text{observed})}{T_{ij}(\text{calculated})}$$

T_{ij}(observed) and T_{ij}(calculated) are determined from Tables E3.1.1 and E3.1.6, respectively. Apply the gravity model [equation (3.1)] to estimate zone-to-zone trips for the horizon year 2020. Friction factors are obtained from Table E3.1.3. The K_{ij} values are used from Table E3.1.7. The final trip

interchange matrix for the horizon year is shown in Table E3.1.8.

3. Mode choice. Use the utility functions to estimate the utilities for auto and transit (Table E3.1.9). The logit model for finding the auto share is

$$P(\text{auto}) = \frac{e^{U_{\text{auto}}}}{e^{U_{\text{auto}}} + e^{U_{\text{transit}}}}$$

Use the logit model to determine the fraction of zone-to-zone trips by auto and transit, as shown in Table E3.1.10. The trip interchange matrix obtained from trip distribution in step 2 and the modal share yield Table E3.1.11.

Table E3.1.9   Utility Values by Mode[a]

	To Zone:		
From Zone:	1	2	3
1	$U_{\text{auto}} = 2.23$	$U_{\text{auto}} = 1.88$	$U_{\text{auto}} = 1.73$
	$U_{\text{transit}} = -0.46$	$U_{\text{transit}} = -0.78$	$U_{\text{transit}} = -0.94$
2	$U_{\text{auto}} = 1.77$	$U_{\text{auto}} = 2.07$	$U_{\text{auto}} = 1.71$
	$U_{\text{transit}} = -0.90$	$U_{\text{transit}} = -0.55$	$U_{\text{transit}} = -1.07$
3	$U_{\text{auto}} = 1.56$	$U_{\text{auto}} = 1.59$	$U_{\text{auto}} = 2.11$
	$U_{\text{transit}} = -1.04$	$U_{\text{transit}} = -0.05$	$U_{\text{transit}} = -0.54$

[a]U_{auto}, auto utility; U_{transit}, transit utility.

Table E3.1.10   Fraction of Trips by Mode

From zone:	To zone:		
	1	2	3
1	P(auto) = 0.94	P(auto) = 0.93	P(auto) = 0.94
	P(transit) = 0.06	P(transit) = 0.07	P(transit) = 0.06
2	P(auto) = 0.94	P(auto) = 0.93	P(auto) = 0.94
	P(transit) = 0.06	P(transit) = 0.07	P(transit) = 0.06
3	P(auto) = 0.93	P(auto) = 0.93	P(auto) = 0.93
	P(transit) = 0.07	P(transit) = 0.07	P(transit) = 0.07

Table E3.1.11   Trip Interchanges by Mode (2020)

From zone:	To Zone:		
	1	2	3
1	Auto trips = 98	Auto trips = 370	Auto trips = 233
	Transit trips = 7	Transit trips = 26	Transit trips = 16
2	Auto trips = 269	Auto trips = 230	Auto trips) = 42
	Transit trips = 19	Transit trips = 17	Transit trips = 3
3	Auto trips = 307	Auto trips = 133	Auto trips) = 8
	Transit trips = 23	Transit trips = 10	Transit trips = 1

4. *Traffic assignment.* The minimum path (all-or-nothing) method is used for loading the trips on each link to yield Table E3.1.12. These trips reflect expected demand for given levels of service. By changing trip time and cost (representing supply functions), demand can be estimated for all or individual links.

Example 3.2 This example illustrates the use of a statewide travel model to estimate the transportation impacts of proposed major corridor improvements on a selected transportation network. The study corridor is the 122-mile corridor (U.S. 31) between Indianapolis and South Bend, Indiana. U.S. 31 is the primary north/south route through north-central Indiana. The proposed major corridor improvement concept for U.S. 31 is for an upgrade of the corridor to Interstate design standards and also includes construction of a new east-side bypass of Kokomo and a new freeway-to-freeway interchange with I-465, as shown in Figure E3.2. The overall study was carried out by Cambridge Systematics, Inc. and Bernardin, Lochmueller & Associates, Inc. (CSI-BLA, 1998).

SOLUTION   The Indiana Statewide Travel Model (ISTM) was used to generate projections of traffic

Table E3.1.12   Auto and Transit Volumes by Link (2020)

Route	Auto and Transit Travel Time[a]	Auto and Transit Trips
1–2	12* (15*)	374 (22)
1–3	7* (12*)	236 (13)
1–2–3	31 (41)	
1–3–2	23 (33)	
2–1	13* (15*)	271 (17)
2–3	19* (26*)	42 (3)
2–3–1	28 (46)	
2–1–3	20* (27*)	
3–1	9* (20*)	309 (20)
3–2	16* (21*)	134 (9)
3–2–1	29 (36)	
3–1–2	21 (35)	

[a]An asterisk indicates the travel time of paths with least travel time. Transit travel time and trips are shown in parentheses.

volumes and travel times on the highway network in the corridor, as well as in the entire state. Developed

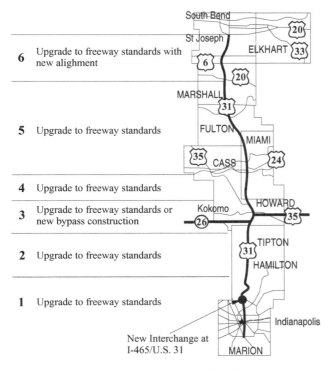

Figure E3.2   Proposed U.S. 31 corridor improvement.

6   Upgrade to freeway standards with new alighment

5   Upgrade to freeway standards

4   Upgrade to freeway standards

3   Upgrade to freeway standards or new bypass construction

2   Upgrade to freeway standards

1   Upgrade to freeway standards

New Interchange at I-465/U.S. 31

U.S. 31, with an average increase of approximately 45% for the entire corridor. In absolute number of trips, the largest increase in trips would be seen at the southern end of the corridor.

Over the past two decades, there have been efforts to improve the TPM using *individual choice* (or *random utility*) *models* and *activity-based models* (Hensher and Button, 2000). Individual choice models try to capture the decision process of individual trip makers given the assumption that the trip maker is rational, has full knowledge, and therefore seeks to choose a transportation alternative mode, route, destination, and so on, that maximizes their utility (utility is measured implicitly or explicitly in terms of travel time, out-of-pocket costs, comfort and security, and other nonmonetary costs). Depending on the number of travel alternatives and the statistical assumptions associated with the demand data, model types include logit, probit, and dogit models. Unlike trip-based demand estimation approaches, activity-based approaches capture the scheduling of and participation in activities that directly generate the need to travel. Also, activity-based methods are considered more responsive to evolving policies oriented toward management rather than facility expansion. A new generation of demand models has been advocated to overcome the limitations of the currently used models (McNally, 2000).

The TPM method has seen wide applications in transportation demand for modes other than highways (such as air and rail) and for flow entities other than passengers (e.g., freight). In freight demand analysis involving spatial interactions of facilities, commodity surpluses and deficits at various geographical points on a transportation network are established and commodities are made to flow from centers of excess supply to those of excess demand. Such flow is governed by trip distribution techniques such as the gravity model.

in 1998 for INDOT to support statewide transportation planning activities, ISTM includes both passenger and freight movements on the 11,300-mile statewide highway network. The model includes 651 internal and 110 external traffic analysis zones. Two future-year (2020) traffic forecasts were developed and compared—one assuming the U.S. 31 improvements are implemented and one assuming they do not occur, as shown in Table E3.2.1.

The transportation network analysis suggested that at the horizon year (2020) the average daily traffic (ADT) is expected to increase significantly along most segments of

Table E3.2.1   Estimated Demand along U.S. 31 (2020)

U.S. 31 Link	Length (miles)	Number of Trips (ADT)		
		No-Build	Build	Difference
I-465 to SR 431	10	78,800	122,200	43,400
SR 431 to SR 26	23	39,800	61,400	21,600
SR 26 to U.S. 35 (north leg)	9	36,400	41,900	5,500
U.S. 35 (north leg) to U.S. 24	11	23,800	37,000	13,200
U.S. 24 to U.S. 30	52	18,500	30,700	12,200
U.S. 30 to U.S. 20 bypass	19	35,200	42,900	7,700
Corridor Total	**122**	**36,100**	**52,600**	**16,500**

Another demand estimation method that involves spatial interaction is *network optimization*, where the demand for each link is determined on the basis of minimized total transportation cost expended in the network. Compared to the traffic assignment method, the optimization method may have serious limitations because (1) it implies that there is only one central decision-making entity for travel in the network and therefore fails to account for the different adaptations of individual users to network changes, and (2) the objective (total cost) function may be rendered concave due to scale economies, and therefore infeasible (Hensher and Button, 2000).

(*b*) *Demand Estimation Based Only on the Attributes of a Corridor or Project or Its Endpoints*   The estimation of aggregate transportation demand for a specific link or node of a network based on the facility data has been discussed extensively. Kanafani (1983) provided models that estimate travel demand for a link between two nodes (ranging from small areas that differ by land use to large cities or major population centers), for a segment within the link, or for a nodal facility. Such estimation can be carried out using either one of two approaches. The first is a *multimodal approach* that recognizes the relationships that exist between modes and thus carries out the estimation in a simultaneous fashion. The structure of models that estimate multimodal intercity demand is similar to that of the TPM approach in that there are several alternative modes, several destinations, and several routes. It has been recommended that because trip distribution analysis (estimation of various demands for various modes at various alternative links) may be limited by the intrinsic characteristics of the cities, demand estimation should be done separately for each pair of cities. The second approach, a *mode-specific approach*, assumes that the modal demands are independent and therefore estimates these demands separately. Steps that could be used for estimating demand between two major population centers based only on the attributes of a corridor or its endpoints are presented next.

Step 1: Establish the Market Segmentation   Demand estimation may be carried out separately for freight and passenger transportation, for work trips and nonwork trips, or for trips that otherwise differ by some attribute. The entire trip-making market could therefore be divided into different segments, the demand estimated for each segment, and the demands summed to yield the overall demand.

Step 2: Select the Demand Function   In this step, data are collected for the project and models are developed to estimate demand as a function of the attributes of the endpoints of the proposed project, such as population

or employment. The analyst could use one of many forms of demand functions, depending on the type of data being used, whether the demand is for a link or a node, whether it is sought merely to estimate demand changes in response to changes in service attributes, and so on. Where only historical data on demand are available, the analyst may estimate future demand on the basis of projections of past trends using time series–based trend lines. Where socioeconomic data are available to derive trip productions and attractions of the endpoints, the gravity models may be more appropriate.

Specific mathematical forms for demand estimation may include the elasticity-based form that is typically used where the analyst is faced with data and time limitations and seeks to estimate changes in demand from an existing or base situation. Common generic mathematical forms for demand estimation are:

$$Linear: \quad V = b_0 + b_1 x$$
$$Multiplicative: \quad V = b_0 x^{b_1}$$
$$Exponential: \quad V = b_0 e^{b_1 x}$$
$$Power: \quad V = (b_0)^x$$
$$Logistic: \quad V = \frac{b_0}{b_1 + e^{b_2 x}}$$
$$Logistic\text{-}product: \quad V = \frac{\alpha}{1 + \gamma x^{\beta}}$$

For simple trend analysis, the x variable simply represents time (years). For other types of demand estimation models, x is a vector of multiple variables, such as socioeconomic system attributes.

Demand Estimation Using Trend Analysis: Future demand can be estimated simply on the basis of past data. The functional form typically selected is one that best fits the historical data (the S-curve has often been used). Obviously, the use of trend analysis to estimate future demand implicitly assumes that the levels of the other factors affecting travel (as well as their relationships) will remain unchanged over time—this can be a rather restrictive assumption. Also, trend analysis does not account for possible future changes in the trip-generating characteristics of the area served by the facility or the wider network areas, or for possible future changes in the service attributes of the facility in question, of other links in the network, or of other competing modes. Given such limitations, trend analysis is generally considered to be more appropriate as a *diagnostic* rather than a *predictive* tool in the estimation of demand (Meyer and Miller, 2001).

Example 3.3   The demand for a certain rail transit system shows stable growth over the past decade, as shown in Table E3.3.1. An analyst seeks to estimate the expected demand at year 2008 when the system is due for improvement. Use the linear and exponential functional forms to predict the expected demand in that year. What assumptions should be made in using the predicted value of demand for evaluation? What are the limitations in using trend analysis for demand estimation?

SOLUTION   The expected demand in the year 2008 can be determined using the mathematical functional forms of the linear and exponential curves as follows:

Linear form:

$$V = 0.089(\text{year} - 1990) + 1.1408 \qquad R^2 = 0.95$$

Thus, the projected demand in 2008 on the basis of linear trends is

$$(0.089)(2008 - 1990) + 1.1408 = 2.74 \text{ million}$$

Exponential form:

$$V = 1.2106e^{0.0499(\text{year}-1990)} \qquad R^2 = 0.98$$

Thus, the projected demand in 2008 on the basis of exponential trends is

$$1.2106e^{0.0499(2008-1990)} = 2.97 \text{ million}$$

While the exponential form gives a higher value of R^2, both forms provide good fits. Consequently, it may be desirable to use both estimates to yield a range of expected demand in 2008.

The underlying assumption in trend analysis is that all demand-contributing factors in the study area are constant over the period of projection. Furthermore, the supply of this mode and that of competing modes (e.g., private automobile or bus transit) are assumed to be constant. A limitation of the trend analysis method of demand estimation is that these assumptions are not always realistic. Changes in socioeconomic characteristics (such as relocation of new businesses, construction of schools,

hospitals, etc.) and improvements or degradations in the supply attributes of this mode or its rival modes are always imminent. Such changes violate the foregoing assumption and can render the demand predictions inappropriate.

Elasticity-Based Models for Demand Estimation: Transportation improvements typically result in changed levels of service, such as trip cost and/or time. Elasticity-based demand models help estimate the new demand levels for a particular transportation mode in response to changes in service attributes, such as trip cost and time. The assumption is that the preimplementation demand level is known. In Section 3.4 we present the concept of elasticity and in Section 3.4.5 we discuss how it can be used for demand estimation.

Gravity-Based Models for Demand Estimation: The concept of gravity model used in TPM (discussed in Section 3.1.4) can be used for direct estimation of demand between two population or employment centers. In its classic formulation the gravity model is analogous to Newton's law of universal gravitation:

$$V_{AB} = N_A N_B I_{AB} \qquad (3.2)$$

where V_{AB} is the demand for transportation between zones A and B; N_A and N_B are the measure of trip attractiveness, such as employment at zones A and B, respectively; I_{AB} and is the travel "impedance" between A and B (i.e., some characteristic or attribute of the transportation system that either impedes or facilitates travel between zones A and B, such as travel distance, time, speed, comfort, security, or out-of-pocket cost). The formulation above shows that the gravity model incorporates demand and supply characteristics by using parameters for trip attractiveness and impedance, respectively.

Most mode-specific travel demand estimation is carried out on the basis of the gravity model. The gravity model used in the traditional four-step transportation planning model (TPM), represents interzonal distribution of trips. Equation (3.1) gives a ratio of the travel propensity for each link relative to the sum of all link travel propensities, in terms of their respective impedances. Thus the gravity

Table E3.3.1   Annual Ridership of a Rail Transit System

Year	1990	1992	1994	1996	1998	2000	2002	2004
Demand (millions of passengers per year)	1.25	1.37	1.45	1.58	1.72	1.95	2.31	2.48

model determines the relative competitiveness of alternative destinations and estimates the shares of travel destinations. Compared to passenger transportation demand, commodity transportation demand is more consistent with economic demand theory and analysis because (1) the reason behind travel decisions are mostly economic (e.g., cost minimization), and (2) the demand for commodity transportation is derived completely from the various demands for the commodities at the points of consumption that are geographically distinct from the points of production—as such, the nature of the demand function can be found by identifying the patterns of production, distribution, and consumption in the network.

Example 3.4 The total air traffic (thousands of passengers per week) between a certain pair of cities, V_{ij}, can be given by

$$V_{ij} = \text{INC}_{ij}^{0.38} \times \text{POP}_{ij}^{0.25} \times \text{TIME}_{ij}^{-1.51}$$

where INC_{ij} is the per capita income averaged across both cities i and j, in tens of thousands; POP_{ij} the average population between the two cities, in millions; and TIME_{ij} the average flying time between the two cities, in hours. Determine the demand when the average per capita income is \$30,000, the average of the two populations is 2 million, and the average flying time is 1.5 hours.

SOLUTION

$$V_{ij} = (3^{0.38})(2^{0.25})(1.5^{-1.51}) = 979 \text{ passengers per week}$$

Other variables that could be used in such models include the distance between the cities, average ticket price, and availability of other modes. However, in developing or using models of this type, the analyst should be careful to ascertain whether the predictive power of the model could be compromised by high correlations between the independent variables. For example, flight distance, ticket price, and flying time may be highly correlated.

(*c*) *General Comments on Demand Estimation Methods*
As with most other real-world models, the main weakness of transportation demand estimation models is that they are often developed on the basis of historical data that may not be adequately representative of the future. Furthermore, transportation planning models are often based on the hypothesized travel patterns of travelers, and such patterns can be validated empirically by observing the trip behavior of passengers. If it were possible to carry out controlled experiments that incorporate specific levels of the transportation system and activity system attributes, the behavior of travelers under each set of conditions could

be ascertained more reliably and used as a basis for future demand prediction. Unfortunately, it is not feasible to carry out such controlled experiments, therefore, past and current transportation and activity system conditions offer the only setting upon which future predictions can be made. As such, demand models are typically most valid when they are applied to future conditions that are not very different from those under which such models were developed. Second, demand models tend to be most reliable in the short term, as they typically fail to incorporate the long-term impacts of changes in trip patterns.

3.2   TRANSPORTATION SUPPLY

3.2.1 Concept of Transportation Supply
The supply of a transportation product or service represents the level of performance of the product or service that a provider is willing to offer at a given level of a service attribute (such as trip price). There are basically two aspects of transportation supply: quantity and quality.

1. *Quantity* refers to the amount of a product or service that the provider makes available or the capacity of a transportation system. For a transit system, for example, quantity may refer to the number of buses or rail cars per hour; and for a highway system, quantity may refer to the number of lanes. In the quantity context, a performance (supply) model estimates the quantity expected to be supplied at a given level of the service attribute, such as trip cost or travel time, at a given period of time.

2. *Quality* refers to the level of service. Examples for transit are cleanliness, security, lack of passenger congestion, and vehicle and track condition. For the highway system, examples are the level of traffic congestion and the pavement surface condition. In the quality context, performance (supply) models typically estimate the rate of deterioration of the transportation product or service over time. For example, the quality of rail tracks decreases with time as accumulated climate and use take their toll.

A specific supply curve represents the supply–price relationship given a set of conditions specific to the transportation product or service in question (referred to as *alternative-specific attributes*, such as travel time, comfort, convenience), and also specific to the producers or service providers (such as technology, policy, and governmental intervention through policies and regulations). Changes in such conditions often result in changes in the levels of transportation supply, even at a fixed price of that service or product. When such changes in conditions (other than price) occur, they are represented as a shift in the supply curve.

In the context of quantity and quality as discussed above, increases in transportation supply may be thought of not only in terms of increasing the fleet size of a transit company, building new roads, or adding lanes to existing roads, but also in terms of investments that are not physical and capital-intensive in nature. For instance, the use of intelligent transportation systems, ramp metering, and managed lanes (high-occupancy vehicle, or high-occupancy and toll lanes, truck-only lanes, etc.) could lead to an increased level of service without any physical enlargements of the road network.

3.2.2 Causes of Shifts in the Transportation Supply Curve

The supply of a transportation service may change even if price remains the same, for reasons such as:

- *Prices of rival transportation services.* The supply of a service may decrease if there is a decrease in the price of a competing transportation service, causing providers to reallocate resources to provide larger quantities of the more profitable service. This may apply more to toll roads, where profit is the primary motive, and to a lesser extent, non-toll roads.
- *Number of transportation modes.* An increased number of modes, such as construction of a subway in a city that already has buses and light rail transit and facilities for autos, indicates an increase in supply, shifting the supply curve to the right (downwards).
- *Prices of relevant inputs.* If the cost of resources used to produce a transportation service increases, the transportation agency would be less capable of supplying the same quantity at a given price, and the supply curve will shift to the left (upwards).
- *Technology.* Technological advances that increase facility capacity or efficiency cause the supply curve to shift to the right (downwards).

3.3   EQUILIBRATION AND DYNAMICS OF TRANSPORTATION DEMAND AND SUPPLY

3.3.1 Demand–Supply Equilibration

At equilibrium conditions, the quantity of trips demanded is equal to the quantity supplied. The equilibrium state is essentially fixed at a given point in time and is often analyzed as such. However, over a period of time, several short- and long-term equilibrium states can occur in response to changes in system supply or system demand, and each equilibrium state can be analyzed separately. The traffic assignment step in TPM discussed in Section 3.1.4

represents the equilibrium state, typically for a peak period.

Example 3.5   The following equations represent the demand and supply functions associated with a given passenger railway route for a particular season.

Demand function:

$$V = 5500 - 22p$$

Supply function:

$$p = 1.50 + 0.0003V$$

where V is the daily passenger trips along the route and p is the fare in dollars. Determine the equilibrium demand and price and comment on the threshold demand and fare.

SOLUTION   Solving the two equations simultaneously yields the following equilibrium values: $V = 5431$ daily passenger trips and $p = \$3.13$. The equilibrium point can also be obtained graphically by plotting the two equations simultaneously and determining the point of intersection. Several other observations can be made: The maximum daily demand along the route is 5500 trips, and the minimum ticket price is \$1.50 per trip.

3.3.2 Simultaneous Equation Bias in Demand–Supply Equilibration

Traditional methods for estimating transportation demand and supply implicitly assume that the supply characteristics are exogenous and fixed, implying that demand and supply functions exist as single independent equations. In reality, one or more of the supply variables may not be exogenous, but rather, may depend on the endogenous variable representing traffic volume, thus introducing a two-way causality problem best known as *simultaneity*. An example is the use of time-series analysis in modeling air travel demand; the issue of simultaneity arises because observed trends in traffic (a representation of demand) and price and capacity (representations of supply) are actually not independent. In such cases, a system of equations needs to be specified to estimate the model parameters reliably. Simultaneity may be ignored if the value of the supply variable at each demand level is assumed to be fixed and exogenous. Where such simultaneity cannot be ignored, it becomes difficult to reliably calibrate the demand and supply models, and the problem of *identification* (which gives rise to such difficulty) needs to be addressed. Several standard econometric texts provide methodologies to identify or address simultaneity (Wooldridge, 2000; McCarthy, 2001; Washington et al., 2003).

3.3.3 Dynamics of Transportation Demand and Supply

Assume that at the same price, there is increased demand due to factors exogenous to the transportation system, such as increasing population, rising employment, or business growth. This causes the demand curve to shift from D_{old} to D_{new} while the supply stays the same; the equilibrium point shifts from (V_0, p_0) to (V_1, p_1). Then, if there is an improvement in the quantity or quality of the transportation system, such as additional highway lanes, congestion mitigation techniques, or an intelligent transportation system (ITS), the supply (performance) function shifts from S_{old} to S_{new} and the system reaches yet another new equilibrium (V_2, p_2). The increase in system performance may then lead to further shifts in demand for the system. For example, the construction of a new interchange or added lanes may be accompanied by increased business activity (such as an increased number of shopping malls or restaurants, or increased sales by existing businesses). There will thus be a new equilibrium point. These demand and supply shifts and resulting changes in equilibrium positions are illustrated in Figure 3.4. In reality, transportation systems undergo such changes constantly, moving from one equilibrium point to another.

3.4 ELASTICITIES OF TRAVEL DEMAND

Analysts involved in transportation system evaluation may often need to adjust their demand forecasts to

reflect changed socioeconomic or transportation system characteristics. Knowledge of demand elasticities enable analysis of the impacts of changes in factors that influence transportation demand. In cases where elasticity values are known, changes in demand from an existing level can be estimated using the methods that are presented in Section 3.4.5.

Elasticity, defined as percentage change in demand for a 1% change in a decision attribute, helps to obviate the dimensionality problems associated with other concepts of demand sensitivity, such as derivatives. The elasticity of travel demand V, with respect to an attribute x can be expressed as.

$$e_x(V) = \frac{x}{V}\frac{\partial V}{\partial x} \qquad (3.3)$$

Table 3.1 presents the elasticity functions for selected mathematical forms of the demand model. The interpretation of elasticity values, methods of computation, and applications are discussed in a subsequent section of this chapter. Demand elasticities can be influenced by factors such as mode type, trip purpose, time of day, trip length, trip-maker characteristics, and existing level of factor. Because the trip maker's decision is typically associated with combined utility maximization, a specific elasticity value cannot be considered while explicitly considering the existing levels of the other factors. As such, the transportation service attributes are important determinants of trip-maker sensitivity to price changes. For example, for a high level of service, the impact of a fare increase will be

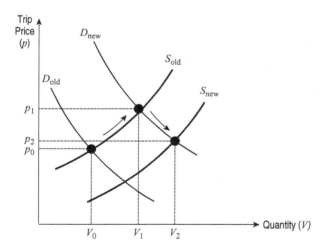

Figure 3.4 Instances of demand and supply equilibration.

Table 3.1 Elasticity Functions for Standard Mathematical Forms of Aggregate Demand

	Elasticity Function: $(x/V)(\partial V/\partial x)$
Linear $V = \alpha + \beta x$	$\dfrac{\beta x}{V} = \dfrac{1}{1 + (\alpha/\beta x)}$
Product $V = \alpha x^\beta$	$e = \beta$
Exponential $V = \alpha e^{\beta x}$	$e = \beta x$
Logistic $V = \dfrac{\alpha}{1 + \gamma e^{\beta x}}$	$\left(1 - \dfrac{V}{\alpha}\right) = -\dfrac{\beta_\gamma x e^{\beta x}}{1 + \gamma e^{\beta x}}$
Logistic-product $V = \dfrac{\alpha}{1 + \gamma x^\beta}$	$\left(1 - \dfrac{V}{\alpha}\right) = -\dfrac{\beta \gamma x^\beta}{1 + \gamma x^\beta}$

Source: Adapted from Manheim (1979).

relatively small (as is the case for the peak-period operations of many rail transit systems). On the other hand, for a poor level of service, a fare increase would probably cause a significant drop in demand.

It has been determined that the overall value of demand elasticity with respect to rail transit fares is much lower than that for bus transit, and suburban bus transit shows higher fare elasticity than bus service. Also, demand-fare elasticities for short trips are likely to exceed those of long trips by a factor of 2. In most cases, the magnitude of demand elasticity for fare decrease is lower than that for fare increase. The elasticities of demand with respect to transportation service attributes (such as travel time) generally exceed those with respect to trip price, and long-run service elasticities typically exceed those of the short run.

Elasticities can be classified by the method of computation, source of the elasticity, and relative direction of response. These are discussed in the next sections.

3.4.1 Classification of Elasticities by the Method of Computation

Two elasticity computation methods can be illustrated using Figure 3.5.

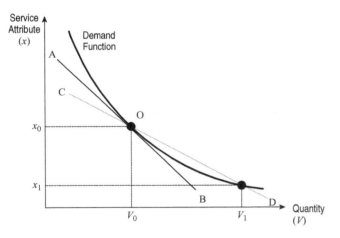

Figure 3.5 Point and arc elasticities.

Point elasticity, expressed as equation (3.3), is proportional to the slope of the tangent (AOB) to the demand curve at (x_0, V_0), where V is the quantity demanded and x is the attribute of the transportation system, such as the out-of-pocket costs associated with a trip.

Arc elasticity, on the other hand, is computed over the arc between (x_0, V_0) and (x_1, V_1) and is proportional to the slope of the line (COD). It is expressed as

$$e_x(V) = \frac{\Delta V/V}{\Delta x/x} = \frac{\Delta V x}{\Delta x V} = \frac{(V_1 - V_0)(x_1 + x_0)/2}{(x_1 - x_0)(V_1 + V_0)/2} \tag{3.4}$$

where V_0 is the quantity demanded when the attribute value is x_0 and V_1 is the quantity demanded when the attribute value is x_1.

It can be seen from the equations above that as Δx approaches zero, the value of arc elasticity becomes equal to that of point elasticity. Typically, specific values of the attribute x and travel demand V can be measured to permit estimation of the arc elasticity, while data for the computation of point elasticities are not so easily available. When the value of elasticity is lower than -1 or greater than 1, the demand is described as being elastic with respect to the attribute (Figure 3.6). However, when elasticity is between -1 and 1, the demand is described as being inelastic or relatively insensitive.

If the demand for a given mode is elastic with respect to the price of travel on that mode, a change in the price is likely to lead to a change in the revenue associated with that mode. This is most readily observed for transit modes and also for highway modes involving a toll. Similarly, significant cross-elasticities across modes influence the level of revenue generated.

Example 3.6: Point Elasticity An aggregate demand function for a rail transit service from a suburb to a downtown area is represented by the equation $V = 500 - 20p^2$, where V is the number of trips made per hour and p is the trip fare. At a certain time when the price was \$1.50, 2000 trips were made. What is the elasticity of demand with respect to price?

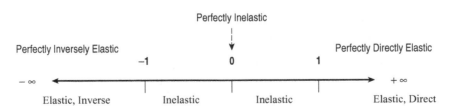

Figure 3.6 Elastic and inelastic regions.

SOLUTION

Point price elasticity, $e_p(V) = \dfrac{\partial V/V}{\partial p/p} = \dfrac{\partial V}{\partial p}\dfrac{p}{V}$

$$= (-20)(2)(1.50)\left(\frac{1.50}{2000}\right)$$

$$= -0.045$$

Example 3.7: Arc Elasticity   Two years ago, the average air fare between two cities was $1000 per trip and 45,000 people made the trip per year. Last year, the average fare was $1200 and 40,000 people made the trip. Assuming no change in other factors affecting trip making (e.g., security, state of the economy), what is the elasticity of demand with respect to the price of travel?

SOLUTION

Arc price elasticity, $e_p(V)$

$$= \frac{\Delta V(p_1 + p_2)/2}{\Delta p(V_1 + V_2)/2}$$

$$= \frac{(40,000 - 45,000)(1,200 + 1,000)/2}{(1200 - 1000)(40,000 + 45,000)/2}$$

$$= -0.647$$

3.4.2 Classification of Elasticities by the Attribute Type

Attributes that affect travel demand include characteristics of the transportation system, such as the price and level of service associated with a given mode, the price and level of service of competing modes, and the characteristics of the socioeconomic system (i.e., income, level of employment, household size, car ownership, etc.). Among these factors, price and income are of particular interest. The elasticities of demand in response to price and income can be termed price elasticity and income elasticity, respectively.

(*a*) *Demand Elasticities with Respect to the Trip Maker's Income*   Transportation planners often seek to predict the impact of changing socioeconomic characteristics on the demand for various modes of transportation. A major indicator of economic trends is income. *Income elasticity* is generally defined as the percent change in travel demand in response to a one percent change in income. In transportation economics, a good or service is considered *normal* if there is a direct relationship between the demand for the transportation service and the income of the consumer (trip maker). Besides, if the demand for a good decreases with increasing income, the good is described as *inferior*. Automobile travel is generally

considered superior, and mass transit is considered to be an inferior good.

(*b*) *Demand Elasticities with Respect to Trip Price*   A study of price elasticities is important because it is often used to assess the impact of the changing prices of a transportation mode (or its rival modes) on the demand for the mode. The level of price elasticity depends on factors such as the price of the rival modes, the income share of the mode, the scope of definition of the mode, and whether the mode is considered a luxury or a necessity.

(*c*) *Demand Elasticities with Respect to Other Attributes*   It is also useful to have knowledge of the elasticity of transportation demand with respect to attributes other than price and income. Such other attributes may include the service reliability of the transportation system and the backgrounds of the system users (for example, household auto availability). This knowledge can help the analyst to make any needed adjustments in future demand in response to changes in such attributes so that more reliable demand predictions can be obtained.

3.4.3 Classification of Elasticities by the Relative Direction of Response: Direct and Cross-Elasticities

Direct elasticity is the effect of the change in an attribute (e.g., price) of a transportation service on the demand for the *same* service. For example, when the transit fare increases, it is likely that transit travel will decrease, depending on the extent of the fare increase. *Cross-elasticity* refers to the effect of a change in an attribute of a transportation service on the demand for an alternative transportation service. Applications of cross-elasticity can be found in the case of substitute services or complementary services. In the case of substitute services, when consumers patronize more of service A in response to an increase in the price of service B, service A is generally described as a perfect substitute for service B. An example is rail freight demand and highway freight demand. An increase in the price of rail transportation causes an increase in the demand for truck transportation. In this case, cross-elasticity is positive. For complementary goods such as auto travel and gasoline, an increase in the price of gasoline results in decreased demand for gasoline and consequently, a decreased demand for auto travel. In this case, the cross-elasticity is negative.

Example 3.8   A 20% increase in downtown parking costs resulted in a 5% reduction in downtown auto trips and a 20% increase in transit patronage for downtown routes. Determine the elasticities of auto and transit demand with respect to parking costs.

SOLUTION Let p_1 and p_2 represent the initial and new parking fee, respectively. A_1 is the auto travel demand before the parking fee increase, A_2 the auto travel demand after the parking fee increase, T_1 the transit travel demand before the parking fee increase, and T_2 the transit travel demand after the parking fee increase. The percent change in auto use with respect to an increase in parking costs is a direct elasticity, while the percent change in transit use with respect to an increase in parking costs is a cross-elasticity:

$$\text{initial parking price} = p_1,$$

$$\text{final parking price} = p_2 = 1.20p_1$$

$$\text{initial transit demand} = V_{T1},$$

$$\text{final transit demand} = V_{T2} = 1.20V_{T1}$$

$$\text{initial auto demand} = V_{A1},$$

$$\text{final auto demand} = V_{A2} = 0.95V_{A1}$$

Arc elasticity of transit demand with respect to parking costs,

$$e_{Tp} = \frac{\Delta V(p_1 + p_2)/2}{\Delta p(V_1 + V_2)/2} = \frac{(V_{T2} - V_{T1})(p_1 + p_2)/2}{(p_2 - p_1)(V_{T1} + V_{T2})/2} = 1$$

Arc elasticity of auto demand with respect to parking costs,

$$e_{Tp} = \frac{\Delta V(p_1 + p_2)/2}{\Delta p(V_1 + V_2)/2} = \frac{(V_{A2} - V_{A1})(p_1 + p_2)/2}{(p_2 - p_1)(V_{A1} + V_{A2})/2}$$
$$= -0.25$$

3.4.4 Examples of Elasticity Values Used in Practice

Demand can be expressed in terms of the amount of travel [vehicle-miles traveled (VMT)], ton-miles of freight, car ownership or vehicle stock, fuel consumption, and so on, and elasticity values have been developed for many of these forms of demand. The concept of elasticities has broad applications in many areas of transportation systems management such as physical changes that increase supply, policy changes that change trip prices and out-of-pocket costs, parking costs, selective road pricing, and so on, as well as changes in the economic environment outside the control of the system planner, such as fuel price changes.

Demand elasticity values may vary by the temporal scope of the analysis and the type of demand measure selected (VTPI, 2006). Short run is typically less than two years, medium run is two to 15 years, and long run is 15 years or more, although these temporal definitions may vary from agency to agency (Litman, 2005). Studies by Button (1993) suggest that long-run elasticities are mostly greater than those of the short run by factors

of 2 to 3. Also, Goodwin et al. (2003) determined that the elasticities of demand expressed in terms of fuel consumption generally exceed elasticities expressed in terms of vehicle travel by factors of 1.5 to 2.

(a) *Demand Elasticity with Respect to General Out-of-Pocket Expenses* Out-of-pocket expenses or the trip price for automobile travel include fuel, road tolls, and parking fees. For transit the trip price includes mainly the fare charged (VTPI, 2006). The elasticity of automobile travel with respect to trip price was found to be -0.23 and -0.28 in the short and long run, respectively (Oum et al., 1992). In another study in Europe (VTPI, 2006), the elasticities for urban peak period travel with respect to trip price were found as follows: -0.384 for automobile and -0.35 for public transit; elasticity values were higher for off-peak travel. Also, elasticity values with respect to out-of-pocket expenses on the basis of automobile trip type are given in Table 3.2.

(b) *Demand Elasticity with Respect to Parking Price* Several studies, such as Clinch and Kelly (2003), Kuzmyak et al. (2003), Pratt (2005), and Vaca and Kuzmyak (2005), provide information on demand elasticities with respect to parking price. Kuzmyak et al. (2003) indicated a range of demand elasticities with respect to parking prices as follows: -0.1 to -0.3, depending on trip type, demographics, location, and other factors. Table 3.3 provides the elasticities of demand for various travel modes with respect to parking price for relatively automobile-oriented urban regions. Hensher and King (2001) determined that a 10% increase in prices at preferred central business district (CBD) parking locations in Sydney, Australia, would cause a 5.41% reduction in demand at those locations, a 3.63% increase in "park-and-ride" trips, a 2.91% increase in public transit trips, and a 4.69% reduction in total CBD trips.

Some researchers have cautioned that the use of parking price elasticities can be misleading, particularly where parking is currently free. It is meaningless to measure a percentage increase from a zero price (VTPI,

Table 3.2 Elasticity of Road Travel with Respect to Out-of-Pocket Expenses

Trip Type	Elasticity
Urban shopping	-2.7 to -3.2
Urban commuting	-0.3 to -2.9
Interurban business	-0.7 to -2.9
Interurban leisure	-0.6 to -2.1

Source: VTPI (2006).

Table 3.3   Demand Elasticities with Respect to Parking Price by Mode

Trip Purpose	Car Driver	Car Passenger	Public Transportation	Walking and Cycling
Commuting	−0.08	+0.02	+0.02	+0.02
Business	−0.02	+0.01	+0.01	+0.01
Education	−0.10	+0.00	+0.00	+0.00
Other	−0.30	+0.04	+0.04	+0.05

Source: TRACE (1999), VTPI (2006).

2006). Policy shifts from free to priced parking typically reduce drive-alone commuting by 10 to 30%, particularly when implemented with improvements in transit service and ride-share programs and other TDM strategies (Litman, 2005).

(*c*) *Demand Elasticity with Respect to Fuel Price*   Road users generally react to increased fuel prices by reducing the amount of driving (typically in terms of vehicle-miles) in the short run, and by purchasing or leasing more-fuel-efficient vehicles in the long run (VTPI, 2006). On the basis of international studies, Goodwin (1992) estimated elasticity values as −0.15 and −0.3 to −0.5 for the short and long run, respectively. Higher values were found by Dargay (1992), who carried out an analysis separately for fuel price increases and decreases. Johansson and Schipper (1997) estimated a long-run car travel demand elasticity of −0.55 to −0.05 with respect to fuel price. Using U.S. data spanning the early 1980s to the mid-1990s, Agras and Chapman (1999) determined that the short- and long-run elasticities of VMT with respect to fuel price were −0.15 and −0.32, respectively. From country to country, there is some variation in demand elasticity with respect to fuel price (Glaister and Graham, 2000).

Some studies have used, implicitly or explicitly, fuel consumption as a surrogate for travel demand. Dahl and Sterner (1991) estimated the elasticity of fuel consumption with respect to fuel price to be −0.18 in the short run and −1.0 in the long run. DeCicco and Gordon (1993) estimated that the medium-run elasticity of vehicle fuel in the United States ranges from −0.3 to −0.5. Hagler Bailly (1999) established fuel consumption elasticities with respect to fuel price in the short run and long run, with separate estimates for various fuel types and transportation modes (Table 3.4).

(*d*) *Demand Elasticity with Respect to Road Pricing and Tolling*   Short-term toll road price elasticities in Spain

Table 3.4   Estimated Fuel Price Elasticities by Mode and Fuel Type

Mode and Fuel Type	Short-Run Elasticity	Long-Run Elasticity
Road gasoline	−0.10 to −0.20	−0.40 to −0.80
Road diesel truck	−0.05 to −0.15	−0.20 to −0.60
Road diesel bus	−0.05 to −0.15	−0.20 to −0.45
Road propane	−0.10 to −0.20	−0.40 to −0.80
Road compressed natural gas	−0.10 to −0.20	−0.40 to −0.80
Rail diesel	−0.05 to −0.15	−0.15 to −0.80
Aviation turbo	−0.05 to −0.15	−0.20 to −0.45
Aviation gasoline	−0.10 to −0.20	−0.20 to −0.45
Marine diesel	−0.02 to −0.10	−0.20 to −0.45

Source: Hagler Bailly (1999), VTPI (2006).

range from −0.21 to −0.83 (Matas and Raymond, 2003.) Litman (2003) reported that the recent congestion pricing fee in downtown London during weekdays led to a 38% and an 18% reduction in private automobile and other traffic (buses, taxis, and trucks), respectively, in that area. Luk (1999) estimated that toll elasticities in Singapore range from −0.19 to −0.58 (average of −0.34).

(*e*) *Demand Elasticity with Respect to Travel Time*   Goodwin (1992) estimated that the elasticity of vehicle travel demand at urban roads with respect to travel time is −0.27 and −0.57 in the short and long run, respectively (the values for rural roads were −0.67 and −1.33, respectively). The elasticities of demand with respect to auto travel times, by trip type and mode, are summarized in Table 3.5. These are long-term elasticities in areas of high vehicle ownership: over 0.45 vehicle per person (TRACE, 1999). Also, demand elasticities with respect to travel time were presented by SACTRA (1994) and separately for auto and

Table 3.5   Elasticity of Demand with Respect to Travel Time by Mode and Trip Purpose

Mode/ Purpose	Auto Driver	Auto Passenger	Public Transport	Walking and Cycling
Commuting	−0.96	−1.02	+0.70	+0.50
Business	−0.12	−2.37	+1.05	+0.94
Education	−0.78	−0.25	+0.03	+0.03
Other	−0.83	−0.52	+0.27	+0.21
Total	−0.76	−0.60	+0.39	+0.19

Source: TRACE (1999), VTPI (2006).

bus in-vehicle time, and for transit-related walking and waiting times, by Booz Allen Hamilton (2003).

(*f*) *Demand Elasticity with Respect to Generalized Travel Costs* The generalized cost of transportation can include the costs associated with travel time, safety, vehicle ownership and operation, fuel taxes, tolls, transit fares, and parking, among others (VTPI, 2006). NHI (1995) provides an elasticity of demand of −0.5 with respect to the generalized cost. Booz Allen Hamilton (2003) estimated the elasticity of demand with respect to the generalized cost of travel in the Canberra, Australia region by time of day: −0.87 for peak, −1.18 for off-peak, and—1.02 overall (peak and off-peak combined). In the United Kingdom, TRL (2004) estimated generalized cost elasticities as follows: 0.4 to −1.7 for urban bus transit, −1.85 for London underground, and −0.6 to −2.0 for rail transport. Lee (2000) estimated the elasticity of vehicle travel demand with respect to generalized cost (fuel, vehicle wear and mileage-related ownership costs, tolls, parking fees and travel time, etc.) as follows: −0.5 to −1.0 in the short run and −1.0 to −2.0 in the long run.

(*g*) *Transit Elasticities* The elasticity of demand with respect to transit fare (Pham and Linsalata, 1991) is generally higher for small cities than for large cities and is also higher for off-peak hours (Table 3.6). Similar values were obtained by TRL (2004), which estimated that (1) metro rail fare elasticities were −0.3 in the short run and −0.6 in the long run; (2) bus fare elasticities were approximately −0.4 in the short run, −0.56 in the medium run, and 1.0 over the long run; and (3) bus fare elasticities were relatively low (−0.24) in the peak period compared to the off-peak period (−0.51).

Kain and Liu (1999) summarized transit demand elasticity estimates from previous studies and determined

Table 3.6 Transit Elasticities by Time of Day and City Size

	Large Cities (more than 1 million population)	Smaller Cities (less than 1 million population)
Average for all hours	−0.36	−0.43
Peak hour	−0.18	−0.27
Off-peak	−0.39	−0.46
Off-peak average	−0.42	
Peak-hour average	−0.23	

Source: Pham and Linsalata (1991), VTPI (2006).

the elasticity values with respect to various attributes as follows: regional employment, 0.25; central city population, 0.61; service (transit vehicle miles), 0.71; and fare, −0.32. For example, a 10% increase in fare would be expected to decrease ridership by 3.2%, all other factors remaining the same.

(*h*) *Freight Elasticities* In a study in Denmark, the price elasticity of highway freight demand was found to be as follows (Bjorn, 1999):

- *Freight volume* (in terms of tonnage distance): −0.47
- *Freight traffic* (in terms of truck trip distance): −0.81

In response to increases in highway freight prices, shippers may utilize existing truck capacity more efficiently or may shift to rail freight modes (Litman, 2005). For freight transportation by rail and road, Hagler Bailly (1999) established the long-run elasticity of demand with respect to price as −0.4, but could be lower or higher, depending on the freight type and other factors. Small and Winston (1999) reviewed various estimates of freight elasticities, a summary of which is provided in Table 3.7.

(*i*) *Final Comments on Elasticity* The value of travel elasticity to be used in any situation depends on the characteristics of the area, the existing level of demand, the trip type, the existing level of the elasticity attribute, the location, and other factors. For example, transit-dependent individuals are generally less sensitive to changes in trip price or other transit service attributes. Litman (2005) found that as the per capita income, drivers, vehicles, and transport options increase, the transit elasticities are likely to increase. Also, in using elasticity values for demand analysis, analysts must consider conditions under which the elasticity values were developed. Elasticity values that are from studies performed many decades ago may be misleading in the current time. For transit demand analysis, for instance, it should be realized that real incomes have increased over the years, and a relatively smaller percentage of the population is transit dependent. Furthermore, the temporal lag of the response must be given due consideration. For example, Dargay and Gately (1997) state that approximately 30% of the response to a price change takes place within one year, and virtually 100% takes place within 13 years.

The common practice of using static rather than dynamic elasticity values overestimates welfare losses from increased user prices and congestion because it ignores society's ability to respond to changes over time (Dargay and Goodwin, 1995). Static elasticities skew investments toward increasing highway capacity and undervalue transit, TDM, and "no build" transportation

Table 3.7   Freight Transportation Elasticities with Respect to Price and Transit Time

Model Type	Attribute	Rail	Truck
Aggregate mode split model	Price	−0.25 to −0.35	−0.25 to −0.35
	Transit Time	−0.3 to −0.7	−0.3 to −0.7
Aggregate model, cost function	Price	−0.37 to −1.16	−0.58 to −1.81
Disaggregate mode choice model	Price	−0.08 to −2.68	−0.04 to −2.97
	Transit time	−0.07 to −2.33	−0.15 to −0.69

Source: Small and Winston (1999), VTPI (2000).

alternatives (Litman, 2005). Evidence of the variation of travel demand elasticities across nations is found in a study by the World Bank (1990) that published values of price elasticities of travel demand in several developing and developed countries.

3.4.5 Application of the Elasticity Concept: Demand Estimation

Elasticity-based demand models help estimate the new demand levels for a particular transportation mode in response to implementation of service attribute changes, such as trip cost increases and travel-time decreases. For this, it is assumed that the preimplementation demand level is known.

(*a*) *Nonlinear Demand Function*   For a demand function of the form $V = kx^a$, where x is an activity or transportation system attribute, the elasticity of demand with respect to the attribute x can be calculated on the basis of two data points (x_1, V_1), and (x_2, V_2) as

$$e_x = a = \frac{\log V_1 - \log V_2}{\log x_1 - \log x_2}$$

The new demand, V_{new}, corresponding to a change in the attribute x, can therefore be estimated as

$$V_{new} = V_1 \left(\frac{x_{new}}{x_1} \right)^{e_x} \qquad (3.5)$$

(*b*) *Linear Demand Function*   A variation to this method of demand estimation is when the demand function is assumed to be linear over the range of interest. In this case, the elasticity can be determined using equation (3.3):

$$e_x = \frac{\partial V/V}{\partial x/x} = \frac{\Delta V/V}{\Delta x/x}$$

$e_x = (\Delta V/V_1)/(\Delta x/x_1)$ when x_1 is used as a base point, and $e_x = (\Delta V/V_2)/(\Delta x/x_2)$ when x_2 is used as a base point. Clearly, the value of elasticity will depend on

which coordinate is used as a base point. If coordinate (x_k, V_k) is used as the base point, the new demand (V_{new}) corresponding to a change in the attribute x can be estimated using equation (3.5):

$$V_{new} = V_k \left(1 + e_x \frac{x_{new} - x_k}{x_k} \right) \qquad (3.6)$$

Example 3.9   A commuter system involves two modes to the downtown area: rail transit and bus transit. When the average bus travel times are 2 and 2.5 hours, respectively, bus riderships are 7500 and 5000, respectively. A new high-occupancy-vehicle lane is being evaluated for implementation, and it is expected that this would reduce the bus travel time to 1 hour from the existing travel time of 2 hours. What is the expected demand of bus transit after the project is implemented assuming (a) a linear demand function and (b) a nonlinear demand function?

SOLUTION   (a) (x_1, V_1) is (2, 7500), and (x_2, V_2) is (2.5, 5000). Assuming that the demand function is linear over the range of interest, the elasticity of demand with respect to travel time can be calculated as follows:

$$e_x = \frac{\partial V/V}{\partial x/x} = \frac{\Delta V/V}{\Delta x/x}$$

$$= \frac{V_2 - V_1}{x_2 - x_1} \frac{x_1}{V_1} \qquad [\text{using}(x_1, V_1) \text{as the base point}]$$

$$= \left(\frac{5000 - 7500}{2.5 - 2.0} \right) \left(\frac{2.0}{7500} \right) = -1.33$$

$$e_x = \frac{\partial V/V}{\partial x/x} = \frac{\Delta V/V}{\Delta x/x}$$

$$= \frac{V_2 - V_1}{x_2 - x_1} \frac{x_2}{V_2} \qquad [\text{using}(x_2, V_2) \text{as the base point}]$$

$$= \left(\frac{5000 - 7500}{2.5 - 2.0} \right) \left(\frac{2.5}{5000} \right) = -2.5$$

Therefore, the new demand can be calculated using equation (3.6) as follows:

$$V_{new} = V_k \left(1 + e_x \frac{x_{new} - x_k}{x_k}\right) = (7500)\left(1 - 1.33\frac{1-2}{2}\right)$$

$$= 12,487 \qquad \text{[using an elasticity value of} -1.33]$$

$$V_{new} = V_k \left(1 + e_x \frac{x_{new} - x_k}{x_k}\right) = (7500)\left(1 - 2.5\frac{1-2}{2}\right)$$

$$= 16,875 \qquad \text{[using an elasticity value of} -2.5]$$

(b) Assuming a nonlinear demand function, the elasticity can be calculated as follows:

$$e_x = a = \frac{\log(V_1/V_2)}{\log(x_1/x_2)}$$

$$= \frac{\log(7500/5000)}{\log(2/2.5)} = \frac{0.1761}{-0.0969} = -1.82$$

In Example 3.9, where the bus travel time is reduced from 2 hours to 1 hour, the new demand (bus ridership) can be calculated using equation (3.5) as follows:

$$V_{new} = V_1 \left(\frac{x_{new}}{x_1}\right)^{e_x} = (7500)\left(\frac{1}{2}\right)^{-1.82} = 26,481$$

Therefore, assuming a nonlinear demand function, it is estimated that the bus ridership will increase' by 253% if the travel time is reduced by 50%.

It has often been cautioned that demand estimation using elasticity-based models are prone to *aggregation bias* because elasticities are typically computed from aggregate data with little segmentation. Also, there are issues of the transferability of models from one area to another, as the elasticity of individual travelers actually depends on the specific characteristics of the activities and transportation systems at each area. Also, the elasticities assume that all other factors besides the factor in question are constant (which may be true only in the short run); therefore, the elasticity-based method may be unsuitable for long-term demand predictions. Furthermore, demand estimation based on elasticity models typically assumes that elasticities are constant or that demand is linear: Both assumptions may be valid only for small changes in the system attributes.

3.4.6 Consumer Surplus and Latent Demand

Analysis of the impact of changes in the market price of a transportation service helps establish whether the consumer's position is better or worse. Such traditional analysis fails to quantify changes in consumer satisfaction due to these price changes. One method used to address this gap is the use of a concept known as *consumer surplus*. This method compares the *value* of each unit of a commodity consumed against its *price*. In other words, consumer surplus is the difference between what consumers are willing to pay for a good or service (indicated by the position on the demand curve) and what they actually pay (the market price). For example, for a certain air transportation route where the average traveler pays $600 per trip but would be willing to pay an average of $650 per trip, the consumer surplus is $50. Consumer surplus measures the net welfare that consumers derive from their consumption of goods and services, or the benefits or satisfaction they derive from the exchange of goods. The total consumer surplus is shown by the area under the demand curve and above the ruling market price ($p^* p_w W$) as shown in Figure 3.7.

Consumer surplus or changes in consumer surplus are typically obtained from structural demand estimation, from which estimates of willingness to pay are derived and compared to expenditures. The total value of willingness to pay is the sum of consumer surplus and consumer expenditure.

Maximization of consumer surplus is the maximization of the economic utility of the consumer. The use of the consumer surplus concept is common in the area of the evaluation of transportation systems. In Figure 3.7, the area enclosed by $p^* O V_w W$ represents the *total community benefit of the transportation service*, and the area enclosed by $p_w O V_w W$ represents the *market value of (or total consumer expenditure for) the service*. It can also be observed that travelers between V_w and V^* do not make

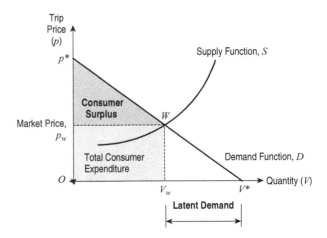

Figure 3.7 Consumer surplus and latent demand.

trips given the prevailing circumstances, but would do so if the price per trip were lower than the equilibrium price. The total of such potential trips is termed *latent demand* (represented by $V^* - V_w$ along the x-axis) and refers to the difference between the maximum possible number of trips and the number of trips that are actually made. An application of the latent demand concept is travel demand management, such as transit fare reduction and other incentives for non–peak hour travel. From Figure 3.7, it is seen that if a zero fare is charged, consumers will demand transit trips up to the point where the demand curve cuts the x-axis.

Figure 3.8 shows how the user impact of a transportation system improvements could be evaluated in terms of consumer surplus by representing such improvement as the resulting area under a transportation demand curve due to a shift in the transportation supply curve. In the figure, the demand curve (as a function of trip price) for a transportation system is depicted by the line D. An improvement in supply, such as increased quantity (e.g., number of guideway lines, highway lanes, transit frequency) or improved quality of service (e.g., increased comfort, safety and security) causes the supply curve to shift from S_{old} to S_{new}. The new consumer surplus is given by the area enclosed by $p^* p_{new} W_{new}$. Thus, change in consumer surplus is represented by the shaded area enclosed by $p_{old} p_{new} W_{new} W_{old}$ and has a magnitude of $(p_{old} - p_{new})(V_{old} + V_{new})/2$.

- *Consumer surplus in cases of perfect elasticity.* When demand for a transportation service is perfectly elastic, the level of consumer surplus is zero since the price that people pay matches the price they are willing to pay. There must be perfect substitutes in the market for this to be the case.

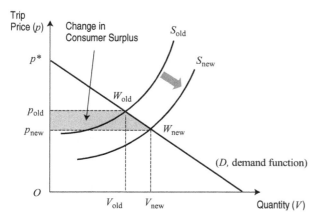

Figure 3.8 Change in consumer surplus.

- *Consumer surplus in cases of perfect inelasticity.* When demand is perfectly inelastic (demand is invariant to changes in price), the amount of consumer surplus is infinite.

Example 3.10 The demand for a transit service between a city and its largest suburb during an off-peak hour, V, is given by $2500 - 350t$ where t is the travel time in minutes. At the current time, the transit trip takes an average of 5 minutes. Determine (a) the time elasticity of demand and (b) the latent demand at this travel time.

SOLUTION

(a) $e_t(V) = \dfrac{\partial V / V}{\partial t / t} = \dfrac{\partial V}{\partial t} \dfrac{t}{V}$

$= (-350) \left[\dfrac{5}{2500 - (350 \times 5)} \right] = -2.3$

(b) At $t = 5$ min, the demand $V = 750$. Therefore, the latent demand is $2500 - 750 = 1750$.

Example 3.11 It is estimated that the demand for a newly constructed parking facility will be related to the price of usage as follows: $V = 1500 - 25P$, where V is the number of vehicles using the parking lot per day and P is the average daily parking fee in dollars. For the first month of operation, parking at the facility is free. (a) How many vehicles would be expected to park at the facility during the first month? (b) After the second month, when a $10 daily fee is charged, how many vehicles would be expected to use the facility, and what would be the loss in consumer surplus?

SOLUTION

(a) During the first month, when $p_1 = 0$, $V_1 = 1500$ vehicles/day.
(b) After the second month, when $p_2 = \$10$, $V_2 = 1500 - (25 \times 10) = 1250$ vehicles/day. Using Figure 3.7, the loss in consumer surplus is given by

$\frac{1}{2} \times (p_2 - p_1)(V_1 + V_2)$

$= (0.5)(10 - 0)(1500 + 1250) = \$13,750$

3.5 EMERGING ISSUES IN TRANSPORTATION DEMAND ESTIMATION

Over the past two decades, increasing availability of detailed travel data has encouraged faster development of disaggregate demand models that seek to predict the travel choices of individual travelers. Developments that have added impetus to such efforts include: (1) the consideration of travelers as *rational* units who seek to maximize their utility associated with the trips they undertake, (2) the quantification of travelers' perceptions of demand and supply, and (3) recognition of the probabilistic nature of travel decisions. Using the disaggregate function directly—given the characteristics of each consumer in the market—the overall demand can be estimated from disaggregate demand models developed for each consumer within each market segment. Further information on disaggregate transportation demand modeling may be obtained from Bhat (2000) and other literature.

Another issue is that of organizational travel demand. Hensher and Button (2000) stated that while demand modeling for passenger travel is important, it is becoming increasingly clear that travel demand by businesses and other organizations needs to be addressed fully. In the past, the latter has received less than the attention deserved because the public sector provided most transportation services, and the purpose of transportation demand modeling had been to allow this component of transportation to interface with users. However, this situation has changed in light of recent and continuing developments, such as deregulation and large-scale privatization. Also, the capacities of transportation networks in the past were defined by peak-volume commuter traffic, but this is no longer the case in the current era.

SUMMARY

An important step in the transportation project development process is the evaluation of alternative policies and regulations for transportation systems operations and use, which depend heavily on transportation demand and supply and interaction between the two parameters. In presenting this material, we recognize that travel demand is not direct but derived, is subject to governmental policies, has a consumption that is unique in time and space, and can be undertaken by several alternative modes that differ by technology, operating and usage policies, and extents of scale economies. We presented a background for transportation demand analysis in the context of transportation supply (or changes thereof). To provide the analyst with some working numbers useful for estimating expected changes in demand in response to changing attributes such as travel time, trip price, income, and parking, we provided recent values of demand elasticities.

EXERCISES

3.1. The demand and supply models for travel between Townsville and Cityburg during a particular season are represented by the following equations:

Demand function:

$$V = 4200 - 29p$$

Supply function:

$$p = 3.10 + 0.02V$$

where V is the number of tickets purchased per month and p is the price of a ticket in dollars. Provide a graphical illustration of the supply and demand functions, and determine the equilibrium demand and price.

3.2. The aggregate demand for a bus transit service serving a newly developed suburban area is represented by the equation $V = 300 - 40p^2$, where V is the number of trips made per month and p is the average price of a ticket for the trip. In a given month, the average price was \$0.75. What is the point elasticity of demand of the bus transit service with respect to price?

3.3. A $w\%$ increase in downtown parking costs resulted in a $f\%$ reduction on downtown auto trips and a $g\%$ increase in transit patronage for downtown routes. Derive expressions for the arc elasticities of auto and transit demand with respect to parking costs.

3.4. The number of automobile trips per hour (V) between two midwestern cities has the following function:

$$V = aT_A^{-2.0}T_T^{0.15}C_A^{-0.5}C_T^{0.6}$$

where T_A and T_T are the travel time for auto and transit, respectively; C_A and C_T are the out-of-pocket costs for auto and transit, respectively; and a is a constant that reflects the size and average income of the population.

(a) At the current time, there are 50,000 automobile trips between the cities every day. If a new parking policy results in an increase of out-of-pocket auto costs from \$5 to \$6, what will be the change in demand?

(b) In addition to part (a), if transit facility improvements lead to a reduction in transit time from 1 hour to 45 minutes, what would be the new demand for automobile travel between the two cities?

3.5. The Kraft demand model can be expressed in the following general form:

$$k \prod_{i=1}^{n} X_i^{c_i}$$

where X is a vector of variables representing the socioeconomic system (such as population and income) and the transportation system (such as travel costs and time) and n is the number of variables. Show that for any variable in the Kraft model, the elasticity of travel demand with respect to each variable is constant.

3.6. For input in evaluation of an improvement project for rail service between cities A and B in a certain state, it is desired to determine the volume of demand. The intercity travel demand is given by the following demand function:

$$Q_{ijm} = 28\text{POP}_i^{0.81}\text{POP}_j^{1.24}\text{PCI}_i^{1.5}\text{PCI}_j^{1.75}\text{PCR}_i^{-0.62}$$
$$\times \text{PCR}_j^{-0.87}\text{RTT}_m^{-1.85}\text{BTT}^{-0.90}\text{RTC}_m^{-2.97}\text{BTC}^{0.57}$$

where POP = average population of the city (millions)

PCI = average per capita income of the city (tens of thousands)

PCR = share of retail in total employment in the city (a fraction)

RTT_m = travel time by mode m relative to the travel time of the fastest mode

BTT = travel time by the fastest mode (min)

RTC_m = travel cost by mode m relative to the cost of the cheapest mode

BTC = travel cost by the cheapest mode (cents)

Estimate the expected level of demand for the rail facility given the following post-implementation data:

City A: population = 1.2 million, average per capita income = \$37,900, share of retail in total employment = 20%

City B: population = 0.8 million, average per capita income = \$45,000, share of retail in total employment = 15%

Table EX3.7.1   Input Information for Exercise 3.7

(a) Dependent Variables in Regression Models

Zone	Cars X_1	Households X_2	Employment X_3	Commercial Area (Acres) X_4
1	370	235	880	5230
2	220	180	495	1200
3	190	136	300	550

(b) Travel Time (min) (2000)

From Zone:	To Zone: 1	2	3
1	10	25	40
2	27	12	29
3	45	24	13

(c) Expected Travel Time (min) (2020)

From Zone:	To Zone: 1	2	3
1	12	28	42
2	29	15	34
3	46	27	16

(d) Trip Interchange Matrix (2000)

From Zone:	To Zone: 1	2	3	P_i
1	680	256	135	1071
2	383	200	121	704
3	210	211	156	577
A_j	1273	667	412	2352

Table EX3.7.2 Travel Times and Travel Costs for Auto and Transit: 2020

(a) Travel Time (Travel Costs) by Auto

Origin	Destination		
	1	2	3
1	12 ($4.5)	28 ($2.9)	42 ($3.5)
2	29 ($5.3)	15 ($2.5)	34 ($4.1)
3	46 ($3.6)	27 ($3.4)	16 ($2.3)

(b) Travel Time (Travel Costs) by Transit

Origin	Destination		
	1	2	3
1	15 ($3.0)	35 ($1.8)	52 ($2.2)
2	38 ($4.5)	22 ($1.1)	40 ($2.7)
3	55 ($2.8)	35 ($2.3)	24 ($1.4)

Expected travel time by rail upon improvement

= 35 minutes

Travel time by fastest mode(auto)

= 28 minutes

Expected travel cost of rail upon improvement

= 75 cents

Travel cost by cheapest mode(bus transit)

= 65 cents

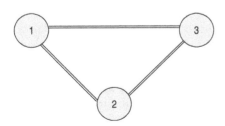

Figure EX3.7

3.7. Use the four-step travel demand modeling procedure to calculate the travel demand on the three links in a three-zone transportation network shown in Figure EX3.7. Use the information in Table EX3.7.1. The following horizon year (2020) trip production and attraction models are given:

$$P_i = 10 + 2.2x_1 + 1.3x_2$$
$$A_j = 30 + 1.27x_3 + 0.035x_4$$

The impedance function is given as $t_{ij}^{-0.5}$, where t_{ij} is the travel time between zones i and j. The calibrated utility functions for auto and transit are given as (time in minutes and cost in dollars)

$$U_{auto} = 3.45 - 0.8\text{Cost} - 0.025\text{Time}$$
$$U_{transit} = 1.90 - 0.26\text{Cost} - 0.028\text{Time}$$

The expected travel times and travel cost for auto and transit in 2020 are given in Table EX3.7.2.

REFERENCES[1]

Agras, J., Chapman, D. (1999). The Kyoto Protocol, CAFE standards, and gasoline taxes, *Contemp. Econ. Policy*, Vol. 17, No. 3.

Bhat, C. R. (2000). Flexible model structures for discrete choice analysis, in *Handbook of Transport Modeling*, ed. Hensher, D. A., Button, K. J., Pergamon Press, Amsterdam, The Netherlands.

Bjorner, T. B. (1999). Environmental benefits from better freight management: freight traffic in a VAR model. *Transp. Res. D.*, Vol. 4, No. 1. pp 45–64.

Booz Allen Hamilton (2003). *ACT Transport Demand Elasticities Study*, Canberra Department of Urban Services, Canberra, Australia, ww.actpla.act.gov.au/plandev/transport/ACTElasticityStudy_FinalReport.pdf.

Button, K. (1993). *Transport Economics*, 2nd ed., Edward Elgar, Aldershot, UK.

[1]References marked with an asterisk can also serve as useful resources for demand estimation.

*Cambridge Systematics (1996). *A Guidebook for Forecasting Freight Transportation Demand*, NCHRP Rep. 388, Transportation Research Board, National Research Council, Washington, DC.

Clinch, P. J., Kelly, A. (2003). *Temporal Variance of Revealed Preference On-Street Parking Price Elasticity*, Department of Environmental Studies, University College, Dublin, Ireland, www.environmentaleconomics.net.

CSI-BLA (1998). *Economic Impacts of U.S. 31 Corridor Improvements*, Cambridge Systematics, Inc., Bernadin, Lochmueller and Associates, Inc., for the Indiana Department of Transportation, Indianapolis, IN.

Dahl, C., Sterner, T. (1991). Analyzing gasoline demand elasticities: a survey, *Energy Econ.*, Vol. 13, pp. 203–210.

Dargay, J. (1992). Demand elasticities, *J. Transp. Econ.* Vol. 26, No. 2, p. 89.

Dargay, J., Gately, D. (1997). Demand for transportation fuels: imperfect price-reversibility? *Transp. Res. B*, Vol. 31, No. 1, pp. 71–82.

Dargay, J. M., Goodwin, P. B. (1995). Evaluation of consumer surplus with dynamic demand, *J. Transp. Econ. Policy*, Vol. 29, No. 2, pp. 179–193.

DeCicco, J., Gordon, D. (1993). *Steering with Prices: Fuel and Vehicle Taxation and Market Incentives for Higher Fuel Economy*, American Council for an Energy-Efficient Economy, Washington, DC, www.aceee.org.

Glaister, S., Graham, D. (2000). *The Effect of Fuel Prices on Motorists*, AA Motoring Policy Unit and the UK Petroleum Industry Association, London, http://195.167.162.28/policyviews/pdf/effect_fuel_prices.pdf.

Goodwin, P. (1992). Review of new demand elasticities with special reference to short and long run effects of price changes, *J. Transp. Econ.*, Vol. 26, No. 2, pp. 155–171.

Goodwin, P., Dargay, J., Hanly, M. (2003). *Elasticities of Road Traffic and Fuel Consumption with Respect to Price and Income: A Review*, ESRC Transport Studies Unit, University College, London, www.transport.ucl.ac.uk.

Hagler Bailly (1999). *Potential for Fuel Taxes to Reduce Greenhouse Gas Emissions from Transport*, transportation table of the Canadian national climate change process, www.tc.gc.ca/Envaffairs/subgroups1/fuel_tax/study1/final_Report/Final_Report.htm.

*Hensher, D. A., Button, K. J. (2000). *Handbook of Transport Modeling*, Pergamon Press, Amsterdam, The Netherlands.

Hensher, D., King, J. (2001). Parking demand and responsiveness to supply, price, location in Sydney Central Business District, *Transp. Res. A*, Vol. 35 No. 3. pp. 177–196.

*ITE (2003). *Trip Generation Handbook*, 2nd ed., Institute of Transportation Engineers, Washington, DC.

Johansson, O., Schipper, L. (1997). Measuring the long-run fuel demand for cars, *J. Transp. Econ. Policy*, Vol. 31, No. 3, pp. 277–292.

Kain, J. F., Liu, Z. (1999). Secrets of success, *Transp. Res. A*, Vol. 33, No. 7–8, pp. 601–624.

*Kanafani, A. (1983). *Transportation Demand Analysis*, McGraw-Hill, New York.

Kuzmyak, R. J., Weinberger, R., Levinson, H. S. (2003). *Parking Management and Supply: Traveler Response to Transport System Changes*, Chap. 18, Rep. 95, Transit Cooperative Research Program, Transportation Research Board, National Research Council, Washington, DC.

Lee, D. (2000). *Demand Elasticities for Highway Travel*, HERS Tech. Doc. Federal Highway Administration, U.S. Department of Transportation, www.fhwa.dot.gov.

Litman, T. (2003). *London Congestion Pricing. Implication for Other Cities*. Victoria Transport Policy Institute, Victoria, BC, Litman, T. (2004). Transit price elasticities and cost elasticities, *J. Public Transp.*, Vol. 7, No. 2, pp. 37–58.

Litman, T. (2005). *Transportation Elasticities: How Prices and Other Factors Affect Travel Behavior*, Victoria Transport Policy Institute, Victoria, BC, Canada.

Luk, J. Y. K. (1999). Electronic road pricing in Singapore, *Road Transp. Res.*, Vol. 8, No. 4, pp. 28–30.

Manheim, M. L. (1979). *Fundamentals of Transportation Systems Analysis*, Vol. 1, *Basic Concepts*, MIT Press, Cambridge, MA.

Matas, A., Raymond, J. (2003). Demand elasticity on tolled motorways, *J. Transp. Stat.*, Vol. 6, No. 2–3, pp. 91–108, www.bts.gov.

McCarthy, P. S. (2001). *Transportation Economics: Theory and Practice—A Case Study Approach*, Blackwell Publishers, Malden, MA.

McNally, M. (2000). The activity-based approach, in *Handbook of Transport Modeling*, ed. Hensher, D. A., Button, K. J., Pergamon Press, Amsterdam, The Netherlands.

*Meyer, M. D., Miller, E. J. (2001). *Urban Transportation Planning: A Decision-Oriented Approach*, McGraw-Hill, Boston, MA.

NHI (1995). *Estimating the Impacts of Urban Transportation Alternatives*, Participant's Notebook, NHI Course 15257, National Highway Institute, Federal Highway Administration, U.S. Department of Transportation, Washington, DC.

Oum, T. H., Waters, W. G., and Yong, J. S. (1992). Concepts of price elasticities of transport demand and recent empirical estimates. *J. Transp. Econ.* Policy, Vol. 26, pp. 139–154.

Pham, L., Linsalata, J. (1991). *Effects of Fare Changes on Bus Ridership*, American Public Transit Association, Washington, DC, www.apta.com.

*Pratt, R. (2005). *Parking Pricing and Fees—Traveler Response to Transportation System Changes*. Transit Cooperative Research Program Report Nr. 95, Transportation Research Board, Washington, DC.

SACTRA (1994). *Trunk Roads and the Generation of Traffic*, Standing Advisory Committee on Trunk Road Assessment, UK Department of Transportation, HMSO, London, www.roads.detr.gov.uk/roadnetwork).

Small, K., Winston, C. (1999). The demand for transportation: models and applications, in *Essays in Transportation Economics and Policy*, Brookings Institute, Washington, DC.

TRACE (1999). Costs of private road travel and their effects on demand, including short and long term elasticities, in *Elasticity Handbook: Elasticities for Prototypical Contexts, TRACE*, prepared for the European Commission, Directorate-General for Transport, Contract RO-97-SC.2035, www.cordis.lu/transport/src/tracerep.htm.

*TRL (2004). *The Demand for Public Transit: A Practical Guide*, Rep. TRL 593 Transportation Research Laboratory Berkshire, UK. (www.trl.co.uk), available at www.demandforpublictransport.co.uk.

Vaca, E., Kuzmyak, J. R. (2005). Parking pricing and fees, Chap. 13 in TCRP Rep. 95, Transit Cooperative Research Program, Transportation Research Board, National Research Council, Washington, DC.

*VTPI (2006). Transportation elasticities, in *Travel Demand Management Encyclopedia*, Victoria Transport Policy Institute, Victoria, BC, Canada. www.vtpi.org/elasticities.pdf. Accessed Jan. 2006.

Washington S., Karlaftis, M., Mannering, F. L. (2003). *Statistical and Econometric Methods for Transportation Data Analysis*, Chapman & Hall/CRC, Boca Raton, FL.

Wooldridge, J. M. (2000). *Introductory Econometrics: A Modern Approach,* South Western College Publishing, Cincinnati, OH.

*World Bank (1990). *A Survey of Recent Estimates of Price Elasticities of Travel Demand, Policy Planning and Research*, Working Papers, World Bank, Washington, DC, www.worldbank.org/transport/publicat/inu-70.pdf.

ADDITIONAL RESOURCES

BTE Transport Elasticities (2005). Database Online (http://dynamic.dotrs.gov.au/bte/tedb/index.cfm) is a comprehensive resource for regularly updated international literature on transportation elasticities.

Chan, Y. (1979). *Review and Compilation of Demand Forecasting Experiences: An Aggregation of Estimation Procedures*, Working Paper, U.S. Department of Transportation, Washington, DC.

Memmott, F. (1983). *Application of Freight Demand Forecasting Techniques*, NCHRP Rep. 283, National Research Council, Washington, DC. Includes a user manual for predicting freight demand using freight generation and distribution, mode choice, and route assignment.

Spielberg, F. (1996). *Demand Forecasting for Rural Passenger Transportation*, Transit Cooperative Research Program Project B-03, Transportation Research Board, National Research Council, Washington, DC. Includes a workbook and spreadsheet template for estimating demand for rural passenger transportation.

U.S. Bureau of the Census (yearly). *Annual Commodity Flow Surveys and Database*. Provides data on freight shipments by industry.

CHAPTER 4

Transportation Costs

Drive thy business or it will drive thee.
                    —*Benjamin Franklin (1706–1790)*

INTRODUCTION

Good decisions at any step of the transportation project development process (PDP) require reliable information on the costs of alternative actions. Each stage of the PDP process involves costs (and benefits) to the agency, facility users, and the community. Certain benefits can be estimated in terms of reductions in user and community costs relative to a given base (typically, do-nothing) alternative.

Transportation costing generally involves estimation of the additional resources needed to increase the quantity or quality of the transportation supply from a given level, and analysts involved in transportation costing often encounter such concepts as economy of scale, price mechanisms, and demand and supply elasticities (McCarthy, 2001).

Typically, the first step in transportation system costing is to describe the physical systems involved and their operations (Wohl and Hendrickson, 1984). The required factors of production (including material, labor, and equipment input), are then identified and their costs determined. Alternatively, an aggregate approach that uses data from several similar past projects can be used to develop average unit costs per facility dimension, usage, or demand. The cost functions and average values presented in this chapter are mostly useful for purposes of sketch planning. For bidding purposes, it is more appropriate to develop precise cost estimates using data from detailed site investigations, engineering designs, and planned policies and operational characteristics of the system.

In this chapter we first present classification systems of the costs encountered in different modes of transportation.

Then the components of agency and user costs are discussed and alternative ways of estimating these costs are presented. We also show how costs can be adjusted to account for differences in implementation time periods, location, and project size (economies of scale). Finally, contemporary costing issues such as cost overruns and vulnerability and risk costs are discussed.

4.1 CLASSIFICATION OF TRANSPORTATION COSTS

Transportation costs may be classified by the source of cost incurrence, the nature of variation with the output, the expression of unit cost, and the point in the facility life cycle at which the cost is incurred.

4.1.1 Classification by the Incurring Party

Transportation costs may be classified by the source of cost incurrence. *Agency costs* are the costs incurred by the transportation facility or service provider; *user costs* are the monetary and nonmonetary costs incurred by the transportation consumers, such as passengers, commuters, shippers, and truckers. Section 4.2 provides a detailed discussion of agency costs. *Community* or *nonuser costs* represent the costs incurred by the community as a whole, including entities not directly involved with use of the facility and are often referred to as secondary costs or *externalities*. Community costs can be nonmonetary (such as disruption of community cohesiveness) or monetary (such as a change in property values). Figure 4.1 shows the various costs categorized by incurring party.

4.1.2 Classification by the Nature of Cost Variation with Output

The costs of transportation systems typically comprise a fixed component, which is relatively insensitive to output volume, and a variable component, which is influenced by output volume, and can be expressed as follows:

$$\text{total cost, TC}(V) = k + f(V)$$

where k is the fixed-cost component (FC), $f(V)$ is the variable-cost component, and V is the output volume.

Agency capital costs can be expressed in terms of the size or number of capacity-enhancing features made available by the proposed project (e.g., the number of lane-miles, line-miles, transit buses or trains); the fixed-cost component comprises the costs of acquiring the right-of-way and relocating or replacing structures and utilities; and the variable-cost component involves cost elements to support the increased operation (e.g., driver and fuel costs for urban bus systems). Agency operating costs are

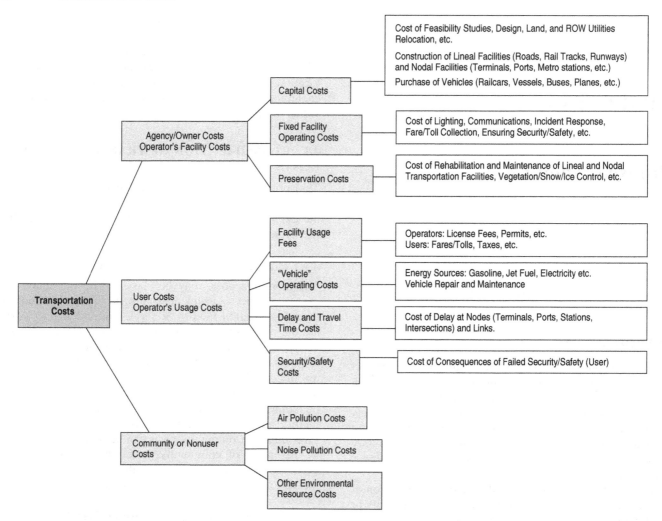

Figure 4.1   Transportation costs categorized by source of cost incurrence.

often more applicable to vehicles using the facility than of the facility itself, which therefore makes these costs a major issue in evaluating transit improvements.

A transportation cost function's mathematical form and the relative magnitudes of its fixed and variable components (variable–fixed cost ratio), and the current output level are all expected to indicate whether or not a transportation system will exhibit scale economies. The variable–fixed cost ratio is in turn influenced by the work scope (construction vs. preservation), the facility dimensions, and the incurring party (facility owner, shipper, or auto user). The ratio is generally low for construction and high for maintenance, low for transportation modes owned and operated by the same entity (such as rail and pipeline transportation), and high for modes where the owner of the fixed asset and the

user/operator are separate entities, such as air, water, and truck transportation. In the last case, the relatively small fixed costs incurred by the operator are those associated with the purchase or lease of vehicles (planes, ships, and trucks) and fixed fees associated with facility use, while the large variable costs arise from fuel use, vehicle maintenance, labor costs, and so on.

4.1.3 Classification by the Expression of Unit Cost

(*a*) *Average Cost*   The average total cost, ATC, is the total cost associated with 1 unit of output. It is calculated as the ratio of the total cost to the output: ATC = TC/V, where TC is the total cost and V is the volume (output). The average fixed cost, AFC, is the fixed cost associated with 1 unit of output and is calculated as the ratio of the fixed cost to the output, $AFC = FC/V$. Similarly, the average variable cost is the cost of 1 unit of output and is calculated as the ratio of the variable cost to the output, $AVC = VC/V$. The concept of average costs is useful in the economic evaluation of transportation system improvements because it helps assess the cost impacts of improvements at a given supply level.

(b) *Marginal Cost* The marginal cost of a transportation good or service is the incremental cost of producing an additional unit of output. The terms of incremental cost, differential cost, and marginal cost have essentially similar meaning but typically are used in contexts that have very subtle differences (Thuesen and Fabrycky, 1964). *Incremental cost* is a small increase in cost. *Differential cost* is the ratio of a small increment of cost to a small increase in production output. Marginal cost analysis is relevant in transportation system evaluation because an agency may seek the incremental cost changes in response to planned or hypothetical production of an additional unit of output with respect to facility construction, preservation, or operations. Marginal cost and average cost can differ significantly. For example, suppose that an agency spends $10 million to build a 10-mile highway and $10.5 million to build a similar 11-mile highway, the average costs are $1 million and $0.954 million, respectively, but the marginal cost of the additional mile is $0.5 million. The expressions related to marginal cost are as follows:

Marginal variable cost:

$$MVC = \frac{\partial VC}{\partial V}$$

Marginal total cost:

$$MTC = \frac{\partial TC}{\partial V} = \frac{\partial FC}{\partial V} + \frac{\partial VC}{\partial V} = \frac{\partial VC}{\partial V} = MVC$$

Like average cost, marginal cost concepts help an agency or shipper to evaluate the cost impacts of various levels of output or the additional cost impact of moving from a certain output level to another.

Example 4.1 A cost function is expressed in the following general form: total cost (TC) $= k + f(V)$, where k is the fixed cost (FC) and $f(V)$ is the variable cost. V is the output. For each of the functional forms shown in Table E4.1 derive expressions for (a) average fixed cost, (b) average variable cost, (c) average total cost, (d) marginal variable cost, and (e) marginal total cost.

SOLUTION   The expressions are shown in Table E4.1.

Example 4.2 The costs of running a metropolitan bus transit system are provided in Table E4.2. Plot the graphs of (a) total cost, variable cost, and fixed costs; (b) average total costs, average variable costs, and average fixed costs; and (c) marginal total costs and average total cost. Show the point at which marginal total cost equals average total cost, and explain the significance of that point.

SOLUTION   The graphs are shown in Figure E4.2. The region on the left of the intersection point (MC < AC) represents scale economies and the region on the right represents scale diseconomies (MC > AC). An agency would prefer to produce goods or provide services in the region where MC < AC. Since revenue is a linear function

Table E4.1   Typical Cost Functions and Expressions for Unit Costs

	$TC = k + aV$ (Linear)	$TC = k + aV^2$ (Quadratic)	$TC = k + ae^V$ (Exponential)	$TC = k + aV^3$ (Cubic)	$TC = k + a \ln V$ (Logarithmic)	$TC = k + ab^V$ (Power)
Average fixed cost = FC/V	k/V	k/V	k/V	k/V	k/V	k/V
Average variable cost = VC(V)/V	a	aV	ae^V/V	aV^2	$(a \log V)/V$	ab^V/V
Average total cost = TC(V)/V	$k/V + a$	$k/V + aV$	$k/V + ae^V/V$	$k/V + aV^2$	$k/V + a \times \log V/V$	$k/V + ab^V/V$
Marginal variable cost = marginal total cost	a	$2aV$	ae^V	$3aV^2$	a/V	$a \ln(b)b^V$

Table E4.2   Transit Agency's Costs

	Annual Ridership (V) in millions							
	1	2	3	4	5	6	7	8
Fixed cost, FC	3	3	3	3	3	3	3	3
Variable cost, VC	1.250	1.375	1.500	1.625	1.750	1.875	2.000	2.125
Total cost, TC	4.250	4.375	4.500	4.625	4.750	4.875	5.000	5.125
Average fixed cost, AFC	0.300	0.150	0.100	0.075	0.060	0.050	0.043	0.038
Average variable cost, AVC	0.125	0.069	0.050	0.041	0.035	0.031	0.029	0.027
Average total cost, AC	0.425	0.219	0.150	0.116	0.095	0.081	0.071	0.064
Marginal variable cost, MVC	—	0.125	0.125	0.125	0.125	0.125	0.125	0.125
Marginal total cost, MC	—	0.125	0.125	0.125	0.125	0.125	0.125	0.125

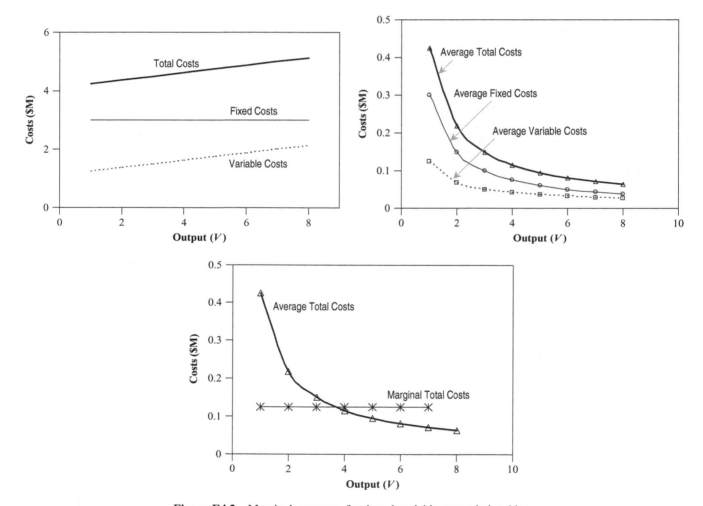

Figure E4.2   Marginal, average, fixed, and variable cost relationships.

of ridership $(R = aV)$, transit agencies are interested in knowing maximum ridership that can be achieved while ensuring that MC is less than or equal to AC. The maximum level of ridership corresponds to the intersection point between the average total cost and the marginal total cost (at that point, revenue is maximized). The generated revenue will most likely not cover the costs incurred by the transit agency. It should be kept in mind, however, that unlike private entities, the primary goal of public agencies is to provide service rather than to maximize profit. As such, for many transit agencies, maximum revenue is less than agency cost and therefore such agencies often operate on subsidies.

Example 4.3 The cost function associated with air shipping operations of a logistics company is Total Cost (in $ millions) $= 1.2 + 150V^2$, where V is the monthly output (volume of goods delivered) in millions of tons. Plot a graph showing the average total cost and marginal total cost.

SOLUTION

The average total cost function is

$$AC = \frac{V}{TC} = \frac{1.2}{V} + 150V$$

The marginal total cost function is

$$MC = \frac{\partial TC}{\partial V} = 300V$$

Plots of these functions are provided in Figure E4.3

(*c*) *General Discussion of the Average and Marginal Cost Concepts* In this section we presented the concepts of

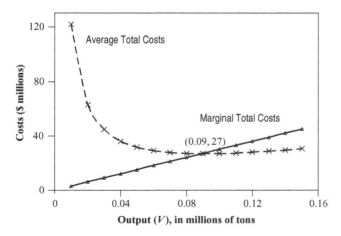

Figure E4.3 Average and marginal cost relationships.

average and marginal effects from a monetary cost perspective. In some transportation problems, the analyst may need to apply these concepts to the consumption of non-monetary resources (e.g. environmental degradation, community disruption) as well as system benefits (e.g., system preservation, congestion mitigation, safety improvement, air quality enhancement).

Another issue is the selection of the appropriate output, V, to be used in the cost analysis. This depends on the transportation mode and the phase of the transportation development process in question. For example, in aggregate costing of rail transit construction, the number of stations and length of the system can be used as output variables. In aggregate costing of rail or airport operations, the number of passenger miles or passenger trips could be used. In the case of highway operations, the traffic volume or vehicle miles of travel could be used. In freight operations costing, ton miles or ton trips could be used.

4.1.4 Classification by Position in the Facility Life Cycle

Life-cycle costs include relevant agency and user costs that occur throughout the life of a transportation asset, including the initial costs. In general, transportation costs over its life cycle may be classified as *initial costs* and as *subsequent costs*. The latter are incurred at later stages of facility life and therefore involve activities such as operations, preservation of the fixed asset or rolling stock, and costs that are associated with salvage or disposal of the physical facility or rolling stock.

4.1.5 Other Classifications of Transportation Costs

Transportation agency costs may also be categorized according to the source of work (activities carried out by an agency's in-house personnel vs. activities let out on contract), the role of the work (activities aimed at preventing deterioration vs. activities geared toward correcting existing defects), or the cycle over which costs are incurred (activities carried out routinely vs. activities carried out at recurrent or periodic intervals).

4.2 TRANSPORTATION AGENCY COSTS

Agency costs refer to the expenditures incurred by the facility owner or operator in providing the transportation service. For fixed assets, agency costs are typically placed into seven major categories: advance planning, preliminary engineering, final design, right-of-way acquisition and preparation, construction, operations, preservation, and maintenance. In some cases, disposal of the fixed asset at the end of its service life involves some costs that are referred to as *salvage costs*. For movable assets (rolling stock), agency costs typically comprise acquisition, vehicle operating preservation, maintenance, and disposal costs.

4.2.1 Agency Costs over the Facility Life Cycle

Several types of agency costs are incurred over the life of a transportation facility. However, not all of these costs may be applicable in a particular evaluation exercise. The analyst must identify the costs that do not vary by transportation alternative and must exclude these costs from the evaluation. Typically, the initial agency costs of planning and preliminary engineering are the same across alternatives. Also, where facility locations have already been decided, location-related expenses, such as right-of-way (ROW) acquisition and preparation, are fixed across alternatives. Furthermore, where it is sought to only evaluate alternative construction practices, preservation strategies, or operational policies, the cost of design can be excluded from the evaluation.

(*a*) *Advance Planning*   These may include the cost of route and location studies, traffic surveys, environmental impact assessments, and public hearings. Advance planning costs are typically estimated as a lump sum based on the price of labor-hours within the transportation agency or from selected consultants. In evaluating alternatives, costs should exclude any costs of advance planning work done prior to development of the alternatives.

(*b*) *Preliminary Engineering*   These may include the costs of carrying out an engineering study of a project, such as geodetic and geotechnical investigations. If some preliminary engineering has been done (especially regarding technical feasibility of competing alternatives), such costs may be excluded from project costs.

(*c*) *Final Design*   These are the costs of preparing engineering plans, working drawings, technical specifications, and other bid documents for the selected design. Final design costs typically are 10 to 20% of construction costs.

(*d*) *Right-of-Way Acquisition and Preparation*   Acquisition costs of ROW land typically include the purchase price, legal costs, title acquisition, and administrative costs of negotiation, condemnation, and settlement. Severance damages are typically significant, and determining the value of remnant acquisitions is often a complex task. In the absence of other information, fees and charges associated with ROW acquisition may be assumed to be 2% of the purchase price. Right-of-way preparation costs include relocation or demolition of structures and utility

relocation. A preliminary estimate of the costs for acquiring and preparing ROW costs can be made by a quick field inventory of the project alignment to determine the volume of structures slated for demolition, and applying the agency's demolition cost rates. For structures that need to be relocated, it is necessary to consider the costs of acquiring new land and reconstructing such structures. The basis for residential relocation payments, including costs of temporary rentals, may be established by existing policy of the transportation agency or government. The relocation cost of existing utility facilities, such as water, gas, telephone, and electricity should be estimated with the assistance of utility companies.

(*e*) *Construction*   At the planning stages, rough approximations of construction cost can be made on the basis of similar past projects. To do this, it may be useful to employ statistical regression to develop such costs as a function of work attributes, location, and so on. Alternatively, the cost may be built up using unit costs of individual constituent work items. Such estimation of transportation project construction costs may seem a relatively easy task but may be complicated by lack of estimating expertise (Dickey and Miller, 1984), a problem that has often led to cost discrepancies in transportation project contracts.

(*f*) *Operations*   These costs may include charges for utility use (e.g., electricity for transit or air terminals, street lighting, and traffic signal systems), safety patrols, traffic surveillance and control centers, ITS initiatives, toll collection, communication equipment, labor, and so on. Given adequate historical data, it may be possible to develop annual operating cost models for estimating future operating costs. Such models are typically a function of facility type and size, age of facility, and level of use.

(*g*) *Preservation and Maintenance*   These are the costs incurred by an agency to ensure that an asset is kept in acceptable physical condition. For a highway agency, for instance, preservation costs include pavement and bridge rehabilitation as well as preventive and routine maintenance, vegetation control, and snow and ice control. Predictions of preservation maintenance costs may be made in the form of simple average cost rates (such as cost per line-mile of rail track or cost per square meter of bridge deck) or statistical models that estimate facility cost as a function of facility dimensions, material type, and other factors.

4.2.2 Techniques for Estimating Agency Costs

Costing of transportation projects and services can generally be carried out in two alternative ways: a disaggregate approach and an aggregate approach. Further details and examples of each approach are provided here.

(a) *Disaggregate Approach* (*Costing Using the Prices of Individual Pay Items or Treatments*) In this approach, the overall cost of an entire project is estimated using the *engineer's estimate* or the contractor's *bid prices* for each specific constituent work activity (also referred to as a *pay item*) of the project. Pay items may be priced in dollars per length, area, or volume, or weight of finished product, and is often reported separately for materials, labor and supervision, and equipment use. This method of costing is more appropriate for projects that have passed the design stage and for which specific quantities of individual pay items are known. It is generally not appropriate for projects whose design details are not yet known.

The use of detailed pay item unit costs for estimating the cost of transportation facilities or services is straightforward but laborious. For a project, there can be several hundreds, sometimes thousands, of pay items that are priced separately. This costing approach typically forms the basis for contract bidding. The first step is the decomposition of a specific work activity (such as rail track installation) into constituent pay items expressed in terms of finished products (such as one linear foot of finished rail guideway) or in terms of specific quantities of material (such as aggregates, concrete, steel beams, formwork), equipment, and labor needed to produce one linear foot of finished guideway. After the various components of the work activity have been identified, a unit price is assigned to them (on the basis of updated historical contract averages or using the engineer's estimates), and the total cost of the work activity is determined by summing up the costs of its constituent pay items. The level of detail of the pay items generally depends on the stage of the transportation project development process at which the cost estimate is being prepared. At the early planning stages, relatively little is known about the prospective design; therefore, the level of identifying the pay items and their costing is quite coarse (Wohl and Hendrickson, 1984). Cost estimators typically refer to four distinct levels of coarseness that reflect the stages at which such estimates are typically required:

1. Conceptual estimate in the planning stage (typically referred to as *predesign estimate* or *approximate estimate*)
2. Preliminary estimate in the design stage (often termed *budget estimate* or *definitive estimate*)
3. Detailed estimate for the final assessment of costs
4. As-built cost estimate that incorporates any cost overruns or underruns

In its coarser form, cost accounting utilizes more aggregated estimates that are for groups of pay items rather than for individual pay items. Average cost values may be used, but a more reliable method would be to develop cost models as a function of facility attributes such as material type, construction type, size, surface or subsurface conditions, and geographical region. There may be other variables, depending on whether the costs being estimated are initial construction costs or whether they are costs incurred over the remaining life of the facility. The time-related variables (such as accumulated environmental and traffic effects) have little or no influence on initial construction cost but significantly affect subsequent costs (i.e., preservation and maintenance costs).

At another level of disaggregation, average cost values and cost models can be developed for each *treatment* (a specific agency activity) that is comprised of multiple pay items or for each pay item.

(b) *Aggregate Approach* (*Costing*) An example of this approach is a model that estimates the overall cost associated with the construction, preservation, or operations per facility output or dimension. In a manner similar to the disaggregate approach, costs developed using the aggregate approach can be in one of two forms:

1. An average rate, where historical costs for each system family are updated to current dollars, averaged, and expressed as a dollar amount per unit output (dimension). *Family* refers to a number of systems placed in one group on the basis of similar characteristics. For example, the estimated average cost of rigid pavement maintenance was determined to be $480/lane-mile per annum (Labi and Sinha, 2003). Average rates may be developed for each subcategory: for example, rigid interstate pavements located in a certain region or certain types of rigid pavements (plain, reinforced, continuously reinforced, etc.).

2. A statistical model, where historical overall costs are modeled as a function of facility characteristics (e.g., facility dimensions, material, construction type, age).

An example of an aggregate cost model (for a heavy rail transit system) is as follows:

$$\text{Unit Cost} = 3.9 \times L^{-0.702} \times U^{1.08} \times ST^{-0.36}$$

where Unit Cost = cost per line-mile-station in $M
L = number of line-model
U = fraction of the system that is underground
ST = number of stations

Also, statistical models can be developed for each subcategory, or differences in subcategories could be included in a broad model as dummy variables. Costs developed using the aggregate approach are typically used for sketch planning and long-range budgeting where the application of specific treatments (and thus their corresponding individual costs) are not known with certainty and only rough approximations of overall costs are sought.

4.2.3 Risk as an Element of Agency Cost

(*a*) *Risk due to Uncertainties in Estimation*   Most cost models that are currently used by transportation agencies treat input variables as deterministic values that do not adequately reflect the uncertainty that actually exists in the real world. Such uncertainty is introduced by factors such as fluctuations in work quality, material and labor prices, climate, etc. (Hastak and Baim, 2001). Risk analysis may be used to address the issue of uncertainty. Risk analysis in transportation costing answers three basic questions about risk (Palisade Corporation, 1997): What are the possible outcomes of cost? What is the probability of each outcome? What are the consequences of decisions based on knowledge of the probability of each outcome? Values of input variables that influence transportation costs are modeled using an appropriate probability distribution that is deemed by the analyst to best fit the data for each variable. Then the expected overall cost outcome is determined. This can be repeated, using Monte Carlo simulation, for several values of the variable within the probability distribution defined.

(*b*) *Risk due to Disasters*   Risk-based transportation costing also involves natural and human-made disasters that can significantly influence the operations and physical structure (and consequently, the costs of physical preservation and operations of such facilities). Natural disasters include floods, earthquakes, and scour, human-made disasters include terrorist attacks and accidental collisions that critically damage transportation infrastructure. The probability of a transportation system failure can be assessed for each vulnerability type. Then the cost of damage or repair in that event can be used to derive a failure cost or vulnerability cost that could be included in the transportation system costing (Chang and Shinozuka, 1996; Hawk, 2003). Vulnerability cost can be defined as follows:

vulnerability cost = probability of disaster occurrence

× cost of damage if the disaster occurs.

Risks are evident in both the probability of the occurrence and the uncertainties of damage cost in the event of disaster. As evidenced from the 2005 Katrina hurricane disaster on the U.S. Gulf coast, estimating the damage cost can be as uncertain as estimating the probability of the event itself.

4.3   TRANSPORTATION USER COSTS

4.3.1 User Cost Categories

User impacts that can be monetized include vehicle operating costs, travel-time costs, and safety costs. Nonuser or community costs (e.g., of air pollution, noise, water pollution, community disruption) are not so easily monetized. Both user and community costs are often related directly to the physical condition as well as the performance of a facility. For example, excessive congestion and poor physical condition of rail lines can translate to high user costs of safety and delay, and high community costs due to noise.

(*a*) *Travel-Time Costs*   Travel time is one of the major items in the evaluation of alternative transportation systems. The cost of travel time is calculated as the product of the amount of travel time and the value of travel time. Methods for assessing the amount and value of travel time (in minutes, hours, etc.) and its monetary value ($ per hour, etc.) are discussed in Chapter 5.

(*b*) *Safety Costs*   The costs of safety can be estimated as either preemptive costs or after-the-fact costs. Preemptive safety costs are incurred mostly by the agency in ensuring that crashes are minimized and may be considered as agency operating costs; after-the-fact safety costs are those incurred by users (through fatality, injury, or vehicle damage), the agency (through damaged facilities such as bridge railings or guardrails), or the community (through damage to abutting property, pedestrian casualties, for example). In Chapter 6 we present unit crash costs and a methodology to estimate safety costs and incremental safety benefits of transportation projects.

(*c*) *Vehicle Operating Costs*   Irrespective of mode, the costs of operating transportation vehicles can be substantial. In Chapter 7 we provide details on VOC components and factors, unit values of VOC, and the methodology for evaluating the impact of transportation system improvements on the operating costs of transportation vehicles.

(*d*) *Noise, Air, and Water Pollution Costs*   Noise, air, and water pollution costs can be estimated in terms of preemptive costs or after-the-fact costs. In any case, there seems to be no universally adopted method of valuation of these effects. Consequently, they are typically not included in economic efficiency analysis of transportation projects but are instead considered in cost-effectiveness framework without monetization.

4.3.2 Impacts of Demand Elasticity, Induced Demand, and Other Exogenous Changes on User Costs

When a transportation system is improved (through enhanced service or physical condition, the resulting decrease in user costs causes a shift of the supply function to the right. This decrease constitutes the user benefits. There are three possible scenarios for which such user benefits can be estimated (Dickey and Miller, 1984): when demand is inelastic, when demand is elastic and there are induced trips, and when demand is elastic and there are generated trips.

The foregoing discussion is presented for a composite user cost but is also applicable to individual user cost types. In some cases where detailed data are unavailable, the analysis of user costs may be simplified by using an overall value for user costs rather than summing up values of individual components of user costs. For each situation, the change in user costs can be calculated using simple geometry: area of a rectangle for Figure 4.2 and area of a trapezoid for Figures 4.3 and 4.4.

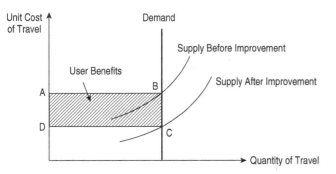

Figure 4.2 Unit user cost when demand is perfectly inelastic.

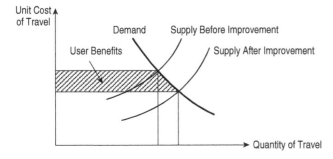

Figure 4.3 Unit user cost when demand is elastic and there are induced trips.

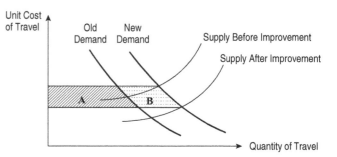

Figure 4.4 Unit user cost when demand is elastic and there are trips generated.

(*a*) *When Demand Is Inelastic*   When demand is inelastic (therefore precluding any induced, generated, or diverted trips), the user benefit occurring from an improved transportation system is taken as the product of the reduction in the unit cost (price) of travel and the number (quantity) of trips (Figure 4.2). For purposes of illustration, a *travel unit* will be taken as a trip. For example, a technological improvement such as electronic tolling that decreases delay and hence reduces the unit cost of each trip would generally cause a downward shift in the supply curve, leading to user benefits. On the contrary, a new transportation policy such as security checks that increases delay (and hence the unit cost of each trip) is reflected by an vertical upward shift in the supply curve (and equilibrium point) indicating negative user benefits in the short run, all other factors remaining the same. In both cases, the number of trips would remain the same because demand is inelastic.

(*b*) *When Demand Is Elastic and There Are Induced Trips*   When demand is elastic, an increase in supply, from classical economic theory, results in lower user cost of transportation and subsequently, increased or induced demand. Thus, the area (shown in Figure 4.3 as user benefits) is trapezoidal in shape and is greater than the rectangular area that corresponds to the product of the unit price reduction and the number of trips. For example, improved transit service through higher service frequency and increased reliability would decrease the user cost of delay. This can be represented by a downward right shift of the supply curve and equilibrium point, all other factors remaining the same. The number of trips and user benefits would increase. On the other hand, an intervention that increases fares would increase the cost of travel, all other factors remaining the same, and would be reflected by an upward left shift of the supply curve and equilibrium point. The number of trips would decrease and the user benefits of such an intervention would be negative.

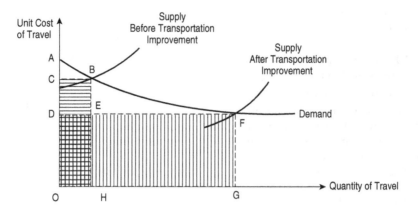

Figure 4.5   Change in total user costs.

(c) When Demand Is Elastic and Trips Are Generated
When demand is elastic and there is a shift in the demand curve (due to increased demand even at the same price), the increase in user benefits (consists of Areas A and B in Figure 4.4) but Area B is due only in part from the improvement. The changes in user benefits for the scenarios discussed above are in response to changes within the transportation system itself, such as nature of demand elasticity, changes in demand (induced or generated), or changes in supply (trip delays, travel times, price). The figures can also help explain the effect of *exogenous changes*, which include:

- Change in prices of VOC components
- Implementation or removal of user subsidies or taxes
- Technological advancements in areas outside (but related to) the transportation system in question

A case in point is the 2005 increase in gasoline prices in the United States. This generally caused an increase in the unit cost of each personal or business trip in the short run. Users with elastic demand reduced their trips, while those with inelastic demand had the same number of trips after the change. In either case, the end result was a negative gain in user benefits. Another example is the user subsidization that typically occurs in some developing countries. When transportation users are subsidized by the government, this lowers the supply curve because the unit cost of each trip is reduced. This leads to increased travel (where demand is elastic) and increased benefits (for either elastic or inelastic demand). The removal of subsidies or the imposition of taxes has the opposite effect.

Transportation projects and services are typically implemented with the objective of lowering congestion, increasing safety, and decreasing travel time—such reductions

in user costs translate into increases in quality of life, business productivity, retention and attraction of investments, increased employment, and so on. However, an increase in transportation supply does not always lead to a decrease in total travel costs. Depending on the shape of the demand and supply functions and the elasticity of demand, a decrease in unit travel costs could lead to a decrease or increase in total user costs (Dickey and Miller, 1984). For example, Figure 4.5 shows that (1) the benefits of the transportation system improvement (the area DCBF) are not necessarily equal to the change in total user costs (the area ODFG − the area OCBH), and (2) the total user costs in this scenario actually increases with the decrease in unit travel costs due to the system improvement (the area represented by rectangle ODFG is much larger than that represented by rectangle OCBH).

4.4 GENERAL STRUCTURE AND BEHAVIOR OF COST FUNCTIONS

A cost function is a mathematical description of the variation of cost with respect to some output variable (typically system dimensions or the level of system use). There are three major aspects of transportation cost functions: the *dependent variable, independent variables* (including the *output dimension*), and the *functional form* of the cost function.

4.4.1 Components of a Transportation Cost Function

(a) Dependent Variable   This is typically the cost of the output, in monetary terms and in a given time period. To adjust for the effects of inflation it is often necessary to express the cost items in constant dollar. For facility construction and improvement projects, construction price indices are used to convert current dollars to

constant dollars, as discussed in Section 4.6.2. Generally, the dependent variable can be a total cost or a unit cost (total cost per unit output). In using the unit cost as the dependent variable, the analyst typically calculates unit costs for each observation (e.g., cost per lane-mile per passenger-mile), or per ton-mile and then develops statistical functions of such costs with respect to output, facility dimensions and/or other characteristics. This approach presupposes that costs are linearly related to the output variable, thus impairing investigation of scale economies. A superior and more flexible approach is to use the total cost as the dependent variable and to use the output variables, among other variables, as the independent variable. Then using calculus, the elasticities of the response variable with respect to each independent variable can be determined, and then the existence and extent of scale economies or diseconomies can be identified.

(*b*) *Independent Variables*   Two types of factors affect cost levels: (1) those related to the output, such as number or frequency of trains or buses, number of trips, tons of material shipped, passenger-miles, vehicle-miles, or ton-miles (these are referred to as *output variables*) and (2) those independent of output, such as spatial location. Output-related variables typically constitute the variable component of a cost function, while the nonoutput variables typically comprise the fixed component. Examples of output variables typically used in cost functions or rates for capital costs of physical transportation infrastructure are shown in Table 4.1.

(*c*) *Functional Form*   Nonlinear functional forms, which include quadratic, cubic, exponential, logarithmic, and power forms, are generally more appropriate than linear forms, as they are capable of accounting for scale economies or diseconomies.

4.4.2 Economies and Diseconomies of Scale

Economy of scale refers to the reduction in average cost per unit increase in output; *diseconomy of scale* refers to the increase in average cost per unit increase in output. Through operational efficiencies (or inefficiencies) or by virtue of inherent features of the facility or its environment, the cost of producing each additional unit may rise or fall as production increases. For a given cost function,

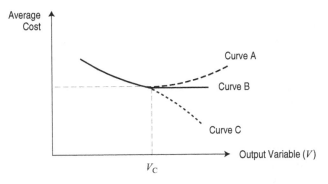

Figure 4.6   Variations of average cost reflecting scale economies and diseconomies.

Table 4.1   Possible Variables for Agency Cost Functions or Rates

	Physical Infrastructure	Operations
Highways	Pavements: cost per lane-mile of new pavement, cost per volume of laid/cast material Bridges: cost per area of new or rehabilitated bridge (measured using deck area)	Congestion/mobility: cost per travel-time reduction, cost per unit resource for incident management Safety: cost per unit reduction in fatal and injury crashes
Bus and rail transit	Cost per bus or railcar, cost per route-mile	Cost per passenger, cost per passenger-mile, cost per revenue vehicle
Rail freight	Track: cost per line-mile Terminals: cost per terminal, cost per floor area (of terminals) Yards: cost per yard area	Cost per passenger, cost per enplanement
Air travel	Cost per area of passenger terminal, cost per runway length, cost per runway area	Cost per ton load of freight, cost per passenger
Marine ports	Cost per area of facility, cost per dock	Cost per passenger-mile, Cost per freight ton-mile

scale economies or diseconomies with respect to any output variable are typically represented by the index of that variable in the cost equation and can be investigated by plotting observed total or average unit cost vs. the output variable (Figure 4.6). Depending on facility type and level of output, the average cost at a certain output level, V_C, may increase (curve A) or decrease (curve C) or may remain the same (curve B).

4.5 HISTORICAL COST VALUES AND MODELS FOR HIGHWAY TRANSPORTATION SYSTEMS

4.5.1 Highway Agency Cost Models

(a) *Cost Models by Improvement Type*   A widely used set of project costs are those developed as part of the Highway Economic Requirements System (Table 4.2). Efforts have been carried out in individual state, provincial, and local highway agencies to derive average costs of capital improvement project types. Wilmot and Cheng (2003) for instance, developed a model to estimate future overall highway construction costs in Louisiana in terms of resource costs (construction labor, materials, and equipment), contract characteristics, and the environment. Also, Labi and Sinha (2003) established average costs for standard pavement preservation treatments and capital improvements in Indiana.

The amounts shown in Table 4.2 are average values, and the cost of a specific project may be less or more than the amount shown, due to such factors as:

- *Number of crossings and ramps (i.e., over water, railway, other highway).* Highway projects with higher numbers of crossings require more bridges, leading to higher overall costs per mile.
- *Right-of-way.* A project that is built within an existing right-of-way has lower unit costs than one that needs additional right-of-way.
- *Environmental impacts.* Projects in environmentally sensitive areas generally have higher unit costs.
- *Existing soil and site conditions.* High variability in soil conditions can translate to higher unit costs.
- *Project size.* Larger projects generally have lower unit costs due to scale economies. For some facilities, however, the need for additional stabilizing structures beyond certain facility dimensions may translate into a greater cost increase per unit increase in dimension, thus reflecting scale diseconomies.
- *Project complexity.* More complex projects typically have higher unit costs.
- *Method of construction delivery.* Projects constructed using traditional contracting processes generally have lower unit costs than those for projects constructed

using alternative processes such as design–build and warranties. It is worth noting, however, that facilities constructed using traditional contracting processes may have higher unit preservation costs over their life cycle.

- *Urban or rural location.* Urban projects generally have higher unit costs than those of their rural counterparts.

Other factors that may affect project costs include the degree of competition for the contract, design standards, labor costs, material and workmanship specifications, and topographic and geotechnical conditions. For the foregoing reasons, comparing or transferring states' construction costs using bid price data should be done with extreme caution. Factors that cause large cost differences should be identified, and unit prices from such contracts may be excluded from the comparison.

(b) *Cost Models for Pay Items and Factors of Production*
A number of state transportation agencies, such as California, Massachusetts, Arizona, Indiana, Texas, and Arkansas, publish their historical transportation construction and maintenance cost data online. In some cases, these data include the prices of individual pay items of the winning bid as well as those of the engineer's estimate. At a national level, pay item data are available through AASHTO's *Trns.prt Estimator*, an interactive Windows-based stand-alone cost estimation system for highway construction. For analysts who are interested in the prices of raw materials, labor, materials, and equipment use, the Federal Highway and Transit Administrations' Web sites, have useful data that track trends in prices. This database is made possible through continual reporting to the FHWA and FTA, cost data from the states that cover key work items and materials. The FHWA publishes bid price data in its quarterly *Price Trends for Federal-Aid Highway Construction* and in its annual *Highway Statistics* series.

4.5.2 Transit Cost Values and Models

Transit agency costs include (1) capital cost items such as land acquisition, construction of tracks (guideways), stations, and ancillary facilities; (2) vehicle (rolling stock) costs, and (3) operating costs. Factors affecting rail transit costs include system length, number of stations, vertical alignment, and fraction of the system underground. Also, it is usually more expensive to build a rail rapid transit line than a bus rapid transit line, partly because rail lines require additional and more expensive facilities, such as power supply, signals, and a safety control system. In a tunnel, however, it may be less expensive to build a rail line because rail cars are smaller than buses,

Table 4.2 Highway Improvement Costs (Thousands of 2005 Dollars per Lane-Mile)

Functional Class	Terrain	Reconstruct and Add High-Cost Lanes	Reconstruct and Add Normal-Cost Lanes	Reconstruct and Widen Lanes	Reconstruct	Major Widening at High Cost	Major Widening at Normal Cost	Minor Widening	Resurface and Improve Shoulders	Resurface
Rural interstate	Flat	759	759	855	713	477	477	386	265	150
	Rolling	888	888	944	734	509	509	415	280	144
	Mountainous	1,023	1,023	1,251	1,042	670	670	570	343	185
Rural other principal arterial	Flat	957	957	729	623	490	490	378	184	94
	Rolling	990	990	820	704	547	547	416	200	94
	Mountainous	1,409	1,409	1,074	880	1,020	1,020	593	273	137
Rural minor arterial	Flat	831	831	562	443	483	483	314	185	79
	Rolling	904	904	707	604	668	668	329	188	84
	Mountainous	1,223	1,223	1,103	791	847	847	435	234	132
Rural major collector	Flat	732	732	641	454	460	460	253	129	45
	Rolling	802	802	776	562	457	457	267	141	52
	Mountainous	1,073	1,073	993	774	781	781	355	181	65
Urban sections	Freeways and expressways	11,227	4,828	3,541	2,169	11,396	4,996	2,102	628	292
	Other divided	6,677	2,667	2,181	1,236	7,139	3,130	1,159	430	196
	Other undivided	4,716	1,724	1,896	1,130	5,327	2,335	1,227	375	222

Source: Costs have been indexed from 1997 dollars shown in the HERS Technical Report Version 3.26, December 2000. The improvement costs in this table include right-of-way.

thus requiring smaller tunnels, and do not emit exhaust gases, whose removal requires special tunnel ventilation facilities (Black, 1995). The various transit types, which are illustrated in Figure 4.7, are defined as follows (TRB, 2003; APTA, 2005):

- *High-speed rail:* a commuter railway primarily for intercity travel. There are several high-speed facilities in Europe and Asia, and recently, a Maglev high-speed transit facility has been constructed in Shanghai, China, to connect the city center and the main airport. Figure 4.8 provides a summary of unit construction cost for high-speed rail.
- *Heavy rail:* an electric railway with the capacity for a heavy volume of traffic, operating on an exclusive right-of-way that is separate from all other vehicular and pedestrian traffic. Heavy rail is often characterized by high-speed and rapid-acceleration

movements, and its passenger railcars operate individually or in multicar trains on fixed rails.

- *Commuter rail:* an electric- or diesel-propelled variation of heavy rail purposely for urban passenger train service, consisting of local short-distance travel operating between a central city and adjacent suburbs. Because of its service characteristics, it is sometimes referred to as *metropolitan rail, regional rail,* or *suburban rail.*
- *Light rail:* lightweight passenger rails system that operates with one- or two-car trains on fixed rails. Unlike heavy-rail service, light rail operates on nonexclusive right-of-way that is mostly not separated from other traffic. Also known as *streetcars, trams,* or *trolley cars,* light-rail vehicles are often operated electrically.

(a)　　　　　　　　　　　　　　　　　(b)

(c)　　　　　　　　　　　　　　　　　(d)

Figure 4.7　Major categories of rail transit and bus transit: (*a*) heavy (rapid) rail (photo courtesy of Doug Bowman, Creative Commons Attribution 2.0 license); (*b*) commuter rail (photo courtesy of LERK, Creative Commons Attribution 2.5 license); (*c*) light rail; (*d*) bus rapid transit (photo courtesy of Shirley de Jong, Creative Commons Attribution ShareAlike 2.5 license).

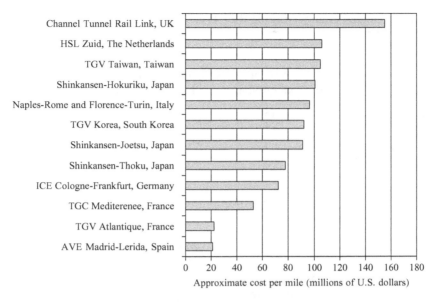

Figure 4.8 Costs of high-speed rail in Europe and Japan. (Adapted from CIT, 2004.)

- *Monorail:* a railway system that uses cars running on a single rail. Typically, the rail is run overhead and the cars are either suspended from it or run above it. Driving power is transmitted from the cars to the track by means of wheels that rotate horizontally, making contact with the rail between its upper and lower flanges.
- *Bus rapid transit:* essentially, a rubber-tired version of light-rail transit with greater operational flexibility. It can include a wide range of facilities, from mixed traffic and curb bus lanes on streets to exclusive busways.
- *Bus transit:* traditional urban bus transit, mostly using city streets.

(*a*) *High-Speed-Rail Capital Costs*   As shown in Figure 4.8, the cost of high-speed rail construction varies from country to country. This is due to variations in availability and prices of factors of production such as land and labor. In the United States, Acela Express high-speed trains operate between Washington, DC and Boston via New York City and Philadelphia along the northeast corridor of the United States.

(*b*) *Heavy (Rapid)-Rail Capital Costs*   On the basis of historical data, the total cost of heavy-rail construction can be decomposed by subsystem as follows (Cambridge Systematics et al., 1992): land, 6%; guideway, 26%; stations, 26%; trackwork, 4%; power, 3%; control, 5%; facilities, 2%; engineering and management/testing, 15%; and vehicles, 13%.

Like high-speed rail, heavy-rail systems typically involve a large capital outlay. For example, the 104-mile 43-station San Francisco Bay Area Rapid Transit, most of which was completed in 1974, cost approximately $3.82 billion in 2005 dollars. In the late 1980s, the cost of building a heavy-rail line was approximately $100 to $300 million per line-mile (in 2005 dollars), depending on the number of stations and the fraction of system constructed underground (Table 4.3). The table shows the capital costs of heavy (rapid)-rail transit systems constructed at four major cities in the United States.

A rough model for estimating the unit cost of heavy (rapid)-rail construction is as follows: For heavy (rapid)-rail systems with 40 to 60% underground, the average cost is $14.4 million per line-mile-station, and the cost model is

$$\text{UC} = 3.906 \times \text{LM}^{-0.702} \times \text{PU}^{1.076}$$
$$\times \text{ST}^{-0.358} \qquad R^2 = 0.94 \qquad (4.1)$$

where UC is the unit cost (cost per line-mile-station), in millions of 2005 dollars, LM the number of line-miles, PU the percentage of system underground, and ST the number of stations. Therefore, given basic information such as the expected system length (miles), average number of lines, number of stations, and surface–underground fraction, the expected overall cost of a heavy (rapid)-rail system can be roughly estimated.

For example, the estimated cost of a two-station 10-lane-mile heavy-rail system with 50% underground is

Table 4.3   Capital Costs of Selected Heavy (Rapid)-Rail Transit Systems

	City	Line-Miles	Percent Underground	Number of Stations	Capital Cost (millions of 2005 dollars)	Cost per Line-Mile (millions of 2005 dollars)	Cost per Line-Mile-Station (millions of 2005 dollars)
Partially	Atlanta	26.8	42	26	4,693	175	6.74
underground	Baltimore	7.6	56	9	2,224	293	32.52
	Washington	60.5	57	57	13,749	227	3.99
Fully above ground	Miami	21	0	20	2,314	110	5.51

Source: Adapted from Cambridge Systematics et al. (1992).

$(3.906)(10)^{-0.702}(50)^{1.076}(2)^{-0.358} = \$40.74$ million per line-mile-station. Also, the cost of a similar 10-station 50-lane-mile system with 50% underground is $(3.906)(50)^{-0.702}(50)^{1.076}(10)^{-0.358} = \$7.40$ million per line-mile-station. Equation (4.1) shows the existence of economies of scale in heavy (rapid)-rail construction costs; the higher the number of stations or line-miles, the lower the cost per station or per line-mile.

Example 4.4   Eighty-five percent of a proposed heavy-rail transit system in the city of Townsville will be located aboveground. The total length is 12 line-miles; four stations are planned. Determine the estimated project cost.

SOLUTION   The cost per line-mile-station = $(3.906)(12^{-0.702})(15^{1.076})(4^{-0.358}) = 7.66$. Therefore, the overall cost of the system is $(7.66)(12)(4) = \$367.68$ million.

While Table 4.3 presents detailed useful information such as the number of line-miles, stations, and the percentage underground, its data are aggregated for all segments of a given city's heavy (metro)-rail systems. On the other hand, Table 4.4 presents the unit construction cost of various segments in each city but does not show details by number of stations, line-miles, and the underground fraction.

(c) Capital Costs of Light-Rail Fixed Facilities   The capital costs of light-rail transit systems vary considerably by construction type. There are generally about six different types of light-rail construction, classified by the extent and manner in which the guideway is buried in the ground. *At-grade structures* are grounded on the surrounding terrain. *Elevated light-rail structures* are installed on columns so that they are above the surrounding terrain. *Fill structures* are constructed on

Table 4.4   Metro-Rail Construction Cost per Mile

Heavy (Metro)-Rail Project	Cost per Mile (millions of 2005 dollars)
Atlanta MARTA	
Phase A	248.0
Phase B	117.2
Phase C	120.5
Baltimore Metro	
Sections A and B	123.0
Section C	357.5
Los Angeles Red Line	
Segment 1	697.8
Segment 2	349.6
Segment 3a	333.3
Washington Metro	
Orange Line	232.5
Red and Blue Lines	203.5
Green Line, Blue Extension	310.7
Average cost per mile	281.2

Source: Adapted from Parsons Brinckerhoff (1996).

an embankment on the existing ground. *Subway light rails* are those located completely below ground. On the basis of the data from light-rail systems in Portland, Sacramento, San Jose, Pittsburgh, and Los Angeles, the distribution of total light-rail construction costs are as follows: guideway elements, 23%; yards and shops, 5%; systems, 10%; stations, 5%; vehicles, 13%; special conditions, 7%; right-of-way, 8%; and soft costs, 29%. *Special conditions* refer mostly to utility relocation; *soft costs* include demolitions, roadway changes, and environmental treatment (Booz Allen Hamilton, 1991). Table 4.5 presents the unit cost of various light-rail projects in the United States.

Table 4.5 Light-Rail Construction Cost per Mile

Light-Rail Project	Cost per Mile (millions of 2005 dollars)
Baltimore Central Line	18.8
Phase 1	
Three extensions	16.4
Dallas DART S&W Oak Cliff	31.3
Park Lane	58.6
Denver RTD	24.4
Central Corridor	
Southwest Extension	20.3
Los Angeles MTA	43.4
Blue Line	
Green Line	49
Portland Tri-Met	26.6
Banfield	
Westside	56.7
Sacramento RTD	12.4
Original Line	
Mather Field Road Extension	15.4
Salt Lake City UTA South Line	21.4
St. Louis MetroLink Phase 1	20.8
San Diego Trolley	31.3
Blue Line	
Orange Line	23.5
Santa Clara County VTA	26.2
Guadalupe Corridor	
Tasman Corridor	43.8
Average cost per mile	36.6

Source: Adapted from U.S. General Accounting Office (2001).

Guideways for Light Rail: Guideway construction typically accounts for 16 to 38% of overall capital costs (Black, 1995). Of the various light-rail construction types, subway guideway construction is by far the most costly, followed by retained-cut guideway systems. Guideways on at-grade levels and elevated fills are the least expensive types of light-rail construction. Table 4.6 presents the capital costs per line-mile (expressed in 2005 dollars) for light-rail guideways constructed at various urban areas in the United States. The average cost is $36.6 million per mile, in 2005 dollars.

For estimating the approximate guideway cost of a light-rail project whose construction type is known, the average cost values shown in the last column of Table 4.6 may be used. However, at the initial planning stage, the type of light-rail construction may not be known. In such cases, the analyst may provide a rough estimation of the project capital costs using the following model, developed using data from 22 U.S. cities where light-rail projects were implemented in the 1992–2005 period (Light Rail Central, 2002):

Total guideway cost

$$= \exp(-1997.92 + 1448.22 \text{ LENGTH}^{0.0005}$$
$$+ 553.55 \text{ STATIONS}^{0.0005}) \qquad R^2 = 0.61 \quad (4.2)$$

where the total guideway cost is in millions of 2005 dollars, LENGTH is the system length in miles, and STATIONS is the number of stations.

Example 4.5 It is proposed to construct a 20-mile light-rail system in the city of Megapolis. The number of stations is not yet known. Given the nature of the terrain, an elevated fill structure is recommended. Determine the estimated cost of the guideway for the system. If the guideway is expected to account for 30% of the capital cost of the overall system, estimate the total capital cost of the project.

SOLUTION Using the average cost for elevated fill structure from Table 4.6, estimated guideway cost = (5.87)(20) = $117.4 million; total system cost = (117.40)(100/30) = $391.33 million.

Example 4.6 A new light-rail system planned for a rapidly growing city will be 21 miles in length and will serve 13 stations. The construction type has not yet been decided. Find the total and average (per mile-station) guideway cost of the system. An alternative being considered is to construct the system to cover 38 miles and to serve 22 stations. Find the average guideway cost of the second alternative, and explain for any differences in average guideway costs between the two alternatives.

SOLUTION Using equation (4.2),

Total cost for alternative 1

$$= \exp[-1997.92 + (1448.22)(21^{0.0005})$$
$$+ (553.55)(13^{0.0005})] = \$868.35 \text{ million}$$

$$\text{Average cost} = \frac{\$868.35}{(21)(13)}$$
$$= \$3.18 \text{ million per mile-station}$$

Total cost for alternative 2

$$= \exp[-1997.92 + (1448.22)(38^{0.0005})$$
$$+ (553.55)(22^{0.0005})] = \$1544.71 \text{ million}$$

Table 4.6   Guideway Cost for Selected Light-Rail Construction Types

Type of Guideway Construction	Guideway Cost[a] (millions of 2005 dollars)					
	Portland	Sacramento	San Jose	Pittsburgh	Los Angeles	Average Cost (per mile)
At-grade	10.74	3.68	5.43	4.10	5.67	5.93
Elevated structure	27.11	3.65		5.67	26.62	15.76
Elevated retained, fill		9.60		8.56	8.41	8.91
Elevated fill				6.23	5.49	5.87
Subway			61.39	64.02	61.82	62.41
Retained cut		44.33	2.36	43.71	27.93	29.58

Source: Adapted from Booz Allen Hamilton (1991).

[a] Design, engineering, right-of-way acquisition, and other administrative costs are excluded.

$$\text{Average cost} = \frac{\$1544.71}{(38)(22)}$$

$$= \$1.84 \text{ million per mile-station}$$

Alternative 2 represents a 42% reduction in average cost. This can be attributed to economy-of-scale effects.

Stations and Yards for Light Rail: Construction of passenger stations and rolling stock maintenance yards often constitutes a significant fraction of overall transit capital costs. Table 4.7(a) presents the unit capital costs of light-rail passenger stations in five cities of the United States, expressed in 2005 dollars. It can be seen that stations for subway station construction are by far the most costly, followed by those for elevated guideway systems.

Table 4.7(b) presents the capital costs of light-rail transit yards and mechanical shops in 2005 dollars. The costs do not include design, engineering, right-of-way acquisition, and other administrative costs. The average cost for the construction of rail transit yards and shops was $600,000 per unit of capacity. *Capacity* represents the maximum number of vehicles that can be held in the maintenance yard.

Example 4.7   It is proposed to construct 20 passenger stations for a planned subway light-rail system. A maintenance yard and shop with a capacity of 60 vehicles is also proposed. Estimate the overall capital cost for stations and yards for the project.

SOLUTION

Average cost of passenger station for subway light-rail

transit system = $26,982,000

Cost of 20 passenger stations

$$= (20)(\$26,982,000) = \$539,640,000$$

Average cost of maintenance yard

$$= \$600,000 \text{ per unit capacity}$$

Cost of 60 capacity units

$$= (60)(\$600,000) = \$36,000,000$$

Total capital cost for stations and yards

$$= \$575,640,000$$

(*d*) *Capital Costs of Monorail Fixed Facilities*   Table 4.8 presents the cost of monorail construction per mile. The average cost is approximately $220 million per mile. This includes the cost of the guideway, stations, and other ancillary structures.

(*e*) *Rolling Stock Capital Costs for the Various Rail Transit Types*   Table 4.9 presents the unit costs of rolling stock for various rail transit system types. Estimated costs for both heavy- and light-rail vehicles exceed $2 million each, expressed in 2005 dollars. Table 4.10 shows the unit costs of rehabilitating rolling stock in 2005 dollars.

Example 4.8   A transit agency wishes to purchase 55 new cars for its heavy-rail system. Also, it is expected that rehabilitation of these cars will be carried out twice in their life cycle. What is the estimated total capital cost of the new fleet over their life cycle?

SOLUTION   From Table 4.9, average purchase cost per car = $2.3 million. Purchase cost of 55 cars = (55)($2.3 million) = $126.5 million. From Table 4.10, average rehabilitation cost per car = $0.84 million. Total rehabilitation

Table 4.7 Light-Rail Transit Capital Costs at Selected Locations

(a) Passenger Station Costs per Station (thousands of 2005 dollars)

Type of Construction	Portland	Sacramento	San Jose	Pittsburgh	Los Angeles	Average
At-grade center platform	831		263		1,656	917
At-grade side platform	910	636	312	3,248	1,401	1,302
Elevated					4,493	4,493
Subway				11,491	42,473	26,982

(b) Maintenance Yards and Shops

Location	Yard and Shop Capital Costs (thousands of 2005 dollars)	Yard and Shop Capacity (vehicles)	Cost per Unit of Capacity (thousands of 2005 dollars)
Portland	22,549	100	226
Sacramento	6,900	50	138
San Jose	31,846	50	637
Pittsburgh	72,323	97	746
Los Angeles	67,817	54	1,256
Average			600

Source: Adapted from Booz Allen Hamilton (1991).

Table 4.8 Monorail Construction Cost per Mile

Monorail Project	Cost per Mile (millions of 2005 dollars)
Las Vegas Extension (planned)	197.6
Newark Airport mini-monorail	274.8
Kitakyushu monorail	179.3
Average cost per mile	217.2

Source: Adapted from Parsons Brinckerhoff (2001), LTK Engineering Services (1999).

cost of 55 cars = (55)($0.84 million)(2) = $92.4 million. Therefore, the estimated total capital cost = $218.9 million.

(*f*) *Bus Rapid Transit Capital Costs* BRT facility development costs depend on the location, type, and complexity of construction. The costs of existing systems were reported to be $7.5 million per mile for independent at-grade busways, $6.6 million per mile for arterial busways located in the road median, and $1 million for mixed traffic and/or curb bus lanes (TRB, 2003). The costs can be many times higher when tunnels and other features for exclusive guideways are included. Table 4.11 shows the costs (in U.S. dollars) of selected bus rapid transit systems at locations around the world.

(*g*) *Rail Transit Operating Costs* Rail transit operating costs consist of salaries, wages, and fringe benefits; utilities (power supplies); and maintenance of rolling stocks, stations, and rail tracks (guideways), while bus transit operating costs include salaries, wages, and fringe benefits; fuel; and vehicle and terminal maintenance. Operating costs may be reported in two ways:

1. As a function of supply-based measures; in other words, operating cost may be expressed as a function of inventory size, system type, or some physical attribute of the system. Examples include operating cost per mile, per vehicle, and per expected vehicle-miles of travel. Note that for rail transit where schedules are not always a reliable indicator of the level of ridership, VMT (unlike passenger-miles of travel) may not be a reliable measure of consumed service demand. Operating cost functions are useful at the facility planning stage where a cost estimate is sought for operating the system.

2. As a function of demand-based measures; in other words, operating cost may be expressed as a function of operating cost per passenger, per vehicle, per passenger-hour, per passenger-mile, and so on. These types of operating cost models are more useful for performance

Table 4.9   Unit Rolling Stock Costs for Various Rail Transit System Types

Type of system	Location	Year	Quantity Ordered	Cost for Total Order[a]	Cost per Car[a]	Average Cost per Car[a]
Heavy (rapid)-rail transit	Chicago	1991	256	350.49	1.37	2.3
	Los Angeles	1989	54	106.70	1.98	
	New York	1990	19	66.35	3.49	
	San Francisco	1989	150	385.44	2.57	
	Washington, DC	1989	68	140.64	2.08	
Light-rail transit	Boston	1991	86	222.86	2.58	2.6
	San Diego	1991	75	205.97	2.75	
	St. Louis	1990	31	76.65	2.46	
Commuter rail	Florida	1990	6	9.96	1.65	2.4
	Los Angeles	1990	40	86.10	2.16	
	New Jersey	1991	50	76.31	1.52	
	New York	1990	39	153.64	3.93	
	Indiana	1991	17	46.43	2.74	

Source: Adapted from Cambridge Systematics et al. (1992).

[a] In millions of 2005 dollars adjusted from actual dollars as of order date. Variations in unit costs are due to type of vehicle, size of order, and options.

Table 4.10   Costs of Rolling Stock Rehabilitation

Type of System	Location	Year	Car Type	Quantity Rehabilitated	Cost for Total Order[a]	Rehabilitation Cost per Car[a]	Average Rehabilitation Cost per Car[a]
Heavy (rapid) rail transit	New York	1991	R33 subway	494	339.35	0.69	0.84
	New York	1991	R44 subway	280	250.54	0.89	
	New York	1990	R44 subway	64	60.78	0.95	
Commuter rail	Maryland	1990		35	11.82	0.34	0.99
	New Jersey	1991		230	376.66	1.64	

Source: Adapted from Cambridge Systematics et al. (1992).

[a] In millions of 2005 dollars adjusted from actual dollars as of 1991. Variations in unit costs are due to type of vehicle, size of order, and options.

assessments than they are for cost estimation of future projects. However, if the future demand is known, these types of operating cost functions can be used to derive operating cost estimates for purposes of future project planning.

Table 4.12 presents average operating costs for various transit modes, in terms of four cost related performance measures. These costs have not been corrected for possible scale economies. Data are for all heavy and light rail systems in the United States and the 20 largest bus systems in terms of average weekday passengers. It is seen that bus transit, as compared to other modes, has lower operating cost per vehicle-hour and per vehicle-mile, slightly higher cost in terms of passenger-mile, and similar costs per passenger-trip. Heavy rail has the lowest operating cost per passenger-mile, followed by light rail. This could be because rail transit cars are larger than those of bus transit, and people tend to make longer trips on rail than on buses. As such, the unit operating costs of rail systems enjoy higher economies of scale than bus transit in terms of passenger-miles. Operating

Table 4.11 Cost of Development for Selected BRT Systems

Facility	Location	Miles	Cost (millions of 2005 dollars)	Cost/Mile (millions of 2005 dollars)	Notes
Bus tunnels	Boston—Silver Line	4.1	1477.09	359.97	Includes bus lanes
	Seattle	2.1	492.36	234.15	
Busway	Hartford	9.6	109.41	10.94	
	Houston—HOV system	98	1072.26	21.88	
	Los Angeles—San Bernardino Freeway	12	82.06	6.56	
	Miami	8.2	64.55	7.66	
	Ottawa	37	320.58	8.75	
	Pittsburgh—South Busway	4.3	29.54	6.56	
	East Busway	6.8	142.24	20.79	
	West Busway	5	300.89	60.18	
	Adelaide (guided bus)	7.4	57.99	7.66	
	Brisbane[a]	10.5	218.83	20.79	
	Liverpool—Parramatta	19	109.41	5.47	
	Runcorn	14	16.41	1.09	
Freeway, reversible	New York—I-495 New Jersey	2.5	0.77	0.33	
Reversible lanes	I-495 New York	2.2	0.11	0.11	
	I-278 Gowanus	5	10.94	2.19	Involves freeway reconstruction
Arterial street median busways	Cleveland	7	240.71	31.73	
	Eugene	4	14.22	3.50	
	Bogota	23.6	201.32	8.75	
	Quito	10	63.02	6.56	
	Belo Horizonte			1.75	
Mixed Traffic–curb bus lanes	Los Angeles	42	9.08	0.22	
	Vancouver—Broadway	11	9.85	1.09	
	Richmond	9.8	48.14	4.49	
	Leeds (guided bus)	2.1	5.47	2.63	
	Rouen (optically guided bus)	28.6	218.83	7.66	

Source: Adapted from TRB (2003).

[a]Excludes costs of downtown bus tunnel built before busway.

costs for bus rapid transit service in Pittsburgh (1989) averaged $0.52 per passenger-trip, and operating costs per vehicle revenue-hour ranged from $50 in Los Angeles to $150 in Pittsburgh (TRB, 2003). A nationwide study by Biehler (1989) showed that bus rapid transit can cost less per passenger trip and per mile than light rail transit, depending on the situation.

Table 4.13 shows the distribution of rail transit operating costs by spending category. Many rail systems involve the use of auxiliary infrastructure such as an automatic train operation (ATO) system, operations control center

(OCC), and an automatic fare collection (AFC) system. In 1979, BART let out an ATO contract for $26.2 million (with subsequent change orders, this amount reached $32.7 million). The cost of installing BART's OCC was $2.9 million, while the AFC cost was $4.96 million in 1968 (change orders brought the contract total to $6.6 million) (BART, 2006).

Example 4.9 A light-rail transit system is proposed for the city of Metroville. From the planned schedule it is estimated that 20 rail vehicles will be needed and that

Table 4.12   Average Operating Costs as a Function of Output, by Transit Mode (2005 Dollars)

Performance Measure	Heavy (Rapid) Rail	Light Rail	Bus[a]
Per revenue vehicle-hour	152.29	150.29	76.50
Per revenue vehicle-mile	6.96	11.02	6.42
Per passenger trip	1.61	1.63	1.61
Per passenger mile	0.33	0.44	0.47

Source: Adapted from Black (1995).

[a]Bus operating costs are presented for comparison purposes only.

each vehicle, on its revenue trips, will travel an average distance of 330 miles a day. Assume that the system will operate all year round. What is the expected annual operating cost of the system?

SOLUTION   From Table 4.12 average operating cost per revenue vehicle-mile = $11.02.

Expected travel for all revenue vehicles in one year

$$= (330)(365)(20) = \$2,409,000 \text{ vehicle-miles}$$

Estimated total operating cost per year

$$= (2,409,000)(11.02) = \$26,547,000$$

(*h*) *Bus Transit Capital Costs*   Bus transit capital costs involve purchase and preservation of buses, construction and preservation of bus facilities (terminals and stations), and sometimes include construction of a bus-only highway lane. The price per bus depends on the size (length or number of seats), type (transit, suburban, or articulated), number of units purchased, and availability of accessories such as air conditioning, automatic transmission, and wheelchair lifts. For small buses, additional cost factors include the chassis type. Tables 4.14(a) and (b) show the range of unit prices for heavy-duty buses and small buses, respectively. Table 4.15 shows the rehabilitation costs for heavy-duty buses 35 ft in length. The cost of constructing bus facilities ranges from $120 to $140 per square foot. The bus transit costs presented in this section are based on historical data, and all costs shown have been adjusted to their 2005 equivalents using FTA cost adjustment factors.

(*i*) *Bus Transit Operating Costs*   As Table 4.16 illustrates, some diseconomies of scale are associated with operating bus transit systems, irrespective of the output variable used for the cost function. For example, the cost per vehicle mile, cost per vehicle hour, and cost per peak vehicle are higher for systems of size exceeding 250 buses than they are for systems of size 100 to 250. It should be noted that *vehicle* refers to *revenue vehicle*, which is a vehicle that is in operation over a route and is available to the public for transport at a given time period.

Using data from Cambridge Systematics et al. (1992), the following operating cost functions were developed:

Cost per vehicle-mile

$$= 2.652 S^{0.184} \text{PBR}^{0.029} \qquad R^2 = 0.92$$

Cost per vehicle-hour

$$= 41.063 S^{0.134} \text{PBR}^{0.247} \qquad R^2 = 0.84$$

Cost per peak vehicle

$$= 11.405 S^{0.020} \text{PBR}^{-0.039} \qquad R^2 = 0.83$$

Table 4.13   Distribution of Rail Transit Operating Costs by Spending Category (Percent)[a]

	Heavy (Rapid) Rail (12 systems)	Light Rail (13 systems)	Commuter Rail (10 systems)
Operator salaries and wages	9.30	18.10	11.0
Other salaries and wages	40.7	34.5	29.6
Fringe benefits	29.2	26.2	28.6
Utilities	8.7	9.4	6.1
Other costs	12.1	11.7	24.7
Total	100	100	100

Source: Adapted from Cambridge Systematics et al. (1992).

[a]Percentages are calculated from average costs in each category for all systems reporting.

Table 4.14   Bus Acquisition Costs

(a) Heavy-Duty Buses

Bus Type	Total Number of Buses Purchased	Average Cost per Bus (2005 dollars)	Range of Cost per Bus (2005 dollars)
60-ft articulated	30	472,555	325,842–501,425
40-ft suburban	162	385,608	NA[a]
40-ft transit	686	300,518	270,128–339,348
35-ft transit	45	294,946	290,388–330,907
30-ft transit	43	288,531	253,245–293,764

Source: Adapted from Cambridge Systematics et al. (1992).

(b) Small Buses

Type	Gross Vehicle Weight Rating (lb)	Cost Range[b] (2005 dollars)
Light duty		
Truck cab type of chassis	9,500–12,500	50,649–101,298
Motor home type of chassis	14,500–18,500	7,5974–126,623
Medium duty (rear engine chassis)	16,500–20,500	109,740–185,713
Heavy duty (integrated body)	22,500–26,000	211,038–295,453

Source: Adapted from Johnson (1991).

[a]NA, not available.

[b]Variations in costs are due to size of order, vehicle configuration, and options.

Table 4.15   Rehabilitation Costs for 35-ft Buses

Location	Year	Quantity Rehabilitated (2005 dollars)	Cost per Bus (2005 dollars)	Average Cost per Bus (2005 dollars)
Dubuque	1990	10	136,822	
Monterey	1990	15	239,438[a]	153, 924
Westchester County	1991	20	85,513	

Source: Adapted from Cambridge Systematics et al. (1992).

[a]Includes addition of wheelchair lift, which added about $20,000 per bus to the rehabilitation cost.

where S is the system size (number of buses operated in maximum service), and cost is in 2005 dollars. The *peak-to-base ratio* (*PBR*) is the number of vehicles operated in passenger service during the peak period (morning and afternoon time periods when transit riding is heaviest) divided by the number operated during the off-peak period. These functions can be used to estimate the future operating costs of a proposed bus transit system if the system size and peak-to-base ratio are known. If the latter variable is unknown, the average cost value can be used. More recent average values of operating costs for buses and other public transportation modes are provided in Tables 4.17 to 4.19 but these do not involve the peak-to-base ratio variable.

Table 4.16   Unit Operating Costs of Bus Transit By System Size and Peak-to-Base Ratio

System Size[a]	Peak-to-Base Ratio[b]	Cost per Vehicle-Mile (2005 dollars)	Cost per Vehicle-Hour (2005 dollars)	Cost per Peak Vehicle (thousands of 2005 dollars)
250 or more buses	Ratio 2.00 (16)[c]	7.88	109.87	253,120
	Ratio < 2.00 (18)	8.24	102.22	314,690
100–249 buses	Ratio 2.00 (20)	6.50	98.87	205,230
	Ratio < 2.00 (30)	6.41	81.44	236,020
50–99 buses	Ratio 1.75 (18)	6.48	89.84	176,160
	Ratio < 1. 75 (15)	6.05	94.54	232,600
25–49 buses	Ratio 1.50 (28)	4.87	65.59	164,190
	Ratio < 1.50 (45)	4.93	65.16	198,390
Fewer than 25 buses	Ratio 1.50 (30)	4.43	61.57	141,950
	Ratio < 1.50 (56)	4.38	59.76	172,740
All sizes	All motor buses (363)[d]	5.28	73.03	191,550
	Trolley buses (5)[e]	9.87	104.28	289,040

Source: Adapted from Cambridge Systematics et al. (1992).

[a]Vehicles operated in maximum service.

[b]Vehicles operated in average p.m. peak divided by vehicles operated in average base period.

[c]Numbers in parentheses are the number of bus systems for which data are available.

[d]The complete motor bus database includes several transit systems for which peak–base ratios are not available. Data are missing for a few transit systems for some of the variables above.

[e]Four of the five trolley bus systems are part of systems in the largest size class above.

Table 4.17   Capital and Operating Costs by Travel Mode for Small Cities[a] (2005 Dollars)

Cost Category	Cost Type	Units of Output	Bus	Demand Responsive
Capital costs	Rolling stock	Per passenger trip	0.48	0.74
		Per passenger-mile	0.15	0.21
	Systems and guideways	Per passenger trip	0.05	0.26
		Per passenger-mile	0.02	0.04
	Facilities and stations	Per passenger trip	0.59	0.04
		Per passenger-mile	0.14	0.01
	Total capital costs	Per passenger trip	1.19	1.56
		Per passenger-mile	0.33	0.35
Operating costs	Total operating costs	Per passenger-mile	1.23	3.39
		Per passenger trip	4.02	16.01
		Per vehicle-mile	4.20	3.55
		Per vehicle-hour	59.44	42.77

Source: Adapted from *FTA (2003)*; ECONorthwest et al. (2002).

[a]Data compiled from 20 randomly selected systems with population <200,000.

Example 4.10   The bus transit agency of a certain medium-sized city plans to augment its current fleet by acquiring 45 new 35-ft buses. The brand of buses specified has a service life of 15 years and will need rehabilitation in the sixth and eleventh years of their service life. (a) How much can the agency expect to spend on the capital cost of the new buses over their service life? (b) Assuming a peak-to-base ratio of 1.4 and an average VMT of 36,500 per year, estimate the annual operating cost of the new fleet.

Table 4.18   Capital and Operating Costs by Travel Mode for Medium-sized Cities[a] (2005 Dollars)

Cost Category	Cost Type	Units of Output	Commuter Rail	Heavy Rail	Light Rail	Bus	Vanpool	Demand Responsive
Capital costs	Rolling stock	Per passenger trip	0.02	0.02	0.51	0.60	0.19	3.10
		Per passenger-mile	<0.01	<0.01	0.10	0.08	0.00	0.26
	Systems and guideways	Per passenger trip	0.15	0.15	5.90	0.15	0.44	0.55
		Per passenger-mile	<0.01	<0.01	5.90	0.03	0.02	0.07
	Facilities and stations	Per passenger trip	0.09	<0.01	0.28	0.15	0.14	0.19
		Per passenger-mile	0.01	<0.01	0.11	0.03	0.01	0.02
	Total capital costs	Per passenger trip	0.27	0.16	6.32	0.71	1.29	3.87
		Per passenger-mile	0.04	<0.01	6.11	0.16	0.04	0.35
Operating costs	Total operating costs	Per passenger-mile	0.45	0.30	1.88	0.82	0.89	2.84
		Per passenger trip	21.19	1.97	2.51	3.70	11.60	27.83
		Per vehicle-mile	18.86	6.24	16.12	5.43	1.58	3.55
		Per vehicle-hour	694.52	164.09	185.71	76.33	57.54	53.27

Source: Adapted from FTA (2003); ECONorthwest et al. (2002).
[a]Data compiled from 20 randomly selected system with population >200,000 and <1,000,000.

SOLUTION   (a) *Capital cost*: From Table 4.14, the average purchase cost per 35-ft bus = $294,946. Therefore, the purchase cost of 45 buses = (45)($294,946) = $13,272,570. From Table 4.15, average rehabilitation cost per bus = $153,924. The total rehabilitation cost of 45 buses = (45)($153,924)(2) = $13,853,160. Therefore, the estimated total capital cost = $27,125,730.

(b) *Operating cost*: From Table 4.16, average operating cost per vehicle-mile = $4.93.

The expected travel for all vehicles in one year = (45)(36,500) = 1,642,500 vehicle-miles. The estimated total operating cost per year = (1,642,500)($4.93) = $8,097,525.

4.5.3 Relationships between Transit Operating Costs, System Size, Labor Requirements, and Technology

Tables 4.17 to 4.19 present the capital and operating costs for transit and other public transportation travel modes for small, medium-sized and large cities in the United States (FTA, 2003). These costs are expressed in terms of operational performance measures. Clear differences in cost are seen across mode types and system size (surrogated by city size). An advantage of capital-intensive transit modes, such as rail, is that the smaller share of labor inputs renders the operating costs of such systems less vulnerable to inflation, a particularly important issue given the frequent and sharp

increases in transit labor costs relative to the cost of living (Black, 1995). For old transit systems, however, this advantage is outweighed by the fact that such rail systems require a relatively large number of nonoperating workers who maintain the vehicles and right-of-way and carry out management and policing duties. Furthermore, the old rail systems are relatively complicated and require considerable attention to prevent failures. On the basis of 1990 data (Booz Allen Hamilton, 1991), labor expenses (including fringe benefits) comprised the following percentages of total operating costs: old heavy-rail systems, 81.9%; new heavy-rail systems, 70.2%; old light-rail systems, 82.7%; new light-rail systems, 62.3%, and bus transit (20 largest systems), 80.2%.

Clearly, in terms of vulnerability of labor (and thus, operating costs) to inflation, old rail systems seem to have little or no advantage over buses. On the other hand, the lower labor cost fraction (and thus lower inflation risk) of new rail systems is evident and may be attributed to use of state-of-the-art technologies for service and fare collection.

4.5.4 Air Transportation Costs

Denver International Airport (DIA) is the only major airport constructed in the United States in the past 20 years. The cost of DIA, including airport planning, land, and construction was approximately $60 million

Table 4.19   Capital and Operating Costs by Travel Mode for Large Cities[a] (2005 Dollars)

Cost Category	Cost Type	Units of Output	Commuter Rail	Heavy Rail	Light Rail	Bus	Vanpool	Demand Responsive
Capital costs	Rolling stock	Per passenger trip	2.51	0.28	0.73	0.34	1.42	3.22
		Per passenger-mile	0.20	0.05	0.20	0.08	0.07	0.59
	Systems and guideways	Per passenger trip	1.64	0.78	17.15	0.15	0.14	0.66
		Per passenger-mile	0.12	0.15	4.28	0.03	0.01	0.11
	Facilities and stations	Per passenger trip	1.60	0.61	12.28	0.15	0.03	0.12
		Per passenger-mile	0.07	0.13	1.52	0.03	0.00	0.01
	Total capital costs	Per passenger trip	5.85	1.76	31.08	0.71	1.65	4.60
		Per passenger-mile	0.38	0.36	6.02	0.16	0.09	0.93
Operating costs	Total operating costs	Per passenger-mile	0.53	0.45	4.80	0.83	0.12	3.68
		Per passenger trip	8.95	2.26	6.16	3.84	3.84	30.55
		Per vehicle-mile	14.46	9.16	27.52	8.18	0.63	4.33
		Per vehicle-hour	413.41	199.84	238.44	102.38	24.37	63.68

Source: Adapted from FTA (2003); ECO Northwest et al. (2002).

[a]Data compiled from 20 randomly selected transit systems at cities with population >1,000,000.

per square mile (GAO, 1995). This excludes the cost of capitalized interest, bond discounts, and costs to other users of airport facilities. The annual (1996) cost of operating that airport was $160 million (GAO, 1996) or $9 per domestic "origin-and-destination" passenger. In 2003, a new runway was added at the cost of $52 per square foot.

4.6   ISSUES IN TRANSPORTATION COST ESTIMATION

The cost estimation of transportation projects is a complex undertaking that requires a great deal of engineering judgment. Due consideration should be given to a number of issues that may significantly influence the reliability of cost estimates. Such issues include methods of cost estimation, spatial or temporal adjustments, adjustments for economies (or diseconomies) of scale, sunk-cost considerations, and other factors. These issues are discussed in the following sections.

4.6.1 Aggregated Estimates for Planning vs. Detailed Engineering Estimates for Projects

Most agencies develop unit cost estimates for construction, preservation, maintenance, and operations activities on the basis of market prices of materials, labor, and equipment use. The overall cost of a project is the sum of the product of the unit costs and the quantities of individual pay items. For the final sum of all items, a percentage may be added for contingencies, such as possible cost overruns or unexpected site conditions. Often, for planning purposes, a quick and approximate estimate is needed. As such, instead of obtaining an estimate based on individual pay items, an aggregate value of cost may be derived using historical data from past contracts.

4.6.2 Adjustments for Temporal and Spatial Variations (How to Update Costs)

(a) *Temporal Variation (Constant vs. Current Dollars)* From a conceptual and computational standpoint, it is easier to prepare cost estimates in constant (and not nominal) dollar amounts, thus removing the effects of inflation from the analysis. Then if cost streams over time are being compared, the necessary discounting or compounding formula can be used to reflect the opportunity cost. This approach assumes that the interest rate does not include inflation effects. Several cost indices are available to adjust cost information across different years. Examples include the FHWA Federal-Aid Highway Construction Price Index, the Federal Capital Cost Index (Schneck et al., 1995), the FHWA Highway Maintenance and Operating Cost Index, the *Engineering News-Record*'s Construction Cost Index, and the R.S. Means City Construction Index. The Federal

Table 4.20 State Cost Factors, 2004

State	Cost Factor	State	Cost Factor	State	Cost Factor
Alabama	1.21	Louisiana	1.32	Oklahoma	0.95
Alaska	1.30	Maine	1.10	Oregon	1.25
Arizona	0.95	Maryland	0.83	Pennsylvania	0.95
Arkansas	0.95	Massachusetts	0.78	Puerto Rico	1.23
California	1.56	Michigan	1.24	Rhode Island	0.98
Colorado	1.26	Minnesota	1.11	South Carolina	1.32
Connecticut	0.88	Mississippi	1.51	South Dakota	1.19
Delaware	1.51	Missouri	0.81	Tennessee	0.90
District of Columbia	0.56	Montana	1.19	Texas	1.19
Florida	1.19	Nebraska	1.15	Utah	1.33
Georgia	1.15	Nevada	1.49	Vermont	1.27
Hawaii	0.76	New Hampshire	1.30	Virginia	0.80
Idaho	1.12	New Jersey	0.70	Washington	1.39
Illinois	0.90	New Mexico	0.69	West Virginia	0.70
Indiana	1.28	New York	0.90	Wisconsin	1.08
Iowa	0.94	North Carolina	0.97	Wyoming	1.24
Kansas	0.59	North Dakota	1.42	United States	1.00
Kentucky	1.39	Ohio	0.85		

Source: FHWA (2005).

Highway Administration's price trends for federal-aid highway construction are based on information received for the contracts that exceed $0.5 million. Effective the first quarter of 1990, the FHWA index was converted to a 1987 = 100 base. The *Engineering News-Record*'s Construction Cost Index uses a 1967 = 100 base. Agency costs can be converted to their current or future values using the price indices from the FHWA price trends (see the General Appendix). Price trend prediction using historical data is useful particularly when long-term economic conditions are predictable. A Web address for a price data source is listed in the Additional Resources section of this chapter.

Broad adjustments of cost to reflect the effect of inflation should be done with caution because inflation rates may be different across components of an overall transportation system. For example, general construction costs typically increase at a faster rate than inflation, whereas ITS and other technology-related costs have seen cost reductions.

(*b*) *Spatial Cost Variations* An analyst may wish to estimate the cost of a proposed project on the basis of similar projects implemented at other states. Given the variation of cost of living and costs of production from state to state, it may be necessary to modify costs

from other states before they are transferred to others. The FHWA (2005) provides state cost factors for capital improvements (Table 4.20).

4.6.3 Adjustments for Economies of Scale

Although economies of scale have long been recognized in cost analysis of transportation systems, there seems to be an inadequate attempt to develop a formal method to duly adjust cost values to account for this effect in transportation systems evaluation. In most past evaluation studies, cost comparisons have traditionally proceeded on the basis of the cost per unit dimension of each facility. For example, the historical costs of flexible vs. rigid pavements and steel vs. concrete bridges have been compared on the basis of their costs per lane-mile and per square foot, respectively, or on the basis of the sum of costs of their individual constituent pay items per some unit quantity. Such an approach implicitly assumes that a linear relationship exists between the cost of each system or pay item and its size. However, relatively few past studies that analyzed infrastructure cost modeling seem to have explicitly recognized and accounted for the nonlinear relationship that typically exists between project cost and project dimension: The greater the project dimension, the lower the unit cost (cost per lane-mile).

Obviously, cost comparison of any two alternative systems must duly account for economy-of-scale effects, because failing to do so may bias the results against the alterative that typically has smaller project dimensions. For example, comparing the unit costs (cost per lane-mile) of a 20-mile warranty pavement to a 3-mile traditionally constructed pavement (all other characteristics remaining the same) would be inappropriate because compared to the traditional pavement the warranty project (by virtue of its greater length) is likely to yield a smaller unit cost and consequently, a higher effectiveness/cost ratio. It is therefore necessary for the different dimensions of competing systems to be adjusted or "brought" to a common dimension. In this way, any differences in their adjusted costs may reflect the differences in their inherent qualities and not their sizes.

Adjustments for economies of scale may be carried out by establishing a correction factor by which unit costs corresponding to a certain dimension can be translated to yield unit costs corresponding to a certain specified standard project dimension. The only information needed for such adjustment is the unit aggregate cost of the project and the unit aggregate cost function for all projects in the same family. The unit cost function may be developed from historical contract data.

Example 4.11 It is sought to construct a 40-line-mile transit system to link the cities of Cityburg and Townsville. Two types of transit systems have emerged as the popular choices: A and B. Systems A and B have the following cost functions: $C_A = -1.05 \ln(X) + 5.2$ and $C_B = 30/X^{0.95}$, respectively, developed on the basis of past projects. C is the cost per line-mile and X is the number of line-miles. The average unit cost of all past projects of types A and B are $207,000 and $285,000 per line-mile, respectively. Would the given unit costs suffice

for the evaluation? If not, give reasons and provide the unit costs that should be used for the evaluation.

SOLUTION   The solution can best be explained using a sketch in Figure E4.11.

Unless there are data for development of cost function, the use of average unit costs for evaluation should be avoided because they correspond to a certain average system dimension that may not be the same as the dimension of the system being proposed. A significant difference in functional forms of cost functions for alternative designs could lead to very different cost estimates for the system, and this difference is influenced by the planned dimension of the system. In the example above, up to 14.2 line-miles, the unit cost of system A is less than that of system B, but beyond 14.2 line-miles, the unit cost of system A exceeds that of system B. For example, for a system dimension of 40 line-miles, systems A and B are expected to cost $133,000 and $90,000, respectively, per line-mile. These values, not the average costs given, should be used for the agency cost aspects of the evaluation of these systems.

4.6.4 Problem of Cost Overruns

At the feasibility and planning stages of the transportation development process, projected capital and operating costs of public transportation projects have typically been underestimated, as studies have shown that project costs have run over their original bid amounts, often by as much as 5 to 14% (Rowland, 1981; Turcotte, 1996; Wagner, 1998; Bordat et al., 2003). It has been argued that the increasing complexity, increased length of communication channels, and distortion of information feedback associated with larger projects translate to higher cost-overrun rates. Nonquantifiable cost-overrun factors include contract document quality, nature of interpersonal

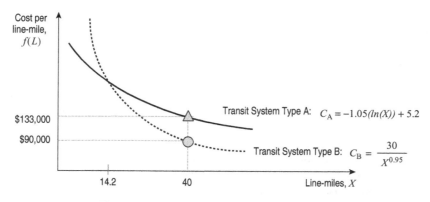

Figure E4.11   Economy-of-scale adjustments.

relations on the project, and contractor policies (Jahren and Ashe, 1990). A FHWA study found that cost overruns were largely attributable to design revisions, difference between the engineer's estimate and the winning bid, and unexpected site conditions, among other reasons (Jacoby, 2001). The causes of overrun costs of transportation projects cited above are attributable to both the contractor and the contracting agency and include inadequate field investigations, unclear specifications, plan errors, design changes, and construction errors (Korman and Daniel, 1998; Wagner, 1998). Also, a FTA study showed that differences between planning and engineering estimates and actual transit construction costs originate from a variety of sources, such as changes in project scope, changes in design standards, unforeseen field conditions, expanded environmental and community requirements, extended implementation periods, underestimation of unit costs, omission of several aspects of project soft costs, and weak estimates of inflation for project capital costs (FTA, 1993). In transportation cost estimation, therefore, sufficient efforts should be made to avoid cost underestimation, such as including a realistic contingency amount or factor to cover possible cost overruns.

4.6.5 Relative Weight of Agency and User Cost Unit Values

An important issue in project economic efficiency analysis or multi-criteria evaluation is the relationship between agency cost and user cost values. Some studies have counted user costs on a dollar-to-dollar basis with agency costs, implying that $1 of agency cost is equivalent to $1 of user cost, therefore adding agency costs directly to user costs to obtain an overall project cost. However, there seems to be a trade-off between agency expenses and user cost; alternative designs and preservation strategies that reduce certain user costs often entail higher agency expenses (FHWA, 2002). Second, agency costs appear in agency budgets, whereas user costs do not but rather, reflect the "pain and suffering" of the facility users (Walls and Smith, 1998). Other researchers have therefore cautioned that only a fraction of user costs should be considered and added to agency costs. But what fraction of the total estimated user cost should be used? In other words, what is the ratio of the value of agency cost to user costs? Currently, there seems to be no consensus on the issue, and evaluation has often been carried out using a direct summation of agency and user costs.

The *societal cost* of a transportation project includes all of the money spent on the construction, preservation, and operation over the service life of the facility and its salvage costs. In addition, societal cost includes user costs (vehicle operation, crashes, and travel time) and nonuser costs (noise, air pollution, etc.), and rehabilitation and maintenance. These costs are incurred by producers, consumers, other affected parties, taxpayers, and, ultimately, community residents.

SUMMARY

Transportation cost analysis is a key aspect of transportation systems evaluation. To avoid bias in the evaluation it is essential to consider all cost aspects (agency, user, and community costs). Benefits are often viewed as the reduction in costs (typically, user and community costs) relative to a base alternative, but may also comprise incoming money streams (such as toll revenue) and noncost attributes such as improved aesthetics and community cohesion. Costs may be classified by the source of cost incurrence (agency, user, and community), the nature of variation with the output (fixed and variable), the expression of unit cost (average and marginal), and the time in the facility life cycle at which the cost is incurred (planning/design, construction, operations, and preservation). Agency costs comprise capital costs, operating costs, and maintenance costs. User costs are due largely to vehicle operation, travel time, delay, and safety. Community or nonuser costs are typically adverse impacts (such as noise, air pollution, etc.) suffered not necessarily by facility users but also by persons living or working near the facility.

Typically, the first step in transportation system costing is to describe the physical systems and their operations, followed by costing of the required factors of production. Alternatively, the cost of providing transportation facilities or using transportation services can be expressed as a mathematical function of facility attributes such as physical dimensions, types, constituent material, use, or physical or institutional environment. The costing process may be carried out using cost accounting methods (a process that is laborious, relatively accurate, and used for contract bidding) or statistical modeling that expresses a unit dimension of finished product as a function of treatment or facility characteristics. For user and community costs, preemptive costs differ from after-the-fact costs, as the former involves costs incurred by the agency in ensuring that adverse user costs are minimized, whereas the latter refers to costs incurred by users due to unfavorable conditions associated with that user cost type.

Issues associated with the estimation of costs for transportation projects include aggregated planning estimates vs. detailed engineering estimates, adjustments for temporal variations (how to update costs), adjustments for economies of scale, sunk-cost considerations, uncertainties in transportation systems costing, the problem of cost overruns, the ratio of values of agency and user costs, and realistic estimation of future maintenance and operating costs. Historical cost values and models for transportation systems are available in project reports, at agency Web sites, and from other sources. However, such costs may be used for sketch planning only, as they are either averaged over several projects or specific to a past project with unique conditions. The actual cost of a future transportation alternative may be less or more than that estimated at the planning stage, due to factors such as the presence of extraneous structures, the need for ROW purchase, possible environmental impacts, existing soil and site conditions, project size, project complexity, and method of construction delivery, among others.

EXERCISES

4.1. Compare the life-cycle costs of the following transit alternatives on the basis of their cost per seat: a railcar that costs $1,500,000 has 70 seats and an expected life of 25 years; and a bus that has an initial cost of $200,000, 40 seats, and an expected life of eight years. Assume an interest rate of 6%.

4.2. The annual fixed costs of operating a transit system between cities A and B is $5 million. Also, every passenger-mile costs the transit agency 80.56. Determine **(a)** the annual variable costs; **(b)** the total annual costs; **(c)** the average total costs; **(d)** the average marginal costs. Plot a graph of the total, average and marginal cost functions for the transit operation.

4.3. It is proposed to construct a suitable cost-effective surface transit system to connect an airport and suburb to downtown. The distance is 5 miles, and a station is planned for each 1-mile interval. Two alternatives are being considered: light rail and heavy rail. For each system, determine:

   (a) The capital costs for guideways, vehicles, and stations.

   (b) The rehabilitation costs of the vehicles (assume rehabilitation intervals of five years). Assume that negligible rehabilitation and maintenance costs of guideway and stations are negligible.

   (c) The operating costs per year. Assume that operating costs are uniform for each year.

   (d) Draw cash flow diagrams to illustrate the cash outflows for each of 10 years.

4.4. In response to growing passenger and freight demand at Lawrenceville City airport, it is proposed to construct an additional runway. Draw a timetable for release of funds for the various categories of agency costs involved and provide specific examples of costs in each category.

4.5. Discuss the essential differences between the cost accounting and aggregate costing approaches. List the merits and demerits of each approach.

4.6. The fixed operating cost of a transit agency is $50,000 per week. Statistical analyses of historical costs have shown that the variable costs are governed by the following cost function: variable costs $= 0.02V^3 - 4V^2 + 750V$, where V is the weekly ridership. If the average fare is $2.75 per rider, determine the ridership that maximizes revenues of the transit agency. Plot a graph of the total costs, fixed costs, and variable costs. Also, plot a graph of the total cost, average total costs, and marginal total costs.

4.7. The operating costs of a package shipper is governed by the cost function $C = 250V^{3.5}$, where V represents the daily output (number of packages transported in millions). Plot the average and marginal cost functions for $V = 1$ to 5 in unit increments.

4.8. A transportation company has a cost function $C = 10 + 2V + 5V^2$, where C represents the annual total operating costs and V is the number of taxicabs. Provide a plot of the total operating cost function, average operating cost function, and marginal operating cost function. Determine and sketch the elasticity function. Comment on the economy-of-scale implications of the operating costs.

REFERENCES[1]

APTA (2005). *Rail Definitions*, American Public Transit Association, Washington, DC, www.apta.com/research/stats/rail/definitions.cfm. Accessed Aug. 19, 2005.

BART (2006). *BART System Facts, Bay Area Rapid Transit*, www.bart.gov.Accessed Oct. 2006.

Biehler, A. D. (1989). Exclusive busways versus light rail transit: a comparison of new fixed-guideway systems, in *Light Rail Transit: New System Successes at Affordable Prices, Spec. Rep. 221*, Transportation Research Board, National Research Council, Washington, DC, pp. 89–97.

Black, A. (1995). *Urban Mass Transportation Planning*, McGraw-Hill, New York.

Booz Allen Hamilton (1991). *Light Rail Transit Capital Cost Study*, Office of Technical Assistance and Safety, Urban Mass Transportation Administration, U.S. Department of Transportation Washington, DC.

Bordat, C., Labi, S., McCullouch, B., Sinha, K. C. (2003). *An Analysis of Cost Overruns, Time Delays and Change Orders in Indiana*, FHWA/JTRP/2004/07, Purdue University, West Lafayette, IN.

*Cambridge Systematics, Urban Institute, Sydec, H., Levinson, Abrams-Cherwony and Associates, Lea and Elliott (1992). *Characteristics of Urban Transportation Systems*,

---

[1]References marked with an asterisk can also serve as useful resources for cost estimation.

rev. ed., Rep. DOT-T-93-07, Federal Transit Administration, Washington, DC.

Chang, S. E., Shinozuka, M. (1996). Life-cycle cost analysis with natural hazard risk, *J. Infrastruct. Syst.*, Vol. 2, No. 3, pp. 118–126.

CIT (2004). *High-Speed Rail: International Comparisons,* Research Report, Commission for Integrated Transport, London, www.cfit.gov.uk/research. Accessed Apr. 2005.

Dickey, J. W., Miller, L. H. (1984). *Road Project Appraisal for Developing Countries*, Wiley, New York.

ECONorthwest, Parsons Brinckerhoff Quade & Douglas (2002). *Estimating the Benefits and Costs of Public Transit Projects: A Guidebook for Practitioners*, TCRP Rep. 78, Transportation Research Board, National Research Council, Washington, DC.

FHWA (2002). *Economic Analysis Primer*, Federal Highway Administration, U.S. Department of Transportation Washington, DC.

_____ (2005). *Price Trends for Federal Aid Highway Construction*, Federal Highway Administration, U.S. Department of Transportation Washington, DC.

FTA (1993). *Transit Profiles: Agencies in Urbanized Areas Exceeding 200,000 Population*, for the 1992 Section 15 Report Year, Federal Transit Administration, Washington, DC.

_____ (2003). *National Transit Database*, Federal Transit Administration, Washington, DC www.ntdprogram.com.

GAO (1995). *Denver International Airport: Information on Selected Financial Issues*, Letter Report GAO-AMD-95-230, General Accounting Office, Washington, DC, www.colorado.edu/libraries/govpubs. Accessed Oct 10, 2006.

GAO (1996). *Denver Airport: Operating Results and Financial Risks*, Letter Report GAO-AMD-96-27, General Accounting Office, Washington, DC. www.colorado.edu/libraries/govpubs. Accessed Oct 10, 2006.

Hastak, M., Baim, E. J. (2001). Risk factors affecting management and maintenance cost of urban infrastructure, *J. Infrastruct. Syst.*, Vol. 7, No. 2, pp. 67–76.

Hawk, H. (2003). *Bridge Life-Cycle Cost Analysis*, NCHRP Rep. 483, Transportation Research Board, National Research Council, Washington, DC.

Jacoby, C. (2001). Report on supplemental agreement reasons, *AASHTO–FHWA Project Cost Overrun Study*, Federal Highway Administration, U.S. Department of Transportation, Washington, DC.

Jahren, C., Ashe, A. (1990). Predictors of cost-overrun rates, *ASCE J. Constr. Eng. Manage.*, Vol. 116, No. 3, pp. 548–551.

Johnson, F. (1991). What about small buses, *Bus Ride*, July.

Korman, R., Daniel, S. (1998). Audit notes "Avoidable" changes, *Eng. News-Rec.*, Vol. 240, No. 13.

Labi, S., Sinha, K. C. (2003). *The Effectiveness of Maintenance and Its Impact on Capital Expenditures*, Tech.

Rep. FHWA/IN/JTRP-2002/27, Purdue University, West Lafayette, IN.

Labi, S., Rodriguez, M. M., Shah, H., Sinha, K. C. (2006). A performance-based approach for estimating bridge preservation funding needs, presented at the 85th Annual Meeting of the Transportation Research Board, Washington, DC.

Light Rail Central (2002). *Status of North American Light Rail Projects*, North American light rail information site, http://www.lightrail.com/. Accessed Dec. 2005.

LTK Engineering Services (1999). Sound transit link light rail project, *Transit Technol. Rev.*, for *Sound Transit* Feb 2, 1999 Issue.

McCarthy, P. S. (2001). *Transportation Economics: Theory and Practice—A Case Study Approach*, Blackwell Publishers, Malden, MA.

Palisade Corporation (1997). *@Risk User's Manual*, Palisade, Newfield, NY, http://www.palisade.com. Accessed Dec. 12, 2003.

Parsons Brinckerhoff (1996). *Characteristics of Light Rail Transit vs. Heavy Rail Transit Which Have Capital Cost Implications*, PB, Norfolk, VA.

Rowland, H. (1981). The causes and effects of change orders on the construction process, Ph.D. dissertation, Georgia Institute of Technology, Atlanta, GA.

Schneck, D. C., Laver, R. S., Threadgill, G., Mothersole, J. (1995). *The Transit Capital Cost Index Study*, report prepared by Booz Allen Hamilton and DRI/McGraw-Hill for the Federal Transit Administration, Washington, DC, www.fta.dot.gov/transit_data_info/reports.

Thuesen, H. G., Fabrycky, W. J. (1964). *Engineering Economy*, 3rd ed., Prentice Hall, Englewood Cliffs, NJ.

TRB (2003). *Bus Rapid Transit, Vol. 2, Implementation Guidelines*, TCRP Rep. 90, Transportation Research Board, National Research Council, Washington, DC.

Turcotte, J. (1996). *Review of the Florida Department of Transportation's Performance in Controlling Construction Cost Overruns and Establishing Accountability for These Problems*, Rep. 96–21, Florida Legislature, Tallahassee, FL, http://www.oppaga.state.fl.us. Accessed Dec. 12, 2002.

U.S. General Accounting Office. (2001). *Mass Transit: Bus Rapid Transit Shows Promise*, GAO, Washington, DC.

Wagner, T. (1998). Highway construction program, http://www.state.de.us/auditor/highway. Accessed Dec. 10, 2003.

Walls J., Smith, M. R. (1998). *Life Cycle Cost Analysis in Pavement Design*, Technical Rep. FHWA-SA-98-079, Federal Highway Administration, Washington, DC.

*Wilmot, C. G., Cheng, G. (2003). Estimating future highway construction costs, *ASCE J. Constr. Eng. Manage.*, Vol. 129, No. 3, pp. 272–279.

Wohl, M., Hendrickson, C. (1984). *Transportation Investment and Pricing Principles*, Wiley, New York.

CHAPTER 5

Travel-Time Impacts

All my possessions for a moment in time.
— *Elizabeth I (1533–1603)*

INTRODUCTION

There is an old adage that "time is money." But can time have a value? The attributes of time make it unexchangeable and therefore, strictly speaking, time cannot be purchased, sold, or bartered. As such, time has no intrinsic value and therefore the term *value of time* actually means "value of goods, services, or some utility that can be produced within a time interval." When the trip is made in less time than before, the reduction in time is considered as "saved" time even though the difference in time was not really saved but was used to perform another activity. This is the conceptual basis upon which transportation analysts consider reductions in travel time to be a "saving" and proceed to measure its benefits in terms of the amount of time saved and the value of each unit of time saved.

Enhancements to a transportation system are often expected to yield increased travel speed or decreased waiting or transfer times, and consequently, reduced travel time. The savings associated with reduced travel time typically constitute the largest component of transportation user benefits. A conference of European Ministers of Transport in Paris in December 2003 concluded that "the valuation standards of time requirements for transport and time savings as a consequence of transport policies are often decisive for the acceptance or rejection of transport policies and transport infrastructure investment projects" (UNESC, 2004).

In this chapter we present issues associated with travel time as a transportation performance measure and methodologies for the assessment of travel-time amounts

and unit monetary values for the purpose of evaluating the travel-time impacts of transportation projects. Given that the values of travel time vary by certain attributes of the trip and the trip-maker, it is important to establish the various ways by which travel-time amounts may be categorized.

5.1 CATEGORIZATION OF TRAVEL TIME

5.1.1 Trip Phase

On the basis of trip phase, components of travel-time amount may be categorized as *in-vehicle travel time* (IVTT) or *out-of-vehicle travel time* (OVTT). IVTT is the time incurred by passengers or freight in the course of their transportation by rail, air, water, or highway vehicles from one point to another. IVTT can be determined as the ratio of the distance traveled to the average operating speed. Operating speed, in turn, is influenced largely by prevailing traffic conditions.

OVTT is the "excess travel time" spent outside a vehicle during the journey. It includes the time spent waiting at terminals or transferring between modes. For auto travel, the excess travel time may include parking search time and walking time to and from parking. For transit travel, the OVTT components are the walking time to and from the transit stop and the waiting time at each end of the trip. For freight transportation, excess travel time includes primarily modal transfer times at ports and terminals. For both passenger and freight transportation, out-of-vehicle travel times can be increased by security concerns or weather problems. For example, in the post-9/11 period, the time spent by passengers at airports increased because of security screening procedures.

The categorization of travel time on the basis of trip phase is important because travelers typically attach different disutilities to different trip phases. Research findings suggest that irrespective of travel mode, people generally attach a higher degree of undesirability (and therefore, higher disutility and greater time value) to the time spent *waiting* for the vehicle compared to that spent *traveling* in it (Mohring et al., 1987). For freight transportation, intermodal transfer times can be critical in the ability to meet the requirement of just-in-time services.

Example 5.1 A work-bound commuter walks from home to a bus stop and takes a bus to reach rail transit station A in 7 minutes. At the station, the person boards the train and undertakes a 13-minute trip to a downtown bus stop, where she boards a bus that takes her to the workplace in 5 minutes. Tabulate the IVTT and OVTT associated with the journey. Assume a waiting time at the transit center and bus stops of 3 minutes and a walk time of 2 minutes.

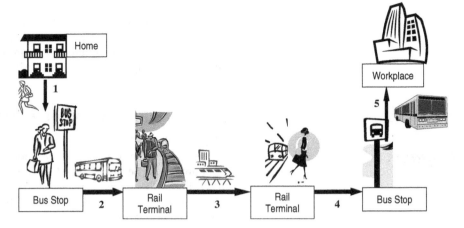

Figure E5.1   Example of trip phases: journey from home to work.

Table E5.1   IVTT and OVTT According to Trip Phase

	Trip Segment	IVTT (min)	OVTT (min)
Journey 1	Walk from home to bus stop	0	2
	Wait at bus stop	0	3
Journey 2	Bus trip from bus stop to rail transit station	7	0
	Wait for rail transit	0	3
Journey 3	Rail transit journey to destination station	13	0
Journey 4	Walk to bus stop	0	2
	Wait at bus stop	0	3
Journey 5	Bus trip from bus stop to workplace	5	0
Total travel time by trip phase		25	13
Total trip travel time		38 min	

SOLUTION   The journey from home to work is illustrated in Figure E5.1, and the IVTTs and OVTTs are tabulated in Table E5.1 according to the trip phase.

5.1.2 Other Bases for Travel-Time Categorization

(*a*) *Traveler Aggregation*   Travel time may be considered with respect to a person or groups of people classified by socioeconomic characteristics, trip origin and destination or trip purpose, vehicle type, and other factors.

(*b*) *Clocking Status*   Travel time is expended by travelers in the course of working (*on-the-clock* travel time) or outside work (*off-the-clock* travel time). Some travel-time estimation procedures treat such travel times separately, as they are likely to have different monetary values.

(*c*) *Flow Entity*   For passenger transportation, hourly travel-time values per dollar are typically expressed per person; for freight transportation, travel time is expressed per ton, cubic foot, gallon, barrel, or other unit.

(*d*) *Time of Day*   Traffic conditions change constantly, and therefore travel speeds and times vary widely from hour to hour. However, two distinct periods of trip-making behavior in a typical day are the peak and off-peak periods, and travel time is typically estimated separately for these two periods.

5.2   PROCEDURE FOR ASSESSING TRAVEL-TIME IMPACTS

The overall framework for assessing travel-time impacts involves the estimation of travel-time amounts, travel-time values, and overall savings in travel-time costs. This is done for two scenarios: a *base-case scenario* (typically, representing the existing situation without intervention) and an *alternative scenario* (typically representing the improved transportation situation after intervention). Specific steps are shown in Figure 5.1 and discussed below.

Step 0: Establish the Base-Case Year   The base case may be for either the current year or a specified future year.

Steps 1 to 3: Estimate the Demand and Capacity Before Intervention   Travel speed and time are the

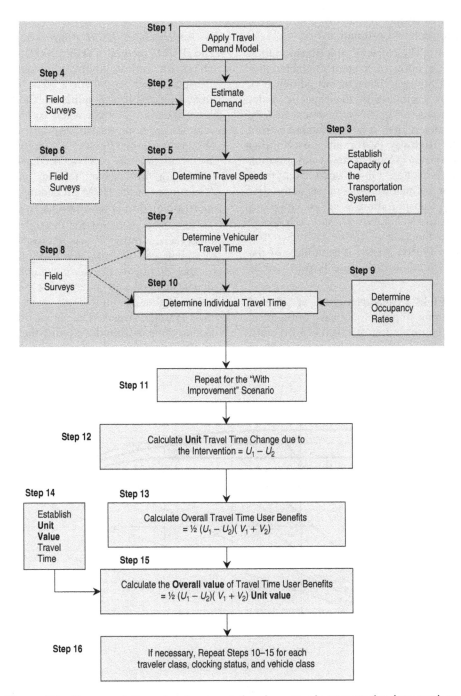

Figure 5.1   Framework for estimating travel time impacts of transportation interventions.

result of both travel demand and capacity of the transportation system. In Steps 1 to 3, therefore, the transportation analyst establishes system demand and capacity so that travel speed and time can be estimated. In base-case scenarios where speed or travel time can be estimated directly from the field, this step can be skipped.

(a) *Demand estimation*   In Chapter 3 we present methods, identify relevant software packages, and provide numerical examples for demand estimation.

(b) *Capacity estimation*   The capacity of a transportation system is typically a function of system characteristics (such as the number of highway lanes or rail guideways). It can be calculated as a product of the capacity under ideal conditions and requisite capacity adjustment factors. Data on system characteristics can be obtained from databases, such as the Highway Performance Monitoring System (HPMS), that currently exist at state transportation agencies in the United States. Given such data, there are methodologies for estimating system capacity. For example, for highway transportation, a set of equations is available in the *Highway Capacity Manual* (HCM) to estimate capacity as a function of traffic characteristics and roadway geometry (TRB, 2000). A summary of the HCM road capacity estimation procedure is provided as Appendix A5.1.

Step 4: Perform Field Measurements of Travel Demand   For the without-improvement case only, as an alternative to (or as a confirmation of results from) steps 1 to 3, it may be necessary to measure the travel demand directly from the field.

Step 5: Determine Travel Speeds before Intervention   Travel speeds may be estimated using approaches provided by the HCM method (TRB, 2000), in which the analyst determines speed as a function of highway class, flow rate, density, and free flow speed (FFS); and the COMSIS method (COMSIS Corporation et al., 1995), in which the analyst determines speed as a function of demand and capacity.

(a) *Approach 1: HCM Approach for Speed Estimation*   The HCM method (TRB, 2000) provides speed–flow curves for various highway classes. Figure 5.2 presents the speed–flow curve for a basic freeway segment with undersaturated flow conditions. The free-flow speed is the mean speed in the field when volumes are less than 1300 vehicles per hour per lane (vphpl). In the absence of field observations, the *Highway Capacity Manual* recommends the calculation of free-flow speed using a set of adjustment factors for traffic characteristics and roadway geometry. A summary of the HCM procedure for roadway operating speed prediction is provided as Appendix A5.2.

The speed of travel for through movements on urban streets where traffic flow is interrupted due to the presence of signals can be estimated using the speed–flow curves in the *Highway Capacity Manual*, as a function of signal density and intersection volume–capacity (*v/c*) ratios. Figure 5.3 shows one such speed–flow curve for class II urban streets. The signal timing and street design assumptions used in developing these curves are provided in the footnotes. Similar curves for different sets of assumptions and classes of urban streets available in the

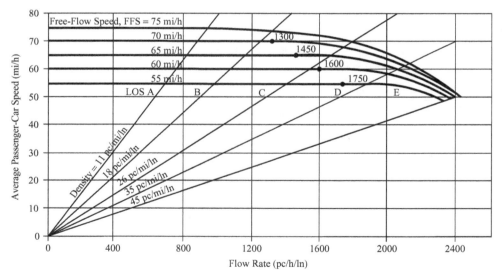

Figure 5.2   Speed flow curves and level of service for basic freeway segments. (From TRB, 2000.)

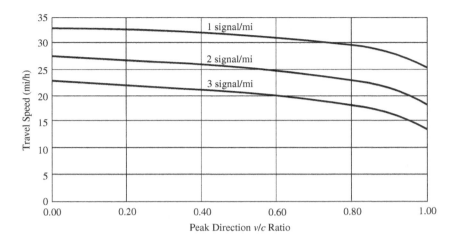

Figure 5.3 Speed flow curves for class II urban streets. Assumptions: 40-mph midblock free-flow speed, 6-mile length, 120-s cycle length, 0.45 g/C. Arrival type 3, isolated intersections, adjusted saturation flow rate of 1700 veh/h, two through lanes, analysis period of 0.25 h, pretimed signal operation. (From TRB, 2000.)

Highway Capacity Manual can be used to determine the average speed at such sections as a function of signal density. For example, using Figure 5.3, the travel speed on a 6-mile urban street with three isolated signalized intersections per mile and peak direction *v/c* ratio of 0.6 is approximately 20 mph.

Example 5.2 Determine the average passenger car speed on a 6-mile urban freeway section during the off-peak period under undersaturated conditions when the flow rate is 1700 vphpl. The free-flow speed is given as 70 mph.

SOLUTION   Using Figure 5.2, corresponding to a free-flow speed of 70 mph and a flow rate of 1700 vphpl, the average passenger car speed is approximately 68 mph under undersaturated conditions.

(b) *Approach II: COMSIS Corporation Method* COM-SIS et al. (1995) provided a procedure for speed estimation under the effects of congestion. Applying traffic simulation model runs with FHWA's FRESIM and NETSIM computer programs, a macroscopic simulation model, QSIM, was developed to examine the effects of queuing on speeds. QSIM produced hourly speed outputs for segments with AWDT/capacity ranging from 1 to 16. Average weekday daily traffic (AWDT) was used instead of annual average daily traffic (AADT) to take into account the effect of varying traffic on weekdays and weekends. Speed look-up tables were developed for the estimation of speed at the end of each hour as a function of AWDT/capacity ratio, depending on the functional

class of the road. Table 5.1 shows the speed look-up table for estimating hourly speed at freeways.

Since the average daily traffic represents the most common traffic demand information for highway networks, the COMSIS approach is well suited for project planning analysis. This method provides an overall measure of the effect of volume changes and capacity improvements on travel time without requiring detailed profiles of volumes by time of day. To use the speed look-up tables, prior determination of the average weekday traffic (AWDT) and roadway capacity is needed. Average weekday traffic (AWDT) can be determined by applying a conversion factor to the AADT. After AWDT and capacity are determined, the hourly speed, daily speed, peak speed, and off-peak speed can be estimated from speed look-up tables such as Table 5.1.

Example 5.3 In 2004, the annual average daily traffic on a 6-mile stretch of Interstate 65 in Indianapolis was 145,210 vehicles. The capacity of the six-lane freeway is 1900 vehicles per hour per lane. Determine the average speed on the freeway during the morning (7:00 to 10:00 a.m.) and afternoon (4:00 to 5:00 p.m.) peak periods using the speed look-up table developed by COMSIS Corporation for urban and rural freeways. Use a factor of 1.0991 for converting AADT to AWDT.

SOLUTION

Annual average daily traffic (AADT) = 145,210 vehicles

Table 5.1   Freeway Speeds on an Average Weekday[a] (Miles per Hour)

Hour Ending	Ratio of Average Weekday Daily Traffic to Capacity															
	1	2	3	4	5	6	7	8	9	10	11	12	13	14	15	16
12 mn. −1 a.m.	59.94	59.89	59.84	59.78	59.72	59.67	59.61	59.55	59.49	59.43	59.37	59.3	59.22	58.96	58.65	58.27
1–2 a.m.	59.97	59.94	59.9	59.87	59.84	59.8	59.77	59.74	59.7	59.66	59.64	59.6	59.55	59.3	59	58.65
2–3 a.m.	59.97	59.95	59.93	59.9	59.87	59.85	59.82	59.8	59.77	59.75	59.72	59.7	59.67	59.42	59.13	58.78
3–4 a.m.	59.97	59.95	59.93	59.91	59.88	59.86	59.84	59.82	59.8	59.78	59.77	59.76	59.73	59.5	59.21	58.87
4–5 a.m.	59.96	59.93	59.89	59.86	59.82	59.78	59.75	59.71	59.69	59.66	59.64	59.63	59.59	59.35	59.06	58.71
5–6 a.m.	59.89	59.8	59.69	59.58	59.47	59.35	59.23	59.12	59.01	58.91	58.8	58.69	58.57	58.29	57.98	57.66
6–7 a.m.	59.7	59.41	59.08	58.73	58.37	57.98	57.56	57.15	56.73	56.25	55.69	54.99	53.83	52.51	50.16	48.57
7–8 a.m.	59.54	59.09	58.56	57.99	57.37	56.73	55.93	54.28	50.56	45.38	40.77	36.86	33.74	30.01	27.34	25.3
8–9 a.m.	59.65	59.33	58.94	58.54	58.11	57.66	57.09	55.52	50.75	43.57	37.21	31.99	27.87	24.56	22.23	20.58
9–10 a.m.	59.74	59.49	59.21	58.92	58.6	58.28	57.94	57.53	56.1	51.18	42.26	33.4	27.54	24.01	21.74	19.98
10–11 a.m.	59.74	59.5	59.22	58.93	58.62	58.3	57.97	57.61	57.2	56.43	53.15	44.21	33.55	27.24	23.88	21.31
11–12 md.	59.72	59.46	59.16	58.84	58.51	58.16	57.79	57.4	56.97	56.51	55.73	52.24	42.13	32.77	26.97	23.04
12–13 p.m.	59.71	59.43	59.12	58.78	58.43	58.06	57.67	57.26	56.82	56.35	55.83	54.14	47.63	38.06	29.75	24.01
13–14 p.m.	59.7	59.42	59.1	58.76	58.39	58.01	57.62	57.19	56.73	56.24	55.69	54.42	50.14	41.55	31.6	24.47
14–15 p.m.	59.67	59.35	58.99	58.6	58.2	57.76	57.31	56.83	56.34	55.79	55.02	53.21	48.32	40.17	30.24	23.18
15–16 p.m.	59.59	59.2	58.74	58.26	57.73	57.17	56.59	56.00	55.32	54.17	51.64	46.85	40.12	32.39	24.88	19.91
16–17 p.m.	59.52	59.06	58.52	57.92	57.29	56.62	55.8	54.49	52.00	47.41	40.97	34.47	28.87	23.98	19.7	17.11
17–18 p.m.	59.52	59.06	58.51	57.91	57.27	56.59	55.54	53.38	48.91	42.11	34.96	28.97	24.31	20.74	17.79	16.12
18–19 p.m.	59.67	59.35	59	58.62	58.2	57.78	57.14	55.59	51.35	43.65	35.04	28.17	23.3	20.01	17.40	15.91
19–20 p.m.	59.77	59.55	59.31	59.05	58.78	58.49	58.2	57.85	56.99	53.65	45.43	34.53	26.26	21.79	18.37	16.34
20–21 p.m.	59.82	59.65	59.46	59.26	59.05	58.84	58.62	58.39	58.15	57.77	55.98	49.27	37.48	28.67	22.29	18.19
21–22 p.m.	59.83	59.68	59.51	59.33	59.14	58.95	58.75	58.54	58.29	58.02	57.71	56.74	52.66	43.71	32.53	23.25
22–23 p.m.	59.86	59.74	59.6	59.46	59.31	59.16	59	58.82	58.61	58.39	58.18	57.92	57.33	54.59	46.24	32.38
23–12 mn.	59.9	59.81	59.71	59.6	59.49	59.38	59.27	59.14	58.99	58.83	58.68	58.52	58.33	57.79	55.68	45.68
Peak[b]	59.59	59.2	58.74	58.24	57.71	57.14	56.39	54.88	51.27	45.16	38.26	32.07	27.27	23.52	20.57	18.69
Off-peak[b]	59.74	59.5	59.21	58.92	58.6	58.27	57.92	57.56	57.12	56.38	54.57	50.31	43.23	36.4	30.20	25.44
Daily	59.68	59.37	59.02	58.64	58.23	57.8	57.28	56.43	54.58	51.24	46.62	41.11	35.3	30.31	25.95	22.71

[a] Free-flow speed of 60 mph assumed in simulation.
[b] Peak period (7:00–10:00 a.m.); off-peak period (4:00–7.00 p.m.)

Therefore,

annual weekday daily traffic (AWDT)

$= (145{,}210)(1.0991) = 159{,}600$ vehicles

per lane capacity $= 1900$ vphpl

two-directional hourly capacity of freeway $= (1900)(6)$

$= 11{,}400$ vehicles/h

Therefore,   AWDT/C $= 159{,}600/11{,}400 = 14$.   From Table 5.1, the average estimated speed during the morning and afternoon peak periods are 26.19 and 23.98 mph, respectively.

Step 6: Perform Field Measurements of Speed   For the base or without-improvement case only, where the travel speed under the existing transportation situation is sought, travel speed can be measured in the field directly as an alternative to (or a way to confirm the results from) step 5. For this there are automated traffic monitoring devices that operate on the basis of laser, radar, infrared, and other technologies. Another way is to drive along with the traffic stream and record the speed of travel.

Step 7: Determine the Vehicular Travel Time before Intervention   Given the simple relationship between travel speed, distance, and time of day, travel time can be found from the speeds estimated using the COMSIS Corporation speed look-up tables. An alternative approach to calculation of travel time is to use the Bureau of Public Roads function (BPR):

travel time (in hours)

$$= t_0 \left[1 + \alpha \left(\frac{\text{traffic flow rate on the link (vphpl)}}{\text{capacity of the link (vphpl)}} \right)^n \right]$$

$$(5.1)$$

where

$$t_0 = \text{free-flow travel time} = \frac{\text{link distance (mi)}}{\text{free-flow speed (mph)}}$$

and α and n are constants.

Example 5.4 Determine the morning and afternoon peak-period travel times on the freeway section in Example 5.3.

SOLUTION The travel speeds during the morning and afternoon peak periods on the freeway were calculated to be 26.19 and 23.98 mph respectively. Therefore, the travel time can be calculated as

$$\text{morning travel time} = \frac{(6)(60)}{26.19} = 13.75 \, \text{min}$$

$$\text{afternoon travel time} = \frac{(6)(60)}{23.98} = 15.0 \, \text{min}$$

Example 5.5 In field studies the traffic flow rate on a four-lane 6-mile section of arterial was reported as 1300 vphpl during the morning peak period. Using the BPR function, determine the travel time on this link during the morning peak period. The capacity of the arterial is 1400 vphpl. Assume that $\alpha = 0.15$ and $n = 4$. The free-flow speed on the arterial is 40 mph.

SOLUTION Using Equation 5.1,

$$\text{travel time} = \left(\frac{6}{40}\right)\left[1 + (0.15)\left(\frac{1300}{1400}\right)^4\right](60)$$
$$= 10 \, \text{min}$$

For the purpose of planning future projects, link or corridor travel times can be obtained from the results of the traffic assignment phase of network-level planning. In cases where network-level assignment data are not available, travel times can be estimated by taking projected traffic volume and capacity as input.

Step 8: Perform Direct Field Measurements of Travel Time For the base case (and for existing transportation conditions in particular), an alternative to the determination of travel time in step 7 (or a way to confirm the results from that step) is to measure travel time directly from the field. For this, the analyst can drive along with the traffic stream and record the time spent on traveling between a specific origin–destination pair. In recent years, the use of license plate recognition, GPS, and other technologies has shown much promise in direct and accurate field measurement of travel time.

Step 9: Determine Occupancy Rates before Intervention This step is needed to convert travel time per vehicle to travel time per vehicle occupant. The vehicle occupancy rates for the base case and the alternative scenarios are generally not expected to differ significantly

except in cases where the transportation intervention is related directly to vehicle occupancies, such as HOV or HOT system implementation and car pooling initiatives.

Step 10: Determine the Average Unit Travel Time without Intervention Unit in-vehicle travel time per traveler,

$$U_1 = \text{OCC} \times \text{TT}_\text{V}$$

where TT_V is the average vehicular operating travel time and OCC is the average vehicle occupancy.

In cases where the travel speeds of trucks and other commercial vehicles are significantly different from passenger vehicles, separate travel time estimates should be made for each vehicle class.

Step 11: Repeat Steps 1 to 10 for the Intervention Scenario Proposed All the steps in the shaded portion of the procedure (with the exception of the field measurements, steps 4, 6, and 8) are repeated for the alternative or intervention scenario. Because this scenario is only hypothetical, no field measurements can be undertaken. Analysts who wish to establish "field" measures of travel demand, travel speeds, or travel times for the intervention scenario (to confirm the values of these parameters) may use available transportation simulation models to accomplish that task.

Step 12: Calculate the Change in Travel Time Expected due to Intervention For most transportation interventions, it is the in-vehicle travel time that is reduced. In a few cases, however, such as the upgrading of freight transfer terminals, construction of additional transit terminals or bus stops, or an increase in transit service frequency, out-of-vehicle travel time is reduced. The change in travel time is given by the expression $U_1 - U_2$, where U_1 and U_2 are the unit travel times without and with the intervention, respectively.

Step 13: Calculate the Travel-Time User Benefits The user benefits of the intervention or improvement, in terms of travel time, are calculated as the change in consumer surplus: $0.5(U_1 - U_2)(V_1 + V_2)$, V_1 and V_2 are the number of trips (or demand) without and with intervention, respectively. In some cases, the intervention may lead to induced travel demand in the long term.

Step 14: Establish the Unit Value of Travel Time In this step, the value of travel time (expressed in terms of dollars/hour/person, for example) is established. This is arguably the most challenging and contentious aspect of travel-time impact analyses. Many transportation agencies have already established travel-time values that can be updated for use in travel-time impact evaluation. Such updating can be carried out using consumer price indices for automobile or transit users and the producer price index for commercial vehicles. Average values of travel time in the United States and other countries are given in Section 5.3.2 and Appendix A.5. The value of travel time varies from place to place and over the years (due to inflation). As such, the use of travel-time value should be carried out with due adjustments made for such considerations.

Example 5.6   In 2000, the value of 1 hour of travel time for automobile users was $16.50. On the basis of CPI trends, determine the value of travel time in 2006.

SOLUTION   From the trends in CPI for passenger transportation,

$$\text{VTT}_{2006} = \text{VTT}_{2000} \times \frac{\text{CPI}_{2006}}{\text{CPI}_{2000}} = (\$16.50)\left(\frac{176.13}{151.58}\right)$$

$$= \$19.17 \text{ per hour}$$

In most countries, it is assumed that the value of time is directly proportional to income, and hence the attributed values of time should change over time in direct proportion to the change in income (typically represented by GDP per capita). Where travel-time values do not exist, the analyst may use one of several available methodologies to establish such values as discussed in Section 5.3.3.

Step 15: Calculate the Value of Travel-Time User Benefits   This is the product of the unit value of travel time (dollars/hour/person) from step 14, and the number of hours represented by the user benefit (from step 13); that is, $0.5(U_1 - U_2)(V_1 + V_2)$(unit value of travel time).

Step 16: If Necessary, Repeat Steps 10 to 13 for Each Traveler Class, Clocking Status, and Vehicle Class   Where the amount and value of travel time is the same for all travelers (or averaged across all travelers), this procedure is carried out only once. However, in cases where travelers and trips are segregated by an attribute such as vehicle class (truck vs. automobile), trip purpose (business vs. personal), type of work-related trip (off-the-clock vs. on-the-clock work), or time of day (peak vs. off-peak), the analysis may be repeated for each attribute and the results are summed up to yield the overall travel-time savings.

5.3   ISSUES RELATING TO TRAVEL-TIME VALUE ESTIMATION

5.3.1 Conceptual Basis of Time Valuation

In allocating time among activities, people implicitly trade off the extra consumption that work earns against the foregone leisure that would be required. There is also the possibility of spending extra money to save travel time and thereby augment the amount of time for working or leisure. This possibility arises in at least three contexts:

1. Choice between a fast and expensive mode or route and a cheaper and slower alternative
2. Choice between costly shortcut routes (often due to tolling) and a free but longer alternative
3. Choice between expensive activity or residences located near a workplace and cheaper activity or residences located far from the workplace

By analyzing the relative sensitivity of such choices to variations in money and time cost, the implicit value of the time of travelers can be estimated. This conceptual framework yields the following important insights into the nature of the value of travel-time savings (Gwilliam, 1997):

- Working time produces goods (which are a direct source of welfare) and therefore has a social value that is independent of the workers' preference values.
- Time vs. money trade-off preferences (and hence the value of travel time) vary from person to person. As such, from a practical viewpoint, some simplifying categorization is vital for travel-time valuation.
- The value of nonwork time could be considered as being equal to the wage rate only in hypothetical situations where persons freely choose how many hours to work and do not consider work to be onerous. As such, nonwork time can only be valued empirically.
- Activity and time are consumed jointly. As such, the value of a time saving is related to the value of its associated activity.
- The value of time savings is a ratio between the marginal utilities of time and money. As such, travel-time value depends on the tightness of the budget constraint (and consequently, income) and the time constraint (and consequently, socioeconomic background and other characteristics of the traveler).

5.3.2 Factors Affecting the Travel-Time Value

Several factors can influence the value of travel time, as shown in Table 5.2. The relative weight of each factor depends on the characteristics of the trip maker and trip, trip length, environmental and seasonal considerations, and mode of travel. Furthermore, given a particular mode of travel, the derived value of travel time depends on the type of approach or model used for the derivation.

Table 5.2   Factors Affecting Value and Amount of Travel Time

Factors Affecting Amount of Travel Time	Factors Affecting Value of Travel Time
How long does it take to travel?	*What is the dollar value of 1 hour of travel?*
Trip length	Mode and vehicle of travel
Vehicle speed	Trip phase (in-vehicle vs. out-of-vehicle)
Vehicle occupancy	Trip purpose and urgency
Other factors	Time of day, day of week, season of year
Weather	Trip location (local vs. intercity)
Security concerns	Traveler's socioeconomic background (age, wage, and occupation)
	Relationship between amount of time used for trip and time used for waiting
	Existing level of legal minimum wage
	Travel-time reduction vs. travel-time extension

(*a*) *Influence of Traveler Income*   Travel-time values have often been estimated as proportions of either personal or household incomes. In general, higher-income travelers value their time more, but the increment in time value is proportionately lower than that of income. Values of time vary between regions within a country as a result of differences in wages and incomes. The evaluation of investments on the basis of travel-time values that reflect such income-related differences (particularly where the users do not pay directly for investment) is likely to yield a vicious cycle: high-income areas yield high project returns, which attract investment and increase income further, whereas the contrary is seen for low-income areas. To avoid this situation, national average wage rates for major categories of labor can be used, and national average income can be applied in the valuation of leisure-time savings, particularly where poverty alleviation or regional redistribution of income is a national objective (Gwilliam, 1997).

(*b*) *Other Traveler Characteristics*   Travelers with higher amounts of free time, such as very young persons and retired elderly persons, are likely to have lower values of time.

(*c*) *Transportation Mode and Vehicle Type*   For a given transportation mode, travel-time factors can play roles that vary from dominant to relatively minor, depending on the class, type, or size of the transportation vehicle. For example, for automobiles and buses, dominant factors include the number of occupants, occupant ages, wages and occupation, trip purpose and urgency, time of day, day of week, season of year, relationship between amount of time used for trip and time used for waiting, and existing legal minimum wage level. For commercial vehicles,

dominant factors include trip purpose, crew wages, and period of travel.

(*d*) *Trip Status (On-the-Clock and Off-the-Clock)*   On-the-clock travel time is associated with work travel, and has values that are based on costs to the employer such as wages and fringe benefits, costs related to vehicle productivity, inventory-carrying costs, and spoilage costs. Off-the-clock trips include trips for commuting to and from work, personal business, and leisure activity. Heavy trucks are assumed to be used only for work, so the value of time for their occupants is the on-the-clock value. Table 5.3 summarizes the estimates of cost components of the value of travel time by vehicle type based on FHWA's HERS software.

(*e*) *Trip Phase (In-Vehicle vs. Out-of-Vehicle)*   The opportunity costs of the time spent inside the vehicle and that spent out of the vehicle may be same but the relative disutility between these two travel-time components may differ from each other. For example, waiting for a bus or train may be more unpleasant than riding in the bus or train, and trip-makers implicitly attach a higher value of travel time for waiting compared to actual traveling. The value of walking and waiting time can be two to three times greater than riding (in-vehicle) (Small, 1992). Recent European studies show that transfer time and waiting time values exceed those of in-vehicle times by a factor of 1.33 to 2, and Chilean studies indicate an even higher ratio. A World Bank publication recommends that where local evidence is unavailable, all "excess" (i.e., out-of-vehicle) travel time should be valued at a premium of 50% above that of in-vehicle travel time (Gwilliam, 1997).

Table 5.3   Distribution of Hourly Travel-Time Values by Vehicle Class (2005 Dollars)

	Vehicle Class						
Category	Small Automobile	Medium-sized Automobile	4-Tire Truck	6-Tire Truck	3- or 4-Axle Truck	4-Axle Combination Truck	5-Axle Combination Truck
Labor/fringe	$32.22	$32.22	$22.10	$26.84	$22.35	$26.92	$26.92
Vehicle productivity	$2.11	$2.48	$2.67	$3.77	$10.78	$9.10	$9.78
Inventory	$0.00	$0.00	$0.00	$0.00	$0.00	$2.02	$2.02
On-the-clock	$34.34	$34.70	$24.77	$30.61	$33.13	$38.04	$38.72
Off-the-clock	$17.54	$17.58	$18.50	$30.61	$33.14	$38.04	$38.73

Source: Updated from Forkenbrock and Weisbrod (2001).

(*f*) *Trip Purpose   Work trips* have usually been valued on the assumption that the value to an employer of the working time of employees must, at the margin, be equal to the wage rate, bumped up by extra costs that are directly associated with employment of labor, such as health benefits, social security taxes, and costs of uniforms. In the United Kingdom, a "bumping-up factor" of approximately 0.33 is typically applied (Gwilliam, 1997). It may be argued that where high levels of unemployment exist, shadow prices below the wage rate could be used.

(*g*) *Trip Length and Size of Travel-Time Reduction*   All other factors remaining the same, differences in trip length may lead to different values of travel time. A recent study (ECONorthwest and Parsons, 2002a) indicated that the value of travel time at peak periods was approximately 45 to 50% of the pretax hourly wage (except for trips of less than 1 mile in length). Time values were determined to range from 8% of the pretax wage rate for trips less than 1 mile, to 49% for trips between 11 and 25 miles, and thereafter dropped to 41%. Off-peak values had the same pattern but were considerably lower than the peak values (generally about two-thirds of the peak values). Also, the unit travel-time value for long trips (travel time exceeding 30 minutes) was 20% higher than that for short trips (travel time less than 20 minutes).

The unit time value for car trips over 50 km in length in Sweden was found to be more than twice that for shorter journeys. For non-car modes travel time value was about 20% higher for long than for short trips. Studies in the UK and the Netherlands showed similar effects, particularly for business travelers. Also, it was determined that the unit value of time was higher when the time savings constituted a larger proportion of the base trip time. The UK and Dutch studies showed very small or zero unit time values for very small time savings (<5 minutes) and

indicated greater unit values for time losses compared to time savings (Gwilliam, 1997).

(*h*) *Direction of Travel-Time Change (Increase vs. Reduction)*   In cases where there is a change in travel time, the value of travel time can also depend on whether the change is favorable (i.e., decreased travel time due to improved conditions) or whether it is adverse (i.e., increased travel time due to worsened travel conditions). In other words, all other factors remaining the same, the value attached to each hour of reduced travel time may be different from that attached to each hour of increased travel time.

(*i*) *Trip Mode*   Some trips (such as park-and-ride trips to work) involve more than one mode. In such cases, the separate effects of changes in aggregate travel times should be identified. Also, empirical evidence suggests that slower modes generally attract low-income travelers who have lower values of time; while faster modes attract travelers with higher incomes and thus higher values of travel time. For example, in-vehicle travel-time values (corrected for income and other factors) were found to be highest for high-speed rail followed by air, car, intercity train, regular train, long-distance bus, and local bus, in that order (VTPI, 2005). Therefore, it has been argued that the time savings for individuals attracted to an improved mode should be valued at the rate appropriate to the mode from which they are transferring.

Travel conditions (which typically are a function of the time of day) significantly influence the value of travel time. Table 5.4 presents the results of a study that investigated travel-time values at Boston and Portland on the basis of transportation mode and time of day. Estimated travel-time costs per passenger mile for peak period and off-peak period travel in areas of high,

Table 5.4   Travel-Time Values in Two Cities (Cents per Passenger-Mile)

City	Urban Density	Expressway Peak	Expressway Off-Peak	Non-expressway Peak	Non-expressway Off-Peak	Commuter Rail Peak	Commuter Rail Off-Peak	Rail Transit Peak	Rail Transit Off-Peak	Bus Peak	Bus Off-Peak	Bicycle Peak	Bicycle Off-Peak	Walk Peak	Walk Off-Peak
Boston	High	24.3	9.6	40.4	23.9	28.9	22.7	40.1	28.6	50.5	39.8	60.6	47.8	243	159
	Medium	15.2	8.0	24.3	15.9	19.8	14.0	28.1	25.3	50.5	39.8	60.6	47.8	202	159
	Low	11.0	8.0	20.2	13.6	19.0	13.3	n/a	n/a	50.5	39.8	60.6	47.8	202	159
Portland, ME	High	11.1	7.8	19.9	13.1	n/a	n/a	n/a	n/a	42.6	33.5	49.8	39.2	166	131
	Medium	10.0	7.1	16.6	11.2	n/a	n/a	n/a	n/a	42.6	33.5	49.8	39.2	166	131
	Low	7.7	6.0	12.4	9.8	n/a	n/a	n/a	n/a	30.2	23.8	49.8	39.2	166	131

Source: VTPI (2005).

medium, and low urban densities are presented. It is clear that the value of travel time (cents per passenger-mile) in congested conditions exceeds that of uncongested conditions, irrespective of travel mode.

5.3.3 Methods for Valuation of Travel Time

Valuation of travel time is typically carried out by comparing travel between two alternative routes or modes, or comparing travel to another economic activity that could have taken place during the travel period. The value of travel time can be found by using the wage rate, revealed preference, or stated preference methods. The basic concept underlying each of these methods can best be explained using time-cost exchange plots.

(*a*) *Exchange Plot*   This involves solicitation of choice preferences of travelers and can be used to explain the behavioral response to travel options varying in terms of time and cost (Hensher and Button, 2000). In this method, the willingness-to-pay concept is considered to be restricted to those who are in a position, and are willing, to trade-off a disadvantage in one attribute to gain an advantage in another. Such persons are referred to as *traders* or *exchangers*. Using this method, the exchange preferences of each person in a group of travelers faced with a choice between two travel options can be obtained. Their respective trade-off values can be plotted on a two-dimensional graph whose axes represent time and cost attributes. Consider two travel options for each traveler such that:

$$\Delta C = \text{cost of option not chosen}$$
$$- \text{cost of option chosen.}$$

$$\Delta t = \text{time for option not chosen}$$
$$- \text{time for option chosen.}$$

$\Delta C > 0$: this indicates that the cost of the chosen option is lower and therefore the traveler is a *cost saver*.

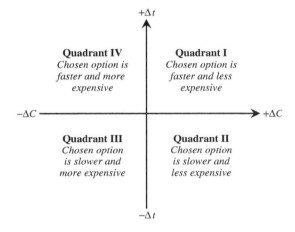

Figure 5.4   Exchange plot for an individual traveler.

$\Delta t > 0$: this indicates that the travel time for the chosen option is less and therefore the traveler is *time saver*. Depending on the sign of ΔC and Δt, an individual traveler can be in one of the four quadrants shown in Figure 5.4.

- *Quadrant I:*   these persons are not exchangers.
- *Quadrant II:*   persons who opt to save cost and spend time, hence $+\Delta C$ and $-\Delta t$. These people are exchangers and cost-savers.
- *Quadrant III:*   these persons are not exchangers.
- *Quadrant IV:*   persons who opt to spend money and save time, hence $-\Delta C$ and $+\Delta t$. These people are exchangers and time-savers.

Exchange plots consider only those people who are faced with a choice situation (i.e., those falling within quadrants II and IV), and involve the following steps:

1. Conduct a survey of travelers by asking how much money they are prepared to pay to gain a certain

amount of time, or how much time they are willing to forego to save a specified amount of money.

2. Plot the trade-off points for various people on an exchange graph.
3. Draw a line through the origin, passing through the two exchange quadrants such that a minimum number of people are misclassified. This can be achieved by making sure that a minimum number of points lie below the line. The line is referred to as the *joint minimum classification* (JMC) *line*.
4. Find the gradient of the JMC line.
5. Compute the reciprocal of the gradient. This is equal to the value of travel time.

Issues associated with exchange plots are as follows:

• This approach is used only when there are equal numbers of observations in quadrant II as in quadrant IV. If there are unequal numbers of observations in the quadrants, a weighting procedure is used for the points in one of the quadrants so that each gradient has an equal weight in determining the location of the JMC line.
• The location of the JMC line is found by manual counting and positioning.
• In this approach, socioeconomic characteristics and other attributes can be considered. Using income levels, for instance, a given sample population can be stratified by income groups, with separate plots made for each income group. Separate values of time can be determined for each group, and the results can be compared for any significant variations.

Exchange plots offer a direct means to explaining the concept of travel-time valuation without resorting to statistical details. When multiple options are involved, this approach is described as *score maximization* to determine the value of travel time (Manski, 1975). The line with the least number of misclassifications provides the maximum score.

Example 5.7 In 2006, a time–cost trade-off survey was conducted among 10 randomly selected commuters along a transportation corridor. People were asked to choose between two alternatives in terms of travel time and cost. Their responses are presented in Table E5.7. Use the exchange plot method to estimate the value of travel time.

SOLUTION   The stated preference data obtained from the survey were used to tabulate Table E5.7. ΔT and ΔC were used to plot the exchange graph shown in Figure E5.7. The JMC line was plotted manually such that the minimum number of people were misclassified and its gradient was calculated as $[15 - (-14)/60]/[2.5 - (-2.5)] = 29/300$. Therefore, the value of travel time = \$10.34/person-hour.

(*b*) *Wage Rate Method*   The wage rate method is the simplest and the most commonly used method to estimate the value of travel time (Forkenbrock and Weisbrod, 2001). In this method, two types of travel time need to be considered: on- and off-the-clock travel time.

Valuation of On-the-Clock Travel Time: Generally, the value of travel time during working periods is

Table E5.7   Travel-Time and Cost Trade-offs

	Time (min)		Cost (dollars)			
Commuter	For the Option Not Chosen (I)	For the Chosen Option II	For the Option Not Chosen (I)	For the Chosen Option II	ΔT (I-II)	ΔC (I-II)
1	68	65	0.65	1.10	3	−0.45
2	45	49	1.36	0.78	−4	0.58
3	57	47	0.42	2.14	10	−1.72
4	55	63	1.73	0.43	−8	1.3
5	55	43	0.89	2.8	12	−1.91
6	56	43	0.90	2.87	13	−1.97
7	58	64	1.83	0.55	−6	1.28
8	53	44	0.80	2.50	9	−1.7
9	50	62	2.44	0.53	−12	1.91
10	56	63	2.45	0.76	−7	1.69

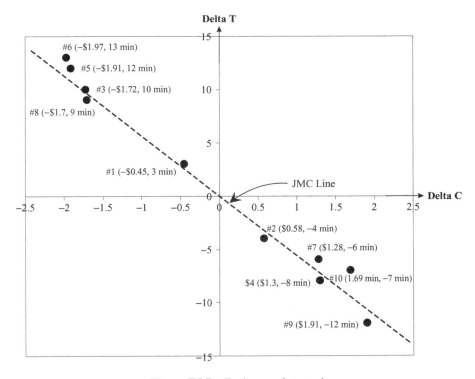

Figure E5.7   Exchange plot graph.

considered equal to the wage rate plus concomitant costs of transportation operations. Particularly, for commercial vehicles, reduced travel time can mean:

- Fewer vehicles are required to haul a given quantity of goods in the same time interval, translating into reduced investment per given output.
- A given vehicle can be used more hours per day or operated more miles during its useful life than it would at greater trip times. Hence, even though depreciation is faster, the rate of depreciation per output is lower.
- Wages are lower for the output achieved.

Examples of on-the-clock travel include technical personnel on their way from office or workshop to attend to a problem or assignment elsewhere, taxi drivers on their usual duty rounds, and roving sales persons, postal and Fedex/UPS delivery workers, and other personnel who advertise, market, or deliver goods and services by moving from one place to another. This includes commercial and industrial haulage.

Work-based travel time may be calculated on the basis of wage rates as follows: Let the wage rate per hour = w (dollars/h), the adjustment for worker benefits = a

(dollars/h), and the value of extra goods and services produced in time interval t (hours) = v_g. Then the value of travel time (dollars/h) = $w + a + v_g/t$.

It is often assumed that any time saving will be converted into additional output by the business traveler or haulage team. In reality, this conversion may not be 100% complete since resources cannot automatically be switched from one task to another. Furthermore, in the case of haulage operations, the maximum use to which travel-time savings may be put depends on the type and size of the crew. Table 5.5 presents the unit work travel-time values as a percentage of wage rate, for various modes.

Valuation of Off-the-Clock Travel Time: HERS considers the value of off-the-clock (nonwork) travel time for drivers as approximately 60% of the wage rate exclusive of benefits, and the value of time for passengers as 45% of the wage rate. Table 5.6 shows the recommended in-vehicle nonwork travel time values as a percentage of the wage rate for various modes of travel. The percentages presented for surface modes apply to all combinations of in- and out-of-vehicle times. The walk access, waiting, and transfer times are valued at 100% of the wage rate.

Table 5.5   Unit Work Travel-Time Values as a Percentage of Wage Rate

	Surface Modes[a]	Air Travel[a]	Truck Drivers
Local travel	100 (80–120)	NA	100
Intercity travel	100 (80–120)	100 (80–120)	100

Source: ECONorthwest and Parsons (2002).
[a]Values in parentheses indicate range. NA, not applicable.

Table 5.6   Unit Nonwork Travel-Time Values as a Percentage of Wage Rate

	Surface Modes[a]	Air Travel[a]	Truck Drivers
Local travel	50 (35–60)	NA	NA
Intercity travel	70 (60–90)	70 (60–90)	NA

Source: ECONorthwest and Parsons (2002).
[a]Values in parentheses indicate range. NA, not applicable.

Table 5.7   Values of Travel Time for Personal and Business Travel

Trip Purpose	Trip Phase	Trip Location	Value of Travel Time (dollars/hour per person)
Personal	In-vehicle	Local	50% of wages
		Intercity	70% of wages
	Out-of-vehicle (waiting, walking, or transfer time)	All locations	100% of wages
Business	In-vehicle	All locations	100% of total compensation
	Out-of-vehicle (waiting, walking, or transfer time)	All locations	100% of total compensation

Source: ECONorthwest and Parsons (2002).

Table 5.7 presents in- and out-of-vehicle travel-time values as a percentage of the wage rate for various modes of travel applicable to both on- and off-the-clock times. According to a World Bank study (Gwilliam, 1997), where it is not possible to derive local values, travel-time values can be estimated using prevailing wage rate and average household income, as shown in Table 5.8.

Example 5.8 It is sought to determine the values of on-the-clock travel time on the basis of the following wage information: hourly wages are $16.25, $12.16, and $16.38 for the users of automobiles, light-duty trucks, and heavy-duty trucks, respectively. Also, the value of fringe benefits (per hour) are $6.44, $6.76, and $9.11, respectively, for the users of these vehicle classes. The average automobile occupancies for on- and off-the-clock

trips are 1.22 and 1.58, respectively. The corresponding average vehicle occupancies for light-duty trucks are 1.03 and 1.18, respectively. The average vehicle occupancy for heavy-duty trucks is 1.04. Assume that the heavy-duty trucks are operated only during working hours. Assume that 10% of all automobile trips and 70% of all light-duty truck trips are made during working hours. These trips include the trips made by rental vehicles and those of automobile trips that are used entirely for work-related travel. The freight inventory value (the time value of the average payload, i.e., the interest cost per hour of the cargo) for heavy-duty trucks is $1.88. Assume that the freight inventory values for light-duty trucks and automobiles are negligible. Determine the value of travel time for personal and work travel.

Table 5.8   Values of Travel Time Based on Wage Rate and Income

Trip Purpose	Rule	Value[a]
Work trip	Cost to employer	$1.33W$
Business	Cost to employer	$1.33W$
Commuting and other nonwork	Empirically observed value	$0.3H$ (for adults), $0.15H$ (for children)
Walking or waiting	Empirically observed value	$1.5 \times$ value for trip purpose
Freight or public transport	Resource cost approach	Vehicle time cost + driver wage cost + occupants' time

Source: Gwilliam (1997).

[a] W, wage rate per hour; H, household income per hour.

SOLUTION   (1) *Computation of the cost of employees per vehicle to employers for 1 hour of travel time*   The cost is computed by multiplying the total compensation of each employee by the average vehicle occupancy of the vehicle:

$$\text{total compensation (dollars/hr)} = \text{wage} + \text{fringe benefits}$$

For automobiles:

$$\text{cost} = \$(16.25 + 6.44)(1.22) = \$27.68/\text{h}$$

For light-duty trucks:

$$\text{cost} = \$(12.16 + 6.76)(1.03) = \$19.49/\text{h}$$

For heavy-duty trucks:

$$\text{cost} = \$(16.38 + 9.11)(1.04) = \$26.51/\text{h}$$

(2) *Computation of the total on-the-clock travel-time value*   This is computed as the sum of the travel-time cost of employees per vehicle to employers and the freight inventory value for the respective vehicle type. The cost of vehicle productivity for each mode is assumed negligible for this case. Table E5.8.1 shows calculated total on-the-clock travel-time values.

(3) *Computation of the weighted average travel-time value for on-the-clock trips based on miles traveled by each mode during working hours*

Weighted travel-time value for automobiles during working hours $= (\$27.68)(0.1) = \$2.77/\text{h}$

Weighted travel-time value for light-duty trucks during working hours $= (\$19.49)(0.7) = \$13.64/\text{h}$

Weighted value of travel time for heavy-duty trucks during working hours $= (\$28.39)(1.0) = \$28.39/\text{h}$

(4) *Total off-the-clock travel-time value*   This is computed as a percentage fraction of wage rates excluding the benefits. It is assumed that heavy-duty trucks do not operate off-the-clock.

For automobiles:

Value of driver's travel time $= 60\%$ of wage rate $= (\$16.25)(0.6)(1) = \$9.75/\text{h}$ (one driver)

Value of passenger's travel time $= 45\%$ of wage rate $= (\$16.25)(0.45)(0.58)(\text{Occupancy} = 1.58) = \$4.24/\text{h}$

Table E5.8.1   Computation of Total On-the-Clock Travel-Time Value (2005 Dollars) for Example 5.8

	Automobiles	Light Trucks	Heavy Trucks
Average vehicle occupancy	1.22	1.03	1.04
Cost of employees	$27.68	$19.49	$26.51
Freight inventory value (per hour)	0.00	0.00	1.88
Total on-the-clock travel-time value	27.68	19.49	28.39

Hence, the total travel time value for automobiles = $9.75 + $4.24 = $13.99/h.

For light-duty trucks:

Value of driver's travel time = 60% of wage rate
  = ($12.16)(0.6)(1) = $7.30/h(one driver)

Value of passenger's travel time = 45% of wage rate
  = ($12.16)(0.45)(0.18)(Occupancy = 1.18)
  = $0.98/h

Hence, the total travel-time value for light-duty trucks = $7.30 + $0.98 = $8.28/h.

(5) *Computation of the weighted off-the-clock travel-time value based on miles traveled by automobiles and light-duty trucks during off-the clock hours.*

Weighted off-the-clock travel-time value for automobiles = ($13.99)(1 − 0.1) = $12.59/h

Weighted off-the-clock travel-time value for light-duty trucks = ($8.28)(1 − 0.7) = $2.48/h

The total weighted average travel time value for each mode is computed by adding the weighted on-the-clock [from Step (3)] and off-the-clock [from Step (5)] travel time values as shown in Table E5.8.2.

The unit travel-time values computed in this example can vary with several other factors (e.g., trip length, income level, traffic density, peak/off-peak hours), as discussed earlier in this chapter.

(*c*) *Revealed Preference Approach (RPA)* In the RPA approach of travel time valuation, actual decisions of travelers regarding the choice of transportation options that differ by travel time and/or travel cost are modeled. Such options could relate to mode choice (fast but costly mode vs. slow but inexpensive mode) or route choice (fast but costly toll route vs. slow but free route).

The underlying principle is that weights (which reflect relative importance) are assigned by travelers to cost and time used for any particular route or mode; the ratio of these weights is a measure of their travel-time value. The proportion of travelers choosing any one of the two alternatives must be known before the ratio can be computed. For two modes or route alternatives *m* and *n*, the proportion of travelers that choose a particular alternative *m* is given as

$$P_m = \frac{e^{U_m}}{e^{U_n} + e^{U_m}} = \frac{1}{1 + e^{U_n - U_m}} \qquad (5.2)$$

where

U_k = satisfaction or utility associated with
       a particular alternative k

$$= \alpha_0 + \sum \alpha_i Z_{ik} \qquad (5.3)$$

Z_{ik} is the *i*th characteristic or service attribute of alternative k (e.g., cost, time, comfort, convenience), and α_0, α_i are coefficients obtained from the revealed behavior of users.

The simplest form of the utility function is when the travel time (t) and travel cost (c) are the only service attributes considered.

$$U_n - U_m = \Delta\alpha_0 + \alpha_1(t_n - t_m) + \alpha_2(C_n - C_m) \qquad (5.4)$$

However, equation (5.4) can account for the circumstances in which the time is spent by including other variables, such as the expected number of crashes and number of speed changes.

Example 5.9   In this example, the two alternatives are a toll route and a non-toll route (free route) from which the traveler must choose. Attributes for each alternative are travel time, out-of-pocket costs (toll and fuel consumption), speed changes (SC) and crash costs (CC). The input data structure for the analysis is shown in Table E5.9 Show how the value of travel time can be estimated.

Table E5.8.2   Weighted Travel-Time Values by Vehicle Class (Dollars/Hour) for Example 5.8

	Automobiles	Light-Duty Trucks	Heavy-Duty Trucks
On-the-clock trips	$2.77	$13.64	$28.39
Off-the-clock trips	12.60	2.48	0.00
Total weighted average	15.37	16.12	28.39

Table E5.9   Input Data Structure for Toll Route vs. Free Route Example

	Travel Time	Toll	Fuel Cost	Crash Cost	Number of Speed-Cycle Changes	Total Out of Pocket Costs	Percentage of Road Users
Alternative 1 (toll route)	T_{toll}	F_{toll}	$Fuel_{toll}$	CC_{toll}	SC_{toll}	$C_{toll} = F_{toll} + Fuel_{toll}$	P_{toll}
Alternative 2 (free route)	T_{free}	$F_{free} = 0$	$Fuel_{free}$	CC_{free}	SC_{free}	$C_{free} = Fuel_{free}$	$1 - P_{toll} = P_{free}$
	ΔT	ΔF	$\Delta Fuel$	ΔCC	ΔSC	ΔC	

Table E5.10   Values for Dependent and Independent Variables Used in Calibration

Sample	P_{toll}	ΔSC (No. of Speed Cycle Changes) (Free−Toll)	ΔT (min) (Free−Toll)	ΔC (Free−Toll)	$\text{Log}_e \dfrac{1 - P_{toll}}{P_{toll}}$
1	0.26	7	15.23	−0.52	1.05
2	0.32	9	13.59	−0.22	0.75
3	0.29	14	12.55	−0.77	0.90
4	0.30	5	19.83	−0.58	0.85
5	0.26	7	15.85	−0.60	1.05
6	0.34	10	19.24	−0.47	0.66
7	0.24	6	16.21	−0.57	1.15
8	0.27	11	13.67	−1.37	0.99
9	0.28	5	18.01	0	0.94
10	0.26	3	19.19	−1.16	1.05

SOLUTION   The differences in utility between the toll and free route can be expressed as follows:

$$U_{free} - U_{toll} = (\alpha_{0\ free} - \alpha_{0\ toll}) + \alpha_1(T_{free} - T_{toll})$$
$$+ \alpha_2(C_{free} - C_{toll}) + \alpha_3(CC_{free} - CC_{toll})$$
$$+ \alpha_4(SC_{free} - SC_{toll})$$

$$P_{toll} = \frac{1}{1 + e^{U_{free} - U_{toll}}}$$

$$U_{free} - U_{toll} = \log_e \frac{1 - P_{toll}}{P_{toll}}$$

$$\log_e \frac{1 - P_{toll}}{P_{toll}} = (\alpha_{0\ free} - \alpha_{0\ toll}) + \alpha_1\ \Delta T + \alpha_2\ \Delta C$$
$$+ \alpha_3\ \Delta CC + \alpha_4\ \Delta SC \qquad \text{(E5.9)}$$

The value of travel time is given by the ratio of the time and cost coefficients, α_1/α_2. The model can also include terms relating to comfort, scenic appeal, and other factors that affect the driving environment.

Example 5.10   Travel choice behavior was observed along 10 locations over a given period during morning peak hours, where commuters had to choose between a toll road and a free road. The differences between trip costs, travel times, and speed-cycle changes are given in Table E5.10 for all the locations. The fraction of commuters choosing the toll road over the free road is also given. Determine the travel-time value (TTV) per vehicle and per person assuming average vehicle occupancy of 1.15.

SOLUTION   The model given in equation (E5.9) can be calibrated using the data.

$$\log_e \frac{1 - P_{toll}}{P_{toll}} = (\alpha_{0\ free} - \alpha_{0\ toll})$$
$$+ \alpha_1\ \Delta T + \alpha_2 \Delta C + \alpha_3 \Delta SC$$

It is assumed that the crash cost is the same on both the routes and is not a consideration in the decision-making process.

The calibrated model using linear regression is as follows:

$$\log_e \frac{1 - P_{\text{toll}}}{P_{\text{toll}}} = \begin{array}{c} (1.97) \\ (4.650) \end{array} - \begin{array}{c} (0.04656) \\ (-2.353) \end{array} \Delta T$$

$$- \begin{array}{c} (0.146) \\ (-1.590) \end{array} \Delta C - \begin{array}{c} (0.047) \\ (-2.986) \end{array} \Delta SC$$

$$R^2 = 0.648$$

The numbers in parentheses (*t*-statistics) indicate that all the variables are significant. Therefore,

$$\text{TTV(per vehicle)} = \frac{\alpha_1}{\alpha_2} = \left(\frac{-0.04656}{-0.146} \right) (60)$$

$$= \$19.12/\text{vehicle-h}$$

$$\text{TTV(per person)} = \frac{\$19.12}{1.15} = \$16.63/\text{person-h}$$

(*d*) *Stated Preference Approach* (SPA)   SPA involves a willingness-to-pay (WTP) survey of individual travelers (by polling or using questionnaires), presenting a series of hypothetical choices closely related to their current modes of travel through repetitive questioning. The change in cost of their present mode or route that would be just sufficient to cause them to switch to the another mode or route can be determined. Such a cost can be termed *switching threshold*.

At relatively little cost and on the basis of a single experiment, SPA can be used in a wide range of contexts offering alternatives designed to give numerous credible trade-off possibilities.

For any two routes or modal alternatives, A and B, the binary logit model can be represented by

$$P_{\text{B}} = \frac{1}{1 + e^{U_{\text{A}} - U_{\text{B}}}} \tag{5.5}$$

where

$$U_{\text{A}} - U_{\text{B}} = \beta_0 + \beta_1(t_{\text{A}} - t_{\text{B}}) + \beta_2(C_{\text{A}} - C_{\text{B}} + \text{ST}_{\text{B}})$$
$$+ \beta_3(\text{CC}_{\text{A}} - \text{CC}_{\text{B}}) \tag{5.6}$$

Here ST_{B} is the switching threshold for alternative B, t_{A} and C_{A} are the time and cost associated with alternative A, and t_{B} and C_{B} are the time and cost associated with alternative B.

CC_{A}, CC_{B} = Crash cost associated with A and B, respectively.

By including the switching threshold ST_{B} in the utility function, the traveler is made indifferent to any specific route or mode choice. The point of indifference (which represents a 50–50 chance of either option being chosen) occurs when $U_{\text{A}} - U_{\text{B}} = 0$. Hence, equation (5.6) can be rewritten as

$$(C_{\text{A}} - C_{\text{B}} + \text{ST}_{\text{B}}) = \lambda_0 + \lambda_1(t_{\text{A}} - t_{\text{B}}) + \lambda_2(\text{CC}_{\text{A}} - \text{CC}_{\text{B}}) \tag{5.7}$$

The value of travel time is given by the coefficient λ_1.

There may be some difficulty in measuring the switching threshold. Some travelers may not be able to envision and properly weigh the options and reliably define what their indifference threshold would be unless they actually experience it. It may be assumed that underestimates and overestimates given by individuals cancel out to produce a reasonably accurate average value of travel time.

Example 5.11   Two travel alternatives are available to commuters traveling between the downtown and suburbs of Metropolis city: rapid rail transit (RRT) and a slower but less expensive surface bus transit (SBT). In a survey, ten SBT users were asked to indicate the amount of money (between zero and five dollars, that would have to be paid to them in order for them to consider RRT as equally attractive as SBT (in other words, the travelers were asked to indicate their switching thresholds). The switching thresholds, and the travel time and cost differentials, are given in Table E5.11. Calculate travel-time value. Assume all other attributes are the same for the two modes.

SOLUTION   Using the Logit Model,

$$P_{\text{RRT}} = \frac{1}{1 + e^{U_{\text{SBT}} - U_{\text{RRT}}}}$$

where P_{RRT} is the probability that an individual travels using RRT and U is the utility attached by an individual to his or her travel choice. The expression can be rewritten as

$$\frac{1 - P_{\text{RRT}}}{P_{\text{RRT}}} = e^{U_{\text{SBT}} - U_{\text{RRT}}}$$

$$\log_e \left(\frac{1 - P_{\text{RRT}}}{P_{\text{RRT}}} \right) = U_{\text{SBT}} - U_{\text{RRT}}$$

When a traveler considers both modes to be equally attractive, $P_{\text{SBT}} = P_{\text{RRT}} = 0.5$. Hence,

$$\log_e \left(\frac{1 - 0.5}{0.5} \right) = U_{\text{SBT}} - U_{\text{RRT}}$$

$$0 = \beta_0 + \beta_1(T_{\text{SBT}} - T_{\text{RRT}}) + \beta_2[C_{\text{SBT}} - (C_{\text{RRT}} - \text{ST}_{\text{RRT}})]$$

$$C_{\text{SBT}} - (C_{\text{RRT}} - \text{ST}_{\text{RRT}}) = \lambda_0 + \lambda_1(T_{\text{SBT}} - T_{\text{RRT}})$$

$$\Delta C + \text{ST}_{\text{RRT}} = \lambda_0 + \lambda_1 \Delta T \tag{5.8}$$

Table E5.11   Time and Cost Data for Model Calibration and Switching Threshold Values

Individual	ΔT (mins/trip) ($\text{TIME}_{\text{SBT}} - \text{TIME}_{\text{RRT}}$)	ΔC ($/trip) ($\text{COST}_{\text{SBT}} - \text{COST}_{\text{RRT}}$)	ST_{RRT} ($/trip)	$\Delta C + \text{ST}_{\text{RRT}}$ ($/trip)
1	3.00	−$2.00	1.00	−1.00
2	8.00	−$3.50	1.50	−2.00
3	6.50	−$3.50	1.75	−1.75
4	5.50	−$2.50	1.00	−1.50
5	4.00	−$2.50	1.50	−1.00
6	7.00	−$5.00	2.75	−2.25
7	5.00	−$4.00	2.75	−1.25
8	1.50	−$3.00	2.25	−0.75
9	7.00	−$4.00	2.00	−2.00
10	8.50	−$5.50	3.00	−2.50

where T_i and C_i represent the travel time and cost associated with mode i. The variable ΔT indicates the additional time taken by "default" alternative (in this case, the surface bus transit) compared to other alternative (in this case, rapid rail transit) for each trip. For each traveler, the variable ΔC represents the additional travel cost of the default alternative relative to the other alternative, and ST_{RRT} represents the traveler's threshold cost value for switching from the default alternative (surface bus transit) to the other alternative (rapid rail). The data for travel time and cost for the two modes and switching threshold values are provided in Table E5.11.

Using any standard statistical software, the regression model shown in Equation 5.8 can be calibrated as follows:

$$\Delta C + \text{ST}_{\text{RRT}} = 0.194 - (0.251)(\Delta T) \qquad R^2 = 0.91$$
$$(1.14) \quad (-8.93)$$

The values in parentheses are the t-statistics of the coefficients. The value of the travel time (per person-hr) TTV can be calculated using the coefficient of ΔT:

$$\text{TTV} = (0.251)(60) = \$15.07/\text{person-hour}$$

The use of logit models to estimate the travel-time value can be generalized further by allowing the parameters in the utility model to vary in the population to account for random taste heterogeneity (Hess et al., 2004). The estimated travel-time value using logit models is sensitive to the model specification. Algers et al. (1998) found that the travel-time value obtained from ordinary logit model specification with fixed model parameters as used here was significantly lower than the value estimated from mixed logit model specification when the coefficients were assumed to be normally distributed in the population.

5.4 CONCLUDING REMARKS

With increased globalization, specialization, and transportation seamlessness, it is expected that travel time, as an evaluation criterion, will play an increasingly important role. As noted in a recent publication by the United Nations (UNESC, 2004), the time costs of international trade have become more important than the resource costs of transportation as evidenced by the strong shift to freight air transport even though air transportation costs, at about 25% of the product value, exceed surface transportation costs. A major reason for this development is the shortening of product cycles. These developments concern not only relatively small high-tech sectors but also labor-intensive sectors, such as the clothing industry. As such, proximity to major market areas seems to be an increasingly important determinant for the location of industries relative to the real wage costs at different locations. The increased importance of transportation times for international and interregional trade indicates the challenge for transport policy to react to, anticipate, and support these developments.

SUMMARY

Transportation provides a means for people and goods to move from one point to another, and travel time is a major resource that is spent in achieving this goal. Transportation system interventions are generally expected to result in increased travel speed (and consequently, reduced travel time). When the trip is made in less time than before, the reduction in time, considered as "saved" time, is used to perform another activity. On the basis of travel time and cost trade-offs, the value of travel time can be estimated and the time-reduction benefits of transportation interventions can be determined. There are countless vital public and private transportation projects of various modes where travel-time savings constitute a large fraction of economic benefits.

In estimating overall travel-time costs or benefits, two important elements are the amount of travel time and the unit value of the travel time. Travel time can be categorized on various bases including trip phase, flow entity, and clocking status. The overall framework for assessing travel-time impacts involves consideration of a base-case scenario and the improvement scenario. The steps involve establishment of the base year; estimation of the demand and capacity of the transportation system with and without intervention; determination of travel speeds and times; field measurements to determine (or confirm) travel demand, speeds, and times; determination of vehicle occupancy rates with/without intervention; calculation of savings (or increase) in travel-time amounts due to the intervention; establishing the unit value of travel time; and calculating the overall cost savings (or increase) in travel-time costs for all traveler classes, clocking status, and vehicle classes.

Behavior exhibited by travelers that enable travel-time valuation are typically in the context of choice between fast and expensive modes or routes and cheaper, slower alternatives, and choice between costly activity or residences located near a workplace and cheaper activity or residences located far from the workplace. By analyzing the relative sensitivity of such choices to variations in time and cost, the implicit value of travel time of travelers can be identified. The valuation of travel time is considered a challenging task and may show some inconsistencies due

to reasons such as difficulty in isolating the relationship between travel-time value and travel characteristics, costliness of data collection, differences between perceived travel costs and actual travel costs, and lack of a consistent explanation of consumer behavior in situations where consumption activities involve the expenditure of time as well as money.

The use of travel time as a transportation investment performance measure (and consequently, as a criterion for impact evaluation) is widespread. In some countries, lack of local information on the value to time savings has led to the exclusion of travel-time savings in economic evaluation.

EXERCISES

5.1. The AADT on a 4-mile stretch of I-70 in Marion County in 2005 was reported as 160,500. The capacity on the eight-lane freeway is 1750 vehicles per hour per lane. Plot the hourly travel time profile for the freeway using the speed look-up table developed by COMSIS Corporation (Table 5.1). Use a conversion factor of 1.12 for converting the AADT to AWDT. A reconstruction project increases the number of lanes on the freeway to 10 and the capacity to 1900 vehicles per hour per lane. Calculate the travel-time savings in the morning peak period between 8:00 and 9:00 a.m. of the opening year (2010). The value of travel time is $14.50 per person per hour in the current year (2005). The CPI index for 2005 is 160.40 and for 2010 is 190.85. Assume that the average vehicle occupancy is 1.07 and that there are 250 working days in the opening year.

5.2. Prove that the value of travel time is given by the ratio of coefficient of travel time and cost in the route choice utility model. Assume that the utility model includes only these two route-specific

Table EX5.3   Input Data for Wage Rate Based Approach

	Trip Purpose	Vehicle Hours Saved	Percent Miles Traveled	Unit Travel-Time (dollars) Value	Average Vehicle Occupancy
Local auto	On-the-clock	300	10		1.22
	Off-the-clock		90		1.58
Intercity auto	On-the-clock	150	15		1.12
	Off-the-clock		85		1.62
Light trucks	On-the-clock	60	100	19.49	1.03
Heavy trucks	On-the-clock	80	100	30.43	1.00

Table EX5.4   Travel Time and Cost Data for Exchange Plot Approach

$T_{toll} - T_{free}$ (min)	4	−6	12	−10	−5	−8	12	10	−5	11
$C_{toll} - C_{free}$ ($)	−0.5	1.75	−1.75	2	1.5	1	−2.25	−2	1.8	−1.85

Table EX5.5   Data for Binary Logit Model to Estimate Travel-Time Value

Location	Fraction Choosing Route B	Travel Time (min)		Travel Cost (dollars)		No. of Speed Changes	
		A	B	A	B	A	B
1	0.68	13.47	31.51	2.35	1.47	12	22
2	0.72	18.06	32.10	2.36	1.80	10	20
3	0.69	18.34	30.86	2.76	1.69	11	27
4	0.69	16.04	33.76	2.72	1.68	15	23
5	0.74	19.09	31.18	2.81	1.73	14	26
6	0.68	18.09	35.44	2.43	1.66	12	24
7	0.72	16.65	29.87	2.51	1.50	15	23
8	0.73	15.68	27.62	3.15	1.48	10	21
9	0.72	15.34	33.35	2.41	1.79	16	22
10	0.73	16.98	37.74	3.43	1.86	16	21

variables. How does the value of travel time change if socioeconomic variables of the traveler are included in the model?

5.3. An economic evaluation has to be performed for a congestion mitigation project implemented on U.S. Route-52 in Indiana. The vehicle hours of travel time saved, unit travel-time value, and the average vehicle occupancy of each mode are given in Table EX5.3. Compute the travel-time savings using the plausible range of travel-time values recommended by USDOT (Tables 5.4 and E5.7). Assume that the wage rate is $16.25 for the automobile passengers and that the fringe benefits are worth $6.44.

5.4. Determine the value of travel time using the exchange plot method for the travel-time and travel-cost data in Table EX5.4, obtained from a stated preference survey of 10 commuters facing the choice of a toll road or a free road.

5.5. Determine the value of travel time using the binary logit model from the route choice data given in Table EX5.5. Assume an average vehicle occupancy of 1.3.

REFERENCES

Algers, S., Bergstrom, P., Dahlberg, M., Dillen, J. L. (1998). Mixed logit estimation of the value of travel time, http://www.nek.uu.se/Pdf/1998wp15.pdf. Accessed Jan. 2006.

COMSIS Corporation, Scientific Applications International Corporation, Garman Associates (1995). *Development of Diurnal Traffic Distribution and Daily, Peak, and Off-Peak Vehicle Speed Estimation Procedures for Air Quality Planning,* Work Order B-94-06, Federal Highway Administration, U.S. Department of Transportation, Washington, DC.

ECONorthwest, Parsons Brinckerhoff Quade & Douglas (2002a). *Estimating the Benefits and Costs of Public Transit Projects: A Guidebook for Practitioners,* TCRP Rep. 78, Transportation Research Board, National Research Council, Washington, DC.

_____(2002b). *A Guidebook for Developing a Transit Performance-Measurement System,* TCRP Tech. Rep. 88, Transportation Research Board, National Research Council, Washington, DC.

FHWA (2002). *The Highway Economic Requirement System (HERS): Technical Manual,* Federal Highway Administration, U.S. Department of Transportation, Washington, DC.

Forkenbrock, D., Weisbrod, G. E. (2001). *Guidebook for Assessing the Social and Economic Effects of Transportation Projects,* NCHRP Rep. 456, Transportation Research Board, National Research Council, Washington, DC.

Gwilliam, K. (1997). *The Value of Time in Economic Evaluation of Transport Projects: Lessons from Recent Research,* Transport No. OT-5, World Bank, Washington, DC.

Hess, S., Bierlaire, M., Polak, J. W. (2004). Estimation of value-of-time using mixed logit models, www.trb-forecasting.org/papers/2005/. Accessed Jan. 11, 2006.

Manski, C. (1975). Maximum score estimation of the stochastic utility model of choice, *J. Econometr.,* Vol. 3, pp. 205–228.

Mohring, H., Schroeter, J., Paitoon, W. (1987). The values of waiting time, travel time, and seat on a bus, *Rand J. Econ.,* Vol. 18, No. 1.

Small, K. (1992). Urban transportation economics, in *Fundamentals of Pure and Applied Economics*, Vol. 51, ed. Lesourne, J., Sonnenschein, H., Harwood Academic Publishers, Chur, Switzerland.

TRB (2000). *Highway Capacity Manual*, Transportation Research Board, National Research Council, Washington, DC.

UNESC (2004). *Transport Trends and Economic Studies on Transport Economics and Track Costs Undertaken by Other Organizations*, TRANS/2005/7/Add.1, United Nations Economic and Social Council, United Nations Organization, New York.

VTPI (2005). *Transportation Cost and Benefit Analysis: Travel Time Costs*, Victoria Transportation Policy Institute, Victoria, BC, Canada, http://www.vtpi.org/tca/tca0502.pdf. Accessed Aug. 19, 2005.

ADDITIONAL RESOURCES

Heggie, I. (1972). *Transport Engineering Economics*, McGraw-Hill, London.

Hensher, D. A. (1977). *The Value of Business Travel Time*, Pergamon Press, Oxford.

Li, J., Gillen, D., Dahlgren, J. (1999). Benefit-cost Evaluation of the Electronic Toll Collection System: *A Comprehensive Framework and Application*, Transp. Res. Rec. 1659, Transportation Research Board, National Research council, Washington, DC.

Stopher, P. R., Meyburg, A. H. (1976). *Transportation Systems Evaluation*, Lexington Books, D. C. Health and Company, Lexington, MA.

USDOT. (1997). Memorandum: Revised departmental guidance for the valuation of travel time in economic analysis, U.S. Department of Transportation, Washington, DC, http:ostpxweb.dot.gov/policy/Data Accessed Jan 2006.

Wardman, M. (1998). The value of travel time: a review of British evidence, *J. Transp. Econ. Policy*, Vol 32, No. 3.

Weisbrod, G., Weisbrod, B. (1997). *Measuring Economic Impacts of Projects and Programs*, Economic Development Research Group, Boston, MA.

APPENDIX A5.1: ESTIMATION OF ROADWAY CAPACITY USING THE HCM METHOD (TRB, 2000)

The primary objective of capacity analysis is to estimate the maximum number of vehicles a facility can accommodate with reasonable safety during a specified time period. The capacity of a roadway segment is highest when all roadway and traffic conditions meet or exceed their base values. These base conditions, which are determined using empirical studies, assume good weather, familiarity of users with transportation facility, good pavement conditions, and uninterrupted traffic flow. In general, the conditions that prevail on most highways are different from the base conditions. As a result, the computations of capacity, service flow rate, and level of service require adjustments.

HCM classifies transportation facilities into two categories of flow: uninterrupted and interrupted. Freeways are an example of an uninterrupted flow facility. The multilane highways and two-lane highways can also have uninterrupted flow in long segments between two points of interruption. This appendix summarizes the HCM capacity analysis methodology for freeways, multilane highways, and two-lane highways.

(*a*) *Basic Freeway Segments* A divided roadway segment having two or more lanes in each direction, full access control, and uninterrupted flow irrespective of traffic merging and diverging from ramps is referred to as a basic freeway segment. The base conditions for basic freeway segments are as follows:

- A minimum lane width of 12 ft
- Minimum right shoulder clearance (between the edge of the travel lane and objects) of 6 ft
- Minimum median lateral clearance of 2 ft
- Traffic stream comprising passenger cars only
- Five or more lanes in each direction of travel (urban areas only)
- Interchange spacing greater than 2 miles
- Driver population comprising of users of high familiarity
- Level terrain (no grades greater than 2%)

As the operating conditions are more restrictive than the base conditions, the base free-flow speed is adjusted according to the extent of deviation from the base conditions, resulting in a reduced free-flow speed. Table A5.1.1 shows the relationship between capacity and free-flow speed for basic freeway segments. It can be noted from Table A5.1.2 that, given a free-flow speed, the capacity of a basic freeway segment is the maximum service flow rate at LOS E. This is because the upper boundary of the LOS E corresponds to a volume/capacity (v/c) ratio of 1.0.

Table A5.1.1 Relationship between Free-Flow Speed and Capacity on Basic Freeway Segments and Multilane Highways

Basic Freeway Segments		Multilane Highways	
Free-flow Speed(mi/h)	Capacity (pc/h/ln)	Free-flow Speed (mi/h)	Capacity (pc/h/ln)
75	2400	60	2200
70	2400	55	2100
65	2350	50	2000
60	2300	45	1900
55	2250		

Table A5.1.2   LOS Criteria for Basic Freeway Segments and Multilane Highways

Criterion	Basic Freeway Segments LOS					Multilane Highways LOS				
	A	B	C	D	E	A	B	C	D	E
	FFS = 75 mi/h					FFS = 60 mi/h				
Maximum Density (pc/mi/ln)	11	18	26	35	45	11	18	26	35	40
Average Speed (mi/h)	75	74.8	70.6	62.2	53.3	60	60	59.4	56.7	55
Maximum v/c	0.34	0.56	0.76	0.9	1	0.3	0.49	0.7	0.9	1
Maximum Service Flow Rate (pc/h/ln)	820	1350	1830	2170	2400	660	1080	1550	1980	2200
	FFS = 70 mi/h					FFS = 55 mi/h				
Maximum Density (pc/mi/ln)	11	18	26	35	45	11	18	26	35	41
Average Speed (mi/h)	70	70	68.2	61.5	53.3	55	55	54.9	52.9	51.2
Maximum v/c	0.32	0.53	0.74	0.9	1	0.29	0.47	0.68	0.88	1
Maximum Service Flow Rate (pc/h/ln)	770	1260	1770	2150	2400	600	990	1430	1850	2100
	FFS = 65 mi/h					FFS = 50 mi/h				
Maximum Density (pc/mi/ln)	11	18	26	35	45	11	18	26	35	43
Average Speed (mi/h)	65	65	64.6	59.7	52.2	50	50	50	48.9	47.5
Maximum v/c	0.3	0.5	0.71	0.89	1	0.28	0.45	0.65	0.86	1
Maximum Service Flow Rate (pc/h/ln)	710	1170	1680	2090	2350	550	900	1300	1710	2000
	FFS = 60 mi/h					FFS = 45 mi/h				
Maximum Density (pc/mi/ln)	11	18	26	35	45	11	18	26	35	45
Average Speed (mi/h)	60	60	60	57.6	51.1	45	45	45	44.4	42.2
Maximum v/c	0.29	0.47	0.68	0.88	1	0.26	0.43	0.62	0.82	1
Maximum Service Flow Rate (pc/h/ln)	660	1080	1560	2020	2300	490	810	1170	1550	1900
	FFS = 55 mi/h									
Maximum Density (pc/mi/ln)	11	18	26	35	45					
Average Speed (mi/h)	55	55	55	54.7	50					
Maximum v/c	0.27	0.44	0.64	0.85	1					
Maximum Service Flow Rate (pc/h/ln)	600	990	1430	1910	2250					

(b) *Multilane Highways*   The base conditions for multi-lane highways are as follows:

- A minimum lane width of 12 ft
- Minimum total lateral clearance of 12 ft from roadside objects (right shoulder and median) in the travel direction
- Traffic stream comprising passenger cars only
- Absence of direct access points along the roadway segment
- Divided highway
- Level terrain (grade less than 2%)
- Driver population comprising of highly familiar roadway users
- Free-flow speed higher than 60 mi/h

The operating free-flow speed is calculated through adjustments to the base free-flow speed according to the prevailing conditions. Procedures for making speed adjustments are discussed in Appendix A5.2.

Table A5.1.1 shows the relationship between free-flow speed and capacity for multilane highways. Again, it is important to note that the values of capacity correspond to the maximum service flow rate at LOS E and a v/c ratio of 1.0.

(c) *Two-Lane Highways*   The base conditions for two-lane highways are as follows:

- A minimum lane width of 12 ft
- Minimum shoulder width of 6 ft
- Highway segment with 0% no passing zones
- Traffic stream comprising of passenger cars only
- No direct access points along the roadway
- Level terrain (grade less than 2%)
- No impediments to through traffic due to traffic control or turning vehicles
- Directional traffic split of 50/50

The capacity for extended lengths of two-lane highway segments under base conditions is 3200 passenger cars per hour combined for both directions. For short lengths of two lane highways, such as bridges or tunnels, the capacity varies from 3200 to 3400 passenger cars per hour for both directions of travel combined.

Example A5.1   Determine the capacity (per lane) on a six-lane divided urban freeway. The free-flow speed was found to be 57.5 mi/h after adjustments for lane width, lateral clearance, number of lanes, and interchange density were made to the base free-flow speed.

SOLUTION   From Table A5.1.1, the capacity corresponding to a free flow speed of 55 mi/h is 2250 pc/h

and corresponding to 60 mi/h is 2300 pc/h. Interpolating linearly, the capacity corresponding to a free-flow speed of 57.5 mi/h will be 2275 pc/h for each lane on the six-lane divided urban freeway.

Alternatively, Exhibit 23-15 on Page 23-14 in HCM (2000) could be used to determine the capacity of the basic freeway segment on the basis of its interchange spacing (in miles) and number of lanes.

APPENDIX A5.2: ESTIMATION OF ROADWAY OPERATING SPEEDS USING THE HCM METHOD (TRB, 2000)

Given the travel demand and system capacity from step 3, the travel speeds can be estimated for both the base case and the case under investigation. This may be done using network-wide travel demand modeling for an overall network (which yields results for each link in the network) or solely for a single link. Even where only a single route or link is under investigation, network-level analyses are typically preferred, because unlike the project-level speed estimation, they typically give due cognizance to trips diverted to or from other routes from or to the facility under the improvement scenario. The vital overall contribution of travel speeds to an evaluation of transportation effects is evidenced in its due consideration to a wide range of impact types, such as vehicle operating costs, vehicular emissions, noise, and energy use. Besides field monitoring, travel speeds may be estimated using approaches provided in the HCM or using the COMSIS method as discussed in Section 5.2. This appendix discusses the HCM method.

The free-flow speed is the mean speed of passenger cars measured under low-to-moderate flows (under 1300 pcphpl). Speeds on a specific freeway section are expected to be virtually constant in this range of flow rates. The free-flow speed can be estimated indirectly on the basis of the physical characteristics of the freeway section under investigation. These physical characteristics include lane width, right-shoulder lateral clearance, number of lanes, and interchange density. The following equation can be used for the estimation of free-flow speed:

For basic freeway sections:

$$FFS = FFS_i - f_{LW} - f_{LC} - f_N - f_{ID}$$

For multilane rural and suburban roads:

$$FFS = FFS_i - F_M - F_{LW} - F_{LC} - F_A$$

where FFS = estimated free-flow speed (mph)
FFS_i = estimated ideal free-flow speed, 70 or 75 mph

f_{LW} = adjustment for lane width
f_{LC} = adjustment for right-shoulder lateral clearance
f_N = adjustment for number of lanes (not applicable to multilane roads)
f_{ID} = adjustment for interchange density (not applicable to multilane roads)
F_M = adjustment for median type (not applicable to freeways)
F_A = adjustment for access points (not applicable to freeways)

HCM recommends that an ideal free-flow speed of 75 mph can be assumed for rural freeways. For urban and suburban freeways, the recommended ideal free-flow speed is 70 mph.

(a) *Adjustment for Median Type*   The first adjustment to free-flow speed relates to the median type. This adjustment is not required for free-flow speed on freeways. For rural and suburban multilane roads, the adjustment factors are given in Table A5.2.1.

(b) *Adjustment for Lane Width*   The ideal lane width is 12 ft. The ideal free-flow speed is reduced when the average width across all lanes within a freeway section is less than 12 ft. Adjustment factors to reflect the effect of narrower average lane widths are provided in Table A5.2.2.

(c) *Adjustment for Right Shoulder Lateral Clearance* According to the HCM, the ideal lateral clearance is 6 ft or greater on the right side and 2 ft or greater on the

median or left side. The ideal free-flow speed has to be adjusted if these requirements are not met. There are no adjustment factors to reflect the effect of median lateral clearance of less than 2 ft. However, lateral clearance of less than 2 ft on either the right or left sides is often rare. The adjustment factors for right shoulder lateral clearance are shown in the Table A5.2.3.

For rural and suburban multilane roads, adjustment factors are given for the total lateral clearance (Table A5.2.4), which is the sum of the lateral clearances of the median (if greater than 6 ft, use 6 ft) and right shoulder (if greater than 6 ft, use 6 ft).

(d) *Adjustment for Number of Lanes*   Freeway sections with five or more lanes in one direction are considered ideal with respect to the free-flow speed. When there are fewer than five lanes, the free-flow speed is less than ideal.

Table A5.2.1   Adjustment Factors for Median Type

Median Type	Reduction in Free-Flow Speed (mph)
Undivided highways	1.6
Divided highways	0

Table A5.2.2   Adjustment Factors for Lane Width

Lane Width (ft)	Reduction in Free-Flow Speed	
	Freeways	Multilane Roads
≥12	0.0	0.0
11	1.9	1.9
10	6.6	6.6

Table A5.2.3   Adjustment Factors for Right Shoulder Lateral Clearance

Right Shoulder Lateral Clearance (ft)	Reduction in Free-Flow Speed (mph)		
	Lanes in One Direction		
	2	3	4
≥6	0.0	0.0	0.0
5	0.6	0.4	0.2
4	1.2	0.8	0.4
3	1.8	1.2	0.6
2	2.4	1.6	0.8
1	3.0	2.0	1.0
0	3.6	2.4	1.2

Table A5.2.4   Adjustment Factors for Total Lateral Clearance

Four-Lane Highways		Six-Lane Highways	
Total Lateral Clearance (ft)	Reduction in Free-Flow Speed (mph)	Total Lateral Clearance (ft)	Reduction in Free-Flow Speed (mph)
12	0	12	0
10	0.4	10	0.4
8	0.9	8	0.9
6	1.3	6	1.3
4	1.8	4	1.7
2	3.6	2	2.8
0	5.4	0	3.9

Table A5.2.5   Adjustment Factors for Number of Lanes

Number of Lanes (One Direction)	Reduction in Free-Flow Speed (mph)
≥5	0.0
4	1.5
3	3.0
2	4.5

Adjustment factors to reflect the effect of the number of lanes on ideal free-flow speed are shown in Table A5.2.5. Only mainline lanes (basic and auxiliary) are considered in the determination of number of lanes. For example, HOV lanes are not included. These adjustment factors were computed on the basis of data collected on urban and suburban freeway sections and do not reflect conditions on rural freeways which typically carry two lanes in each direction. Hence, the value of the adjustment factor for rural freeways is taken as zero.

(*e*) *Adjustment for Interchange Density*   The ideal interchange density according to the HCM is 2-mile interchange spacing. If the density of interchanges is greater, the ideal free-flow speed is reduced. The HCM-recommended adjustment factors for interchange density are given in Table A5.2.6. An interchange is defined as having at least one on-ramp. Hence, interchanges with only off-ramps are not considered in determining interchange density. Interchanges considered should include typical interchanges with arterials or highways and major freeway to freeway interchanges.

(*f*) *Adjustment for Access Point Density*   This adjustment factor is applicable to rural and suburban multilane roads. It is not applicable to freeways. When the data on the

Table A5.2.6   Adjustment Factors for Interchange Density

Interchanges per Mile	Reduction in Free-Flow Speed (mph)
≤0.50	0.0
0.75	1.3
1.00	2.5
1.25	3.7
1.50	5.0
1.75	6.3
2.00	7.5

Table A5.2.7   HCM-Recommended Access Point Density for Different Types of Developments

Type of Development	Access Points per Mile (One Side of Roadway)
Rural	0–10
Low-density suburban	11–20
High-density suburban	21 or more

Table A5.2.8   Adjustment Factors for the Effects of Access Point Density on Free-Flow Speed

Access Points per Mile	Reduction in Free-Flow Speed (mph)
0	0.0
10	2.5
20	5.0
30	7.5
40 or more	10.0

number of access points on the highway section is not available, the HCM recommends the use of the values shown in Tables A5.2.7 and A5.2.8, depending on the type of development.

Example A5.2   Determine the ideal free-flow speed on a 6-mile urban freeway section with three lanes in each direction, a lateral clearance of 4 ft on the right and left sides and with a lane width of 11 ft over the entire section. There are six interchanges within the section.

SOLUTION   Assuming an ideal free-flow speed of 70 mph on the urban freeway under consideration, the free-flow speed on the freeway section can be calculated using the equation

$$FFS = FFS_i - f_{LW} - f_{LC} - f_N - f_{ID}$$

where

Factor due to lane width, f_{LW}	2.0 mph (refer to Table A5.2.2)
Factor due to right shoulder lateral clearance, f_{LC}	0.8 mph (refer to Table A5.2.3)
Factor due to number of lanes, f_N	3.0 mph (refer to Table A5.2.5)
Interchange density, ID	6 interchanges over 6 miles of freeway 1 interchange per mile

Factor due to interchange   2.5 mph (refer to
  density, f_{ID} Table A5.2.6)

Hence, the free-flow speed on the given freeway section is

$$FFS = 70 - 2 - 0.8 - 3.0 - 2.5 = 61.7 \text{ mph}$$

APPENDIX A5.3: TRAVEL TIMES USED IN WORLD BANK PROJECTS

Tables A5.3.1 and A5.3.2 list the values of passenger and crew travel times, respectively, that have been used in World Bank projects.

Table A5.3.1   Values of Passenger Travel Time ($/h)

Year	Country	Motor-cycle	Car	Pick-up	Bus	Truck	Rail	Project
1992	Venezuela		2.72	2.14	1.66			Urban Transport (Caracas only)
1996	Uruguay		1.10	1.10	0.29			National Road Network Analysis (1996–1999 Plan)
1996	Ukraine				0.15			Urban Transport
1993	Tunisia		1.07		0.48		0.48	Urban Transport II
1983	Tunisia	0.33	0.33	0.33	0.33	0.33		Urban Transport II
1975	Thailand	1.00	1.50		0.50			Bangkok Traffic Management
1990	Sri Lanka	0.41	0.82		0.16	0.16		Colombo Urban Transport
1993	St. Lucia		1.14	1.49	0.91	1.10		West coast road study
1994	Russia				0.35			Urban Transport
1993	Perú	0.69	0.69	0.69	0.69	0.69	0.69	Transport Rehabilitation (Road Component)
1995	Lebanon		1.72	2.59	1.24			National Roads
1995	Latvia				1.80			Municipal Services Development (Riga UT component)
1994	South Korea		2.57		1.70			Pusan Urban Transport Management
1987	South Korea	\$0.50 to \$1.5 per passenger-hour for work-related trips						Kyonggi Regional Transport
1984	South Korea		1.65		0.45	0.90		Seoul Urban Transportation
1995	Kenya		1.24	0.24	0.24			Urban Infrastructure Transport III
1993	Jordan							
1992	Jordan							Swaileh–Queen Alia International Airport Road
1983	Jordan	1.26	1.26	1.26	1.26			Amman Transport and Municipal Development
1985	Indonesia		2.06	2.06	0.42			Regional Cities Urban Transport
1996	India		1.00		0.75			Andra Pradesh State Highway

(continued overleaf)

Table A5.3.1   (*continued*)

Year	Country	Motor-cycle	Car	Pick-up	Bus	Truck	Rail	Project
1994	India	0.58	0.62		0.56/0.24			National Highway III
1991	Honduras		0.60	0.60	0.14			Road Rehabilitation and Maintenance
1992	Guatemala		0.80	1.00	0.28			Road Maintenance Program
1995	Ghana		0.05	0.05	0.05	0.05		Highway Sector Investment Program
1996	Dominican Republic		0.73					National Highway
1981	Côte d'Ivoire	0.67	0.67	0.67	0.67	0.67		Urban II
1995	Colombia		1.72		0.32	0.32		Bogota Urban Transport
1996	China	0.12	0.12	0.12	0.12	0.12		Tianjin Urban Development and Environment
1993	China	0.33	0.33	0.33	0.33	0.33		Shanghai Metropolitan Transport II
1990	China		Working time at $0.20/h and nonpaid time at $0.05/h					Medium-Sized Cities Development
1989	China	0.20	0.20	0.20	0.20	0.20		Shanghai Metropolitan Transport I
1987	Cameroon		1.47		1.47			Urban II (Douala Infrastructure Component)
1989	Burkina Faso		0.63		0.63			Urban II
1995	Brazil		4.46		1.28		0.78	Recife Metropolitan Transport Decentralization
1979	Brazil		0.71		0.15		0.22	Urban Transport II (Porto Alegre)
1993	Bangladesh		0.91	0.91	0.35			Jamuna Bridge
1990	Bangladesh		0.57	0.43	0.23	0.23		Road Rehabilitation and Maintenance II

Source: Gwilliam (1997).

Table A5.3.2   Values of Crew Travel Time ($/h)

Year	Country	Car	Pick-up	Mini-bus	Bus	2-Axle Truck	3-Axle Truck	>3-Axle Truck	Project
1992	Venezuela	1.39	1.39	1.39	1.39	1.39	1.39	1.39	Urban Transport (Caracas only)
1993	Spain		42.29		21.14		25.36	22.86	Catalunya Highway Maintenance and Rehabilitation

(*continued overleaf*)

Table A5.3.2 (*continued*)

Year	Country	Car	Pick-up	Mini-bus	Bus	2-Axle Truck	3-Axle Truck	>3-Axle Truck	Project
1991	Sierra Leone				0.47	0.47	0.47	0.47	Road Rehabilitation and Maintenance
1993	St. Lucia		2.49	2.49		2.99	3.46	3.94	
1993	Nigeria		0.25	0.25	1.41	0.47	0.47	0.98	Multistate Roads II
1987	Niger		1.05	1.05		1.73	1.73	2.79	National Transport Investment Program
1994	Nepal		0.40		0.84	0.54			Road Maintenance and Rehabilitation
1992	Mexico	1.33			3.87	1.67	3.33	3.33	Trunk Roads Network Maintenance Strategy
1995	Lebanon				2.79	2.67	2.67	2.67	National Roads
1995	Kenya	0.51	0.65		0.98	1.31	1.93		Urban Infrastructure
1993	Jordan				1.02	1.81	1.81	1.81	Transport III
1992	Jordan	1.02	1.02	1.02	1.02	1.81	1.81	1.81	Swaileh–Queen Alia International Airport Road
1996	India	0.40			1.80	1.80	1.80		Andhra Pradesh State Highway
1994	India		0.44		1.02	0.87	1.04	1.04	National Highway III
1991	Honduras		0.39		0.96		0.96	1.35	Road Rehabilitation and Maintenance
1992	Guatemala		1.00	1.50	1.90	1.25	1.25	1.25	Road Maintenance Program
1995	Ghana	0.28	0.28		0.56	0.56	0.56	0.56	Highway Sector Investment Program
1996	Dominican Republic				1.09	0.93	1.09	1.09	National Highway
1989	Chile		1.00		3.00	1.20	1.80		Road Sector II
1985	Chile		1.00		3.00	1.20	1.80		Road Sector I
1987	Cameroon	5.52							Urban II (Douala Infrastructure Component)
1994	Brazil				3.29	2.32	2.32	2.81	State Highway Management II
1993	Bangladesh				0.84	0.70			Jamuna Bridge
1990	Bangladesh	0.46	0.46	1.03	1.03	0.83	0.83	0.83	Road Rehabilitation and Maintenance II
1994	Algeria				2.96	2.76	3.57	3.37	Highway VI

Source: Gwilliam (1997).

CHAPTER 6

Evaluation of Safety Impacts

I am prepared for the worst, but hope for the best.
　　　　　　　　—*Benjamin Disraeli (1804–1881)*

INTRODUCTION

Transportation projects generally have a direct or indirect safety component that reduces the rate or severity of crashes. As such, safety enhancement is considered a key aspect of user benefits associated with physical or policy changes in a transportation system. In the period 1992–2002, approximately 40,000 to 45,000 fatalities per year were experienced on the U.S. transportation system. Of this, 90 to 95% was highway-related (USDOT, 2004). As seen in Figure 6.1, for every 100,000 residents in 2002, highways had a fatality rate of approximately 15 deaths, while railroads had 0.33. In the figure, the fatality statistics for air transportation include air carrier service, commuter service, air taxi service, and general aviation; for the highway mode, fatalities include all types of highway motor vehicles, bicycles, and pedestrians. Railroad fatalities include deaths from railroad highway–rail grade-crossing incidents. For transit fatality statistics, the modes considered include: motor bus, heavy rail, light rail, commuter rail, trolley bus, aerial tramway, automated guideway transit, cablecar, ferry boat, and monorail. Waterborne fatalities include those due to vessel- or non-vessel-related incidents on commercial and recreational vessels. Pipeline facilities include hazardous liquid and gas pipelines.

For people under 65 years of age, the Center for Disease Control has ranked transportation accidents as the third-leading cause of death in the United States (after cancer and heart disease) each year from 1991 to 2000 (USDHHS, 2003). During those years, an annual average of nearly 36,000 people under 65 lost their lives due to transportation accidents. A far larger number of people are injured than killed; an estimated 3.0 million people suffered some type of injury involving passenger and freight transportation in 2002, and a majority of these injuries (98%) resulted from highway crashes (USDOT, 2004).

The economic cost of transportation crashes, which is borne by individuals, insurance companies, and government, consists of loss of market productivity, property damage, loss of household productivity and workplace costs. Intangible costs include pain and suffering, and loss of life. The costs of crashes can be very high. For instance, motor vehicle crashes in the United States cost an estimated $230 billion in 2000, representing approximately $820 per person or 2% of the gross domestic product (USDOT, 2004).

Within the highway mode, safety problems are most pernicious at roads in rural areas and at roads that have only one lane in each direction. Most of these roads were designed and built many decades ago using standards that have become outdated. As such, they are generally characterized by operational and safety deficiencies arising from inadequate road geometry, driver information deficiencies, lack of passing opportunities, and traffic conflicts due to driveways. Transportation projects typically include interventions to upgrade these and other facilities to acceptable standards.

In this chapter we present a procedural framework that can be used by analysts to assess the safety impacts of transportation investments. Much of the discussion focuses on the highway mode, because compared to all other modes, highway safety continues to be the major transportation safety problem. Nevertheless, the general concepts discussed here are applicable to other modes of transportation. We first present the basic taxonomy associated with transportation safety, briefly discuss the factors that affect crashes, identify possible safety projects, and present evidence of the agency costs and effectiveness (user benefits) of various project types. Then the procedural framework for safety evaluation is presented. This essentially comprises the product of two elements: change in *crash frequency* after the proposed transportation intervention, and unit *crash monetary costs*. Crash frequency or its reduction can be estimated using crash relationships (rates, equations), developed from national data or preferably, recent local data. We also identify existing software packages that may be used or customized for safety evaluation of highway projects and list some current resources for safety evaluation.

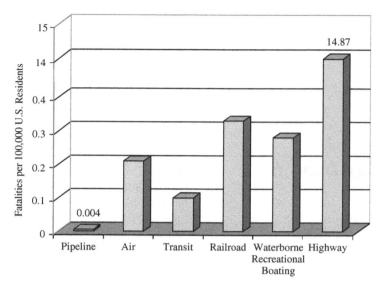

Figure 6.1 Transportation fatality distribution by mode 2002. (From USDOT, 2004.)

6.1 BASIC DEFINITIONS AND FACTORS OF TRANSPORTATION SAFETY

6.1.1 Definition of a Crash

The most basic unit for measuring transportation safety is a crash. A *crash* can be defined as a collision involving at least one moving transportation vehicle (car, truck, plane, boat, railcar, etc.) and another vehicle or object. Transportation crashes are typically caused by factors such as driver, pilot, or operator error, mechanical failure, and poor design of the guideway, roadway, waterway, or runway. A crash can also involve noncollision off the transportation path, such as a vehicle rollover.

6.1.2 Transportation Crashes Classified by Severity

On the basis of severity, transportation crashes are broadly classified into three categories:

1. A *fatal crash* is one where the highest casualty level is a fatality.
2. An *injury crash* is one where the highest casualty level is a nonfatal injury.
3. A *property-damage-only crash* is one that involves a loss of all or part of the transporting vehicle and/or property, but no injury or fatality.

Transportation crashes can also be scaled on the basis of the extent of injury. For example, for highway crashes, two commonly used injury scales are the abbreviated injury scale (AIS) and the KABCO injury scale.

(*a*) *Abbreviated Injury Scale for Crash Severity* Introduced in 1969 by the Association for the Advancement of Automotive Medicine, the AIS is an anatomical scoring system and ranks injuries on a scale that represents the "threat to life" associated with an injury (Table 6.1). The AIS score of the most life-threatening injury [i.e., the maximum AIS or (MAIS)] is often used to describe the type and extent of injury sustained by one or more persons involved in the crash.

(*b*) *KABCO Injury Scale* Established by the American National Standards Institute, the KABCO injury scale (Table 6.2) is designed for police coding of crash details at a crash scene. The coding does not require medical expertise—the police officer at the crash scene assesses the sustained injuries and assigns a code depending on the level of severity. The KABCO system has faced some criticism because it does not always classify injuries classification in a consistent manner (e.g., the code assigns equal severity to a broken arm and a severed spinal cord). Therefore, in a bid to reduce the variability in reporting, the National Highway Traffic Safety Administration (NHTSA) uses both AIS and KABCO scales to describe transportation injuries.

6.1.3 Categories of Factors Affecting Transportation Crashes

Figure 6.2 shows the categories of factors that affect the frequency and severity of transportation crashes. This is followed by a brief discussion of each factor category.

Table 6.1 Abbreviated Injury Scale

Code	Severity	Description
AIS 6	Fatal	Loss of life due to decapitation, torso transaction, massively crushed chest, etc.
AIS 5	Critical	Spinal chord injury, excessive second- or third-degree burns, cerebral concussion (unconscious more than 24 hours)
AIS 4	Severe	Partial spinal cord severance, spleen rupture, leg crush, chest wall perforation, cerebral concussion (unconscious less than 24 hours)
AIS 3	Serious	Major nerve laceration; multiple rib fracture, abdominal organ contusion; hand, foot, or arm crush/amputation
AIS 2	Moderate	Major abrasion or laceration of skin, cerebral concussion finger or toe crush/amputation, close pelvic fracture
AIS 1	Minor	Superficial abrasion or laceration of skin, digit sprain, first-degree burn, head trauma with headache or dizziness
AIS 0	Uninjured	No injury

Source: Blincoe et al. (2002).

Table 6.2 KABCO Scale for Crash Severity

Code	Severity	Injury Description
K	Fatal	Any injury that results in death within 30 days of crash occurrence
A	Incapacitating	Any injury other than a fatal injury which prevents the injured person from walking, driving, or normally continuing the activities the person was capable of performing before the injury occurred (e.g., severe lacerations, broken limbs, damaged skull)
B	Injury evident	Any injury other than a fatal injury or an incapacitating injury that is evident to observers at the scene of the crash in which the injury occurred (e.g., abrasions, bruises, minor cuts)
C	Injury possible	Any injury reported that is not a fatal, incapacitating, or nonincapacitating evident injury (e.g., pain, nausea, hysteria)
O	Property damage only	Property damage to property that reduces the monetary value of that property

Source: NSC (2001).

(a) Environmental Factors Environmental conditions such as poor visibility, high winds, rain and snow storms, ice on a roadway or runway or on airplane wings, animals that cross vehicle paths, and birds that get sucked into plane engines are significant factors of transportation crashes.

(b) Engineering Factors Unfavorable roadway or guideway geometry (e.g., dimensions, alignment, sight distances) and topography (e.g., steep grades, mountain passes) are often associated with frequent crashes. Also, the poor condition of roadway or runway pavement surfaces (surface defects, low skid resistance, and so on) and of the guideway (deteriorated, deformed, or cracked guideway elements) can lead to crashes. Furthermore, for surface transportation, the absence of crash barriers at high embankments and other hazardous sites contribute to crash occurrence. The operational or usage characteristics of the transportation facility also influence the crash experience. For example, crash rates may be expressed as a function of the congestion level of the transportation facility (AASHTO, 2003). The analysis of safety impacts of transportation investments proceeds on the premise that such investments, besides their primary objective of facility preservation or capacity expansion, also enhance user

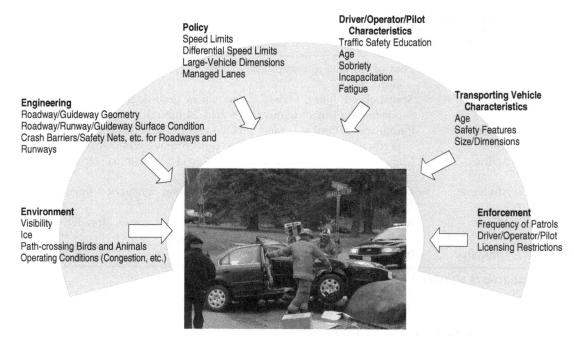

Figure 6.2   Factors affecting transportation crash occurrence and severity. (Photo courtesy of Peter Gene, Creative Commons Attribution-ShareAlike 2.0.)

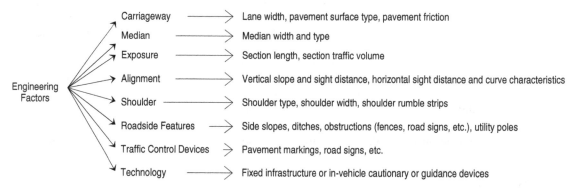

Figure 6.3   Engineering factors of highway transportation crashes.

safety. Interventions typically result in improved physical characteristics and dimensions and enhanced operational performance of the transportation facility, and the safety benefits of interventions are more visible particularly where the preintervention features are below established standards. The engineering factors that affect highway traffic safety are shown in Figure 6.3.

The safety impacts of changes in engineering factors are typically expressed in terms of crash reduction factors or accident modification factors. A *crash reduction factor* indicates the extent by which crashes are reduced in response to a specific intervention or improvement

that enhances the safety-related engineering features of the facility. For example, if the crash reduction factor of shoulder widening is 10%, a road section that currently has narrow shoulders and experiences 50 crashes per year can be expected to have a reduction of 5 crashes per year after shoulder widening. An *accident modification factor* for a certain safety condition (e.g., addition of shoulders) is a factor that is multiplied with the number of crashes predicted for a base situation (e.g., absence of shoulders) to obtain the number of crashes that can be expected for the alternative situation (presence of shoulders). For highway transportation, improvements include enhancements to the carriageway, shoulder, median, alignment, roadside hazard elimination, and traffic control devices. Also technological devices may be embedded in the facility or placed in vehicles to serve as warning devices in case of hazardous situations. In many cases, the extent of crash reduction is not fixed but varies, depending on the extent of the improvement and the defect severity (e.g., widening a narrow lane by 2 ft may yield a higher crash reduction than widening the same lane by 1 ft; also, widening a narrow lane by 1 ft may yield a higher crash reduction than widening a wide lane by the same margin). Typically, crash reduction functions are discussed from the perspective of engineering improvements, but the concept could be extended to improvements in other crash factors, such as policy, enforcement, vehicle, and operator characteristics.

From the perspective of transportation systems evaluation, engineering factors are considered particularly pertinent because (a) enhancements in such factors can help reduce the crash contributions of the other crash factors (for example, enhanced facility condition or alignment renders the overall transportation operating environment more forgiving of operator error or limitations, vehicle inadequacies, and poor environmental conditions) and (b) engineering factors, to a greater extent compared to other crash factors, are within the direct control of transportation agencies.

(c) Policy Factors Recent years have seen increased attention to national policies such as sobriety laws for airline pilots, truck and transit operators, a 10-hour driving limit for truck drivers, seat belt use, and helmet use (for motorcycles). The most visible, yet probably most contentious policy factor in highway safety is that of speed limits. Policies that result in changed speed limits or establishment of speed differentials by vehicle class may lead to changes in crash rates and severities, depending on highway functional class, crash severity type, existing speeds, and other factors. Other policy factors that may influence safety include the managed lanes concept, which reduces the size heterogeneity of traffic—a traffic stream that is comprised of vehicles of uniform size may be safer than one that consists of vehicles of different sizes.

(d) Driver Characteristics Crashes are also influenced by characteristics of drivers, operators, and pilots of transportation vehicles, such as age and gender (Islam and Mannering, 2006), experience, and alcohol or drugs. Kweon and Kockleman (2003) showed that in road transportation, for example, young and middle-aged men are slightly more likely to have a crash than their female counterparts, but the opposite is true for older

age groups. Also, younger and older drivers tend to have relatively high crash rates per vehicle-mile. Furthermore, professional drivers (operators of trucks, buses, taxis, etc.) generally have low "per mile" crash rates but relatively high "per vehicle-year" crash rates because of their relatively large amounts of travel. Intoxicated drivers tend to have crash rates (crashes per vehicle-mile) that far exceed those of sober drivers; approximately one-third of all traffic fatalities involve at least one intoxicated driver.

(e) Vehicle or Mode Characteristics Vehicle design features affect crash frequency and severity. Differences in size, weight, and shape of vehicles in a traffic stream can increase the likelihood of collisions. Also, occupants in passenger cars are twice as likely to have fatalities as those in larger and heavier vehicles. Newer vehicles tend to have design features and safety equipment that provide greater crash protection than that of older models, thus reducing crash severity, if not frequency. Recent research suggests that some drivers in vehicles with more safety features tend to drive more aggressively thus offsetting the intended benefits of safety features (Winston et al., 2006). Buses and other transit vehicles tend to have low crash rates per mile and have low injury rates for their occupants. Sport utility vehicles and large vans tend to have a high rate of rollover crashes, and motorcyclists, bicyclists, and pedestrians tend to have greater injuries when involved in a crash.

(f) Enforcement Factors The frequency of patrols and the establishment of effective driver education and licensing restrictions generally help to improve safety. Also, the higher severity of penalties for traffic infractions generally tends to encourage operator responsibility and thus can increase traffic safety.

6.2 PROCEDURE FOR SAFETY IMPACT EVALUATION

For purposes of evaluating the safety impacts of transportation projects (by comparing the "with" and "without improvement" scenarios), this chapter focuses on the engineering factors. The overall framework (Figure 6.4) revolves around three tasks:

1. Estimating the extent to which relevant engineering factors (or aggregated combination thereof) would be changed (such as lane-width increase)
2. Ascertaining the impact of each unit change of the engineering factor on crash reduction
3. From the results of tasks 1 and 2, computing the overall change in crashes expected due to the given intervention

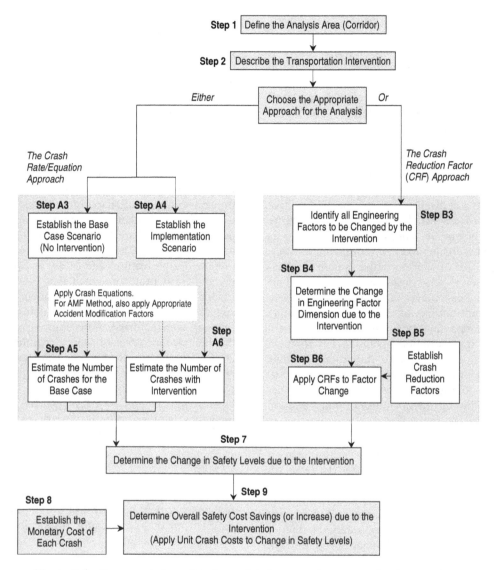

Figure 6.4 Framework for estimating safety impacts of transportation interventions.

The alternative to the use of crash reduction factors is one that involves an implicit or explicit combination of factors (such as road class) where existing crash rates or equations are used to determine the safety levels (number of crashes) for the "with improvement" and "without improvement" states of the facility. The steps of the framework for evaluating the safety impacts of transportation improvements are presented next.

Step 1: Define the Analysis Area Typically, only a specific transportation facility (e.g., road section or intersection) is analyzed. At the network level, the safety impacts of a systemwide transportation policy or other intervention can be evaluated by dividing the network into individual facility (or families of facilities) and carrying out the analysis for each facility.

Step 2: Describe the Intervention

(a) *Transportation Intervention* A transportation intervention or improvement may expand the capacity of the transportation system; improve the operational performance of the system; preserve the fixed assets by improving, for instance, roadway, runway, or guideway condition; upgrade the transportation facility to a higher class; preserve rolling stock (to improve the condition of mobile assets, thus lessening the likelihood of mechanical failure); or a policy-related intervention.

(b) *Approach for the Evaluation* There are two alternative approaches to determining the safety impacts of an intervention: *crash rate/crash equation approach*, and the *crash reduction factor approach*.

The choice of approach is dictated by the type of data and models that are available. Where only crash rates or crash equations are available, using the crash rate/crash equation approach (see the left-hand shaded box in Figure 6.4) may be preferable. Where detailed crash reduction factors for each engineering factor are available, the crash reduction factor approach can be used (see the right-hand shaded box in Figure 6.4).

Steps 3 to 6: Estimate the Crash Frequency Steps 3 to 6 involve estimation of the number of crashes with and without the improvement. There are a number of ways of doing this (see step 2): Using crash rates, crash equations with and without accident modification factors, or crash reduction factors (Figure 6.5). For the crash rates, the constant *a* is the crash rate for each category of facility. For the crash equations, the variable VMT is a measure of exposure in terms of traffic volume (AADT) and section length, and the vector X_i refers to various engineering features, such as the width of a lane, shoulder, or median; shoulder type; horizontal and vertical curve characteristics; and left-turn provisions. Most engineering features have an associated factor for crash reduction or accident modification (Appendix A6).

(a) *Crash Rate–Crash Equation Approach* Details of this approach are as follows:

1. Establish the function that gives the expected safety levels of each family of facilities. This may be in the form of *average crash rate values* (crashes per VMT, crashes per mile, or crashes per AADT) (examples provided in Table 6.3), or *regression equations that estimate crash frequencies or rates* as functions of the operating and physical characteristics of the facility (examples provided in Table 6.4).

2. Determine the values of the independent variables (representing the state of each engineering factor) as they pertain to the facility in question. If the crash rate method is being used, this step involves determination of the exposure or usage. For example, Figure 6.5 shows the determination of the number of crashes if VMT is used as a measure of exposure. If a regression equation is being used, determine the values of each variable in the regression equation, such as section VMT, lane width, shoulder type, and so on. This is done for both the base case (without the improvement) and the intervention case (with the improvement).

3. Substitute the given levels of the independent variables or exposure into the crash equation or crash rates to determine the total safety levels (number of crashes). This is done for both the without-improvement and with-improvement situations. For the existing without-improvement situation, the actual number of crashes, if known, may be used instead of estimating it from the table or the equation. Due to data aggregation, the crash rate approach may yield less precise estimates of safety impacts than the crash equation approach.

Example 6.1 A 6-mile urban "minor arterial" highway section is to receive major upgrading that will improve the design standards to the freeway and expressway category. Assume that crash reduction factors for the individual treatments associated with the upgrade are unknown, and crash prediction equations for both facility types are not available. Estimate the number of crashes with and without the upgrade. Assume traffic volumes of 7520 and 7800 vehicles per day (vpd) before and after the upgrade, respectively.

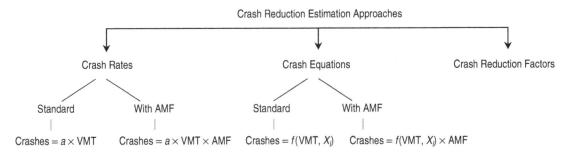

Figure 6.5 Approaches for estimating reduction of crash frequency (for steps 3 to 6).

Table 6.3   Motor Vehicle Traffic Fatality and Injury Rates by Functional Class

| Area Class | Functional Class | Number of Crashes (per 100 million VMT) | |
		Fatal	Non-Fatal
Rural	Interstate	1.05	25.08
	Other principal arterial	1.96	50.87
	Minor arterial	2.33	70.52
	Major collector	2.51	86.79
	Minor collector	3.16	106.02
	Local	3.52	147.79
Urban	Interstate	0.56	46.56
	Other freeway & expressway	0.75	68.60
	Other principal arterial	1.30	124.69
	Minor arterial	1.08	126.89
	Collector	1.00	104.95
	Local	1.33	194.40

Source: FHWA (1998).

Table 6.4   Selected Crash Estimation Functions

Facility	Equation
Urban freeways (AASHTO, 2003)	$\%\Delta C = 100 \left[\dfrac{3.0234\,(V_1/C_1) - 1.11978\,(V_1/C_1)^2}{3.0234\,(V_0/C_0) - 1.11978\,(V_0/C_0)^2} - 1 \right]$

$\%\Delta C$ = percentage change in crash rate (crashes per VMT)
V_0, C_0 = volume and capacity of highway without improvement (pcphpl)
V_1, C_1 = volume and capacity of highway with improvement (pcphpl)

Urban, four-leg signalized intersections (Bauer and Harwood, 2000)

Total crashes

$$Y = e^{-3.428}(X_1)^{0.224}(X_2)^{0.503} \exp(0.063X_{19} + 0.622X_{20} - 0.2X_{21} - 0.310X_5 - 0.13X_{22} - 0.053X_{16} - 0.115X_{11} - 0.225X_3 - 0.13X_{17})$$

Fatal + injury crashes

$$Y = e^{-5.745}(X_1)^{0.215}(X_2)^{0.574} \exp(-0.051X_{19} + 0.4X_{20} - 0.240X_{21} - 0.290X_5 - 0.155X_{22} - 0.163X_3 - 0.151X_{17} + 0.005X_4)$$

Y = expected number of total multiple-vehicle accidents in a three-year period
X_1 and X_2 = average daily traffic (veh/day) on minor and major road, respectively
X_{19} = pretimed signal timing design
X_{20} = fully actuated signal timing design
X_{21} = 1 if multiphase (>2) signal timing, 0 otherwise
X_5 = 1 if no access control on major road; 0 otherwise
X_{22} = number of lanes on minor road
X_3 = 1 if major road has $\leq$ 3 through lanes in both directions of travel combined; 0 otherwise
X_{17} = 1 if major road has 4 or 5 through lanes in both directions of travel combined; 0 otherwise
X_4 = design speed on major road (mph)

Table 6.4   *(continued)*

Facility	Equation

Urban, four-leg intersections with stop control on the minor road (Bauer and Harwood, 2000)

Total crashes

$$Y = e^{-4.664}(X_1)^{0.281}(X_2)^{0.620}\exp(-0.941X_{15} - 0.097X_{16} + 0.401X_3 + 0.120X_{17}$$
$$-0.437X_5 - 0.384X_{11} - 0.160X_8 - 0.153X_6 - 0.229X_7)$$

Fatal + injury crashes

$$Y = e^{-4.693}(X_1)^{0.206}(X_2)^{0.584}\exp(-0.747X_{15} - 0.081X_{16} - 0.382X_5 + 0.282X_3$$
$$+0.049X_{17} - 0.020X_{14} - 0.3X_{11} - 0.079X_6 - 0.401X_7)$$

Y = expected number of total multiple-vehicle accidents in a three-year period
X_1 and X_2 = average daily traffic (veh/day) on minor and major road, respectively
X_{15} = 1 if left turns are prohibited; 0 otherwise
X_{16} = average lane width on major road (ft)
X_3 = 1 if major road has ≤3 through lanes in both directions of travel combined; 0 otherwise
X_{17} = 1 if major road has 4 or 5 through lanes in both directions of travel combined; 0 otherwise
X_5 = 1 if no access control on major road; 0 otherwise
X_{11} = 1 if there is no free right-turn lane; 0 otherwise
X_8 = 1 if the intersection has no lighting; 0 otherwise
X_6 = 1 if minor arterial; 0 otherwise
X_7 = 1 if major collector; 0 otherwise
X_{14} = outside shoulder width on major road (ft)

Urban, three-leg intersections with stop control (Bauer and Harwood, 2000)

Total crashes

$$Y = e^{-5.557}(X_1)^{0.245}(X_2)^{0.683}\exp(-0.559X_{11} - 0.402X_{15} + 0.019X_{12} + 0.210X_{13}$$
$$-0.006X_4 - 0.147X_{18} - 0.037X_{16})$$

Fatal + injury crashes

$$Y = e^{-6.618}(X_1)^{0.238}(X_2)^{0.696}\exp(-0.581X_{11} - 0.393X_{15} - 0.057X_{12} + 0.209X_{13}$$
$$-0.182X_{18} - 0.048X_{16} + 0.094X_{18})$$

Y = expected number of total multiple-vehicle accidents in a three-year period
X_1 and X_2 = average daily traffic (veh/day) on minor and major road, respectively
X_{11} = 1 if there is no free right-turn lane; 0 otherwise
X_{15} = 1 if left turns are prohibited; 0 otherwise
X_{12} = 1 if there is no left-turn lane; 0 otherwise
X_{13} = 1 if there is a curbed left-turn lane; 0 otherwise
X_{18} = presence of median of major road; 0 otherwise
X_{16} = average lane width on major road
X_8 = 1 if the intersection has no lighting; 0 otherwise

Highway segments (Forkenbrock and Foster, 1997)

$$Y = e^{0.517} \times 0.972^{\text{PSR}} \times 1.068^{\text{TOPCURVE}} \times 1.179^{\text{PASSRES}} \times 1.214^{\text{ADTLANE}} \times 0.974^{\text{RIGHTSH}} \times 0.933^{\text{LANES}} \times 1.051^{\text{TOPGRAD}}$$

Y = Crash rate in millions of VMT
PSR = present serviceability rating of the pavement surface ranging from 0 (failed) to 5 (excellent)
TOPCURV = the severity of the worst horizontal curve ranging from 0 (no curve) to 12 (sharpest curve)

(continued overleaf)

Table 6.4   (*continued*)

Facility	Equation
	PASSRES = dummy variable representing the presence/absence of passing restrictions (1/0, respectively) ADTLANE = hourly traffic volume in thousands per lane RIGHTSH = right shoulder width (ft) LANES = dummy variable representing the number of lanes (1 for 4 lanes, 0 for 2 lanes) TOPGRAD = measure of the average vertical grade ranging from 0 (no grade) to 12 (severe grade)

Rural FLSC (four-leg stop-controlled) intersections at rural two-lane highways (Bauer and Harwood, 2000)

Total crashes

$$Y = e^{-10.025}(X_1)^{0.532}(X_2)^{0.758}\exp(0.321X_3 + 0.009X_4 + 0.2X_5 + 0.181X_6 + 0.173X_7 + 0.122X_8 + 0.053X_9 - 0.159X_{10} + 0.157X_{11})$$

Fatal + injury crashes

$$Y = e^{-10.294}(X_1)^{0.546}(X_2)^{0.680}\exp(0.385X_3 + 0.013X_4 + 0.183X_9 - 0.234X_{10} + 0.261X_6 + 0.170X_7 + 0.219X_8)$$

Y = expected number of total multiple-vehicle accidents in a three-year period
X_1 and X_2 = average daily traffic (veh/day) on minor and major road, respectively
X_3 = 1 if major road has $\leq$3 through lanes in both directions of travel combined; 0 otherwise
X_4 = design speed on major road (mph)
X_5 = 1 if no access control on major road; 0 otherwise
X_6 = 1 if minor arterial; 0 otherwise
X_7 = 1 if major collector; 0 otherwise
X_8 = 1 if the intersection has no lighting; 0 otherwise
X_9 = 1 if surrounding terrain is flat; 0 otherwise
X_{10} = 1 if surrounding terrain is mountainous; 0 otherwise
X_{11} = 1 if there is no free right-turn lane; 0 otherwise

Rural TLSC (three-leg stop-controlled) intersections at rural two-lane highways (Bauer and Harwood, 2000)

Total crashes
$$Y = e^{-9.178}(X_1)^{0.383}(X_2)^{0.830}\exp(0.213X_{12} + 0.124X_{13} + 0.225X_5 + 0.145X_6 + 0.211X_7 - 0.017X_{14} - 0.045X_9 + 0.095X_{10})$$

Fatal + injury crashes

$$Y = e^{-9.141}(X_1)^{0.384}(X_2)^{0.781}\exp(-0.03X_{14} + 0.169X_8 + 0.180X_{12} + 0.062X_{13} + 0.164X_6 + 0.192X_7 - 0.219X_{11})$$

Y = expected number of total multiple-vehicle accidents in a three-year period
X_1 and X_2 = average daily traffic (veh/day) on minor and major road, respectively
X_{12} = 1 if there is no left-turn lane; 0 otherwise
X_{13} = 1 if there is a curbed left-turn lane; 0 otherwise
X_5 = 1 if no access control on major road; 0 otherwise
X_6 = 1 if minor arterial; 0 otherwise
X_7 = 1 if major collector; 0 otherwise
X_{14} = outside shoulder width on major road (ft)
X_9 = 1 if surrounding terrain is flat; 0 otherwise
X_{10} = 1 if surrounding terrain is mountainous; 0 otherwise
X_8 = 1 if the intersection has no lighting; 0 otherwise
X_{11} = 1 if there is no free right-turn lane; 0 otherwise.

SOLUTION As no safety information is available for the highway section or the local region, national crash rates associated with highway classes can be used. From Table 6.3, the average crash rates for the initial highway class (urban minor arterial) as well as for the class to which it will be upgraded (other freeway and expressway), an approximation of expected crashes for each scenario can be determined as follows:

Without improvement:

For urban minor arterials, rate of fatal crashes

$$= 1.08 \text{ per } 10^8 \text{ VMT}$$

Annual VMT $= (7520)(6)(365) = 16{,}468{,}800$

Number of fatal crashes expected per annum

$$= (1.08)(10^{-8})(16{,}468{,}800) = 0.18$$

With improvement:

For urban freeways and expressways, rate of fatal crashes

$$= 0.75 \text{ per } 10^8 \text{ VMT}$$

Annual VMT $= (7800)(6)(365) = 17{,}082{,}000$

Number of fatal crashes expected per annum

$$= (0.75)(10^{-8})(17{,}082{,}000) = 0.13$$

Example 6.2 The monthly PDO crash frequency prediction equation for rural principal arterials in a certain state is

$$\text{PDO crashes} = 0.8921 + 0.7097 \ln(\text{LENG})$$
$$+ 0.2409 \ln(\text{AADT}) - 0.1128\text{LW}$$
$$- 0.0676\text{SW} - 0.0624\text{PSI}$$
$$- 0.0553\text{ARAD} + 0.0646\text{AGRAD}$$

where $\ln(\text{LENG}) =$ the natural logarithm of section length (miles), $\ln(\text{AADT}) =$ the natural logarithm of section traffic volume, LW $=$ the lane width (feet), SW $=$ shoulder width (ft), PSI $=$ present serviceability index (a measure of pavement condition), ARAD $=$ average radius (tens of ft) of all horizontal curves, and AGRAD $=$ average grade of vertical curves (%).

Table EX6.2 shows the improvement of specific road factors after a major rehabilitation of a major rural principal arterial.

Assume that all other roadway factors are not changed significantly by the improvement (section length $= 20$ miles, traffic volume $= 75{,}254$ vpd, average vertical

Table E6.2 Change in Road Factors

	Without Improvement	With Improvement
Lane width (ft)	8	10
Shoulder width (ft)	2	4
Pavement condition (PSI)	3	4
Horizontal alignment (average curve radius, ft)	500	600

grade $= 1.3\%$). Estimate the expected number of crashes with and without the improvement.

SOLUTION Without the improvement, the number of property-damage crashes is

$$0.8921 + (0.7097 \times \ln 20) + (0.2409 \times \ln 75{,}254)$$
$$- (0.1128 \times 8) - (0.0676 \times 2) - (0.0624 \times 3)$$
$$- (0.0553 \times 500/10) + (0.0646 \times 1.3) = 1.65$$

With the improvement, the number of property-damage crashes is

$$0.8921 + [0.7097 \times \ln 20) + (0.2409 \times \ln 75{,}254)$$
$$- (0.1128 \times 10) - (0.0676 \times 4) - (0.0624 \times 4)$$
$$- (0.0553 \times 600/10) + (0.0646 \times 1.3) = 0.67$$

Example 6.3 In a bid to reduce congestion, it is proposed to add a lane to an existing urban freeway that currently has a volume–capacity (v/c) ratio of 1.15. It is expected that after the capacity expansion, the v/c ratio would fall to 0.75. Determine the percentage change in crash rate.

SOLUTION

$$\frac{V_0}{C_0} = \text{volume–capacity ratio without improvement} = 1.15$$

$$\frac{V_1}{C_1} = \text{volume–capacity ratio with improvement} = 0.75$$

Using the equation in Table 6.4, the reduction in crash rate is given by

$$\%\Delta C = (100)\left[\frac{(3.0234)(0.75) - (1.11978)(0.75)^2}{(3.0234)(1.15) - (1.11978)(1.15)^2} - 1\right]$$
$$= 17.95\%$$

(b) *The Accident Modification Factor Approach*   In this approach, the established crash rates or equations, such as those shown in Tables 6.3 and 6.4, are multiplied by a factor [the accident modification factor (AMF)] that represents the safety improvement to yield a new frequency of crashes. AMFs are the incremental effects of safety of specific elements of traffic control and highway design. The AMF for a nominal or base element is 1.00. A set of elements associated with a higher crash experience than the nominal condition has an AMF exceeding 1.00, and another set that has a lower crash experience than the nominal has an AMF of less than 1.00.

For a transportation improvement under evaluation, AMF is given by the ratio of the AMF of the with-intervention scenario to the AMF without intervention. Thus, for a project that has an AMF of 90%, one can expect crashes to be reduced by 10%.

The use of crash rates with AMF is relatively straightforward—the accident modification factor represents all the safety impacts associated with improvement related to the various engineering features. If the AMF applies only to certain crash types or patterns (also referred to as *related crashes*), certain adjustments are necessary to obtain the AMF on all crashes (Harwood et al., 2003). Example 6.4 shows how AMF values could be used to adjust the number of crashes predicted on the basis of crash rates. The general procedure is similar to that for crashes predicted using crash equations. A caution: The specific road feature whose AMF factor is being used must not be present as an independent variable in the crash prediction model—doing so would mean double-counting its effects. NCHRP's *Research Results Digest 229* (Harkey et al., 2004) provides a comprehensive list of AMFs for various traffic engineering and ITS improvements (some of these are presented in Table A6.3.).

Example 6.4   A rural 6-mile-long minor arterial road segment has a traffic volume of 10,000 per day. As part of a corridor improvement project, the existing shoulder width is widened from 2 ft to 6 ft. Estimate the number of fatal crashes with and without improvement. Use the crash rates in Table 6.3 and the accident modification factors in Appendix Table A6.4. Assume that the VMT remains the same.

SOLUTION   From Table 6.3, the fatal crash rate for rural minor arterials = 2.33 per 100 million VMT.
Without improvement:

Expected number of fatal crashes
$$= \frac{(2.33)(10{,}000)(365)(6)}{100 \times 10^6} 0.57 = 0.51$$

Accident modification factor for 2-ft shoulders
$$= (1.30)$$
Modified expected number of fatal crashes
$$= (0.51)(1.30) = 0.66$$

With improvement:

Expected number of fatal crashes
$$= \text{same as above} = 0.51$$
Accident modification factor for 6-ft shoulders = 1.00

Modified expected number of fatal crashes
$$= (0.51)(1.00) = 0.51$$

(c) *Crash Reduction Factor Approach*

(c1) Identify all engineering factors that are likely to be changed by the intervention. For example, highway improvements may add lanes, increase lane width, improve pavement surface friction, remove road side obstacles, and so on.

(c2) Establish the extent to which each relevant engineering factor (identified in step c1) will be changed by the intervention.

(c3, c4) Obtain the crash reduction factors for improvements in individual crash factors. The crash reduction factor (CRF) for each improvement is a measure of the efficacy of that improvement in reducing crashes associated with deficient levels of the corresponding engineering factor. It is calculated simply as the percentage decrease in the number of crashes:

$$\text{CRF} = \frac{C_{\text{WO}} - C_{\text{W}}}{C_{\text{WO}}} \times 100 = \left(1 - \frac{C_{\text{W}}}{C_{\text{WO}}}\right) \times 100$$

where C_{WO} is the number of crashes *without* the improvement and C_{W} is the number of crashes *with* the improvement.

Alternatively, C_{WO} and C_{W} can be defined as follows: C_{WO} is the average number of crashes at all sites that *lack* the improved feature at a given time and C_{W} is the average number of crashes at all otherwise similar sites that *have* the improved feature at the same time. C_{WO} and C_{W} are given or are estimated from crash prediction models.

For example, a CRF of 0.2 for shoulder paving means that if an unpaved shoulder were to be paved, a 20% reduction in crashes is expected. Obviously, most crash reduction factors are only average values, because the efficacy of the improvement would depend on the extent of the treatment (widening an 8-ft lane to 10 ft and widening a 8-ft lane to 12 ft will have different crash reduction effects) as well as the existing severity of the factor deficiency (widening a 8-ft lane to 10 ft will yield a crash reduction that is different from that of widening a 10-ft lane to 12 ft).

Many highway agencies have established a set of crash reduction factors for each safety countermeasure and extent thereof. When local or national data on crash reduction factors are not available, the analyst can collect field data or use an existing relevant data set to develop crash prediction equations from which crash reduction factors can be established using the procedures described in Section 6.3.

Example 6.5   An intersection improvement project in a certain city is proposed. It involves the provision of left-turn lanes at the signalized intersection between two major urban arterials. Also, the signal timing was redesigned to include a dedicated green phase for left turns. Currently, there are 6 fatal, or injury crashes per year at the intersection over a three-year period. What reduction in fatal or injury crashes can be expected due to the project? Assume that the effects of such improvements on safety are mutually exclusive and complementary.

SOLUTION   If C_W and C_{WO} are the number of crashes at similar sites that are with improvement and without improvement, respectively, at a given time, the crash reduction can be given by

$$\mathrm{CRF} = \frac{C_{WO} - C_W}{C_{WO}} \times 100$$

From Table A6.1, the appropriate CRF is 0.53.

$$\Rightarrow C_{WO} - C_W = \frac{\mathrm{CRF} \times C_{WO}}{100} = \frac{(53)(6)}{100} = 3$$

Estimated number of crashes saved due to improvement = 3 crashes per year.

Example 6.6   As part of a major corridor expansion project to facilitate international freight and passenger travel, a stretch of an existing multilane urban minor arterial highway is to have a median installed (full restriction of access between opposing lanes) and full control of access from local roads. Also, the pavement is to be resurfaced to improve its skid resistance. Determine the safety impacts of the corridor improvement project in terms of total crashes. Without the improvement, the total number of all crashes over a three-year period is 23.

SOLUTION   From Table A6.1, the crash reduction factors are as follows:

Median installation : 25% → 6 crashes saved

Resurfacing(to improve surface friction) : 10%
   → 2 crashes saved

Total reduction in total crashes = 6 + 2 = 8

Number of crashes after improvements
   = 23 − 8 = 15

Therefore, there are 23 and 15 crashes without and with improvement, respectively, over a three year period.

Final Comments on Steps 3 to 6:   In these steps, the analyst estimates the expected number of crashes using one of many alternative approaches. Although a few aspects deal with predictions of frequencies of specific crash types (Table 6.4), the discussion is generally for total crashes. In cases where separate models for different crash severities are unavailable and where the analyst needs to segregate all predicted crashes by severity type (for purposes of costing or reporting), approximate distributions from past crash histories may be used. Such distributions are expected to vary from region to region and also across transportation facilities that differ by class, location, and so on. For highway facilities, a rough guide for the distribution of total crashes, for planning purposes, is as follows (Labi, 2006): fatal crashes, 0.5 to 1%; injury crashes, 20 to 30%; PDO crashes, 70 to 80%.

Step 7: Determine the Safety Benefits   Crash cost is one of the several categories of user costs that decrease with improved facility or safer roadway. When demand is elastic, there will be an increase in demand due to the shift in the supply curve, reflecting improved safety, that is, reduced safety cost of transportation (Figure 6.6). Therefore, in case of elastic demand, the safety benefits of a transportation intervention can be calculated as follows: safety savings = $(0.5)(U_1 - U_2)(V_1 + V_2)$, where U_1 and U_2 are the unit safety rates or "costs" (number of crashes per million VMT per year, for example) without and with the improvement, and V_1 and V_2 are the travel demand values (millions of VMT) without and with the improvement, respectively. When demand is inelastic, user safety benefit occurring from an improved transportation system is taken as the product of the reduction in the unit safety cost of travel and the (quantity) of travel demand (millions of VMT per year).

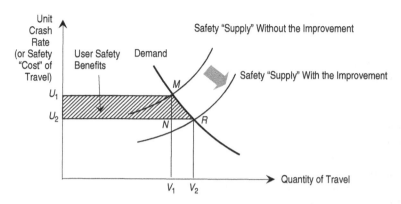

Figure 6.6   User benefits of increased safety due to a transportation intervention.

Step 8: Establish the Unit Monetary Crash Cost
When safety benefits are expressed in terms of the number of reduced crashes per VMT, the corresponding monetary cost savings is determined as the product of the *crash reduction per VMT* and the *unit monetary crash cost* to yield the dollars saved per VMT. The unit monetary cost of crashes is a function of (1) market or economic costs, which include property damage, insurance and legal costs, medical costs, and lost productivity, and (2) nonmarket costs, the emotional and social costs of casualties resulting from road crashes (Lindberg and Borlänge, 1999; Miller et al., 2000). To estimate the cost of a road crash. Blincoe et al. (2002) examined the economic cost of motor vehicle crashes to society using the human capital approach by discounting to present value the victim's income that is foregone due to the victim's premature death or injury. Loehman et al. (2000) applied the willingness-to-pay (WTP) approach to estimate the value of pain, grief, suffering, and uncompensated lost time resulting from crash-related injuries. Lindberg and Borlänge (1999) used the concept of marginal external costs to estimate the cost of road crashes. The marginal external costs are the incremental costs of a crash borne by society at large, including family and friends, and can also include costs borne by the victims of the crash. Using the WTP approach, Lindberg and Borlänge (1999) concluded that the nonmarket cost component was the dominant component and overshadows all other cost components of road crashes: the nonmarket costs account for 90% for fatal, 80% for severe injury, and 60% for light injury crash costs.

The two commonly used sources for the dollar value estimates are the annual publication of the National Safety Council Estimates and the 1988 FHWA memorandum. Also, the cost of road crashes can be based on a weighted injury scale by using indices to the level of severity of

the road crash. The unit costs of each crash severity type are available for injury scales such as the KABCO rating scale (NSC, 2001) and the abbreviated injury scale (Blincoe et al., 2002). Table 6.5 shows the unit crash cost values for KABCO crash coding scheme, updated using consumer price indices from the U.S. Department of Labor (USDL, 2006).

Step 9: Determine the Overall Safety Cost Savings Due to the Intervention   Given the expected number of crashes reduced due to the improvement (from step 7) and the unit cost per crash (from step 8), the analyst can calculate the dollar value of the overall crash cost savings.

Example 6.7   The injury crash rate with and without the improvement project at a rural two-lane highway is 2.87 and 3.5 per million VMT, respectively. Determine the user safety benefits in monetary terms due to the reduction in injury crashes. Assume an average vehicle occupancy rate of 1.00. The annual VMT is 1.5 and 1.8 millions for the without- and with-improvement scenarios, respectively.

Table 6.5   Unit Crash Costs on the Basis of the KABCO Injury Scale

Code	Severity	Unit Cost (2005 dollars)
K	Fatal	3,654,299
A	Incapacitating	181,276
B	Injury Evident	46,643
C	Injury Possible	22,201
O	Property Damage Only	2,116

Source: Updated from NSC (2001).

SOLUTION

Crash rate without improvement = 3.5 per million VMT

Crash rate with improvement = 2.87 per million VMT

Safety savings = $0.5(U_1 - U_2)(VMT_1 + VMT_2)$

$= (0.5)(3.5 - 2.87)(1.5 + 1.8) = 1.04$

From Table 6.5,

Average cost of incapacitating injury crash = \$181,276

Injury crash cost savings = $(1.04)(\$181,276)$

$= \$188,527$

due to the improvement project in the first year.

6.3 METHODS FOR ESTIMATING CRASH REDUCTION FACTORS

In the methodology presented in Section 6.2, a critical part of the CRF approach for crash reduction prediction is the establishment of crash reduction factors. Many state highway agencies have established crash reduction factors and functions associated with various improvements or interventions using their local data. These may be used by the analyst. However, in cases where crash reduction factors or crash prediction functions for other jurisdictions may not be applicable to a specific evaluation problem, the analyst should develop CRF values using local data. Generally, two types of studies can be used to develop crash reduction functions or factors: *before-and-after studies* and *cross-sectional (with-and-without) studies*.

6.3.1 Before-and-After Studies

A vital requirement in before-and-after studies is the recognition that some other extenuating factors besides the safety intervention may be partly responsible for the safety improvement and hence the crash frequency, or number of crashes per year, n_B, in the before period B without improvement may not be the same as the crash frequency, n_{A*}, in the after period A without improvement. Such extenuating factors may include random trends in crash occurrence or changes in other engineering factors, such as pavement friction factor, slopes, and VMT. In such a scenario, the crash frequency n_B for the before period B cannot be used as a reference in estimation of the crash reduction factor. Hence, the crash frequency n_B for the without-intervention scenario is adjusted for the change in annual exposure (VMT, AADT, etc.), and the crash reduction factor is calculated as follows:

$$\text{CRF} = \left(1 - \frac{n_A}{n_{A*}}\right) \times 100 \qquad \text{where} \quad n_{A*} = \frac{E_A}{E_B}n_B$$

where n_A is the crash frequency with the improvement, and E_A and E_B represent the exposure (VMT, AADT, etc.) in the after and before periods, that is, with and without the improvement, respectively.

Example 6.8 At a certain site, 30 crashes were reported over three years before a lane-widening project. The number of crashes reduced to 22 when observations were made over three years after the improvement project. The AADT on the 4.5-mile section changed from 12,260 before the improvement to 13,430 after the improvement. Calculate the crash reduction factor. Assume that all the other engineering factors remain constant over time.

SOLUTION   The crash frequencies before and after the improvement project are $n_B = 30/3 = 10$ crashes per year and $n_A = 22/3 = 7.333$ crashes per year. Since the AADT changed when the number of crashes was observed after the improvement, the crash frequency in the before period is adjusted for the change in exposure as follows:

$$n_{A*} = \frac{(10)(13,430)(4.5)}{(12,260)(4.5)} = 10.954$$

Therefore, the CRF can be calculated as

$$\text{CRF} = 100\left(1 - \frac{7.333}{10.954}\right) = 33.05\%$$

Conventional before-and-after studies use crash frequency data from several years before and after an intervention, from single or several control sites (where no improvement has been made), to estimate the CRF. The crash reduction factor at the control site is determined to estimate the change in the number of crashes due to factors other than those in the improvement project, such as random trends in crash occurrence or changes in VMT or any other engineering factors affecting safety (Figure 6.3). Detailed steps for computing crash reduction factors using the control site method are available in standard texts (Hauer, 1997).

Shortcomings of the Before-and-After Approach: Before-and-after studies, which involve a one-to-one match of improved sites with control sites, can suffer from the *regression-to-the-mean* (RTM) phenomenon (Hauer, 1997). RTM simply means that if a location has been selected for implementing a transportation improvement or intervention based on a short-span crash history, it is likely that in the ensuing years, crash experience would decrease (i.e., would regress to the long-term average crash rate) even if no interventions are made. As such, a decrease of crash experience (or part thereof) could mistakenly be attributed to the intervention thus overestimating the effectiveness of the intervention.

To adjust observed crash data to account for the RTM effect, the *empirical Bayesian* (EB) *procedure* can be used (Hauer, 1997; Harwood, et al., 2000). The EB method is applicable where there are data on historical crash frequency and estimated crash frequency. EB adjusts the predicted number of crashes by assigning weights to the crash frequencies predicted and observed (C_P and C_O, respectively) and utilizes these parameters to determine the number of crashes that can be expected (C_E). The weight is calculated on a parameter that is designed to account for overdispersion. The formula used to estimate the expected number of crashes is as follows:

$$C_E = w_P C_P + w_O C_O$$

where w_P, the weight for predicted crashes, $= 1/(1 + kC_P)$; w_O, the weight for observed crashes, $= 1 - w_P = kC_P/(1 + kC_P)$; and k is the overdispersion parameter. Suggested k values are as follows (AASHTO, 2003): 0.31 for roadway segments, 0.54 for three-leg stop-controlled intersections, 0.24 for four-leg stop-controlled intersections, and 0.11 for four-leg signalized intersections.

Example 6.9 For a certain roadway segment, six crashes, over a three year period, are observed after a roadway geometry improvement project. It was predicted that the section will have five crashes. Using the EB procedure, find the number of crashes expected for the segment after the improvement project. Assume no changes in engineering factors over time other than those due to the improvement project.

SOLUTION   For roadway segments, the overdispersion factor, $k = 0.31$. Therefore,

weight of crashes predicted, w_P

$$= \frac{1}{1 + kC_P} = \frac{1}{1 + (0.31)(5)} = 0.392$$

weight of crashes observed, $w_O = (1 - w_P) = 0.608$

number of crashes expected, C_E

$$= (0.392)(5) + (0.608)(6) = 5.608$$

For a safety improvement effectiveness evaluation at a road section (a function of the difference in before and after crash values), the EB value should preferably be used. If the number of crashes predicted (five) is used, the effect of the improvement would be underestimated.

Also, using the number of crashes observed (six) would lead to overestimation of effectiveness.

6.3.2 Cross-Sectional Studies

Cross-sectional analyses may involve a straightforward comparison of crashes at sections with and without the crash factor under investigation. Such analyses may also involve an approach where models are developed using data from several sections during a given time period, which differ by the crash factor under investigation. This approach was used by Tarko et al. (2000) to estimate crash reduction factors from given crash equations. Considering that the expected number of crashes with and without an improvement are $n_W = f(X_W)$ and $n_{WO} = f(X_{WO})$, respectively, where X is a vector of crash factors, the general formulation was stated as follows:

$$\text{CRF} = \left[1 - \frac{f(X_W)}{f(X_{WO})} \right] \times 100$$

Depending on the functional form for $f(X)$, the crash reduction function may take one of several forms. In the Tarko et al. (2000) study, the functional form was exponential:

$$f(X) = kYQ^\gamma e^{\beta X}$$

where k is a constant, Y and Q are exposure variables representing the temporal span of data and indicate the section length and traffic volume, respectively, and β is the slope parameter associated with the variable X. It is often assumed that crash reductions of roadway factor improvements are independent of each other, but some research studies have established composite crash reduction factors for specific combinations of multiple crash factors.

Example 6.10 Tarko et al. (2000) developed the following crash prediction model for signalized intersections:

$$C = e^k Y Q^\gamma e^{\beta_1 X_1 + \beta_2 X_2 + \cdots + \beta_n X_n}$$

where C is the number of crashes over a period of Y years; Q is the traffic volume entering the intersection (AADT); $X_1, X_2, \ldots, X_n$ are independent variables representing various roadway factors; k, γ, and β are constants. (a) Derive an expression for the crash reduction function for any roadway factor X_j. (b) Using cross-sectional data collected for several signalized intersections in a certain city, a crash prediction equation was developed based on the functional form above (after the natural logarithm is taken for both sides). Derive the crash prediction equation if the estimated values of the parameter coefficients are

Table E6.10

Variable Code	Variable Description	Coefficient
Constant	Constant term	−6.3771
ln Y	Natural log of the number of years	1.0000
ln Q	Natural log of traffic volume	0.7821
X_1	Number of lanes, including turning lanes	0.0673
X_2	Separation between directions by adding median with divisional islands on approaches	−0.5499
X_3	Number of raised separation at the intersection	0.4627
X_4	Average width of the separation	−0.0257

given in Table E6.10. (c) Using the results above, develop the crash reduction function for each roadway factor.

SOLUTION

(a) Let X_j^B and X_j^A be the values of the roadway factor before and after the improvement. Then the crash reduction function with respect to this roadway factor can be derived as follows:

$$\text{CRF} = 1 - \frac{e^k Y Q^\beta e^{\alpha_1 X_1 + \alpha_2 X_2 + \cdots + \alpha_j X_j^A + \cdots + \alpha_n X_n}}{e^k Y Q^\beta e^{\alpha_1 X_1 + \alpha_2 X_2 + \cdots + \alpha_j X_j^B + \cdots + \alpha_n X_n}}$$

$$= 1 - \frac{e^{\alpha_j X_j^A}}{e^{\alpha_j X_j^B}}$$

$$= 1 - e^{\alpha_j (X_j^A - X_j^B)}$$

(b) Taking the natural logarithm on both sides of the functional form

$$C = e^k Y Q^\beta e^{\alpha_1 X_1 + \alpha_2 X_2 + \cdots + \alpha_j X_j + \cdots + \alpha_n X_n}$$

$$\ln C = k + \ln Y + \beta \ln Q + \alpha_1 X_1 + \alpha_2 X_2 + \cdots + \alpha_j X_j + \cdots + \alpha_n X_n$$

Substituting the value of the coefficients yields

$$\ln C = -6.3771 + \ln Y + 0.7821 \ln Q + 0.0673 X_1$$
$$- 0.5499 X_2 + 0.4627 X_3 - 0.0257 X_4$$

(c) Using the results from (a) and (b), the crash reduction functions with respect to each of the roadway factors above is given as

$$\text{CRF}(X_1) = 1 - e^{0.0673(X_1^A - X_1^B)}$$

$$\text{CRF}(X_2) = 1 - e^{-0.5499(X_2^A - X_2^B)}$$

$$\text{CRF}(X_3) = 1 - e^{0.4627(X_3^A - X_3^B)}$$

$$\text{CRF}(X_4) = 1 - e^{-0.0257(X_4^A - X_4^B)}$$

6.3.3 Comparison of the Before-and-After and Cross-Sectional Methods

The key difference between the before-and-after and cross-sectional studies is that the former uses data pertaining changes in safety over time, whereas the latter uses data on the differences in safety between locations at a given point in time. The main advantage of the before-and-after approach is that it is more conformable to the concept of controlled experimentation. Its main shortcoming is the great amount of effort or resources needed to ensure a proper experimental design and execution of such studies, particularly over the desired range of levels of each roadway factor. The main advantage of cross-sectional models is that they make use of data that is often readily available at highway agencies and are much less expensive in terms of time and effort compared to before-and-after studies. The main disadvantage of the cross-sectional approach is that it requires an extensive amount of data to ensure proper specification and is often subject to estimation problems related to data quality. However, with ongoing automation of roadway inventory data at highway agencies, the effect of specification-related problems is increasingly being mitigated, and the number and range of crash factors that can be included in cross-sectional models is being broadened. A combination of before-and-after analysis and a cross-sectional analysis using negative binomial regression was proposed by Poch and Mannering (1996).

6.3.4 Elasticity of Crash Frequency

Crash reduction efficacy of safety-related transportation projects can be expressed in terms of the marginal effects (such as elasticities) on crash frequency of unit changes in levels of each engineering variable. However, this is applicable only if the change is small.

$$E_{x_j} = \frac{\partial f}{f} \frac{x_j}{\partial x_j}$$

where E is the elasticity of crash frequency with respect to the jth independent variable, x_j is the magnitude of the variable X_j under consideration, and f is the crash prediction function.

Table 6.6   Common Functional Forms and Elasticity Functions

	Functional Form of the Crash Prediction Equation, $f(X)$	Elasticity Function $[X_j/f(X)](\partial f/\partial X_j)$	References
Linear	$\beta_0 + \beta_1 X_1 + \beta_2 X_2 + \cdots + \beta_n X_n$	$\dfrac{\beta_j X_j}{\beta_0 + \beta_1 X_1 + \cdots + \beta_n X_n}$	
Product	$\beta_0 \times X_1^{\beta_1} X_2^{\beta_2} \cdots X_n^{\beta_n}$	β_j	Forkenbrock and Foster (1997); Tarko et al. (2000)
Exponential	$\beta_0 e^{\beta_1 X_1 + \beta_2 X_2 + \cdots + \beta_n X_N}$	$\beta_j X_j$	Forkenbrock and Foster (1997)

Table 6.6 presents the elasticity functions corresponding to three common functional forms of crash prediction equations. In many cases, the elasticity function is not a constant but is a function of the value of the X_j variable. In the context of crash reduction, this implies that the effectiveness of a safety improvement often depends on the level of the existing engineering factor or deficiency.

6.4   SAFETY-RELATED LEGISLATION

Safety has long been a key consideration in transportation-related federal legislation such as transportation funding reauthorizations. Initial requirements set forth by the 1991 Intermodal Surface Transportation Efficiency Act (ISTEA) set the stage for the establishment of safety management systems in various states and therefore helped establish the databases and knowledge bases needed for systematic safety impact evaluation of transportation projects. The Transportation Equity Act for the 21st Century (TEA-21) of 1998 focused on five deployment goals designed to improve the efficiency, safety, reliability, service life, environmental protection, and sustainability of the nation's surface transportation system. In 2005, the Safe, Accountable, Flexible, Efficient Transportation Equity Act: A Legacy for Users (SAFETEA-LU) was signed to reaffirm the national emphasis on transportation safety. SAFETEA-LU established a new core highway safety improvement program that is structured and funded to make significant progress in reducing highway fatalities. It created an agenda for increased highway safety by doubling the funds for safety infrastructure and by requiring results-driven strategic highway safety planning.

6.5   SOFTWARE PACKAGES FOR SAFETY IMPACT EVALUATION OF TRANSPORTATION INVESTMENTS

6.5.1   Interactive Highway Safety Design Model

IHSDM is a suite of software analysis and evaluation tools for assessing the safety impacts of geometric design

decisions. For a given highway project, IHSDM checks existing or proposed designs against relevant design policy values and estimates the expected safety and operational performance of the design (FHWA, 2003). IHSDM therefore helps transportation planners to incorporate safety considerations in project selection. The overall IHSDM contains modules for safety evaluation tasks and concepts such as crash prediction, design consistency monitoring, driver–vehicle interaction, and intersection safety diagnostics. The current version of IHSDM focuses on rural two-lane highways, and future versions are expected to include other road classes.

6.5.2 Indiana's Safety Management System

Several safety management systems have been developed at the state level. In Indiana, the system has been automated to form a software package that consists of several evaluation modules for assessing project- or network-level safety impacts of transportation projects (Lamptey et al., 2006). By determining the safety impact of individual treatments associated with transportation projects, *SMSS-IN* helps planners in quantifying and monetizing the reductions in fatal, injury, and PDO crashes and produces outputs that can be used for economic efficiency analysis of transportation projects.

6.6   CONSIDERATIONS IN SAFETY IMPACT EVALUATION

The procedural evaluation framework presented in Section 6.2 can be used for assessing the safety impacts of transportation projects. This generally involves an estimation of crash frequencies with and without an intervention using crash rates, crash equations, or crash reduction factors. Choosing an appropriate method to estimate crash frequency depends on the availability of data. The crash rate method is the least data intensive but may provide the least reliable estimates of future crash frequency; the crash reduction factor method generally yields more reliable crash estimates but is data intensive and may be plagued with problems of overlapping (where the project involves multiple safety interventions). Furthermore, regardless of which estimation approach is chosen, the analyst will have to decide whether the given crash relationships (crash rates, equations, or reduction factors) are sufficiently representative of the given problem. In many cases, such relationships exist only at a more aggregate level (such as regional or national) or may be local but outdated. As such, recent local data may need to be collected to develop such relationships so that they can be used for crash prediction for specific projects.

Another issue is that of the influence of other crash factors. Prediction of future crashes on the basis of current relationships (rates, equations, or reduction factors) proceeds on the implicit assumption that the status of the other crash factors (such as enforcement levels, operator characteristics, education, and policy) will remain the same in the future. Crash occurrence is a complex interaction of the various crash factors; as such, it is not very certain how future changes in the nonengineering factors will affect the expected number of future crashes that were estimated on the basis of only the engineering factors. Elvik and Vaa (2004) cataloged over a hundred road safety measures associated with highway engineering, traffic control, vehicle design, public information, and police enforcement that have been tried and tested at locations all over the world and have provided some discussion of the interrelationships between factors.

The issue of equity arises in the context of safety impact evaluation of transportation projects. The analyst must ascertain whether a transportation intervention yields greater safety benefits to certain population groups while other groups get significantly lower (or even negative) safety benefits. For example, upgrading a local minor collector street to major arterial status may improve the safety of through traffic but may pose a hazard for residents (particularly children) of the area (Forkenbrock and Weisbrod, 2001).

There is also the issue of crash cost sources and responsibilities. The largest components of the total motor vehicle crash cost are market productivity (the cost of foregone paid labor due to death and disability) and property damage, each accounting for about 26% of the total costs. The loss of household productivity (the cost of foregone household labor) accounted for 9% of the total cost. Workplace cost (2% of the total cost) is the disruption due to the loss or absence of an employee such that it requires training a new employee, overtime to accomplish the work of the injured employee, and administrative costs to process personnel changes. Other costs are

associated with insurance administration (7%), legal (5%), and emergency services (less than 1%). Ultimately, all citizens, whether or not they are involved in a crash, pay a part of motor vehicle crash costs through insurance premiums, taxes, out-of-pocket expenses, and so on. Data from 2000 indicate that approximately one-fourth of the total crash cost is paid directly by those involved, while society in general pays the rest. Insurance companies, which are funded by all insured drivers (whether or not they are involved in a crash) paid about 50% of the cost and the government paid 9% (NHTSA, 2002). These are the economic costs only and therefore do not include the intangible consequences of these events to individuals and families, such as pain and suffering and loss of life.

SUMMARY

In this chapter we presented a procedural framework for assessing the safety impacts of transportation projects. While the safety issue remains a key consideration in evaluation of projects for all transportation modes, in this chapter we focus on the highway mode because of the overwhelming dominance of the highway safety problem. The general evaluation framework, however, is applicable to projects associated with other transportation modes.

Even with highway transportation, it is only the engineering factors that typically are mostly affected by improvements to the system. The overall framework presented in this chapter may be applicable to impact evaluation of increased enforcement levels or regulatory initiatives, such as increased patrols, changed speed limits, stricter driver under influence (DUI) laws, and so on.

In the past, safety evaluation included primarily those projects that were directly safety related, such as guardrail installation, treatments of freeway gore areas, and so on. As such, safety considerations were not included for projects such as pavement preservation. In the case of federal 3R projects, for instance, safety engineers did not participate in the design of such projects. At a later time when it was necessary to accommodate safety-related improvements (such as reconstructing sharp curves, replacing or extending bridges with narrow decks) in 3R projects, safety evaluation of such projects was stymied. In recent years, it has been duly recognized that there are safety impacts associated with most projects and state agencies have subsequently reshaped their 3R design procedures. New practices for 3R projects include various safety-related tasks grouped in the following categories: safety-conscious design practices, design practices for key highway features, planning and programming 3R projects, safety research and training, and other design procedures and assumptions.

EXERCISES

6.1. For each mode of transportation, the factors that affect crashes may be categorized broadly as follows: system engineering features, environment (weather), operator characteristics (age, education, etc.), vehicle characteristics, policy, and so on. Against this background, explain why crashes are still by far highest for the highway mode of transportation compared to the other modes.

6.2. Mention some initiatives that have helped reduce the high rate of highway crashes over the past 20 years. Even at their current rates, highway crashes are unacceptably high. What can be done to further reduce the rate of highway crashes?

6.3. What is the difference between "safety impacts of transportation projects" and "impacts of transportation safety projects"? Give three examples of highway transportation projects for which safety impacts are typically evaluated in addition to other impact types. Also, give three examples of highway transportation safety projects.

6.4. Two-lane rural and urban roads experience unique operational difficulties and safety problems, such as the lack of passing opportunities due to oncoming traffic and/or poor sight distance. As part of a proposed major corridor improvement of a two-lane highway near Brunswick Town, it is intended to construct a passing lane at a certain crash-prone stretch of the highway. This would enable left-turners to seek refuge in an island as they wait for a gap to make the turn, and would also enable passing traffic to bypass the waiting left turners. Currently, all 70 crashes per year at that T-intersection are due to rear-ending of waiting left-turners. Of all crashes, 2 are fatal crashes, 20 are injury crashes, and the rest are PDO crashes. What will be the safety impact of the transportation project in terms of (a) crash frequency and (b) crash costs? Use Table A6.2 to obtain the appropriate crash reduction factor and Table 6.5 for the unit crash costs.

6.5. To reduce severe congestion and intolerable travel times for commuters using State Road 555, a two-lane highway connecting the City of Light to its fast-growing western suburbs, it is proposed to upgrade the highway to a four-lane facility. The project will also involve pavement resurfacing, shoulder widening, and passing opportunities. It is expected that there will be a 5% increase in traffic due to the project. Values of the roadway factors with

Table EX6.5   Values of Roadway Factors

	Without Improvement	With Improvement
Pavement condition (PSR)	Fair (2.5)	Very good (4.4)
Horizontal alignment (TOPCURV)	Good (4)	Good (4)
Passing restrictions (PASSRES)	2	0
Traffic volume per lane (ADTLANE)	2.5	Determine this value
Lane class	0	1
Road shoulder width (RIGHTSH)	2 ft	4 ft
Vertical alignment (TOPGRAD)	Good (4)	Good (4)

and without the improvement project are given in Table EX6.5.

Using the crash prediction model developed by Forkenbrock and Foster (1997) in Table 6.4, determine the safety impact of the project in terms of crash reduction on the basis of:

(a) The aggregate approach. Here, use the crash prediction equation to directly determine the number of crashes with and without the improvement project.

(b) The disaggregate approach. Here, apply marginal effects analysis to derive the crash reduction function (for each affected roadway factor) from the crash prediction equation. Then using the data given, determine the reduction in crashes associated with each factor and sum them up to get the overall crash reduction.

(c) Compare the results from **(a)** and **(b)**. Comment on the relative ease of each approach. Under what circumstances is it more appropriate to use the disaggregate approach?

6.6. An existing rural two-lane county road has a lane width of 6 ft and unpaved shoulders of 1 ft width. It is proposed to upgrade the road to higher standards.

(a) On the basis of safety impacts only, which of the following alternative schemes would have the greatest impact?

(1) Widen the lane to 8 ft and do nothing to the shoulder (technically, this means adding the shoulder to the lane and constructing new 2-ft-wide shoulders). Use Table A6.2(a).

(2) Do nothing on the lane and widen shoulder width to 3 ft. Use Table A6.2(b).

(3) Pave the shoulder and do nothing else. Use Table A6.1.

(b) What other decision parameter beside effectiveness (expected crash reduction) of each action would be needed to make a final decision?

6.7. An existing urban freeway currently has a volume–capacity ratio of 1.05. It is planned to add a lane to accommodate increasing traffic growth at this highway. It is expected that the volume–capacity ratio after the capacity expansion will be 0.82. Determine the safety impact of the improvement.

6.8. For a four-leg stop controlled intersection in a certain city, seven crashes were observed in a 3-year period. Also, it has been predicted that the section will have five injury crashes over the next three-year period. Using the EB procedure, find the expected number of injury crashes for the intersection over that period.

REFERENCES

AASHTO (2003). *A Manual of User Benefit Analysis for Highways*, 2nd Edition American Association of State Highway and Transportation Officials, Washington, DC.

Bauer, K. M., Harwood, D. W. (2000). *Statistical Models of At-Grade Intersection Accidents*, Addendum, FHWA-RD-99-094, Federal Highway Administration, U.S. Department of Transportation, Washington, DC.

Blincoe, L., Seay, A., Zaloshnja, E., Miller, T., Romano, E., Luchter, S., Spicer, R. (2002). *The Economic Impact of Motor Vehicle Crashes, 2000*, DOT HS 809 446, National Highway Traffic Safety Administration, U.S. Department of Transportation, Washington, DC.

Elvik, R., Vaa, T. (2004). *The Handbook of Road Safety Measures*, Elsevier, Amsterdam, The Netherlands.

FHWA (1998). *Highway Statistics, 1997*. Office of Highway Information Management, Federal Highway Administration, Washington, DC.

FHWA (2003). IHSDM Preview CD-ROM, Version 2.0, FHWA-SA-03-005, Federal Highway Administration, U.S. Department of Transportation, Washington, DC.

Forkenbrock, D. J., Foster, N. S. J. (1997). *Accident cost saving and highway attributes. Transportation*, Vol. 24, No. 1, pp. 79–100.

Forkenbrock, D. J., Weisbrod, G. E. (2001). *Guidebook for Assessing the Social and Economic Effects of Transportation Projects*, NCHRP Rep. 456, Transportation Research Board, National Research Council, Washington, DC.

Gan, A., Shen, J., Rodriguez, A., Brady, P. (2004). Crash Reduction Factors: A State-of-the-Practice Survey of State Departments of Transportation. Procs., Annual Meeting of the Transportation Research Board, Washington, DC.

Harwood, D. W., Hobar, C. J. (1987) *Low-Cost Methods for Improving Traffic Operations on Two-Lone Roads*. Technical Rep. FHWA-IP-87-2. Federal Highway Administration, Washington, DC.

Harwood, D. W., Council, F. M., Hauer, E., Hughes, W. E., Vogt, A. (2000). *Prediction of the Expected Safety Performance of Rural Two-Lane Highways*, FHWA-RD-99-207, Federal Highway Administration, U.S. Department of Transportation, Washington, DC.

Harwood, D. W., Kohlman Rabbani, E. R., Richard, K. R., McGee, H. W., Gittings, G. L. (2003). *Systemwide Impact of Safety and Traffic Operations Design Decisions of Resurfacing, Restoration, or Rehabilitation (RRR) Projects*, NCHRP Rep. 486, Transportation Research Board, National Research Council, Washington, DC.

Harkey, D. L., Srinivasan, R., Zegeer, C. V., Persaud, B., Lyon, C., Eccles, K., Council, F. M., McGee H. (2004). *Crash Reduction Factors for Traffic Engineering and Intelligent Transportation System Improvements—State of Knowledge Report* NCHRP Results Digest 229, Transportation Research Board, NRC, Washington, DC.

Hauer, E. (1997). *Observational Before–After Studies in Road Safety*, Pergamon Press, Oxford, UK.

Islam, S., Mannering, F. (2006). Driver aging and its effect on male and female single-vehicle accident injuries: Some additional evidence. *J. Safety Res.* Vol. No. 37, No. 3, pp. 267–276.

Kweon, Y., Kockleman, K. (2003). Driver attitudes and choices: Speed Limits, seat Belt use, and drinking-and-driving, Presented at 82nd Annual meeting of the Transportation Research Board, Washington, DC.

Labi, S. (2006). *Effects of Geometric Characteristics of Rural Two-Lane Roads on Safety, Vol. I, Analysis of Engineering Factors*, Tech. Rep. FHWA/IN/JTRP-2006/2, Purdue University, West Lafayette, IN.

Lamptey, G., Labi, S., Sinha, K. C. (2006). *Development of a Safety Management System and Software for the State of Indiana*, Tech. Rep. FHWA/IN/JTRP-2006/5, Purdue University, West Lafayette, IN.

Lindberg, G., Borlänge, V. (1999). *Calculating Transport Accident Costs: Final Report of the Expert Advisors to the High Level Group on Infrastructure Charging* (Working Group 3). Swedish National Road and Transportation Research Institute, Borlärye, Sweden.

Loehman, E., Islam, S., Sinha K. C. (2000) Willingness to pay to avoid statistical risk of injuries and death from motor vehicle accidents, Presented at the Economic Science Association Regional Conference, Tucson, AZ.

Miller, T. R., Viner, J., Rossman, S., Pindus, N., Gellert, W., Dillingham, A., Blomquist, G. (1991). *The Costs of Highway Crashes*, Urban Institute, Washington DC.

Miller, T. R., Lawrence, B., Jensen, A., Waehrer, G., Spicer, R., Lestina, D., Cohen, M. (2000). *The Consumer Products Safety Commission's Revised Injury Cost Model*, U.S. Consumer Product Safety Commission, Claverton, MD.

NHTSA (2002). *The Economic Impact of Motor Vehicle Crashes, 2000*, National Highway Traffic Safety Administration, U.S. Department of Transportation, Washington, DC, www.nhtsa.dot.gov/people/economic/. Accessed Dec. 2005.

NSC (2001). *Estimating the Costs of Unintentional Injuries, 2000* National Safety Council, Spring Lake Drive, Itasca, IL, www.nsc.org/lrs/statinfo/estcost0.htm. Accessed Dec. 19, 2005.

Poch, M., Mannering, F. (1996). Negative binomial analysis of intersection-accident frequencies, *J. Transp. Eng.*, Vol. 122, No. 2, pp. 105–113.

Tarko, A., Sinha, K. C., Eranky, S., Brown, H., Roberts, E., Scinteie, R., Islam, S. (2000). *Crash Reduction Factors for Improvement Activities in Indiana*, Joint Transportation Research Program, Purdue University, West Lafayette, IN.

Turner, D. S. (1984). Prediction of bridge accident rates, *J. Transp. Eng.*, Vol. 110, No. 1.

USDHHS (2003). *National Vital Statistics Reports: Deaths, 1991–2000* Centers for Disease Control, National Center for Health Statistics, U.S. Department of Health and Human Services, Washington, DC, www.cdc.gov/nchs/products.htm. Accessed Mar. 14, 2003.

USDL (2006). *Consumer Price Index*, Bureau of Labor Statistics, U.S. Department of Labor, Washington, DC, ftp://ftp.bls.gov/pub/special.requests/cpi/cpiai.txt. Accessed Jan. 11, 2006.

USDOT (2004). *Transportation Statistics Annual Report, 2004*, Bureau of Transportation Statistics, U.S. Department of Transportation, Washington, DC.

Winston, C., Maheshri, V., Mannering, F. (2006). An exploration of the offset hypothesis using disaggregate data. The case of air bags and antilock brakes. *J. Risk Uncertainty*, Vol. 32, No. 2, pp. 83–99.

Zegeer, C. V., Stewart, J. R., Reinfurt, D. W., Herf, L., Huntes, W. (1987). *Safety Effects of Cross-Section Design for Two-lane Roads*, Vols I and II, Tech. Rep. FHWA-RD-87-008, Federal Highway Administration, Washington, DC.

Zegeer, C. V., Stewart, R., Reinfurt, D., Council, F., Newman, T., Hamilton, E., Miller, E., Hunter, W. (1991). *Cost-Effective Geometric Improvements for Safety Upgrading of Horizontal Curves* Tech. Rep. FHWA-RD-90-021, Federal Highway Administration, Washington, DC.

ADDITIONAL RESOURCES

Crash Rates

Annual Reports, National Highway Traffic Safety Administration, Washington, DC.

Highway Statistics Annual Reports, Table VM-1, Federal Highway Administration, U.S. Department of Transportation, Washington, DC, http://www.fhwa.dot.gov/policy/ohim/hs02/index.htm.

National Transportation Statistics Annual Reports, Table 2-2, Bureau of Transportation Statistics, U.S. Department of Transportation, revised, Washington, DC, http://www.bts.gov/.

Crash Equations (Safety Performance Functions)

Bauer, K. M., Harwood, D. W. (1996) Statistical Models of At-Grade Intersection Accidents, FHWA-RD-96-125, Federal Highway Administration, U.S. Department of Transportation, Washington, DC.

Harwood, D. W., Council, F. M., Hauer, E., Hughes, W. E., Vogt, A. (2000). *Prediction of the Expected Safety Performance of Rural Two-Lane Highways*, FHWA-RD-99-207, for the Federal Highway Administration, U.S. Department of Transportation, Washington, DC.

Hauer, E., Harwood, D. W., and Griffith, M. S. (2002). Estimating Safety by the Empirical Bayes Method: A Tutorial, *Transp. Res. Rec. 1784*, Transportation Research Board, National Research Council, Washington, DC.

Vogt, A. (1999). *Crash Models for Rural Intersections: 4-Lane by 2-Lane Stop-Controlled and 2-Lane by 2-Lane Signalized*, FHWA-RD-99-128, Federal Highway Administration, U.S. Department of Transportation, Washington, DC.

Vogt, A., Bared, J. G. (1998). *Accident Models for Two-Lane Rural Roads: Segments and Intersections*, Transp. Res. Rec. 1635, Transportation Research Board, National Research Council, Washington, DC.

Crash Reduction Factors

Council, F. M., Stewart, J. R. (1999). *Safety Effects of the Conversion of Rural Two-Lane to Four-Lane Roadways based on Cross Sectional Models*, Transp. Res. Rec. 1665, Transportation Research Board, National Research Council Washington, DC, pp. 35–43.

Shen, J., Gan, A. (2003). Development of Crash Reduction Factors: Methods, Problems, and Research Needs, *Transp. Res. Rec. 1840*, Transportation Research Board, National Research Council, Washington DC, pp. 50–56.

Crash Costs

Miller, T. R., Lawrence, B., Jensen, A., Waehrer, G., Spicer, R., Lestina, D., and Cohen, M. (2000). *The Consumer Products Safety Commission's Revised Injury Cost Model*, U.S. Consumer Product Safety Commission, Claverton, MD.

Lindberg, G. (1999). *Calculating Transport Accident Costs*, Final Report of the Expert Advisors to the High Level Group on Infrastructure Charging (Working Group 3), Swedish National Road and Transportation Research Institute Borlänge, Sweden.

NSC (2001). *Estimating the Costs of Unintentional Injuries, 2000*, National Safety Council Spring Lake Drive, Itasca, IL, http://www.nsc.org/lrs/statinfo/estcost0.htm.

Database for Crash Analysis

The Highway Safety Information System (HSIS) is a multistate database containing crash, roadway inventory, and traffic volume data. It is operated by the University of North Carolina Highway Safety Research Center (HSRC) and LENDIS Corporation under contract with the FHWA. Analysts can use this database to develop crash rates, crash equations, or crash reduction functions for subsequent use in predicting crashes for projects under investigation.

Resources for Overall Safety Impact Evaluation

FHWA (2000). *Highway Economic Requirements System: Technical Report*, Federal Highway Administration, U.S. Department of Transportation, Washington, DC.

Forkenbrock, D. J., Weisbrod, G. E. (2001). *Guidebook for Assessing the Social and Economic Effects of Transportation Projects*, NCHRP Rep. 456, Transportation Research Board, National Research Council, Washington, DC.

Harwood, D. W., Kohlman Rabbani, E. R., Richard, K. R., McGee, H. W., Gittings, G. L. (2003). *Systemwide Impact of Safety and Traffic Operations Design Decisions of Resurfacing, Restoration, or Rehabilitation (RRR) Projects*, NCHRP Rep. 486, Transportation Research Board, National Research Council, Washington, DC.

Fitzpatrick, K., Balke, K., Harwood, D. W., Anderson, I.B. (2000). *Accident Mitigation Guide for Congested Rural Two-Lane Highways*, NCHRP Rep. 440, Transportation Research Board, National Research Council, Washington DC.

APPENDIX A6: CRASH REDUCTION AND ACCIDENT MODIFICATION FACTORS

Table A6.1   Crash Reduction Factors: All Highways

Activity Category	Specific Activity	Crash Reduction Factor (%) All Crashes
Channelization	Channelize intersection	23
	Provide left-turn lane (with signal)	24
	Provide left-turn lane (without signal)	40
	Install two-way left turn in median	34
	Add mountable median	15
	Add nonmountable median	25
	Provide right-turn-lane	28
	Increase turn-lane length	28
	Horizontal alignment changes	50
Geometric improvements	Gentler horizontal curve	
	Change in horizontal curvature	
	20 to 10°	48
	15 to 5°	63
	10 to 5°	45
	Improve vertical curve	43
	Improve sight distance at intersection	31
	Superelevation	46
Median device installation	Install median barrier (general)	25
	Install raised median	23
	Add flush median	52
	Add flush median with refuge for left turns	44
Widening of lane/shoulder, shoulder paving	Widen lane	28
	Widen paved shoulder	29
	Widen unpaved shoulder	22
	Pave shoulder	17
	Stabilize shoulder	24
Lane additions	Add acceleration/deceleration lane	16
	Add lanes	23
	Add turning lane	17
Bridge improvements	Bridge replacement	46
	Bridge widening	48
	Bridge deck repair	14
	Bridge rail upgrade	20
Intersection improvements	Increase turning radii	13
	improve sight distance	33
Freeway improvements	Construct interchange	57
	Modify entrance/exit ramp	25
	Construct frontage road	35

(*continued overleaf*)

Table A6.1   (*continued*)

Activity Category	Specific Activity	Crash Reduction Factor (%) All Crashes
Traffic signal improvements	Install sign	27
	Change 2WSC to signal	28
	Change 2WSC to signal and add lane	36
	General upgrade of existing signal system	25
	Replace lenses with larger ones (12 in.)	12
	Improve signal phasing	25
	Improve signal timing	12
	Add exclusive left-turn phase (protected)	29
	Install/improve pedestrian signal	23
	Remove unwarranted signal	66
Guardrail improvements	Install guardrail	20
	Upgrade guardrail	10
	Install guardrail at bridge	24
	Install guardrail at outer lane in curve	63
	Install guardrail at culverts	27
Pavement improvements	General pavement treatment	25
	Groove pavement	19
	Resurface with skid-resistant material	10
	Resurfacing (general)	20
	Install rumble strips	30
	Groove shoulder	25
Roadside improvements	Relocate fixed objects	40
	Install impact attenuators	30
	Flatten side slope	25

Source: Harkey et al. (2004).

Table A6.2   Crash Reduction Factors: Rural Two-Lane Highways

(a) Factors for Lane Widening

Amount of Lane Widening (ft)	% Reduction in Crashes
1	12
2	23
3	32
4	40

Source: Zegeer et al. (1987).

(b) Factors for Shoulder Widening[a]

Amount of Lane Widening (ft)	% Reduction in Crashes	
	Paved	Unpaved
2	16	13
4	29	25
6	40	35
8	49	43

Source: Zegeer et al. (1987).

[a]Values are for run-off-road, head-on, opposite-direction sideswipe crashes.

(c) Factors for Increasing Roadside Recovery Distance[a]

Amount of Increased Roadside Recovery Distance (ft)	5	8	10	12	15	20
% Reduction in "Related" Crash Types	13	21	25	29	35	44

Source: Zegeer et al. (1987).

[a]Values are for run-off-road, head-on, opposite-direction sideswipe crashes.

(d) Factors for Side-Slope Improvements

	Side Slope After Flattening							
	1:4		1:5		1:6		1:7 or Flatter	
Side Slope Before Flattening	Single Vehicle	Total	Single Vehicle	Total	Single Vehicle	Total	Single Vehicle	Total
1:2	10	6	15	9	21	12	27	15
1:3	8	5	14	8	19	11	26	15
1:4	0	—	6	3	12	7	19	11
1:5	—	—	0	—	6	3	14	8
1:6	—	—	—	—	0	—	8	5

Source: Zegeer et al. (1987).

(e) Factors for Bridge Shoulder Widening[a]

Bridge Shoulder Width on Each Side before Widening	Bridge Shoulder Width after Widening					
	2 ft	3 ft	4 ft	6 ft	7 ft	8 ft
0	23	42	57	78	83	85
1	—	25	45	72	78	80
2	—	—	27	62	71	74
3	—	—	—	48	60	64
4	—	—	—	44	44	50

Source: Turner (1984).

[a]Width of bridge lanes assumed constant.

(f) Factors for Providing Passing Opportunities

	% Reduction in Crashes	
Countermeasure	Total Crashes	Fatal + Injury Crashes
Passing lanes	16	13
Short four-lane section	29	25
Turnout	40	35
Shoulder use section	49	43

Source: Harwood and Hoban (1987).

(g) Factors for Increased Roadside Recovery Distance at Curve Sections

Increase in Roadside Clear Recovery Distance (ft)	Percent Reduction in Total Curve Crashes
5	9
8	14
10	17
12	19
15	23
20	29

Source: Zegeer et al. (1991).

(h) Factors for Flattening Side Slopes on Curves

	Percent Reduction in Total Curve Crashes			
Initial Side Slope of	Side Slope After Treatment			
Curve (Before Treatment)	1:4	1:5	1:6	1:7 or flatter
1:2	6	9	12	15
1:3	5	8	11	15
1:4	—	3	7	11
1:5	—	—	3	8
1:6	—	—	—	5

Source: Zegeer et al. (1991).

(i) Factors for Curve Widening

Total Amount of Lane or Shoulder Widening at Curve (ft)		% Reduction in Crashes		
Total	Per Side	Lane Widening	Paved-Shoulder Widening	Unpaved-Shoulder Widening
2	1	5	4	3
4	2	12	8	7
6	3	17	12	10
8	4	21	15	13
10	5	—	19	16
12	6	—	21	18
14	7	—	25	21
16	8	—	28	24
18	9	—	31	26
20	10	—	33	29

Source: Zegeer et al. (1991).

Table A6.3   Accident Modification Factors: All Highways

(a) General Improvements

		AMF	
Activity	Facility Type	All Crashes	Fatal + Injury Crashes
Add shoulder rumble strips (effect on single-vehicle run-off road crashes)	Urban and rural freeways	0.82	—
	Other highways	0.79	—
Install roundabout	Urban and rural freeways	0.87	—
	Other highways	0.93	—
	Urban single lane (prior control—stop sign)	0.28	0.12
	Rural single lane (prior control—stop sign)	0.42	0.18
	Urban Multilane (prior control—stop sign)	0.95	—
	Urban single/multilane (prior control—signal)	0.65	0.26
Install guardrails	All facilities	0.56 (all injury crashes)	0.56
Install traffic signal	Three-leg intersections		
	All crash patterns	—	0.86
	Right-angle crashes	—	0.66
	Rear-end crashes	—	1.50
	Four-leg Intersections		
	All crash patterns	—	0.77
	Right-angle crashes	—	0.33
	Rear-end crashes	—	1.38

(b) Exclusive Turning Lanes

Activity	Facility Type	AMF for One Approach		AMF for Two Approaches	
		All Crashes	Fatal + Injury Crashes	All Crashes	Fatal + Injury Crashes
Add exclusive left-turn lane	Four-leg rural stop-controlled intersection	0.72	0.65	0.52	0.42
	Three-leg rural stop-controlled intersection	0.56	0.45	—	—
	Four-leg rural signalized intersection	0.82	—	0.67	—
	Three-leg rural signalized intersection	0.85	—	—	—
	Four-leg urban stop-controlled intersection	0.73	0.71	0.53	0.50
	Three-leg urban stop-controlled intersection	0.67	—	—	—
	Four-leg urban signalized intersection	0.90	0.91	0.81	0.83
	Three-leg urban signalized intersection	0.93	—	—	—
Add exclusive right-turn lane	Four-leg rural stop-controlled intersection	0.86	0.77	0.74	0.59
	Four-leg urban signalized intersection	0.96	0.91	0.92	0.83

Source: Harkey et al. (2004).

Table A6.4   Accident Modification Factors: Rural Two-Lane Highways

(a) Factors for Providing Superelevation at Horizontal Curves

Existing Superelevation Deficiency	Accident Modification Factor
0.00	1.00
0.01	1.00
0.02	1.06
0.03	1.09
0.04	1.12

Source: Zegeer et al. (1991).

(b) Factors for Shoulder Widening

Shoulder Width (ft)	Accident Modification Factor[a]
0	1.50
2	1.30
4	1.15
6	1.00
8	0.87

Source: Harwood et al. (2000).
[a]For run-off-road, head-on, opposite-direction sideswipe crashes.

(c) Factors for Shoulder Surface Improvement[a]

Shoulder Type	Shoulder Width (ft)							
	0	1	2	3	4	6	8	10
Paved	1.00	1.00	1.00	1.00	1.00	1.00	1.00	1.00
Gravel	1.00	1.00	1.00	1.01	1.01	1.02	1.02	1.03
Composite	1.00	1.01	1.02	1.02	1.03	1.04	1.06	1.07
Turf	1.00	1.01	1.03	1.04	1.05	1.08	1.11	1.14

Source: Harwood et al. (2000).
[a]For run-off-road, head-on, opposite-direction sideswipe crashes.

CHAPTER 7

Vehicle Operating Cost Impacts

Better to be wise by the misfortunes of others than by your own.

—*Aesop (560 B.C.)*

INTRODUCTION

Vehicle costs are direct expenses that comprise the costs of vehicle ownership (fixed) and vehicle operation (variable). The latter category, typically referred to as *vehicle operating costs* (VOCs), varies with vehicle use and is typically expressed in cents per mile traveled by a vehicle. For most transportation modes, VOC involves energy use, tires, maintenance, repairs, and mileage-dependent depreciation. Fixed vehicle costs are those that are largely independent of vehicle use and are generally unaffected by transportation improvements; examples are insurance costs, time-dependent depreciation, financing, and storage. Such costs are therefore typically excluded from VOC impact evaluation of projects.

VOC *savings* or *benefits* of a transportation improvement or intervention simply refer to *the reduction in vehicle operating costs compared to an existing situation or a base-case alternative.*

For areawide or corridor-level projects involving multimodal systems, an improvement in any part of the system can affect VOCs of the other parts or other modes. For example, service improvement in commuter rail or provision of a bus rapid transit along a corridor can affect the level of service on highway facilities in the same corridor because the shift of some travelers from automobile to transit would lead to improved highway level of service due to reduced congestion and thus, lower vehicle operating costs at the highway section.

In this chapter we identify VOC components and factors and present a procedural framework for assessing the

VOC impacts of transportation improvements. Then a comparison of various VOC estimation methodologies and software for the highway mode, is presented.

7.1 COMPONENTS OF VEHICLE OPERATING COST

The components of vehicle operating cost are the individual items associated with vehicle operation on which expenses are directly incurred. These include the costs of energy needed to propel the vehicle, fluids, and other light consumables associated with mechanical working of the drivetrain, occasional replacement of the vehicle's contact surfaces with the guideway, vehicle repair and maintenance, and vehicle depreciation.

7.1.1 Fuel

Fuel is a key component of vehicle operating costs. For highway vehicles for instance, fuel costs can account for 50 to 75% of usage-related costs. Fuel cost can be estimated on the basis of fuel efficiency and unit fuel price. Fuel efficiency, in turn, depends primarily on vehicle class, type, age, and speed. Automobile associations, petroleum institutes, and government energy agencies publish fuel prices (dollars per gallon) on a regular basis. In the United States, the average prices of gasoline and diesel in 2005 were \$2.2 and \$2.4, respectively (USDOE, 2005b). Fuel prices for VOC computation purposes should be derived by subtracting the federal and state gasoline taxes from retail prices. On a mileage basis, the unit costs of fuel (including oil) in 2003–2004 ranged from approximately 7 cents per vehicle-mile for small autos to over 21 cents per vehicle-mile for large trucks (Barnes and Langworthy, 2003; AAA, 2005). Generally, very low speeds, steep uphill grades, and curves lead to higher fuel consumption rates and hence higher overall fuel costs. In the Highway Economic Requirements System (HERS) model (FHWA, 2002), the change in vehicle fuel efficiencies across the years is accounted for in VOC estimation using an adjustment factor.

7.1.2 Shipping Inventory

The inventory cost of cargo (freight transportation) is a special category of user cost. The entity that ships the cargo (the *client*) is a user of a shipping service made available by a *carrier*. In the course of transporting perishable or valuable cargo, the client incurs holding costs that represent an opportunity cost: If at the beginning of the shipment, the client had a cash amount worth the cargo being shipped, such an amount would have earned some interest by the time the cargo reaches its destination. So by having the cargo transported, the client is foregoing some benefits. Higher inventory costs are generally directly related to cargo value, greater cargo perishability, higher prevailing opportunity cost of money, and slower speed of the shipping vehicle. To compute the inventory cost for a given vehicle class, an hourly discount rate is typically determined and multiplied by the average value of shipments undertaken by that vehicle class (FHWA, 2002). AASHTO (2003) recommends that the inventory costs of cargo per vehicle-mile should be applied to the unit user cost attributed to cargo-carrying transportation vehicles. The most significant VOC factors that affect the shipping inventory costs are speed and delay, but cargo value and interest rate also can be influential. Higher cargo value and interest rates and greater travel or transfer delay translate to higher unit costs of shipping inventory, and higher speeds lead to lower inventory costs. For example, at a 10% interest rate, two trucks each shipping $100,000 cargo, one traveling at 60 mph and the other at 50 mph, incur inventory costs of approximately 2.5 and 6 cents per mile, respectively (AASHTO, 2003).

7.1.3 Lubricating Oils for Mechanical Working of the Drivetrain

The lubricating oil cost includes the cost of engine oil, transmission fluids, brake fluids, and other similar consumables associated with the operation of vehicle engine and drive train. Oil cost is a product of unit price (dollars/quart) and consumption rates (quarts/mile). The consumption rates depend on the amount of use as well as characteristics of the guideway and vehicle, and operational conditions such as speed, delay, grade, and curves. Typically, the cost of this set of VOC components is reported together with fuel costs, but some sources report them separately. In 2005 dollars, oil costs ranged from $1.73 to $4.32 per quart (Appendix A7.2).

7.1.4 Preservation of the Vehicle–Guideway Contact Surface

At their points of contact, both the vehicle and guideway experience deterioration due to wear and tear. For highways and runways, the vehicle contact is a tire; for railways, the contact is typically a steel wheel. Updated tire costs (2005 dollars) from the HERS technical report (Appendix A7.2), are as follows: $54.71 per tire for small autos, $86.54 for medium-sized to large autos, $95.39 for four-tire single-unit trucks, $95.38 for six-tire single-unit trucks, $230.10 for single-unit trucks of three or more axles, and $569.74 for combination trucks. Of the various VOC factors, pavement condition, grade, curvature, and speed changes are those that most influence the rate of wear of contact surfaces (Thoresen and Roper, 1996).

7.1.5 Vehicle Repair and Maintenance

Repair and maintenance costs are incurred on vehicle parts that need replacement or replenishment after some amount of use. For gasoline-powered vehicles, these include the cost of batteries, alternators, fuel pumps, air pump, tire rims, electrical parts such as bulbs and fuses, and so on. These costs also include costs of replacing parts due to crashes, misuse, or other adversarial factors. In some methodologies, the cost of vehicle repair and maintenance is not reported separately but is added to other nonfuel costs. In Year 2005 dollars, the unit cost of vehicle repair and maintenance generally ranged from 4.7 cents per vehicle-mile for small to medium-sized vehicles to 9.3 cents per vehicle-mile for trucks (AAA, 2005). Vehicle repair and maintenance are influenced by pavement condition, curvature, and to a lesser extent, speed, grade, and speed change.

7.1.6 Depreciation

Vehicle depreciation is a function of vehicle usage (miles of travel) and vehicle age (years since manufacture). Table 7.1 presents the depreciation costs of selected vehicle classes and types. It can be seen that mileage-based depreciation rates are similar across vehicle classes: This seems reasonable because the lower initial cost of cars is balanced by their shorter service lives compared with trucks, so the net effect is that rates of mileage-based depreciation are similar across vehicle types (Barnes and Langworthy, 2003). Mileage-based depreciation costs can account for a significant fraction of overall vehicle operating costs. In some literature, the cost of vehicle depreciation is reported together with other nonfuel costs.

The values presented in Table 7.1 are average values. Depreciation rates actually vary by factors such as grade, curves, surface condition, and speed. An improvement in the transportation facility can produce a smoother pavement and improved driving conditions (through reduced stop-and-go situations). Also, all other factors remaining the same, increased speed can lead to reduced depreciation rates, as illustrated by Figure 7.1 for straight constant-speed sections (FHWA, 2002).

7.1.7 VOC Data Sources and Average National VOC Rates

Data on the trends in VOC component prices and consumption rates are available from published and online national resources. These are produced by a number of

Table 7.1 Average Vehicle Depreciation Costs (2005 Dollars)

	Total Depreciation (cents/h)	Average Travel (mi/y)	Mileage-Related Depreciation		Time-Related Depreciation (cents/h)
			(cents/mi)	(cents/h)	
Small autos	219	11,575	14	80	139
Medium-sized to large autos	257	11,575	12	73	185
Four-tire Single-unit trucks	278	12,371	6	36	242
Six-tire	393	10,952	10	55	338
3+ axles Combination trucks	1,122	15,025	22	209	913
3 or 4 axles	946	35,274	7	129	817
5+ axles	1,017	66,710	8	232	785

Source: Cost values are updated from their 1995 values in FHWA (2002).

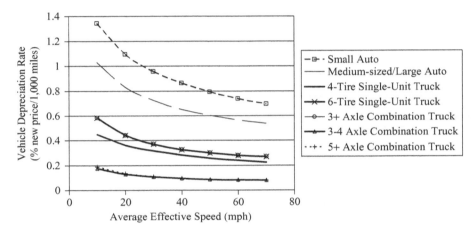

Figure 7.1 Depreciation rate by speed for straight sections (from FHWA, 2002.).

organizations, such as the International Energy Agency, Oak Ridge National Laboratory, national automobile associations, energy agencies, petroleum institutes, and private organizations, including Runzheimer International. Also, national agencies such as the Bureau of Labor Statistics in the United States provide monthly reports on changes in the prices paid by consumers for commodities including vehicle oil and tires, using the consumer and producer price indices. Table 7.2 presents the prices of selected VOC components by vehicle type.

7.2 FACTORS THAT AFFECT VEHICLE OPERATING COST

For all modes of transportation, vehicle operating costs are affected by factors such as vehicle–operator characteristics, economic factors, condition and other charact-

eristics of the fixed transportation facility, and policy–institutional factors. Although we focus on highway transportation in this section, the principles and concepts can be adapted to other transportation modes. Figure 7.2 shows the categories of highway VOC factors.

7.2.1 Vehicle Type

Vehicle operating costs are influenced by size, class, and other vehicle characteristics. Trucks and buses generally have higher operating costs than automobiles, as they consume more fuel and oil and have higher prices for their vehicle parts. Even for a given vehicle type, there could be changes in VOC over time due to improved vehicle technology and fuel efficiency. If the analyst seeks to carry out long-term VOC impact evaluation, future levels of fuel efficiency could be extrapolated from past trends and duly factored in the VOC computation process.

Table 7.2   Average Vehicle Operating Costs (Cents/Vehicle Mile)

	Fuel and Oil	Maintenance and Repair	Tires	Mileage-Dependent Depreciation	Total
Small autos	5.4	3.5	0.5	13.9	20.59
Medium-sized autos	6.44	4.12	1.58	12.5	20.59
Large autos	7.50	4.33	1.90	12.5	22.17
SUVs	8.34	4.33	1.58	12	22.70
Vans	7.50	4.12	1.69	12	21.75
Trucks	21.41	11.09	3.70	10.6	44.64

Source: Costs are updated to 2005 from the following: nontruck fuel, maintenance and repair, and tires, AAA (2005); truck fuel, maintenance and repair, and tires, Barnes and Langworthy (2003); and, depreciation estimations and projections are on the basis of data from FHWA (2002).

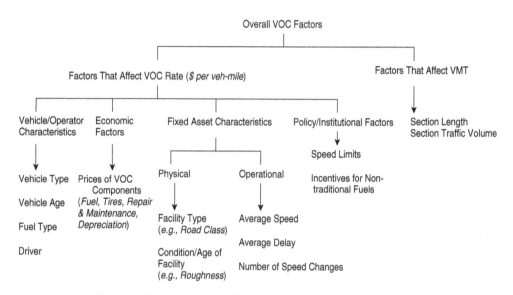

Figure 7.2   Factors that affect highway vehicle operating costs.

In some cases, analysts may seek the operating costs associated with bicycling and walking to facilitate a more comprehensive comparison of transportation alternatives that include these modes. A standard bicycle with basic accessories can cost $100 to $500 with annualized maintenance costs of $20 to $40 for tire replacement, tire pumping, and security; for walking, the main consumable is that of footwear, which typically lasts 500 to 5000 miles of walking distance (VTPI, 2004). The human energy use associated with walking and cycling may be considered a benefit rather than a cost, particularly if traveling using these transportation modes substitute for other exercise activities.

7.2.2 Fuel Type

The uncertainties in supply and increasing costs of fossil fuels coupled with their adverse environmental effects have led to growing use of alternative energy sources for transportation. In evaluating the impacts of transportation improvements, therefore, analysts need to account for the increasing percentage of alternative-fuel vehicles in the traffic stream. At the current time, electric and hybrid vehicles have relatively high purchase costs (150 to 200% of the price of a comparable gasoline car). Electric cars require new battery sets every 20,000 to 30,000 miles costing $2000 to $3000 (averaging 6 to 15 cents per vehicle-mile), and consume 0.25 to 0.5 kWh per mile, so energy costs average 2 to 5 cents per kWh based on typical residential energy rates (USDOE, 2005a). The maintenance costs, including battery replacements, are significantly higher for electric cars (over four fold) compared to hybrid or conventional cars (VTPI, 2005). Even with traditional fuels, there are differences in cost across fuel types: in 2005, the average price of diesel was approximately 10% higher than that of regular leaded gasoline. Also, there are price differences across the three standard grades of gasoline.

7.2.3 Longitudinal Grade

Uphill movements impose additional loads on vehicle engines and therefore require greater consumption of energy compared to downhill or level movements. For downhill trips, fuel consumption is lower than for uphill or level trips, but increased brake applications may lead to increased wear and tear of brake linings and therefore to increased cost of the brake maintenance component of VOC. Figure 7.3 illustrates the general relationships between grade and VOC at various speeds. Generally, overall VOC is lowest for sections with gentle downward slopes (0 to −4%). Table 7.4 shows how the vehicle operating cost for medium-sized automobiles can be determined for a given speed and longitudinal grade. Detailed equations that indicate the effect of grade on the consumption of fuel and for other VOC components, are provided in Appendix A7.1 and the HERS manual (FHWA, 2002). The rate of consumption of each VOC component is subsequently multiplied by the unit price of the component and appropriate adjustment

factors for fuel efficiency and pavement condition to determine the overall cost of the component.

Example 7.1 A 2.15-mile section of State Road 25 on rolling terrain received major improvements in vertical alignment. The average grade of the section was reduced from 3.2% to 2.5%. Traffic volume and composition, and speed were the same after the improvement. Assume that the traffic stream has a 50:50 directional split and is composed primarily of medium-sized automobiles, and the traffic volume is 43,340 vpd. In both cases, the average speed is 50 mph. What is the first year user benefit in terms of VOC?

SOLUTION
Before improvement:

Uphill traffic: VOC at +3.2% grade = $275/1000 VMT

Downhill traffic: VOC at −3.2% grade

    = $190/1000 VMT

Average: $232.5/1000 VMT

After improvement:

Uphill traffic: VOC at +2.5% grade = $260/1000 VMT

Downhill traffic: VOC at −2.5% grade

    = $200/1000 VMT

Average: $230/1000VMT

Change in unit costs:

$(\text{VOC}_{\text{before}} - \text{VOC}_{\text{after}})$ or $(U_1 - U_2)$

    = $2.5/1000 VMT = $0.0025/VMT

First-year user benefits

    $= (0.5)(U_1 - U_2)(\text{VMT}_1 + \text{VMT}_2)$

    $= (0.5)(0.0025)(2)(43,340 \times 2.15 \times 365) = \$85,028$

7.2.4 Vehicle Speed

Vehicle operating speed is the dominant factor in determining VOC (Bennett, 1991; Thoresen and Roper, 1996; Bennett and Greenwood, 2001; FHWA, 2002). Transportation improvements influence travel speeds and therefore can profoundly affect VOC. For some vehicles, fuel consumption decreases with increasing speed to a certain point, after which there is little significant change (or sometimes, an increase) in fuel consumption with increasing speed. Factors that affect operating speeds, and subsequently influence fuel VOC, are speed limits (set by

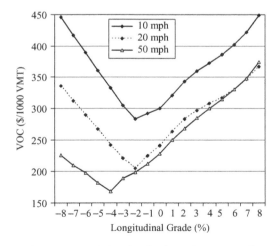

Figure 7.3 Impact of longitudinal grade on medium-sized automobile VOC at various speeds. (Based on data from Zaniewski, 1982.)

policy) and traffic conditions (which vary by the time of day—peak vs. nonpeak). In this section we discuss the impact of speed on shipping inventory costs and present some VOC models based on speed and other factors.

(a) *Inventory Shipping*   Inventory cost is affected by vehicle speed and is calculated as follows (AASHTO, 2003):

$$U_{IC} = (100) \frac{r}{(365)(24)} \frac{1}{S} P \qquad (7.1)$$

where U_{IC} is the user inventory cost in cents per vehicle-mile, r the annual interest rate, P the cargo value in dollars, and S the vehicle speed in miles per hour.

Example 7.2   Due to a new speed limit policy, the average truck operating speed on a certain interstate freeway increased from 56.5 mph to 61.2 mph. Find the decrease in shipping inventory costs per year for trucks that comprise 22% of the overall traffic stream of 82,500 vehicles per day (vpd). Each truck hauls an average of $1.5 million worth of goods daily. Assume an 8% interest rate.

SOLUTION   Using equation (7.1), the daily changes in inventory costs per truck due to the change in travel speed, ΔU_{IC}, can be estimated as follows:

$$\Delta U_{IC} = (100) \frac{r}{(365)(24)} \left(\frac{1}{S_0} - \frac{1}{S_1} \right) P$$

$$= (100) \left(\frac{0.08}{8760} \right) \left(\frac{1}{56.5} - \frac{1}{61.2} \right) (\$1,500,000)$$

$$= 1.9178 \text{ cents/vehicle-mile}$$

number of trucks per year

$$= (0.22)(82,500)(365) = 6,624,750$$

total reduction in inventory cost for all trucks per year

$$= (1.9178/100)(6,624,750) = \$127,050 \text{ per mile}$$

(b) *VOC Models and Look-up Table Based on Speed and Vehicle Class*   Hepburn (1994) developed a VOC model for urban roadways that considers the sum of four VOC components (tires, vehicle depreciation, maintenance, and fuel) as a function of two VOC factors: speed and vehicle class. The model is particularly useful for evaluating VOC impacts of transportation interventions that mostly yield a change in average operating speeds or policies that cause a shift in vehicle class distribution. The Hepburn function is as follows:

For "low" average travel

speeds (<50 mph) : $\text{VOC} = C + \dfrac{D}{S}$

Table 7.3   Parameters for Hepburn's VOC–Speed Model (2005 Cents)

Vehicle Type	C	D	a_0	a_1	a_2
Small automobile	24.8	45.5	27.2	0.035	0.00021
Medium-sized automobile	28.5	95.3	33.5	0.058	0.00029
Large automobile	29.8	163.4	38.1	0.093	0.00033

For "high" average travel

speeds (>50 mph) : $\text{VOC} = a_0 - a_1 S + a_2 S^2$

where VOC is in cents/mile, S is speed (mph) and C, D, a_0, a_1, and a_2 are coefficients that are functions of vehicle class. The coefficient values are provided in Table 7.3.

The Hepburn model assumes that depreciation depends entirely on vehicle use and that the depreciation rate is constant throughout vehicle life. Furthermore, the model is for tangent, level, and urban road sections with pavement roughness assumed to remain constant over time, and all VOC component costs assumed to vary with distance, with the exception of fuel cost, which varies with speed. It does not explicitly consider the consumption rates and prices of individual VOC components for each vehicle class but is nevertheless useful for quick estimation of VOC.

Example 7.3   A straight and level urban arterial has an average operating speed of 35 mph. What is the unit VOC of medium-sized automobiles that use this highway?

SOLUTION   Knowing the values of C and D from Table 7.3,

$$\text{VOC} = C + \frac{D}{S} = 28.5 + \frac{95.3}{35}$$

$$= 31.22 \text{ cents/vehicle-mile}$$

(c) *VOC Models Based on Speed, Grade, and Vehicle Class*   Zaniewski (1982) provided a VOC model as a function of speed, grade, and vehicle class. Table 7.4 presents the VOCs for medium-sized autos, with updated cost values. If the project section consists of several segments with different grades or VMTs, the unit VOC (dollars/vehicle-mile) is estimated separately for each segment. It should be noted, however, that the vehicles at the time of the Zaniewski study (ca. 1980) had 17% lower fuel efficiency than vehicles in 1997 (FHWA, 2002), and even lower compared to vehicles in 2005. As

Table 7.4   VOC by Vehicle Speed and Roadway Grade[a]

Grade(%)	Speed (mph)													
	5	10	15	20	25	30	35	40	45	50	55	60	65	70
8	591	507	451	414	403	395	398	406	414	422	444	467	477	492
7	552	476	424	391	379	369	369	376	385	393	417	444	454	467
6	526	454	406	372	361	347	346	352	362	372	398	422	430	444
5	499	435	389	358	346	333	332	335	346	354	376	395	410	429
4	481	421	379	347	333	319	317	322	329	338	352	367	387	412
3	459	406	364	335	322	309	307	310	317	322	332	340	367	395
2	435	387	347	319	307	297	292	297	301	302	314	319	346	377
1	403	362	325	297	288	279	272	272	279	282	292	301	322	346
0	376	338	302	272	264	255	247	247	254	257	273	287	301	319
−1	367	329	288	254	243	235	232	235	237	239	254	265	282	299
−2	357	319	273	231	212	217	219	223	225	225	237	246	264	284
−3	385	344	292	249	228	209	197	191	212	213	225	235	250	270
−4	422	376	322	273	250	227	212	202	195	190	217	225	239	255
−5	461	407	350	301	276	249	231	217	212	205	204	197	228	243
−6	499	439	379	327	301	273	250	235	228	223	231	213	210	228
−7	537	470	406	352	325	299	273	255	247	237	232	227	223	219
−8	914	503	437	379	350	324	297	279	265	255	249	239	235	228

[a]Cost/1000 VMT for medium-sized autos in 2005 dollars.

such, Table 7.4 should be used after stating the necessary assumptions regarding fuel efficiency, or after making due adjustments for fuel efficiency changes over the years.

Example 7.4   A highway section consists of two segments A and B that have the characteristics listed in Table E7.4 Determine the total vehicle operating costs for each segment. Assume that all vehicles are medium-sized automobiles, and assume further that the values in Table 7.4 reflect current fuel consumption rates.

Table E7.4   Highway Segment Data

	Segment A	Segment B
Traffic volume (ADT)	5320	8580
Average grade (%)	+4.0	+1.5
Speed (mph)	30	50
Length (miles)	5.7	2.6
Directional split	68% on upward slope, 32% on downward slope	45% on upward slope, 55% on downward slope

SOLUTION   Given the average speeds and grades, the unit vehicle operating cost is determined from Table 7.4 as follows:

Segment A:

Unit VOC = $(319)(0.68) + (227)(0.32)$

   = $289.56 per 1000 VMT

VMT = $(5.7)(5320) = 30,324$ vehicle-miles daily

Overall VOC = $(\$289.56)(30,324) = \$8,781$ per day

Segment B:

Unit VOC = $(292)(0.45) + (232)(0.55)$

   = $259 per 1000 VMT

VMT = $(2.6)(8580) = 22,308$ vehicle-miles daily

Overall VOC = $(259)(22,308)$

   = $5,778 per day

(*d*) *VOC Models Based on Speed, Gradient, Curvature, and Pavement Condition*   Some VOC models, such as the World Bank's HDM (Bennett and Greenwood, 2001) and the HERS model (FHWA, 2002), estimate the unit cost of each VOC component as a function of speed, grade, and pavement condition. This is done for basic sections (straight sections with constant speed), and then excess vehicle operating costs due to speed changes and curvature are calculated. The excess VOC is added to the basic costs to yield the overall VOC for the section.

7.2.5 Delay

Nodes and links in the networks of various transportation modes may often experience delay, which translates into higher vehicle operating costs. In evaluating transportation improvements at such facilities, VOC costs, particularly for fuel and inventory, can be expressed as a function of time delay. On highway links, for instance, delay can involve decelerating to a stop, idling, and accelerating from a stopped position. Such stop-and-go traffic leads to additional strain on a vehicle, which is translated into higher use of fuel and oil. All three phases involve fuel consumption rates that generally exceed that of constant-speed travel. The primary share of overall delay costs is attributed to acceleration of vehicles after being slowed or stopped rather than fuel consumed in decelerating or idling during delay periods (AASHTO, 2003). The impact of travel delay on VOC (fuel and inventory shipping cost components) can be estimated using a methodology provided by AASHTO (2003). In the methodology, the analyst estimates the delay with and without improvement using field measurements (applicable only to the existing situation), simulation, or analytical travel delay models. Using the estimated change in delay, fuel consumption rates per minute of delay (Table 7.5) and fuel price (Appendix A7.2), the total cost

of delay can be calculated. This is repeated for each vehicle class. Example calculations are provided below.

(a) *Change in Fuel Costs due to Delay Change* For a given vehicle class, the change in fuel costs due to a change in travel delay is found as follows (AASHTO, 2003):

$$\text{change in fuel VOC} = g(D_0 - D_1)p$$

where: g is the fuel consumption in gallons per minute of delay (from Table 7.5), $D_0 - D_1$ = change in delay (minutes) due to the transportation improvement, and p is the price of fuel. The parameters g and p are specific to vehicle class.

Example 7.5 Modernization and optimization of the traffic signal system at a busy urban arterial yielded, on average, a 9-minute reduction in delay per trip for users of the arterial. The traffic volume is 4300 vph and is composed of 25% small autos, 30% large autos, 25% SUVs, 10% two-axle single-unit trucks, 5% three-axle single-unit trucks, and 5% multiple-unit trucks. After improvement, average free-flow speed increases from 45 mph to 50 mph, and traffic volume and composition remain unchanged. Determine the reduction in fuel costs during peak hours due to the decrease in delay. Assume that fuel cost is $2.20 per gallon. Use the fuel consumption rates provided in Table 7.5, and assume simple averages across vehicle classes.

Table 7.5   Fuel Consumption (Gallons) per Minute of Delay by Vehicle Type

Free-Flow Speed (mph)	Small Automobile	Large Automobile	SUV	Two-Axle Single-Unit Truck	Three-Axle Single-Unit Truck	Multiple-Unit Truck
20	0.011	0.022	0.023	0.074	0.102	0.198
25	0.013	0.026	0.027	0.097	0.133	0.242
30	0.015	0.030	0.032	0.122	0.167	0.284
35	0.018	0.034	0.037	0.149	0.203	0.327
40	0.021	0.038	0.043	0.177	0.241	0.369
45	0.025	0.043	0.049	0.206	0.280	0.411
50	0.028	0.048	0.057	0.235	0.321	0.453
55	0.032	0.054	0.065	0.266	0.362	0.495
60	0.037	0.060	0.073	0.297	0.404	0.537
65	0.042	0.066	0.083	0.328	0.447	0.578
70	0.047	0.073	0.094	0.360	0.490	0.620
75	0.053	0.080	0.105	0.392	0.534	0.661

Source: Adapted from AASHTO (2003).

Table E7.5 Estimation of Change in Fuel Consumption Costs due to Delay

	Small Auto	Large Auto	SUV	Two-Axle Single-Unit Truck	Three-Axle Single-Unit Truck	Multiple-Unit Truck	Total
Traffic volume (vph)	1075	1290	1075	430	215	215	4300
Fuel consumption rate (gals/min)	0.025	0.043	0.049	0.206	0.280	0.411	
Fuel price ($/gal)	$2.2 (average for all vehicle classes)						
Change in delay due to the improvement, $D_0 - D_1$	9 min (average for all vehicle classes) per peak hour						
Change in fuel consumption costs	$532	$1098	$1043	$1754	$1192	$1750	$7369

SOLUTION   The traffic volume for each vehicle class is determined by multiplying the percentage composition by the total traffic volume. Using Table 7.5, the fuel consumption rates are determined, and the change in fuel consumption cost is presented in Table E7.5.

Therefore, the total reduction in fuel costs during the peak hours due to the decrease in delay is $7,369/hr

(b) *Change in Shipping Inventory Costs due to Delay* Example 7.2 provided a method to evaluate the impact of a change in shipping speed (due to a transportation improvement) on inventory costs. AASHTO (2003) provides a methodology for estimating the impact of time delay on shipping operating costs, as follows: The change in inventory cost per shipping vehicle due to a change in delay is given by $\Delta I(D) = I(D)\Delta D$, where ΔD is the change in delay (in minutes) and

$$I(D) = \text{inventory costs (cents per vehicle-minute)}$$
$$= (100)\frac{r}{(365)(24)(60)}P$$

where r is the interest rate (per annum) and P is the dollar value of the cargo being transported by the shipping vehicle.

In some cases, the analyst is provided with an estimate of the expected change in delay, but in other cases, change in delay will need to be estimated (by calculating the delay before and after the improvement). Delay can be estimated on the basis of prevailing traffic conditions and road inventory. Methodologies for estimating delay are found in available literature, such as the *Highway Economics Requirements System* (HERS) *Technical Manual* (FHWA, 2002, pp. 4-7 to p. 4–10).

Example 7.6 A freeway was constructed in 2005 to bypass a city center. This improvement led to a 10-minute reduction in travel delay per trip for shippers who transport goods across the city. If the average value of cargo is $265,000 per truck and the interest rate is 6%, determine (a) the shipping inventory costs per vehicle before the construction, (b) the reduction in shipping inventory costs due to the construction in 2005, and (c) the change in user benefits accrued to shippers in 2005 compared to pre-construction conditions. The pre-construction period daily truck traffic (ADTT) was 33,000, and the trip time was 1.5 hours. Assume a 5% ADTT increase due to induced demand.

SOLUTION
(a) *Shipping Inventory Cost per Truck*
The unit inventory cost of the shipment before improvement can be calculated as follows:

$$I(D) = P \times [r/(365 \times 24 \times 60)] = (\$265,000) \times$$
$$[0.06/(365 \times 24 \times 60)] = \$0.03025/\text{truck-minute}$$

The unit inventory cost after the improvement is the same as that before the improvement because there is no change in the total cargo value and the annual interest rate. Since the travel time reduces by 10 minutes after the improvement, the total inventory cost saved due to the improvement is

Change in unit inventory cost = $0.03025/truck-minute
$$\times \text{ 10 minutes} = \$0.3025/\text{truck}$$

(b) *Reduction in Shipping Inventory Cost*
The unit shipping inventory cost in dollars/truck-mile, U, can be calculated as follows:
Before Improvement:

$$U_{\text{before}} = \$0.03025/\text{truck-minute} \times (60/S_{\text{before}})$$
$$= \$1.815T/L \text{ per truck-mile}$$

where S_{before} = average speed before improvement (mph) = L/T, L = average truck trip length (miles) and T = average truck trip time (hours).

Total yearly inventory cost

$$= (\$1.815T/L)(ADTT)(L)(365)$$

After Improvement:

$$U_{after} = \$0.03025/\text{truck-minute} \times (60/S_{after})$$

$$= \$1.815[T - (10/60)]/L \text{ per truck-mile}$$

Total yearly inventory cost

$$= \{\$1.815[T - (10/60)]/L\}/\{1.05ADTT(L)(365)\}$$

Therefore, the reduction in total shipping inventory cost = Total yearly inventory cost before the improvement−Total yearly inventory cost after the improvement = 1.815 $ADTT(365)\,[T - 1.05\{T - (10/60)\}] = 1.815ADTT(365)$ $[0.175 - 0.05T] = \$2,186,168$ in the first year.

(*iii*) *Change in User Benefits (or Change in Consumer Surplus)*

The change in user benefits can be calculated based on the change in consumer surplus, which is given by the following formula:

User benefits

$$= \frac{1}{2}(U_{before} - U_{after}) \times (VMT_{before} + VMT_{after})$$

$$= \frac{1}{2}([(1.815(10/60)/L)][2.05ADTT(L)(365)]$$

$$= \$3,734,703.$$

7.2.6 Speed Changes

Vehicles travel at different speeds due to geometric and/or traffic conditions. It has been shown that the more frequent the speed change of a vehicle, the higher the associated operating cost, particularly its fuel component. When vehicles slow down or pick up speed, they experience additional strain that is translated into a higher use of fuel and oil. As such, highway projects that smoothen traffic flow by reducing the frequency and intensity of speed changes ultimately reduce the costs of vehicle operation.

An extreme case of speed change is stop−start conditions, which are usually typical of city driving. Barnes and Langworthy (2003) showed that for maintenance, repair, and depreciation, worsening stop−start conditions will increase costs of fuel consumption and to a smaller extent, the costs of maintenance, repair, and depreciation.

Table 7.6   Percent Decrease in VOC from City to Highway Driving Conditions[a]

VOC Component	Automobile	Pick-up/ Van/SUV	Commercial Truck
Fuel	29%	23%	24%
Maintenance/repair	16%	14%	13%
Depreciation	16%	14%	13%
Total	20%	17%	18%

Source: Data from Barnes and Langworthy (2003).
[a]Pavement in good condition (PSI = 3.5 or above) assumed for both cases.

Table 7.6 can be used to estimate the impact of a change in driving conditions on VOC, for each vehicle class.

VOC Model Based on Speed-Change Frequency: For each vehicle class, VOC equations in the *HERS Technical Manual* (FHWA, 2002) can be used to estimate the total unit costs of speed changes per thousand vehicle-miles, as the sum of speed change costs due to five VOC components (fuel, oil, tires, maintenance and repair, and vehicle depreciation). The template for VOC computations is shown in Table 7.7. The set of equations for computing the rate of consumption of each VOC component, including at speed-change frequency, is presented in Appendix A7.1.

7.2.7 Horizontal Curvature

A vehicle negotiating a horizontal curve requires extra energy to counter centrifugal forces in order to stay in a radial rather than a tangential path. Furthermore, the side friction increases tire wear and tear, and the

Table 7.7   Percent Decrease in VOC from Poor to Good Pavement Conditions[a]

VOC Component	Automobile	Pickup/ Van/SUV	Commercial Truck
Fuel	0%	0%	0%
Maintenance/repair	20%	21%	20%
Tires	18%	17%	20%
Depreciation	21%	20%	20%
Total	15%	13%	11%

Source: Data from Barnes and Langworthy (2003).
[a]Good pavement, PSI = 3.5 or above (i.e., $IRI = 85$ or below); poor pavement, PSI = 2.5 or below (i.e., IRI = 170 or above); highway driving (not city driving) assumed for both cases.

frequency and cost of maintenance and replacement. The VOC due to curves involves fuel, tire, and maintenance and repair, and is typically expressed as a function of the rate of consumption and unit prices of these VOC components, vehicle type, and average speed. In the HERS methodology, VOC for curve negotiation speed is estimated separately for sections with low speeds (<55 mph) and those with high speeds (>55 mph).

(*a*) *Low-Speed Sections*   VOC due to curve negotiation at these sections is estimated using VOC vs. curve–degree tables from Zaniewski (1982). These tables show the costs due to curves for each vehicle type as a function of curvature and speed.

(*b*) *High-Speed Sections*   In the HERS manual, the VOC due to curves for each vehicle type is calculated using the rate of consumption of VOC component for curve sections, by vehicle class, the unit prices of VOC components, and the adjustment factor for VOC component. The set of equations for VOC consumption rates at curve sections is presented in Appendix A7.1. An example computation is provided in Section 7.3.2.

7.2.8 Road Surface Condition

To some extent, pavement roughness, often measured in terms of the present serviceability index (PSI) or international roughness index (IRI), can affect the maintenance, tire, repair, and depreciation cost components of VOC. This is because the motion of vehicle tires on a rough pavement surface is associated with greater resistance to movement, which can lead to higher levels of fuel consumption compared to traveling at a similar speed on a smooth surface; and a bumpy ride, which leads to increased vibration and wear and tear of vehicle parts. Also, an indirect effect of poor pavement conditions is that road users may be forced to drive at lower speeds, leading to higher fuel consumption. Transportation projects such as resurfacing that improve pavement surfaces can therefore lead to reductions in VOCs.

Zaniewski (1982) suggested that there can be significant impacts of pavement roughness on nonfuel vehicle operating cost components, particularly for rough pavements. Most other research on the relationship between pavement condition and VOC has been conducted outside the United States by the World Bank and other international agencies. Examples include a New Zealand study (Opus Central Laboratories, 1999) which suggests that at superior levels of pavement condition (low roughness), increments in condition have relatively little incremental effect on vehicle operating cost (Figure 7.4), and that additional costs of vehicle operation start to accrue only when the IRI exceeds approximately 100 in/mi (3.33 m/km). For paved roads in poor condition and for gravel roads, changes in road surface condition can lead to significant reductions in VOC. Barnes and Langworthy (2003) reported on a previous study that suggested that a unit increase in IRI (in m/km) can generally lead to an increase of $200 (1.67 cents/vehicle-mile, assuming 12,000 annual mileage) in vehicle maintenance and repair costs alone. Also, Barnes and Langworthy (2003) developed adjustment factors for all VOC components combined, as a function of pavement condition (Figure 7.5). They assumed a baseline PSI of 3.5 or better (an IRI of

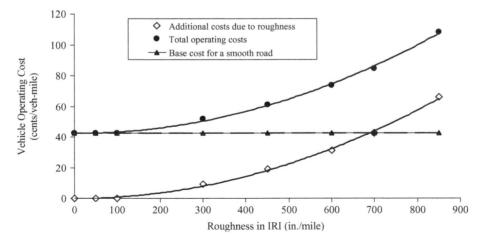

Figure 7.4   Relationship between VOC and pavement roughness. (Adapted from Opus Central Laboratories, 1999.)

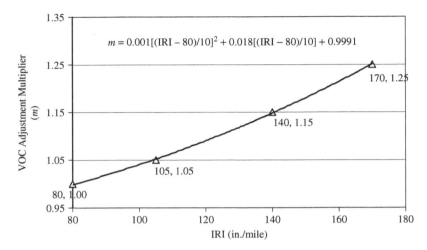

Figure 7.5 VOC adjustments for pavement roughness levels.

about 85 in/mi or 1.35 m/km), at which further increases in pavement condition would have no impact on operating costs, and then adjusted for three levels of rougher pavement as shown in the figure. The figure can be used to estimate the VOC corresponding to a given pavement condition. For the depreciation component, there seem to be relatively few studies that have explicitly shown a relationship with pavement roughness. However, it seems obvious that in the long term, a vehicle which is operated on a rough pavement surface is likely to lose its value faster than one that is operated on a smooth-surfaced pavement.

Example 7.7 A warranty HMA resurfacing project on Interstate 599 yielded a performance jump of 40 IRI (in/mi). If the base vehicle operating cost is $143 per 1000 vehicle-miles, (a) determine the change in unit VOC due to resurfacing. Use the Barnes and Langworthy relationship. The IRI before improvement was 110 in/mi. (b) If the traffic volume is 67,500 vpd and the section is 6.5 miles in length, determine the overall change in VOC.

SOLUTION (a) *Before improvement*: IRI = 110 in/mi, and the VOC adjustment multiplier is given by

$$m = (0.001)\left(\frac{110-80}{10}\right)^2 + (0.018)\left(\frac{110-80}{10}\right)$$
$$+ 0.9991 = 1.06$$

$$\text{VOC} = (1.06)(143) = \$151.58/1000 \text{ VMT}$$

After improvement: IRI = 110 − 40 = 70 in/mi, $m = 1.00$ since 70 is less than 80, and therefore VOC =

$143/1000 VMT. Change in unit VOC = 151.58 − 143 = \$8.58/1000 VMT.

(b) Overall change in VOC = ($8.58)(67,500)(365) (6.5)/1000 = \$1.374 million.

Table 7.7 presents the percentage changes in VOC for pavements that transition from poor condition (PSI of 2 or higher) to good condition (PSI of 3.5 or lower) under the given highway conditions. VOCs corresponding to intermediate values of pavement condition can be determined by interpolation. This table can therefore be used to estimate the impact of changes in pavement condition (within the PSI range given) on VOC for each VOC component or for all components combined, in response to a pavement improvement project.

Example 7.8 After replacing a highway pavement section that had 2.5 PSI and a total VOC value of $152/1000 VMT for automobiles, the pavement now has a PSI value of 3.7. Estimate the new automobile VOC.

SOLUTION From Table 7.7, the average adjustment in VOC upon pavement improvement from poor to good condition is a 15% reduction, that is, (0.85)(152) = \$129.2/1000 VMT.

FHWA's HERS methodology duly incorporates the effect of pavement conditions on the individual VOC components of oil consumption, tire wear, and depreciation, and this can be done for each of six vehicle classes. Also, HERS utilizes a pavement condition adjustment factor to account for differences in pavement condition relative to a reference pavement condition. These factors are provided in Appendix A7.3.

7.2.9 Other VOC Factors

Other factors that can influence the cost of vehicle operation include driver behavior, condition of vehicle, vehicle weight, prices of vehicle maintenance (reflected in costs of labor, vehicle consumables, and spare parts), and weather severity. Operating costs for transit vehicles (such as buses and trolleys) are also affected by other factors, such as transit schedules (which typically depend on passenger demand) and vandalism.

7.3 PROCEDURE FOR ASSESSING VOC IMPACTS

The framework for assessing VOC impacts of transportation interventions (Figure 7.6) revolves around three tasks:

1. Estimating the unit VOC rates (i.e., dollars/vehicle-mile) with and without intervention
2. Estimating the amounts of travel (VMT) with and without the intervention
3. Calculating the user VOC benefits of intervention

7.3.1 Steps for Assessing the Impacts

Step 1: Define the Analysis Area This involves identification of the project and its limits and provides vital benchmarks for collecting project data such as grades, section length, average operating speeds, and other data needed for VOC estimation.

Step 2: Describe the Transportation Intervention Different interventions have different impacts on VOC factors and consequently, on VOC components. For example, for evaluating a highway project or policy change that influences only vehicle speed, there may be no need to collect data on grades. Therefore, this step helps the analyst to select the appropriate VOC factors for the analysis and could therefore guide in selecting the appropriate methodology or software to be used. Fuel cost typically dominates VOC, and fuel price and vehicle fuel efficiency values are determinants of vehicle fuel costs.

Step 3: Consider the Base-Case Scenario The base case typically refers to the current condition without intervention or improvement. It may also refer to a future condition without an intervention.

Steps 4 to 6: Establish the Values of Relevant VOC Factors. Use Models, Look-up Tables, or Graphs to Determine VOC per Vehicle-Mile Data on average speeds, grades, pavement condition, vehicle-type distribution, and/or other relevant VOC factors are collected for the "with" and "without" intervention scenarios. By applying data from the selected models or look-up

tables, the unit vehicle operating cost (dollars/vehicle-mile) is estimated. This is done for both the base case and the intervention scenario. Depending on the appropriate VOC factors, the analyst may choose from a variety of models, such as those described below.

(a) *Hepburn (1994) VOC Model* As described in Section 7.2.4(b), this model estimates VOC as a function of speed and vehicle class only. The VOC components considered are: tires, vehicle depreciation, maintenance, and fuel. This model is useful for highway transportation interventions that mostly yield a change in average operating speeds only.

(b) *1982 FHWA VOC Model* This model, developed by Zaniewski (1982), is based on two VOC factors: speed and grade (Table 7.4). For highway transportation projects that involve significant changes in grade through extensive vertical realignment, this method can be used instead of the Hepburn model. Details are provided in Section 7.2.4(c).

(c) *FHWA HERS (1999) Model* This is probably the most comprehensive of all VOC models currently in use in the United States. It considers a wide array of VOC factors: speed, speed changes, curvature, pavement condition, and vehicle class and five VOC components: fuel, oil, tires, vehicle depreciation, and maintenance and repair. To facilitate the analysis, a software package is available (FHWA, 2002). Use of the HERS methodology for steps 4, 5, and 6 of the framework, is described in Section 7.3.2 and Example 7.11.

(d) *AASHTO (2003) Model* This model presents for each automobile class, a single VOC (aggregated for all VOC components), in reference to 1999 unit vehicle operating costs reported by AAA. Updated values of the AAA's unit VOC values can be found on the AAA Web site and in its publications. These values are generally not decomposed by VOC factors or their levels, and it may be assumed that they represent the unit costs under average conditions of the VOC factors In AASHTO (2003), the VOC components for which VOC is reported individually are fuel and inventory. Fuel costs are presented as a function of speed and vehicle class and inventory costs are presented as a function of the value of the inventory, interest rate, and shipping delay.

(e) *The World Bank's HDM Model* This model helps to estimate VOC for motorized and nonmotorized vehicles that operate on paved or unpaved roads (Bennett and Greenwood, 2001). It considers a wide range of VOC factors: speed, speed changes, curvature, pavement condition, and vehicle class; and VOC components; fuel, oil, tires, vehicle depreciation, and maintenance and repair.

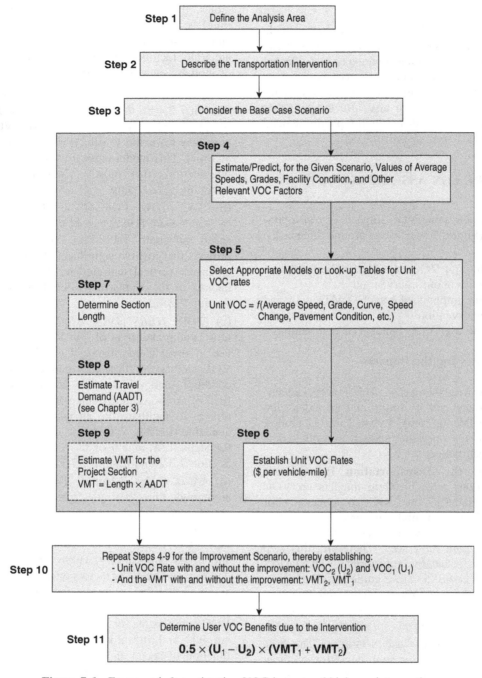

Figure 7.6   Framework for estimating VOC impacts of highway interventions.

(f) *Australia's NIMPAC Model*   This model helps to estimate VOC for motorized and nonmotorized vehicles (Thoresen and Roper, 1996). Detailed estimates are provided for different VOC components, and their effects on the various VOC factors are described.

Steps 7 to 9: Determine the Section Length and Travel Demand (AADT), and Estimate the VMT   These parameters are computed for the base case as well as the intervention scenario. The section length generally remains the same after intervention unless the project entails significant vertical or horizontal alignment or length extension. Traffic volumes rarely remain the same after improvement—they increase due to induced travel and attracted traffic from other routes. Chapter 3 discussed how travel demand can be estimated for a link in a transportation network.

Step 10: Repeat Steps 4 to 9 for the Intervention Scenario By the end of this stage of the framework, the analyst should have established the values of the following parameters: VOC_1 or U_1, VOC_2 or U_2, VMT_1, and VMT_2. This is done for each vehicle class and facility segment under consideration.

Step 11: Determine the User Overall VOC Savings Due to Intervention The overall VOC savings are determined for each vehicle class or road segment under consideration, and the sum of all savings is computed. In the latter case, computation of the user benefits of VOC will depend on whether demand is elastic or inelastic. When demand is inelastic (therefore precluding any induced, generated, or diverted trips), the user VOC benefit occurring from an improved transportation system is taken as the product of the reduction in the unit VOC and the number of vehicle-miles. When demand is elastic and there are induced trips, the increase in supply results in a lower cost of transportation and subsequently, increased demand. Then the user VOC benefit is given by $(0.5)(U_1 - U_2)$ $(VMT_1 + VMT_2)$.

Example 7.9 An ambitious portfolio of ITS programs at a certain 14.5-mile urban freeway is expected to increase average operating speed from 35 mph to 50 mph. Determine the change expected in the total VOC of medium-sized automobiles due to the improvement. Assume a traffic volume of 76,250 vpd, that medium-sized vehicles constitute 35% of the traffic stream, and that the traffic volume increases by 5% after improvement.

SOLUTION
Before improvement:

$$U_1 \text{ or } VOC_1 = C + \frac{D}{S} = 28.5 + \frac{95.3}{35}$$
$$= 31.22 \text{ cents/vehicle-mile}$$
$$VMT_1 = (14.5)(0.35)(76,250) = 386,969 \text{ per day}$$

After improvement:

$$U_2 \text{ or } VOC_2 = 28.5 + \frac{95.3}{50}$$
$$= 30.41 \text{ cents/vehicle-mile}$$
$$VMT_2 = (14.5)(0.35)(76,250)(1.05)$$
$$= 406,317 \text{ per day}$$

User VOC benefit due to the improvement
$$= (0.5)(VOC_1 - VOC_2)(VMT_1 + VMT_2)$$
$$= (0.5)(0.3122 - 0.3041)(386,969 + 406,317)$$
$$= \$3212 \text{ per day}$$

Example 7.10 As part of its upgrade to a state highway, a certain section of county road is to receive vertical realignment through massive cut-and-fill earthwork operations. The project comprises two segments with the characteristics given in Table E7.10. Determine the user VOC benefit per year due to the improvement.

SOLUTION *Segment 1:*

Before improvement: From Table 7.4, unit VOC,

$$U_1 = (332 \times 0.52) + (231 \times 0.48) = \$283.52/1000 \text{ VMT}$$
$$VMT_1 = (0.87)(65,200) = 56,720 \text{ vehicle-miles}$$

Table E7.10 Characteristics of the Base Case and Improvement Scenarios

	Before Improvement				After Improvement			
	Length (miles)	Traffic Volume (ADT)	Grade(%)	Speed	Length (miles)	Traffic Volume (ADT)	Grade(%)	Speed
Segment 1: 52% uphill, 48% downhill	0.87	65,200	5.0	35	0.86	68,000	2.0	50
Segment 2: 60% uphill, 40% downhill	1.2	53,200	2.0	45	1.2	54,300	1.5	50

After improvement: Similarly, unit VOC

$$U_2 = (302 \times 0.52) + (225 \times 0.48) = \$265.04/1000 \text{ VMT}$$

$$\text{VMT}_2 = (0.86)(68,000) = 58,480 \text{ vehicle-miles}$$

$$\begin{aligned}
\text{VOC benefit} &= (0.5)(U_1 - U_2)(\text{VMT}_1 + \text{VMT}_2) \\
&= (0.5)(283.52 - 265.04) \\
&\quad (56,720 + 58,480)(365/1000) \\
&= \$388,537
\end{aligned}$$

Segment 2:

Before improvement: Unit VOC,

$$U_1 = (301 \times 0.6) + (225 \times 0.4) = \$270.60/1000 \text{ VMT}$$

$$\text{VMT}_1 = (1.2)(53,200) = 63,840 \text{ vehicle-miles}$$

After improvement: Unit VOC,

$$U_2 = (292 \times 0.6) + (232 \times 0.4)$$
$$= \$268/1000 \text{ VMT}$$

$$\text{VMT}_2 = (1.2)(54,300) = 65,160 \text{ vehicle-miles}$$

$$\begin{aligned}
\text{VOC benefits} &= (0.5)(U_1 - U_2)(\text{VMT}_1 + \text{VMT}_2) \\
&= (0.5)(270.6 - 268)(63,840 + 65,160)(365/1000) \\
&= \$62,210
\end{aligned}$$

Total VOC benefits due to the improvement = \$(388,527 + 61,210) = \$449,747 per year.

7.3.2 Implementation of Steps 4 to 6 Using the HERS Method

In HERS, VOC (CSOPCST) is estimated as the sum of operating cost per thousand vehicle-miles due to following

VOC components: fuel, oil, tires, maintenance and repair, and vehicle depreciation, separately assuming a *basic section* (straight and level section with no speed change). Then the *excess* VOC is computed separately for *speed-change sections* and *curved sections* and added to the basic section costs to obtain the overall VOC for the entire highway segment.

The cost is calculated for each vehicle type and VOC component. This is done for basic sections (Table 7.8a), speed-change sections (Table 7.8b), and curved sections (Table 7.8c).

- *Rates of consumption of VOC components*. Appendix A7.1 presents equations for estimating the rate of consumption of various VOC components at constant speed, excess consumption for speed changes and curved sections, for different vehicle classes.
- *Unit costs of VOC components*. These are provided in Appendix A7.2 or in the HERS manual (FHWA, 2002).
- *Pavement condition adjustment factors for VOC components*. These are provided in Appendix A7.3 or in the HERS manual (FHWA, 2002).
- *Component adjustment factors*. These factors reflect reductions in consumption rates of various VOC components between 1980 and 1997. As these are the most recent values available, they are included in Appendix A7.4 (FHWA, 2002).

Example 7.11   It is proposed to improve a certain 5.2-mile urban arterial section. The section is straight with no speed changes. Assume that the current volume is 41,000 vpd of small autos and a 6% traffic growth after the improvement. Estimate the constant-speed operating

Table 7.8   HERS VOC Computation Templates
(a) Basic Sections

VOC Component	(a) Rate of Consumption per 1000 VMT	(b) Pavement Condition Adjustment Factor	(c) Unit Cost	(d) Component Adjustment Factor	(e) = (a) × (b) ×(c/d) Total Cost ($/1000 VMT)
Fuel (FC)	CSFC gallons	PCAFFC	COSTF	FEAF	
Oil (OC)	CSOC quarts	PCAFOC	COSTO	OCAF	
Tire (TW)	CSTW % worn	PCAFTW	COSTT	TWAF	
Maintenance and repair (MR)	CSMR % of average cost	PCAFMR	COSTMR	MRAF	
Depreciation (VD)	CSVD % of new price	PCAFVD	COSTV	VDAF	

$$\sum e = \text{CSOPCST}$$

Table 7.8 (*continued*)

(b) Excess VOC due to Speed Change

VOC Component	(a) Excess Rate of Consumption per 1000 VMT	(b) Unit Cost ($)	(c) Component Adjustment Factor	(d) = (a) × (b/c) Total Cost ($/1000 VMT)
Fuel (FC)	VSFC gallons	COSTF	FEAF	
Oil (OC)	VSOC quarts	COSTO	OCAF	
Tire (TW)	VSTW % worn	COSTT	TWAF	
Maintenance and repair (MR)	VSMR % of average cost	COSTMR	MRAF	
Depreciation (VD)	VSVD % of new price	COSTV	VDAF	
				$\sum d$ = VSCOPCST

(c) Excess VOC at High-Speed Curved Sections

VOC Component	(a) Excess Rate of Consumption per 1000 miles	(b) Unit Cost ($)	(c) Component Adjustment Factor	(d) = (a) × (b/c) Total Cost ($/1000 VMT)
Fuel, FC	CSFC gallons	COSTF	FEAF	
Tire, TW	CSTW % worn	COSTT	TWAF	
Maintenance and Repair, MR	CSMR % of average cost	COSTMR	MRAF	
				$\sum d$ = COPCST

cost per thousand vehicle-miles (CSOPCST) due to each of five VOC components (fuel, oil, tires, maintenance and repair, and vehicle depreciation). The road conditions without the improvement (expected values with the improvement are shown in parentheses) are as follows: average grade, 3% (2.5%); pavement condition, 3.1 (4.2) PSR; and average speed, 23 (35) mph.

SOLUTION (*1*) *Unit costs of VOC components* These are given in Table E7.11.1 along with component adjustment factors from the HERS Manual.

(*2*) *Pavement condition adjustment factors before and after improvement* The equations in Appendix A7.3 are used to compute the pavement condition adjustment factors (PCAFOC) for small autos:

Fuel: PCAFFC = 1, for before and after improvement.

Oil: PCAFOC$_{before}$

$$= 2.64 + 0.0729 PSR^2 - 0.722 PSR$$
$$= 2.64 + (0.0729)(3.1^2) - (0.722)(3.1) = 1.102$$

Tire wear: PCAFTW$_{before}$

$$= 2.40 - 1.111 \ln(PSR)$$
$$= 2.40 - 1.111 \ln(3.1) = 1.143$$

Maintenance and repair: PCAFMR$_{before}$

$$= 3.19 + 0.0967 PSR^2 - 0.96 PSR$$
$$= 3.19 + (0.0967)(3.1^2) - (0.96)(3.1) = 1.143$$

Vehicle depreciation: PCAFVD$_{before}$

$$= 1.136 - 0.106 \ln(PSR)$$
$$= 1.136 - 0.106 \ln(3.1) = 1.016$$

Pavement condition adjustment factors after improvement are computed using the same equations and on the basis of improved PSR (4.2).

(*3*) *Rates of consumption of VOC components before improvement* The equations in Appendix A7.1 are used, as shown below:

Table E7.11.1 Adjustment Factors and Unit Costs by VOC Component

VOC Component	Unit Cost, ($)	Component Adjustment Factor
Fuel (FC)	COSTF = 0.871	FEAF = 1.536
Oil (OC)	COSTO = 3.573	OCAF = 1.05
Tire (TW)	COSTT = 45.20	TWAF = 1.0
Maintenance and repair (MR)	COSTMR = 84.10	MRAF = 1.0
Depreciation (VD) price/vehicle	COSTV = 18,117	VDAF = 1.30

[a]From FHWA (2002).

Fuel: $\text{CSFC}_{\text{before}}$

$$= 100.82 - 4.9713S + 0.11148S^2 - 0.0011161S^3$$
$$+ (5.1089 \times 10^{-6})S^4 + 3.0947G$$
$$= 100.82 - (4.9713)(23) + (0.11148)(23)^2$$
$$- (0.0011161)(23)^3 + (5.1089 \times 10^{-6})(23)^4$$
$$+ (3.0947)(3)$$
$$= 42.587 \text{ gals/1000 VMT}$$

Oil: $\text{CSOC}_{\text{before}}$

$$= \exp[2.7835 - 0.79034\ln(S) - 1.1346/S^{1.5}$$
$$+ 0.65342G^{0.5}]$$
$$= \exp[2.7835 - 0.79034\ln(23) - 1.1346/23^{1.5}$$
$$+ (0.65342)(3^{0.5})] = 4.165 \text{ quarts/1000 VMT}$$

Tire wear: $\text{CSTW}_{\text{before}}$

$$= \exp[-2.55 + 0.0001621S^2 + 0.01441S$$
$$+ 1.473\ln(G) - 0.001638SG]$$
$$= \exp[-2.55 + (0.0001621)(23^2) + (0.01441)(23)$$
$$+ 1.473\ln(3) - (0.001638)(23)(3)]$$
$$= 0.534\% \text{ worn/1000 VMT}$$

Maintenance and repair needs: $\text{CSMR}_{\text{before}}$

$$= 48.3 + 0.00865S^2 + 0.0516SG$$
$$= 48.3 + (0.00865)(23^2) + (0.0516)(23)(3)$$
$$= 56.436 \text{ (\% average cost/1000 VMT)}$$

Depreciation rate: $\text{CSVD}_{\text{before}}$

$$= 2.2 + 0.001596S - 0.38\ln(S)$$
$$= 2.2 + (0.001596)(23) - 0.38\ln(23)$$
$$= 1.045\% \text{ new price/1000 VMT}$$

(4) VOC computation for the before-improvement case Results are summarized in Table E7.11.2.

In this example, excess VOC computation is not needed because the section is a straight constant-speed section, and therefore involves only basic section costs. Furthermore, there is no need to repeat the computation for other vehicle classes because the traffic stream is comprised primarily of small autos.

(5) Rates of consumption of VOC components after improvement The rates of consumption of VOC components for small autos at constant speed at the new average effective speed (S) and new gradient (G %) are:

Fuel: $\text{CSFC}_{\text{after}}$

$$= 100.82 - 4.9713S + 0.11148S^2 - 0.0011161S^3$$
$$+ (5.1089 \times 10^{-6})S^4 + 3.0947G$$
$$= 100.82 - (4.9713)(35) + (0.11148)(35)^2$$
$$- (0.0011161)(35)^3 + (5.1089 \times 10^{-6})(35)^4$$
$$+ (3.0947)(2.5)$$
$$= 30.938 \text{ gals/1000 VMT}$$

Oil consumption rate: $\text{CSOC}_{\text{after}}$

$$= \exp[2.7835 - 0.79034\ln(S) - 1.1346/S^{1.5}$$
$$+ 0.65342G^{0.5}]$$
$$= \exp[2.7835 - 0.79034\ln(35) - 1.1346/35^{1.5}$$
$$+ (0.65342)(2.5^{0.5})]$$
$$= 2.722 \text{ quarts/1000 VMT}$$

Tire wear: $\text{CSTW}_{\text{after}}$

$$= \exp[-2.55 + 0.0001621S^2 + 0.01441S$$
$$+ 1.473\ln(G) - 0.001638SG]$$
$$= \exp[-2.55 + (0.0001621)(35^2) + (0.01441)(35)$$

Table E7.11.2   VOC Computation for the Before-Improvement Case

VOC Component	(a) Rate of Consumption, per 1000 VMT	(b) Pavement Condition Adjustment Factor	(c) Unit Cost, ($)	(d) Component Adjustment Factor	(e) = (a) × (b) × (c/d) Total Cost ($/1000 VMT)
Fuel (FC)	42.587	1	0.871	1.536	24.149
Oil (OC)	4.165	1.102	3.573	1.05	15.619
Tire (TW)	0.00534	1.143	45.2	1	0.276
Maintenance and repair (MR)	0.564	1.143	84.1	1	54.215
Depreciation (VD)	0.01045	1.016	18,117	1.3	147.963
				$\sum e$ = CSOPCST =	242.222

Table E7.11.3   VOC Computation for the After-Improvement Case

VOC Component	(a) Rate of Consumption per 1000 VMT	(b) Pavement Condition Adjustment Factor	(c) Unit Cost, ($)	(d) Component Adjustment Factor	(e) = (a) × (b) × (c/d) Total Cost ($/1000 VMT)
Fuel (FC)	30.938	1	0.871	1.536	17.544
Oil (OC)	2.722	0.894	3.573	1.05	8.281
Tire (TW)	0.00527	0.805	45.2	1	0.192
Maintenance and repair (MR)	0.6341	0.864	84.1	1	46.075
Depreciation (VD)	0.00905	0.984	18,117	1.3	124.104
				$\sum e$ = CSOPCST =	196.196

Table E7.11.4   Total User VOC Benefits

	Before Improvement	After Improvement
VOC ($/1,000 VMT)	242.222	196.196
VMT (in thousands)	(41,000/1,000)(5.2) = 213.6	226
Total User Benefits = 0.5(242.222 − 196.196)(213.6 + 226) = $10,107 per day		

$+ 1.473 \ln(2.5) - (0.001638)(35)(2.5)]$

$= 0.527\%$ tire wear/1000 VMT

Maintenance and repair: $\text{CSMR}_{\text{after}}$

$= 48.3 + 0.00865S^2 + 0.0516SG$

$= 48.3 + (0.00865)(35^2) + (0.0516)(35)(2.5)$

$= 63.41\%$ average cost/1000 VMT

Depreciation rate: $\text{CSVD}_{\text{after}}$

$= 2.2 + 0.001596S - 0.38 \ln(S)$

$= 2.2 + (0.001596)(35) - 0.38 \ln(35)$

$= 0.905\%$ new price/1000 VMT

(6) VOC computation for the after-improvement case The results are summarized in Table E7.11.3.

(7) Evaluation The total user VOC benefits are estimated in Table E7.11.4.

7.4 SPECIAL CASE OF VOC ESTIMATION: WORK ZONES

Transportation improvements aim at providing a better level of service. A paradoxical twist, however, is that during the implementation (construction) period (typically lasting a few days to one or more years) users may suffer a severe deterioration of levels of service along the affected transportation corridor. For example, work zones for highway rehabilitation and maintenance can significantly raise the costs of safety, travel time, and vehicle operation. Fortunately, the duration of the work zone is typically very small compared with the service life of the facility, so the reduction in operating costs due to the improved facility over its service life far outweighs the increase in operating costs due to the associated work zones. When traffic demand exceeds work-zone capacity, traffic flow becomes constricted and a queue is formed. The stop-and-go conditions at work-zone queues translate into increased frequency of speed changes and delay, and consequently, higher user costs. In some cases, voluntary or involuntary traffic detours by road users, in a bid to avoid a highway work zone, add to travel distance. In such cases, even if the VOC rates remain the same, VMT increases and therefore the overall VOC increases. Most agencies make conscious efforts to reduce the extra costs of vehicle operation, safety and travel time by adopting optimal values of work zone configuration parameters such as timing, duration, scope, and traffic control.

7.5 SELECTED SOFTWARE PACKAGES THAT INCLUDE A VOC ESTIMATION COMPONENT

The multiplicity of VOC components and factors lend a considerable degree of complexity to the process of VOC estimation. There are a number of software packages that can help in estimating vehicle operating cost, as part of typically an overall economic efficiency analysis. A brief discussion of the features of each selected package is herein provided.

7.5.1 AASHTO Method

AASHTO's User Benefit Analysis for Highways (AASHTO, 2003) spreadsheet package estimates safety, travel time, and VOC impacts as part of overall economic analysis. VOC is estimated separately for two categories of VOC components: fuel and "other" (tires, maintenance, etc.). For other VOCs, AASHTO (2003) refers to a 1999 AAA table that presents average automobile operating costs irrespective of VOC factor levels. On the other hand, fuel VOC is given as a function of vehicle type and speed, a relationship adopted from the Cohn et al. (1992) study.

7.5.2 HERS Package: National and State Versions

The HERS methodology (FHWA, 2002) helps estimate the impacts of VOC (as well as safety and travel time) associated with highway widening, pavement improvement, and alignment enhancement at a network level but can also be used for evaluating individual projects. HERS calculates unit VOCs (i.e., VOC per VMT) on the basis of five VOC components (fuel consumption, oil consumption, tire wear, maintenance and repair, and depreciable value) as a function of several VOC factors (speed, speed change, horizontal curves, and vehicle type). The process involves three stages:

1. Constant-speed VOC calculated as a function of average speed, grade, and pavement condition
2. Excess VOC due to speed-change cycles
3. Additional excess VOC as a function of road curvature

The HERS VOC estimation models are derived from consumption rates and prices that were established by Zaniewski (1982), with some adjustments made on the basis of information from Claffey and Associates (1971) and Daniels (1974). All five VOC components are included in the calculation of VOCs at constant-speed sections and also in determination of the excess VOC due to speed-change cycles. In calculating the excess costs due to curves, only the following VOC components are considered: fuel, tire wear, and maintenance and repair. Data needed to run the HERS package include retail prices of gasoline and diesel, cost of oil and tires, mileage-based maintenance costs per mile for new automobiles, new vehicle prices for medium-weight and heavy trucks, and price indices. These data are largely available on the Internet and in published reports. The national version of HERS contains data that generally reflect an average of national conditions, while the state version allows users to input state-specific data.

7.5.3 HDM-4 Road User Effects

HDM-4 road user effects (HDM-RUE) is a special module of the World Bank's Highway Development and Management, Version 4 (HDM-IV) that estimates vehicle operating cost for road segments, duly incorporating the effects of operating speed and congestion. The software can generate output data relating VOC to each of several factors, such as road grade, pavement roughness, surface texture, speed change, congestion, and speed (Bennett and Greenwood, 2001). Also, HDM-RUE can incorporate "willingness to pay" costs in the VOC estimation. In place of the default data, local data can be used to recalibrate the VOC model. The model includes detailed analyses such as

VOC estimation for heavy truck trailers and mechanistic modeling of tire consumption.

7.5.4 Surface Transportation Efficiency Analysis Model

In the VOC module of the surface transportation efficiency analysis model (STEAM), vehicle operating cost is estimated separately for fuel and nonfuel components (DeCorla-Souza and Hunt, 1999). Fuel consumption is considered variable (with operating speed) for autos and trucks but is considered fixed for local and express buses and for light and heavy rail. For auto and truck only, nonfuel costs are considered fixed per mile regardless of speed. Nonfuel costs include tires and maintenance but exclude mileage-based depreciation and oil consumption. For speed estimation, STEAM uses demand data directly from the traffic assignment step of the four-step transportation demand model. An analyst can quickly investigate the impact of different fuel costs without having to adjust each of the speed–consumption estimates by multiplying the fuel consumption amount at a given average speed by the fuel cost. STEAM enables the user to carry out uncertainty analysis and yields confidence intervals for its VOC outputs. Besides vehicle operating costs, other performance measures analyzed by STEAM are the user costs of safety and delay, and emissions.

7.5.5 Other Models That Include a VOC Estimation Component

The packages described above, some to a greater extent than others, enable the analyst first to estimate the unit cost of vehicle operation and then use the estimated VOCs and other data to obtain the overall VOC (for all vehicles using a system within a given time period). On the other hand, there are other packages that do not estimate unit VOCs but rather, utilize prior established unit VOCs for overall VOC estimation, which is then entered into an overall economic analysis procedure. These include:

1. *The California life-cycle benefit–cost analysis model (Cal-B/C)*, which uses a look-up table for fuel consumption of three vehicle types (autos, trucks, and buses) in gallon/mile estimates. To calculate total fuel costs, fuel consumption is multiplied by fuel cost per gallon, excluding taxes. Nonfuel costs are estimated on a per-mile basis (System Metrics Group and Cambridge Systematics, 2004).
2. *MicroBencost*, which enables the calculation of unit (per vehicle-mile) and overall vehicle operating cost (as well as safety and delay user costs) for a wide range of projects, such as new highway and bridge construction, bypasses, pavement preservation, lane addition, safety projects, railroad crossing projects, and HOV lanes. Default values include the unit operating cost of each vehicle class. The program estimates these costs for eight vehicle types and each hour of the day and duly accounts for traffic growth over time. It should be noted that the MicroBencost procedures for estimating the VOC of intersections differs significantly from the 1977 AASHTO procedures (McFarland et al., 1993).

7.6 COMPARISON OF VOC ESTIMATION METHODS AND SOFTWARE

7.6.1 Levels of Detail

Most models use fuel, oil, tire, and maintenance cost components. Many models assume that VOC factors are primarily vehicle type and speed. Other factors, such as curves, grades, and speed changes, have been used in only a few models. Inclusion of all components and factors in VOC estimation poses significant data and modeling challenges that may not be justified by the relatively small gain in the overall accuracy of results (Forkenbrock and Weisbrod, 2001). The VOC estimation models in HERS, MicroBencost, and HDM-RUE are relatively more comprehensive than are other models. Bein (1993) reviewed VOC models in use worldwide and suggested that VOC estimation methodologies that are based on fuel consumption and speeds may be more appropriate than those that use a wider array of VOC components or factors because fuel consumption rates and speeds are easily measurable. From these perspectives, the STEAM methodology (Cambridge Systematics, 2000) which separates the fuel component from other VOC components, can be expected to provide good estimates of VOC, particularly where detailed data are unavailable.

7.6.2 Data Sources

Most models for fuel consumption utilize results from a common source. The MicroBencost model derived its data from the FHWA 1982 study (Zaniewski, 1982). Also, for unit costs and consumption rates of VOC components, StratBencost (a companion package of MicroBencost) utilized data from HERS. In turn, data on the consumption rates of VOC components data used in HERS were obtained from the 1982 FHWA study. The STEAM model relies on several sources, including the 1982 FHWA study. It should be noted, however, that STEAM uses fixed-cost-per-mile unit VOC costs and does not include mileage-based depreciation.

SUMMARY

Most VOC methodologies estimate unit vehicle operating costs (dollars/VMT) as a function of travel speeds. Other VOC factors are road class, vehicle type, prevailing traffic condition, roadway gradient, roadway curvature, and road surface type and condition. Some methodologies, such as HERS and HDM provide vehicle operating cost equations that account for these factors. While the use of such equations typically yields more reliable VOC estimates, data for such detailed VOC estimation methods may not always be available.

In evaluating the future VOC impacts of a proposed project, the analyst must contend with the inherent uncertainties in the VOC estimation process. Future projections of VOC component types, rates, and unit costs are based on current (or at most foreseeable) trends of vehicle technology, and the economy. With the advent of vehicles that operate on electricity, natural gas, hybrids, and improved gasoline engines, the fuel component of VOC is subject to future uncertainties. Also, increased longevity of vehicles translates into reduced depreciation rates and therefore affect that cost of mileage-related depreciation. Furthermore, existing VOC methodologies cover only the key aspects of VOC components. Other relatively minor components that are excluded at the current time may play a more visible role in future, and may need to be accounted for at that time. Another area of uncertainty is that of VOC factors. Urban growth, changes in speed limits, implementation of managed lanes, and redesignation of highways (typically upgrades to higher classes) are likely to result in operating speeds and speed-change frequencies that are different from their respective levels envisaged at the time of evaluation. Increased or decreased economic development may also affect the expected inventory costs of vehicle operation. Furthermore, facility physical deterioration may result in pavement, runway, or guideway surface conditions that differ from projected trends and may therefore result in VOC values that differ from the expected values.

EXERCISES

7.1. Sectional improvements are proposed for an existing highway to ease traffic flow between a suburb and a downtown area. For the existing road, the overall length is 8 miles, the average grade is 4.5%, the AADT is 13,500, the v/c ratio is 0.7, and the average operating speed is 35 mph. There are three sharp curves (10 degrees and 1.5 miles each), at which the speed limit is 25 mph. There are 30% trucks in the traffic stream. After the project is completed, the new road will be 6.5 miles long with an average

grade of 2%. It is expected that the traffic volume will increase by 10% due to induced and diverted traffic, and a v/c ratio of 0.5 and operating speed of 55 mph are expected. One of the curves will be eliminated and the other two curves will be rendered less sharp: 4 degrees). Determine the annual benefits of the new project from the perspective of vehicle operating costs only.

7.2. Before physical and operational improvements at a 12-mile urban interstate freeway, the traffic volume was 100,000 per day and the unit VOC was 0.41 cent/VMT. After the improvements, it is estimated that the traffic volume increased to 125,000 per day and the VOC reduced to 0.31 cent/VMT. Determine the user VOC benefits of the project.

7.3. The speed limit at a 4.2-mile highway section was recently changed from 35 mph to 45 mph. Using (**a**) Hepburn's VOC model and (**b**) FHWA's model, determine the expected change in the unit and overall VOC of medium-sized automobiles due to the improvement. Assume that operating speeds are generally 5 mph higher than the speed limits. Also, assume a flat terrain and a traffic volume of 6,250 vpd comprised of medium-sized vehicles. The traffic volume increases by 5% after the change in speed limit. Compare your answers from parts (a) and (b).

7.4. In upgrading a 12-mile four-lane state highway section to an Interstate highway, it is proposed to carry out geometric improvements. The changes in grade, travel demand, and speed predicted to occur after the upgrade are presented in Table EX7.4. Determine the overall VOC savings due to the upgrade. Use the FHWA 1982 VOC estimation model.

7.5. A 7.5-mile two-lane urban arterial received a series of traffic flow improvements, such as construction of passing lanes, deceleration lanes, and channelization. Due to these improvements, the average traffic flow speed on a roadway increased from 26 mph to 34 mph. The average grade of the roadway is 2%, and the direction split is 55% uphill and 45% downhill. The traffic volume of 9000 vpd comprises: small automobiles, 25%; medium-sized automobiles, 60%; large automobiles, 15%. Assume zero truck traffic. After the improvement, the traffic volume increases by 8% due to induced demand, and the share of small vehicles decreases by 6% while those of medium-sized and large vehicles increase by 3% each.

Table EX7.4[a]   Traffic and Geometric Characteristics[a]

	Before Improvement			After Improvement		
	Traffic Volume (ADT)	Grade (%)	Speed (mph)	Traffic Volume (ADT)	Grade (%)	Speed (mph)
Segment A: 50% traffic uphill, 50% downhill	18,500	−2.5	50	6,800	−1.0	70
Segment B: 55% traffic uphill, 45% downhill	34,320	+3.0	50	5,430	+1.5	70

[a]Assume that directional splits and road length remain the same after the upgrade. Assume medium-sized vehicles only.

(a) Determine the VOC benefits of the improvement using Hepburn's model.

(b) Determine the VOC benefits of medium-sized vehicles only using FHWA's 1982 methodology. Compare your results with that of the Hepburn model and comment on any differences.

(c) How could the Hepburn model be improved?

7.6. A change in speed limit policy resulted in decreased average truck operating speed on the Brandon Expressway from 69.7 mph to 57.2 mph. What is the overall percentage increase in shipping inventory costs per year? Shipping trucks make up 35% of the traffic stream. The AADT is 121,540. The average value of daily cargo per truck is $0.32 million. Assume that the prevailing interest rate is 7.5%.

REFERENCES[1]

*AAA (2005). Your driving costs, American Automobile Association and Runzheimer International, ww2.aaa.com and www.runzheimer.com. Accessed Dec. 2005.

*AASHTO (2003). *User Benefit Analysis for Highways*, American Association of State Highway and Transportation Officials, Washington, DC.

Al-Omari, B., Darter, M. I. (1994). Relationship between International Roughness Index and Present Serviceability Rating, *Transportation Research Record 1435*, Transportation Research Board, Washington, DC.

*Barnes, G. Langworthy, P. (2003). *The Per-Mile Costs of Operating Automobiles and Trucks*, Rep. MN/RC 2003-19, University of Minnesota, Minneapolis, MN.

Bein, P. (1993). VOC model needs of a highway department, 1993. *Road Transp. Res.*, Vol. 2, No. 2, pp. 40–54.

Bennett, C. R. (1991). *Calibration of Vehicle Operating Cost: Final Report, Thailand Road Maintenance Project*, Rep. 1106-THA, N.I., prepared for the Asian Development Bank by Lea International, Vancouver, BC, Canada.

Bennett, C. R., Greenwood, I. D. (2001). *Modeling Road User and Environmental Effects in HDM-4*, HDM-4 Ref. Vol. 7, World Road Association, Paris.

Cambridge Systematics (2000). *Surface Transportation Efficiency Analysis Model (STEAM 2.0): User Manual*, Federal Highway Administration, U.S. Department of Transportation, Washington, DC, http://www.fhwa.dot.gov/steam/20manual.htm. Accessed Dec. 2005.

Claffey, P. J., and Associates (1971). *Running Costs of Motor Vehicles as Affected by Road Design and Traffic*, NCHRP Rep. 111, Highway Research Board, National Research Council, Washington, DC.

Cohn, L., Wayson, R., Roswell, A. (1992). Environmental and energy considerations, in *Transportation Planning Handbook*, Institute of Transportation Engineers, Washington, DC.

Daniels, C. (1974). *Vehicle Operating Costs in Transportation Studies*, E.S.U. Technical Series, No. 1, Spencer House, London.

DeCorla-Souza, P., Hunt, J. T. (1999). *Use of STEAM in Evaluating Transportation Alternatives*, Federal Highway Administration, U.S. Department of Transportation, Washington, DC, http://www.fhwa.dot.gov/steam/steam-it.pdf. Accessed Dec. 2005.

*FHWA (2002). *Highway Economic Requirements System Technical Manual*, Federal Highway Administration, U.S. Department of Transportation, Washington, DC.

Forkenbrock, D. J., Weisbrod, G. E. (2001). *Guidebook for Assessing the Social and Economic Effects of Transportation Projects*, NCHRP Rep. 456, Transportation Research Board, National Research Council, Washington, DC.

Hepburn, S. (1994). A simple model for estimating vehicle operating costs in urban areas, *Road Transp. Res.*, Vol. 3, No. 2, pp. 112–118.

McFarland, W. F., Memmott, J. L., Chui, M. K. (1993). *Microcomputer Evaluation of Highway User Benefits*, NCHRP Rep. 7–12, Transportation Research Board, National Research Council, Washington, DC.

Opus Central Laboratories (1999). *Review of VOC–Pavement Roughness Relationships Contained in Transfund's Project Evaluation Manual*, Rep. 529277.00, Opus, Lower Hutt, New Zealand.

System Metrics Group and Cambridge Systematics (2004). *The California Life-Cycle Benefit/Cost Analysis Model (Cal-B/C): User's Guide*, California Department of Transportation, Sacramento, CA.

[1]References marked with an asterisk can also serve as useful resources for VOC estimation.

*Thoresen, T., Roper, R. (1996). *Review and Enhancement of Vehicle Operating Cost Models: Assessment of Non-urban Evaluation Models, Transp. Res. Rep.* 279, *Australian Road Research Board*, Victoria, Australia.

USDOE (2005a). Alternative Fuels Data Center (www.afdc.doe. gov) and Fuel Economy Web site, www.fueleconomy.gov/feg/ hybrid_sbs.shtml. Accessed Jan. 15, 2006.

_____(2005b). *A Primer on Gasoline Prices*, Energy Information Administration, U.S. Department of Energy, Washington, DC, http://www.eia.doe.gov. Accessed Jan. 15, 2006.

*VTPI (2004). The cost of driving and the savings from reduced vehicle use, in *TDM Encyclopedia*, Victoria Transport Policy Institute, Victoria, BC, Canada, http://www.vtpi.org/tdm/ tdm82.htm. Accessed June 4, 2004.

_____(2005). *Transportation Cost and Benefit Analysis: Vehicle Costs*, Victoria Transportation Policy Institute, Victoria, BC, Canada.

Zaniewski, J. P. (1982). *Vehicle Operating Costs, Fuel Consumption, and Pavement Type and Condition Factors*, prepared by the Texas Research and Development Foundation, for the U.S. Department of Transportation, Washington, DC.

ADDITIONAL RESOURCES

Chesher, A., Harrison, R. (1987). *Vehicle Operating Costs: Evidence from Developing Countries*, a World Bank Publication, Johns Hopkins University Press, Baltimore, MD.

New Jersey DOT (2001). *Road User Cost Manual*, Sec. 2, *Road User Cost Components*, New Jersey Department of Transportation, NJ.

The national automobile associations (of the United States, Canada, UK, etc.) generally present the average cost per mile of operating a passenger vehicle in terms of fuel, oil, maintenance, and tire replacement and maintenance costs on the basis of vehicle type, model, driving style, and origin of travel. They may also provide the costs of routine servicing, repairs and replacements, and warranties.

The U.S. Department of Energy's brochure *A Primer on Gasoline Prices*, an Energy Information Administration brochure, is available at http://www.eia.doe.gov/pub/oil_gas/ petroleum/analysis_publications/primer_on_gasoline_prices/ html/petbro.html.

National petroleum institutes (of the United States, Canada, UK, etc.) provide summaries of the retail price of gasoline and their price trends in terms of month. These prices are also categorized into regions, subregions, states or provinces, and cities.

The U.S. Bureau of Labor Statistics provides monthly reports on changes in the prices paid by urban consumers for a representative basket of goods and services, including oil and tires for four-tire vehicles, accessible at www.bls.gov/cpi. For larger vehicles, the appropriate index for tires is the producer price index, which is accessible at www.bls.gov/ppi/home.htm.

APPENDIX A7.1:   FHWA (2002) HERS MODELS FOR VOC COMPUTATION[1]

[1]The following abbreviations are used in this appendix:

S, average effective speed (mph)

G, gradient (%)

D, degree of curvature

GR, grade (%)

CSMAX, maximum speed during speed-change cycle

CSFC, constant-speed fuel consumption (gal/1000 VMT)

CSOC, constant-speed oil consumption (qt/1000 VMT)

CSTW, constant-speed tire wear (% worn/1000 VMT)

CSMR, constant-speed maintenance and repair (% average cost/1000 VMT)

CSVD, constant-speed depreciation (% new vehicle price/1000 VMT)

SCCFC, excess fuel consumption for speed-change cycles (gal/1000 cycles)

SCCOC, excess oil consumption for speed-change cycles (qt/1000 cycles)

SCCMR, excess maintenance and repair for speed-change cycles (% average cost/1000 cycles)

SCCD, excess depreciation for speed-change cycles (% new price/1000 cycles)

CFC, excess fuel consumption due to curves (gal/1000 VMT)

CTW, excess tire wear due to curves (% worn/1000 VMT)

CMR, excess maintenance and repair due to curves (% average cost/1000 VMT)

Table A7.1.1   VOC for Fuel Consumption at Constant Speed on Straight Sections (CSFC)

(a) Small Automobile

$G \geq 0$ $\qquad$ $\text{CSFC} = 100.82 - 4.9713S + 0.11148S^2 - 0.0011161S^3 + (5.1089 \times 10^{-6})S^4 + 3.0947G$

$G < 0$ and $S \leq 40$ $\qquad$ $\text{CSFC} = (91.045 - 4.0552S + 0.060972S^2 + 4.0504G + 0.4227G^2)/(1 - 0.014068S$
$\qquad \qquad +0.0004774S^2 - 0.045957G + 0.0054245G^2)$

$G < 0$ and $S > 40$ $\qquad$ $\text{CSFC} = 23.373 + 3.6374G + 0.21681G^2 + (72.562/[1 + \exp(-[(S - 81.639)/7.4605)])]$

(b) Medium-Sized/Large Automobile

$S \leq 40$ $\qquad$ $\text{CSFC} = 88.556 - 3.384S + 1.7375G + 0.053161S^2 + 0.18052G^2 + 0.076354SG$

$S > 40$ $\qquad$ $\text{CSFC} = 85.255 - 2.2399S + 2.7478G + 0.028615S^2 + 0.041389G^2 + 0.046242SG$

(c) Four-Tire Truck

$G \geq 0$ and $20 < S < 55$ $\qquad$ $\text{CSFC} = 115.41 - 3.6397S + 7.0832G + 0.050662S^2 - 0.34401G^2 + 0.096956SG$

$G \geq 0$ and $S \leq 20$ $\qquad$ $\text{CSFC} = 120.7 + -5.0201S + 0.1088S^2 + 9.8816G - 1.3755G^2 + 0.11582G^3$

$G < 0$ and $S \leq 10$ $\qquad$ $\text{CSFC} = 161.2 - 6.622S - 87.758\ln(S)/S - 1.0889G^2 - 0.13217G^3$

$G < 0$ and $10 < S \leq 20$ $\qquad$ $\text{CSFC} = 106.31 - 2.7456S + 5.0147G - 0.001281S^2 + 0.94555G^2 + 0.19499SG$

$G < 0$ and $20 < S < 55$ $\qquad$ $\text{CSFC} = 351.5 - 184.42\ln(S) + 0.71838G + 28.297[\ln(S)]^2 + 1.0105G^2 + 2.8947G\ln(S)$

$G \geq 1.5$ and $S \geq 55$ $\qquad$ $\text{CSFC} = 110.4 + 0.000249S^3 - 18.93\ln(S) + 8.06G$

$-2.5 \leq G < 1.5$ and $S \geq 55$ $\text{CSFC} = (28.77 + 0.183655S + 3.34032G)/(1 - 0.0074966S - 0.049703G)$

Otherwise $\qquad$ $\text{CSFC} = \exp(2.784 + 0.02014S + 0.06881G)$

(d) Six-Tire Truck

$G \geq 0$ and $S < 55$ $\qquad$ $\text{CSFC} = 298.60 - 13.131S + 53.987G + 0.30096S^2 - 4.7321G^2 - 0.88407SG$
$\qquad \qquad - 0.0020906S^3 + 0.22739G^3 + 0.02875SG^2 + 0.0045428S^2G$

$G < 0$ and $S < 55$ $\qquad$ $\text{CSFC} = 273.05 - 9.2427S + 58.195G + 0.14718S^2 + 6.7665G^2 - 1.3785SG$
$\qquad \qquad - 0.00046068S^3 + 0.13884G^3 - 0.079555SG^2 + 0.012622S^2G$

$G \geq 1.5$ and $S \geq 55$ $\qquad$ $\text{CSFC} = 361.11 - 8.1978S + 11.186G + 0.077607S^2 - 0.27665G^2 - 0.035211SG$

$G < 1.5$ and $S \geq 55$ $\qquad$ $\text{CSFC} = 101.5 + 0.000186S^3 + 1.102G^2 + 18.22G$

(e) 3+ Axle Single-Unit Truck

$G \geq 3$ and $S \leq 20$ $\qquad$ $\text{CSFC} = 68.536 + 12.823S + 122.45G + 0.023896S^2 + 0.36758G^2 - 6.2014SG$

$3 \geq G \geq 0$ and $S \leq 20$ $\qquad$ $\text{CSFC} = 254 - 3.0854S - 2.177G - 0.063346S^2 + 24.848G^2 + 4.3101SG$
$\qquad \qquad + 0.0012816S^3 - 1.2432G^3 - 1.6437SG^2 + 0.0013556S^2G$

$G < 0$ and $S \leq 20$ $\qquad$ $\text{CSFC} = (259.66 - 19.925S + 0.49931S^2 - 0.0045651S^3 - 1.5876G)/(1 - 0.058535S$
$\qquad \qquad + 0.00077356S^2 - 0.14916G + 0.024241G^2)$

$G > 3$ and $S > 20$ $\qquad$ $\text{CSFC} = 290.45 - 2.598S + 25.823G + 0.024983S^2 - 2.2654G^2 + 0.21897SG$

$3 \geq G \geq 0$ and $S > 20$ $\qquad$ $\text{CSFC} = 1208.8 - 586.87\ln(S) + 80.955[\ln(S)]^2 + 93.99G - 13.477G^2$

$0 > G \geq -3$ and $S > 20$ $\qquad$ $\text{CSFC} = \exp(6.0673 - 0.1139S + 0.023622S\ln(S) + 0.79191G - 0.022171G^3)$

$G < -3$ and $S > 20$ $\qquad$ $\text{CSFC} = (-1.3978/(1 + (((S - 40.215)/ - 11.403)^2))) + (47.024/$
$\qquad \qquad (1 + (((G + 0.01611)/5.4338)^2))) + (-26.724)/(1 + (((S - 40.215)/$
$\qquad \qquad -11.403)^2)))(1/(1 + (((G + 0.01611)/5.4338)^2)))$

(f) 3–4 Axle Combination-Unit Truck

$S > 20$ and $3 \geq G \geq -3$, $\text{CSFC} = (1087.9 - 576.71\ln(S) + 82.039[\ln(S)]^2 + 22.325G)/(1 - 0.17121\ln(S)$
   or $S \leq 20$ and $G \geq -3$ $\qquad - 0.035147G)$

$G < -3$ $\qquad$ $\text{CSFC} = -239.17 + 61.115\ln(S) + 2221.9/S - 4411.6\exp(-S)$

$S > 20$ and $G > 3$ $\qquad$ $\text{CSFC} = \exp(4.5952 + 0.0049349S\ln(S) + 0.31272G)$

(g) 5+ Axle Combination-Unit Truck

$3 \geq G \geq -3$ $\qquad$ $\text{CSFC} = (1618.8 - 864.83\ln(S) + 124.88[\ln(S)]^2 + 32.087G)/(1 - 0.16247\ln(S)$
$\qquad \qquad - 0.07074G + 0.011717G^2 - 0.0011606G^3)$

$G < -3$ $\qquad$ $\text{CSFC} = -305.94 + 76.547\ln(S) + 2737.7/S - 5493.1\exp(-S)$

$G > 3$ $\qquad$ $\text{CSFC} = (1607 - 986.23\ln(S) + 149.01[\ln(S)]^2 + 84.747G)/(1 - 0.17168\ln(S)$
$\qquad \qquad - 0.021455G)$

Source: FHWA (2002).

Table A7.1.2   VOC for Oil Consumption at Constant Speed (CSOC)

(a) Small Automobile

$G > 0$ and $S < 55$ $\mathrm{CSOC} = \exp(2.7835 - 0.79034\ln(S) - 1.1346/S^{1.5} + 0.65342G^{0.5})$

$G \geq 0$ and $55 \leq S \leq 70$ $\mathrm{CSOC} = -170.4 + 34.02\ln(S) + 1939/S + 0.4747G - 0.003296SG$

$G \geq 0$ and $S \geq 70$ $\mathrm{CSOC} = -170.4 + 34.02\ln(S) + 1939/S + 0.27G$

$G \leq 0$ and $S < 55$ $\mathrm{CSOC} = 1.0435 + (327.89/((1 + (((S + 7.1977)/3.0141)^2))$
$\times (1 + (((G + 8.0484)/2.8984)^2)))))$

Otherwise $\mathrm{CSOC} = -170.4 + 34.02\ln(S) + 1939/S$

(b) Medium-sized/Large Automobile

$G > 0$ and $S < 55$ $\mathrm{CSOC} = \exp(-1.5698 + 9.8768/S^{0.5} - 7.6187S + 0.70702G^{0.5})$

$G \geq 0$ and $55 \leq S < 70$ $\mathrm{CSOC} = 9.5234 - 0.29873S + 0.0026913S^2 + 0.28997G^{1.00129}$

$G \geq 0$ and $S \geq 70$ $\mathrm{CSOC} = -173.3 + 34.6\ln(S) + 1973/S + 0.29G$

$-3 \leq G \leq 0$ and $15 \leq S < 55$ $\mathrm{CSOC} = 0.42295 + 0.35839S - 0.029984S^2 + 0.0010392S^3$
$- 0.000016196S^4 + 9.3539 \times 10^{-8}S^5 - 0.0024G$

$G < -3$ and $15 \leq S < 55$ $\mathrm{CSOC} = 1/(-0.18739 + 0.0014953S^{1.5} - 1.7461/G)$

$G \leq 0$ and $S < 15$ $\mathrm{CSOC} = \exp(1.7713 - 0.12178S^{0.5}\ln(S) + 0.14636G + 0.11002G^2$
$+ 0.0082804G^3)$

Otherwise $\mathrm{CSOC} = -173.3 + 34.6\ln(S) + 1973/S$

(c) Four-Tire Truck

$G > 0$ and $S < 50$ $\mathrm{CSOC} = \exp(2.47 - 0.604\ln(S) - 0.00994GR^2 + 0.277G - 0.001248SG)$

$G > 0$ and $50 \leq S \leq 70$ $\mathrm{CSOC} = 16.41 + 0.004424S^2 - 0.5255S + 1.296G - 0.2664\ln(S)G$

$G > 0$ and $S > 70$ $\mathrm{CSOC} = 16.41 + 0.004424S^2 - 0.5255S + 0.19G$

$\min(-3.5, -S/6.0) < G \leq 0$ and $S < 50$ $\mathrm{CSOC} = 8.45 + 0.0000352S^3 - 0.00567S^2 + 0.370S - 4.12\ln(S)$

$G < \min(-3.5, -S/6.0)$ and $S < 50$ $\mathrm{CSOC} = \exp(0.92 - 0.000295S^2 - 0.751\ln(S) - 0.0269GR^2 - 0.584G)$

Otherwise $\mathrm{CSOC} = 16.41 + 0.004424S^2 - 0.5255S$

(d) Six-Tire Truck

$G > 0$ and $S < 55$ $\mathrm{CSOC} = \exp(3.8424 - 0.93964\ln(S) - 1.7418/S + 0.80327G^{0.5})$

$G > 0$ and $S \geq 55$ $\mathrm{CSOC} = 51.76 + 0.002513S^2 - 14.29\ln(S) + 0.7485G$

$-1.5 < G \leq 0$ and $S < 55$, or $-S/10 \leq G \leq 0$ and $S < 55$ $\mathrm{CSOC} = 13.98 + 0.0000603S^3 - 0.00857S^2 + 0.523S - 6.17\ln(S)$

$G < -S/10$ and $S \geq 70$ $\mathrm{CSOC} = \exp(1.41 + 0.000519S^2 - 0.0845S - 0.0344G^2 - 0.649G)$

Otherwise $\mathrm{CSOC} = 51.76 + 0.002513S^2 - 14.29\ln(S)$

(e) 3+ Axle Single-Unit Truck

$G > 0$ and $S < 55$ $\mathrm{CSOC} = \exp(4.36 + 0.00711S - 0.869\ln(S) - 0.01712GR^2 + 0.338G)$

$\min(-1.5, -S/12.5) < G \leq 0$ and $S < 55$ $\mathrm{CSOC} = 20.2 + 0.0000724S^3 - 0.0103S^2 + 0.662S - 8.52\ln(S)$

$G \leq \min(-1.5, -S/12.5)$ and $S < 55$ $\mathrm{CSOC} = \exp(1.77 + 0.00055S^2 - 0.0769S - 0.0343GR^2 - 0.646G)$

$G > 0$ and $S \geq 55$ $\mathrm{CSOC} = 22.85 + 0.006514S^2 - 0.7188S + 1.615G$

$-S/12.5 \leq G \leq 0$ and $S \geq 55$, or $G \leq 0$ and $S \geq 90$ $\mathrm{CSOC} = 22.85 + 0.006514S^2 - 0.7188S$

Otherwise $\mathrm{CSOC} = \exp(1.77 + 0.00055S^2 - 0.0769S - 0.0343GR^2 - 0.646G)$

(f) 3–4 Axle Combination-Unit Truck

$G > 0$ and $S < 45$ $\mathrm{CSOC} = \exp(3.92 - 0.661\ln(S) - 0.01718GR^2 + 0.361G - 0.000640SG)$

$G > 0$ and $45 \leq S \leq 70$ $\mathrm{CSOC} = 78.59 + 0.003813S^2 - 21.76\ln(S) + 2.1254G - 0.0109SG$

$G > 0$ and $S > 70$ $\mathrm{CSOC} = 78.59 + 0.003813S^2 - 21.76\ln(S) + 1.41G$

$\min(-1.5, -S/12.5) < G \leq 0$, or $G \leq 0$ and $S \geq 70$ $\mathrm{CSOC} = 20.2 + 0.0000724S^3 - 0.01034S^2 + 0.662S - 8.52\ln(S)$

Otherwise $\mathrm{CSOC} = \exp(1.85 + 0.000458S^2 - 0.0746S - 0.0336GR^2 - 0.638G)$

Table A7.1.2   (*continued*)

(g) 5+ Axle Combination-Unit Truck	
$G > 0$ and $S < 55$	$\text{CSOC} = \exp(4.60 - 0.668 \ln(S) - 0.01879 GR^2 + 0.394G - 0.000873SG)$
$G > 0$ and $S \geq 55$	$\text{CSOC} = 9.383 + 0.003478S - 0.271S + 3.040G$
$\min(-1.5, -S/15.0) < G \leq 0$ and $S < 55$	$\text{CSOC} = 42.6 + 0.000189S^3 - 0.0273S^2 + 1.633S - 18.96 \ln(S)$
$\min(-1.5, -S/15.0) < G \leq 0$ and $S \geq 55$	$\text{CSOC} = 9.383 + 0.003478S^2 - 0.271S$
$G \leq \min(-1.5, -S/15.0)$ and $S < 55$	$\text{CSOC} = \exp(2.52 + 0.000397S^2 - 0.0675S - 0.0353GR^2 - 0.652G)$
$G \leq \min(-1.5, -S/15.0)$ and $S \geq 55$	$\text{CSOC} = 115.8 + 0.5094S - 37.27 \ln(S) - 3.064G$

Table A7.1.3   VOC for Tire Wear at Constant Speed (CSTW)

(a) Small Automobile	
$G \geq 2.5$ and $S < 55$	$\text{CSTW} = \exp(-2.55 + 0.0001621S^2 + 0.01441S + 1.473 \ln(G) - 0.001638SG)$
$G \geq 2.5$ and $S \geq 55$	$\text{CSTW} = 1.314 + 0.000733S^2 - 0.05758S + 0.01514G^2 + 0.003997SG$
$0 < G < 2.5$ and $S < 15$, or $-S/20 < G < 2.5$ and $S \geq 15$	$\text{CSTW} = 0.1959 + 2.51 \times 10^{-6}S^3 - 0.0352 \ln(S) + 0.01754G^2 + 0.00348SG$
$-1.5 < G \leq 0$ and $S < 15$, or $-S/10 < G \leq -S/20$ and $S \leq 15$	$\text{CSTW} = 0.0604 + 2.92 \times 10^{-8} \times S^4 + 0.0000796S^2 + 0.0274G^2 + 0.074G + 0.0000568S^2G$
Otherwise	$\text{CSTW} = \exp(-5.39 - 0.000895S^2 + 0.0962G + 2.83 \ln(-G) - 0.00397SG)$

(b) Medium-sized/Large Automobiles	
$G \geq 2.5$ and $S < 55$	$\text{CSTW} = \exp(-2.39 + 0.0001564S^2 + 0.01367S + 1.475 \ln(G) - 0.001586SG)$
$0 < G < 2.5$ and $S < 15$, or $-S/20 \leq G < 2.5$ and $15 \leq S < 55$	$\text{CSTW} = 0.229 + 2.65 \times 10^{-6}S^3 - 0.0403 \ln(S) + 0.0214G^2 + 0.00392SG$
$-1.5 < G \leq 0$ and $S < 15$, or $-S/10 < G < -S/20$ and $15 \leq S < 55$	$\text{CSTW} = 0.08 + 3.0 \times 10^{-6}S^3 + 0.029G^2 + 0.0828G + 0.000056S^2G$
$G \leq -1.5$ and $S < 15$, or $G \leq -S/10$ and $15 \leq S < 55$	$\text{CSTW} = \exp(-5.22 - 0.000771S^2 + 0.0843G + 2.81 \ln(-G) - 0.00323SG)$
$G \geq 0.5$ and $S \geq 55$	$\text{CSTW} = 1.318 + 0.000743S^2 - 0.05661S + 0.01941G^2 + 0.00417SG$
$-S/10 + 1 < G < 0.5$ and $S \geq 55$, or $G < 0.5$ and $S \geq 80$	$\text{CSTW} = -0.2022 + 0.000237S^2 + 0.0213G^2 - 1.0322G + 0.3099 \ln(S)G$
Otherwise	$\text{CSTW} = -0.2613 + 0.000164S^2 + 0.02065G^2 + 0.005452SG - 0.03975 \ln(S)G$

(c) Four-Tire Truck	
$G \geq 2.5$ and $S < 55$	$\text{CSTW} = \exp(-2.08 + 0.0001517S^2 + 0.012S + 1.367 \ln(G) - 0.001389SG)$
$0 < G < 2.5$ and $S < 15$ or $-S/20 < G < 2.5$ and $15 \leq S < 55$	$\text{CSTW} = 0.297 + 2.9 \times 10^{-6}S^3 - 0.0421 \ln(S) + 0.0234G + 0.00429SG$
$-2.5 < G \leq 0$ and $S < 15$ or $-S/10 < G \leq -S/20$ and $15 \leq S \leq 55$	$\text{CSTW} = 0.1294 + 3.64 \times 10^{-6}S^3 + 0.0324G^2 + 0.1085G + 0.0000631S^2G$

(*continued overleaf*)

Table A7.1.3   (*continued*)

$G \leq -2.5$ and $S < 15$ or $G < -S/10$ and $15 \leq S \leq 55$	$CSTW = \exp(-5.45 - 4.13 \times 10^{-6}S^3 - 0.01377S + 2.79\ln(-G))$
$G \geq 0.5$ and $S \geq 55$	$CSTW = 1.365 + 0.000736S^2 - 0.05471S + 0.0197G^2 + 0.004395SG$
$(-S/10 + 1) < G < 0.5$ and $S \geq 55$ or $G < 0.5$ and $S \geq 80$	$CSTW = abs(-0.1554 + 0.000258S^2 + 0.0205G^2 - 0.05138G + 0.005058SG)$
Otherwise	$CSTW = \max(0.01, -0.2177 + 0.000208S^2 + 0.02376G^2 + 0.005895SG$ $-0.03288\ln(S)G)$

(d) Six-Tire Truck

$G \geq 2.5$ and $S < 55$	$CSTW = \exp(-1.572 + 0.0000943S^2 + 0.01509S + 1.65\ln(G) - 0.001535SG)$
$G \geq 2.5$ and $S \geq 55$	$CSTW = 2.206 + 0.001267S^2 - 0.09683S + 0.07733GR^2 + 0.01096SG$
$0 < G < 2.5$ and $S < 15$, or $-S/25 < G < 2.5$ and $S \geq 15$	$CSTW = 0.353 + 4.5 \times 10^{-6}S^3 - 0.0556\ln(S) + 0.0855GR^2 + 0.01012SG$
$-1.5 < G \leq 0$ and $S < 15$; or $-S/14 < G < -S/25$ and $S \geq 15$	$CSTW = 0.104 + 5.37 \times 10^{-8}S^4 + 0.0001578S^2 + 0.1282GR^2 + 0.222G$ $+0.000168S^2G$
Otherwise	$CSTW = \exp(-3.16 - 3.35 \times 10^{-6}S^3 - 0.0308S + 2.28\ln(-G) - 0.00377SG)$

(e) 3 + Axle Single-Unit Truck

$G \geq 2.5$ and $S < 55$	$CSTW = \exp(-1.71 + 0.0000511S^2 + 0.01134S + 1.575\ln(G) - 0.001038SG)$
$G \geq 2.5$ and $S \geq 55$	$CSTW = 1.085 + 0.000405S^2 - 0.03274S + 0.05955G^2 + 0.00577SG$
$-0.5 < G < 2.5$ and $S < 15$, or $-S/30 < G < 2.5$ and $S \geq 15$	$CSTW = 0.0896 + 0.0001308S^2 + 0.0552G^2 + 0.1181G + 0.00402SG$
$-S/20 < G \leq -S/30$ and $S \geq 15$	$CSTW = 0.0345 + 0.000387S^2 + 0.257G^2 + 0.01988SG$
Otherwise	$CSTW = \exp(-3.30 - 0.0275S + 0.1868G + 2.92\ln(-G) - 0.00275SG)$

(f) 3–4 Axle Combination-Unit Truck

$G > 3$	$CSTW = 0.27453 - 0.016411S + 0.090845G + 0.00035502S^2 + 0.047978G^2$ $+0.0042709SG$
$G < -3$	$CSTW = abs(-0.14758 + 0.01337S + 0.0040158G - 0.000053182S^2 + 0.052391G^2$ $+0.0044432SG)$
$-3 \leq G \leq 3$	$CSTW = 0.15566 - 0.0058457S + 0.041763G + 0.00021374S^2 + 0.056992G^2$ $+0.0050156SG$

(g) 5+ Axle Combination-Unit Truck

$G \geq 2.5$ and $S < 55$	$CSTW = \exp(-1.6 + 0.0000684S^2 + 0.00608S + 1.567\ln(G) - 0.000762SG)$
$G \geq 2.5$ and $S \geq 55$	$CSTW = 1.122 + 0.000357S^2 - 0.03264S + 0.06295G^2 + 0.005081SG$
$-0.5 < G < 2.5$ and $S < 15$ or $-S/35 < G < 2.5$ and $S \geq 15$	$CSTW = 0.1432 + 1.248 \times 10^{-6}S^3 + 0.0639G^2 + 0.1167G + 0.00332SG$
$G \geq -S/35$ and $S \geq \max(15, -25G)$	$CSTW = -0.1283 + 1.442 \times 10^{-6}S^3 + 0.01044S + 0.208G^2 + 0.01337SG$
Otherwise	$CSTW = \exp(-3.05 - 1.5 \times 10^{-6}S^3 - 0.01358S + 2.13\ln(-G) - 0.001779SG)$

Table A7.1.4   VOC for Maintenance and Repair at Constant Speed (CSMR)

(a) Small Automobile

$G \geq 0$	$CSMR = 48.3 + 0.00865S^2 + 0.0516SG$
$-1.5 \leq G < 0$ and $S \leq 25$, or $G < 0$ and $25 < S < 55$ and $S \geq -12.2G + 4$	$CSMR = 45.1 + 0.00582S^2 + 0.23S + 0.0502SG$
$G < -1.5$ and $S \leq 25$, or $G < 0$ and $25 < S < 55$ and $S < -12.2G + 4$	$CSMR = -5.83 - 0.01932S^2 - 23.4G$
$-0.14S + 3.6 < G < 0$ and $S \geq 55$	$CSMR = 73.35 + 0.01397S^2 - 0.7398S + 0.04994SG$
Otherwise	$CSMR = 4.27 - 0.0208S^2 - 23.63G$

(b) Medium-Sized/Large Automobile

$G \geq 0$	$CSMR = 48.4 + 0.00867S^2 + 0.0577SG$
$-1.5 \leq G < 0$ and $S \leq 25$, or $G < 0$ and $-12.2G + 4 \leq S$ and $25 < S < 55$	$CSMR = 45.19 + 0.00584S^2 + 0.229S + 0.0562SG$
$G < -1.5$ and $S \leq 25$, or $G < 0$ and $-12.2G + 4 > S$ and $25 < S < 55$	$CSMR = -6.67 - 0.018S^2 - 23.4G$
$-0.14S + 3.6 < G < 0$ and $S \geq 55$	$CSMR = 72.46 + 0.01373S^2 - 0.7081S + 0.05597SG$
Otherwise	$CSMR = -5.415 - 0.01912S^2 - 23.51G$

(c) Four-Tire Truck

$G \geq 0$	$CSMR = 49.2 + 0.00881S^2 + 0.0545SG$
$-1.5 \leq G < 0$ and $S \leq 20$, or $G < 0$ and $20 < S < 55$ and $S \geq -10G + 6$	$CSMR = 46.0 + 0.00595S^2 + 0.231S + 0.0531SG$
$G < -1.5$ and $S \leq 20$, or $G < 0$ and $20 < S < 55$ and $S < -10G + 6$	$CSMR = -12.43 - 0.019S^2 - 23.5G$
$G < 0$ and $S \geq 55$ and $G > -0.1S + 0.75$, or $G < 0$ and $S > 70$	$CSMR = 72.36 + 0.01373S^2 - 0.6841S + 0.0532SG$
Otherwise	$CSMR = -13.83 - 0.0197S^2 - 24.01G$

(d) Six-Tire Truck

$-4 \leq G \leq -1$ and $S > -1.6667G^3 - 17.5G^2$ $-70.833G - 45$ and $S < -1.6667G^3$ $-17.5G^2 - 70.833G - 40$	$CSMR = 1/(0.96223 + 2.3017 \times 10^{-6}S^3 - 0.33129\exp(S/44.4878)$ $+0.48203/G - 0.00029083\exp(-G))$
$G \geq -1$, or $G < -1$ and $S \geq -1.6667G^3$ $- 17.5G^2 - 70.833G - 40)$	$CSMR = 44.2 + 0.01147S^2 + 0.1462SG$
Otherwise	$CSMR = -0.722 - 0.00697S^2 - 15.9G$

(e) 3+ Axle Single-Unit Truck

$-4 \leq G \leq -1$ and $S > -1.6667G^3 - 17.5G^2$ $-75.833G - 45)$ and $S < -1.6667G^3$ $-17.5G^2 - 75.833G - 40)$	$CSMR = 1046.8 - 499.21\ln(S) + 106.76[\ln(S)]^2 + 601.98G$ $+154.36G^2 + 15.039G^3$
$G \geq -1$, or $G < -1$ and $S \geq -1.6667G^3$ $- 17.5G^2 - 75.833G - 40$	$CSMR = 46 + 0.008S^2 + 0.146SG$
Otherwise	$CSMR = 1.6996 + 0.094776S - 0.016324S^2 + 0.00037673S^3$ $-4.0767 \times 10^{-6}S^4 + 1.4984 \times 10^{-8}S^5 - 14.684G$

(f) 3–4 Axle Combination-Unit Truck

$-3 \leq G \leq -1$ and $S \geq (-7.5G^2 - 52.5G$ $- 25)$ and $S < -7.5G^2 - 52.5G - 20$	$CSMR = 169.6 + 6.4867S + 333.98G + 48.825G^2$

(*continued overleaf*)

Table A7.1.4   (*continued*)

| $G > -1$, or $G < -1$ and $G \geq -3$ and $S \geq -7.5G^2 - 52.5G - 20$ | $\text{CSMR} = 46 + 0.008S^2 + 0.146SG$ |
| Otherwise | $\text{CSMR} = 2.44881 - 0.0404901S^{1.5} - 15.8112G$ |

(g) 5+ Axle Combination-Unit Truck

$G \geq 0$	$\text{CSMR} = 44.9 + 0.01148S^2 + 0.254SG$
$G < 0$ and $S > 25$ and $S \geq -40G - 15$	$\text{CSMR} = 78.7 + 1.545S - 20.6\ln(S) + 0.254SG$
Otherwise	$\text{CSMR} = 0.996 - 0.00149S^2 - 15.8G$

Table A7.1.5   VOC for Depreciation at Constant Speed (CSVD)

(a) Small Automobile
$\text{CSVD} = 2.2 + 0.001596S - 0.38\ln(S)$

(b) Medium-Sized/Large Automobile
$\text{CSVD} = 1.725 + 0.001892S - 0.311\ln(S)$

(c) Four-Tire Truck
$\text{CSVD} = 0.742 + 0.000589S - 0.1307\ln(S)$

(d) Six-Tire Truck

| $S < 55$ | $\text{CSVD} = 1.126 + 0.0028S - 0.247\ln(S)$ |
| $S \geq 55$ | $\text{CSVD} = 0.2006 + 4.936/S$ |

(e) 3 + Axle Single-Unit Truck

| $S < 55$ | $\text{CSVD} = 1.126 + 0.00279S - 0.247\ln(S)$ |
| Otherwise | $\text{CSVD} = 0.2006 + 4.936/S$ |

(f) 3–4 Axle Combination-Unit Truck

| $S < 55$ | $\text{CSVD} = 0.354 + 0.000974S - 0.0806\ln(S)$ |
| Otherwise | $\text{CSVD} = 0.05657 + 1.598/S$ |

(g) 5+ Axle Combination-Unit Truck

| $S < 55$ | $\text{CSVD} = 0.395 + 0.001215S - 0.0941\ln(S)$ |
| Otherwise | $\text{CSVD} = 0.05657 + 1.598/S$ |

Table A7.1.6   VOC for Excess Fuel Consumption at Curved Sections (CFC)

(a) Small Automobiles

$S \geq 1/(0.001147 + 0.008062D^{0.5} + 0.008862/D$	$\text{CFC} = \max(0, 18387.7115(1/(1 + (D/39.459)^{-3.0419}))$ $(1)/(1 + (S/104.38)^{-6.2768}))$
$D \geq 6$ and $S > 10$ and $S \leq -0.6807D + 30.944$	$\text{CFC} = \max(0, -0.046905 + 0.95904\ln(D) - 0.02218S$ $- 0.17662[\ln(D)]^2 + 0.000957S^2 - 0.021388S\ln(D))$
$D \geq 6$ and $S > 10$, or $D < 6$ and $S \leq 25$	$\text{CFC} = \max(0, -1.9503 + 1.0112\ln(D) + 0.31328S$ $-0.16763[\ln(D)]^2 - 0.012903S^2 - 0.031507S\ln(D))$
Otherwise	$\text{CFC} = 0$

(b) Medium-Sized/Large Automobile

$D \leq 5$ and $S \leq 1/(-0.0137 + 0.0123D^{0.5}$ $+0.0299/D^{0.5})$	$\text{CFC} = \max(0, -0.34211 + 0.28291D + 0.014828S$ $-0.016971D^2 - 0.00024465S^2 - 0.0047869DS)$
$D > 5$ and $S \leq 1/(-0.0137 + 0.0123D^{0.5}$ $+0.0299/D^{0.5})$	$\text{CFC} = \max(0, -0.79434 + 1.1403\ln(D) + 0.052408S$ $-0.1933[\ln(D)]^2 - 0.00060403S^2 - 0.028889S\ln(D))$
$S > 1/(-0.0137 + 0.0123D^{0.5} + 0.0299/D^{0.5})$	$\text{CFC} = \max(0, \exp(-18.864 - 0.02183D^{1.5}$ $+ 2.6113D^{0.5} + 1.80792S^{0.5}))$

Table A7.1.6   (*continued*)

(c) Four-Tire Truck

$D < 6$ and $S \geq -0.5682D^2 + 0.75D + 55.818$, $CFC = \max(0, \exp(779.63 - 1.2743D + 3.1889D^{0.5}\ln(D)$
    or $D \geq 6$ and $S \geq -0.0055D^2 - 0.7634D$ $-2.9306D^{0.5} - 25.106S^{0.5} - 10,108.5/S^{0.5}$
    $+43.597$ $+10,588.5\ln(S)/S))$

$D \geq 10$ and $S \leq 57.993 + 1.1162D - 13.963D^{0.5}$, $CFC = \max(0, -0.45381 + 0.98231\ln(D) + 0.10049S$
    or $8 \leq D < 10$ and $S \leq 25$ $-0.15[\ln(D)]^2 - 0.0011603S^2 - 0.046122S\ln(D))$

$4 \leq D < 8$ and $S \leq -2.5D + 45$ $CFC = \max(0, -2.0296 + 2.4402\ln(D) + 0.087398S$
                                            $- 0.50234[\ln(D)]^2 - 0.0012841S^2 - 0.036879S\ln(D))$

$2 \leq D < 4$ and $S \leq 35$, or $D < 2$ $CFC = \max(0, \exp(0.0010091 - 5.4673/D^2 - 0.082805S$
    and $S \leq 2.5D + 30$ $+0.011991S^2 - 0.0018375S^{2.5}))$

Otherwise $CFC = 0$

(d) Six-Tire Truck

$D \geq 10$ and $S \geq 27.9 - 0.0144D^2 + 300/D^2$, $CFC = \max(0, \exp(-50.349 - 0.98363D - 0.05974D^2$
    or $2 \leq D < 10$ and $S \geq 1/(0.0127 + 0.00484D$ $+59.476\exp(D/31.649) - 90.158/S^{0.5}))$
    $-0.000675D^2 + 3.97 \times 10^{-5}D^3)$, or $D < 2$
    and $S \geq 1/(0.0286 + 0.00429D - 0.00429D^2)$

$D \geq 5$ and $S \leq 27.9 - 0.0144D^2 + 300/D^2$ $CFC = \max(0, -9.7649 + 7.88\ln(D) + 6.036\ln(S)$
                                          $- 1.0423[\ln(D)]^2 - 1.053[\ln(S)]^2 - 1.464\ln(D)\ln(S))$

$1 < D < 5$ and $S \leq 30 + 70/D - 100/D^2$ $CFC = \max(0, \exp(1.604 - 4.6423/D^{1.5} - 0.000062414S^3))$

Otherwise $CFC = 0$

(e) 3+ Axle Single-Unit Truck

$D \geq 6$ and $S \geq 44.375 - 1.8236D + 0.02044D^2$ $CFC = \exp(371.346 + 5.1878\ln(D) + 10.1521/D^{0.5}$
    $+0.0018571D^2 - 0.000053954D^2$ or $6 > D \geq 2$ $-12.1424S^{0.5} - 4915.79/S^{0.5} + 5093.19\ln(S)/S))$
    and $S \geq 39.5 + 4.1667D - 0.83333D^2$

$D < 2$ and $S \geq 24.5 - 10D + 10D^2$ $CFC = \max(0, 1.3873 + 8.977\exp(-0.5(((D - 2.2124)/1.071)^2$
                                         $+ ((S - 102.44)/16.633)^2)))$

$2 \leq D \leq 16$ and $S < 10$ $CFC = \max(0, -4.0824 + 1.833D - 0.15946D^2 + 0.0044245D^3$
                                         $+ 0.56919S - 0.038513S^2 + 0.00079158S^3)$

$D \leq 1$ and $S < 25$ $CFC = 0$

Otherwise $CFC = \max(0, -8.9743 - 0.099969D + 16.366\ln(S)$
                                         $+ 0.0052265D^2 - 3.6805(\ln(S))^2 - 0.11371D\ln(S))$

(f) 3–4 Axle Combination-Unit Truck

$D < 6$ and $S \leq 20$ $CFC = \max(0, (-0.069855 + 0.4852\ln(D)$
                                         $+ 0.029223\ln(S))/(1 - 0.30752\ln(D)$
                                         $+ 0.10364(\ln(D))^2 - 0.52169\ln(S) + 0.10545(\ln(S))^2))$

$20 < S \leq 64 + 0.93749D - 13.928D^{0.5}$ and $D < 6$ $CFC = \max(0, -36.549 + 8.3919D + 19.444\ln(S)$
                                         $- 0.19172DCA^2 - 2.6623[\ln(S)]^2 - 1.9932D\ln(S))$

$20 < S \leq 64 + 0.93749D - 13.928D^{0.5}$ and $D \geq 6$ $CFC = \max(0, -44.639 + 15.079\ln(D) + 31.738\ln(S)$
                                         $- 1.734[\ln(D)]^2 - 5.305[\ln(S)]^2 - 3.6061\ln(D)\ln(S))$

$D > 1$ and $S \geq 67 + 0.93749D - 13.928D^{0.5}$ $CFC = \max(0, \exp(948,774.18 + 1.056802(\ln(D))^2$
                                         $+ 11,715.15S^{0.5}\ln(S) + 54,041.58(\ln(S))^2$
                                         $- 133,443.12S^{0.5} - 268,395.66\ln(S) + 309,522.12/\ln(S)$
                                         $- 1,311,374.33/(S^{0.5})))$

$D \leq 1$ and $S \geq 67 + 0.93749D - 13.928D^{0.5}$ $CFC = \max(0, (-13.559 - 1.1956D + 0.37772DCA^2$
                                         $+ 3.5166\ln(S))/(1 - 0.37771D + 0.1152DCA^2$
                                         $- 0.1529\ln(S)))$

Otherwise $CFC = 0$

(*continued overleaf*)

Table A7.1.6   (*continued*)

(g) 5 + Axle Combination-Unit Truck

$D < 6$ and $S \le 20$ $\mathrm{CFC} = \max(0, \exp(4.892 - 5.8015/D^2 - 0.070341S - 6.612/S^{0.5}))$

$20 < S \le 64 + 0.93749D - 13.928D^{0.5}$ and $D < 6$ $\mathrm{CFC} = \max(0, -0.76579 + 9.3637\ln(D) + 0.025171S$
$\qquad\qquad - 0.75491[\ln(D)]^2 + 0.00010068S^2 - 0.22116S\ln(\mathrm{D}))$

$S \le 64 + 0.93749D - 13.928D^{0.5}$ and $D \ge 6$ $\mathrm{CFC} = \max(0, -44.672 + 15.308\ln(D) + 31.804\ln(S)$
$\qquad\qquad - 1.8472[\ln(D)]^2 - 5.3075[\ln(S)]^2 - 3.5085\ln(D)*\ln(S))$

$D > 1$ and $S \ge 67 + 0.93749D - 13.928D^{0.5}$ $\mathrm{CFC} = \max(0, \exp(-37.185 - 0.0034062D^2 + 1.262(\ln(D))^2$
$\qquad\qquad - 0.00000046205S^3 + 8.9915\ln(S)))$

$D \le 1$ and $S \ge 67 + 0.93749D - 13.928*D^{0.5}$ $\mathrm{CFC} = \max(0, -3.3518 + 58.52/((1 + ((D - 3.8448)/3.142)^2)$
$\qquad\qquad (1 + ((S - 99.792)/14.486)^2)))$

Otherwise $\mathrm{CFC} = 0$

Table A7.1.7   VOC for Excess Tire Wear at Curved Sections (CTW)

(a) Small Automobile

$D \ge 16$ and $S \ge -0.031746D^2 + 0.74603D + 21.19$, $\mathrm{CTW} = \max(0, 351{,}887\exp(-\exp(-(D - 51.408)/19.756)$
or $16 \ge D \ge 6$ and $S \ge 45 - 1.9167D + 0.041667D^2$, $\qquad - (D - 51.408)/19.756 + 1)$
or $D < 6$ and $S \ge (-442.3 + 2959.4/D^{0.5}$ $\qquad \times \exp(-\exp(-(S - 122.22)/38.201)$
$- 6735.1/D + 6810.6/D^{1.5} - 2582.5/D^2$ $\qquad - (S - 122.22)/38.201 + 1))$

$D \ge 16$ and $S < -0.031746D^2 + 0.74603D + 21.19$, $\mathrm{CTW} = \max(0, -21.508 + 13.474\ln(D) + 19.67\ln(S)$
or $16 > D \ge 8$ and $S < 45 - 1.9167D + 0.041667D^2$ $\qquad - 1.5206(\ln(D))^2 - 3.5315[\ln(S)]^2$
$\qquad - 0.6298\ln(D)\ln(S))$

$8 > D \ge 6$ and $S < 45 - 1.9167D + 0.041667D^2$, $\mathrm{CTW} = \max(0, -3.3578 + 3.5095D + 0.080638S$
or $D < 6$ and $S < -442.3 + 2959.4/D^{0.5}$ $\qquad - 0.18665D^2 - 0.00054297S^2 - 0.061173DS)$
$- 6735.1/D + 6810.6/D^{1.5} - 2582.5/D^2$

(b) Medium-Sized-Large Automobile

$D \ge 16$ and $S \ge -0.031746D^2 + 0.74603D + 21.19$, $\mathrm{CTW} = \max(0, 519{,}464\exp(-\exp(-(D - 48.665)/18.647)$
or $6 \le D < 16$ and $S \ge 45 - 1.9167D + 0.041667D^2$, $\qquad - (D - 48.665)/18.647 + 1)$
or $D < 6$ and $S \ge -442.3 + 2959.4/D^{0.5}$ $\qquad \times \exp(-\exp(-(S - 127.84)/39.862)$
$- 6735.1/D + 6810.6/D^{1.5} - 2582.5/D^2$ $\qquad - (S - 127.84)/39.862 + 1))$

$D \ge 16$ and $S < -0.031746D^2 + 0.74603D + 21.19$, $\mathrm{CTW} = \max(0, -31.7 + 20.767\ln(D) + 22.783\ln(S)$
or $6 \le D < 16$ and $S < 45 - 1.9167D + 0.041667D^2$ $\qquad - 2.5841[\ln(D)]^2 - 3.9522[\ln(S)]^2$
$\qquad - 4.4831\ln(D)\ln(S))$

$D < 6$ and $S < -442.3 + 2959.4/D^{0.5} - 6735.1/D$ $\mathrm{CTW} = \max(0, -4.4955 + 4.542D + 0.088792S$
$+ 6810.6/D^{1.5} - 2582.5/D^2$ $\qquad - 0.27253D^2 - 0.00042329S^2 - 0.07399DS)$

(c) Four-Tire Truck

$D \ge 16$ and $S \ge 0.02381D^2 - 1.4524D + 42.143$, $\mathrm{CTW} = \max(0, 450{,}515\exp(-\exp((D - 49.07)/18.816)$
or $6 \le D < 16$ and $S \ge 35 - 125/D + 750/D^2$, $\qquad -(D - 49.07)/18.816 + 1)$
or $(D < 6$ and $S \ge 23.334 + 112.5/D$ $\qquad \times \exp(-\exp(-(S - 124.84)/38.88)$
$-150.83/D^2 + 25/D^3$ $\qquad -(S - 124.84)/38.89 + 1))$

$D \le 16$ and $S < 0.02381D^2 - 1.4524D + 42.143$ $\mathrm{CTW} = \max(0, -13.126 + 79.095/D + 254.26/S$
$\qquad - 39.567/D^2 - 694.97/S^2 - 217.62/DS)$

$6 \le D < 16$ and $S < 35 - 125/D + 750/D^2$, $\mathrm{CTW} = \max(0, -2.743 + 3.5215D + 0.077273S$
or $D < 6$ and $S < 23.334 + 112.5/D$ $\qquad - 0.16376D^2 - 0.00069592S^2 - 0.064592DS)$
$- 150.83/D^2 + 25/D^3$

(d) Six-Tire Truck

$D \ge 16$ and $S \ge 0.02381D^2 - 1.4524D + 42.143$, $\mathrm{CTW} = \max(0, 377{,}675\exp(-\exp(-(D - 51.703)/19.791)$
or $6 \le D < 16$ and $S \ge 35 - 125/D + 750/D^2$, $\qquad - (D - 51.703)/19.791 + 1)$
or $D < 6$ and $S \ge 23.334 + 112.5/D$ $\qquad \exp(-\exp(-(S - 120.93)/37.611)$
$-150.83/D^2 + 25/D^3$ $\qquad - (S - 120.93)/37.611 + 1))$

Table A7.1.7   (*continued*)

$D \geq 16$ and $S < 0.02381D^2 - 1.4524D + 42.143$, or $6 \leq D < 16$ and $S < 35 - 125/D + 750/D^2$	$CTW = \max(0, -26.586 + 17.42\ln(D) + 19.303\ln(S) - 2.1482[\ln(D)]^2 - 3.3487[\ln(S)]^2 - 3.81\ln(D)\ln(S))$
$D < 6$ and $S < 23.334 + 112.5/D - 150.83/D^2 + 25/D^3$	$CTW = \max(0, -4.0066 + 3.8372D + 0.11043S - 0.2262D^2 - 0.0011358S^2 - 0.064529DS)$

(e) 3+ Axle Single-Unit Truck

$D \geq 16$ and $S \geq 0.02381D^2 - 1.4524D + 42.143$, or $16 > D \geq 6$ and $S \geq 35 - 125/D + 750/D^2$, or $D < 6$ and $S \geq 23.334 + 112.5/D -150.83/D^2 + 25/D^2$	$CTW = \max(0, 707,192\exp(-\exp(-(D - 44.524)/16.77) - (D - 44.524)/16.77 + 1) \times \exp(-\exp(-(S - 132.23)/40.729) - (S - 132.23)/40.729 + 1))$
$D \geq 16$ and $S < 0.02381D^2 - 1.4524D + 42.143$, or $16 > D \geq 6$ and $S < (35 - 125/D + 750/D^2$	$CTW = \max(0, 7.4369 + 29.473/D + 6.5816 * \ln(S) - 541.46/D^2 - 3.8133[\ln(S)]^2 + 45.797(\ln(S))/D)$
$D < 6$ and $S < 23.334 + 112.5/D - 150.83/D^2 + 25/D^3$	$CTW = \max(0, -4.6194 + 4.5401D + 0.10837S - 0.26588D^2 - 0.00099725S^2 - 0.076619DS)$

(f) 3–4 Axle Combination-Unit Truck

$D \geq 16$ and $S \geq (0.02381D^2 - 1.4524D + 42.143$, or $16 > D \geq 6$ and $S \geq 35 - 125/D + 750/D^2$, or $D < 6$ and $S \geq 23.334 + 112.5/D - 150.83/D^2 + 25/D^3$	$CTW = \max(0, 578,653\exp(-\exp(-(D - 54.618)/20.44) - (D - 54.618)/20.44 + 1) \times \exp(-\exp(-(S - 120.41)/37.427) - (S - 120.41)/37.427 + 1))$
$D \geq 16$ and $S < 0.02381D^2 - 1.4524D + 42.143$, or $(16 > D \geq 6$ and $S < 35 - 125/D + 750/D^2$	$CTW = \max(0, -26.305 + 16.264\ln(D) + 20.114\ln(S) - 1.7217[\ln(D)]^2 - 3.4077[\ln(S)]^2 - 4.0945\ln(D)\ln(S))$
$D < 6$ and $S < 23.334 + 112.5/D - 150.83/D^2 + 25/D^3$	$CTW = \max(0, -3.8937 + 3.8291D + 0.092128S - 0.22412D^2 - 0.00082522S^2 - 0.064764DS)$

(g) 5+ Axle Combination-Unit Truck

$D \geq 16$ and $S \geq 0.02381D^2 - 1.4524D + 42.143$, or $16 > D \geq 6$ and $S \geq 35 - 125/D + 750/D^2$, or $1 \leq D < 6$ and $S \geq (23.334 + 112.5/D - 150.83/D^2 + 25/D^3$	$CTW = \max(0, \exp(-40.193 + 14.371\exp(D/ - 53.803) + 1.2303(\ln(D))^2 - 1.8886S/\ln(S) + 7.0737S^{0.5}))$
$D < 1$ and $S \geq 23.334 + 112.5/D - 150.83/D^2 + 25/D^3$	$CTW = \max(0, 1/(1.1442 - 0.015388D^3 - 9704.3/S^{1.5} + 27,917\ln(S)/S^2 - 42,372/S^2))$
$D \geq 16$ and $S < 0.02381D^2 - 1.4524D + 42.143$, or $16 > D \geq 6$ and $S < 35 - 125/D + 750/D^2$	$CTW = \max(0, -27.686 + 18.235\ln(D) + 24.103\ln(S) - 2.2305[\ln(D)]^2 - 4.3932[\ln(S)]^2 - 4.4593\ln(D)\ln(S))$
$D < 6$ and $S < (23.334 + 112.5/D - 150.83/D^2 + 25/D^3)$	$CTW = \max(0, -4.9124 + 4.8372D + 0.12051S - 0.2845D^2 - 0.0011691S^2 - 0.08169DS)$

Table A7.1.8   VOC for Excess Maintenance and Repair at Curved Sections (CMR)

(a) Small Automobile

$D > 10$ and $S \geq -0.65D + 34.5$, or $D < 10$ and $S \geq -2.4444D + 52.444$	$CMR = \max(0, \exp(-19.624 - 1.0614D^{0.5}\ln(D) + 6.4853D^{0.5} + 0.033374S^{1.5} - 0.00046284S^2\ln(S)))$
$5 \leq D \leq 10$ and $(-1D + 20) \leq S \leq (-1D + 25)$	$CMR = 0.1$
Otherwise	$CMR = 0$

(b) Medium-sizes/Large Automobile

$D > 12$ and $S \geq -0.5D + 30$, or $D < 12$ and $S \geq -2.3636D + 52.364$	$CMR = \max(0, \exp(-37.927 + 3.2935\ln(D) + 1.8096/D + 7.8477\ln(S)))$

(*continued overleaf*)

Table A7.1.8   (*continued*)

(c) Four-Tire Truck

$5 \leq D \leq 10$ and $-1D + 20 \leq S \leq -1D - 25$ — CMR $= 0.1$

Otherwise — CMR $= 0$

$D \geq 12$ and $S \geq -0.45D + 30.4$, or $D < 12$
and $S \geq -2.2727D + 52.273$ — CMR $= \max(0, \exp(594.56 - 0.021279D^{1.5} + 2.6656D^{(0.5)}$
$-19.444S^{0.5} - 7777/S^{0.5} + 8121.8\ln(S)/S))$

$3.5 < D < 8.5$ and $17.5 \leq S < 22.5$,
or $4.5 < D < 10.5$ and $12.5 \leq S \leq 17.5$,
or $7.5 < D < 12.5$ and $7.5 < S \leq 12.5$ — CMR $= 0.1$

Otherwise — CMR $= 0$

(d) Six-Tire Truck

$D \geq 8$ and $S \geq -0.0038D^2 - 0.3106D + 27.272$ — CMR $= \max(0, \exp(9.6157 + 0.12975D - 157.95/D^2$
$+ 7095.5\exp(-D) - 106.49\ln(S)/S))$

$D < 8$ and $S \geq -0.625D^2 + 3.125D + 40$, — CMR $= \max(0, \exp(-314.6 + 2.5973D\ln(D) - 1.4569D^2$
$+ 0.30227D^{2.5} + 2642/\ln(S) - 2565.9/S^{0.5}))$

$1 \leq D \leq 3$ and $-10D + 37.5 \leq S \leq 10D + 2.5$, or
$3 < D \leq 5$ and $S < 32.5$, or $5 < D < 8$ and
$S < 0.8333D^2 - 14.167D + 85$ or $D \geq 8$ and
$S < -0.0038D^2 - 0.3106D + 24.5$ — CMR $= 0.1$

$D > 4.5$ and $12.5 < S < -0.35D^2 + 3.85D + 12$ — CMR $=$ CMR $+ 0.1$

Otherwise — CMR $= 0$

The fourth equation for six-tire truck excess maintenance and repair due to curves is incremental; the condition is true when certain other conditions are true and the equations adds value to the existing CMR value.

(e) 3+ Axle Single-Unit Truck

$D \geq 10$ and $S \geq -0.75D + 40$,
or $10 > D \geq 2$ and $S \geq -2.1875D + 54.375$
or $D < 2$ and $S \geq (5D^2 - 2.5D + 35$ — CMR $= \max(0, \exp-50.038 + 0.71092[\ln(D)]^2$
$+ 0.50522\ln(D) - 0.08522S + 13.02\ln(S)))$

$1 \leq D \leq 3$ and $10D + 2.5 \geq S \geq -10D + 37.5$, or
$3 < D \leq 5$ and $S \leq 32.5$, or $D \geq 14$ and
$S \leq -0.3125D + 21.875$, or $5 < D < 14$ and
$S \leq -1.66667D + 40.8333$ — CMR $= 0.1$

$4.5 \leq D \leq 10.5$ and $S > 39.3 - 13.497D$
$+ 2.215D^2 - 0.11833D^3$ and $S < 17.5 + 5/$
$(1 + \exp(-((D - 7.0222)/ - 0.07845))))$ — CMR $=$ CMR $+ 0.1$

Otherwise — CMR $= 0$

The third equation for 3+ axle single-unit truck excess maintenance and repair due to curves is used to increment the value for CMR derived from the second equation under certain conditions. The conditions for this may be true when the condition for the second equation is true.

(f) 3–4 Axle Combination-Unit Truck

$D \geq 17.5$ and $S \geq -0.4D + 29.5$,
or $17.5 > D \geq 2.5$ and $S \geq -1.5D + 48.75$,
or $D < 2.5$ and $S \geq 4D + 35$ — CMR $= \max(0, \exp(304.96 - 0.90108D + 2.0321D^{0.5}\ln(D)$
$- 0.70003\ln(D) - 41.773\ln(S)$
$- 1312.1/S^{0.5} + 2080.7/S))$

$D \leq 3$ and $-6.667D^2 + 38.33D + -22.5 \geq S \geq$
$6.667D^2 - 38.33D + 62.5$, or $3 < D \leq 6$ and
$S \leq 32.5$, or $6 < D < 10$ and $S \leq -2.5D + 47.5$,
or $D \geq 10$ and $S \leq -0.5D + 27.5$ — CMR $= 0.1$

$3.5 < D < 6.5$ and $12.5 < S < 22.5$,
or $5.5 < D < 12.5$ and $7.5 < S < 17.5$ — CMR $=$ CMR $+ 0.1$

Otherwise — CMR $= 0$

Table A7.1.8   (*continued*)

The third equation for 3–4 axle combination-unit truck excess maintenance and repair due to curves is, under certain conditions, used to increment the CMR value derived from the second equation. Under some circumstances, the conditions for both the second and third and equations will be true.

(g) 5+ Axle Combination-Unit Truck

$D \geq 10$ and $S \geq (-0.5D + 32.5)$,
   or $10 > D > 3$ and $S \geq -2.8571D + 56.071$,
   or $D < 3$ and $S \geq 6.6667D + 27.5$

$$\text{CMR} = \exp(703.2 + 0.75135[\ln(D)]^2 - 1.3433/D^{0.5} - 62.464\ln(S) - 2045.3/\ln(S) + 3128.1/S))$$

$1.5 \leq D \leq 3$ and $5D + 22.5 > S > -5D + 17.5$, or $16 \leq D \leq 25$ and $S \leq 17.5$, or $D > 25$ and $S \leq -1D + 42.5$, or $3 < D < 16$ and $S \leq -1.5385D + 42.115$

$$\text{CMR} = 0.1$$

$2.5 \leq D \leq 6.5$ and $12.5 \leq S \leq 27.5$,
   or $3.5 \leq D \leq 10.5$ and $7.5 \leq S \leq 22.5$,
   or $9.5 \leq D \leq 15$ and $2.5 < S \leq 17.5$,
   or $15 \leq D \leq 17$ and $7.5 \leq S \leq 17.5$
   or $17 \leq D < 22.5$ and $7.5 \leq S \leq 12.5$

$$\text{CMR} = \text{CMR} + 0.1$$

$3.5 < D < 6.5$ and $17.5 \leq S < 22.5$,
   or $4.5 < D < 10.5$ and $12.5 \leq S \leq 17.5$,
   or $(7.5 < D < 10.5$ and $7.5 < S \leq 12.5)$

$$\text{CMR} = \text{CMR} + 0.1$$

Otherwise

$$\text{CMR} = 0$$

The third and fourth equations for 5+ axle combination-unit truck excess maintenance and repair due to curves are, under certain conditions, used to increment the CMR value derived from the second equation. These conditions for one or both of these equations may be true when the condition for the second equation is true.

Table A7.1.9   **Excess VOC for Fuel Consumption due to Speed Variability (SCCFC)**

(a) Small Automobile
$\text{CSMAX} < 5$ $\text{SCCFC} = 0.00424\text{CSMAX}^3$
$\text{CSMAX} \geq 5$ $\text{SCCFC} = 0.04547 + 0.08559\text{CSMAX} + 3677 \times 10^{-8}\text{CSMAX}^3$

(b) Medium-Sized/Large Automobile
$\text{CSMAX} < 5$ $\text{SCCFC} = 0.008\text{CSMAX}^3$
$\text{CSMAX} \geq 5$ $\text{SCCFC} = 0.03401 + 0.1902\text{CSMAX} + 4491 \times 10^{-8}\text{CSMAX}^3$

(c) Four-Tire Truck
$\text{CSMAX} < 5$ $\text{SCCFC} = 0.00904\text{CSMAX}^3$
$\text{CSMAX} \geq 5$ $\text{SCCFC} = 0.8137 + 0.1576\text{CSMAX} + 7327 \times 10^{-8}\text{CSMAX}^3$

(d) Six-Tire Truck
$\text{CSMAX} < 5$ $\text{SCCFC} = 0.1184\text{CSMAX}^2$
$\text{CSMAX} \geq 5$ $\text{SCCFC} = 3.09 + 0.02843\text{CSMAX}^2$

(e) 3+ Axle Single-Unit Truck
$\text{CSMAX} < 5$ $\text{SCCFC} = 0.174\text{CSMAX}^2$
$\text{CSMAX} \geq 5$ $\text{SCCFC} = 4.477 + 0.03862\text{CSMAX}^2$

(f) 3–4 Axle Combination-Unit Truck
$\text{CSMAX} < 5$ $\text{SCCFC} = 0.324\text{CSMAX}^2$
$\text{CSMAX} \geq 5$ $\text{SCCFC} = 6.342 + 0.5855\text{CSMAX} + 0.03191\text{CSMAX}^2$

(g) 5+ Axle Combination-Unit Truck
$\text{CSMAX} < 5$ $\text{SCCFC} = 0.3584\text{CSMAX}^2$
$\text{CSMAX} \geq 5$ $\text{SCCFC} = 2.052 + 1.167\text{CSMAX} + 0.03292\text{CSMAX}^2$

Table A7.1.10   Excess VOC for Oil Consumption due to Speed Variability (SCCFC)

(a) Small Automobile
CSMAX < 5 $SCCOC = 0.00004CSMAX^3$
CSMAX $\geq$ 5 $SCCOC = 0.000879 + 0.000934CSMAX - 1612 \times 10^{-8}CSMAX^2 + 193 \times 10^{-9}CSMAX^3$

(b) Medium-sized/Large Automobile
CSMAX < 5 $SCCOC = 0.00004*CSMAX^3$
CSMAX $\geq$ 5 $SCCOC = 0.000801 + 0.000869CSMAX - 1617 \times 10^{-8}CSMAX^2 + 197 \times 10^{-8}CSMAX^3$

(c) Four-Tire Truck
CSMAX < 5 $SCCOC = 0.0002CSMAX^2$
CSMAX $\geq$ 5 $SCCOC = \exp(-6.242 + 0.5935\ln(CSMAX) + 0.000131CSMAX^2)$

(d) Six-Tire Truck
CSMAX < 5 $SCCOC = 0.00068*CSMAX^2$
CSMAX $\geq$ 5 $SCCOC = \exp(-5.069 + 0.6392\ln(CSMAX) + 0.000169CSMAX^2)$

(e) 3+ Axle Single-Unit Truck
CSMAX < 5 $SCCOC = 0.00136*CSMAX^2$
CSMAX $\geq$ 5 $SCCOC = \exp(-4.408 + 0.6632\ln(CSMAX) + 0.000148CSMAX^2)$

(f) 3–4 Axle Combination-Unit Truck
CSMAX < 5 $SCCOC = 0.00136*CSMAX^2$
CSMAX $\geq$ 5 $SCCOC = \exp(-4.408 + 0.6632\ln(CSMAX) + 0.000148CSMAX^2)$

(g) 5+ Axle Combination-Unit Truck
CSMAX < 5 $SCCOC = 0.0028*CSMAX^2$
CSMAX $\geq$ 5 $SCCOC = \exp(-3.735 + 0.6849\ln(CSMAX) + 0.000112CSMAX^2)$

Table A7.1.11   Excess VOC for Tire Wear due to Speed Variability (SCCFC)

(a) Small Automobile
CSMAX < 5 $SCCTW = 0.0008CSMAX^2$
CSMAX $\geq$ 5 $SCCTW = \exp(-7.112 + 1.999\ln(CSMAX) - 8384 \times 10^{-8}CSMAX^2)$

(b) Medium-Sized/Large Automobile
CSMAX < 5 $SCCTW = 0.0012CSMAX^2$
CSMAX $\geq$ 5 $SCCTW = \exp(-6.64 + 1.947\ln(CSMAX) - 9909 \times 10^{-8}CSMAX^2)$

(c) Four-Tire Truck
CSMAX < 5 $SCCTW = 0.0012CSMAX^2$
CSMAX $\geq$ 5 $SCCTW = \exp(-6.568 + 1.906\ln(CSMAX) - 7502 \times 10^{-8}CSMAX^2)$

(d) Six-Tire Truck
CSMAX < 5 $SCCTW = 0.0016CSMAX^2$
CSMAX $\geq$ 5 $SCCTW = \exp(-6.387 + 1.984\ln(CSMAX) - 988 \times 10^{-7}CSMAX^2)$

(e) 3+ Axle Single-Unit Truck
CSMAX < 5 $SCCTW = 0.0012CSMAX^2$
CSMAX $\geq$ 5 $SCCTW = \exp(-6.595 + 1.918\ln(CSMAX) - 6855 \times 10^{-8}CSMAX^2)$

(f) 3–4 Axle Combination-Unit Truck
CSMAX < 5 $SCCTW = 0.0008CSMAX^2$
CSMAX $\geq$ 5 $SCCTW = \exp(-7.111 + 2.0276\ln(CSMAX) - 0.000102CSMAX^2)$

(g) 5+ Axle Combination-Unit Truck
CSMAX < 5 $SCCTW = 0.0012CSMAX^2$
CSMAX $\geq$ 5 $SCCTW = \exp(-6.643 + 1.947\ln(CSMAX) - 721 \times 10^{-7}CSMAX^2)$

Table A7.1.12   Excess VOC for Maintenance and Repair due to Speed Variability (SCCFC)

(a) Small Automobile
CSMAX < 5 $SCCMR = 0.0016CSMAX^2$
CSMAX ≥ 5 $SCCMR = \exp(-6.284 + 0.006889CSMAX + 1.881\ln(CSMAX) - 7388 \times 10^{-8}CSMAX^2)$

(b) Medium-Sized/Large Automobile
CSMAX < 5 $SCCMR = 0.0016CSMAX^2$
CSMAX ≥ 5 $SCCMR = \exp(-6.277 + 0.007347CSMAX + 1.876\ln(CSMAX) - 7275 \times 10^{-8}CSMAX^2)$

(c) 4-Tire Truck
CSMAX < 5 $SCCMR = 0.0016CSMAX^2$
CSMAX ≥ 5 $SCCMR = \exp(-6.39 + 1.958\ln(CSMAX) - 1781 \times 10^{-8}CSMAX^2)$

(d) 6-Tire Truck
CSMAX ≤ 5 $SCCMR = 0.0012CSMAX^2$
CSMAX > 5 $SCCMR = \exp(-6.427 + 0.01826CSMAX + 1.758\ln(CSMAX) - 0.000103CSMAX^2)$

(e) 3+ Axle Single-Unit Truck
CSMAX < 5 $SCCMR = 0.0008CSMAX^2$
CSMAX ≥ 5 $SCCMR = \exp(-7.446 - 0.005514CSMAX + 2.212\ln(CSMAX) + 5075 \times 10^{-8}CSMAX^2)$

(f) 3–4 Axle Combination-Unit Truck
CSMAX < 5 $SCCMR = 0.0012CSMAX^2$
CSMAX ≥ 5 $SCCMR = \exp(-6.639 + 0.006003CSMAX + 1.912\ln(CSMAX))$

(g) 5+ Axle Combination-Unit Truck
CSMAX < 5 $SCCMR = 0.0012CSMAX^2$
CSMAX ≥ 5 $SCCMR = \exp(-6.705 + 0.008136CSMAX + 1.94\ln(CSMAX))$

Table A7.1.13   Excess VOC for Depreciation due to Speed Variability (SCCD)

(a) Small Automobile
CSMAX < 60 $SCCD = 0.0004CSMAX$
CSMAX ≥ 60 $SCCD = \exp(-4.327 + 0.000168CSMAX^2)$

(b) Medium-Sized/Large Automobile
CSMAX < 5 $SCCD = 0.0004CSMAX$
5 ≤ CSMAX < 50 $SCCD = 0.001 + 0.0002CSMAX$
CSMAX ≥ 50 $SCCD = \exp(-4.973 + 0.000228CSMAX^2)$

(c) 4-Tire Truck
CSMAX < 60 $SCCD = 0.0002CSMAX$
CSMAX ≥ 60 $SCCD = \exp(-5.0007 + 0.000162CSMAX^2)$

(d) 6-Tire Truck
CSMAX < 5 $SCCD = 0.0004CSMAX$
5 ≤ CSMAX < 40 $SCCD = 0.001429 + 0.000221CSMAX$
CSMAX ≥ 40 $SCCD = \exp(-4.957 + 0.000294CSMAX^2)$

(e) 3+ Axle Single-Unit Truck
CSMAX < 5 $SCCD = 0.0006CSMAX$
5 ≤ CSMAX < 55 $SCCD = 0.001 + 0.0004CSMAX$
CSMAX ≥ 55 $SCCD = \exp(-4.439 + 0.000231CSMAX^2)$

(*continued overleaf*)

Table A7.1.13   (*continued*)

(f) 3–4 Axle Combination-Unit Truck	
CSMAX < 60	SCCD = 0.0002CSMAX
CSMAX ≥ 60	SCCD = exp($-5.007 + 0.000162$CSMAX^2)
(g) 5+ Axle Combination-Unit Truck	
CSMAX < 60	SCCD = 0.0002CSMAX
CSMAX ≥ 60	SCCD = exp($-5.007 + 0.000162$CSMAX^2)

APPENDIX A7.2: VOC COMPONENT UNIT COSTS

Table A7.2.1   Unit VOC Component Costs in 2005 Dollars

	Fuel ($/gal)	Oil ($/qrt)	Tires ($/tire)	Maintenance and Repair ($/1000 mi)	Depreciation Value ($)
Automobiles					
Small	1.89	4.32	54.71	101.80	21,929
Medium-Sized/Large	1.89	4.32	86.54	123.58	25,865
Trucks					
Single unit, four-tires	1.05	4.32	95.38	157.11	27,873
Single unit, six-tires	1.05	1.73	230.10	294.01	41,650
Single unit, 3+ axles	0.92	1.73	569.74	415.77	91,630
Combination, 3–4 axles	0.92	1.73	569.74	430.66	106,140
Combination, 5+ axles	0.92	1.73	569.74	430.66	115,411

Source: Updated from HERS (FHWA, 2002).

APPENDIX A7.3: PAVEMENT CONDITION ADJUSTMENT FACTORS

Table A7.3.1   Constant-Speed Operating Costs, Pavement Condition Adjustment Factors[a]

(a) General Equation[b]

	Vehicle Class		
		Trucks	
VOC Component	Four-Tire Vehicles	Single-Unit	Combination
Maintenance and repair (PCAFMR)	$3.19 + 0.0967$PSR^2 − 0.961PSR	$1.724 + 0.00830$PSR^2 − 0.661 ln(PSR)	$2.075 + 0.273$PSR − 1.622 ln(PSR)
Depreciation (PCAFVD)	$1.136 − 0.106$ ln(PSR)	$1.332 − 0.262$ ln(PSR)	$1.32 − 0.254$ ln(PSR)
Oil consumption (PCAFOC)	$2.64 + 0.0729$PSR^2 − 0.722PSR	$1.176 − 0.1348$ ln(PSR)	
Tire wear (PCAFTW)	$2.40 − 1.111$ ln(PSR)	$1.668 + 0.001372$PSR^3 − 0.581 ln(PSR)	

Source: FHWA (2002).

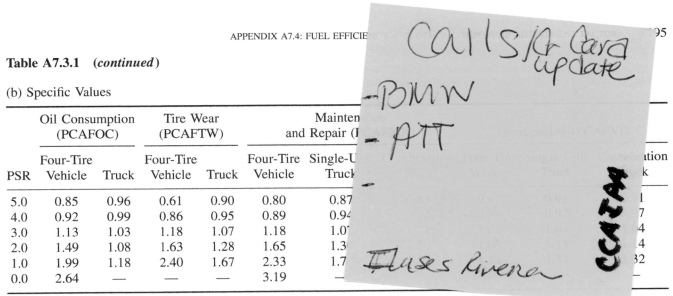

Table A7.3.1 (*continued*)

(b) Specific Values

	Oil Consumption (PCAFOC)		Tire Wear (PCAFTW)		Mainten... and Repair (...		
PSR	Four-Tire Vehicle	Truck	Four-Tire Vehicle	Truck	Four-Tire Vehicle	Single-U... Truck	
5.0	0.85	0.96	0.61	0.90	0.80	0.87	
4.0	0.92	0.99	0.86	0.95	0.89	0.94	
3.0	1.13	1.03	1.18	1.07	1.18	1.0	
2.0	1.49	1.08	1.63	1.28	1.65	1.3	
1.0	1.99	1.18	2.40	1.67	2.33	1.7	
0.0	2.64	—	—	—	3.19	—	

Source: FHWA (2002).

[a]PCAFOC, pavement condition adjustment factor for oil consumption; PCAFTW, pavement condition adjustment factor for tire wear; PCAFMR, pavement condition adjustment factor for maintenance and repair; PCAFVD, pavement adjustment factor for depreciation expenses.

[b]PSR $= 5e^{-0.26\text{IRI}}$ when IRI is in mm/m, or $= 5e^{-0.0041\text{IRI}}$ when IRI is in in/mi (Al-Omari and Darter, 1994).

APPENDIX A7.4: FUEL EFFICIENCY AND OTHER COMPONENT ADJUSTMENT FACTORS

Table A7.4.1 Fuel Efficiency Adjustment Factors

Vehicle Class	Factor
Automobiles (All Sizes)	1.536
4-Tire Trucks	1.596
6-Tire Trucks	1.207
3+ Axle and Combination Trucks	1.167

Source: FHWA (2002).

Table A7.4.2 Adjustment Factors for Other VOC Components

VOC Component	Factor
Fuel (FEAF)	1.536
Oil (OCAF)	1.050
Tire (TWAF)	1.000
Maintenance and Repair (MRAF)	1.000
Depreciation (VDAF)	1.300

Source: FHWA (2002).

CHAPTER 8

Economic Efficiency Impacts

Hatred of costs can often be more intense than love of benefits.

—*Lord Bertrand Russell (1872–1970)*

INTRODUCTION

In Chapter 4 we presented procedures for planning-level estimation of the costs of construction, preservation, and fixed-facility operations, as well as other project costs typically borne by a transportation agency or operator. In Chapters 5 to 7 we discussed procedures for evaluating the monetary benefits of transportation investments from the perspectives of specific performance measures (travel time, safety, and vehicle operation). For a given transportation problem, there are typically several alternative decisions or actions, each with its unique set of costs and benefits. The combined monetary cost and benefit impact of each alternative can be represented by a performance measure known as *economic efficiency*, which is derived using the principles of economic analysis. Economic analysis is a decision-making tool that assesses the efficiency of investments from a monetary standpoint and incorporates the monetized costs and benefits associated with alternative decisions and actions. Across alternatives, differences in the amounts and timings of costs and benefits are likely to influence the relative attractiveness of such alternatives even if the initial investment requirements are not very different. Decisions to select the best of several alternative actions are encountered at every stage of the transportation project development process, and such choices are often made on the basis of economic considerations. As such, economic efficiency analysis (often referred to as *benefit–cost analysis*) can help guide transportation decision making in the various areas of design, construction, preservation, and operations.

8.1 INTEREST EQUATIONS AND EQUIVALENCIES

The fundamental principle underlying all engineering economic efficiency analyses is that the value of money is related directly to the time at which the value is considered. A given amount of money at the current time is not equivalent to the same amount at a past or future year, due to the combined forces of inflation and opportunity cost that erode the value of money over time. *Inflation* refers to the increase in prices of goods and services with time and is reflected by a decrease in the purchasing power of a given sum of money with time. *Opportunity cost* is the income that is foregone at a later time by not investing a given sum of money at a current period. In an engineering economic analysis of alternatives, all monetary amounts are in constant dollars. Inflation is not considered on the assumption that all costs and benefits of various alternatives are affected equally by inflation. If there is reason to believe that future component prices will be affected differently, appropriate adjustments should be made to reflect the differential impact of inflation (AASHTO, 1977).

8.1.1 Cash Flow Illustrations

The time stream of amounts of money that occur within a given period can be displayed either as a *cash flow table* or a *cash flow diagram*. On a cash flow table, there are two columns: one for time and the other for amount. On a cash flow diagram, time is represented on a horizontal axis, while vertical arrows depict the inflow or outflow of money at various points in time. The sign of the amount and the direction of the arrows in cash flow tables and figures, respectively, indicate the movement of the amount. A popular convention is to represent money "coming in" (i.e., returns or benefits) by positive signs and upward arrows pointing away from the horizontal time line in the cash flow table and diagram, respectively. This convention also stipulates that money "going out" (i.e., disbursements or costs) is represented by negative signs and downward arrows in the cash flow table and diagram, respectively. The entire *payment period* (often referred to as the *analysis period, planning horizon*, or *planning period*) is represented by the interval between the present time (often denoted by time $= 0$) and the end of the period (denoted by time $= N$). The planning period is typically divided into a number of equal periods called *compounding periods*. Each period is typically taken as one year.

8.1.2 The Concept of Interest

The amount by which a given sum of money differs from its future value is typically represented as *interest*.

Borrowed money to be paid back to a financial institution at a future time must comprise the initial amount (principal) plus interest. This reflects the fact that the value of the initial amount is not the same as the value of the amount at the time of payback. Interest is therefore described as the price of borrowing money, or simply, the time value of money, and the change in interest over time is referred to as the *interest rate*. The interest rate is used to determine the future value of a present sum or cash flow and the *discount rate* is the interest rate used in determining the present value of a future sum or cash flow.

A 10% annual interest rate indicates that for every dollar borrowed in the initial year, 10 cents must be paid as interest at the end of each year. Central banks typically control interest rates to remedy current or expected economic problems. For instance, in a sluggish economy, the U.S. Federal Reserve Board decreases interest rates to discourage saving and encourage individual spending and business investments; interest rates are increased when the economy is overheated. In a stable economy such as those of developed countries, interest rates are typically lower than those in economies with high inflation and a low certainty of investment returns.

8.1.3 Types of Compounding and Interest Rates

The interest rate associated with a borrowed sum of money could take many forms, such as being simple or compound, discrete, continuous, fixed, or variable (Figure 8.1). In *simple interest* computations, the amount of interest at the end of each period is the same, as each of such amounts are a *fixed* percentage of the initial amount. The amount of compound interest in a given period is the interest charged on the total amount owed at the end of the preceding period (i.e., the sum of the principal and the previous period's amount of interest). Therefore, amounts borrowed on compound interest involve higher payments for amortization. In the current business environment, interest is typically computed using compound interest rates.

Values of interest that are computed only at the end of each compounding period and with a constant interest rate are typically referred to as *fixed periodic rates*. In such instances, there is only one compounding period (e.g., a fixed annual rate refers to an interest rate with a one-year compounding period). In many cases, the compounding period is less than one year (quarterly, monthly, or weekly). In the financial environment, it is customary to quote interest rates on an annual basis followed by the compounding period if different from one year in length. For example, a case where the interest rate is 5% per interest period and the interest period is six months may be described as "10% compounded semiannually." In this case, the annual rate of interest (10%) is referred to as the *nominal interest rate*. Close examination of

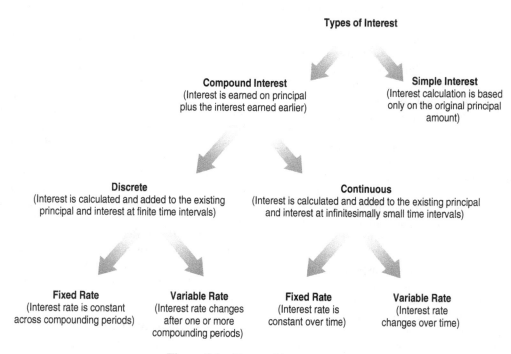

Figure 8.1   Types of interest rates.

this case would show that the actual annual rate on the principal is not 10% but a rate that exceeds 10% because compounding occurs twice during the year. The actual or exact rate of interest earned on the principal during one year is known as the *effective interest rate*, and can be computed using the equation $r_e = (1 + r_n/m)^m - 1$, where r_n is the nominal interest rate per year, r_e the effective interest rate (when compounding occurs m times during the year), and m the number of interest periods in a year (because $m > 0$, $r_e \geq r_m$). For example, if compounding occurs every four months (i.e., three times a year) and the nominal annual interest rate is 5%, then the effective annual interest rate is: $(1 + 0.05/3)^3 - 1 = 5.08\%$.

8.1.4 Interest Equations and Key Variables

Interest equations, also referred to as *equivalency equations*, are relationships between amounts of money that occur at different points in time and are used to estimate the worths of a single amount of money or a series of monetary amounts from one time period to another to reflect the time value of money. The key components of such relationships are the interest factors which are functions of the interest rate and the payment period. Interest factors are expressed as a formula (see Tables 8.1 and A8.1) or as a table of values derived from such a formula (see Table A8.3) and are provided separately for discrete compounding and continuous compounding of the interest rate. Table A8.1 presents interest factors for an annual series that follow an arithmetic gradient pattern for the case of discrete compounding. In some cases the analyst may be faced with nonuniform but systematic annual payments that will need conversion to a uniform annual series to facilitate the analysis. The functions for converting a few selected nonuniform series to their equivalent uniform annual series are presented in Table A8.2. Interest equations typically involve the following five key variables: P, the initial amount (at time = 0); F, the amount at a specified future period (at time = N); A, a periodic (typically at the end of each year) amount; i, the effective interest rate for the compounding period; and N, a specified number of compounding periods, or the *analysis period*.

(*a*) *Analysis Period*   The analysis period, which is often referred to as the *project time horizon*, needs to be determined prior to the economic efficiency analysis of a project. The selection of an appropriate analysis period involves trade-off between two considerations (Dickey and Miller, 1984).

1. On one hand, a long analysis period often seems appropriate because transportation facilities are typically designed to provide service for generations, and it is often preferable to select an analysis period that is equal or close to the service life; otherwise, the often problematic issue of accounting for remaining service lives or residual values of the facility and of its user costs may arise.

2. On the other hand, when the analysis period selected is too long, the effect of discounted facility preservation and maintenance costs and user benefits over the analysis period would overwhelm the initial costs and may render initial cost amounts insignificant. Second, long analysis periods may be unrealistic in cases where the regional or national economy is prone to a high degree of uncertainty due to fluctuating economic trends (which invalidate the interest rate values used for the analysis), political upheavals (particularly in some developing countries), delays in starting or completing the project construction, and technological changes in facility rehabilitation and maintenance that change the benefits and costs associated with annual and periodic expenditure streams from previously established values used for the initial analysis.

As such, in selecting an analysis period, factors that need to be taken into account include the project type (and consequently, the length of service life), the variation in service lives of alternative investment options, the nature of the regional or national economy (developed vs. developing), the forecast uncertainties, the social discount rate and its stability, the rate of technological change, possible competing/complementary facilities, and the likelihood of construction or implementation time delays.

In cases where competing alternatives have different service lives, the best economic efficiency criterion to use is the equivalent uniform annual values of costs and benefits. It may also be possible to express the analysis period as a common multiple of the service lives of the investment alternatives (a similar replacement is assumed when each alternative reaches the end of its service life, this continuing until the end of the analysis period). Where the analysis period does not equal (or cannot be expressed as) a common multiple of the service lives of competing investment options, a replacement cycle to perpetuity may be assumed. Also, in the case where the same analysis period must be used for alternatives with different service lives, some alternatives would involve a residual value at the end of the analysis period, and such values would need to be translated into monetary values so that they can be considered fully in the economic efficiency analysis.

Table 8.1 Interest Equations for Discrete and Continuous Compounding[a]

Description	Cash Flow Diagram	Computational Formula	Factor Computation
Finding the future compounded amount (F) at the end of a specified period given the initial amount (P) and interest rate		$F = P \times$ SPCAF Single payment compound amount factor, SPCAF $(i\%, N)$	$\text{SPCAF} = (1 + i)^N$ $\text{SPCAF} = e^{Ni}$
Finding the initial amount (P) that would yield a given future amount (F) at the end of a specified period given the interest rate		$P = F \times$ SPPWF Single payment present worth factor, SPPWF $(i\%, N)$	$\text{SPPWF} = \dfrac{1}{(1 + i)^N}$ $\text{SPPWF} = \dfrac{1}{e^{Ni}}$
Finding the uniform yearly amount (A) that would yield a given future amount (F) at the end of a specified period given the interest rate		$A = F \times$ SFDF Sinking fund deposit factor, SFDF $(i\%, N)$	$\text{SFDF} = \dfrac{i}{(1 + i)^N - 1}$ $\text{SFDF} = \dfrac{e^i - 1}{e^{Ni} - 1}$
Finding the future compounded amount (F) at the end of a specified period due to annual payments (A) given the interest rate		$F = A \times$ USCAF Uniform series compound amount factor, USCAF $(i\%, N)$	$\text{USCAF} = \dfrac{(1 + i)^N - 1}{i}$ $\text{USCAF} = \dfrac{e^{Ni} - 1}{e^i - 1}$
Finding the initial amount (P) that is equivalent to a series of uniform annual payments (A) given the interest rate		$P = A \times$ USPWF Uniform series present worth factor, USPWF $(i\%, N)$	$\text{USPWF} = \dfrac{(1 + i)^N - 1}{i(1 + i)^N}$ $\text{USPWF} = \dfrac{1 - e^{-Ni}}{e^i - 1}$
Finding the amount of uniform yearly payments (A) that would completely recover an initial amount (P) at the end of a specified period, given the interest rate		$A = P \times$ CRF Capital recovery factor, CRF $(i\%, N)$	$\text{CRF} = \dfrac{i(1 + i)^N}{(1 + i)^N - 1}$ $\text{CRF} = \dfrac{e^i - 1}{1 - e^{-Ni}}$

[a] In the fourth column, upper and lower equations are for discrete and continuous compounding, respectively. For fixed discrete compounding yearly, $i =$ nominal interest rate and N represents the number of years. When there is more than one compounding period per year, the equations and tables can be used as long as there is a cash flow at the end of each interest period. In that case, i represents the interest rate per period and N is the number of periods. When the compounding is more frequent than a year, but the cash flows are annual, the equation can be used with N as number of years and i as the effective annual interest rate. Interest factors may be computed using equations provided on this page or read from Table A8.3.

(b) *Interest Rate*   Use of low interest rates for economic efficiency analysis tends to favor alternatives with high initial costs or with benefits occurring far off in the later years of the analysis period because benefits and costs that lie farther into the future receive more weight than do those that are more imminent. For a similar reason, using high interest rates tends to favor alternatives with low initial costs and/or have benefits that mostly occur early in the analysis period. As such, the value chosen as the interest rate has a profound influence on the outcome of economic efficiency evaluation.

The "real" (constant-dollar) interest rates used for economic efficiency impact evaluation of transportation investments typically range from 4 to 8%. The U.S. Office of Management and Budget recommends the use of a rate of 7% to represent the private-sector rate of return on capital investment. However, other agencies typically use lower rates to take cognizance of the social rate of time preference. A recent survey found that a 4% rate has been used for many years by the Army Corps of Engineers (which had the effect of favoring projects with long service lives or with net benefits occurring many years into the "future"), a 5% rate for several states, a 7% rate by the United Kingdom Department of Transport, and an 8% rate by the British Columbia Ministry of Transportation and Highways (Weisbrod, 2000).

(c) *Residual Value*   In some economic efficiency studies, the analysis period is not equal to the service life of the facility. As such, there is some finite *residual value* of the facility remaining at the end of the analysis period, and such a "benefit," or negative cost, needs to be taken into consideration in the analysis. In some cases, an agency may incur net residual costs at the end of the analysis period, often due to salvage or disposal expenses. The two fundamental components of residual value are the *remaining service life* and the *salvage value*.

The remaining service life (RSL) of a facility at the end of the analysis period is the additional time during which the facility can still provide acceptable levels of service. Failure to account for different RSL values across alternatives can result in bias in the evaluation. Salvage value is the value of recovered or recycled materials and assumes that the transportation facility (or component thereof) is removed from service or replaced at the end of the analysis period. A difference between RSL and salvage value is that the former is used for evaluation in cases where the transportation facility continues to operate at the end of the analysis period, whereas the latter is used when the end of the analysis period coincides with the termination of the facility.

Example 8.1   Five years from now, an airport authority intends to rehabilitate its runways at a cost of $1.5 million. Ten years from now, the runways will be replaced. At that time, the salvage value of reclaimable materials will be $0.75 million. Assuming an interest rate of 8%, find the combined present worth of these costs.

SOLUTION

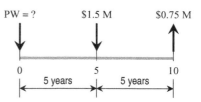

Figure EX8.1

$$PW = 1.5M \times SPPWF\ (8\%, 5)$$
$$- 0.75M \times SPPWF\ (8\%, 10) = \$347,400.$$

Example 8.2   A major corridor investment is expected to yield $50,000 per year in reduced crash costs, $20,000 per year in reduced vehicle operating costs, and $405,000 per year in reduced travel-time costs. What is the combined present worth of these benefits? Assume that the interest rate is 5%; the analysis period is 20 years; and salvage value is $1.0 million

SOLUTION

$$PW = (50,000 + 20,000 + 405,000) \times USPWF\ (5\%, 20)$$
$$+ 1,000,000 \times SPPWF\ (5\%, 20) = \$6.30M$$

(d) *Interest Equations for Continuous Compounding of the Interest Rate*   In some cases of economic evaluation, not only is the interest rate compounded several times within a year, but it is possible for the frequency of compounding of such rates (and consequently, their periods) to be so many that the number of compounding periods can be considered infinite. Consider the general case of an investment where i is the nominal interest rate per year and m is the number of interest periods in a year. This means that the interest rate per compounding period is given by i/m. The continuously compounded value of a single amount P after n years is given by

$$F = P\left(1 + \frac{i}{m}\right)^{mn} = P\left(1 + \frac{i}{m}\right)^{(m/i)in}$$

But $\lim_{m\to\infty}(1+i/m)^{m/i}=e$. Therefore, $F=Pe^{in}$, where e^{in} is defined as the continuously compounded amount factor for the case of infinitely multiple compounding periods. Similarly, the continuously discounted value of a single future amount F, after n years, is given by $P=(F/e^{in})$. The factor $1/e^{in}$ is defined as the continuously discounted factor for the case of infinitely multiple discounting periods. For the case of an infinite number of compounding periods in a year, the effective annual interest rate is given by

$$\frac{F-P}{P}=\frac{P(e^i)-P}{P}=e^i-1$$

Other special cases of interest equations are presented in Section 8.1.5.

8.1.5 Special Cases of Interest Equations

Most problems encountered in economic efficiency analysis for transportation decision making can be solved using the interest equations presented in Table 8.1. However, there are some variations of the problem, such as when periodic payments are being made to perpetuity, when there are infinite compounding periods in a year or when payments are not only compounded continuously but are being made with interest that is also continuously compounded. These special cases are discussed below.

(a) Present Worth of Periodic Payments in Perpetuity Consider the case of a transportation facility with a life cycle of N years, as shown in Figure 8.2. All postconstruction investments made during the life cycle can be compounded into a single amount, R. If it is assumed that the facility will be kept in service to perpetuity, then the life-cycle investment, R, will be repeated at every N-year period. The period N is assumed to be constant for this discussion, but N could be increasing or decreasing

with time, depending on the level of use and technological advances. Increasing levels of use would generally translate to decreasing values of N with time, and vice versa. Also, increasing quality of construction or preservation materials and other inputs would generally lead to increasing values of N. As with most transportation facilities, it is assumed that the initial investment (P) is not the same as the periodic investments (R), as the latter typically involves reconstruction, rehabilitation, and maintenance. A case in point is water port construction (where the initial investment includes right-of-way acquisition, geotechnical treatments, deck construction, dredging, etc., while recurring investments may involve dock structural rehabilitation and dredging). Another example is in highway construction, where the initial investment includes right-of-way acquisition, embankment construction, relocation of utilities, wetlands restoration, and other costs that are typically not found in the recurring investments of pavement resurfacing or reconstruction.

The present worth of all payments in perpetuity is given by

$$
\begin{aligned}
\mathrm{PW}_\infty \\
&= P+\frac{R}{(1+i)^N}+\frac{R}{(1+i)^{2N}}+\frac{R}{(1+i)^{3N}}+\cdots\\
&= P+R\left[\frac{1}{(1+i)^N}+\frac{1}{(1+i)^{2N}}+\frac{1}{(1+i)^{3N}}+\cdots\right]\\
&= P+R\left[\frac{1}{1-1/(1+i)^N}-1\right]=P+\frac{R}{(1+i)^N-1}
\end{aligned}
$$

In cases where the facility already exists, P is a sunk cost, and the present worth is then equal to $R/\big((1+i)^N-1\big)$.

Example 8.3 A new airport runway will cost $7.2 million to construct, including design, land acquisition, and other initial costs. It is expected that every 40

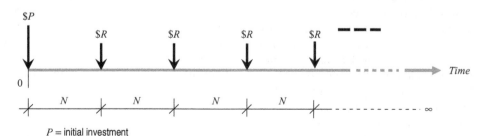

P = initial investment
R = compounded amount of all cash flows within a replacement life cycle
N = length of replacement life cycle of the facility

Figure 8.2   Present worth of periodic payments in perpetuity.

years, the runway will be reconstructed at a cost of $3.1 million. Calculate the present worth of the initial and all reconstruction costs to perpetuity. Assume $i = 4\%$.

SOLUTION

$$\text{PW}_\infty = 7.2\text{M} + \frac{3.1}{(1 + 0.04)^{40} - 1} = \$7.46\text{M}$$

Example 8.4 After several decades of service, a railway bridge is slated for reconstruction. The estimated service life of the structure is 60 years. The reconstruction cost is $600,000. During its replacement cycle, the bridge will require two rehabilitation events, each costing $200,000, at the twentieth and fortieth years and the average annual cost of maintenance is $5000. At the end of the replacement cycle, the bridge will again be reconstructed and the entire cycle is assumed to recur to perpetuity. What is the present worth of all bridge agency costs in perpetuity? Assume an interest rate of 5%.

SOLUTION   All costs within the life cycle of the bridge are illustrated as follows:

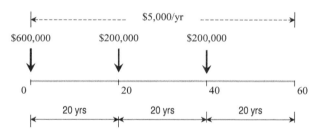

R = compounded life-cycle cost

$= 600{,}000\text{SPCAF}(5\%, 60) + 200{,}000\text{SPCAF}(5\%, 40)$

$\quad + 200{,}000\text{SPCAF}(5\%, 20)$

$\quad + 5000\text{USCAF}(5\%, 60) = \$14{,}914{,}087$

Present worth of all costs in perpetuity ($\text{PW}_{5\%,\infty}$):

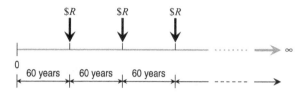

$$\text{PW}_\infty = \frac{R}{(1 + i)^N - 1} = \frac{14{,}914{,}087}{(1 + 0.05)^{60} - 1} = \$843{,}596$$

(*b*) *Present Worth of Continuously Compounded Payments with Continuously Compounded Interest*   Another special case of economic evaluation involves exponentially increasing costs or benefits with continuously compounded interest (Figure 8.3). Consider a general case where the initial amount, R_0, grows exponentially at a rate of r expressed as a percentage per year. Bringing all future streams to the present gives the present worth of $R_0 = R$, the present worth of $R_1 = Re^r/e^i$, the present worth of $R_2 = Re^{2r}/e^{2i}$, and the present worth of $R_n = Re^{nr}/e^{ni}$. Summing up the values of all present worth yields

$$\begin{aligned} \text{PW}_{\text{CCP,CCI}} &= R\left[1 + \frac{e^r}{e^i} + \frac{e^{2r}}{e^{2i}} + \cdots + \frac{e^{nr}}{e^{ni}}\right] \\ &= R\left[\frac{e^{n(r-i)} - 1}{(r - i)}\right] \qquad r < i \end{aligned}$$

the present worth of continuously compounded payments with continuously compounded interest.

Example 8.5 The average annual cost of operating the physical infrastructure of a small airport facility is currently $100,000. Due to the growth in air traffic, the annual costs are compounded continuously at 3% per annum. What is the present worth of the operating costs

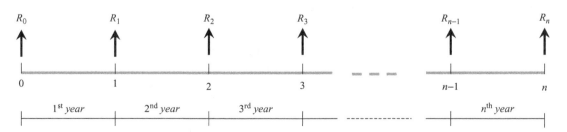

Figure 8.3 Present worth of continuously compounded payments with continuously compounded interest.

over a period of 10 years? Assume a 10% interest rate that is compounded continuously.

SOLUTION

$$PW_{CCP,CCI} = 100,000 \left[\frac{e^{10(0.03-0.1)} - 1}{0.03 - 0.1} \right] = \$719,160$$

8.2 CRITERIA FOR ECONOMIC EFFICIENCY IMPACT EVALUATION

After the present and future sums representing the benefits and/or costs of the relevant performance measures (expenses for facility construction preservation, and operation, and savings in safety, travel time, and vehicle operation) have been brought to their present worth or annualized, the question then is: How are they used to assess the economic efficiency of a proposed project? There are several criteria for doing this:

- Present worth of costs (PWC)
- Equivalent uniform annual cost (EUAC)
- Equivalent uniform annual return (EUAR)
- Net present value (NPV)
- Internal rate of return (IRR)
- Benefit–cost ratio (BCR)

These criteria are also sometimes referred to as *indicators* or *measures of economic efficiency*. The first two criteria are applicable only when all alternatives are associated with a similar level of benefits and cost minimization is therefore the sole evaluation criterion.

8.2.1 Present Worth of Costs
This method converts all costs of a transportation project into an equivalent single cost assumed to occur at the beginning of the analysis period.

Example 8.6 An airplane purchase is proposed by an airline. The initial cost of airplane type A is $50 million, the average annual maintenance cost is $0.25 million, and the salvage value will be $8 million. For airplane type B, the initial cost is $30 million, the average annual maintenance cost is $0.75 million, and the salvage value will be $2 million. Both types have a useful life of 15 years. Which alternative should be selected? Assume a 7% interest rate.

SOLUTION

$$PW_{C_A}(\text{in millions}) = 50 + 0.25USPWF(7\%, 15)$$
$$- 8SPPWF(7\%, 15) = \$49.38M$$

$$PW_{C_B}(\text{in millions}) = 30 + 0.75USPWF(7\%, 15)$$
$$- 2SPPWF(7\%, 15) = \$36.11M$$

Alternative B is more desirable because it has a lower present worth of life-cycle costs.

8.2.2 Equivalent Uniform Annual Cost
This method combines all of the costs of a transportation project into an equivalent annual cost over an analysis period of n years. This method is useful when alternatives have different analysis periods or when they have similar levels of effectiveness.

Example 8.7 Bus transit services in MetroCity can be performed satisfactorily using any one of two alternative bus types, A and B. Type A has an initial cost of $100,000, an estimated life of six years, annual maintenance and operating costs of $8000, and $20,000 salvage value. Type B has an initial cost of $75,000, an estimated life of five years, annual maintenance and operating costs of $8000 for the first two years and $12,000 for the remaining four years, and $10,000 salvage value. Find the equivalent annual cost of each alternative, and decide which option is more desirable. Assume a 6% interest rate.

SOLUTION

$$EUAC_A(\text{thousands}) = 100CRF(6\%, 6) + 8USPWF(6\%, 6)$$
$$\times CRF(6\%, 6) - 20SFDF(6\%, 6)$$
$$= \$25.47$$
$$EUAC_B(\text{thousands}) = 75CRF(6\%, 5) + 8USPWF(6\%, 2)$$
$$\times CRF(6\%, 6) + 12USPWF(6\%, 4)$$
$$\times SPPWF(6\%, 2) \times CRF(6\%, 6)$$
$$- 40 SFDF(6\%, 6) = \$22.57$$

Alternative B is more desirable because it has a lower value of equivalent uniform annual cost.

8.2.3 Equivalent Uniform Annual Return
The EUAR method combines all costs and benefits or returns associated with a transportation project into a single annual value of return (benefits less costs) over the analysis period. This method can be used when the alternatives have different levels of costs and different levels of benefits, or different analysis periods.

Example 8.8 Two alternative designs are proposed for renovating a water port. Alternative A involves an initial project cost of $200 million, an estimated life of 25 years, a salvage value of $22 million, annual maintenance and operating costs of $15 million, and annual benefits of $50 million in terms of monetized savings in inventory delay, safety and security, and vessel operations. Alternative B has an initial project cost of $175 million, an estimated life of 25 years, annual maintenance and operating costs of $16 million, a salvage value of $15 million, and annual benefits of $40 million. Find the equivalent uniform annual return of each alternative and identify the alternative that should be undertaken. Assume a 4% interest rate.

SOLUTION

$$EUAR_A(\text{millions}) = 50 - 200CRF(4\%, 25)$$
$$- 15 + 22SFDF(4\%, 25) = \$22.73M$$
$$EUAR_B(\text{millions}) = 40 - 175CRF(4\%, 25)$$
$$- 16 + 15SFDF(4\%, 25) = \$13.6M$$

Alternative A is more desirable because it has a higher equivalent annual return.

8.2.4 Net Present Value

The NPV of an investment is the difference between the present worth of benefits and that of costs. NPV reflects the value of the project at the time of the base year of the analysis, which may be considered the year of decision making. NPV is often considered as the most appropriate of all economic efficiency indicators because it provides a magnitude of net benefits in monetary terms. If a project involves borrowing or obtaining equity capital, then the the interest required to obtain the funds should be considered a cost. Among competing transportation projects or policies, the alternative with the highest NPV is considered the most "economically efficient."

Example 8.9   For the problem in Example 8.8, determine the net present value for each alternative.

SOLUTION

$$NPV_A(\text{millions}) = 50USPWF(4\%, 25)$$
$$- 200 - 15USPWF(4\%, 25) + 22SPPWF(4\%, 25)$$
$$= \$355M$$
$$NPV_B(\text{millions}) = 40USPWF(4\%, 25)$$
$$- 175 - 16USPWF(4\%, 25) + 15SPPWF(4\%, 25)$$
$$= \$206M$$

Alternative A is more desirable because it has a higher net present value.

8.2.5 Internal Rate of Return

Agencies that seek to invest money in a project ask themselves whether their investment will pay back a net rate of return that is greater than some minimum acceptable rate or whether it will yield a net profit before within a given period of time. The smaller the acceptable rate of return, the longer investors are willing to wait to see a net profit, and vice versa. The *minimum attractive rate of return* (MARR) is the lowest rate of return that investors will accept before they invest, considering the likely investment risks or the opportunity to invest elsewhere for possibly greater returns. MARR is related (inversely) to the *payback period* (the time taken for an investment to pay back to the investors a particular outlay such as their initial investment).

An economic rate of return is defined as the *vestcharge*, that is, the interest rate at which the net present worth or equivalent uniform annual return is equal to zero. The internal rate of return (IRR) method determines the interest rate that is associated with a zero net present value (NPV) and is consequently associated with an equivalency of the present worth of benefits and present worth of costs. Then the IRR is compared to the minimum attractive rate of return (MARR). If the IRR exceeds the MARR, the investment is considered worthwhile. Considering the general case discussed earlier, the IRR value is found by equating the present worth of benefits to the present worth of costs, or by equating the equivalent uniform annual benefits with the equivalent uniform annual costs.

Example 8.10   An urban rail transit agency is considering the purchase of a new $30,000 ticketing system that will reduce travel time. The estimated life of the system is 10 years, at which time the value of the system will be $15,000. The expected travel-time savings per year is $5000 per year, and the average annual maintenance and operating cost is $2000. Is the project economically more desirable than the do-nothing alternative? The minimum attractive rate of return is 5%.

SOLUTION   Equating the net cash flow on both sides, we have:

$$5000USPWF(i\%, 10) + 15,000SPPWF(i\%, 10)$$
$$\approx 30,000 + 2000USPWF(i\%, 10)$$

Solving this equation by trial and error yields $i = 6.25\% > 5\%$. It is, therefore, economically more efficient to undertake the project than the do-nothing alternative.

8.2.6 Benefit–Cost Ratio

The *benefit–cost ratio* (BCR) is a ratio of the equivalent uniform annual value (or net present value) of all benefits to that of all costs incurred over the analysis period. An investment with a BCR exceeding 1 is considered to be economically feasible, and the alternative with the highest BCR value is considered the best alternative.

Example 8.11 For the port problem in Example 8.8, determine the benefit–cost ratio for each alternative.

SOLUTION

$$BCR_A = \frac{PWB_A}{PWC_A}$$
$$= \frac{50USPWF(4\%, 25) + 22SPPWF(4\%, 25)}{200 + 15USPWF(4\%, 25)}$$
$$= 1.93$$
$$BCR_B = \frac{PWB_B}{PWC_B}$$
$$= \frac{40USPWF(4\%, 25) + 15SPPWF(4\%, 25)}{175 + 16USPWF(4\%, 25)}$$
$$= 1.48$$

Alternative A is economically more efficient because it has a higher benefit cost ratio.

Certain procedures recommend that maintenance costs be considered as negative benefits. Using such an approach, the maintenance costs appear as a negative value added to the numerator of a benefit–cost ratio function. On the other hand, certain agencies such as the U.S. Office of Management and Budget (OMB) recommend a different treatment whereby maintenance costs are recognized as an element of the total life-cycle cost and implicitly appear as costs added to the denominator of the overall costs in the benefit–cost ratio (BCR) function.

The U.S. Flood Control Act of 1936 was probably the first instance where reference was explicitly made to the BCR concept in public project evaluation. By definition, any project with a positive NPV will also have a B/C ratio exceeding 1. However, projects with relatively high levels of benefits and costs have a higher NPV than those with smaller benefits and costs but may have higher or lower B/C ratios. Because of its inherent ambiguities, the BCR method is generally not recommended for transportation evaluation unless all B/C ratios are accompanied with explicit values of benefits and costs.

8.2.7 Evaluation Methods Using Incremental Attributes

The foregoing discussion pertained to determining the values of an economic efficiency performance criterion (benefit, cost, IRR, benefit–cost ratio, etc.) associated with each individual investment. The best investment is which yields the "most desired" value of the performance criterion. In public projects, benefits represent savings in user costs. Consequently, mutually exclusive projects require an incremental approach including pairwise comparisons. In the incremental approach, a particular investment, generally the least cost or do-nothing, is taken as the base case or base alternative. The approach in this pairwise comparison method is to determine if the incremental gain in benefit justifies the additional cost.

Example 8.12 Three alternative congestion mitigation projects are being considered for an urban freeway corridor. The costs associated with the alternatives are given in Table E8.12. Assume an interest rate of 5% and an analysis period of 20 years. Which of the alternatives would you recommend on the basis of economic efficiency?

Table E8.12 Project Cost Data

Alternative Costs	A: Road Widening	B: HOV Facility	C: ITS (Ramp Metering and Incident Management)
Initial cost ($1000s)	8352	8400	4500
Annual maintenance and operation ($1000s)	20	563	1000
Annual user cost ($1000s)	1670	1100	1750

SOLUTION   Comparing B with A yields

NPV_{B-A}(thousands)

= present worth of user cost savings

– present worth of additional costs

= $(1,670 - 1,100)USPWF(5\%, 20)$

$- [(8400 - 8352) + (563 - 20)USPWF(5\%, 20)]$

= $7103 - (48 + 6766) = \$289$

Therefore Alternative B is a better than alternative A. Comparing B with C yields

$\text{NPV}_{\text{B-C}}$(thousands)

$$= (1750 - 1100)\text{USPWF}(5\%, 20)$$

$$- [(8400 - 4500) + (563 - 1000)\text{USPWF}(5\%, 20)]$$

$$= 8100 - 3900 + 5446 = \$9{,}646$$

Alternative B is a better alternative than alternative C. Therefore, of the three alternatives, B is the most economically efficient.

8.2.8 General Discussion of Economic Efficiency Criteria

Each economic efficiency analysis method has its unique logic, merits, and demerits. The equivalent uniform annual cost and present value of costs methods are applicable only when all competing alternatives are associated with the same level of service, and therefore the monetary value of benefits are similar across all alternatives. The equivalent uniform annual return and net present value methods consider the benefits and are therefore appropriate where competing alternatives have significant and very different levels of service. NPV, which is expressed as a monetary value and not a rate, ratio, or index, provides a readily comprehensible magnitude of the net benefit of an investment. For this reason, NPV is the method recommended by many agencies. Like the NPV and EUAR, the IRR method considers both benefits and costs. Also, no assumption is needed about the interest rate, although the minimum attractive rate of return must be specified. The main disadvantage of the IRR method is that a unique solution may not always be guaranteed. Many multilateral agencies, including the World Bank, have used the IRR method for project appraisal. The benefit–cost ratio duly considers both benefits and costs but is susceptible to the problems of any ratio-based index: different values of BCR may be obtained depending on the definition, units, and dimensions of the benefits and costs. Most important, BCR does not provide any indication of the total extent of benefit. Another issue associated with economic efficiency is that of relative weights: most analyses proceed on the assumption that all monetary values have the same weight regardless of source. Some agencies, however, assign weights on monetary amounts depending on the source, such as agency costs and user costs (see Chapter 18).

8.3 PROCEDURE FOR ECONOMIC EFFICIENCY ANALYSIS

The overall framework for carrying out an economic efficiency analysis on the basis of benefit and cost considerations is illustrated as Figure 8.4, which is a

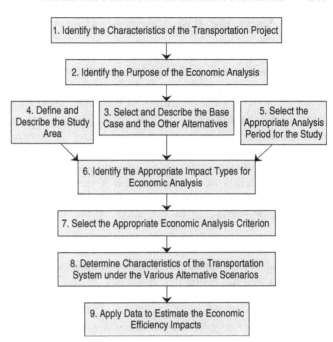

Figure 8.4 Framework for economic efficiency impact evaluation.

synthesis of the procedures presented by Booz Allen Hamilton (1999), Cambridge Systematics (2000), and Forkenbrock and Weisbrod (2001).

Step 1: Identify the Characteristics of the Transportation Project   This step involves preliminary work for the analysis, such as identifying the scope of the project (passenger or freight or both), the mode (highway, transit, rail, etc.), the flow unit (trains, buses, trucks, airplanes, boats, etc.), the network feature (terminals or routes), and the scale (i.e., specific site location, specific strip or corridor, or an entire systemwide area such as city or county). The type of transportation improvement for each alternative is then identified. Examples include upgrading or expanding the existing facility, maintaining its services or providing a new facility, mode, or service. Finally, the purpose of the project should be identified, such as addressing an existing congestion problem, meeting expected future demand, and generating new economic development.

Step 2: Identify the Purpose of the Analysis   There could be several reasons for an agency's efforts to evaluate the economic efficiency impacts of transportation projects. The specific purpose of the analysis is largely influenced by the type and purpose of the overall project itself. The purpose of the analysis, in turn, influences the framework and methodology for the analysis.

Step 3: Select the Base Case and Transportation Alternatives

(a) *Base Case* The base case should be chosen such that it allows a realistic representation of past, current, or possible future conditions. A base case can be "do-nothing" or the current condition, or the least-cost option under future conditions.

(b) *Alternative Scenario* The base case and alternative scenarios may differ by travel mode, facility type, service type facility location (in terms of setting, routing, or alignment), facility or service size (in terms of capacity and cost), area served or the expected change in the level or quality of service to be provided by the facility or service.

Step 4: Definite and Describe the Study Area The factors that should be considered in selecting a primary study area for an economic efficiency analysis are as follows:

- *The area of jurisdiction for the sponsoring agency.* This could be the agency responsible for project funding, project spending (implementation), project evaluation, or a combination thereof. The study area could be a corridor, neighborhood, city, county, state, or province.
- *The area of direct project influence.* Whether the project involves a route/line or a specific terminal facility, the "area of direct influence" includes the area in which facility users or the community are affected.

Step 5: Select the Appropriate Analysis Period for Study Section 8.1.4 discussed factors that need to be taken into account in selecting the analysis period. The analysis period selected should primarily be long enough to distinguish between the costs and benefits between alternatives.

Step 6: Select the Appropriate Impact Types for the Evaluation Economic efficiency impacts are evaluated on the basis of the monetized equivalents of individual impact types, including (1) direct revenue (toll receipts and other out-of-pocket costs, etc.), and (2) impact types that can be monetized (e.g., travel time, vehicle operating costs, safety). The selection of impact types may also be influenced by other factors, such as the preferences of the sponsoring agency and the availability of data.

Step 7: Select the Appropriate Economic Efficiency Criterion for the Evaluation Factors that affect the selection of an appropriate economic efficiency criterion include the size of the project (and consequently, the levels of benefits and costs), the variation between the levels

of service (therefore, the benefits) of competing projects, the preferences of the sponsoring agency, the relationship between the facility service life, and the analysis period. In carrying out an economic efficiency analysis, it is vital to identify correctly the benefits and costs associated with each alternative so that the benefits and costs are not unduly over- or understated (Wohl and Hendrickson, 1984).

Step 8: Determine the Characteristics of the Transportation System under Various Alternative Scenarios In this step, the analyst investigates how each alternative investment will affect users in terms of travel time, vehicle operating cost, and safety, besides other direct costs incurred and benefits accrued. Existing transportation planning software packages, such as QRS-II, Tranplan, EMME2, and MINUTP, may be used to construct transportation network simulations and analysis, and to forecast the demand and user impacts of changes to the transportation system. These packages consider both the supply and demand for transportation.

(a) *Supply-Side Modeling* The user impacts of transportation projects are typically analyzed using either a full simulation model or a sketch planning model that includes the affected project or corridor. Such a model should cover all travel modes of interest, and there should be data on the transportation system supply under alternative scenarios, such as capacity, projected vehicle volumes and trip distribution patterns, and system performance (i.e., the resulting travel times and costs for users of the affected travel modes, routes or links, and terminals or transfer points).

1. *Full simulation models* estimate traffic patterns, volumes, and travel times for each link and node of the network. Simulation models are applicable to situations where trip diversion and rerouting are components of user impacts and are also useful for separately estimating user impacts for various user categories (e.g., different types of businesses) that have significantly different origin–destination patterns.
2. *Sketch planning models* are generally used where there are relatively few routes or modal diversion alternatives to be considered, such as a transportation alternative that affects access in a downtown area. Typically capable of implementation using a simple spreadsheet, sketch planning models are used to estimate the volumes and travel time and cost impacts of localized transportation improvements.

Another aspect of user cost estimation under variations in supply-side conditions relates to work zones associated with the construction or preservation activities of the facility.

(b) *Demand-Side Analysis* To complement data obtained from the supply-side analysis, information on travel demand (such as trip volumes by origin–destination combination) at the base year and at future years are necessary for each alternative transportation investment scenario. Using data from the supply and demand sides, the simulation or sketch planning model is used to estimate changes in volumes, travel times, travel costs, and volume–capacity ratios for the affected portions of the transportation system for each scenario. In Chapter 3, which deals with transportation demand and supply, we discuss how this step could be carried out.

Step 9: Apply the Data to Calculate the Economic Efficiency Impacts The calculation of project costs and user benefits involves the use of average values from available databases or models.

(a) *Project Costs* These costs typically consist of (1) right-of-way, rolling stock, and construction costs; (2) operating costs; and (3) maintenance costs. They are also typically referred to as *agency costs* and are discussed in Chapter 4.

(b) *Project User Benefits* The primary user benefit components are (1) savings in vehicle operating cost (2) travel-time savings, and (3) increased safety. Such benefits are realized over the entire project life and generally grow with the increasing travel volumes. In certain circumstances of the system operation, such as construction/rehabilitation work zones or congestion conditions, however, there may actually be an increase in certain user costs compared to normal operations, but these are typically short-lived.

The cost, time, and safety benefits for existing and diverted trips should be estimated directly. For induced trips (which would otherwise not occur), there is typically no relevant travel time or safety benefit. It is naturally expected that there would be some benefit for diverted and induced trips (trip makers enticed to switch to the new facility), which, nevertheless, does not exceed the benefits associated with the existing trips (otherwise, people would have switched in the preimplementation situation). Within these extremes, the exact magnitude of benefits for induced trips may vary depending on the nature of the improvement alternative relative to existing alternatives. When it is not possible to estimate accurately the benefits of induced trips, it is recommended that they be estimated to be roughly one-half of the per-trip benefit accruing for existing trips (Forkenbrock and Weisbrod, 2001; AASHTO, 2003). In economic terms, that is equivalent to a consumer surplus concept, in which

there is a linear demand response in terms of willingness to pay for increasing benefit levels.

8.4 SOFTWARE PACKAGES FOR ECONOMIC EFFICIENCY ANALYSIS

8.4.1 Surface Transportation Efficiency Analysis Model

FHWA's STEAM package assesses the economic efficiency of physical investments in multimodal urban transportation infrastructure as well as policy alternatives such as pricing and demand management measures (Cambridge systematics, 2000). Modes that can be analyzed are auto, carpool, truck, local bus, express bus, light rail, and heavy rail. The temporal scope of application includes total average weekday traffic, peak, or off-peak periods. Also, multiple trip purposes may be considered. The model is closely linked to outputs from the four-step urban transportation planning process: The study area is partitioned into traffic analysis zones and aggregated to districts where separate benefit and cost factors may be specified. Costs and benefits can be reported at the corridor or regional level. The benefit categories include reductions in vehicle operating costs, travel times, crash costs, emissions, energy consumption, and noise; and the agency costs include capital (infrastructure investment) and operating costs. The economic efficiency of each alternative is expressed as a net present worth or a benefit–cost ratio. Other quantitative impacts that can be considered include congestion, access to jobs, revenues and transfers (revenues) from fares, tolls, and fuel taxes, as well as the levels of risk in the results estimated (probability distributions for key outputs). Other related FHWA benefit–cost analysis tools include IMPACTS, SMITE, SCRITS, and SPASM. These software packages and their supporting documentation are available on-line at: http://www.fhwa.dot.gov/steam/links.htm.

8.4.2 MicroBenCost Model

In the United States, from the late 1970s through the early 1990s, the most widely accepted benefit–cost analysis model highways and transit was that presented in the 1977 AASHTO *Manual for User Benefit Analysis of Highway and Bus-Transit Improvements* (AASHTO, 1977). To facilitate use of the model, the Texas Transportation Institute and the National Cooperative Highway Research Program developed the MicroBencost computer software package (McFarland et al., 1993). The MicroBencost program compares the costs of an existing situation to that of a planned transportation improvement. It selects the best improvement alternative from several candidate projects and presents an objective ranking of projects in the order of their potential benefits. MicroBencost can analyze a broad spectrum of projects, including new location or bypass facilities, pavement rehabilitation and bridge rehabilitation and replacement.

8.4.3 Highway Development and Management Standards Model

The highway development management (HDM) model was sponsored by the World Bank, and several other international organizations. HDM evaluates highway projects, standards, and programs mostly in developing countries and makes comparative economic evaluations of alternative construction and preservation scenarios either for a given road section or for an entire road network. The HDM model duly considers that the costs of construction, maintenance, and vehicle operation are functions of road characteristics such as vertical alignment, horizontal alignment, and road surface condition (University of Birmingham, 2005).

8.4.4 Highway Economic Requirements system

FHWA's HERS is an economic efficiency analysis tool that uses technical standards to identify highway deficiencies applies incremental benefit–cost analysis to select the most economically efficient portfolio of improvements for systemwide implementation, and predicts the system condition and user cost levels resulting from a given level of investment (FHWA, 2002a). The amounts and total costs of travel time, safety, and vehicle operation and the cost of emissions associated with each alternative improvement are used to assess the economic efficiency. In cases where funding is not available to achieve "optimal" spending levels, HERS prioritizes economically worthwhile improvement options according to relative merit (benefit–cost ratios are used) and then selects the best set of projects for systemwide implementation. HERS minimizes the expenditure of public funds while simultaneously maximizing highway user benefits given funding constraints or user-specified performance objectives. The HERS software is used not only for economic efficiency impact evaluation, but also for program development and needs assessments.

8.4.5 California DOT'S Cal-B/C System

The California life-cycle benefit–cost (Cal-B/C) analysis model carries out economic efficiency evaluation for planned highway and transit improvement projects. The model is capable of analyzing the impacts of lane additions, HOV lanes, passing/truck climbing lanes, and intersection improvements. Transit modes that can be analyzed include passenger rail, light rail, and bus transit. Cal-B/C calculates the savings in travel time, vehicle operating cost, accident cost, and emissions. Performance measures include life-cycle costs and benefits, net present value, benefit–cost ratio, rate of return on investment, and project payback period. The model enables quick economic efficiency analysis, and comparison and ranking of roadway and transit alternatives that have similar benefits (Booz Allen Hamilton, 1999).

8.5   LIFE-CYCLE COST ANALYSIS

Life-cycle cost analysis (LCCA) is a special case of economic efficiency analysis where the streams of a facility's benefits and costs extend over an appreciable length of time such as one *life cycle* of the facility. LCCA can be used at the project level or the network level. Project-level LCCA focuses on a specific facility, such as a runway, bridge, or road segment, whereas network-level analysis considers an entire inventory of facilities. In both cases, LCCA helps in evaluating the overall long-term economic efficiency between competing alternative investment options by evaluating the benefits and costs of various alternative preservation and improvement strategies or funding levels over the life cycle(s), or part thereof, of a facility or facilities. The monetized costs and benefits associated with each alternative *activity profile* (planned actions over a facility life or part thereof) are determined and the alternative with the highest net present value is typically selected. The Federal Highway Administration (FHWA) and the Federal Transit Administration (FTA) encourage the use of LCCA in analyzing all major investment decisions. Studies conducted in the United States and abroad strongly suggest that cost-effective long-term investment decisions could be made at lower costs if LCCA were adopted properly (Darter et al., 1987; Peterson, 1985; Mouaket and Sinha, 1990; Al-Mansour and Sinha, 1994; FHWA, 2002b).

8.6   CASE STUDY: ECONOMIC EFFICIENCY IMPACT EVALUATION

As part of a major corridor expansion project, it is proposed to improve a 12-mile stretch of an existing 4 lane urban arterial highway. The improvement will involve lane and shoulder widening, median closings (full restriction of access between opposing lanes), and full control of access from local roads. Other details about the "do-nothing" and "improvement" scenarios are presented in Table 8.2. Also, the analysis period is 10 years, the interest rate is 5% per year, and the travel time value is $14 for autos. Also assume that the traffic stream is comprised only of medium-size automobiles and that the delay due to congestion is negligible. For estimating the annual number

Table 8.2   Corridor Expansion Project Data

	Do Nothing	Improvement
Initial construction cost ($)[a]	0	$70,000,000
Average annual preservation cost ($/lane-mile)[a]	25,000	1,500
Residual value at end of 10 years ($)[a]	0	500,000
Geometric features		Median closings, full control of access from local roads
Average annual daily traffic, assume constant	25,280	46,100 (expected)
Lane width (ft)	6	8
Shoulder width (ft)	2	4
Average operating speed (mph)	50	65

[a]Assume that all monetary amounts are in 2005 dollars.

of PDO and fatal-injury crashes, the following equations can be used:

$$N_{\text{PDO}} = 0.9211 + 0.8817 \ln(\text{length})$$
$$+ 0.3812 \ln(\text{AADT}) - 0.1375\text{LW} - 0.0717\text{SW}$$

$$N_{\text{FI}} = -0.1211 + 0.0610 \ln(\text{length})$$
$$+ 0.022 \ln(\text{AADT}) - 0.006\text{LW} - 0.006\text{SW}$$

Determine whether the improvement is economically feasible.

1. *Determine the vehicle operating cost impacts.* From Chapter 7, the VOC at urban roadways for four VOC components (tires, vehicle depreciation, maintenance, and fuel), combined, can be estimated as follows:

$$\text{VOC} = a_0 + a_1 v_s + a_2 v_s^2 \qquad \text{for } v_s > 50 \text{ mph}$$

where speed is in mph and VOC is in cents per vehicle-mile, and a_0, a_1, and a_2 are coefficients that dependent on vehicle type. Using the parameters provided in Table 7.3 the VOC for each alternative can be found as follows:

Do-nothing alternative: Unit VOC,

$$U_{\text{DN}} = 33.5 - (0.058)(50)$$
$$+ (0.00029)(50)^2 = 31.33 \text{ cents per veh-mile}$$

$$\text{VMT}_{\text{DN}} = (25,280)(12 \text{ mi})$$
$$= 303,360 \text{ vehicle-miles}$$

Improvement alternative: Unit VOC,

$$U_{\text{IMP}} = 33.5 - (0.058)(65) + (0.00029)(65)^2$$
$$= 30.96 \text{ cents per veh-mile}$$

$$\text{VMT}_{\text{IMP}} = (46,100)(12 \text{ mi})$$
$$= 553,200 \text{ vehicle-miles}$$

Therefore, the annual VOC benefits to be derived by the improvement are

$$0.5(U_{\text{DN}} - U_{\text{IMP}})(\text{VMT}_{\text{DN}} + \text{VMT}_{\text{IMP}})$$
$$= (0.5)(31.33 - 30.96)(303,360 + 553,200)/100$$
$$= \$1584.60 \text{ per day} = \$578,392 \text{ per year}$$

2. *Determine the safety impacts.*

Do-nothing alternative: Estimated annual number of PDO crashes,

$$N_{\text{PDO}} = 0.9211 + 0.8817 \ln(12) + 0.3812 \ln(25,280)$$
$$- (0.1375)(6) - (0.0717)(2) = 6.01$$

Therefore, the PDO crash rate,

$$U_{\text{PDO,DN}} = \frac{N_{\text{PDO}}}{\text{VMT}} = \frac{6.01}{(12)(25,280)(365)}$$
$$= 5.43 \text{ per 100 million VMT}$$

The estimated annual number of fatal/injury crashes,

$$N_{\text{FI}} = -0.1211 + 0.0610 \ln(12) + 0.022 \ln(25,280)$$
$$- (0.006)(6) - (0.006)(2) = 0.21$$

$$\text{VMT}_{\text{DN}} = (12)(25,280)(365) = 1.107 \text{ million/year}$$

Therefore, the fatal/injury crash rate,

$$U_{\text{FI,DN}} = 0.21/1.107 = 0.19 \text{ per million VMT}$$

Improvement alternative:

$$N_{\mathrm{PDO}} = 0.9211 + 0.8817 \ln(12)$$
$$+\ 3812 \ln(46,100) - (0.1375)(8)$$
$$-\ (0.0717)(4) = 5.82$$

$$\mathrm{VMT}_{\mathrm{IMP}} = (12)(46,100)(365) = 2.019 \text{ million VMT}$$

Therefore, the PDO crash rate,

$$U_{\mathrm{PDO,IMP}} = 5.82/2.019$$
$$= 2.88 \text{ per 100 million VMT}$$

$$N_{\mathrm{FI}} = -0.1211 + 0.0610 \ln(12) + 0.022 \ln(46,100)$$
$$-\ (0.006)(8) - (0.006)(4)$$
$$= 0.195$$

Therefore, the fatal-injury crash rate, $U_{\mathrm{FI,IMP}} = 0.195/2.019 = 0.097$ per 100 million VMT.

Safety savings of the improvement over do-nothing:

For PDOs:

$$\text{Benefits} = 0.5(U_{\mathrm{DN}} - U_{\mathrm{IMP}})(\mathrm{VMT}_{\mathrm{DN}} + \mathrm{VMT}_{\mathrm{IMP}})$$
$$= (0.5)(5.43 - 2.88)(1.107 + 2.019)$$
$$= 3.99 \text{ crashes}$$

Assuming average cost per PDO crash

$$= \$5936$$

Therefore,

$$\text{PDO safety benefits per year} = (3.99)(\$5936)$$
$$= \$23,662$$

For fatal/injury:

$$\text{Benefits} = 0.5(U_{\mathrm{DN}} - U_{\mathrm{IMP}})(\mathrm{VMT}_{\mathrm{DN}} + \mathrm{VMT}_{\mathrm{IMP}})$$
$$= (0.5)(0.19 - 0.097)(1.107 + 2.019)$$
$$= 0.15 \text{ crashes}$$

Assuming average cost of a fatal/injury crash

$$= \$999,958$$

Therefore,

$$\text{fatal-injury safety benefits per year}$$
$$= (0.15)(\$999,958) = \$150,000$$

Total annual safety benefits

$$= \$23,662 + \$150,000 = \$173,662$$

3. *Determine the travel-time impacts.*

Do-nothing alternative:

$$\text{Average travel time, } U_{\mathrm{DN}} = \frac{\text{distance}}{\text{speed}}$$
$$= \frac{12}{50} = 0.24 \text{ h per trip}$$

Improvement alternative:

Average travel time, U_{IMP}

$$= \frac{\text{distance}}{\text{speed}} = \frac{12}{65} = 0.185 \text{ h per trip}$$

Travel-time savings (hours) of the improvement over do-nothing $= 0.5(U_{\mathrm{DN}} - U_{\mathrm{IMP}})(V_{\mathrm{DN}} + V_{\mathrm{IMP}})$

$$= (0.5)(0.24 - 0.185)(25,280 + 46,100)$$
$$= 1963 \text{ h/day} = 716,495 \text{ h/year}$$

Assuming the value of travel time $= \$14/\mathrm{h}$, the annual travel-time benefits of the improvement scenario $= (716,477)(\$14) = \$10,030,930$.

4. *Calculate the net present value of the improvement scenario over the do-nothing scenario.*

Assuming no change in future year traffic volumes:

$$\text{NPV} = \text{Present worth of benefits} - \text{Present worth of costs}$$
$$= (578,392 + 173,662 + 10,030,930)\mathrm{USPWF}(5\%, 10 \text{ yrs})$$
$$-\ \{70,000,000 - [(25,000 - 1,500)12 \times 4]$$
$$\times\ \mathrm{USPWF}(5\%, 10 \text{ yrs}) - 500,000 \, \mathrm{SPPWF}(5\%, 10 \text{ yrs})\}$$
$$= 83,266,202 - [70,000,000 - 8,710,416 - 306,950]$$
$$= \$22 \text{ M.}$$

Therefore, the improvement is economically feasible. If the increase in traffic is considered, the NPV will be ever higher.

8.7   FINAL COMMENTS ON ECONOMIC EFFICIENCY ANALYSIS

Consistent with the pursuit of social justice, equitable distribution of project benefits is increasingly being considered by many transportation agencies. In striving to ensure such distribution of benefits, transportation analysts may consider the concept of *Pareto efficiency*, which states that an allocation of goods or services is Pareto efficient if no alternative allocation can better the condition for at least one person without negatively affecting anyone else. Boardman et al. (2001) discussed the concept of Pareto efficiency and its relationship to economic efficiency analysis, stating that the latter utilizes a decision rule with less conceptual appeal but greater feasibility than the actual Pareto efficiency rule. This decision rule is based on the well-known *Kaldor–Hicks criterion*, which states that a policy should be adopted if and only if those who would gain are capable of fully compensating those who would lose and yet remain better off. This criterion provides the basis for the *potential Pareto efficiency rule*, or more commonly, the *net benefits criterion*, which states that only policies that have positive net benefits should be adopted. Besides its feasibility, the potential Pareto efficiency rule is often considered justified for the following reasons: (1) by always choosing policies with positive net benefit, society maximizes aggregate wealth and therefore helps the relatively poor individuals in society; (2) different policies will have different gainers and losers, and costs and benefits tend to average out among individuals so that each individual is likely to realize positive net benefits from the full set of policies; and (3) it appropriately guards against unduly granting excessive weight to the preferences of organized groups (stakeholders) that have relatively large influence on political systems, and or granting insignificant weights to the perspectives of unorganized or uninfluential individuals in the society.

Another issue in economic analysis is the valuation of certain components of economic efficiency. Critics argue that economic efficiency debases the terms of public discourse by assigning monetary values to intangibles such as human life. An opposing school of thought contends that it is appropriate to place a value on intangibles, at least in the statistical sense, in order to assess properly the policies and projects that have a profound impact on these intangibles. In other areas of economic efficiency analysis, critics charge that such economic efficiency evaluations undermine democracy by imposing a single goal (efficiency) in the evaluation of public projects and policies. This would be true if public policy were determined strictly via benefit–cost analysis results compared to public policy being determined solely via democratic processes that give equal weight to all interests. In the real world, however, these extreme situations do not exist: Economic efficiency analyses rarely serve as a single decisive yardstick for policy making. Besides, it can be argued that by using economic efficiency analysis as one of the factors for decision making, less organized and less vocal constituencies who

have little electoral clout often have their interests better represented. In Chapter 18 we provide a methodology through which an analyst can incorporate a wide array of performance measures, including economic efficiency, socioeconomic impacts, and environmental effects to arrive at transportation decisions that achieve specified performance targets within established constraints.

SUMMARY

Decisions to select the best of several alternative actions are encountered at various stages of the transportation development process and economic efficiency (benefit–cost) is one of the most widely used and objective performance measures used to compare such alternatives. Economic efficiency analysis proceeds on the assumption that all significant benefits and costs can be expressed in monetary values. We reviewed the concepts of economic efficiency analysis, such as the fundamental principle that the value of money is related directly to the time at which the value is considered. Special cases of interest equations include the present worth of periodic payments in perpetuity and the effect of an infinite number of compounding periods in a year.

Criteria for economic efficiency evaluation, which utilize various forms of interest equations, include the equivalent uniform annual cost, present value of costs, equivalent uniform annual return, net present value, internal rate of return, and benefit–cost ratio. Each criterion has its unique logic, merits, and shortcomings, and agencies have their preferred choice of criteria.

The overall framework for carrying out economic efficiency analysis involves identification of the characteristics of the transportation project and the purpose of the analysis; selection of the base case and transportation alternatives; selection of the appropriate geographic study area, analysis period, impact types, and choice of an economic efficiency criterion; determination of the characteristics of the transportation system under various alternative scenarios; and application of the data to calculate the economic efficiency impacts. In reality, there is a great deal of variation associated with such input parameters, which consequently can makes it difficult to predict outcomes with absolute certainty. A number of software packages are available for evaluating transportation alternatives on the basis of economic efficiency.

EXERCISES

8.1. List two possible transportation interventions (projects, practice, or policy changes) in your locality where the results of an economic efficiency analysis would be useful in deciding whether to go ahead with that intervention. For each example, indicate (1) the motivation for assessing the economic efficiency and (2) the stage of the PDP with which the project/policy is concerned.

8.2. Explain the relationship between opportunity cost, interest rate, and inflation.

8.3. If you borrow $4000, what single payment must you make after five years to repay the principal and interest at 10%? Alternatively, what uniform annual payment would be required?

8.4. Solve the following using a 7% interest rate compounded annually: (**a**) What is the amount that will be accumulated in a sinking fund at the end of 15 years if $200 is deposited in the fund at the beginning of each of the 15 years? (**b**) Uniform deposits were made on January 1 of 1991, 1992, 1993, and 1994 into a fund intended to provide $1000 on January 1 of 2005, 2006, and 2007. What were the size of these deposits?

8.5. (**a**) What annual expenditure for 12 years is equivalent to spending $1000 at the end of the first year, $1500 at the end of the fifth year, and $2000 at the end of the ninth year if the interest rate is 8% per annum?

(**b**) What single amount paid at the beginning of the first year is equivalent to the series of unequal payments in part (a), with an interest rate of 8%?

8.6. What uniform annual payment for 30 years is equivalent to spending $10,000 at the present time (year 0), $10,000 at the end of 10 years, $10,000 at the end of 20 years, and $2000 a year for 30 years? Assume an interest rate of 8%.

8.7. (**a**) What is the difference between effective and nominal interest rate?

(**b**) What is the effective annual interest rate if the nominal interest rate is 12% and there are two compounding periods in the year?

(**c**) What nominal annual interest rate, compounded quarterly, yields an effective annual interest rate of 22%?

(**d**) What is the effective annual interest rate for a nominal annual interest rate of 10% compounded continuously?

8.8. Explain the following terms: (**a**) continuously compounded interest and (**b**) periodic payments to perpetuity.

8.9. The rate of growth of traffic on a newly constructed bridge is 3% per year. By the end of the first year,

500,000 vehicles will have traveled it. Determine the number of vehicles using the bridge in the tenth year of service. Assuming that a toll of $0.75 is collected per vehicle, calculate the present worth of the total toll collections during the 10-year period. Assume an interest rate of 10% with continuous compounding. For simplicity, cash flows can be considered discrete, occurring at the end of each year.

8.10. Assuming that interest is compounded monthly, determine the present worth of the cash streams illustrated in Figure EX8.10.

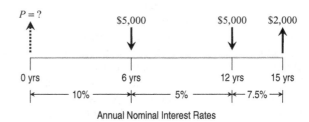

Figure EX8.10

8.11. A proposed transportation project has two alternatives with the costs and benefits shown in Table EX8.11. The user benefits refer to reductions in user costs for vehicle operation, delay, and crashes relative to a do-nothing alternative. Assuming a 10% interest rate per year and a MARR of 8%, and assuming that the do-nothing alternative is not feasible, indicate which alternative you would recommend, on the basis of any of the following economic efficiency criteria: (1) equivalent uniform annual return, (2) internal rate of return, or (3) benefit–cost ratio method.

Table EX8.11   Costs and Benefits of Alternatives

	Alternative 1	Alternative 2
Initial construction costs	$50 million	$20 million
Annual operating costs	$0.5 million	$1 million
Frequency of rehabilitation	Every 5 years	Every 3 years
Rehabilitation cost	$1 million	$1.2 million
Average annual maintenance cost	$0.75 million	$1.5 million
Annual user benefits	$20 million	$12 million
Service life	15 years	9 years
Salvage value	$2.5 million	$0.5 million

8.12. Three alternative congestion mitigation strategies are being considered to enhance mobility at a congested bypass freeway. The prospective net cash flows for these alternatives are shown in Table EX8.12. The MARR is 15% per year. Using the NPV method and incremental analysis procedure, determine which is the best congestion mitigation alternative from the economic efficiency standpoint.

Table EX8.12   Cash Flows under Alternatives

End of year	A Implement-ation Cost	B Implement-ation Cost	C Implement-ation Cost
0	−$6 million	−$8 million	−$7 million
1	+$2 million	+$3 million	+$2.5 million
2	+$3 million	+$4.1 million	+$3.4 million
3	+$3.8 million	+$4.5 million	+$4.1 million
4	+$4.1 million	+$4.7 million	+$4.4 million
5	+$4.2 million	+$4.8 million	+$4.5 million

8.13. The initial investment for constructing a median and guardrails for a four-lane rural highway is $1,500,000. Maintaining these facilities is expected to cost $2000 annually for the first five years of service and $8000 for the next five years of service. It is expected that these facilities will be rehabilitated at the end of the tenth year at a cost of $50,000, after which the maintenance costs are expected to decrease to $5000 per year. What is the equivalent uniform annual cost over a 15-year period of service if the interest rate is 6% per year? Assume that zero salvage value is 20% of the initial construction cost. If the investment is scheduled five years from now, what amount should be set aside now to provide for these improvements?

8.14. An improvement in all the terminals on a transit network is scheduled every five years in perpetuity. Each improvement costs $25,000. What is the present worth of all the costs of the improvement project if the interest rate is 8% per year?

8.15. At the current year (end of year 0), the user benefits from a road capacity enhancement project is $10 million and the total project cost is $100 million. What must be the minimum rate of growth of benefit (in percent per year) for the project to be feasible economically? Assume that the project life is 30 years and that the benefit grows continuously.

The interest rate is 5% per year compounded continuously.

REFERENCES[1]

AASHTO (1977). *A Manual for User Benefit Analysis of Highway and Bus-Transit Improvements*, American Association of State Highway and Transportation Officials, Washington, DC.

———— (2003). *User Benefit Analysis for Highways*, Amer. Assoc. of State and Transportation Officials, Washington, DC.

Al-Mansour, A. I., Sinha, K. C. (1994). *Economic Analysis of Effectiveness of Pavement Preventive Maintenance*, Transp. Res. Rec. 1442, Transportation Research Board, National Research Council, Washington, DC.

*Boardman, A. E., Greenberg, D. H., Vining, A. R., Weimer, D. L. (2001). *Cost–Benefit Analysis: Concepts and Practice*, Prentice Hall, Upper Saddle River, NJ.

Booz Allen Hamilton (1999). *The California Life-Cycle Benefit/Cost Analysis Model (Cal-B/C)*, Technical Supplement to User's Guide, California Department of Transportation, Sacramento, CA.

*Cambridge Systematics (2000). *Surface Transportation Efficiency Analysis Model (STEAM 2.0):- User Manual*, Federal Highway Administration, U.S. Department of Transportation, Washington, DC, http://www.fhwa.dot.gov/steam/20manual.htm. Accessed Jan. 2004.

Darter, M. I., Smith, R. E., Shahin, M. Y. (1987). Use of life-cycle costing analysis as a basis for determining the cost-effectiveness of maintenance and rehabilitation treatments for developing a network level assignment procedure, *Proc. North American Pavement Management Conference*, Toronto, ON, Canada.

*Dickey, J. W., Miller, L. H. (1984). *Road Project Appraisal for Developing Countries*, Wiley, New York.

FHWA (2002a) *Highway Economic Requirements System*, Technical Report, Federal Highway Administration, U.S. Department of Transportation. Washington DC.

FHWA (2002b). *Life-Cycle Cost Analysis Primer*, Federal Highway Administration, U.S. Department of Transportation, Washington, DC.

Forkenbrock, D. J., Weisbrod, G. E. (2001). *Guidebook for Assessing the Social and Economic Effects of Transportation Projects*, NCHRP Rep. 456, Transportation Research Board, National Research Council, Washington, DC.

GASB (1999). *Basic Financial Statements—and Management's Discussion and Analysis—for State and Local Governments*, Statement 34, Governmental Accounting Standards Board, Norwalk, CT.

McFarland, W. F., Memmott, J. L., Chui, M. L. (1993). *Microcomputer Evaluation of Highway User Benefits*, NCHRP Rep. 7–12, Transportation Research, Board, National Research, Council, Washington, DC.

Mouaket, I. M., Sinha, K. C. (1990). *Cost Effectiveness of Rigid and Composite Highway Pavement Routine Maintenance*, FHWA/JHRP-90-15, Joint Highway Research Project, School of Civil Engineering, Purdue University, West Lafayette, IN.

[1]References marked with an asterisk can also serve as useful resources for economic efficiency evaluation.

Peterson, D. E. (1985). *Life Cycle Cost Analysis of Pavement*, NCHRP Synth. Hwy. Pract. 122, Transportation Research Board, National Research Council, Washington, DC.

University of Birmingham (2005). *HDM Technical User Guide*, UB, Birmingham, UK.

Weisbrod, G. (2000). *Current Practices for Assessing Economic Development Impacts from Transportation Investments*, NCHRP Synth. Hwy. Pract. 290, Transportation Research Board, National Research Council, Washington, DC.

Wohl, M. Hendrickson, C. (1984). *Transportation Investment and Pricing Principles: An Introduction for Engineers, Planners and Economists*, Wiley, New York.

ADDITIONAL RESOURCES

British Columbia Ministry of Transport and Highways, Planning Services Branch (1992). *Economic Analysis Guidebook*, BC Ministry, Victoria, BC, Canada.

Cambridge Systematics, Bernardin Lochmueller (1996). *Major Corridor Investment–Benefit Analysis System*, Indiana Department of Transportation, Indianapolis, IN.

Campen, J. T. (1986). *Benefit, Cost and Beyond*, Ballinger Books, Cambridge, MA.

Gramlich, E. M. (1990). *A Guide to Benefit–Cost Analysis*, Prentice Hall, Upper Saddle River, NJ.

Hawk, H. *Bridge Life-Cycle Cost Analysis*, NCHRP Rep. 483, Transportation Research Board, National Research Council, Washington, DC, 2002.

Lewis, D. (1996). *Primer on Transportation, Productivity and Economic Development*, NCHRP Rep. 342, Transportation Research Board, National Research Council, Washington, DC.

Nas, T. (1996). *Cost–Benefit Analysis: Theory and Application*, Sage Publications, Newbury Park, CA.

PIARC (2004). *Economic Evaluation Methods for Road Projects in PIARC Member Countries*, PIARC Committee C9 on Economic and Financial Evaluation, World Road Association, Cedex, France.

Stokey, E., Zeckhauser, R. (1978). *A Primer for Policy Analysis*, W.W. Norton, New York.

Transportation Association of Canada (1994). *A Primer on Transportation Investment and Economic Development*, TAC, Ottawa, Canada.

Transport Canada (1994). *Guide to Benefit–Cost Analysis*, Rep. TP 11875E, TC, Ottawa, ON, Canada.

TRB (1997). *Transportation Research Circular 477*, Transportation Research Board, Washington, National Research Council, DC.

U.S. Office of Management and Budget (1992). *Guidelines and Discount Rates for Benefit–Cost Analysis of Federal Programs*, Circ. A-94, *Fed. Reg.*, Vol. 57, No. 218.

Wilbur Smith Associates (1993). *Guide to the Economic Evaluation of Highway Projects*, Iowa Department of Transportation, Des Moines, IA.

Zerbe, R. O., Dively, D. (1994). *Benefit–Cost Analysis in Theory and Practice*, HarperCollins, New York.

APPENDIX A8

Table A8.1 Interest Formulas for Arithmetic Gradient Series with Discrete Compounding

Description	Cash Flow Diagram	Computational Formula	Factor Computation
Finding the future compounded amount (F) at the end of a specified period due to linearly increasing annual payments (G), given the interest rate.		$F = G \times \text{GSCAF}$ The gradient series compounded amount factor, GSCAF ($i\%$, N), may be computed as shown.	$\text{GSCAF} = \dfrac{(1 + i)^N - 1}{i^2} - \dfrac{N}{i}$

Table A8.1   (*continued*)

Description	Cash Flow Diagram	Computational Formula	Factor Computation
Finding the amount of linearly increasing annual payments (G) that would yield a future compounded amount (F) at the end of a specified period, given the interest rate.	$(N-1)G=?$ $(N-2)G=?$ $1G=?$ $0G=?$ 0 1 2 $\cdots$ $N-1$ N F	$G = F \times \text{GSSFDF}$ The gradient series sinking fund deposit factor, GSSFDF ($i\%$, N), may be computed as shown.	$\text{GSSFDF} = \dfrac{i^2}{(1+i)^N - 1 - Ni}$
Finding the initial amount (P) that would be equivalent to specified linearly increasing annual payments (G), given the interest rate.	$(N-1)G$ $(N-2)G$ $P=?$ $1G$ $0G$ 0 1 2 $\cdots$ $N-1$ N	$P = G \times \text{GSPWF}$ The gradient series present worth factor, GSPWF ($i\%$, N), may be read from Table A.8.2 or may be computed as shown.	$\text{GSPWF} =$ $\left[\dfrac{(1+i)^N - 1}{i^2} - \dfrac{N}{i} \right] \Big/ ((1+i)^N)$
Finding the amount of linearly increasing annual payments (G) that would completely recover an initial amount (P) at the end of a specified period, given the interest rate.	$(N-1)G=?$ $(N-2)G=?$ P $1G=?$ $0G$ 0 1 2 $\cdots$ $N-1$ N	$G = P \times \text{GSCRF}$ The gradient series capital recovery factor, GSCRF ($i\%$, N), may be computed as shown.	$\text{GSCRF} = \dfrac{(1+i)^N \times i^2}{(1+i)^N - 1 - Ni}$

Table A8.2   Conversion Factors between Uniform Annual Series and Selected Nonuniform Series

Type of Compounding	Direction of Conversion	Computational Formula
Discrete compounding	Linear gradient series (G) to equivalent uniform series (A)[a]	$A = G\left[\dfrac{1}{i} - \dfrac{N}{(1+i)^N - 1}\right] = G \times \text{GSUAF}$
	Geometric series (M) to equivalent uniform series (A)[b]	$A = M\dfrac{[(1+t)/(1+i)]^N - 1}{t - i}\dfrac{i(1+i)^N}{(1+i)^N - 1}$
Continuous compounding	Linear gradient series (G) to equivalent uniform series (A)	$A = G\left(\dfrac{1}{e^i - 1} - \dfrac{N}{e^{Ni} - 1}\right)$
	Geometric series (M) to equivalent uniform series (A)	$A = M\dfrac{(1+t)^N - e^{Ni}}{1 + t - e^i}\dfrac{e^i - 1}{e^{Ni} - 1}$

[a]GSUAF, the gradient series uniform amount factor, can also be read off from standard equivalency tables.
[b]The cash flow patterns are changing at a constant rate of $t\%$ per period. The initial cash flow in this series, M, occurs at the end of period 1. The cash flow at the end of period 2 is $M(1 + t)$ and at the end of period N is $M(1 + t)^{N-1}$.

Table A8.3   Compound Interest Factors

	Single Payment		Uniform Payment Series				Arithmetic Gradient	
	Compound Amount Factor: Find F Given P	Present Worth Factor: Find P Given F,	Sinking Fund Factor: Find A Given F,	Capital Recovery Factor: Find A Given P,	Capital Amount Factor: Find F Given A,	Present Worth Factor: Find P Given A,	Gradient Uniform Series: Find A Given G,	Gradient Present Worth: Find P Given G,
N	F/P	P/F	A/F	A/P	F/A	P/A	A/G	P/G
				2%				
1	1.020	0.9804	1.0000	1.0200	1.000	0.980	0.000	0.000
2	1.040	0.9612	0.4950	0.5150	2.020	1.942	0.495	0.961
3	1.061	0.9423	0.3268	0.3468	3.060	2.884	0.987	2.846
4	1.082	0.9238	0.2426	0.2626	4.122	3.808	1.475	5.617
5	1.104	0.9057	0.1922	0.2122	5.204	4.713	1.960	9.240
6	1.126	0.8880	0.1585	0.1785	6.308	5.601	2.442	13.680
7	1.149	0.8706	0.1345	0.1545	7.434	6.472	2.921	18.903
8	1.172	0.8535	0.1165	0.1365	8.583	7.325	3.396	24.878
9	1.195	0.8368	0.1025	0.1225	9.755	8.162	3.868	31.572
10	1.219	0.8203	0.0913	0.1113	10.950	8.983	4.337	38.955
11	1.243	0.8043	0.0822	0.1022	12.169	9.787	4.802	46.998
12	1.268	0.7885	0.0746	0.0946	13.412	10.575	5.264	55.671
13	1.294	0.7730	0.0681	0.0881	14.680	11.348	5.723	64.948
14	1.319	0.7579	0.0626	0.0826	15.974	12.106	6.179	74.800
15	1.346	0.7430	0.0578	0.0778	17.293	12.849	6.631	85.202

Table A8.3 (*continued*)

	Single Payment		Uniform Payment Series				Arithmetic Gradient	
	Compound Amount Factor: Find F Given P	Present Worth Factor: Find P Given F,	Sinking Fund Factor: Find A Given F,	Capital Recovery Factor: Find A Given P,	Capital Amount Factor: Find F Given A,	Present Worth Factor: Find P Given A,	Gradient Uniform Series: Find A Given G,	Gradient Present Worth: Find P Given G,
N	F/P	P/F	A/F	A/P	F/A	P/A	A/G	P/G
16	1.373	0.7284	0.0537	0.0737	18.639	13.578	7.080	96.129
17	1.400	0.7142	0.0500	0.0700	20.012	14.292	7.526	107.555
18	1.428	0.7002	0.0467	0.0667	21.412	14.992	7.968	119.458
19	1.457	0.6864	0.0438	0.0638	22.841	15.678	8.407	131.814
20	1.486	0.6730	0.0412	0.0612	24.297	16.351	8.843	144.600
21	1.516	0.6598	0.0388	0.0588	25.783	17.011	9.276	157.796
22	1.546	0.6468	0.0366	0.0566	27.299	17.658	9.705	171.379
23	1.577	0.6342	0.0347	0.0547	28.845	18.292	10.132	185.331
24	1.608	0.6217	0.0329	0.0529	30.422	18.914	10.555	199.630
25	1.641	0.6095	0.0312	0.0512	32.030	19.523	10.974	214.259
26	1.673	0.5976	0.0297	0.0497	33.671	20.121	11.391	229.199
27	1.707	0.5859	0.0283	0.0483	35.344	20.707	11.804	244.431
28	1.741	0.5744	0.0270	0.0470	37.051	21.281	12.214	259.939
29	1.776	0.5631	0.0258	0.0458	38.792	21.844	12.621	275.706
30	1.811	0.5521	0.0246	0.0446	40.568	22.396	13.025	291.716
36	2.040	0.4902	0.0192	0.0392	51.994	25.489	15.381	392.040
40	2.208	0.4529	0.0166	0.0366	60.402	27.355	16.889	461.993
48	2.587	0.3865	0.0126	0.0326	79.354	30.673	19.756	605.966
50	2.692	0.3715	0.0118	0.0318	84.579	31.424	20.442	642.361
52	2.800	0.3571	0.0111	0.0311	90.016	32.145	21.116	678.785
60	3.281	0.3048	0.0088	0.0288	114.052	34.761	23.696	823.698
70	4.000	0.2500	0.0067	0.0267	149.978	37.499	26.663	999.834
72	4.161	0.2403	0.0063	0.0263	158.057	37.984	27.223	1,034.056
80	4.875	0.2051	0.0052	0.0252	193.772	39.745	29.357	1,166.787
84	5.277	0.1895	0.0047	0.0247	213.867	40.526	30.362	1,230.419
90	5.943	0.1683	0.0040	0.0240	247.157	41.587	31.793	1,322.170
96	6.693	0.1494	0.0035	0.0235	284.647	42.529	33.137	1,409.297
100	7.245	0.1380	0.0032	0.0232	312.232	43.098	33.986	1,464.753
4%								
1	1.040	0.9615	1.0000	1.0400	1.000	0.962	0.000	0.000
2	1.082	0.9246	0.4902	0.5302	2.040	1.886	0.490	0.925
3	1.125	0.8890	0.3203	0.3603	3.122	2.775	0.974	2.703
4	1.170	0.8548	0.2355	0.2755	4.246	3.630	1.451	5.267
5	1.217	0.8219	0.1846	0.2246	5.416	4.452	1.922	8.555
6	1.265	0.7903	0.1508	0.1908	6.633	5.242	2.386	12.506
7	1.316	0.7599	0.1266	0.1666	7.898	6.002	2.843	17.066
8	1.369	0.7307	0.1085	0.1485	9.214	6.733	3.294	22.181
9	1.423	0.7026	0.0945	0.1345	10.583	7.435	3.739	27.801
10	1.480	0.6756	0.0833	0.1233	12.006	8.111	4.177	33.881

(*continued overleaf*)

Table A8.3 (*continued*)

N	Single Payment		Uniform Payment Series				Arithmetic Gradient	
	Compound Amount Factor: Find F Given P F/P	Present Worth Factor: Find P Given F, P/F	Sinking Fund Factor: Find A Given F, A/F	Capital Recovery Factor: Find A Given P, A/P	Capital Amount Factor: Find F Given A, F/A	Present Worth Factor: Find P Given A, P/A	Gradient Uniform Series: Find A Given G, A/G	Gradient Present Worth: Find P Given G, P/G
11	1.539	0.6496	0.0741	0.1141	13.486	8.760	4.609	40.377
12	1.601	0.6246	0.0666	0.1066	15.026	9.385	5.034	47.248
13	1.665	0.6006	0.0601	0.1001	16.627	9.986	5.453	54.455
14	1.732	0.5775	0.0547	0.0947	18.292	10.563	5.866	61.962
15	1.801	0.5553	0.0499	0.0899	20.024	11.118	6.272	69.735
16	1.873	0.5339	0.0458	0.0858	21.825	11.652	6.672	77.744
17	1.948	0.5134	0.0422	0.0822	23.698	12.166	7.066	85.958
18	2.026	0.4936	0.0390	0.0790	25.645	12.659	7.453	94.350
19	2.107	0.4746	0.0361	0.0761	27.671	13.134	7.834	102.893
20	2.191	0.4564	0.0336	0.0736	29.778	13.590	8.209	111.565
21	2.279	0.4388	0.0313	0.0713	31.969	14.029	8.578	120.341
22	2.370	0.4220	0.0292	0.0692	34.248	14.451	8.941	129.202
23	2.465	0.4057	0.0273	0.0673	36.618	14.857	9.297	138.128
24	2.563	0.3901	0.0256	0.0656	39.083	15.247	9.648	147.101
25	2.666	0.3751	0.0240	0.0640	41.646	15.622	9.993	156.104
26	2.772	0.3607	0.0226	0.0626	44.312	15.983	10.331	165.121
27	2.883	0.3468	0.0212	0.0612	47.084	16.330	10.664	174.138
28	2.999	0.3335	0.0200	0.0600	49.968	16.663	10.991	183.142
29	3.119	0.3207	0.0189	0.0589	52.966	16.984	11.312	192.121
30	3.243	0.3083	0.0178	0.0578	56.085	17.292	11.627	201.062
31	3.373	0.2965	0.0169	0.0569	59.328	17.588	11.937	209.956
32	3.508	0.2851	0.0159	0.0559	62.701	17.874	12.241	218.792
33	3.648	0.2741	0.0151	0.0551	66.210	18.148	12.540	227.563
34	3.794	0.2636	0.0143	0.0543	69.858	18.411	12.832	236.261
35	3.946	0.2534	0.0136	0.0536	73.652	18.665	13.120	244.877
40	4.801	0.2083	0.0105	0.0505	95.026	19.793	14.477	286.530
45	5.841	0.1712	0.0083	0.0483	121.029	20.720	15.705	325.403
50	7.107	0.1407	0.0066	0.0466	152.667	21.482	16.812	361.164
55	8.646	0.1157	0.0052	0.0452	191.159	22.109	17.807	393.689
60	10.520	0.0951	0.0042	0.0442	237.991	22.623	18.697	422.997
65	12.799	0.0781	0.0034	0.0434	294.968	23.047	19.491	449.201
70	15.572	0.0642	0.0027	0.0427	364.290	23.395	20.196	472.479
75	18.945	0.0528	0.0022	0.0422	448.631	23.680	20.821	493.041
80	23.050	0.0434	0.0018	0.0418	551.245	23.915	21.372	511.116
85	28.044	0.0357	0.0015	0.0415	676.090	24.109	21.857	526.938
90	34.119	0.0293	0.0012	0.0412	827.983	24.267	22.283	540.737
95	41.511	0.0241	0.0010	0.0410	1,012.785	24.398	22.655	552.731
100	50.505	0.0198	0.0008	0.0408	1,237.624	24.505	22.980	563.125

Table A8.3   (*continued*)

	Single Payment		Uniform Payment Series				Arithmetic Gradient	
	Compound Amount Factor: Find F Given P	Present Worth Factor: Find P Given F,	Sinking Fund Factor: Find A Given F,	Capital Recovery Factor: Find A Given P,	Capital Amount Factor: Find F Given A,	Present Worth Factor: Find P Given A,	Gradient Uniform Series: Find A Given G,	Gradient Present Worth: Find P Given G,
N	F/P	P/F	A/F	A/P	F/A	P/A	A/G	P/G
				5%				
1	1.050	0.9524	1.0000	1.0500	1.000	0.952	0.000	0.000
2	1.103	0.9070	0.4878	0.5378	2.050	1.859	0.488	0.907
3	1.158	0.8638	0.3172	0.3672	3.153	2.723	0.967	2.635
4	1.216	0.8227	0.2320	0.2820	4.310	3.546	1.439	5.103
5	1.276	0.7835	0.1810	0.2310	5.526	4.329	1.903	8.237
6	1.340	0.7462	0.1470	0.1970	6.802	5.076	2.358	11.968
7	1.407	0.7107	0.1228	0.1728	8.142	5.786	2.805	16.232
8	1.477	0.6768	0.1047	0.1547	9.549	6.463	3.245	20.970
9	1.551	0.6446	0.0907	0.1407	11.027	7.108	3.676	26.127
10	1.629	0.6139	0.0795	0.1295	12.578	7.722	4.099	31.652
11	1.710	0.5847	0.0704	0.1204	14.207	8.306	4.514	37.499
12	1.796	0.5568	0.0628	0.1128	15.917	8.863	4.922	43.624
13	1.886	0.5303	0.0565	0.1065	17.713	9.394	5.322	49.988
14	1.980	0.5051	0.0510	0.1010	19.599	9.899	5.713	56.554
15	2.079	0.4810	0.0463	0.0963	21.579	10.380	6.097	63.288
16	2.183	0.4581	0.0423	0.0923	23.657	10.838	6.474	70.160
17	2.292	0.4363	0.0387	0.0887	25.840	11.274	6.842	77.140
18	2.407	0.4155	0.0355	0.0855	28.132	11.690	7.203	84.204
19	2.527	0.3957	0.0327	0.0827	30.539	12.085	7.557	91.328
20	2.653	0.3769	0.0302	0.0802	33.066	12.462	7.903	98.488
21	2.786	0.3589	0.0280	0.0780	35.719	12.821	8.242	105.667
22	2.925	0.3418	0.0260	0.0760	38.505	13.163	8.573	112.846
23	3.072	0.3256	0.0241	0.0741	41.430	13.489	8.897	120.009
24	3.225	0.3101	0.0225	0.0725	44.502	13.799	9.214	127.140
25	3.386	0.2953	0.0210	0.0710	47.727	14.094	9.524	134.228
26	3.556	0.2812	0.0196	0.0696	51.113	14.375	9.827	141.259
27	3.733	0.2678	0.0183	0.0683	54.669	14.643	10.122	148.223
28	3.920	0.2551	0.0171	0.0671	58.403	14.898	10.411	155.110
29	4.116	0.2429	0.0160	0.0660	62.323	15.141	10.694	161.913
30	4.322	0.2314	0.0151	0.0651	66.439	15.372	10.969	168.623
31	4.538	0.2204	0.0141	0.0641	70.761	15.593	11.238	175.233
32	4.765	0.2099	0.0133	0.0633	75.299	15.803	11.501	181.739
33	5.003	0.1999	0.0125	0.0625	80.064	16.003	11.757	188.135
34	5.253	0.1904	0.0118	0.0618	85.067	16.193	12.006	194.417
35	5.516	0.1813	0.0111	0.0611	90.320	16.374	12.250	200.581

(*continued overleaf*)

Table A8.3   (*continued*)

	Single Payment		Uniform Payment Series				Arithmetic Gradient	
	Compound Amount Factor: Find F Given P	Present Worth Factor: Find P Given F,	Sinking Fund Factor: Find A Given F,	Capital Recovery Factor: Find A Given P,	Capital Amount Factor: Find F Given A,	Present Worth Factor: Find P Given A,	Gradient Uniform Series: Find A Given G,	Gradient Present Worth: Find P Given G,
N	F/P	P/F	A/F	A/P	F/A	P/A	A/G	P/G
40	7.040	0.1420	0.0083	0.0583	120.800	17.159	13.377	229.545
45	8.985	0.1113	0.0063	0.0563	159.700	17.774	14.364	255.315
50	11.467	0.0872	0.0048	0.0548	209.348	18.256	15.223	277.915
55	14.636	0.0683	0.0037	0.0537	272.713	18.633	15.966	297.510
60	18.679	0.0535	0.0028	0.0528	353.584	18.929	16.606	314.343
65	23.840	0.0419	0.0022	0.0522	456.798	19.161	17.154	328.691
70	30.426	0.0329	0.0017	0.0517	588.529	19.343	17.621	340.841
75	38.833	0.0258	0.0013	0.0513	756.654	19.485	18.018	351.072
80	49.561	0.0202	0.0010	0.0510	971.229	19.596	18.353	359.646
85	63.254	0.0158	0.0008	0.0508	1,245.087	19.684	18.635	366.801
90	80.730	0.0124	0.0006	0.0506	1,594.607	19.752	18.871	372.749
95	103.035	0.0097	0.0005	0.0505	2,040.694	19.806	19.069	377.677
100	131.501	0.0076	0.0004	0.0504	2,610.025	19.848	19.234	381.749
				6%				
1	1.060	0.9434	1.0000	1.0600	1.000	0.943	0.000	0.000
2	1.124	0.8900	0.4854	0.5454	2.060	1.833	0.485	0.890
3	1.191	0.8396	0.3141	0.3741	3.184	2.673	0.961	2.569
4	1.262	0.7921	0.2286	0.2886	4.375	3.465	1.427	4.946
5	1.338	0.7473	0.1774	0.2374	5.637	4.212	1.884	7.935
6	1.419	0.7050	0.1434	0.2034	6.975	4.917	2.330	11.459
7	1.504	0.6651	0.1191	0.1791	8.394	5.582	2.768	15.450
8	1.594	0.6274	0.1010	0.1610	9.897	6.210	3.195	19.842
9	1.689	0.5919	0.0870	0.1470	11.491	6.802	3.613	24.577
10	1.791	0.5584	0.0759	0.1359	13.181	7.360	4.022	29.602
11	1.898	0.5268	0.0668	0.1268	14.972	7.887	4.421	34.870
12	2.012	0.4970	0.0593	0.1193	16.870	8.384	4.811	40.337
13	2.133	0.4688	0.0530	0.1130	18.882	8.853	5.192	45.963
14	2.261	0.4423	0.0476	0.1076	21.015	9.295	5.564	51.713
15	2.397	0.4173	0.0430	0.1030	23.276	9.712	5.926	57.555
16	2.540	0.3936	0.0390	0.0990	25.673	10.106	6.279	63.459
17	2.693	0.3714	0.0354	0.0954	28.213	10.477	6.624	69.401
18	2.854	0.3503	0.0324	0.0924	30.906	10.828	6.960	75.357
19	3.026	0.3305	0.0296	0.0896	33.760	11.158	7.287	81.306
20	3.207	0.3118	0.0272	0.0872	36.786	11.470	7.605	87.230
21	3.400	0.2942	0.0250	0.0850	39.993	11.764	7.915	93.114
22	3.604	0.2775	0.0230	0.0830	43.392	12.042	8.217	98.941
23	3.820	0.2618	0.0213	0.0813	46.996	12.303	8.510	104.701
24	4.049	0.2470	0.0197	0.0797	50.816	12.550	8.795	110.381
25	4.292	0.2330	0.0182	0.0782	54.865	12.783	9.072	115.973

Table A8.3   (*continued*)

	Single Payment		Uniform Payment Series				Arithmetic Gradient	
	Compound Amount Factor: Find F Given P	Present Worth Factor: Find P Given F,	Sinking Fund Factor: Find A Given F,	Capital Recovery Factor: Find A Given P,	Capital Amount Factor: Find F Given A,	Present Worth Factor: Find P Given A,	Gradient Uniform Series: Find A Given G,	Gradient Present Worth: Find P Given G,
N	F/P	P/F	A/F	A/P	F/A	P/A	A/G	P/G
26	4.549	0.2198	0.0169	0.0769	59.156	13.003	9.341	121.468
27	4.822	0.2074	0.0157	0.0757	63.706	13.211	9.603	126.860
28	5.112	0.1956	0.0146	0.0746	68.528	13.406	9.857	132.142
29	5.418	0.1846	0.0136	0.0736	73.640	13.591	10.103	137.310
30	5.743	0.1741	0.0126	0.0726	79.058	13.765	10.342	142.359
31	6.088	0.1643	0.0118	0.0718	84.802	13.929	10.574	147.286
32	6.453	0.1550	0.0110	0.0710	90.890	14.084	10.799	152.090
33	6.841	0.1462	0.0103	0.0703	97.343	14.230	11.017	156.768
34	7.251	0.1379	0.0096	0.0696	104.184	14.368	11.228	161.319
35	7.686	0.1301	0.0090	0.0690	111.435	14.498	11.432	165.743
40	10.286	0.0972	0.0065	0.0665	154.762	15.046	12.359	185.957
45	13.765	0.0727	0.0047	0.0647	212.744	15.456	13.141	203.110
50	18.420	0.0543	0.0034	0.0634	290.336	15.762	13.796	217.457
55	24.650	0.0406	0.0025	0.0625	394.172	15.991	14.341	229.322
60	32.988	0.0303	0.0019	0.0619	533.128	16.161	14.791	239.043
65	44.145	0.0227	0.0014	0.0614	719.083	16.289	15.160	246.945
70	59.076	0.0169	0.0010	0.0610	967.932	16.385	15.461	253.327
75	79.057	0.0126	0.0008	0.0608	1,300.949	16.456	15.706	258.453
80	105.796	0.0095	0.0006	0.0606	1,746.600	16.509	15.903	262.549
85	141.579	0.0071	0.0004	0.0604	2,342.982	16.549	16.062	265.810
90	189.465	0.0053	0.0003	0.0603	3,141.075	16.579	16.189	268.395
95	253.546	0.0039	0.0002	0.0602	4,209.104	16.601	16.290	270.437
100	339.302	0.0029	0.0002	0.0602	5,638.368	16.618	16.371	272.047
				7%				
1	1.070	0.9346	1.0000	1.0700	1.000	0.935	0.000	0.000
2	1.145	0.8734	0.4831	0.5531	2.070	1.808	0.483	0.873
3	1.225	0.8163	0.3111	0.3811	3.215	2.624	0.955	2.506
4	1.311	0.7629	0.2252	0.2952	4.440	3.387	1.416	4.795
5	1.403	0.7130	0.1739	0.2439	5.751	4.100	1.865	7.647
6	1.501	0.6663	0.1398	0.2098	7.153	4.767	2.303	10.978
7	1.606	0.6227	0.1156	0.1856	8.654	5.389	2.730	14.715
8	1.718	0.5820	0.0975	0.1675	10.260	5.971	3.147	18.789
9	1.838	0.5439	0.0835	0.1535	11.978	6.515	3.552	23.140
10	1.967	0.5083	0.0724	0.1424	13.816	7.024	3.946	27.716
11	2.105	0.4751	0.0634	0.1334	15.784	7.499	4.330	32.466
12	2.252	0.4440	0.0559	0.1259	17.888	7.943	4.703	37.351
13	2.410	0.4150	0.0497	0.1197	20.141	8.358	5.065	42.330
14	2.579	0.3878	0.0443	0.1143	22.550	8.745	5.417	47.372
15	2.759	0.3624	0.0398	0.1098	25.129	9.108	5.758	52.446

(*continued overleaf*)

Table A8.3 (*continued*)

	Single Payment		Uniform Payment Series				Arithmetic Gradient	
	Compound Amount Factor: Find F Given P	Present Worth Factor: Find P Given F,	Sinking Fund Factor: Find A Given F,	Capital Recovery Factor: Find A Given P,	Capital Amount Factor: Find F Given A,	Present Worth Factor: Find P Given A,	Gradient Uniform Series: Find A Given G,	Gradient Present Worth: Find P Given G,
N	F/P	P/F	A/F	A/P	F/A	P/A	A/G	P/G
16	2.952	0.3387	0.0359	0.1059	27.888	9.447	6.090	57.527
17	3.159	0.3166	0.0324	0.1024	30.840	9.763	6.411	62.592
18	3.380	0.2959	0.0294	0.0994	33.999	10.059	6.722	67.622
19	3.617	0.2765	0.0268	0.0968	37.379	10.336	7.024	72.599
20	3.870	0.2584	0.0244	0.0944	40.995	10.594	7.316	77.509
21	4.141	0.2415	0.0223	0.0923	44.865	10.836	7.599	82.339
22	4.430	0.2257	0.0204	0.0904	49.006	11.061	7.872	87.079
23	4.741	0.2109	0.0187	0.0887	53.436	11.272	8.137	91.720
24	5.072	0.1971	0.0172	0.0872	58.177	11.469	8.392	96.255
25	5.427	0.1842	0.0158	0.0858	63.249	11.654	8.639	100.676
26	5.807	0.1722	0.0146	0.0846	68.676	11.826	8.877	104.981
27	6.214	0.1609	0.0134	0.0834	74.484	11.987	9.107	109.166
28	6.649	0.1504	0.0124	0.0824	80.698	12.137	9.329	113.226
29	7.114	0.1406	0.0114	0.0814	87.347	12.278	9.543	117.162
30	7.612	0.1314	0.0106	0.0806	94.461	12.409	9.749	120.972
31	8.145	0.1228	0.0098	0.0798	102.073	12.532	9.947	124.655
32	8.715	0.1147	0.0091	0.0791	110.218	12.647	10.138	128.212
33	9.325	0.1072	0.0084	0.0784	118.933	12.754	10.322	131.643
34	9.978	0.1002	0.0078	0.0778	128.259	12.854	10.499	134.951
35	10.677	0.0937	0.0072	0.0772	138.237	12.948	10.669	138.135
40	14.974	0.0668	0.0050	0.0750	199.635	13.332	11.423	152.293
45	21.002	0.0476	0.0035	0.0735	285.749	13.606	12.036	163.756
50	29.457	0.0339	0.0025	0.0725	406.529	13.801	12.529	172.905
55	41.315	0.0242	0.0017	0.0717	575.929	13.940	12.921	180.124
60	57.946	0.0173	0.0012	0.0712	813.520	14.039	13.232	185.768
65	81.273	0.0123	0.0009	0.0709	1,146.755	14.110	13.476	190.145
70	113.989	0.0088	0.0006	0.0706	1,614.134	14.160	13.666	193.519
75	159.876	0.0063	0.0004	0.0704	2,269.657	14.196	13.814	196.104
80	224.234	0.0045	0.0003	0.0703	3,189.063	14.222	13.927	198.075
85	314.500	0.0032	0.0002	0.0702	4,478.576	14.240	14.015	199.572
90	441.103	0.0023	0.0002	0.0702	6,287.185	14.253	14.081	200.704
95	618.670	0.0016	0.0001	0.0701	8,823.854	14.263	14.132	201.558
100	867.716	0.0012	0.0001	0.0701	12,381.662	14.269	14.170	202.200
				8%				
1	1.080	0.9259	1.0000	1.0800	1.000	0.926	0.000	0.000
2	1.166	0.8573	0.4808	0.5608	2.080	1.783	0.481	0.857
3	1.260	0.7938	0.3080	0.3880	3.246	2.577	0.949	2.445
4	1.360	0.7350	0.2219	0.3019	4.506	3.312	1.404	4.650
5	1.469	0.6806	0.1705	0.2505	5.867	3.993	1.846	7.372

Table A8.3   (*continued*)

N	Single Payment		Uniform Payment Series				Arithmetic Gradient	
	Compound Amount Factor: Find F Given P F/P	Present Worth Factor: Find P Given F, P/F	Sinking Fund Factor: Find A Given F, A/F	Capital Recovery Factor: Find A Given P, A/P	Capital Amount Factor: Find F Given A, F/A	Present Worth Factor: Find P Given A, P/A	Gradient Uniform Series: Find A Given G, A/G	Gradient Present Worth: Find P Given G, P/G
6	1.587	0.6302	0.1363	0.2163	7.336	4.623	2.276	10.523
7	1.714	0.5835	0.1121	0.1921	8.923	5.206	2.694	14.024
8	1.851	0.5403	0.0940	0.1740	10.637	5.747	3.099	17.806
9	1.999	0.5002	0.0801	0.1601	12.488	6.247	3.491	21.808
10	2.159	0.4632	0.0690	0.1490	14.487	6.710	3.871	25.977
11	2.332	0.4289	0.0601	0.1401	16.645	7.139	4.240	30.266
12	2.518	0.3971	0.0527	0.1327	18.977	7.536	4.596	34.634
13	2.720	0.3677	0.0465	0.1265	21.495	7.904	4.940	39.046
14	2.937	0.3405	0.0413	0.1213	24.215	8.244	5.273	43.472
15	3.172	0.3152	0.0368	0.1168	27.152	8.559	5.594	47.886
16	3.426	0.2919	0.0330	0.1130	30.324	8.851	5.905	52.264
17	3.700	0.2703	0.0296	0.1096	33.750	9.122	6.204	56.588
18	3.996	0.2502	0.0267	0.1067	37.450	9.372	6.492	60.843
19	4.316	0.2317	0.0241	0.1041	41.446	9.604	6.770	65.013
20	4.661	0.2145	0.0219	0.1019	45.762	9.818	7.037	69.090
21	5.034	0.1987	0.0198	0.0998	50.423	10.017	7.294	73.063
22	5.437	0.1839	0.0180	0.0980	55.457	10.201	7.541	76.926
23	5.871	0.1703	0.0164	0.0964	60.893	10.371	7.779	80.673
24	6.341	0.1577	0.0150	0.0950	66.765	10.529	8.007	84.300
25	6.848	0.1460	0.0137	0.0937	73.106	10.675	8.225	87.804
26	7.396	0.1352	0.0125	0.0925	79.954	10.810	8.435	91.184
27	7.988	0.1252	0.0114	0.0914	87.351	10.935	8.636	94.439
28	8.627	0.1159	0.0105	0.0905	95.339	11.051	8.829	97.569
29	9.317	0.1073	0.0096	0.0896	103.966	11.158	9.013	100.574
30	10.063	0.0994	0.0088	0.0888	113.283	11.258	9.190	103.456
31	10.868	0.0920	0.0081	0.0881	123.346	11.350	9.358	106.216
32	11.737	0.0852	0.0075	0.0875	134.214	11.435	9.520	108.857
33	12.676	0.0789	0.0069	0.0869	145.951	11.514	9.674	111.382
34	13.690	0.0730	0.0063	0.0863	158.627	11.587	9.821	113.792
35	14.785	0.0676	0.0058	0.0858	172.317	11.655	9.961	116.092
40	21.725	0.0460	0.0039	0.0839	259.057	11.925	10.570	126.042
45	31.920	0.0313	0.0026	0.0826	386.506	12.108	11.045	133.733
50	46.902	0.0213	0.0017	0.0817	573.770	12.233	11.411	139.593
55	68.914	0.0145	0.0012	0.0812	848.923	12.319	11.690	144.006
60	101.257	0.0099	0.0008	0.0808	1,253.213	12.377	11.902	147.300
65	148.780	0.0067	0.0005	0.0805	1,847.248	12.416	12.060	149.739
70	218.606	0.0046	0.0004	0.0804	2,720.080	12.443	12.178	151.533
75	321.205	0.0031	0.0002	0.0802	4,002.557	12.461	12.266	152.845
80	471.955	0.0021	0.0002	0.0802	5,886.935	12.474	12.330	153.800
85	693.456	0.0014	0.0001	0.0801	8,655.706	12.482	12.377	154.492

(*continued overleaf*)

Table A8.3   (*continued*)

N	Single Payment		Uniform Payment Series				Arithmetic Gradient	
	Compound Amount Factor: Find F Given P F/P	Present Worth Factor: Find P Given F, P/F	Sinking Fund Factor: Find A Given F, A/F	Capital Recovery Factor: Find A Given P, A/P	Capital Amount Factor: Find F Given A, F/A	Present Worth Factor: Find P Given A, P/A	Gradient Uniform Series: Find A Given G, A/G	Gradient Present Worth: Find P Given G, P/G
90	1,018.915	0.0010	0.0001	0.0801	12,723.939	12.488	12.412	154.993
95	1,497.121	0.0007	0.0001	0.0801	18,701.507	12.492	12.437	155.352
100	2,199.761	0.0005	0.0000	0.0800	27,484.516	12.494	12.455	155.611

				9%				
1	1.090	0.9174	1.0000	1.0900	1.000	0.917	0.000	0.000
2	1.188	0.8417	0.4785	0.5685	2.090	1.759	0.478	0.842
3	1.295	0.7722	0.3051	0.3951	3.278	2.531	0.943	2.386
4	1.412	0.7084	0.2187	0.3087	4.573	3.240	1.393	4.511
5	1.539	0.6499	0.1671	0.2571	5.985	3.890	1.828	7.111
6	1.677	0.5963	0.1329	0.2229	7.523	4.486	2.250	10.092
7	1.828	0.5470	0.1087	0.1987	9.200	5.033	2.657	13.375
8	1.993	0.5019	0.0907	0.1807	11.028	5.535	3.051	16.888
9	2.172	0.4604	0.0768	0.1668	13.021	5.995	3.431	20.571
10	2.367	0.4224	0.0658	0.1558	15.193	6.418	3.798	24.373
11	2.580	0.3875	0.0569	0.1469	17.560	6.805	4.151	28.248
12	2.813	0.3555	0.0497	0.1397	20.141	7.161	4.491	32.159
13	3.066	0.3262	0.0436	0.1336	22.953	7.487	4.818	36.073
14	3.342	0.2992	0.0384	0.1284	26.019	7.786	5.133	39.963
15	3.642	0.2745	0.0341	0.1241	29.361	8.061	5.435	43.807
16	3.970	0.2519	0.0303	0.1203	33.003	8.313	5.724	47.585
17	4.328	0.2311	0.0270	0.1170	36.974	8.544	6.002	51.282
18	4.717	0.2120	0.0242	0.1142	41.301	8.756	6.269	54.886
19	5.142	0.1945	0.0217	0.1117	46.018	8.950	6.524	58.387
20	5.604	0.1784	0.0195	0.1095	51.160	9.129	6.767	61.777
21	6.109	0.1637	0.0176	0.1076	56.765	9.292	7.001	65.051
22	6.659	0.1502	0.0159	0.1059	62.873	9.442	7.223	68.205
23	7.258	0.1378	0.0144	0.1044	69.532	9.580	7.436	71.236
24	7.911	0.1264	0.0130	0.1030	76.790	9.707	7.638	74.143
25	8.623	0.1160	0.0118	0.1018	84.701	9.823	7.832	76.926
26	9.399	0.1064	0.0107	0.1007	93.324	9.929	8.016	79.586
27	10.245	0.0976	0.0097	0.0997	102.723	10.027	8.191	82.124
28	11.167	0.0895	0.0089	0.0989	112.968	10.116	8.357	84.542
29	12.172	0.0822	0.0081	0.0981	124.135	10.198	8.515	86.842
30	13.268	0.0754	0.0073	0.0973	136.308	10.274	8.666	89.028
31	14.462	0.0691	0.0067	0.0967	149.575	10.343	8.808	91.102
32	15.763	0.0634	0.0061	0.0961	164.037	10.406	8.944	93.069
33	17.182	0.0582	0.0056	0.0956	179.800	10.464	9.072	94.931
34	18.728	0.0534	0.0051	0.0951	196.982	10.518	9.193	96.693
35	20.414	0.0490	0.0046	0.0946	215.711	10.567	9.308	98.359

Table A8.3   (*continued*)

	Single Payment		Uniform Payment Series				Arithmetic Gradient	
	Compound Amount Factor: Find F Given P	Present Worth Factor: Find P Given F,	Sinking Fund Factor: Find A Given F,	Capital Recovery Factor: Find A Given P,	Capital Amount Factor: Find F Given A,	Present Worth Factor: Find P Given A,	Gradient Uniform Series: Find A Given G,	Gradient Present Worth: Find P Given G,
N	F/P	P/F	A/F	A/P	F/A	P/A	A/G	P/G
40	31.409	0.0318	0.0030	0.0930	337.882	10.757	9.796	105.376
45	48.327	0.0207	0.0019	0.0919	525.859	10.881	10.160	110.556
50	74.358	0.0134	0.0012	0.0912	815.084	10.962	10.430	114.325
55	114.408	0.0087	0.0008	0.0908	1,260.092	11.014	10.626	117.036
60	176.031	0.0057	0.0005	0.0905	1,944.792	11.048	10.768	118.968
65	270.846	0.0037	0.0003	0.0903	2,998.288	11.070	10.870	120.334
70	416.730	0.0024	0.0002	0.0902	4,619.223	11.084	10.943	121.294
75	641.191	0.0016	0.0001	0.0901	7,113.232	11.094	10.994	121.965
80	986.552	0.0010	0.0001	0.0901	10,950.574	11.100	11.030	122.431
85	1,517.932	0.0007	0.0001	0.0901	16,854.800	11.104	11.055	122.753
90	2,335.527	0.0004	0.0000	0.0900	25,939.184	11.106	11.073	122.976
95	3,593.497	0.0003	0.0000	0.0900	39,916.635	11.108	11.085	123.129
100	5,529.041	0.0002	0.0000	0.0900	61,422.675	11.109	11.093	123.234
				10%				
1	1.100	0.9091	1.0000	1.1000	1.000	0.909	0.000	0.000
2	1.210	0.8264	0.4762	0.5762	2.100	1.736	0.476	0.826
3	1.331	0.7513	0.3021	0.4021	3.310	2.487	0.937	2.329
4	1.464	0.6830	0.2155	0.3155	4.641	3.170	1.381	4.378
5	1.611	0.6209	0.1638	0.2638	6.105	3.791	1.810	6.862
6	1.772	0.5645	0.1296	0.2296	7.716	4.355	2.224	9.684
7	1.949	0.5132	0.1054	0.2054	9.487	4.868	2.622	12.763
8	2.144	0.4665	0.0874	0.1874	11.436	5.335	3.004	16.029
9	2.358	0.4241	0.0736	0.1736	13.579	5.759	3.372	19.421
10	2.594	0.3855	0.0627	0.1627	15.937	6.145	3.725	22.891
11	2.853	0.3505	0.0540	0.1540	18.531	6.495	4.064	26.396
12	3.138	0.3186	0.0468	0.1468	21.384	6.814	4.388	29.901
13	3.452	0.2897	0.0408	0.1408	24.523	7.103	4.699	33.377
14	3.797	0.2633	0.0357	0.1357	27.975	7.367	4.996	36.800
15	4.177	0.2394	0.0315	0.1315	31.772	7.606	5.279	40.152
16	4.595	0.2176	0.0278	0.1278	35.950	7.824	5.549	43.416
17	5.054	0.1978	0.0247	0.1247	40.545	8.022	5.807	46.582
18	5.560	0.1799	0.0219	0.1219	45.599	8.201	6.053	49.640
19	6.116	0.1635	0.0195	0.1195	51.159	8.365	6.286	52.583
20	6.727	0.1486	0.0175	0.1175	57.275	8.514	6.508	55.407
21	7.400	0.1351	0.0156	0.1156	64.002	8.649	6.719	58.110
22	8.140	0.1228	0.0140	0.1140	71.403	8.772	6.919	60.689
23	8.954	0.1117	0.0126	0.1126	79.543	8.883	7.108	63.146
24	9.850	0.1015	0.0113	0.1113	88.497	8.985	7.288	65.481
25	10.835	0.0923	0.0102	0.1102	98.347	9.077	7.458	67.696

(*continued overleaf*)

Table A8.3 (*continued*)

N	Single Payment		Uniform Payment Series				Arithmetic Gradient	
	Compound Amount Factor: Find F Given P F/P	Present Worth Factor: Find P Given F, P/F	Sinking Fund Factor: Find A Given F, A/F	Capital Recovery Factor: Find A Given P, A/P	Capital Amount Factor: Find F Given A, F/A	Present Worth Factor: Find P Given A, P/A	Gradient Uniform Series: Find A Given G, A/G	Gradient Present Worth: Find P Given G, P/G
26	11.918	0.0839	0.0092	0.1092	109.182	9.161	7.619	69.794
27	13.110	0.0763	0.0083	0.1083	121.100	9.237	7.770	71.777
28	14.421	0.0693	0.0075	0.1075	134.210	9.307	7.914	73.650
29	15.863	0.0630	0.0067	0.1067	148.631	9.370	8.049	75.415
30	17.449	0.0573	0.0061	0.1061	164.494	9.427	8.176	77.077
31	19.194	0.0521	0.0055	0.1055	181.943	9.479	8.296	78.640
32	21.114	0.0474	0.0050	0.1050	201.138	9.526	8.409	80.108
33	23.225	0.0431	0.0045	0.1045	222.252	9.569	8.515	81.486
34	25.548	0.0391	0.0041	0.1041	245.477	9.609	8.615	82.777
35	28.102	0.0356	0.0037	0.1037	271.024	9.644	8.709	83.987
40	45.259	0.0221	0.0023	0.1023	442.593	9.779	9.096	88.953
45	72.890	0.0137	0.0014	0.1014	718.905	9.863	9.374	92.454
50	117.391	0.0085	0.0009	0.1009	1163.909	9.915	9.570	94.889
55	189.059	0.0053	0.0005	0.1005	1880.591	9.947	9.708	96.562
60	304.482	0.0033	0.0003	0.1003	3034.816	9.967	9.802	97.701
65	490.371	0.0020	0.0002	0.1002	4893.707	9.980	9.867	98.471
70	789.747	0.0013	0.0001	0.1001	7887.470	9.987	9.911	98.987
75	1,271.90	0.0008	0.0001	0.1001	12708.954	9.992	9.941	99.332
80	2,048.40	0.0005	0.0000	0.1000	20474.002	9.995	9.961	99.561
85	3,298.97	0.0003	0.0000	0.1000	32979.690	9.997	9.974	99.712
90	5,313.02	0.0002	0.0000	0.1000	53,120.226	9.998	9.983	99.812
95	8,556.68	0.0001	0.0000	0.1000	85,556.760	9.999	9.989	99.877
100	13,780.61	0.0001	0.0000	0.1000	137,796.123	9.999	9.993	99.920

CHAPTER 9

Economic Development Impacts

The chief business of the American people is business.
—Calvin Coolidge (1872–1933)

INTRODUCTION

Transportation plays a vital role in the economy of any nation. On the whole, this is reflected in its large contribution to national gross domestic product (GDP), its consumption of a large amount of goods and services, employment of a large number of people, and the revenue it makes available to federal, state, and local governments. Summary statistics indicate strong relationships between gross domestic product and travel (Figure 9.1). Since the 1930s, growth in the GDP and vehicle-miles of travel (VMT) have exhibited similar patterns, even during the period of energy disruptions of the 1970s (USDOT, 2005). The economy and transportation have a bidirectional relationship: increased economic output leads to an increased amount of travel, and increased travel leads to higher economic output. Such a relationship suggests that the econometric phenomenon known as *simultaneity* exists between transportation and the economy.

Studies have demonstrated that investments in highways and other public transport capital reduce the costs of transportation and production, and consequently, contribute to economic growth and productivity. The USDOT (2005) reported that every $1 billion invested in transportation infrastructure generates more than $2 billion in economic activity and creates up to 42,000 jobs. It has been estimated that highway construction directly generates an average of 7.9 jobs per $1 million spent (1996 dollars) on construction (Keane, 1996); public transportation directly supports an average of 24.5 jobs per million passenger-miles; and air transportation supports as

many as 1000 on-site jobs per 100,000 annual passengers, depending on site-specific factors (Weisbrod and Weisbrod, 1997).

In general, economic development impacts should be considered when the transportation project requires substantial investment and/or when public concerns are significant. In this chapter we present the concept of economic development as a performance criterion for transportation system evaluation and we provide a methodology for assessing the economic development impacts of transportation projects.

9.1 ECONOMIC DEVELOPMENT IMPACT TYPES

9.1.1 Economic Development Impact Types

Economic development impact types or performance measures can generally be categorized as follows (Bendavid-Val, 1991; De Rooy, 1995; McConnell and Brue, 1999; Weisbrod, 2000):

1. Impact types relating to overall area economy, such as economic output, gross regional product, value added, personal income, and employment
2. Impact types relating to specific aspects of economic development such as productivity, capital investment, property appreciation, and fiscal impacts that include tax revenues and public expenditure

Economic development impact types are strongly related to each other, and in some cases, two or more impact types present different perspectives of the same type of economic development changes. For example, increased number of jobs in a region is often strongly correlated with higher wages and higher income tax revenue. Increased capital investment in a region is also often associated with increased property values and higher levels of tax revenue from businesses and property tax. As such, evaluation by simple addition of the individual impacts may lead to double-counting. For example, the benefits of truck travel time savings should not be counted separately from increased industrial competitiveness (due to lower transportation costs) resulting from time savings. It seems therefore, reasonable for transportation agencies to utilize only a few economic development impact types in evaluating transportation projects or programs, and the selection of these impact types should be made on the basis of project or program objectives and data availability. Weisbrod and Beckwith (1990) presented an evaluation technique that helps to avoid double-counting. In that technique, economic development benefits are measured in terms of changes in disposable income,

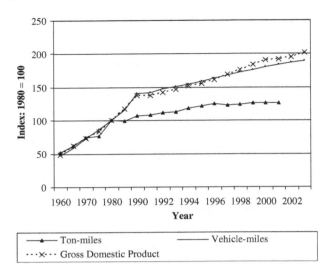

Figure 9.1   U.S. travel and GDP trends, 1960–2003). (From US DOT, 2005.)

and all other impacts not embodied in that performance measure, such as travel time and safety, are estimated separately.

9.1.2 Economic Development Impact Mechanisms

The mechanisms by which transportation projects can impact the economy can be broadly classified as follows (Forkenbrock and Weisbrod, 2001):

(*a*) *Direct Mechanism*   The most significant impact of transportation investments on the economy is the reduction of transportation costs. With increased direct benefits (reduction in crashes, travel time, and vehicle operating costs) offered to users of improved transportation facilities, businesses in the region are afforded improved accessibility to markets and resources (labor, materials, and equipment) and consequently, reap the benefits of reduced business costs and enhanced productivity. Other direct effects include temporary impacts such as short-term wealth and job creation from spending on construction and ongoing operations. Construction-period impacts can be important, especially if they are large in relation to the economy affected, as in some developing countries.

(*b*) *Indirect Mechanism*   Any significant change in business activity due to direct effects will in turn have impacts on "secondary" entities such as local businesses that supply materials and equipment to businesses that are affected directly. Detailed guidelines for estimating the indirect effects of proposed transportation projects are presented by NCHRP (1998).

(*c*) *Induced Mechanism*   Increased personal wages in a region may induce increased spending. This would lead to induced benefits to businesses that provide utilities, groceries, apparel, communications, and other consumer services in the region.

(*d*) *Dynamic Mechanism*   This involves long-term changes in economic development and related parameters such as business location patterns, workforce, labor costs, prices, and resulting land-use changes. These changes in turn affect income and wealth in the area. In some cases, such changes in economic development invites growth that would have occurred elsewhere if the transport investment did not take place. Thus, the geographic scope of the evaluation is an important aspect of such analyses, as discussed in the next section.

The total impact on the economy is estimated as the sum of benefits accrued through all four mechanisms. The ratio of total effect and direct effect is generally termed an *economic multiplier*. Effects that are not direct are often referred to as *multiplier effects*. Figure 9.2 illustrates the functional interrelationships between different economic development impacts types that are typically used in calculating economic multipliers.

9.1.3 Selection of Appropriate Measures of Economic Impact

The outcome of economic development impact assessment of transportation projects can be influenced by the *spatial scope (geographic scale)* selected for the evaluation. For relatively small study areas, the location movements of businesses will probably be perceived as "new activities," while for relatively larger areas, such movements will probably be seen as "internal redistributions" of business activity within the study area (Weisbrod, 2000). In large study areas, it has been found that internal redistributions of activity typically have little or no impact on total regional economic activity.

Closely related to the spatial scope of the analysis is the *project/program scope*. Available literature suggests that the nature and magnitude of economic development impacts of transportation investments depend on whether the transportation stimulus is just a means of providing access (typically, a microlevel stimulus that is relatively small in impact area), a program-level stimulus (typically, affecting a network of transportation facilities in a relatively large area), or a project-level stimulus that falls between these two extremes. Construction of a new interstate highway interchange is an example of microlevel stimulus.

In addition to the spatial scope of the economic development impacts of the transportation investment, the

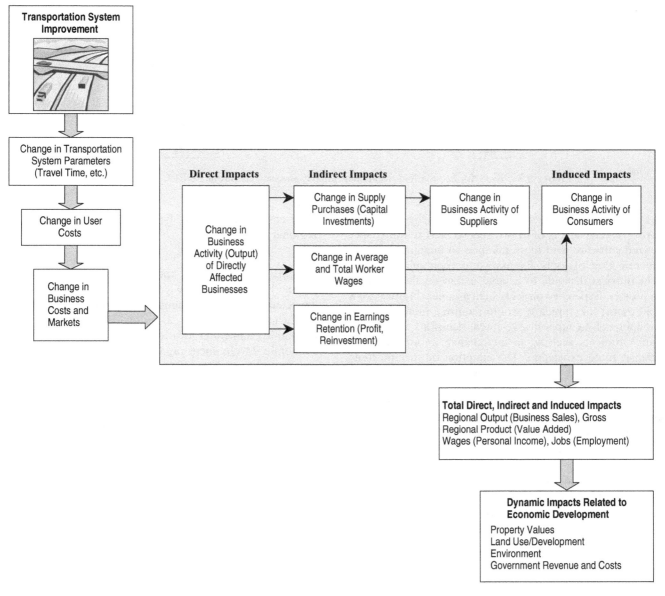

Figure 9.2 Types and mechanisms of economic development impacts.

relative maturity of a transportation system at the time of the investment needs to be considered (Forkenbrock and Foster, 1996). The introduction of new transportation infrastructure into an area with a relatively undeveloped transportation system will generally have a larger impact than when it is implemented in an area with a mature system. The same could be said regarding the relative size and "maturity" of the underlying economy itself. Impacts of transportation investments in poor sustenance economies are likely to differ from those in wealthy

industrialized economies. In the latter case, impacts can be marginal (CUBRC et al., 2001).

9.2   TOOLS FOR ECONOMIC DEVELOPMENT IMPACT ASSESSMENT

The tools that have typically been used to assess economic impacts range from highly qualitative and less data intensive (i.e., surveys and interviews) to highly quantitative (i.e., economic simulation models), as shown

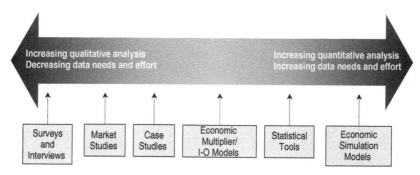

Figure 9.3 Tools for economic development impact assessment. (From CUBRC et al., 2001.)

in Figure 9.3. The latter group of approaches typically involve greater levels of effort, special staff training, specialized software, and more reliance on quantitative data. The selection of tools for assessing economic development impacts depends to a large extent on the scope of the project. Expensive projects such as a new highway or a rapid transit line typically would require a more quantitative approach to support investment decisions than would routine projects such as an interchange improvement or a transit route expansion. The adoption of specific tools is also influenced by the type and amount of resources available, including the level of analytical expertise. A discussion of the tools presented in Figure 9.3 is provided below. Further information on these tools is available in NCHRP Report 456 (Forkenbrock and Weisbrod, 2001) and Weisbrod (2000).

9.2.1 Surveys and Interviews

One method of assessing the expected economic development impacts from transportation investments is to conduct interviews with local businesses, local governmental officials, and community or neighborhood leaders. Survey-type methods used for economic development impact analysis include:

- Expert interviews
- Business surveys
- Shopper origin–destination surveys
- Corridor inventory methods using vehicle origin–destination logs

The first two tools (interviews and surveys of personnel involved in economic development analysis, affected businesses, and facility users such as trucking operators) typically provide valuable insights to potential impact types and mechanisms. Such tools also provide a direct and practical basis for establishing impact scenarios of the physical stimulus or policy change proposed. Also, these tools can indicate whether there will be increased

local competition among businesses or improved overall regional competitiveness.

(*a*) *Expert Interviews*   Expert interviews involve soliciting the judgment of knowledgeable persons regarding the expected impacts of a change in the transportation system on business activities in a region. Experts may include economic planners at local or state government level and economic development organizations who have acquired accumulated experience in business conditions at a particular locality or region. This tool has been used widely, for example, in Florida (Cambridge Systematics et al., 1999b) and in Scotland (Halcrow Fox, 1996). The application of both interview-based methods and forecasting models (Cambridge Systematics, 1996; EDRG and Bernardin, 1998; N-Y Associates and EDRG, 1999) allows an agency to cross-check impacts predicted using either tool, thereby increasing confidence in study findings.

How to Carry Out an Expert Interview: The analysis begins with the development of one or more scenarios representing how travel conditions, business costs, and market access may change after implementation of a transportation project. Key business representatives, developers, and planners are then asked about their perceptions of existing transportation needs, existing barriers, constraints, or threats to economic growth in the community, and how the project under consideration would likely affect economic growth prospects of existing businesses and new businesses that might be attracted to the area. The discussion can also include economic development transfer effects, such as long-term population gains, long-term employment gains, and long-term property value increases. Building or improving a highway corridor may reduce the benefits derived from existing highways in the transportation network (Forkenbrock and Foster, 1996). Also, if any of the economic activity attracted to the corridor is shifted from other sites within the state, that activity cannot be viewed as new economic development but rather as a transfer from one location to another within the same study region. Identification of such transfer effects are critical in the evaluation process (Forkenbrock, 1991). Expert interviews typically focus on specific topics which may include locations (such as particular neighborhoods in a city or different communities in a region) and industries (which represent existing dominant sectors in the economy or special growth opportunities). The interviews can take the form of one-on-one conversations, written surveys, or focus-group discussions that bring a range of participants together to exchange ideas on the likely impacts of an investment. Inconsistency of survey results, a disadvantage of group interviews, can be reduced with use of a Delphi process (Dickey and Watts, 1978). In this process, experts are interviewed individually, informed of the initial results obtained from the group of experts, and given the opportunity to revise their responses in light of the responses from the other experts. This process can help the experts to achieve a consensus. Impact types commonly used in expert interview methods are business sales and property values. Changes in employment and wages are typically assumed to be proportional to changes in business sales. A drawback of the expert interview method is that expectations for change, adverse or beneficial, can be subject to misrepresentation due to local political or other agenda (Forkenbrock and Weisbrod, 2001).

(b) *Business Surveys* Business surveys are typically designed to collect quantitative and qualitative data regarding the potential short-term impacts during construction as well as long-term effects of a proposed project on business activities. Business surveys can be implemented using questionnaires that are mailed to target groups or by interviews conducted in person or by telephone. Also, the Internet is increasingly being used to post such surveys. The target groups for business surveys include local business leaders, representatives of business organizations, and transportation-related organizations such as individual or corporate truckers. Advantages of the questionnaire survey tool over the more personal interview tool include a larger number of respondents, but a disadvantage is that it generally requires greater follow-up efforts to mitigate selection bias and to achieve sample-size targets. Some business surveys may also involve a panel of experts. Interview methods are probably more effective than questionnaires in avoiding panel attrition and therefore are often used for panel surveys. Available literature contains several examples of business surveys (Peat Marwick Main & Co., 1988; W.S. Atkins et al.,

1990; Bechtel Corporation, et al., 1994; Gillis and Casavant, 1994; N-Y Associates and EDRG, 1999).

How to Carry Out a Business Survey: A group of business establishments along the transportation corridor can be selected through stratified random sampling on the basis of size and type of business establishment. The survey can include employees, customers, business owners, and managers. Survey participants can be asked about their current commuting patterns (i.e., transportation mode and residential location) and how the proposed transportation improvement could impact their commutes. Business owners and managers can be asked about their customer and delivery markets and the possible business cost savings associated with the transportation project proposed. The responses should be analyzed and interpreted with caution as some survey respondents may tend to provide unsubstantiated opinions motivated by parochial interests.

A business survey was conducted using a questionnaire to assess the economic impact on businesses during the reconstruction of I-65/70 in downtown Indianapolis (Sinha et al., 2004). Affected businesses in the study area were comprised mostly of restaurants, retail stores, entertainment-related establishments, motels, and hotels. The questionnaire was mailed to 504 businesses on a list furnished by the Indianapolis Downtown Business Association. The extent of financial impacts was assessed through a five-point-scale question. In a retrospective study, Palmer et al. (1986) used business surveys to estimate the effects of road construction on adjacent economic activities.

(c) *Shopper Origin–Destination Surveys* The impact of a proposed transportation facility on a community's economy can be estimated by surveying shoppers either by interviews or by self-administered questionnaires. Surveys of shoppers can also provide information on how their trip-making characteristics could be affected by changes in cost, convenience, or time involved in accessing various shopping areas during or after the project implementation. Examples of application include Cambridge Systematics (1989a), Yeh et al. (1998), EDRG and SRF Consulting Group (1999), and Lichtman (1999).

How to Carry Out a Shopper Origin–Destination Survey: A bus rapid transit line is proposed to serve passengers traveling from a suburb to a new shopping mall 10 miles away. A household survey can be conducted in the suburb to identify current shopping locations, frequency of shopping trips, origin and shopping trip time, and how the proposed new transit line may affect shopping patterns.

(*d*) *Corridor Inventory Methods*   Corridor inventory methods include windshield surveys, vehicle origin–destination logs, and business activity data collection. Windshield surveys are inventories of business activity types and levels (such as sales volume), and conditions existing along a transportation route. These are typically conducted by traveling through the corridor where changes are proposed and using origin–destination logs of trucks to describe how the existing transportation network is used by businesses and to identify the type and value of shipments. This tool can yield general information on economic vitality in a corridor or area. Vehicle origin–destination logs can be used to gather data on shipment types and values. Geographical information systems (GISs) are used increasingly to store business activity data and vehicle origin–destination logs in a geo-coded format, and to map the travel patterns of business suppliers, customers, and "on-the-clock" workers.

After the data on local businesses have been collected and collated, the dependence of each type of business establishment on the mode of travel is assessed, and the potential reduction in transportation-related business costs due to the proposed transportation project, can be estimated. Spreadsheet-based models have been developed to assess separately business dependence on traffic flow changes that either (1) inhibit businesses' local access, (2) bypass them, or (3) take their property (Cambridge Systematics, 1996; Weisbrod and Neuwirth, 1998). A number of studies have utilized business vehicle logs for assessing the economic development impacts of proposed transportation facilities (Cambridge Systematics, 1989b; EDRG and Bernardin, 1998). Windshield surveys have been conducted as part of the Wisconsin Highway 29 Study (Cambridge Systematics, 1989b) and the Southwest

Indiana Highway Corridor Study (Cambridge Systematics, 1996).

How to Develop a Spreadsheet-Based Model: For purposes of illustration, consider a highway corridor improvement that increases traffic throughput and total volumes in an area network. The project also reduces direct driveway access to some area businesses. A four-step spreadsheet-based model can be used to assess the vulnerability of local business establishments to future accessibility losses associated with the proposed transportation change, as follows (Weisbrod and Neuwirth, 1998):

1. Compile an inventory of businesses along the affected route. This corresponds to column A of Table 9.1.
2. Use business or customer interview data to estimate the extent to which each business along the route depends on the volume of area traffic. Alternatively, professional judgment based on direct observations or results from prior studies may be used to estimate the degree of business sensitivity to area traffic. This corresponds to column B in Table 9.1. In this example it is assumed that a business's customers are persons with an original intent of visiting that business or persons who were attracted to that business only because they perceived the business sign from a distance and thus decided to visit that business.
3. Obtain estimates of the change in traffic levels expected along the corridor and of accessibility losses of businesses due to changes such as new median islands that block access to from the other side of the road and left-turn restrictions. This corresponds to columns C and D in Table 9.1.
4. Use a spreadsheet collation of the data (such as shown in Table 9.1) to calculate the overall effect on business sales. The basic formula for estimating

Table 9.1   Sample Spreadsheet Computation of Pass-by Traffic Effect

(A) Inventory of Business	(B) Percent of Customers from Pass-by Traffic	(C) Expected % Change in Total (Both Directions) Pass-by Traffic	(D) Expected % Change in Bypass Traffic Unable to Access Store	(E) Overall % Change in Retail Sales
Double X Gas Station	100	35	55	−39
Big Bun Fast Food	70	35	0	25
Comfort Hotel	15	35	5	4
Fishbone Restaurant	100	35	30	−6

the overall change in retail sales is

$$
\text{col. E} = \left\{ \frac{\text{col. B}}{100} \left[\left(1 + \frac{\text{col. C}}{100} \right) \left(1 - \frac{\text{col. D}}{100} \right) \right] \right.
$$
$$
\left. - \frac{\text{col. B}}{100} \right\} \times 100
$$

9.2.2 Market Studies

Market studies are typically smaller-scale analyses that typically relate to redistribution within a region, not across regions. Market studies can help estimate the existing levels of supply (i.e., land or business locations) and demand (i.e., sales) for key business activities in the analysis area, and typically are able to forecast potential future growth in specific business markets and to estimate how much business growth could be expected with improved transportation services and reduced costs. An inherent assumption is that the proposed project changes the size of the customer market (i.e., change in the level of pass-by traffic or in the breadth of the market area). Such markets for key business activities in an area include those for offices, tourism, and real estate. Using market studies for retail businesses, for instance, an analyst can predict the area (in square feet) of new retail development likely to occur after implementation of the transportation project, and the increase in property values, increase in tax revenues from the new retail businesses, and job increases due to the new development. Market studies generally use site analysis tools or corridor-specific tools, such as the windshield survey discussed in Section 9.2.1(d), to complement other evaluation tools. *Gravity models* are also used in market studies to predict effects on business activities by estimating changes in accessibility to market opportunities, represented as residential access to workplaces or shopping centers, or business access to labor markets or customer markets. Changes in business activities are assumed to be proportional to changes in accessibility resulting from the proposed transportation project. A gravity measure of accessibility to a business location can be obtained by weighting market opportunities by the impedance (e.g., travel cost or travel time) to reach the markets, as follows (CUBRC et al., 2001):

$$
A_i = \sum_j \frac{D_j}{t_{ij}^\alpha} \tag{9.1}
$$

where A_i is the accessibility of location i, D_j is the number of market opportunities of a particular type (shopping, business, or other commercial) at location j,

t_{ij} is the generalized time or cost of travel from i to j, and α is a calibrating factor, typically between 1.5 and 2.0.

The market study approach has been used in economic development impact assessments in New York (Clark Patterson Associates et al., 1998), Maryland (Maryland DOT, 1998), San Diego (SDAG, 1996), Connecticut (Bechtel Corporation et al., 1994), and Massachusetts (Cambridge Systematics, 1988).

How to Conduct a Market Study: Consider the situation described in Section 9.2.1(c). A market study can be conducted to assess the likely change in market sales due to changes in access to the shopping center associated with the new transit service in a manner consistent with the procedure described by Forkenbrock and Weisbrod (2001). The extent of customer attraction is a simple calculation that relates (1) the market share observed for shopping centers in the study area to (2) the relative travel-time and cost of accessing them from different parts of the study area, compared with the time and cost of accessing competing shopping areas. The potential shopper base in each major part of the study area can be estimated using shopping surveys. Table 9.2, an example of survey results, assumes that shopping center i is the only shopping center that gains increased accessibility and market share due to the new transit line. Accessibility indices associated with the shopping center from each of the residential market areas, as well as a composite index of accessibility to the shopping center, can be computed using the gravity model formula. The composite index is a weighted average of the area accessibility values weighted by the number of households in each area. This index can be interpreted as a proportional change in retail sales for the shopping center resulting from the proposed transportation project. For the illustrative example given, the computations are shown in Table 9.3.

9.2.3 Comparative Analysis Tools: Case Studies

In comparative analysis (or case studies), it is assumed that the impact of the proposed transportation improvement on the area economy will be a close reflection of the impacts of a past similar intervention elsewhere. This approach is appropriate in situations where the study area is small, available economic data are limited, and where parallels to experiences elsewhere can be established easily and confidently (Weisbrod, 2000). Transportation projects for which this tool has been used include community bypasses, interchanges, added transit stations, and airports. This tool is particularly compatible with public hearings because case studies facilitate understanding and appreciation by lay people compared to complex economic analyses. The primary drawback to this tool is that the

Table 9.2   Gravity Model Input Data on Accessibility to the Shopping Area

Market Area: Place of Residence, j	Total Market Opportunities at Market Area, j^a (D_j)	Average Travel Time Between Shopping Area i and Market Area, j (t_{ij})	
		Base Case (Current), mins	With New Transit Line (Proposed), mins
Downtown	4,000	35	25
Northern suburb	7,000	30	30
Southern suburb	10,000	55	45
Eastern suburb	6,000	45	35
Western suburb	9,000	15	15

[a] Assumed to be same as the number of households. A more detailed model may incorporate market area average income or other demographic features.

Table 9.3   Gravity Model Calculations of Accessibility to the Shopping Area[a]

Market Area: Place of Residence, j	Gravity Model Market Index, D_j/t_{ij}^2		Percent Change
	Base Case (Current)	With New Transit Line (Proposed)	
Downtown	3.3	6.4	96
Northern suburb	7.8	7.8	0
Southern suburb	3.3	4.9	49
Eastern suburb	3.0	4.9	65
Western suburb	40.0	40.0	0
Composite index	13.3	14.4	8
Market share	18%[b]	19.4%[c]	+8

[a] In situations where there are other shopping areas that would become more accessible due to the new transit facility, the index values for all shopping areas in the region should be calculated and assessed for relative changes to determine the actual shifts in market shares.

[b] Initial market share measured before the new transit service.

[c] New market share estimated given the initial market share and the percent change in composite index due to the proposed transportation project.

selection of appropriate case studies to use for comparison purposes can be fairly subjective, and it is impossible to control for all the influential variables.

How to Carry Out a Comparative Case Study: Steps involved in a comparative case study are as follows:

1. *Identify case studies of similar transportation changes.* Identify similar projects in recent years and determine whether there are any existing case studies. If no such study exists, studies can be undertaken to assess postimplementation effects. For example, to estimate the impacts of the proposed Denver Airport (Colorado

National Banks, 1989), case studies of the economic effects of constructed or expanded airports at Dallas–Fort Worth, Atlanta, and Kansas City were conducted.

2. *Determine the factors affecting the local context.* The local setting of the proposed transportation project may be a small town, a downtown area, a suburban area, or a rural region. Its economy may be focused on tourism, manufacturing, commerce, or agriculture, etc., or a mix thereof. The local situation for the project under investigation should be adequately described to assess the appropriateness of available case studies. For example, the Denver Airport study (Colorado National Banks, 1989) examined similarities and differences between that airport and

three previously constructed airports in terms of business mix and growth, the timing of growth, key supporting infrastructure, airport site development policies, supportive public policies, and international flights.

3. *Assess the implications of case study findings for the project proposed.* Depending on the degree of match in terms of project type and context, case studies can offer predictions of economic development impacts that may turn out to be good estimates or may deviate substantially from the true impacts. Adjustments to the predictions from case studies may be necessary.

9.2.4 Economic Multiplier/Input–Output Models

The economic multiplier approach is a quantitative impact assessment method that is most applicable to investment-driven transportation projects that impact business attraction, expansion, retention, or tourism directly. Typical multipliers are expressed in terms of regional economic output, employment, or income. Their magnitudes vary depending on the type of transportation investment and its relationship to other investments in the regional economy, and the size of the existing regional economy. As a rule of thumb, the economic output multiplier values for most transportation investments are: 2.5 to 3.5 for national impacts, 2.0 to 2.5 for state impacts, and 1.5 to 2.0 for local area impacts (Weisbrod and Weisbrod, 1997). For example, if a $10 million highway improvement takes place along a corridor, it can be expected that the net impact on the local level of economic activity in the study area would increase by 15–20 million dollars. Assessing the economic development impacts of transportation projects on the basis of economic multipliers should be carried out with caution because multipliers typically involve attractions from other regions.

The economic multiplier approach is based largely on *input–output modeling.* Input–output models are essentially accounting frameworks that track interindustry transactions such as the number of units of purchases (inputs) that each industry requires from all industries to produce 1 unit of sales (output). These models provide a means for calculating the indirect and induced effects on business sales and spending, given a set of direct project effects on business sales, employment, or wages. A limitation of this methodology is that interindustry relationships are derived from national forecasts, which are not necessarily applicable to lower levels of the analysis. Furthermore, input–output models are static. They must be used in conjunction with a broader set of techniques to forecast the effects of long-term economic development. In the United States, three major software packages have been used for input–output modeling: Implan (Minnesota Implan Group, 2004), RIMS II (US DOC, 1997), and PC input–output (Reg. Science Research Corporation, 1996).

IMPLAN and PC input–output ask the user to provide a description of the direct effects of an investment and then automatically generate estimates of the indirect, induced, and total economic development effects of the project. On the other hand, RIMS II provides a default set of input–output multipliers that users may apply to their own data. Some state transportation agencies have customized input–output models (Babcock, 2004). An example of the input–output methodology is presented below. This follows the steps given by the Minnesota Implan Group (2004) and Babcock et al. (2003).

How to Conduct an Input–Output Analysis: To illustrate the I/O analysis, a simplified transactions matrix is provided in Table 9.4 to describe the flow of goods and services among three sectors of the economy in a given region. The columns show purchases (input) for each industry, and the rows show sales (output) from each industry to others. For example, to produce $35 million output, the transportation sector purchased $3 million from construction sector, $8 million from manufacturing sector, $12 million from transportation sector, and made $12 million of payments to the *final payments* sector. Final payments are made by industries to households (workers), gross savings (interest, profit), government (taxes), and imports. In addition, the transportation

Table 9.4   Illustrative Input–Output Transactions Matrix

Sector	Construction	Manufacturing	Transportation	Final Demand	Total Output
Construction	7	9	3	21(5)	40
Manufacturing	8	20	8	24(7)	60
Transportation	6	6	12	11(5)	35
Final payments[a]	19(7)	25(7)	12(5)	0	56
Total inputs	40	60	35	56	191

[a]Values in parentheses refer to households.

sector sold $6 million to construction, $6 million to manufacturing, $12 million to transportation, and $11 million to the final demand sector (households, investment, government, and exports). Final demand consists of purchases of goods and services for final consumption in contrast to an intermediate purchase where the goods will be remanufactured further (Minnesota Implan Group, 2004).

Information on the region's economy is provided in Table 9.5. Given the input–output transactions matrix shown in Table 9.4 and the information provided in Table 9.5, the total (direct, indirect, and induced) output, employment, and income multipliers can be determined by applying the input–output methodology, as discussed below.

First, the *direct requirements matrix* (also known as the *A matrix*) is determined. This indicates the input (purchase) requirements of each industry to produce an average $1 of output (sales). The *purchase coefficients*, or *input ratios*, are obtained by dividing purchase data in each industry column of the transactions matrix by the corresponding output value for that industry (Table 9.6). The columns represent production functions that indicate where an industry spends (and in what proportions) to generate each dollar of its output. In the example provided, the third column (transportation) shows that to produce an average $1 of output, the transportation sector buys $0.09 (= 3/35) from construction firms, $0.23 (= 8/35) from manufacturing industries, $0.34 (= 12/35) from

transportation establishments, and makes $0.34 (= 12/35) of payments to the final payments sector ($0.14 of these payments are made to households).

Then the *total (direct, indirect, and induced) requirements matrix* is estimated. This includes the direct and multiplier effects (i.e., effects of household income and spending in addition to the interindustry interaction) in the economy. These effects are defined in Section 9.1.2. The total requirements matrix (Table 9.7) derives from the direct requirements matrix A (Table 9.6) by estimating the $(I - A)$ inverse[1] (known as the *Leontief inverse*), where I is the identity matrix.

For example, for the transportation sector to increase its output by $1, it would eventually require an output of $1.765 (including the initial $1 increase). At the same time, the construction sector must increase its output by $0.375, and the manufacturing sector must increase its output by $0.797. In this grossly simplified economy, the total economic output increase due to a $1 increase in transportation sector output is the sum of these three values, or 3.347 times larger than the initial output expansion in transportation. This is the *output multiplier* concept. Consider an investment of $100 million for a highway construction project along a corridor. Assume that the construction sector will be the only beneficiary of this investment. Through the ripple effects in the economy, the investment would be expected to increase the total level of economic output by an estimated $296 million [($100)(2.96)].

Employment multipliers can be obtained by combining the information in Table 9.7 with the industry employment–output ratios provided in Table 9.5. To obtain the total (direct, indirect, and induced) employment multipliers for each industrial sector, each of the entries in the column of the Leontief inverse

[1] In matrix notation, the A matrix can be written as a series of linear equations, as follows $X = A \cdot X + Y$. This notation simply states that output X is equal to transactions (AX) plus final payments (Y). Then we have $(I - A) \cdot X = Y$ or $X = (I - A)^{-1} \cdot Y$ (Minnesota Implan Group, 2004).

Table 9.5   Illustrative Employment–Output Ratios for the Three Sectors of the Economy in the Region

Sector	Employment	Output ($\times 10^6$)	Employment/ Output Ratio
Construction	10,000	$1000	0.00001
Manufacturing	6,500	300	0.00002
Transportation	8,000	800	0.00001

Table 9.6   Illustrative Direct Requirements Matrix[a]

Input	Construction	Manufacturing	Transportation	Final Demand	Total Output
Construction	0.18	0.15	0.09	0.38 (0.08)	0.21
Manufacturing	0.20	0.33	0.23	0.43 (0.13)	0.31
Transportation	0.15	0.10	0.34	0.20 (0.08)	0.18
Final payments	0.48 (0.18)	0.42 (0.12)	0.34 (0.14)	0	0.29
Total inputs	1.00	1.00	1.00	1.00	1.00

[a]Values in parentheses refer to households.

Table 9.7   Illustrative Total (Direct, Indirect, and Induced) Requirements Matrix

A matrix	Construction	Manufacturing	Transportation	Household Demand
Construction	0.18	0.15	0.09	0.08
Manufacturing	0.20	0.33	0.23	0.13
Transportation	0.15	0.1	0.34	0.08
Household payments	0.18	0.12	0.14	0
$(I - A)$ *matrix*				
Construction	0.82	−0.15	−0.09	−0.08
Manufacturing	−0.2	0.67	−0.23	−0.13
Transportation	−0.15	−0.1	0.66	−0.08
Household payments	−0.18	−0.12	−0.14	1
$(I - A)^{-1}$ *matrix*				
Construction	1.427	0.407	0.375	0.179
Manufacturing	0.664	1.821	0.797	0.339
Transportation	0.467	0.406	1.765	0.209
Household payments	0.402	0.349	0.410	1.102
TOTAL	2.960	2.983	3.347	

in Table 9.7 is multiplied by its employment–output ratio and then the column is summed, as shown in Table 9.8. The value $(37.4)(10^{-6})[= (0.375)(0.00001) + (0.797)(0.00002)+(1.765)(0.00001)]$ is the total employment change due to a dollar of investment in the transportation sector or 38 jobs per million dollars of transportation output. The employment multipliers for the construction and manufacturing sector would be 33 jobs per million dollars of construction output and 45 jobs per million dollars of manufacturing output, respectively.

Finally, the income multipliers are calculated by dividing the value in the household row of the total (direct, indirect, and induced) requirements matrix (Table 9.7) by their corresponding values in the household row of the direct requirements matrix (Table 9.6), as shown in Table 9.9. The total income generated due to the investment of \$1 in transportation would be \$2.870 $(= 0.402/0.14)$. This concept is known as the *income multiplier*. In the given example, the income multipliers

Table 9.8   Employment Multipliers

Employment	Construction $(\times 10^{-6})$	Manufacturing $(\times 10^{-6})$	Transportation $(\times 10^{-6})$
Construction	14.3	4.1	3.8
Manufacturing	13.3	36.4	15.9
Transportation	4.7	4.1	17.7
Total	32.3	44.6	37.4

Table 9.9   Income Multipliers for Each Sector

Income	Construction	Manufacturing	Transportation
Total	2.297	2.991	2.870

for the construction and manufacturing sector are \$2.297 per dollar of construction output and \$2.991 per dollar of manufacturing output, respectively.

9.2.5 Statistical Analysis Tools

Statistical models, typically using regression analysis, are developed on the basis of either historical time series or cross-sectional data on transportation investment, public infrastructure levels, and economic indicators (e.g., employment, wages, and land values). This methodology has been used in the past to identify the relationship between transportation investment levels and accompanying changes in business location and regional development patterns (Evers et al., 1988; Duffy-Deno and Eberts, 1991; Lombard et al., 1992). In other studies, the issue addressed is how the existing stock (and not changes thereof) of transportation infrastructure has affected national economic productivity and the level of national economic growth over time (Aschauer, 1990; Munnell, 1990; Pinnoi, 1993; Toen-Gout and van Sinderen, 1994; Boarnet, 1995; Arsen, 1997; Bell and McGuire, 1997; Nadiri and Mamuneas, 1998; Fraumeni, 1999). An advantage of this approach is its ability to analyze the simultaneous effect of a large number of variables, time lag effects, and functional forms.

There are several examples of statistical models. Queiroz and Gautam (1992) used time-series regression analysis of U.S. data from 1950 to 1988 to investigate the relationship between per capita GNP and road density:

$$PGNP = \begin{cases} -3.39 + 1.24LPR & \text{(no time lag)} \\ R^2 = 0.93 \\ -2.9 + 1.22LPR & \text{(time lag of one year)} \\ R^2 = 0.93 \\ -2.5 + 1.2LPR & \text{(time lag of two years)} \\ R^2 = 0.92 \end{cases}$$

where PGNP is the per capita GNP ($1000 in 1982 constant dollars/inhabitant) and LPR is the density of paved roads (km/1000 inhabitants).

Many factors contribute to GNP growth, and the strong correlation does not necessarily mean road expansion results in GNP growth. There is also a possibility of simultaneity between road expansion and GNP growth; road expansion contributes to GNP growth and GNP growth leads to road expansion. With more recent and expanded data, the issues of causality, correlation, and simultaneity can be addressed.

Lombard et al. (1992) developed county-level cross-sectional regression models for assessing the economic impact of highway expenditure in Indiana. Another example of statistical models are those developed by Gkritza et al. (2006) to investigate the relationship between statewide changes in economic development and investments in expanded highway capacity in Indiana over a 20-year period:

$$REMIEMP = -156 + 10.56NEWLNMI$$
$$- 168.40URBAN + 347.21I$$
$$+ 43.75ACCAIRP - 90.86CENTRAL$$
$$\text{adjusted } R^2 = 0.55$$

$$REMINCMI = -8.71 + 0.51NEWLNMI$$
$$- 4.51RESTURBAN + 14.08I$$
$$+ 2.04ACCAIRP - 3.78CENTRAL$$
$$+ 0.022PRCOSTMI$$
$$\text{adjusted } R^2 = 0.47$$

$$REMIOUTMI = -77 + 3.00NEWLNMI$$
$$- 17.93URBAN + 65.85I$$
$$+ 15.97ACCAIRP$$
$$\text{adjusted } R^2 = 0.47$$

$$REMIGRPMI = -27.21 + 2.18NEWLNMI$$
$$- 16.16RESTURBAN + 21.43I$$
$$- 19.25ST + 8.13ACCAIRP$$
$$- 22.44CENTRAL$$
$$\text{adjusted } R^2 = 0.40$$

where REMIEMP is the net change in employment (jobs), REMINCMI the net change in real disposable income (millions of 1996 dollars), REMIOUTMI the net change in output (millions of 1996 dollars), REMIGRPMI the net change in gross regional product (millions of 1996 dollars), NEWLNMI the new (added) lane-miles, URBAN (1 for a project located in an urban area, 0 for rural projects), RESTURBAN [1 for a project located in an urban area (excluding Marion county with Indianapolis), 0 otherwise], I (1 for interstate highway improvements, 0 otherwise), ST (1 for improvements to a state highway, 0 otherwise), ACCAIRP the degree of accessibility to major airports (1, low to 5, high), CENTRAL (1 for a project located in central Indiana, 0 otherwise), and PRCOSTMI the project investment (millions of 1996 dollars).

9.2.6 Economic Simulation Models

(a) Regional Economic Simulation Models Economic simulation models, which predict economic growth in a given region in response to changes in transportation policies or projects, are extensions of the I/O model discussed in a preceding section. These models typically have four components: (1) a base-case forecast of future economic growth or decline in the region; (2) a model to estimate growth in business sectors in response to direct changes in their relative operating costs and markets; (3) estimation of overall changes in the flow of money in the regional economy, including indirect and induced effects using input–output tables or charts; and (4) a mechanism to predict the future economic growth or decline relative to the base case if the project were implemented. Modeling tools typically generate outputs that reflect changes in employment, personal income, business output, and gross regional product (value added) over a relatively long period of time, typically 20 to 30 years. A common economic simulation tool is the Regional Economic Models, Inc. (REMI) dynamic input–output model (Treyz et al., 1992). Another model is the Regional Economic Impact Model for Highway Systems (Politano and Roadifer, 1989). For long-range planning, these models are preferred over simple input–output modes due to their dynamic nature and ability to account for productivity changes that may develop as a result of transportation decisions over a

20- to 30-year planning horizon (CUBRC et al., 2001). However, data collection and analysis for such models can require considerable effort and expertise.

(*b*) *Hybrid Economic Simulation Models* A number of simulation-based models have been developed to include significant economic factors to provide more reliable prediction of how markets respond to changes in land use and transportation access. An example of such land-use/economic hybrid models is the TELUS (Transportation, Economic, and Land-Use System) developed for the North New Jersey Transportation Planning Authority (NJIT, 1998). Other examples are the METROSIM model (Anas, 1999) and the MEPLAN model (Echenique, 1994). Also, a number of state DOTs have developed integrated traffic and economic models to estimate economic impacts of their major highway corridor projects. For example, Indiana DOT developed the Major Corridor Investment–Benefit Analysis System (MCIBAS), a five-step integrated modeling system that includes a travel demand model, a user-benefit calculation model, a macroeconomic simulation model, and a benefit–cost framework (Cambridge Systematics, 1998). MCIBAS has been applied in several studies (Cambridge Systematics and Bernardin, 1998a, b; Cambridge Systematics et al., 2003). A procedure similar to MCIBAS is the Highway Economic Analysis Tool (HEAT) developed for the Montana Department of Transportation (Cambridge Systematics and EDRG, 2005). Finally, the Mid-Ohio Regional Planning Commission developed a Freight Transportation Investment Model, which utilized a REMI macroeconomic simulation model component for estimating the economic development impacts of the city of Columbus inland port (Cambridge Systematics et al., 1999a).

9.3 ESTIMATION OF LONG-TERM REGIONAL ECONOMIC DEVELOPMENT IMPACTS

A common approach for calculating the regional economic development effects of a transportation improvement is to use a regional economic simulation model in combination with a traffic (network simulation) model. For a given set of project alternatives, the traffic model estimates direct impacts of the transportation system improvement on traffic patterns, volumes, and speeds, and calculates travel cost savings by flow type (passengers and goods) and by trip purpose (business and nonbusiness). The travel cost savings (i.e., reductions in travel time, safety-related costs, or vehicle operating costs) are then translated into user benefits and expressed in terms of monetary values (as described in Chapters 5 to 7). Economic development benefits are then estimated in terms of business savings from market economies of scale, productivity, logistic

opportunities for just-in-time production economies, and shift in business growth and locational factors. User benefits associated with nonbusiness trips are excluded from the economic impact analysis, as they do not affect directly the cost or productivity of doing business and are assumed to be incapable of producing any secondary economic impact. The estimated efficiency benefits of business auto, truck, and other travel modes, over the analysis period, are first translated into financial consequences and then allocated to various types of existing businesses located in the study area. These direct business cost savings are allocated among industries based on: (1) relative sensitivity to transportation cost changes, and (2) each industry's share of economic activity in the study area, in terms of employment. This methodology is discussed in detail in Weisbrod and Grovak (2001) and Cambridge Systematics (1998). The estimated business expansion impacts by the business sector are used as direct impacts for input into the regional economic simulation model.

In addition to the direct cost savings for businesses, transportation projects can potentially enhance strategic connections between specific locations and activities and can expand the size of market reach to customers and labor, thus attracting out-of state business activity and investment. Business attraction impacts are typically estimated as changes in employment by industry. The net business attraction or expansion impacts or tourism impacts can be estimated exogenously by conducting surveys of area firms or interviewing owners or operators of tourism or recreation businesses (Weisbrod and Beckwith, 1992). Also, the method of *location quotients* (*LQ*) can be applied to quantify the magnitude of business attraction effects. A location quotient is an indicator of regional specialization, or a region's competitiveness for a specific industry, measured in terms of employment (Glickman, 1977). A *LQ* of 1 means that an industry has the same share of a regional economy as it does of the national economy. The higher the *LQ*, the greater the competitive advantage of a region for the specific industry. Location quotients can be calculated using the location quotient calculator, a tool produced by the Bureau of Labor Statistics (BLS, 2005). Past studies have applied *LQ* analysis to estimate potential business attraction associated with highway investments (Cambridge Systematics, 1998a, b). It was assumed that if the location experienced strong growth without the transportation improvement (indicated by *LQ* > 1), new business attractions would be limited. However, it may be argued that an LQ exceeding 1 could, on the contrary, spur growth: the potential for business attraction might be higher in a region with competitive advantages such as a skilled labor force or agglomeration economies. In general, it is difficult to accurately predict business attraction impacts because transportation investments constitute only one of several factors in business location decisions. It is possible to make broad estimates about the types and sizes of businesses that may be attracted to a region as a result of a major transportation project; however, this should be done with caution so as not to include business attraction impacts that represent net transfers among regions within the study area.

The next step is to input the results of the preceding step into the regional economic simulation model, run a simulation for the long-term impacts, and evaluate the results. The economic effects of potential business and/or tourism attraction, in terms of business sales by industry, employment by industry, personal income, population, and other variables, can be estimated separately from the direct business expansion impacts. The model is run twice and the total (direct, indirect, and induced) effects associated with the project alternative is calculated on a year-by-year basis over the analysis period with and without the business and/or tourism attractions. In general, the construction period benefits are not included because it is assumed that construction expenditures are short-term and temporary in nature, and would have been spent anyway by state and local governments—either on the

project in question or on other similar projects that would yield comparable capital expenditure benefits (Weisbrod and Beckwith, 1992). The overall analysis procedure is illustrated in Figure 9.4.

How to Conduct a Long-Term Regional Economic Development Impact Analysis: To illustrate how the analytical framework presented in Figure 9.4 is used in practice, consider the case of an urban interstate widening project with geometric and operational characteristics presented in Table 9.10.

The project is scheduled for construction in 2006. The state government seeks to predict the estimated statewide economic development impact of this project 20 years after its implementation (2008–2027). It is assumed that benefits will begin to accrue in the first year of highway operation. Assume an interest rate of 5%.

First, the user benefits due to the proposed project, travel-time savings, vehicle operating cost changes, and crash cost savings are estimated. To assess the broader economic impact of these benefits, estimates on the distribution of two categories of vehicle trips, truck trips and automobile trips for business purposes, are developed. The estimation results for year 1 and year 20 of the analysis period are presented in Figure 9.5. A key issue is the use of an appropriate value of travel time. This is addressed in Chapter 5.

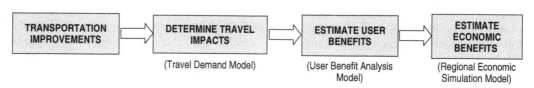

Figure 9.4   Procedure for analyzing regional economic development impact in the long term.

Table 9.10   Project Data

Type of project	Added travel lanes	Base-case average daily traffic in 2005	117,244
Functional class	Urban interstate	Base-case average daily traffic in 2025	173,843
Length of construction period (years)	2	Proposed system average daily traffic in year 1	122,635
Project costs (millions of 2003 dollars)	167	Proposed system average daily traffic in year 20	181,836
Start lanes	6	Base-case capacity (veh/h)	6,224
End lanes	10	Proposed system capacity (veh/h)	10,373
Project length (miles)	7.3	SU/combination-unit trucks (%)	5.9/5.3

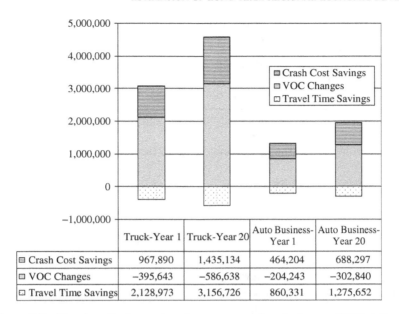

	Truck-Year 1	Truck-Year 20	Auto Business-Year 1	Auto Business-Year 20
⊟ Crash Cost Savings	967,890	1,435,134	464,204	688,297
▢ VOC Changes	−395,643	−586,638	−204,243	−302,840
▢ Travel Time Savings	2,128,973	3,156,726	860,331	1,275,652

Figure 9.5 User benefits by mode, trip purpose, and analysis year (2003 dollars).

Table 9.11   Estimation of Business Cost Savings from User Benefits

User Benefit	Corresponding Cost Savings to Business
Time savings business travel (on-the-clock worker time)	Value of additional productive labor hours (for nonsalaried portion of workers)
Other trips (includes commuting)	(May lead to additional spending or affects wages for recruiting workers)
Operating cost savings business travel (pickups and deliveries)	Direct cost savings
Other travel (includes commuting)	Increase in disposable personal income (may also affect wage rates)
Safety improvements business travel (on-the-clock worker time)	Reduction in insurance costs and worker absenteeism
Other travel	Reduction in insurance cost, raising disposable income

Source: Weisbrod and Weisbrod (1997).

The estimated cumulative user benefits in 2003 dollars, for truck trips over the 20-year analysis period is $24.6 million, and that for auto business trips over the same period is $17.1 million. Consistent with the analysis procedure presented in the preceding section, these benefits are first translated into financial consequences (i.e., direct business cost savings) and then allocated to various types of existing businesses located in the study area. The direct business cost savings from travel efficiency improvements are estimated from user benefits as shown in Table 9.11.

After the estimated business expansion impacts by business sector (industry) have been quantified, they are used as direct impact input data for the REMI dynamic simulation model. The REMI model is then run and the simulation results are compared to the baseline to determine the total (including direct, indirect, and induced) economic effects associated only with business cost savings resulting from the proposed highway project over a 20-year period, as shown in Table 9.12. The economic effects of potential business and/or tourism attraction are estimated separately in the next analysis step.

Table 9.12   Economic Effects Associated with Cost Savings

Net change in employment (jobs)	112
Net change in GRP or value added[a] (millions of 2003 dollars)	46.1
Net change in real personal disposable income[b] (millions of 2003 dollars)	26.9
Net change in output[c] (millions of 2003 dollars)	88.1

[a]Change in the sum of wage income and corporate profit generated in the region; reflects the overall economic activity in a region.
[b]Change in wage income earned by workers within the region (adjusted for inflation and net of taxes).
[c]Change in business sales in the region.

Table 9.13   Results of Simulation for Highway Project

Net change in employment (jobs)	620
Net change in GRP or value added[a] (millions of 2003 dollars)	105.8
Net change in real personal disposable income[b] (millions of 2003 dollars)	55.5
Net change in output[c] (millions of 2003 dollars)	218.4

[a]Change in the sum of wage income and corporate profit generated in the region; reflects the overall economic activity in a region.
[b]Change in wage income earned by workers within the region (adjusted for inflation and net of taxes).
[c]Change in business sales in the region.

The LQ method is applied to quantify the magnitude of business attraction (Cambridge Systematics, 1998a, b). A simplified metric of the magnitude of business attraction is applied (this is assumed to be proportional to that of business expansion by a factor of $1/LQ$). The LQ for the manufacturing industry located in the urban area where the highway improvement takes place, calculated using the location quotient calculator (BLS, 2005), is 0.61. The magnitude of business attraction is proportional to that of business expansion by a factor of $1/LQ = 1/0.61 = 1.64$. Therefore, $(112)(1.64) = 184$ jobs are estimated to be attracted as a result of the highway improvement. For analytical purposes, it is assumed that the additional jobs created would be in equal increments over the analysis period. It is also assumed that the significant increases in accessibility due to the project would benefit manufacturers most, because manufacturers are particularly dependent on reliable truck transportation. Other industries that are expected to produce statewide attraction benefits include the wholesale trade, transportation, and warehousing industries. Finally, tourism attraction impacts associated with the highway project are anticipated to be limited at the state level.

The REMI model is then run and the simulation results are compared to the baseline to determine the total economic development impacts resulting from both cost savings for businesses and business attraction impacts associated with the highway project. The project is predicted to generate 620 (direct, indirect, and induced) jobs that would accrue to industries that benefit most from increased access to buyer and supplier markets and accrue multiplier effects from increased business and consumer spending, such as manufacturing, retail trade, and services industries. The results of the simulation performed for the proposed highway project over the 20-year analysis period are summarized in Table 9.13.

9.4   CASE STUDY: ECONOMIC DEVELOPMENT IMPACT ASSESSMENT

To illustrate the analytical steps involved in the procedural framework, the case of a highway corridor related to the I-69 project in the southwestern part of Indiana is considered. A set of five alternatives was identified for study in the environmental impact statement (EIS). The project involves a 142-mile expressway from Evansville to Indianapolis through multiple, mostly rural counties. The project cost is estimated at $1.8 billion (2003 dollars). It is expected that the new highway, which is a part of the NAFTA corridor from Mexico to Canada, will provide improved accessibility and spur economic growth. The steps followed by the consultants (Cambridge Systematics and Bernardin, 2003) in the assessment of economic development potentials of the five alternatives are discussed below.

1. *Delineate the impact area.* The study area included five regions along the corridor. The impacts of the proposed highway on the rest of the state, state of Illinois, state of Kentucky, and the rest of the United States were also assessed.

2. *Select the analysis period.* Since the purpose of the analysis is an impact assessment for a proposed new facility, a future year or period of time after the new facility opens is selected. A 20-year analysis period was selected for this project.

3. *Select the economic development measures.* Economic performance measures related to economic development considered in the I-69 corridor study included: (a) net change in employment, (b) net change in real disposable income, and (c) net change in real output or business sales.

4. *Determine the existing economic conditions.* Economic data were collected from the publications of the U.S. Bureau of the Census, the U.S. Bureau of Economic Analysis, and other agencies to assess current economic conditions in terms of per capita income, population, and employment growth. Interviews of local businesses and officials were conducted to help evaluate the validity of the economic forecasts and to corroborate the economic data.

5. *Select the analysis methods.* The procedure used for assessing long-term economic impacts followed primarily the methods discussed in Section 9.3 and was based on the use of an integrated traffic and regional simulation model (MCIBAS) that includes a travel demand model, a user-benefit calculation model, a macroeconomic simulation model, and a benefit–cost framework. The travel demand model was used to get estimates of systemwide changes in vehicle-miles of travel and vehicle-hours of travel. Accessibility factors were developed for labor, customer, supplier, and tourism markets. Access of businesses to labor and customer markets was measured as the population within 45 minutes of travel time to the average business, while access of freight customer and supplier markets was measured as the number of employees within 3 hours of travel time to the average business.

6. *Estimate the benefits.* Reductions in travel time, crash costs, and vehicle operating costs for trucks and automobiles used for business purposes were estimated using the user-benefit calculation module of MCIBAS mentioned in step 5.

7. *Estimate the economic impacts.* Direct regional economic impacts, including business cost savings, business attraction benefits, and increased tourism associated with each alternative were estimated using the MCIBAS modules, as shown in Table 9.14. The size of impacts would depend primarily on traffic volumes and increases in accessibility to labor, customer, supplier, and tourism markets.

8. *Estimate the secondary economic impacts.* The macroeconomic simulation model (REMI), a MCIBAS component, was used to estimate the total economic effects (including direct, indirect, and induced) with respect to such changes as business sales (output), employment, and income due to the direct economic impacts estimated in step 7. The resulting information is given in Table 9.15.

9. *Compare the alternatives.* Tables and graphs were generated showing the economic impacts for each alternative as a result of the highway project. Alternatives were ranked against their potential economic development effects. In this example, Alternative III appears to be the

Table 9.14   I-69 Corridor Study: Direct Impacts

Direct Economic Impact	Alternative				
	I	II	III	IV	V
Reduction in annual production costs in 2025 (millions of 2001 dollars)	1.7	10.5	20.2	9.0	16.5
Increase in annual business sales in 2025 (millions of 2001 dollars)	202.8	391.7	631.8	430.1	538.8
Increase in tourism visitor-days in 2025 (thousands of days)	42.4	88.0	168.8	94.1	143.2

Source: Cambridge Systematics and Bernardin (2003).

Table 9.15   I-69 Corridor Study: Secondary Impacts

Secondary Economic Impact	Alternative				
	I	II	III	IV	V
Net change in employment (jobs) in 2025	1400	2500	4300	2700	3800
Net change in real disposable income (millions of 2001 Dollars) in 2025	52	99	165	106	142
Net change in output (millions of 2001 dollars) in 2025	245	495	808	537	679

Source: Cambridge Systematics and Bernardin (2003).

most desirable from the perspective of economic development impacts.

10. *Benefit-cost computation.* To avoid double counting, the components of benefits included in the benefit-cost ratio computation are taken only as the user benefits for non-business travelers (i.e., savings in travel time, vehicle operating cost, and crash costs, for personal travel) plus real personal income impacts (i.e., net change in real disposable income).

SUMMARY

Economic development impacts of transportation projects represent the effects on economic activity in a project area or region. These differ from economic efficiency impacts (which involve the valuation of individual user benefits) or broader social impacts. Economic development impacts occur through mechanisms that can be broadly classified as direct, indirect, induced, or dynamic. The sum of all these effects represents the total effect on economic growth. Common performance measures for economic development impacts are employment, business output (sales), value added, wealth or personal income, and property values. These measures typically overlap and should not be added to yield the total impact. The selection of appropriate performance measures and data collection techniques for the evaluation depends on the purpose of the transportation project, the project type, size of impact area, usefulness of information available for public information and for decision making, and motivation for the evaluation. The motivation for assessing economic development impacts include forecasting the future impacts of proposed projects, to estimating the current economic role of existing systems and facilities, and measuring the actual impact of projects already completed. Analytical techniques for economic development impact studies range in complexity from simple case studies to complex economic simulation models. The key is to match the analytical tool to the purpose and level of desired sophistication of the analysis, within the given resources.

EXERCISES

9.1. What is the difference between economic efficiency impacts and economic development impacts?

9.2. Discuss some typical economic benefits of transportation investments.

9.3. Discuss the detailed measurement process, merits and limitations of any performance measure used to quantify the effect of transportation investments on manufacturing productivity.

Table EX9.5.1   Transactions Matrix

Input	Manufacturing	Utilities	Trade	Final Demand	Total Output
Manufacturing	8	7	10	25(4)	50
Utilities	14	10	6	12(3)	42
Trade	12	5	4	12(3)	33
Final payments[a]	16(6)	20(7)	13(5)	0	49
Total inputs	50	42	33	49	174

[a]Values in parentheses refer only to households.

9.4. Discuss how you would establish a statistical relationship for predicting the overall impacts of transportation on economic development. Identify the statistical and economic issues that are likely to arise and suggest ways by which they could be addressed.

9.5. Input–output analysis: The transactions matrix shown in Table Ex9.5.1 describes the flows of goods and services between three individual sectors of a highly simplified economy in a given region. The illustrative employment–output ratios for the three sectors of the economy in the region that provide information on labor productivity for each sector are provided in Table Ex9.5.2. Estimate the direct, indirect, and induced output, employment, and income multipliers by applying the input–output methodology described in this chapter.

Table EX9.5.2   Employment–Output Ratios

Sector	Employment	Output	Employment/ Output Ratio
Manufacturing	50,000	$50,000,000	0.001
Utilities	40,000	20,000,000	0.002
Trade	100,000	30,000,000	0.003

9.6. A 44-mile four-lane freeway is planned for construction in 2006. The investment required includes the following costs:

- The cost associated with purchasing the land where the highway will be built (including the real estate cost): $55,060,000
- The cost of engineering services involved in project design and study: $9,730,000
- The cost of constructing the highway: $194,540,000

The total investment required is $259,330,000. All costs are in constant 2001 dollar values. Assess the economic impacts of the investment in terms of employment, earnings, and output. Use each of the following two approaches:

1. Input–output analysis: Use the RIMS II and IMPLAN software packages to estimate the output, income, and employment economic multipliers.

2. Regional economic modeling: Use the REMI software package to evaluate the regional economic

impacts of the new highway construction in the long run (year 2020).

Study the outputs to evaluate the impacts and discuss the results. State any assumptions made.

REFERENCES[1]

Anas, A. (1999). Application of the METROSIM model in New York, presented at the Transportation Research Board Annual Meeting, Transportation Research Board, National Research Council, Washington, DC.

Arsen, D. (1997). Is there really an infrastructure/economic development link? in *Dilemmas of Urban Economic Development: Issues in Theory and Practice*, ed. Bingham, R. D., Mier, R., Sage Publications, Newbury Park, CA.

Aschauer, D. A. (1990). Highway capacity and economic growth, *Econ. Perspect.*, (Federal Reserve Bank of Chicago), Vol. 14, No. 5.

Babcock, M. W. (2004). *Approximation of the Economic Impacts of the Kansas Comprehensive Transportation Program*, Kansas Department of Transportation, Topeka, KS, Dec.

*Babcock, M. W., Emerson, M. J., Prater, M. (2003). A model-procedure for estimating economic impacts of alternative types of highway improvement, *Transp. J.*, Vol. 34 No. 4 (Summer), pp. 30–44.

*Bechtel Corporation, Bartram & Cochran, Clarke Tamaccio Architects, Fitzgerald & Halliday, Basile Baumann & Prost, FAS Consultant Services (1994). *Griffin Line Corridor, Economic Development Assessment*, Greater Hartford Transit District, Hartford, CT.

Bell, M. E., McGuire, T. J. (1997). *Macroeconomic Analysis of the Linkages Between Transportation Investments and Economic Performance*, NCHRP Rep. 389, Transportation Research Board, National Research Council, Washington, DC.

*Bendavid-Val, A. (1991). *Regional and Local Economic Analysis for Practitioners*, 4th ed., Praeger Publishers, Westport, CT.

*BLS (2005). *Location Quotient Calculator*, Bureau of Labor Statistics, U.S. Department of Labor, http://data.bls.gov/LOCATION_QUOTIENT/servlet/lqc.ControllerServlet, Nov.

Boarnet, M. (1995). *Highways and Economic Productivity: Interpreting Recent Evidence*, University of California Transportation Center, Berkeley, CA, Oct.

Cambridge Systematics (1988). *Economic and Social Impacts of Orange Line Replacement Transit Service*, Massachusetts Bay Transportation Authority Boston, MA, May.

———— (1989a). *Applied Development Economics and Research Unlimited, Downtown Economic Study*, Sacramento Department of Public Works, Sacramento, CA, June.

———— (1989b). *Highway 29/45/10 Corridor Study: Economic Benefits and Cost–Benefit Evaluation, Final Report*, Wisconsin Department of Transportation, Madison, WI.

———— (1996). *Economic Impacts of the Southwest Indiana Highway Corridor*, Indiana Department of Transportation, Indianapolis, IN.

[1]References asterisked can also serve as useful resources for estimating the economic development impacts of transportation projects.

_____ (1998). *Economic Impact Analysis System: Major Corridor Investment–Benefit Analysis System*, Indiana Department of Transportation, Indianapolis, IN.

Cambridge Systematics; Bernardin, Lochmueller and Associates (1998a). *Economic Impacts of SR 26 and U.S. 35 Corridor Improvements*, Indiana Department of Transportation, Indianapolis, IN.

_____ (1998b). *Economic Impacts of U.S. 31 Corridor Improvements*, Indiana Department of Transportation, Indianapolis, IN.

Cambridge Systematics, EDRG (Economic Development Research Group) (2005). *Montana Highway Reconfiguration Study*, Montana Department of Transportation, Helena, MT.

Cambridge Systematics, IMPEX Logistics, Eng-Wong & Taub Associates, Tioga Group (1999a). *Summary Report: MORPC Inland Port III*, Mid-Ohio Regional Planning Commission, Columbus, OH, Mar.

Cambridge Systematics, Economic Development Research Group Economic Competitiveness Group (1999b). *Transportation Cornerstone Florida*, Florida Chamber of Commerce, Tallahassee, FL, May.

Cambridge Systematics; Bernardin, Lochmueller and Associates; Dyer Environmental Services (2003). *SR 101 Corridor Improvement Feasibility/NEPA Study*, Indiana Department of Transportation, Indianapolis, IN.

Clark Patterson Associates et al., (1998). *Airport Corridor Major Investment Study (Monroe County)*, Tech. Memo., CPA, Rochester, NY, Nov. 18.

Colorado National Banks (1989). *Ready for Takeoff: The Business Impact of Three Recent Airport Developments in the U.S.*, CNB, Denver, CO.

*CUBRC (Calspan-University of Buffalo Research Center), State University of New York, Cambridge Systematics (2001). *Handbook of Economic Development*, New York State Department of Transportation, Albany, NY, Sept.

De Rooy, J. (1995). *Economic Literacy*, Three Rivers Press, New York.

Dickey, J. W., Watts, T. W. (1978). *Analytic Techniques in Urban and Regional Planning*, McGraw-Hill, New York.

Duffy-Deno, K. T., Eberts, R. W. (1991). Public infrastructure and regional economic development: a simultaneous equations approach, *J. Urban Econ.* Vol. 30, pp. 329–343.

Echenique, M. H. (1994). Urban and regional studies at the martin Centre: its origins, its present, its future, *Environ. Plan. B*, Vol. 21, pp. 517–567.

EDRG (Economic Development Research Group), Bernardin, Lochmueller & Associates (1998). *Economic Impact Assessment of KY69 Improvements*, Kentucky Transportation, Commission, Frankfurt, KY.

EDRG (Economic Development Research Group), SRF Consulting Group (1999). *Economic Impact Analysis: St. Croix River Crossing*, Minnesota Department of Transportation St. Paul, MN, and Wisconsin Department of Transportation, Madison, WI.

Evers, G. H. M., Van Der Meer, P. H., Oosterhaven, J., Polak, J. B. (1988). Regional impacts of new transport infrastructure: a multi-sectoral potentials approach, *Transportation*, Vol. 14, pp. 113–126.

Forkenbrock, D. J. (1991). *Putting Transportation and Economic Development in Perspective*, Transp. Res. Rec. 1274, Transportation Research Board, National Research Council, Washington, DC, pp. 3–11.

*Forkenbrock, D. J., Foster, N. S. J. (1996). Highways and business location decisions, *Econ. Devel. Q.*, Vol. 10, No. 3, pp. 239–248.

*Forkenbrock, D. J., Weisbrod, G. (2001). *Guidebook for Assessing the Social and Economic Effects of Transportation Projects*, NCHRP Rep. 456, Transportation Research Board, National Research Council, Washington, DC.

Fraumeni, B. M. (1999). *Productive Highway Capital Stocks, Policy Forum, Transportation Investment and New Insights from Economic Analysis*, Background Paper, Eno Transportation Foundation, Washington, DC, Feb. 23.

Gillis, W. R. Casavant, K. (1994). *Lessons from Eastern Washington: State Route Mainstreets, Bypass Routes and Economic Development in Small Towns*, EWITS Res. Rep. #2, Department of Agricultural Economy, Washington State University, Pullman, WA, http://www.bts.gov/ntl/data/732pdf.

Gkritza, K., Labi, S., Mannering, F. L., Sinha, K. C. (2006). Economic development effects of highway added-capacity projects, presented at the Transportation and Economic Development Conference, Little Rock, AR, Mar.

Glickman, N. J. (1977). *Econometric Analysis of Regional Systems: Explorations in Model Building and Policy Analysis*, Academic Press, New York.

Halcrow Fox (1996). *Strathclyde Tram: Development Impacts, Phase III Report*, Strathclyde PTE, Glasgow, Scotland.

Keane, T. (1996). *Highway Infrastructure Investment and Job Generation*, FHWA-PL-96-015, Federal Highway Administration, U.S. Department of Transportation, Washington, DC. Related version available at http://www.tfhrc.gov/pubrds/spring96/p96sp16htm.

Lichtman, L. (1999). *Durand U.S. Highway 10 Relocation Alternatives Economic Impact Analysis*, Economic Planning and Development Section, Bureau of Planning; Wisconsin Department of Transportation, Madison, WI, Nov.

Lombard, P. C., Sinha, K. C., Brown, D. (1992). *Investigation of the Relationship Between Highway Infrastructure and Economic Development in Indiana*, Transp. Res. Rec. 1359, Transportation Research Board, National Research Council, Washington, DC, pp. 76–81.

Maryland DOT (1998). *Middle River Employment Center Access Study: Land Use Analysis Committee Market Analysis Report*, Maryland Department of Transportation, Baltimore, MD.

McConnell, C., Brue, S. (1999). *Economics: Principles, Problems and Policies*, 14th ed., McGraw-Hill, Boston, MA.

Minnesota Implan Group (2004). *Implan Professional User's Guide*, MIG, Stillwater, MN.

Munnell, A. (1990). Infrastructure investment and economic growth, *J. Econ. Perspect.*, No. 6, pp. 189–198.

Nadiri, M. I., Mamuneas, T. P. (1998). *Contribution of Highway Capital to Output and Productivity Growth in the U.S. Economy and Industries*, Federal Highway Administration, U.S. Department of Transportation, Washington, DC.

*NCHRP (1998). *Guidance for Estimating the Indirect Effects of Proposed Transportation Projects*, Rep. 403, Transportation Research Board, National Research Council, Washington, DC.

NJIT (1998). *TELUS: The Transportation, Economic, and Land-Use System: A State-of-the-Art Transportation Information System for the 21st Century*, New Jersey Institute of Technology, National Center for Transportation Technology, Rutgers University Center for Urban Policy Research, and North New Jersey Transportation Planning Authority, Rutgers, NJ, Apr.

N-Y Associates, EDRG (Economic Development Research Group) (1999). *Economic Impact Study (Phase II) of the Proposed Mississippi River Bridge at St. Francisville, Louisiana and the Zachary Taylor Parkway, Alexandria, Louisiana to Poplarville, Mississippi*, Louisiana Department of Transportation and Development, Baton Rouge, LA, May.

Palmer, J., Cornwell, J. P., Black, W. (1986). *Effects of Road Construction on Adjacent Economic Activities: A Retrospective Study*, Indiana University, Bloomington, IN, May.

Peat Marwick Main & Co. (1998). *Economic Impacts of Upgrading U.S. Route 219 to Interstate Standards from Springville, New York to Du Bois, Pennsylvania*, New York State Department of Transportation, Albany, NY.

Pinnoi, N. (1993). *Transportation and Manufacturing Productivity (Transportation Productivity and Multimodal Planning)*, Texas Transportation Institute, College Station, TX, Oct.

Politano, A. L., Roadifer, C. (1989). *Regional Economic Impact Model for Highway Systems*, Transp. Res. Rec. 1229, Transportation Research Board, National Research Council, Washington, DC.

Queiroz, C., Gautam, S. (1992). *Road Infrastructure and Economic Development: Some Diagnostic Indicators*, Western Africa and Infrastructure and Urban Development Departments, World Bank, Washington, DC.

Regional Science Research Corporation (1996). *PC I-O User's Manual*, RSRC, Hightstown, NJ.

SDAG (1996). *Economic Feasibility Study of the San Diego & Arizona Eastern Railway*, San Diego Association of Governments, San Diego, CA.

Sinha, K. C., McCulloch, B., Bullock, D., Konduri, S., Fricker, J. D., Labi, S. (2004). *An Evaluation of the Hyperfix Project for the Reconstruction of I65/70 in Downtown Indianapolis*, Tech. Rep. FHWA/IN/JTRP-2004/2, Purdue University, West Lafayette, IN, Sept.

*Toen-Gout, M., van Sinderen, J. (1994). *The Impact of Investment in Infrastructure on Economic Growth*, Res. Rep. 9503, Research Centre for Economic Policy, Erasmus University, Rotterdam, The Netherlands.

*Treyz, G., Rickman, D., Shao, G. (1992). The REMI economic–demographic forecasting and simulation model, *Int. Regional Sci. Rev.*, Vol. 14.

USDOC (1997). *Regional Multipliers: A User Handbook for the Regional Input–Output Modeling System (RIMS II)*, 3rd ed., Bureau of Economic Analysis, U.S. Department of Commerce, Washington, DC, Mar.

USDOT (2005). Based on *Transportation Statistics Annual Report, 2004*, plus additional estimates from Bureau of Transportation Statistics, U.S. Department of Transportation, Washington, DC.

Weisbrod, G. (2000). *Current Practices for Assessing Economic Development Impacts from Transportation Investments*, NCHRP Synth. Pract. 290, Transportation Research Board, National Research Council, Washington, DC.

Weisbrod, G., Beckwith, J. (1992). Measuring economic development benefits for highway decision making: the Wisconsin case, *Transp. Q.*, Jan., pp. 57–79.

Weisbrod, G., Grovak, M. (2001). Alternative methods for valuing economic benefits of transportation projects, prepared for the Transportation Association of Canada, Benefit Cost Analysis Symposium, Feb.

Weisbrod, G., Neuwirth, R. (1998). *Economic Effects of Restricting Left Turns*, NCHRP Res. Results Dig. 231, Transportation Research Board, National Research Council, Washington, DC.

Weisbrod, G., Weisbrod, B. (1997). *Assessing the Economic Impact of Transportation Projects: How to choose the Appropriate Technique for Your Project*, Transp. Res. Circ. 477, Transportation Research Board, National Research Council, Washington, DC.

W.S. Atkins, Steer Davies & Gleave, ECOTEC (1990). *Midland Main Line Strategy Study*, tech. rep., Department for Transport, UK, 1990.

Yeh, D., Leong, D., Russell, C. R. (1998). *Allouez Bridge Alternatives Analysis: Impacts on the Local Economy, Economic Planning and Development Section*, Wisconsin Department of Transportation Madison, WI.

ADDITIONAL RESOURCES

Economic Development Research Group (2006). Representative projects-Aviation and economic development; Ground transportation and economic development, www.edrgroup.com/edr1/consulting. Assessed Nov 2006.

Eberts, R. (2004). Understanding the impact of transportation on economic development, Transportation in the New Millenium, Transportation Research Board, National Research Council, Washington, DC.

Litman, T. (2002). Economic development impacts of transportation demand management, TRB 2002 Conference on Transportation and Economic Development, Portland, OR.

CHAPTER 10

Air Quality Impacts

Clearing the air, literally or metaphorically, does tons of good to all.

—Anonymous

INTRODUCTION

An *air pollutant* is a gas, liquid droplet, or solid particle which, if dispersed in the air with sufficient concentration, poses a hazard to flora, fauna, property, and climate. Air pollution, a visible environmental side effect of transportation, has become a public health concern for millions of urban residents worldwide (TRB, 1997). Transportation or "mobile" sources of air pollution, particularly motor vehicles, are a primary source of local carbon monoxide problems and are considered the main cause of excess regional photochemical oxidant concentrations. Transportation vehicles typically emit carbon monoxide, nitrogen oxides, small particulate matter, and other toxic substances that can cause health problems when inhaled. Air pollution also has adverse effects on forests, lakes, and rivers. The contribution of transportation vehicle use to global warming remains a cause for much concern as anthropogenic impacts on the upper atmosphere become increasingly evident. Airports, for instance, are a major source of local violations of ambient carbon monoxide standards and contribute to regional photochemical oxidant problems. In the current era, rail travel is increasingly being powered by electricity and is therefore typically not associated with significant air pollution, except in cases where the source of rail energy generation is associated with significant pollution, such as coal-based electrical power generation.

In this chapter we discuss the transportation sources and adverse impacts of air pollution and factors that affect pollutant emissions and concentrations. We also describe how to estimate pollutant emissions and concentrations using various models and present a general methodology to estimate the air quality impacts of transportation projects. In addition, possible measures to mitigate air pollution impacts, and air quality legislation, are discussed.

10.1 AIR POLLUTION SOURCES AND TRENDS

10.1.1 Pollutant Types, Sources, and Trends

Primary air pollutants are those emitted directly into the atmosphere and include carbon monoxide, hydrocarbons, sulfur oxides, nitrogen oxides, and particulate matter. *Secondary air pollutants* such as ozone and acidic depositions, are those formed in the atmosphere as a result of physical and chemical processes (such as hydrolysis, oxidation, and photochemistry) on primary pollutants. *Greenhouse gases*, such as carbon dioxide, are also direct emissions although not as yet included in USEPA list primary air pollutants.

Natural sources of air pollution include forest fires and volcanoes; *anthropogenic sources* include power generation, fuel use, slash-and-burn agricultural practices, and transportation. Table 10.1 describes the types, sources, effects, and scales of transportation pollutants.

Total air pollution increased from 1960 to 1970 but decreased thereafter despite a great increase in vehicular travel (Figure 10.1). Emissions of volatile organic compounds and particulate matter have declined steadily over the years, while there has been only a slight increase in sulfur dioxide emissions. Also, lead emissions have dropped sharply following the development of lead-free gasoline. The drop in pollutant emissions over the years is often attributed to governmental intervention through the establishment of increasingly restrictive federal emission standards. For example, between 1980 and 1995, the allowable level of carbon monoxide emissions from a passenger car was reduced from 7.0 to 3.4 g/mi.

In the last decade, transportation contributed about 83% of the carbon monoxide (CO), 45% of the volatile organic compounds (VOCs), and 53% of the nitrogen oxide (NO_x) emissions in the United States (USEPA, 2005). Tailpipe emission rates have declined significantly over the past few decades. However, the actual reductions may be smaller because the standard tests do not reflect real driving conditions; and vehicles producing harmful emissions are typically not measured in these tests (BTS, 1997; Homburger et al., 2001). Also, increased vehicle mileage has offset much of the reduction in per-mile emissions, so vehicle emissions continue to be a major

Table 10.1   Air Pollutants from Transportation Sources

Pollutant	Description	Source	Effects	Scale
Carbon monoxide (CO)	Colorless and odorless toxic gas formed by incomplete combustion of fossil fuels. The most plentiful of mobile-source air pollutants.	Vehicle and aircraft engines	Human health (undermines oxygen-carrying ability of blood), climate change.	Very local
Fine particulates (PM_{10}; $PM_{2.5}$)	Inhalable solid particles emitted by mobile sources: droplets of unburned carbon, bits of rubber, metal, material from brake pads, lead particles, etc.	Diesel engines and other sources	Human health (causes respiratory problems), aesthetics.	Local and regional
Nitrogen oxides (NO_x)	Primarily, NO and NO_2, caused by oxidation of atmospheric nitrogen. Some are toxic, all contribute to ozone formation.	Engine	Helps formation of corrosive acids that damage materials; kills plant foliage, impairs respiratory system; absorbs light and reduces visibility; contributes to ozone formation.	Regional
Volatile organic compounds	Includes hydrocarbons (HC) such as methane (CH_4). Emitted from unburned fuel from fuel tanks and vehicle exhausts. Smog is a haze of photochemical oxidants caused by the action of solar ultraviolet radiation on HC and NO_x.	Fuel production and engines	Human health, ozone precursor.	Regional
Lead	Formed by burning leaded fuel.	Fuel production and engines	Affects circulation, reproductive, nervous, and kidney systems; suspected of causing hyperactivity and lowered the learning ability in children.	Regional
Airborne toxins (e.g., benzene)	Pollutants that are carcinogenic or have effects on human reproductive or developmental systems.	Fuel production and engines	Human health risks.	Very local
Ozone (O_3)	Highly reactive photochemical oxidizer formed in atmosphere through reactions involving NO_x, VOCs, and sunlight.	NO_x and volatile organic compounds	Human health (respiratory), plants, aesthetics; ground-level O_3 is a primary component of smog, which impairs visibility.	Regional

Table 10.1 (*continued*)

Pollutant	Description	Source	Effects	Scale
Sulfur oxides (SO_x)	Formed by burning of sulfur-containing fossil fuels and oxidation of sulfur; SO_2 is a colorless water-soluble pungent and irritating gas.	Diesel engines	Human health risks, causes acid rain that harms plants and property; lung irritant; causes acid rain.	Regional
Carbon dioxide (CO_2)	By-product of combustion.	Fuel production and engines	Climate change.	Global
Chlorofluorocarbons (CFCs)	Nontoxic, nonflammable chemicals containing atoms of carbon, chlorine, and fluorine. Classified as halocarbons, a class of compounds that contain atoms of carbon and halogen atoms.	Air conditioners manufactured before the 1980s	Climate change (depletion of outer ozone layer).	Global
Road dust	Dust particles created by vehicle movement.	Vehicle use	Human health, aesthetics.	Local

Source: Carpenter (1994), Faiz et al. (1996), USEPA (1999), Holmen and Niemeier (2003).

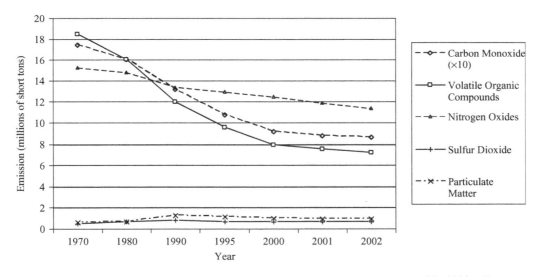

Figure 10.1   Trends in pollutant emissions from transportation sources, 1970–2002. (From USEPA, 2005.)

problem. The overall level of emissions depends heavily on traffic flow characteristics, such as the average flow speed, the frequency and intensity of vehicle acceleration and deceleration, the number of stops, and the vehicle operating mode.

Although highways continue to be the major contributor of transportation air pollution, contributions from other modes should not be underestimated. More than 120 million people live in areas with unhealthy air due to high levels of smog, and most of the busiest airports in the United States are located in, and contribute pollution to, urban areas where air quality is already a problem. Furthermore, it is anticipated that the relative contribution of airport activities to overall emissions will increase over time. Airport emissions are becoming the largest *point sources* in many urban areas, emitting as much NO_x as a large power plant. Sources of air pollution at airports are aircraft (main engines, auxiliary power units), ground service equipment (aircraft tugs, baggage tractors, etc.), and ground access vehicles at airports.

10.1.2 Categories of Air Pollution

There are generally two categories of air pollutants: criteria air pollutants and greenhouse gases.

(*a*) *Criteria Air Pollutants*   This category consists of carbon monoxide (CO), nitrogen oxide (NO_x), volatile organic compounds (VOCs), particulate matter of size 10 μm or less, particulate matter of size 2.5 μm or less, sulfur dioxide (SO_2), and ammonia (NH_3). Gasoline-powered light vehicles continue to be the source of most carbon monoxide emissions from highway vehicles. In the United States, heavy diesel-powered vehicles account for 46% of NO_x emissions from highway vehicles, and light gasoline vehicles are responsible for about 48%. With regard to volatile organic compounds, the transportation sector accounted for just over 54% of total emissions in 2002, and gasoline-powered vehicles were responsible for 91% of highway vehicle VOC emissions. In 2002, the transportation sector also accounted for just over 54% of particulate matter emissions of size 10 μm or less. Most of these were from gasoline vehicles. A similar distribution was seen for particulate matter of smaller size (2.5 μm or less). With regard to lead, the transportation sector (highway vehicles in particular) has long been identified as a dominant source of lead emissions, but its share has dwindled over the years from 82% in 1970 to about 13% in 1999. This is due largely to a 1978 regulatory action calling for reduced lead content of gasoline fuels. Only a small share of transportation lead emissions is now attributed to highway fuel use (USEPA, 2005). In some developing countries, however, lead continues to be a major air pollutant from transportation sources.

(*b*) *Greenhouse Gases*   The atmosphere serves as a blanket for retaining and redistributing heat to maintain Earth's mean surface temperature at levels that are conducive for life. This role is played by certain gases in the atmosphere known as *greenhouse gases*, which include carbon dioxide (CO_2), methane (CH_4) nitrous oxides, hydrofluorocarbons, perfluorocarbons, and sulfur hexafluoride. These gases, released by anthropogenic sources, have reached levels that threaten to expand the natural layer of greenhouse gases, thus leading to greater retention of radiation energy, accelerated global warming, and consequent damage to the global ecology and

development of extreme weather patterns. Transportation sources are significant in this regard: Most CO_2 emissions are from petroleum fuels, particularly motor gasoline, and CO_2 accounts for 80% of the total greenhouse gas emissions.

10.2   ESTIMATING POLLUTANT EMISSIONS

10.2.1 Some Definitions

- *Emission*. This is the discharge of pollutants into the atmosphere. The overall magnitude of emissions depends on the number of emitting sources, the diversity of source types, the nature and scale of activity at the polluting source, and the emission characteristics. For instance, more pollutants are emitted by motor vehicles at higher altitudes, due to inefficient combustion caused by air thinness.
- *Mobile emission*. A mobile source of air pollution is one that is capable of moving from one place to another under its own power, such as a motorized vehicle. Emissions from mobile sources are described as mobile emissions. The total air quality in an area is measured in terms of the ambient concentration of pollutants that are emitted by mobile and stationary sources.
- *Emission factors*. An *emission factor* is an average estimate of the rate at which a pollutant is released into the atmosphere as a result of some activity (such as motor vehicle operation) in terms of activity level such as VMT (vehicle-miles of travel) or VHT (vehicle-hours traveled) for motor vehicles.

10.2.2 Factors Affecting Pollutant Emissions from Motor Vehicles

The major factors that affect the level of vehicle emissions can generally be classified as follows: travel-related, driver-related, highway-related, vehicle-related, fuel type, and environmental (Figure 10.2). An NCHRP study (Report 394) provides information on the sensitivity of vehicle emissions in response to changes in these factors (Chatterjee et al., 1997). We discuss the factors below.

(*a*) *Travel-Related Factors*   Travel-related factors include vehicle engine operating modes, speeds, and accelerations and decelerations. Three *operating modes* are typically considered in estimating exhaust emissions: cold start, hot start, and hot stabilized period. Emission rates differ significantly across these modes. The EPA defines a *cold-start* as any start of a vehicle engine occurring 4 hours or later following the end of the preceding trip for non-catalyst-equipped vehicles, and 1 hour or later following the end of the preceding trip for catalyst-equipped

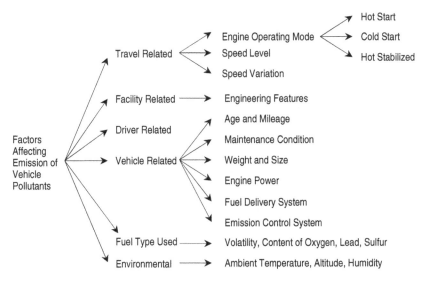

Figure 10.2 Factors affecting vehicle emissions.

vehicles. *Hot starts* are those that occur less than 4 hours after the end of the preceding trips for non-catalyst-equipped vehicles, and less than 1 hour after the end of the preceding trip for catalyst-equipped vehicles. The time between the start and the end of a trip is called the *hot-stabilized period*. Emission rates of HC and CO are higher during cold starts than during hot starts and are lowest during hot-stabilized operation. The difference in vehicle emission rates between operating modes are due to their different air-to-fuel ratios and catalytic conversion rates. In the cold-start mode of vehicle engine operation, the catalytic emission control system is not fully functional and the low air-to-fuel ratio leads to the high HC and CO emissions. The emission of NO_x is, however, low during cold-start modes.

The type, speed, and acceleration of a vehicle and the load on its engine have significant impacts on the level of emissions. HC and CO emissions are highest at low speeds. Figure 10.3 shows the effect of speed on CO and NO_x emissions by vehicle type and fuel type. It is seen, for example, that for most vehicle types, CO and NO_x emissions generally are high at low speeds, decrease with increasing speed to their minimum rates, and then stay flat or increase slightly depending on the vehicle or fuel type, or the pollutant in question. The smoothness and consistency of vehicle speed, traffic conditions, and driving behavior can influence emissions. Sharp acceleration at a high speed and heavy load on an engine require more fuel to feed the engine, thus generating more HC and CO emissions but cause little change in NO_x emissions.

(*b*) *Facility-Related Factors* Certain facility designs can encourage transportation vehicles to operate at low-emitting speeds or modes. For highway transportation, examples include low grade, existence of ramps and signals, acceleration and deceleration lanes, and channelization. It has been shown, for example, that traffic signal coordination can result in up to a 50% reduction in emissions under certain circumstances (Rakha et al., 1999).

(*c*) *Driver-Related Factors* Driver behavior varies significantly by person and by traffic condition, and can influence emission rates. For example, aggressive drivers typically exert more frequent and severe accelerations and decelerations than do their less aggressive counterparts. Such abrupt changes in velocity impose heavy loads on the engine and thus result in higher levels of emissions.

(*d*) *Vehicle-Related and Other Factors* Vehicle emissions are influenced by vehicle age, mileage, condition, weight, size, and engine power. Older model vehicles typically emit more pollutants than do newer ones and heavier and larger vehicles emit more pollutants than are emitted by lighter and smaller vehicles (Ding, 2000). Fuel type also affects emission levels significantly. Furthermore, there is a difference in the combustion processes of the two major engine types that translate into different pollutant emissions rates. Table 10.2 shows pollutant emissions by highway vehicle type and transit mode under average operating conditions.

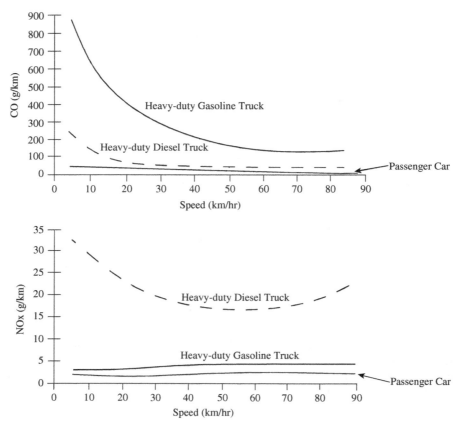

Figure 10.3 Variation in CO and NO$_x$ emission rates by speed, vehicle, and fuel type. (From Faiz et al., 1996.).

(*e*) *Environmental Factors*   At low temperatures, more time is required to warm up the engine and the emission control system, thus increasing the level of cold-start emissions. At higher temperatures, on the other hand, combustive emissions are low, but evaporative emissions are high, due to the increased fuel evaporation rate.

10.2.3 Approaches for Estimating Pollutant Emissions from Highways

In evaluating the impact of transportation improvements on air quality, the first step is to estimate the change in emissions as a result of changes in the average speed of vehicles, increases in motor vehicle trips, and increases in VMT due to these improvements. The second step is to determine the resulting change in pollutant concentrations due to the change in emissions. For highway transportation, emission models can be grouped as follows: speed-based, modal, microscopic, and fuel-based models (Ding, 2000).

(*a*) *MOBILE 6.0 Mobile Source Emission Factor Model*
The EPA MOBILE6 is a speed-based model that estimates highway transportation emission factors in gms/vehicle-mile for three pollutants: hydrocarbons (HC), carbon monoxide (CO), and oxides of nitrogen (NO$_x$), for gasoline- and diesel-fueled highway motor vehicles and certain specialized vehicles, such as natural gas–fueled or electric vehicles. MOBILE6 estimates the emission factors for 28 individual vehicle types under various conditions, such as ambient temperature, travel speed, operating mode, fuel volatility, and mileage accrual rates, and considers four vehicle roadway facilities: freeways, arterial and collectors, local roadways, and freeway ramps (USEPA, 2002). The fleet average emission factor (EF) for a vehicle class, calendar year, pollutant, and emission-producing process is given as follows (Koupal and Glover, 1999):

$$\text{EF}_{ijk} = \sum_{m=1}^{n} [\text{FVMT}_{im}(E_{ijkm}C_{ijkm})] \qquad (10.1)$$

Table 10.2(a) Pollution Emissions by Mode (g/VMT)

	VOC	CO	NO_x	CO_2
Automobile	1.88	19.36	1.41	415.49
SUVs, light truck	2.51	25.29	1.84	521.63
Bus	2.3	11.6	11.9	2386.9
Diesel-powered rail	9.2	47.6	48.8	9771.0

Source: TCRP (2003).

Table 10.2(b) Pollutant Emissions by Truck Type (g/VMT)[a]

Truck Type	Road Class	VOC	CO	NO_x	PM-10	PM-10 Exhaust Only
Single-unit gasoline truck	Local	7.06	144.07	5.94	0.13	0.11
	Arterial	2.29	59.87	7.18	0.13	0.11
	Urban freeway	1.31	51.39	8.12	0.13	0.11
	Rural freeway	1.31	75.87	8.84	0.13	0.11
Single-unit diesel truck	Local	1.18	6.86	14.95	0.42	0.38
	Arterial	0.59	2.86	15.34	0.42	0.38
	Urban freeway	0.42	2.21	22.69	0.42	0.38
	Rural freeway	0.41	2.8	30.39	0.42	0.38
Combination-unit diesel truck	Local	1.22	7.64	16.07	0.41	0.37
	Arterial	0.61	3.18	17.02	0.41	0.37
	Urban freeway	0.43	2.48	25.65	0.41	0.37

Source: http://www.fhwa.dot.gov/ENVIRonment/freightaq/appendixb.htm.

[a]Emission estimates may differ somewhat from EPA National Emission Inventory (NEI) heavy-duty truck estimates, due to differences in aggregation methods for vehicle class and speed.

where EF_{ijk} is the fleet-average emission factor for calendar year i, pollutant type j, and emission-producing process k (e.g., exhaust, evaporative); $FVMT_{im}$ the fractional VMT attributed to model year m for calendar year i ($n = 28$ in MOBILE6); E_{ijkm} the basic emission rate for calendar year i, pollutant j, process k, and model year m; and C_{ijkm} the correction factor (e.g., for temperature, speed) for calendar year i, pollutant j, process k, and model year m.

The MOBILE6 model produces separate emission factors for the start- and running-modes. The running-mode emission factors are based only on hot-stabilized operating conditions; the start emissions represent the additional emissions that result from a vehicle start. The model provides daily and hourly emission factors for each hour of day. In addition, it incorporates enhancements such as update of fuel effects on emissions, use of diurnal evaporative emissions based on real-time diurnal testing, update of hot-soak evaporative emission factors,

update of heavy-duty engine emission conversion factors, update of fleet characterization data, and a provision for distinct emission factor calculations for a wider range of vehicle categories. To facilitate implementation, a software package has been developed for the MOBILE6 model (see Section 10.2.5).

(b) Emission Models Based on Vehicle Operating Modes The term *engine operating mode* refers to engine temperature (hot start, cold start, etc.), while *vehicle operating mode* refers to speed change (or lack thereof), such as cruise, acceleration, deceleration, and idling. Barth et al. (1996) and An et al. (1997) developed modal emission models for light-duty cars and trucks. These models predict the engine power, engine speed, air-to-fuel ratio, fuel use, engine-out emissions, and catalyst pass fraction and finally estimate tailpipe emissions and fuel consumption. The vehicle power demand is modeled as a function of the operating variables (i.e., vehicle acceleration and speed),

specific vehicle parameters (e.g., vehicle mass, transmission efficiency, effects of accessories), and road conditions. The fuel use rate is a function of the power demand, engine speed, and air/fuel ratio, and the engine-generated emissions are estimated using the fuel rate and other factors, as follows:

$$E = \text{FR} \times g \times \text{CPF} \qquad (10.2)$$

where E is the tailpipe emission in g/s, FR the fuel-use rate in g/s, g the grams of engine-out emissions per gram of fuel consumed, and CPF the catalyst pass fraction (the ratio of tailpipe emissions to engine-out emissions).

Another modal emission model is MEASURE (Mobile Emissions Assessment System for Urban and Regional Evaluation) (Guenslar et al., 1998). The emission rates estimated by MEASURE are dependent on both vehicle mode variables (vehicle speed, acceleration profile, idle times, and power demand) and vehicle technology variables (fuel metering system, catalytic converter type, availability of supplemental air injection, and transmission speed). Also, the models estimate the emission rates for each pollutant type.

(c) *Microscopic Emission Models*   Microscopic emission models are used in traffic operations software packages to estimate emissions at highway segments, interchanges, and intersections. The emission rates are estimated incrementally as a function of the instantaneous vehicle fuel consumption, speed, acceleration, and engine power. The Transportation Analysis and Simulation System (TRANSIMS), for example, does this by multiplying the fractional power change at a given time and the emission difference for the given speed and power, and then adding the result to the emissions at constant power. The Traffic Simulation and Dynamic Assignment Model (INTEGRATION) accounts for vehicle stops and accelerations and decelerations at freeways and arterials and estimates emissions by computing fuel consumption for each vehicle on a second-by-second basis for three operation modes (constant-speed cruise, velocity change, and idling) as a function of travel speed (USEPA, 1998). Vehicle emissions are then estimated as a function of fuel consumption, ambient air temperature, and the extent to which a particular vehicle's catalytic converter has already been warmed up during an earlier portion of the trip (Rouphail et al., 2001). INTEGRATION also has the ability to capture congestion effects on emissions (Sinha et al., 1998). FHWA's TRAF-NETSIM tracks the movements of individual vehicles on a second-by-second basis at single intersections and at freeway segments

and ramps, and estimates hot-stabilized emissions of CO, HC, and NO_x as a function of vehicle travel speed and acceleration.

(d) *Fuel-Based Emission Models*   Fuel-based models estimate vehicle emissions on the basis of fuel consumed as vehicles operate in various operating modes. An example is the SYNCHRO traffic model, which first predicts fuel consumption as a function of vehicle-miles, total delay in vehicle-hours/hour, and total stops in stops/hour. Then, to estimate vehicle emissions, the fuel consumption is multiplied by an adjustment factor based on the emission type (Rouphail et al., 2001).

(e) *Greenhouse Gas Emission Models* CO_2, one of the biggest by-products of engine combustion (USEPA, 2006), is a significant greenhouse gas. For every gallon of motor fuel burned, approximately 20 pounds of CO_2 are emitted into the atmosphere. The USEPA has developed a score-based model for estimating the amount of this greenhouse gas. The score is determined on the basis of a vehicle's fuel economy and fuel type, because each type of fuel contains a different amount of carbon per gallon. The scale used ranges from 0 (maximum CO_2 emission) to 10 (least CO_2 emission), and the average score for model year 2005 was 5. Table 10.3 shows the score that corresponds to fuel efficiency rates (mpg) and fuel type. The fuel efficiency rate is a combination of rates from city and highway driving condition as follows:

Combined fuel economy(mpg)

$$= 1/(0.55/\text{city mpg} + 0.45/\text{highway mpg}).$$

10.2.4 Procedure for Estimating Highway Pollutant Emissions

A transportation agency may seek to evaluate either (1) the existing air quality *situation* at a given time (with no intent of any transportation intervention) or (2) the estimated air quality (using models) or actual air quality (using field measurements) after a planned or past transportation intervention. Air quality is typically measured in terms of emissions and/or resulting concentrations of selected air pollutants. Transportation interventions first lead to changes in traffic flow patterns (operating speeds, speed change frequencies, traffic composition); in the medium term, such interventions cause changes in travel demand patterns (trip purposes, route, frequency, mode, etc.); and in the long term, they lead to changes in land-use patterns (locations of residences and businesses). The short-term effects lead to changes in emission rates,

Table 10.3 Greenhouse Gas Score Model

Fuel Type and Fuel Economy[a]					CO_2 Emissions (pounds/mile)	Greenhouse Gas Score
Gasoline	Diesel	E85	LPG	CNG		
44 and higher	50 and higher	31 and higher	28 and higher	33 and higher	Less than 0.45	10
36 to 43	41 to 39	26 to 30	23 to 27	27 to 32	0.45 to 0.54	9
30 to 35	35 to 40	22 to 25	20 to 22	23 to 26	0.55 to 0.45	8
26 to 29	30 to 34	19 to 21	17 to 19	20 to 22	0.65 to 0.74	7
23 to 25	27 to 29	17 to 18	15 to 16	18 to 19	0.75 to 0.84	6
21 to 22	24 to 26	15 to 16	14	16 to 17	0.85 to 0.94	5
19 to 20	22 to 23	14	13	14 to 15	0.95 to 1.04	4
17 to 18	20 to 21	13	12	13	1.05 to 1.14	3
16	18 to 19	12	11	12	1.15 to 1.24	2
15	17	11	10	11	1.25 to 1.34	1
14 and lower	16 and lower	10 and lower	9 and lower	10 and lower	1.35 and higher	0

Source: USEPA (2006).
[a]E85 = 85% ethanol and 15% gasoline, LPG = liquefied petroleum gas, CNG = compressed natural gas.

while the medium- and long-term effects lead to changes in travel amounts (vehicle-miles of travel); thus, the short-, medium-, and long-term effects all lead to a change in overall emissions. For example, a lane-widening project may reduce congestion and improve traffic flow by reducing speed-change cycles (subsequently, reducing pollution) in the short term but may attract induced demand in the long run thus increasing pollution. Also, transportation interventions such as ramp metering and HOV lanes may have adverse air quality effects in the short term (due to queuing and congestion in certain areas) but beneficial air quality impacts in the long term due to overall decreased travel delay. Figure 10.4 illustrates the sequence of impacts of transportation intervention on air quality. The stages discussed in step 1 are for the intervention scenario, while the stages in step 2 are for the no-intervention scenario (base case), which is the do-nothing situation at the current time or at a future time.

Step 1: Determine the Transportation Intervention This may be a policy change or physical enhancement, such as improvements in alignment design, traffic management, or transit operations.

Step 1.1: Identify the Short-Term Effect Most transportation interventions typically lead to changes in operational characteristics, often in the form of increased vehicle operating speeds and fewer speed-change (acceleration and deceleration) events. These operational changes that happen in the short term, also termed *first-order effects* (Dowling et al., 2005), have two impacts: (a) changes in the emission rates of vehicles using the facility, and (b) changes in travel demand patterns.

Step 1.2: Identify the Operational Changes The operational changes that affect travel demand patterns in the medium term (8 to 14 months of the intervention) are typically referred to as *second-order effects* (Dowling et al., 2005). The higher speeds (and hence lower travel times) due to the intervention may induce travelers to undertake more frequent trips, change their current mode to one that benefits most from the intervention, or change their trip schedules. Second-order effects may result in new travel amounts and frequencies (and ultimately, increased total emissions even if emission rates decrease or remain the same).

Step 1.3: Identify the Locational Shifts of Residences and Businesses Operational improvements and changes in demand patterns may lead in the long term to changes in home and business location patterns. Improved traffic flow and reduced congestion tend to attract new businesses and residences or retain existing ones. These can be considered as *third-order effects* (Dowling et al., 2005).

Step 1.4: Establish the New Travel Amounts and Frequencies In investigating the air quality impacts of

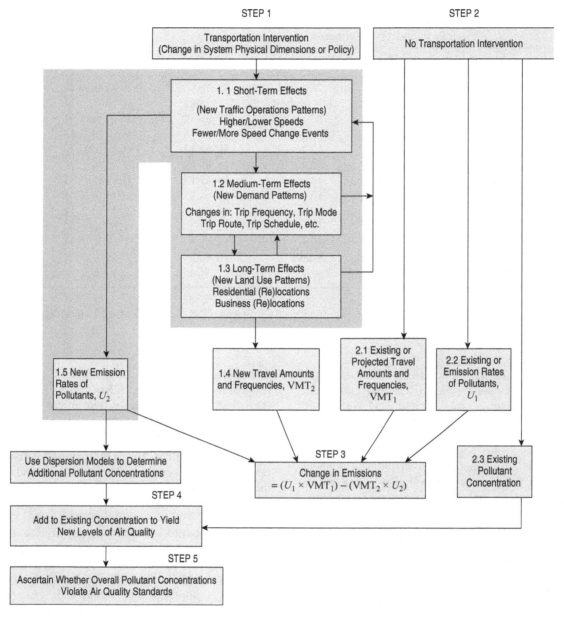

Figure 10.4 Procedure for assessing air quality impacts of transportation interventions.

transportation interventions, most analytical procedures implicitly exclude this step by stopping at the short-term effects (see the shaded area in Figure 10.4.). As such, these procedures assume implicitly that the second- and third-order effects are negligible.

Step 1.5: Establish the New Emission Rates of Pollutants Changes in the emission rates of pollutants may be due to (a) transportation intervention policies that directly affect the rates of pollutant emissions, such

as new emission standards, restriction of vehicles with excessive pollutants, and enforcement of vehicle exhaust inspections and laws, and (b) changes in speed and acceleration–deceleration events arising from physical improvements such as channelization and lane addition. MOBILE6 can be used to estimate the new emission rates due to both types of interventions. The MOBILE6 model and software are described in Sections 10.2.3(a) and 10.2.5, respectively.

Steps 1.6 and 1.7: Estimate the New Total Pollutant Emissions and Concentrations

(a) *Change in Emissions* Knowing the emission rate per travel activity (from step 1.5) and the amount of travel activity (step 1.4), estimating the total emissions (also referred to as *emissions inventory*) is a straightforward task. The differences in various approaches for estimating total emissions stem largely from their respective definitions of the term *travel activity*. In approaches where MOBILE6 is used to establish emission rates (step 1.5), travel activity is defined in terms of vehicle-miles of travel. Total emissions are estimated as follows:

$$\text{total emissions} = \text{emission per vehicle-mile of travel}$$
$$\times \text{ total vehicle-miles of travel for the project} \quad (10.3)$$

In other approaches, such as the comprehensive modal emission model (Dowling, 2005), which define travel activity in terms of vehicle-hours of travel, total emissions are estimated as follows:

$$\text{total emissions} = \text{emission per vehicle-hour of travel}$$
$$\times \text{ total vehicle-hours of travel for the project} \quad (10.4)$$

(b) *Change in Ambient Concentrations* Given the new level of emissions due to the transportation intervention, the associated concentration can be estimated if the levels of dispersion factors (wind speed and direction, mixing height, etc.) are known (see Section 10.3).

Steps 2.1 and 2.2: Analyze the Existing Situation (at the Current or Some Future Year)

This step analyzes a base-case scenario against which the air pollution impacts of intervention can be assessed. If the base case is taken as the current year, the base-case air quality impacts (pollutant concentrations) are established by one of the following methods: (a) measuring the pollutant concentrations directly using air quality monitoring equipment, or (b) carrying out steps similar to steps 1.1 to 1.6 using current-year data on emission rates, traffic operations, and so on. If the base case is for some future year projected from a current do-nothing situation, then estimates of the base case air quality can be predicted by carrying out steps similar to steps 1.1 to 1.6 using data projected for emission rates, traffic operations, and so on, at the future year of interest.

Step 3: Determine the Difference in Total Pollutant Emissions

If the intent of the analysis is to ascertain whether the transportation intervention had (or will have) an impact on the existing air quality of the area, step 3 should be included. Step 3 simply expresses the new emission relative to the base-case emission and quantifies the extent to which the intervention contributes to the improvement or degradation of air quality.

Steps 4 and 5: Estimate the Overall Pollutant Concentrations

If the intent is to determine whether the transportation intervention would lead to a violation of air quality standards or ameliorate existing levels to acceptable levels, it is necessary to carry out step 4. In this step, pollutant concentrations due to mobile sources are added to those from stationary sources. In step 5, the overall concentration for each pollutant is compared with established air quality thresholds to ascertain whether any standards have been violated.

10.2.5 Software for Estimating Pollutant Emissions

The most common software used for estimating pollutant emissions is MOBILE6, whose theoretical procedure is discussed in Section 10.2.3(a). This package utilizes inputs, such as the frequency of starts per day and their distribution by hour and the enforcement of inspection maintenance programs, and incorporates external conditions such as temperature and humidity. MOBILE6 also requires a temporal distribution of traffic during the day for major traffic indicators. Hourly distributions can be input instead of 24-hour averages. Also, the fleet characterization projections of future vehicle fleet size and the fraction of travel are based on considerations that include vehicle age, mileage accumulation rate, and vehicle class. (Vehicles classes are shown in Table 10.4.) Data on the key traffic-related variables (vehicle registration distribution, annual mileage accumulation rate, and the distribution of vehicle miles of travel) are input by vehicle class and roadway type. Local data on mileage accumulation are typically more difficult to obtain because odometer readings are typically not recorded on an annual basis unless an inspection and maintenance program is operational in the region under study. MOBILE6 outputs the emission rates (g/vehicle-mile) for three pollutants: HC, NO_x, and CO. Description of the general MOBILE6 input file and a sample output file are provided in Appendix A-10.

Example 10.1 A portfolio of transportation projects, including traffic signal optimization, lane widening, and channelization, has been undertaken in the city of Townsville. These projects have helped to reduce the traffic congestion in the city and have increased the average speeds of vehicles. However, these improvements have been accompanied by an increase in the amount of travel, due partly to city residents taking advantage of lower congestion and induced demand from nearby towns. It is sought to evaluate the impact of the transportation projects on vehicle emissions in the city. Amounts of travel (by vehicle class) and the speeds corresponding to the without- and with-improvement scenarios are presented in Table E10.1.1. Use MOBILE6 to evaluate the impact of the portfolio of transportation improvements on air quality in terms of emission rate changes of the three key pollutants (HC, CO, and NO$_x$). Assume a 10-mile road length; consider 2004 and 2006 as the without and with-improvement years, respectively; for other air quality parameters, use the default values, provided in MOBILE6.

SOLUTION

1. Input data: VMT fractions (for the vehicle classes listed in the first column of Table E10.1.1) are

entered in the input file. Other input data including speeds and analysis years are entered into the input file as described in Appendix A10.

(1) *Calculation of VMT fractions* (see Table E10.1.1):

$$\text{VMT fraction} = \frac{\text{VMT}}{\text{total VMT}}$$

For example, VMT Fraction for LDV in the without-improvement scenario = 9,975/25,000 = 0.399

(2) *Estimation of emissions*

The vehicle classes (and combinations thereof) that appear in the default descriptive outputs are listed as follows:

Table 10.4   Vehicle Classes in MOBILE6

Number	Abbreviation	Description
1	LDGV	Light-Duty Gasoline Vehicle (Passenger Cars)
2	LDDV	Light-Duty Diesel Vehicles (Passenger Cars)
3	LDGT1	Light-Duty Gasoline Trucks 1 (0-6,000 lbs. GVWR: 0-3,750 lbs. LVW)
4	LDGT2	Light-Duty Gasoline Trucks 2 (0-6,000 lbs. GVWR: 3,751-5,750 lbs. LVW)
5	LDDT12	Light-Duty Diesel Trucks 1 and 2 (0–6,000 lbs. GVWR)
6	LDGT3	Light-Duty Gasoline Trucks 3 (6,001–8,500 lbs. GVWR: 0–5,750 lbs. ALVW[a])
7	LDGT4	Light-Duty Gasoline Trucks 4 (6,001–8,500 lbs. GVWR: 5,751 lbs. and greater ALVW)
8	LDDT34	Light-Duty Diesel Trucks 3 and 4 (6,001–8,500 lbs. GVWR)
9	HDGV2B	Class 2b Heavy-Duty Gasoline Vehicles (8,501–10,000 lbs. GVWR)
10	HDDV2B	Class 2b Heavy-Duty Diesel Vehicles (8,501–10,000 lbs. GVWR)
11	HDGV3	Class 3 Heavy-Duty Gasoline Vehicles (10,001–14,000 lbs. GVWR)
12	HDDV3	Class 3 Heavy-Duty Diesel Vehicles (10,001–14,000 lbs. GVWR)
13	HDGV4	Class 4 Heavy-Duty Gasoline Vehicles (14,001–16,000 lbs. GVWR)
14	HDDV4	Class 4 Heavy-Duty Diesel Vehicles (14,001–16,000 lbs. GVWR)
15	HDGV5	Class 5 Heavy-Duty Gasoline Vehicles (16,001–19,500 lbs. GVWR)
16	HDDV5	Class 5 Heavy-Duty Diesel Vehicles (16,001–19,500 lbs. GVWR)
17	HDGV6	Class 6 Heavy-Duty Gasoline Vehicles (19,501–26,000 lbs. GVWR)
18	HDDV6	Class 6 Heavy-Duty Diesel Vehicles (19,501–26,000 lbs. GVWR)
19	HDGV7	Class 7 Heavy-Duty Gasoline Vehicles (26,001–33,000 lbs. GVWR)
20	HDDV7	Class 7 Heavy-Duty Diesel Vehicles (26,001–33,000 lbs. GVWR)
21	HDGV8A	Class 8a Heavy-Duty Gasoline Vehicles (33,001–60,000 lbs. GVWR)
22	HDDV8A	Class 8a Heavy-Duty Diesel Vehicles (33,001–60,000 lbs. GVWR)
23	HDDV8B	Class 8b Heavy-Duty Gasoline Vehicles (>60,000 lbs. GVWR)
24	HDGV8B	Class 8b Heavy-Duty Diesel Vehicles (>60,000 lbs. GVWR)
25	HDGB	Gasoline Buses (School, Transit and Urban)
26	HDDBT	Diesel Transit and Urban Buses
27	HDDBS	Diesel School Buses
28	MC	Motorcycles (Gasoline)

[a]ALVW = Alternative Vehicle Weight: The adjusted loaded vehicle weight is the numerical average of the vehicle curb weight and the gross vehicle weight rating (GVWR).

Table E10.1.1   VMT Fractions by Vehicle Class for Each Scenario[a]

Vehicle Class		Percentage %	AADT Without Improvement	AADT With Improvement	VMT Fraction Without Improvement	VMT Fraction With Improvement
LDV	LDGV	80	9975	13120	0.399	0.403
	LDDV	20				
LDT1	LDGT1	80	1425	1856	0.057	0.057
	LDDT1	20				
LDT2	LDGT2	80	4750	6112	0.19	0.188
	LDDT2	20				
LDT3	LDGT3	80	450	608	0.018	0.019
	LDDT3	20				
LDT4	LDGT4	80	225	288	0.009	0.009
	LDDT4	20				
HDV2B	HDGV2B	50	1900	2464	0.076	0.076
	HDDV2B	50				
HDV3	HDGV3	50	450	576	0.018	0.018
	HDDV3	50				
HDV4	HDGV4	50	400	480	0.016	0.015
	HDDV4	50				
HDV	HDGV5	50	200	256	0.008	0.008
	HDDV5	50				
HDV6	HDGV6	50	850	1120	0.034	0.034
	HDDV6	50				
HDV7	HDGV7	25	1200	1536	0.048	0.047
	HDDV7	75				
HDV8A	HDGV8A	25	700	928	0.028	0.028
	HDDV8A	75				
HDV8B	HDGV8A	10	1625	2080	0.065	0.064
	HDDV8A	90				
HDBS	HDGB	5	75	128	0.003	0.004
	HDDBS	95				
HDBT	HDDBT	100	475	640	0.019	0.020
MC			300	384	0.012	0.012

[a]Average speed without improvement = 28 mph. Predicted average speed with improvement = 35 mph.

For LDGV: LDGT 1 and 2 combined—LDGT 12
LDGT 3 and 4 combined—LDGT 34
LDGT 1, 2, 3, and 4 combined—LDGT

For LDDV: LDDT 1, 2, 3 and 4 combined—LDDT
For all HDGV and HDGB combined—HDG
For all HDDV and HDDB combined—HDD

For all 28 sub-classes combined—All Vehicles

In the descriptive output file of MOBILE6, emissions for all 28 vehicle sub-classes can be reported by using the following commands: EXPAND LDT EFS, EXPAND HDGV EFS, EXPAND HDDV EFS, and EXPAND BUS EFS.

Emission estimates with and without the improvement are given in Table E10.1.2.

The VMT distributions and emission values for the "with improvement" scenario are shown in parentheses. Emission rates are shown for each vehicle type and pollutant type. The exhaust HC, CO, and NO_x emissions of heavy-duty vehicles are reported only as "composite" exhausts, not as either start or running. Figure E10.1 shows the levels of major pollutants for the "with improvement" and "without improvement" scenarios.

Table E10.1.2   Emission Values for Without- and With-Improvement Scenarios

Vehicle Type		LDGV	LDGT12	LDG34	LDGT	HDGV	LDDV	LDDT	HDDV	MC	All Veh
GVWR			< 6000	> 6000	(All)						
VMT Distribution		0.3968	0.2459	0.0267		0.0922	0.0022	0.0014	0.2228	0.012	1
		(0.4005)	(0.2435)	(0.0272)		(0.0918)	(0.0023)	(0.0014)	(0.2216)	(0.0118)	(1)
Composite emission factors (g/mi)	Composite VOC	1.141	1.373	2.221	1.456	4.198	0.989	2.287	1.083	3.01	1.519
		(0.763)	(0.932)	(1.684)	(1.007)	(3.207)	(0.929)	(1.724)	(0.882)	(3)	(1.108)
	Composite CO	6.4	10.08	14.35	10.5	45.36	2.306	3.786	6.297	25.02	11.299
		(4.3)	(6.87)	(10.45)	(7.23)	(32.19)	(2.302)	(2.931)	(5.438)	(25.06)	(8.144)
	Composite NO$_x$	0.651	0.823	1.111	0.852	4.405	1.591	2.511	12.847	0.77	3.775
		(0.444)	(0.611)	(0.923)	(0.643)	(3.697)	(1.534)	(1.973)	(10.452)	(0.77)	(3.022)
Exhaust emissions (g/mi)	VOC start	0.145	0.208	0.354	0.222		0.179	0.685		0.389	
		(0.086)	(0.129)	(0.244)	(0.141)		(0.169)	(0.473)		(0.389)	
	VOC running	0.276	0.487	0.854	0.523		0.81	1.602		2.083	
		(0.135)	(0.254)	(0.524)	(0.281)		(0.76)	(1.251)		(2.087)	
	VOC total exhaust	0.421	0.695	1.208	0.745	2.52	0.989	2.287	1.083	2.47	0.879
		(0.221)	(0.384)	(0.768)	(0.422)	(1.715)	(0.929)	(1.724)	(0.882)	(2.48)	(0.59)
	CO start	1.83	3.19	4.51	3.32		0.499	1.286		2.898	
		(1.44)	(2.46)	(3.72)	(2.59)		(0.501)	(0.924)		(2.9)	
	CO running	4.58	6.89	9.84	7.18		1.807	2.5		22.12	
		(2.86)	(4.41)	(6.73)	(4.64)		(1.802)	(2.006)		(22.161)	
	CO total exhaust	6.4	10.08	14.35	10.5	45.36	2.306	3.786	6.297	25.02	11.299
		(4.3)	(6.87)	(10.45)	(7.23)	(32.19)	(2.302)	(2.931)	(5.438)	(25.06)	(8.144)
	NO$_x$ start	0.11	0.136	0.179	0.14		0.046	0.128		0.318	
		(0.065)	(0.089)	(0.128)	(0.093)		(0.045)	(0.097)		(0.318)	
	NO$_x$ running	0.541	0.688	0.932	0.712		1.546	2.383		0.453	
		(0.379)	(0.523)	(0.795)	(0.55)		(1.49)	(1.876)		(0.454)	
	NO$_x$ total exhaust	0.651	0.823	1.111	0.852	4.405	1.591	2.511	12.847	0.77	3.775
		(0.444)	(0.611)	(0.923)	(0.643)	(3.697)	(1.534)	(1.973)	(10.452)	(0.77)	(3.022)
Non-exhaust emissions (g/mi)	Hot soak loss	0.108	0.085	0.124	0.089	0.32	0	0	0	0.131	0.098
		(0.096)	(0.082)	(0.135)	(0.088)	(0.278)	(0)	(0)	(0)	(0.132)	(0.089)
	Diurnal loss	0.031	0.032	0.056	0.034	0.103	0	0	0	0.027	0.031
		(0.024)	(0.027)	(0.05)	(0.029)	(0.092)	(0)	(0)	(0)	(0.025)	(0.026)
	Resting loss	0.111	0.109	0.195	0.118	0.345	0	0	0	0.376	0.112
		(0.087)	(0.092)	(0.177)	(0.101)	(0.306)	(0)	(0)	(0)	(0.368)	(0.095)
	Running loss	0.446	0.405	0.564	0.42	0.729	0	0	0	0	0.359
		(0.318)	(0.31)	(0.49)	(0.328)	(0.645)	(0)	(0)	(0)	(0)	(0.275)
	Crankcase loss	0.006	0.009	0.011	0.009	0.012	0	0	0	0	0.006
		(0.005)	(0.009)	(0.01)	(0.009)	(0.011)	(0)	(0)	(0)	(0)	(0.005)
	Refueling loss	0.018	0.038	0.064	0.041	0.17	0	0	0	0	0.034
		(0.012)	(0.028)	(0.055)	(0.03)	(0.16)	(0)	(0)	(0)	(0)	(0.028)
	Total non-exhaust	0.719	0.678	1.013	0.766	1.679	0	0	0	0.534	0.64
		(0.542)	(0.548)	(0.916)	(0.645)	(1.492)	(0)	(0)	(0)	(0.525)	(0.518)

The default HC specification is VOC. However, the analyst can select the HC pollutant(s) for which emissions should be reported by including one of the five optional run-level commands in the "command" input file. The HC pollutants are total hydrocarbons (THC), nonmethane hydrocarbons (NMHC), volatile organic compounds (VOC), total organic gases (TOG), and nonmethane organic gases (NMOG).

(3) *Estimation of air quality impacts*

Table E10.1.3 presents the results of the analysis. Air quality impacts are experienced not only by users, but the society as a whole. Their impacts can be estimated as the difference between emissions with and without the improvement. The impact values in Figure E10.1 and Table E10.1.3 are for all vehicles, and computations are:

$$\text{Emission impact} = U_1(\text{VMT}_1) - U_2(\text{VMT}_2)$$

where U_1, U_2 are emission rates, and VMT_1, VMT_2 are vehicle-miles of travel without and with the improvement, respectively. For example,

$$\text{VOC impact} = (1.519 \times 250{,}000 - 1.108 \times 325{,}760)$$
$$= 18{,}800 \text{ g/day}$$

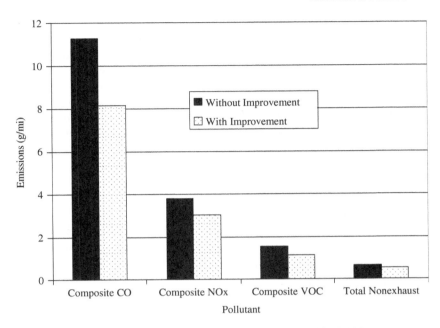

Figure E10.1   Estimates of Emissions from MOBILE6.

Table E10.1.3   Air Quality Impacts of the Improvement

Pollutants	Without Improvement (g/mile)	With Improvement (g/mile)	Impacts (g/day)
Composite VOC	1.519	1.108	18,800
Composite CO	11.299	8.144	171,760
Composite NO$_x$	3.775	3.022	40,700

10.3   ESTIMATING POLLUTANT CONCENTRATION

Steps 1.6(b) and 1.7(b) of the procedure for air quality impact assessment (see Section 10.2) involve an estimation of pollutant concentrations. Details of this task are presented in the present section.

Pollutants emitted from their sources disperse into the atmosphere, where they are transformed or diluted. The resulting amount (mass or volume) of a pollutant per unit volume of air is described as the *concentration* of the pollutant in the air. The atmospheric concentration of a pollutant is affected by the level

of emissions, topographical features, altitude, meteorological conditions, and physical mixing and chemical reactions in the atmosphere. The harmful effects of air pollutants are typically measured in terms of their concentrations.

The dispersion of transportation pollutant emissions in an area or space can be likened to a small hypothetical box into which a specific amount of gas is instantaneously emitted. In the real world, however, the situation is made more complex by the fact that (1) the emission occurs continuously; (2) dispersion of the pollutant occurs not only by diffusion but is aided (and thus rendered more complex from the analytical standpoint) by laminar or turbulent advection (movement) of wind, deposition, chemical reactions, confinement of air masses through the effects of topography, and/or the *inversion* phenomenon (trapping of polluted air due to differences in temperature of air masses); and (3) the pollutant emitted is really not confined to the box but is released from that enclosed space at a certain varying rate that depends on dynamic factors such as ambient temperature and wind speed.

10.3.1 Factors Affecting Pollutant Dispersion

(*a*) *Meteorological Factors*   The atmosphere is the typical medium for pollutant transfer from emission sources to receptors (humans, vegetation, etc.). Atmospheric conditions, which can be expressed in terms of temperature, atmospheric stability, precipitation, wind speed and direction, humidity, and intensity of solar radiation, govern the temporal (hourly, daily, and seasonal) and spatial variation of the transmission and therefore the concentration of air pollutants. Atmospheric stability is related to the change in temperature or wind speed or direction with height (also referred to as *temperature gradient* and *wind shear*, respectively). A stable atmosphere suppresses vertical motion within its domain and therefore generally leads to higher pollutant concentrations, while an unstable atmosphere enhances motion and ultimately lowers pollutant concentration. *Thermal inversion* is a phenomenon characterized by an increase in temperature with height (a reversal of the normal condition) leading to the entrapment of cold air layers by a higher layer of warm air. Such conditions lead to the accumulation of pollutants in the underlying layer of cold air. Wind speed is also a significant factor; the greater the wind speed, the higher the dispersion of air pollutants. Another meteorological factor is surface roughness; the movement of air near Earth's surface is resisted by frictional effects proportional to the surface roughness. *Ceiling height*, which is defined as the height above which relatively rigorous vertical mixing occurs, varies by day and by season. Ceiling heights may reach several thousand feet during summer daylight hours but only a few hundred feet on winter nights. As such, nighttime and winter conditions are associated with a relatively small volume of air available for dispersion and are therefore generally characterized by higher pollutant concentrations.

(b) Topography and Urban Spatial Form   Through the phenomena of air drainage and radiation, the topography of a region affects the wind speed and direction and the atmospheric temperature and subsequently affects the dispersion (and concentration) of pollutants. Air pollution problems are aggravated in metropolitan areas that experience the street "canyon" effect created by tall buildings. Assessing the causes and magnitude of air pollution in metropolitan areas can be a complex undertaking, due to the range and diversity of polluting sources, meteorological conditions, topographic features, and urban spatial forms.

10.3.2 Pollutant Dispersion Models

Pollutants emitted into the atmosphere are dispersed by molecular diffusion, eddy diffusion, and random shifts (Wayson, 2002). Dispersion factors include meteorological conditions such as the wind speed and temperature

gradient, the number of emission sources, and the emission rates of these sources. Atmospheric stability is the resistance to vertical motion of wind. High atmospheric stability as in flat terrain, retards dispersion, whereas low stability (high turbulence) facilitates dispersion. The three most common methods for assessing the impact of emissions on pollutant concentration are the box model, the Gaussian plume model, and the numerical model.

(a) Box Model   This model assumes uniform dispersion of pollutants to fill a single large boxlike space. Two key factors that control pollutant dispersion (and thus concentration) in the local environment are wind speed and mixing height, and the *ventilation factor* is the product of these two factors. Increasing either the mixing height or the wind speed increases the effective volume in which pollutants are allowed to mix. Consider a city with an area A $(a \times b)$ square miles, mixing height H miles, and an average wind speed of v mph (Figure 10.5).

For a pollutant particle emitted at one corner of the city:

1. The maximum distance for transport across the city (i.e., the distance necessary to reach the upwind edge of the box) is $\sqrt{a^2 + b^2 + H^2}$ miles.
2. The maximum time taken to be transported across the city to the upwind edge,

$$t_{\max} = \frac{\text{distance}}{\text{speed}} = \frac{\sqrt{(a^2 + b^2 + H^2)}}{v} \text{ hours}$$

3. For all particles emitted throughout the city, average time taken to be transported across the city to the

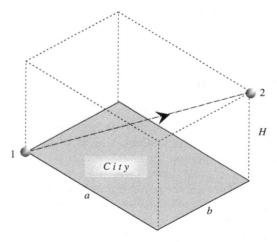

Figure 10.5   Box model for pollutant dispersion.

upwind edge,

$$t_{avg} = \frac{\sqrt{(a^2 + b^2 + H^2)}}{2v} \text{ hours}$$

Assuming that M grams of pollutant are released every t_{max} hour, the concentration of pollutant every t_{max} hours is given by

$$\frac{M}{abH} \text{ g/mi}^3 \qquad (10.5)$$

Example 10.2 The city of Santa Mateo is approximately rectangular in shape with dimensions of 3.5 miles by 2.1 miles. The topographical nature of the area is such that the effective mixing height is 1.2 miles. A particle of a certain pollutant is emitted at the southeastern corner of the city.

(a) Find the maximum distance taken by the particle to travel out of the box.
(b) If the wind speed is 3.5 mph in a SE–NW direction, find the (1) maximum time and (2) the average time taken by a particle of the pollutant emitted from any section of the city to clear the mixing box.
(c) If 1000 g of the pollutant is released in bursts every 2 hours, find the maximum concentration of the pollutant at any given time.

SOLUTION   A mixing box is defined with the following dimensions (in miles): $3.5 \times 2.1 \times 1.2$

(a) The maximum distance for transport across the city is

$$\sqrt{a^2 + b^2 + H^2} = (3.5^2 + 2.1^2 + 1.2^2)^{0.5}$$
$$= 4.25 \text{ mi}$$

(b) For all particles emitted throughout the city:

(1) Maximum time taken to be transported across the city and out of the box,

$$t_{max} = \frac{\text{distance}}{\text{speed}} = \frac{4.25}{\text{wind speed}}$$
$$= \frac{4.25}{3.5} = 1.21 \text{ h}$$

(2) Average time taken to be transported across the city and out of the box,

$$t_{avg} = \frac{1.21}{2} = 0.61 \text{ h}$$

(c) From (b) (1), all pollutant emissions would disperse out of the mixing box completely in 1.21 hours (the residual concentration after 1.21 hours is zero). Two hours after release, therefore, the residual concentration of the pollutant is zero. Therefore, if 1000 g of the pollutant are released in bursts every 2 hours, the maximum concentration will be

$$\frac{1000}{(3.5)(2.1)(1.2)} = 113.38 \text{ g/mi}^3$$

Clearly, the reliability of the results from the box model approach depends on a number of assumptions such as uniformity of dispersion. At any specific receptor site within the box, this assumption is typically violated, particularly when the averaging time is very small. The box model has also been applied to nonhighway modes. Cohn and McVoy (1982) cited an example of the FAA box model that can be used to assess CO emissions at airports. In the case of airports, the receptors are passenger loading areas (where emissions are from ground aircraft and service vehicles) and passenger pickup and drop-off areas (where emissions are from highway vehicles dropping or picking up passengers). Whereas the short-term maximum concentrations in such areas may be unbearable to persons (receptors) at such points, the overall average concentration throughout the entire airport box space may be too little to be of concern. The box model therefore may underestimate air pollution severity, particularly at localized but sensitive receptors.

(b) *Gaussian Plume Model*   This model is based on the random wafting of plumes side to side and up and down, resulting in the increased plume size with time. At any point in the plume, pollutant concentration can be described using a normal distribution, with the plume center having the highest concentration. As one moves away from the source, the maximum concentration level decreases while the concentration standard deviation increases (Figure 10.6).

The Gaussian plume model assumes that (1) there is continuous emission from the source and that diffusion in the direction of travel is negligible, (2) diffused material is a stable gas that remains suspended in the air for long periods and therefore no material is deposited from the plume as it moves downwind, (3) at any point in the plume (cross-sectional plane perpendicular to the direction of dispersion), the distribution of pollutant concentration (from the crosswind and vertical directions) is normal, and (4) the spread of the plume can be represented by the standard deviation of the pollutant concentration, which is consistent with the averaging time of the concentration estimate.

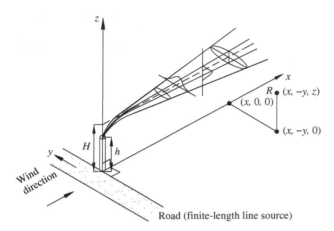

Figure 10.6 Gaussian model for plume formation.

The Gaussian equation is used by most dispersion models to estimate the dispersion of nonreactive pollutants released from an emitting source at a steady rate. The steady-state pollutant concentration, C ($\mu g/ft^3$), at a point specified by the x, y, and z coordinates in the vicinity of the transportation facility is given by

$$C = \frac{Q}{2\pi U \sigma_y \sigma_z} \exp\left(-\frac{y^2}{2\sigma_y^2}\right) \exp\left[-\frac{1}{2}\left(\frac{z+H}{\sigma_z}\right)^2\right]$$
$$\times \exp\left[-\frac{1}{2}\left(\frac{z-H}{\sigma_z}\right)^2\right] \tag{10.6}$$

where Q is the emission rate of the pollutant ($\mu g/s$), U the average wind speed at stack height (ft/s), σ_y, and σ_z the standard deviation of dispersion in the y and z directions, respectively; y the horizontal distance from the plume centerline, z the vertical distance from ground level, and H the effective stack height in ft (= physical stack height + vertical rise of plume).

A uniform average emission rate, Q, is defined for the finite-length line source (FLLS) in weight units of pollutant emissions per unit distance per unit time (e.g., $\mu g/ft\text{-}s$). The x-axis is parallel to the wind direction and the y-axis is parallel to the FLLS. In Figure 10.6, the road (FLLS) is perpendicular to the wind, but this is not always true. In configurations where the road is not perpendicular to the wind, an equivalent FLLS that is perpendicular to the wind can be established. Equation (10.6) is for a single point source. Where there are multiple sources, the concentration at a receptor due to emissions from each source can be calculated separately, and the total concentration is the sum of such concentrations from pollutants moving along the line, in the direction of the

y-axis. Cooper and Alley (2002) showed that the sum of concentrations experienced at the receptor due to an emission source moving between limits y_1 and y_2 along the finite-length line source is given by

$$C = \frac{K}{\sqrt{2\pi}}(G_U - G_L) \tag{10.7}$$

where

$$K = \frac{Q}{U \sigma_z}\left[\exp\left(\frac{-(z-H)^2}{2\sigma_z^2}\right) + \exp\left(\frac{-(z+H)^2}{2\sigma_z^2}\right)\right]$$

and G_U and G_L are Gaussian distribution functions (see Appendix A10.2) corresponding to the upper and lower values of y_1/σ_{y1} and y_2/σ_{y2}, respectively, where σ_{y1} and σ_{y2} are the variances of pollutant concentration at endpoints 1 and 2 of each FLLS.

The Gaussian plume model is widely used to assess pollutant dispersion and concentration, but its assumptions may not always hold, particularly in cases of fluctuating wind directions. Also, the assumption of stable gases may not always be appropriate where the pollutants themselves undergo chemical reactions as they are being dispersed. Furthermore, deposition can and does occur in the case of certain pollutants, such as lead particles and hydrocarbon droplets. Also, the model can lead to misleading results in nonhomogeneous terrain. There are other point-specific models that can overcome some of these limitations (Kretzschmar et al., 1994).

Example 10.3 A busy highway passes near a nursing home for elderly persons. A plan view of the road at that location is shown in Figure E10.3.1. Determine the expected CO concentration at ground level at the nursing home. The CO emission factor is 20 g/mi per vehicle. Wind speed is 2 ft/s, $H = 0$ ft, and traffic volume is 15,000 veh/h. Assume that when $x = 50$ ft, σ_y and σ_z are 20 and 12 ft, respectively; and when $x = 67.5$ ft, σ_y and σ_z are 22 and 14 ft, respectively. Assuming that the road is the sole CO source, determine whether the concentration at the nursing home violates the standard of 35 ppm.

SOLUTION   The emission rate of 20 g/mile per vehicle is expressed in temporal terms as follows:

$$Q = \left(\frac{20 \text{ g}}{\text{mile-veh}}\right)\left(\frac{15,000 \text{ veh}}{1 \text{ h}}\right)\left(\frac{1 \text{ h}}{3600 \text{ s}}\right)$$
$$= 83 \text{ g/mile-s}$$

This means that for each mile of the study segment, 83 g of CO is emitted every second. For consistency with the

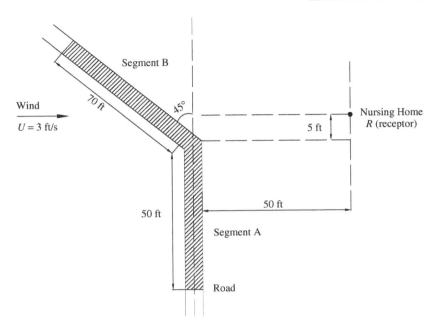

Figure E10.3.1   Road layout near nursing home.

dispersion equation, this can be expressed in micrograms and feet:

$$Q = (83)\left(\frac{1 \text{ mile}}{5280 \text{ ft}}\right)\left(\frac{10^6 \text{ }\mu\text{g}}{1 \text{ g}}\right) = 15,720 \text{ }\mu\text{g/ft-s}$$

Then, the emission rates are adjusted to account for the relative direction between the wind and road traffic. To do this, finite-length line source lines are established perpendicular to the wind direction and passing through the midpoints of the two segments (Figure E10.3.2).

For road segment A: The centerline is perpendicular to the wind source, so there is no need for any adjustment. The emission rate on FLLS A (the finite-length line source) due to traffic on segment A is simply equal to 15,720 μg/ft-s. The length of FLLS-A is 50 ft.

For road segment B: The distance P shown in Figure E10.3.2 is $(70 \sin 45°)/2 = 24.75$ ft.

Length of FLLS-B = $(70 \text{ ft})(\cos 45°) = 49.45$ ft

Equivalent emission rate at FLLS-B

= $(15,720)(70/49.5) = 22,230$ μg/ft-s

The x, y, and z coordinates of each endpoint of the FLLS lines are determined as follows:

FLLS-A: start point $x = 50$ ft; $y = -55.0$ ft, $z = 0$ ft; endpoint $x = 50$ ft, $y = -5.0$ ft, $z = 0$

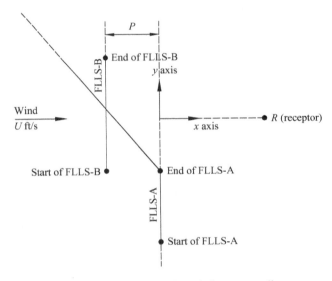

Figure E10.3.2   Finite length line source line.

FLLS-B: start point $x = 74.7$ ft; $y = -5$ ft, $z = 0$ ft; endpoint $x = 74.7$ ft; $y = 44.5$ ft; $z = 0$

Consider FLLS-A: $y_1/\sigma_{y1} = -55/20 = -2.75$ and $y_2/\sigma_{y2} = -5.0/20 = -0.25$. From Appendix A10.2, $G_1 = G(-0.25) = 0.4013$ and $G_2 = G(-2.75) = 0.0030$. Thus, $G_1 - G_2 = 0.3983$.

$$K = \frac{15,720}{(2)(12)} \left[\exp\left(\frac{-(0-0)^2}{(2)(12^2)} \right) \right.$$

$$\left. + \exp\left(\frac{-(0+0)^2}{(2)(12^2)} \right) \right] = 1310.00$$

$$C_{FLLS-A} = \left(\frac{1310}{\sqrt{2\pi}} \right) (0.3983) = 208.16 \ \mu g/ft^3$$

Consider FLLS-B: $y_1/\sigma_{y1} = -5/22 = -0.227$, and $y_2/\sigma_{y2} = +44.5/22 = 2.023$. From Appendix A10.4, $G_1 = G(2.023) = 0.9785$ and $G_2 = G(-0.227) = 0.4102$. Thus, $G_1 - G_2 = 0.5683$.

$$K = \frac{22,230}{(2)(14)} \left[\exp\left(\frac{-(0-0)^2}{(2)(12^2)} \right) \right.$$

$$\left. + \exp\left(\frac{-(0+0)^2}{(2)(12^2)} \right) \right] = 1587.86$$

$$C_{FLLS-B} = \left(\frac{1587.86}{\sqrt{2\pi}} \right) (0.5683) = 359.98 \ \mu g/ft^3$$

Therefore, the total concentration at the receptor is $208.16 + 359.98 = 568.14 \mu g/ft^3$ or 17.54 ppm. This does not exceed the threshold concentration of 35 ppm, so the estimated air quality level at the nursing home does not violate established standards.

(c) *Numerical Models* A numerical air quality model involves a three-dimensional grid of conceptual boxes that occupy the space above a transportation corridor. Emissions from the highway vehicles are considered as a pollutant source "feeding" the series of boxes immediately overlying the highway. Within each box, the pollutant particles diffuse to fill the box at some given rate. Then the pollutant diffuses into the immediately outlying boxes at some given rate. The movement of pollutant from box to box is aided further by local wind effects. Assuming that the local wind effects, diffusion, and emissions are reasonably represented with well-behaved functions of time, the movement of pollutant particles across the boxes can be predicted with successive time increments. The smaller the boxes, the more valid is the assumption that there is uniform concentration within each box. As such, estimates of pollutant concentration can be made at any spatial point within the region, represented by the three-dimensional box grid. When the numerical model is used, restrictive assumptions in the case of the Gaussian plume or box models regarding nondeposition, nonreactions, and so on, are overcome: It is possible to simulate the deposition of pollutants or chemical reactions involving pollutants in each box, as had been done successfully for photochemical oxidant models for the city of Los Angeles

(Cohn and McVoy, 1982). The computational effort and data collection associated with the numerical approach can be very challenging, but the advent of faster computers has helped make this approach very attractive for use.

10.3.3 Software for Estimating Pollutant Dispersion and Concentrations

A number of air dispersion models have been developed for highway and transportation projects. These include the HYROAD, ADMS, California Line Source (CALINE 4), HIWAY, PAL, TEXIN 2, and CAL3QHC models. The HIWAY and PAL models can only be used for free-flow conditions (Wayson, 2002). Models recommended by the EPA, such as TEXIN 2, CALINE 4, and CAL3QHC, account for queueing delays and excess emissions due to variations in engine modes and cruise.

(a) *HYROAD (HYbrid ROADway Model)* HYROAD analyzes intersections and predicts their ambient carbon monoxide concentrations. The model, which is equipped with a graphical user interface, comprises three modules: traffic, emissions, and dispersion. First, the traffic module microscopically simulates the traffic flow by modeling the movement of each vehicle at the intersection. This module yields speed distribution information that is used in the emission module to establish composite emission factors and spatial and temporal distribution of emissions. For each 10-m roadway segment and for each signal phase, vehicle speed and acceleration distributions are observed, and flow and turbulence are analyzed. The last module establishes pollutant dispersion characteristics near the intersection. The model gives hourly concentration of pollutants, including carbon monoxide and other gas-phase pollutants, particulate matter, and air toxins, at specific distances from the intersection (System Application International, 2002).

(b) *ADMS-3 (Atmospheric Dispersion Modeling System)* Developed by Cambridge Environmental Research Consultants of UK, ADMS-3 is an advanced model for calculating the concentrations of pollutants that are emitted continuously from point, line, volume, and area sources, or discretely from point sources. The model includes algorithms which take into account the terrain, wet deposition, gravitational settling, dry deposition, chemical reactions, plume rise as a function of distance, and meteorological conditions, among others (Carruthers et al., 1994).

(c) *CALINE Version 4* The California Line Source Dispersion Model version 4 (CALINE4), predicts air pollution concentrations near lineal transportation facilities.

Developed by the California Department of Transportation, this model is based on the Gaussian diffusion equation and employs a mixing zone concept to characterize pollutant dispersion from the roadway. Given the source strength (emissions), meteorology, and site geometry, CALINE4 can predict pollutant concentrations at receptors located within 500 m of the facility. It also has special options for modeling air quality near highway intersections, street canyons, and parking facilities.

Example 10.4 A certain interstate highway section in the U.S. Midwest consists of three links; A, B, and C (Figure E10.4.1). The highway section passes through a suburban area, and the mean elevation is sea level. The coordinates of each link are as follows: A start (4000, 4000); A end (4200, 4000); B start (4200, 4000); B end (4500, 3500); C start (4500, 3500), C end (5000, 3500). Assume a background CO concentration of 0 ppm, a wind direction standard deviation of 5, and a width of the pollutant mixing zone of 20 m. The link activity and running conditions are provided in Table E10.4.1. Determine the mean concentration of CO at the following receptor sites: site 1 (4100, 3950, 1.8); site 2 (4300, 3700, 1.8), and site 3 (4750, 3550, 1.8).

SOLUTION A sample of the CALINE4 output is provided in Figure E10.4.2. Multiple runs of the model for the various time periods yield the results shown in Table E10.4.2. CO concentrations are in ppm.

10.4 AIR POLLUTION FROM OTHER MODES

Figure 10.7 shows comparative pollutant emission rates from various transportation modes (Holmen and Niemeier,

2003). Compared to other modes, diesel trains and trucks emit relatively low pollutants per passenger-mile or per ton-mile. Electric trains do not cause local pollution except when their power sources are fossil-burning electricity plants that lack pollution controls. With regard to NO_x, the greatest polluters are automobiles and trucks. Diesel trains and buses also emit some NO_x, and the least-emitting sources are electric trains (at the points of power generation). For SO_x, the most significant source is electric trains (at the points of power generation). CO_2 emission is largely due to the use of fossil fuels. Automobiles and trucks are the most significant sources of CO_2; electric trains are the least.

10.4.1 Air Transportation

Air transportation pollution comes from two sources: airport activities and aircraft emissions. With regard to air pollutant emissions due to airport activities it is estimated that aircraft engines contribute approximately 45%; ground access vehicle operations, including passenger drop-offs and pickups, contribute 45%; and ground support equipment contributes 10% (Holmen and Niemeier, 2003). Future aviation trends seem to involve high-flying subsonic and supersonic aircraft, and such travel is expected to cause further depletion of ozone in the stratosphere.

Air Quality Impact Analysis for Air Transportation: The Emissions and Dispersion Modeling System (EDMS) is an FAA-approved model specifically developed for the aviation community to assess the air quality impacts of proposed airport development projects. EDMS is designed to assess the air quality impacts of airport emission

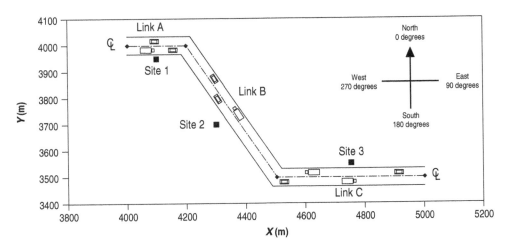

Figure E10.4.1 Site layout for CALINE4 run.

Table E10.4.1   Link Activity and Running Conditions for CALINE4 Run

Time (h)	No. of Cars/Hour, Four Lanes	CO Emission Factor (g/mi)	Wind Speed (m/s)	Stability	Mixing Height (m)	WDIR (deg)	Temperature (°C)
1	2263	12.06	1.94	5	520	10	19.4
2	1670	12.06	1.87	5	526	12	18.9
3	1711	12.06	2.05	5	527	15	18.6
4	1962	12.06	1.44	5	527	10	18.3
5	3173	9.06	2.04	5	561	15	18.0
6	4816	9.06	2.66	4	642	40	17.8
7	5579	9.06	1.41	4	738	70	19.0
8	5938	6.19	1.71	3	834	50	21.0
9	6160	6.19	2.56	3	930	50	22.9
10	6305	6.19	2.96	3	1026	20	24.3
11	6400	9.57	2.29	3	1122	0	25.3
12	6550	14.66	2.78	3	1218	355	26.1
13	6700	18.03	3.01	3	1314	350	26.3
14	6550	14.66	3.02	3	1410	0	26.5
15	6400	9.57	2.30	3	1410	20	26.9
16	6350	7.14	2.90	4	1410	10	26.8
17	6320	7.14	1.96	4	1410	25	26.5
18	5774	6.19	2.06	4	1407	40	26.0
19	5399	6.19	1.73	4	1375	60	25.1
20	5325	6.19	1.60	5	1243	110	23.7
21	4838	6.19	1.65	5	1059	150	22.0
22	4253	6.19	1.72	5	882	70	21.0
23	3785	6.19	1.25	5	689	55	20.3
24	3160	9.06	1.95	5	527	40	19.9

```
                  CALINE4: CALIFORNIA LINE SOURCE DISPERSION MODEL
                                        PAGE  1
                                   JOB: CL4 Example
                                   RUN: Hour 1
                             POLLUTANT: Carbon Monoxide
        I. SITE VARIABLES
        U      = 2.1 M/S              Z0      = 50. CM          ALT=  0. (M)
        BRG    = 15.0 DEGREES         VD      = .0 CM/S
        CLAS   =  5 (E)               VS      = .0 CM/S
        MIXH   = 527. M               AMB     = .0 PPM
        SIGTH  =  5. DEGREES   TEMP   = 18.6 DEGREE (C)
        II. LINK VARIABLES
        LINK                 * LINK COORDINATES (M)  *    EF  H    W
        DESCRIPTION          * X1   Y1   X2   Y2     * TYPE VPH (G/MI) (M) (M)
        -----------------    *----------------------     *-------------------------------
        A.   Link A          * 4000 4000 4200 4000       * AG  1711 12.1   .0 20.0
        B.   Link B          * 4200 4000 4500 3500       * AG  1711 12.1   .0 20.0
        C.   Link C          * 4500 3500 5000 3500       * AG  1711 12.1   .0 20.0
        III. RECEPTOR LOCATIONS
        *  COORDINATES (M)
        RECEPTOR       *  X   Y   Z
        -----------  *---------------------
        1. Site 1  *  4100 3950 1.8
        2. Site 2  *  4300 3700 1.8
        3. Site 3  *  4750 3550 1.8
        IV. MODEL RESULTS (PRED. CONC. INCLUDES AMB.)
        * PRED          *  CONC/LINK
        * CONC *   (PPM)

        RECEPTOR       * (PPM)           *  A  B  C
        ------------ *---------------    *---------------
        1. Site 1  *    .2             *  .2  .0  .0
        2. Site 2  *    .2             *  .0  .2  .0
        3. Site 3  *    .0             *  .0  .0  .0
```

Figure E10.4.2   Sample output of a standard CALINE4 run.

Table E10.4.2 Estimated CO Concentrations (ppm)

Time (h)	Receptor Site 1	Receptor Site 2	Receptor Site 3
1	0.20	0.20	0.00
2	0.20	0.20	0.00
3	0.20	0.20	0.00
4	0.30	0.30	0.00
5	0.20	0.20	0.00
6	0.30	0.20	0.00
7	0.30	0.40	0.00
8	0.40	0.20	0.00
9	0.30	0.20	0.00
10	0.20	0.20	0.00
11	0.40	0.40	0.00
12	0.50	0.50	0.00
13	0.60	0.60	0.00
14	0.50	0.50	0.00
15	0.40	0.30	0.00
16	0.20	0.20	0.00
17	0.30	0.30	0.00
18	0.30	0.20	0.00
19	0.30	0.20	0.00
20	0.20	0.40	0.40
21	0.00	0.00	0.30
22	0.20	0.20	0.00
23	0.30	0.20	0.00
24	0.30	0.20	0.00

sources, particularly aviation sources, which consist of aircraft, auxiliary power units, and ground support equipment. EDMS offers a limited capability to model other airport emission sources that are not aviation-specific, such as ground access vehicles and stationary sources. EDMS performs emission and dispersion calculations and uses updated aircraft engine emission factors from the International Civil Aviation Organization's engine exhaust emissions data bank and vehicle emission factors from the EPA's MOBILE6 model.

10.4.2 Rail Transportation

Rail pollution depends on the power source, which includes coal and steam, diesel, and electricity. In the United States and Western Europe, steam traction has been phased out almost entirely. In other parts of the world, steam is still one source of rail power. Coal-powered steam locomotives consume coal to build up steam that is used to power the vehicles. In doing so, they emit heavy spurts of smoke containing CO_2, SO_x, and NO_x into the atmosphere and pollute the areas near rail lines with smoke particulates. Because steam engines are far less thermally efficient than gasoline, diesel, or electric vehicles, they emit higher amounts of pollutants per energy produced than the other power types. Diesel-powered locomotives and highway trucks produce similar pollutants: CO, NO_x, HC, and carbon-based particulates. In terms of emission per ton-mile, however, diesel rail locomotives are approximately three times cleaner than trucks (Holmen and Niemeier, 2003). For electric-powered rail, the only contribution to air pollution may come from the power sources that generate the electricity used to power such vehicles, particularly where the fuel used is coal or other fossil fuels. Other atmospheric effects of electric railways are emissions resulting from high-speed contact of pantographs on wires, but these are considered negligible (Carpenter, 1994).

10.4.3 Marine Transportation

Commercial marine vessels are responsible for only 2% of the global fossil fuel consumption, but constitute a significant source of ocean air pollution. In terms of emissions per ton of fuel consumed, vessel engines are the least clean combustion sources. These engines produce 14% of the global nitrogen emissions from fossil fuels and 16% of all sulfur emissions from petroleum (Talley, 2003). Marine transportation causes emission of reactive organic gases (ROGs), CO, and NO_x but there have been relatively few studies to quantify the levels of such emissions. Compared to highway sources, waterborne vessels emit relatively small amounts of HC and CO, but their relative contribution to overall pollution is expected to increase with increasing enforcement of pollution standards of other modes (Holmen and Niemeier, 2003).

10.4.4 Transit (Various Modes)

Potter (2003) presented information regarding typical emissions from various transit types in Germany and the United Kingdom (Table 10.3) and established that urban public transit is significantly cleaner than automobiles in terms of NO_x and CO emissions per passenger-distance. For electric rail, indirect SO_2 emissions (i.e., from power-generating plants that produce such electricity) are high but emissions of other pollutants are low, relative to other transit types.

Table 10.5 provides emission rates for both newly manufactured and remanufactured locomotives built originally after 1972. These values are expressed in grams per brake horsepower-hour (g/bhp-hr) and grams of pollutant emitted per gallon of fuel consumed (g/gal). The latter emission rates are obtained by multiplying the emission rates

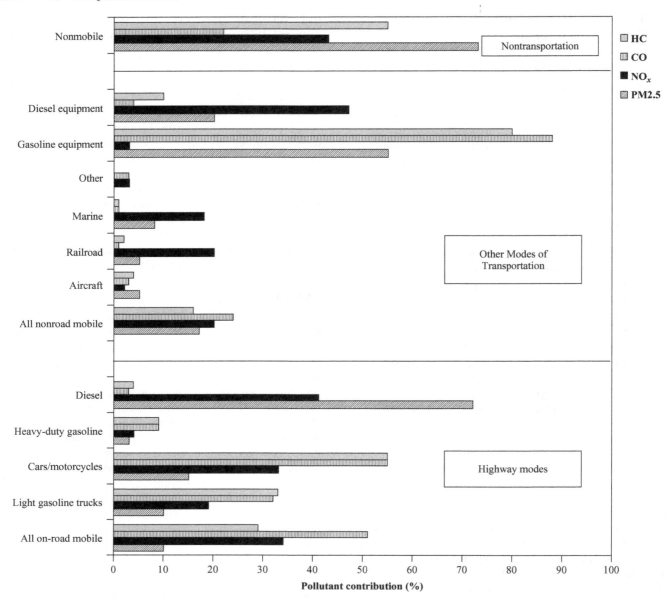

Figure 10.7 Pollutant contributions by mode. (From Holmen and Niemeier, 2003.).

in g/bhp-h with an appropriate conversion factor of 20.8 bhp-h/gal set by the EPA.

10.5 MONETARY COSTS OF AIR POLLUTION

The cost of environmental or resource degradation can be measured in one of three ways: (1) as the cost of cleaning up the air near the source of degradation, (2) as the cost associated with addressing the effects of degradation, and (3) as the willingness of persons to pay to avoid the degradation of their residences or businesses.

10.5.1 Methods of Air Pollution Cost Estimation

(*a*) *Cost Based on Cleaning up the Air at or near the Polluting Source* The costs of cleaning up the air before or after its dispersion involves the installation of air scrubbers at intervals along the polluting line source to clean the air before or as it disperses to adjoining populated areas, a measure which may be less feasible in rural areas than in urban areas. The installation intervals would depend on the characteristics of the traffic (volume, % trucks, speed, etc.), the environment (temperature,

Table 10.5   Emission Rates for Tier 0, Tier 1, and Tier 2 Locomotives[a]

Tier	Type of Haul	HC g/bhp-h	HC g/gal	CO g/bhp-h	CO g/gal	NO_x g/bhp-h	NO_x g/gal	PM g/bhp-h	PM g/gal
Tier 0 (locomotives manufactured 1973–2001)	Line-haul	0.48	10	1.28	26.6	8.6	178	0.32	6.7
	Switch	1.01	21	1.83	38.1	12.6	262	0.44	9.2
Tier 1 (locomotives manufactured 2002–2004)	Line-haul	0.47	9.8	1.28	26.6	6.7	139	0.32	6.7
	Switch	1.01	21	1.83	38.1	9.9	202	0.44	9.2
Tier 2 (locomotives manufactured after 2004)	Line-haul	0.26	5.4	1.28	26.6	5.0	103	0.17	3.6
	Switch	0.52	11	1.83	38.1	7.3	152	0.21	4.3

Source: USEPA (1997).
[a] Estimated controlled values.

wind speed, direction, etc.), and the scrubber capacities. Air pollution costs, if quantified in this manner, can be rather excessive, as the costs of purchasing, operating, and maintaining scrubbers are very high.

(b) Cost Based on Addressing the Effects of Pollution This cost could be described as the *social damage* effect of air pollution. It includes the health care expenses involved with treating respiratory illnesses engendered or exacerbated by an air pollution problem and the cost to repair physical infrastructure and compensation for or remediation of destroyed or degraded crops, forests, and groundwater by acidic depositions formed by chemical reactions between pollutants and atmospheric gases.

(c) Costs Based on the Willingness-to-Pay Approach The costs of air pollution can be estimated by assessing the extent to which affected persons and businesses are willing to pay to avoid an air pollution problem. The assumption is that people are perfectly aware of the

adverse impacts of air pollution on their health and property, and that their stated preferences closely reflect their actual or revealed preference.

10.5.2 Air Pollution Cost Values

The European Economic Commission has supported a great amount of research aimed at valuing the pollution costs of transportation; and air pollution cost estimates have been developed for various pollutant types, transportation modes, and operating speeds. For example, it is estimated that at 1999 conditions, the cost of CO_2 emissions was $26/ton, a value considered consistent with other estimates of global abatement costs for meeting the Kyoto Protocol (Friedrich and Bickel, 2001).

Delucchi (2003) provided external cost estimates of direct motor vehicle use in urban areas of the United States in 1990 (Table 10.6). The marginal costs for health, visibility, and crops were estimated for each kilogram of pollutants, emitted as shown in the table, and are for each 10% change in motor vehicle use.

Table 10.6   Incremental External Costs of Direct Auto Use in Urban Areas[a]

	PM_{10}	NO_x	SO_x	CO	VOCs
Health	13.7–187	1.6–23.3	9.6–90.9	0.0–0.1	0.1–1.5
Visibility	0.4–3.9	0.2–1.1	0.9–4.0	0.0	0.0
Crops	NE[b]	NE	NE	0.0	0.0
Total	14.1–191	1.8–24.5	10.5–94.9	0.0–0.1	0.1–1.5

Source: Delucchi (2003).
[a] Dollars/kilogram for a 10% change in auto use.
[b] NE means not established.

Table 10.7   U.S. National Ambient Air Quality Standards

Pollutant	Measure	Standard Value				Standard Type
Carbon monoxide (CO)	8-h average	9	ppm	(10	mg/m^3)	Primary
	1-h average	35	ppm	(40	mg/m^3)	Primary
Nitrogen dioxide (NO$_2$)	Annual arithmetic mean	0.053	ppm	(100	μg/m^3)	Primary and secondary
Ozone (O$_3$)	1-h average	0.12	ppm	(235	μg/m^3)	Primary and secondary
	8-h average	0.08	ppm	(157	μg/m^3)	Primary and secondary
Lead (Pb)	Quarterly average			1.5	μg/m^3	Primary and secondary
Particulate PM 10 (particles with diameters of 10 μm or less)	Annual arithmetic mean			50	μg/m^3	Primary and secondary
	24-h average			150	μg/m^3	Primary and secondary
PM 2.5 (particles with diameters of 2.5 μm or less)	Annual arithmetic mean			15	μg/m^3	Primary and secondary
	24-h average			65	μg/m^3	Primary and secondary
Sulfur dioxide (SO$_2$)	Annual arithmetic mean	0.03	ppm	(80	μg/m^3)	Primary
	24-h average	0.14	ppm	(365	μg/m^3)	Primary
	3-h average	0.50	ppm	(1300	μg/m^3)	Secondary

Source: USEPA (2002).

10.6   AIR QUALITY STANDARDS

Environmental agencies in most countries have established air quality standards. The U.S. ambient air quality standards are shown in Table 10.7. *Primary standards* represent the minimum requirements to maintain public health. *Secondary standards* are set to protect public welfare, which includes the prevention of soiling of buildings and other public infrastructure, restriction of visibility, and degradation of materials. Other definitions are as follows:

- *Specified concentration level:* the maximum concentration of air pollutant specified.
- *Averaging time:* the time duration that an area is subjected to an air pollutant.
- *Return period:* the maximum frequency or minimum interval with which the maximum concentration specified can be exceeded.

An example of an air quality standard is as follows: *The eight-hour average ambient CO standard is nine ppm (ten mg/m^3) not to be exceeded more than once in a year.* Many urban areas experience occasional violations of the 8-hour standard. On the other hand, violations of the 1-hour standard are rare and occur when there is unusually heavy traffic lasting for only a few hours of the day due to, for example, peak-hour travel or freeway incidents.

Emission standards can also be expressed as the weight of pollutants emitted per unit of power generated. Table 10.8, for example, shows the emission standards in

Table 10.8   Emission Standards for Heavy-Duty Diesel Vehicles (g/kWh)

	Europe (2005)	Japan (2004)	United States (1998)
CO	1.5	2.22	15.5
HC	0.46	0.87	1.3
NO$_x$	3.5	3.38	4.0
PM$_{10}$	0.02	0.18	0.1

Source: Stanley and Watkiss (2003).

g/kWh for heavy-duty diesel vehicles in Europe, Japan, and the United States.

Under international agreements, aircraft emission standards are set through the United Nations' International Civil Aviation Organization (ICAO). In the United States, the EPA establishes emission standards for aircraft engines and the FAA enforces these standards. The EPA regulates NO$_x$, hydrocarbon (HC), carbon monoxide (CO), and smoke emissions from aircraft.

10.7   MITIGATING AIR POLLUTION FROM TRANSPORTATION SOURCES

The reduction of automotive air pollution can be achieved through a variety of measures, including legislation and enforcement, vehicle engine standards, promotion of less polluting modes of transportation, improved fuel quality, alternative fuels, transportation planning and traffic management, and economic instruments (Faiz et al. 1996). In the United States the Congestion Mitigation and Air Quality (CMAQ) program funds projects designed to help metropolitan areas with poor air quality to reach the national air quality standards. Eligible projects are listed below:

- *Traffic flow improvements:* signal modernization and traffic management/control such as incident management and ramp metering and intersection improvements.
- *Transit improvements:* system or service expansion, replacement of buses with cleaner vehicles, and marketing strategies such as shared ride services: park-and-ride facilities, establishment of vanpool or carpool programs, and programs to match drivers and riders.
- *Demand management strategies:* promotion of employee trip reduction programs and development of transport management plans, including improved commercial vehicle operations in urban areas.
- *Nonmotorized transportation:* development of bicycle trails, storage facilities, and pedestrian walkways, as well as promotional activities.
- *Inspection and maintenance:* updating vehicle inspection and maintenance quality assurance programs, construction of advanced diagnostic facilities or equipment purchases, conversion of a public fleet to alternative fuel vehicles, and other projects.
- *Other activities:* outreach activities, experimental pilot projects and innovative financing and fare and fee subsidy programs.

Other Modes: Airlines are investing significant amounts of resources and taking steps aimed at ensuring improved levels of environmental performance (Somerville, 2003). These include development of performance indicators, open reporting of environmental performance, participation in ICAO initiatives, and sponsoring research projects. With regard to marine air pollution, it has been proposed that to reduce the polluting effects at ports, transiting vessels should be required to stop their engines and receive power from shore-side sources of electricity (Talley, 2003).

10.8 AIR QUALITY LEGISLATION AND REGULATIONS

10.8.1 National Legislation

The Air Pollution Control Act of 1955 was the first in a long chain of federal legislation related to the air quality impacts of transportation. In 1963, the Clean Air Act (CAA) was passed (subsequently amended in 1965 and several times later) to enforce emission standards for new vehicles. The Air Quality Control Act of 1967 led to the establishment of air quality criteria. The CAA amendments of 1970 provided federal controls in individual states for regulating and reducing motor vehicle and aircraft emissions. To achieve this goal, the CAA established the National Ambient Air Quality Standards (NAAQS) for pollutants considered harmful to public health and the environment, whereby states were required to prepare state implementation plans (SIPs), a document that outlines how a state intends to deal with air pollution problems. Also, the NAAQS were established for six principal pollutants, called *criteria pollutants* (Section 10.1.2). Regions that do not meet these standards are classified as *nonattainment areas*. Depending on the severity of the air quality problem, nonattainment areas are classified as marginal, moderate, serious, and severe and/or extreme. Also, passed in 1970, the Federal Aid Highway Act required the U.S. Department of Transportation and the EPA to develop and issue guidelines governing the air quality impacts of highways and required the development of transportation control plans and measures for air quality improvement. In a 1977 amendment to the CAA, penalties were established for areas that failed to carry out good faith efforts to meet air quality standards. The 1990 CAA strengthened conformity requirements that require metropolitan planning organizations in nonattainment and maintenance areas to use the most recent mobile source emission estimate models to show that (a) all federally funded and "regionally significant projects," including nonfederal projects in regional transportation improvement programs (TIPs) and plans will not lead to emissions higher than those in the 1990 baseline year, and (2) by embarking on these projects, emissions will be lower than in the no-build scenario. If a transportation plan, program, or project does not meet conformity requirements, it must be modified to offset the negative emission impacts or the EPA will need to work with the appropriate state agency to modify the SIP. If any of the foregoing actions is not accomplished, the transportation plan, program, or project cannot be implemented. The Intermodal Surface Transportation Efficiency Act of 1991 (ISTEA) reinforced the CAA90 requirement that transportation plans conform to air quality enhancement initiatives and provided state and local governments with the funding and flexibility to improve air quality through development of a balanced, environmentally sound intermodal transportation program. In the SAFETEA-LU act of 2005, the air quality conformity process was improved with changes in the frequency of conformity determinations and conformity horizons.

10.8.2 Global Agreements

On the global level, there have been efforts to regulate the extent of the global warming phenomenon (of which transportation sources are a major contributor). The Kyoto Protocol is an agreement negotiated in 1997 in Kyoto, Japan as an amendment to the United Nations Framework Convention on Climate Change (an international treaty on global warming that was adopted at the Earth Summit in Rio de Janeiro in 1992). By ratifying this protocol, countries committed to a reduction in their emissions of carbon dioxide, methane, nitrous oxide, sulfur hexafluoride, HFCs, and PFCs, or to engage in emissions trading if they maintain or increase emissions of these pollutants. In the agreement, industrialized countries are expected to reduce their collective emissions of greenhouse gases by approximately 5% (over 1990 levels). At the treaty's implementation in February 2005, the agreement was ratified by 141 countries whose collective emissions represent over 60% of the total global levels. Several countries including the U.S. have not ratified the Kyoto Protocol, citing economic reasons. However, the evidence on the possible cataclysmic effect of global warming is mounting (Gore, 2006). A recent study commissioned by the British government indicated that the costs related to climate change due to carbon emissions could seriously affect the world's economy, reducing as much as 20% of the total gross domestic product (Timmons, 2006).

SUMMARY

Transportation, particularly the highway mode, continues to be a major contributor to air pollution. It has adverse effects not only on a local and regional scale but also on a global scale by contributing to global warming. The major factors affecting pollutant emissions are travel related, and the EPA-sponsored software MOBILE6 is the common emission estimation tool. Factors that affect dispersion of air pollutants include meteorological conditions, topographical features, and the number and rate of emission sources. Methods for pollutant concentration estimation include the Gaussian plume, numerical, and box models. CALINE 4 is the most commonly used software package for estimating the concentration of pollutants.

The cost of air pollution can be measured by assessing the cost of cleaning the air near the pollution source, the cost of restoring the health and condition of affected persons and property, and the willingness of persons to pay to avoid degradation of air quality at their residences or businesses. Air quality standards, established to preserve public health and welfare from air pollution damage, involve specified concentration levels, averaging times, and return periods.

Efforts to reduce automotive air pollution has been spearheaded by industrialized countries through a variety of measures, including legislation and enforcement, vehicle engine standards, promotion of less polluting modes of transportation, improved fuel quality, use of alternative fuels, and transportation planning and traffic management. The Congestion Mitigation and Air Quality Improvement program provides funds to states for projects designed to help metropolitan areas to attain and maintain the national ambient air quality standards. The Clean Air Act provided strong governmental control in regulating and reducing motor vehicle and aircraft emissions. At the global level, the Kyoto Protocol ratified in 2005 signifies a genuine effort to regulate the anthropogenic causes of the global warming phenomenon.

EXERCISES

10.1. An increase in gasoline prices led to the following changes in VMT on the local street network of Cityville: light-duty vehicles, 5% reduction; motorcycles, 10% increase; heavy-duty vehicles, 8% reduction. If the average speed is expected to increase from 20 mph to 22 mph and all other default data in MOBILE6 remain the same, estimate the impact of the change in gas price on the emissions of CO, HC, and NO_x.

10.2. A series of CMAQ programs in a certain metropolitan area led to a 7% reduction in VMT for all vehicle classes and an increase in average speed from 16 mph to 25 mph. Using MOBILE6, assess the impact of the CMAQ programs on emissions of key air pollutants. Assume that all other data are the same as the data used in Example 10.1.

10.3. A state increased its rural interstate speed limit from 65 mph to 70 mph. Assuming that all other factors are the same, what will be the impact on air pollution emissions? Use MOBILE6. Assume that all other data are the same as the data used in Example 10.1.

10.4. A number of road-widening, intersection improvement, and curve-straightening projects on Interstate 778 led to an increased average speed from 45 mph to 60 mph. What was the net impact of the improvements on emissions? Assume that all other data are the same as the data used in Example 10.1. Use MOBILE6.

10.5. A freeway passes near a school. Determine the expected CO concentration at a height of 2 ft at the school. The CO emission factor is 25 g/mi. Wind speed is 3.5 ft/s, $H = 1$ ft, and traffic volume is 9000 veh/h. Assume that when $x = 50$ ft, σ_y and σ_z are 30 and 15 ft., respectively; and when $x = 67.5$ ft., σ_y and σ_z are 25 and 16 ft, respectively. Assume the same configuration as shown for Example 10.2.

REFERENCES[1]

An, F., Barth, M. J., Norbeck, J., Ross, M. (1997). *Development of Comprehensive Modal Emissions Model: Operating Under Hot-Stabilized Conditions*, Transp. Res. Rec. 1587, Transportation Research Board, National Research Council, Washington, D.C.

Barth, M. J., Johnston, E., Tadi, R. R. (1996). *Using GPS Technology to Relate Macroscopic and Microscopic Traffic Parameters*, Transp. Res. Rec. 1520, Transportation Research Board, National Research Council, Washington, D.C.

BTS, (1997). Mobility and access, in *Transportation Statistics Annual. Report, 1997*, Bureau of Transportation Statistics, U.S. Department of Transportation, Washington, DC.

Carpenter, T. G. (1994). *The Environmental Impact of Railways*, Wiley, Chichester, UK.

Carruthers, D. J., Holroyd, R. J., Hunt., J. C. R., Weng, W.-S., Robins, A. G., Apsley, D. D., Thompson D. J., Smith, F. B. (1994). UK-ADMS: a new approach to modeling dispersion in the Earth's atmospheric boundary layer, *J. Wind Eng. Ind. Aerodynam.*, Vol. 52, pp. 139–153.

Chatterjee, A., Miller, T. L., Philpot, J. W., Wholley, T. F., Guensler, R., Hartgen, D., Margiotta, R. A., Stopher, P. R. (1997). *Improving Transportation Data for Mobile Source Emission Estimates*, NCHRP Rep. 394, Transportation Research Board, National Research Council, Washington, DC.

Cohn, L. F., McVoy, G. R. (1982). *Environmental Analysis of Transportation Systems*, Wiley, New York.

*Cooper, C. D., Alley, F. C. (2002). *Air Pollution Control: A Design Approach*, 3rd ed., Waveland Press, Prospect Heights, IL.

Delucchi, M. (2003). Environmental externalities of motor vehicle use, in *Handbook of Transport and the Environment*, ed. Hensher, D. A., Button, K. J., Elsevier, Amsterdam, The Netherlands.

Ding, Y. (2000). Quantifying the impact of traffic-related and driver-related factors on vehicle fuel consumption and emissions, M.S. thesis, Virginia Polytechnic Institute and State University, Blacksburg, VA.

Dowling, R., Ireson, R., Skabardonis, A., Gillen, D., Stopher, P. (2005). *Predicting Air Quality Effects of Traffic-Flow Improvements: Final Report and User's Guide*, NCHRP Rep. 535, Transportation Research Board, National Research Council, Washington, DC.

*Faiz, A., Weaver, C. S., Walsh, M. P. (1996). *Air Pollution from Motor Vehicles*, World Bank, Washington, DC.

Friedrich, R., Bickel, P., Eds. (2001). *Environmental External Costs of Transport*, Springer-Verlag, Berlin.

Glover, E. L., Koupal, J. W. (1999). *Determination of CO Basic Emission Rates, OBD and I/M Effects for Tier 1 and Later LDVs and LDTs*, EPA 420-P-99-017, U.S. Environmental Protection Agency, Washington, DC.

Gore, A. (2006). *An Inconvenient Truth: The Planetary Emergency of Global Warming*, Rodale, Emmaus, PA.

*Guenslar, R., Washington, S., Bachman, W. (1998). Overview of the MEASURE modeling framework, *Proc. ASCE Conference on Transportation Planning and Air Quality III*, Lake Tahoe, CA.

Holmen, B. A., Niemeier, D. A. (2003). Air quality, in *Handbook of Transport and the Environment*, ed. Hensher, D. A., Button, K. J., Elsevier, Amsterdam, The Netherlands.

Homburger, W. S., Hall, J. W., Reilly, W. R., Sullivan, E. C. (2001). *Fundamentals of Traffic Engineering*, 13th ed., Institute of Transportation Studies, University of California, Berkeley, CA.

Koupal, J. W., Glover, E. L. (1999). *Determination of NO_x and HC Basic Emission Rates, OBD and I/M Effects for Tier 1 and Later LDVs and LDTs*, EPA 420-P-99-009, U.S. Environmental Protection Agency, Washington, DC.

Kretzschmar, J. G., Maes, G., Cosemans, G. (1994). *Operational Short Range Atmospheric Dispersion Models for Environmental Impact Assessment in Europe*, Vols. 1 and 2, E&M.RA9416, Flemish Institute for Technological Research, Mol, Belgium.

Potter, S. (2003). Transport energy and emissions: urban public transport, in *Handbook of Transport and the Environment*, ed. Hensher, D. A., Button, K. J., Elsevier, Amsterdam, The Netherlands.

Rakha, H., Van Aerde, M., Ahn, K., Trani, A. A. (1999). *Requirements for Evaluating the Environmental Impacts of Intelligent Transportation Systems Using Speed and Acceleration Data*, Transp. Res. Rec. 1664, Transportation Research Board, National Research Council, Washington, DC.

Rouphail, N. M., Frey, C. H., Colyar, J. D., Unal, A. (2001). Vehicle emissions and traffic measures: exploratory analysis of field observations at signalized arterials, *Proc. Transportation Research Board 80th Annual Meeting*, Washington, DC.

Sinha, K. C., Peeta, S., Sultan, M. A., Poonuru, K., Richards, N. (1998). *Evaluation of the Impacts of ITS Technologies on the Borman Expressway Network*, Tech. Rep. FHWA/IN/JTRP-98/5, Joint Transportation Research Program, West Lafayette, IN.

Somerville, H. (2003). Transport energy and emissions: aviation, in *Handbook of Transport and the Environment*, ed. Hensher, D. A., Button, K. J., Elsevier, Amsterdam, The Netherlands.

Stanley, J., Watkiss, P. (2003). Transport energy and emissions: buses, in *Handbook of Transport and the Environment*, ed. Hensher, D. A., Button, K. J. Elsevier, Amsterdam, The Netherlands.

System Application International (2002). *User's Guide to HYROAD: The Hybrid Roadway Intersection Model*, Tech. Rep. SYSAPP-02-073d, National Cooperative Highway Research Program, National Research Council, Washington, DC.

Talley, W. K. (2003). Environmental impacts of shipping, in *Handbook of Transport and the Environment*, ed. Hensher, D. A., Button, K. J. Elsevier, Amsterdam, The Netherlands.

Timmons, H. (2006). Britain warns of high costs of global warming, *New York Times*, October 31, 2006 issue, New York, NY.

---

[1]References marked with an asterisk can also serve as useful resources for air quality impact estimation.

TCRP (2003). *Travel Matters: Mitigating Climate Change with Sustainable Surface Transportation*, Transit Cooperative Research Program, Transportation Research Board, National Research Council, Washington, DC.

TRB (1997). *Toward a Sustainable Future: Addressing the Long-Term Effects of Motor Vehicle Transportation on Climate and Ecology*, Spec. Rep. 251, Transportation Research Board, National Research Council, Washington, DC.

USEPA (1997). *Emission Factors for Locomotives: Technical Highlights*, EPA 420-F-97-051, Air and Radiation, Office of Mobile Sources, U.S. Environmental Protection Agency, Washington, DC.

———— (1998). *Assessing the Emissions and Fuel Consumption Impacts of Intelligent Transportation Systems (ITS)*, EPA 231-R-98-007, U.S. Environmental Protection Agency, Washington, DC.

———— (1999). *Indicators of the Environmental Impacts of Transportation*, EPA 230-R-99-001, U.S. Environmental Protection Agency, Washington, DC.

*———— (2002). *User's Guide to MOBILE6.0 Mobile Source Emission Factor Model*, EPA 420-R-02-001, U.S. Environmental Protection Agency, Washington, DC.

———— (2005). *National Emissions Inventory (NEI) Air Pollutant Emissions Trends Data*, U.S. Environmental Protection Agency, Washington, DC, http://www.epa.gov/ttn/chief/trends/index.html. Accessed Feb. 26, 2006.

USEPA (2006). *Green Vehicle Guide*, U.S. Environmental Protection Agency, Washington, DC. www.epa.gov/emissweb/about.htm. Accessed Nov 15, 2006.

Wayson, R. L. (2002). Environmental considerations during transportation planning, in *The Civil Engineering Handbook*, 2nd ed., ed. Chen, W. F., Liew J. Y. R., CRC Press, Boca Raton FL.

ADDITIONAL RESOURCES

Bennett, C. R., Greenwood, I. D. (2001). *Modeling Road User and Environmental Effects in HDM-4*, Vol. 7, *HDM-4 Documentation*, Highway Development and Management Series, World Bank, Washington, DC.

ECMT (1998). *Efficient Transport for Europe: Policies for Internalization of External Costs*, European Conference of Ministers of Transport, Organization for Economic Cooperation and Development, Paris.

EEA (1996). *Guidance Report on Preliminary Assessment Under EC Air Quality Directives*, Tech. Rep. 11, European Environment Agency, Copenhagen, Denmark.

Forckenbrock, D. J., Sheeley, J. (2004). *Effective Methods for Environmental Justice Assessment*, NCHRP Rep. 532, Transportation Research Board, National Research Council, Washington, DC.

Gorham, R. (2002). *Air Pollution from Ground Transportation: An Assessment of Causes, Strategies and Tactics, and Proposed Actions for the International Community*, Global Initiative on Transport Emissions, a partnership of the United Nations and the World Bank Division for Sustainable Development, Department of Economic and Social Affairs, United Nations, New York.

Horowitz, J. L. (1982). *Air Quality Analysis for Urban Transportation Planning*, MIT Press, Cambridge, MA.

USEPA (2000). *AP-42: Compilation of Air Pollutant Emission Factors*, office of Air Quality Planning and Standards, U.S. Environmental Protection Agency, Research Triangle Park, NC.

USDOT (1996). *The Congestion Mitigation and Air Quality Improvement (CMAQ) Program of the ISTEA: Guidance Update*, FHWA and FTA, U.S. Department of Transportation, Washington, DC.

APPENDIX A10.1: USING MOBILE6 TO ESTIMATE EMISSIONS

A10.1.1 Details of the MOBILE6 Input File

The input file used by MOBILE6 comprises the control file, which manages the input data, program execution, and output; the basic files, which contain input data, are common to all scenarios at each program run; and the scenario files, which provide information on individual scenarios under investigation. The basic files enable the input of any emission-related parameters that differ from the default values available in MOBILE6. For several emission parameters, MOBILE6 utilizes default values that are representative of national averages but can be substituted by local data to yield more reliable emission estimates. A typical MOBILE6 *output file* consists of total exhaust and nonexhaust emissions by vehicle type and composite emission factors.

(*a*) *Basic Data File*   This basic date file contains information (Figure A10.1) that is input only once (at the first use) of the MOBILE6 for a particular program run. Inputs in this file, which are specific to the location, are substitutes for the default national average values in MOBILE6.

Engine Starts per Day and Distribution by Hour: The frequency of starts per day influences engine exhaust start emission estimates for light-duty gasoline cars, diesel passenger cars, trucks, and motorcycles but does not affect the emission estimates for heavy-duty diesel-fueled vehicles and buses. For gasoline-fueled vehicles, including heavy-duty vehicles and buses, this parameter also affects the extent of evaporative hot-soak losses that occur at trip ends. MOBILE6 assigns a separate default value for the number of engine starts per day to each of 25 vehicle classes and for each of 25 vehicle age categories. These values differ by the day of week. The analyst needs to input (1) values for engine starts per day for all vehicle classes affected by the **Starts per Day** command; and (2) average fraction of all engine starts that occur in each hour of a 24-hour day, for both weekdays and weekends.

Inspection Maintenance Program Status: The user can specify the status of any existing I/M program using the **I/M Program** command. If this command is not used,

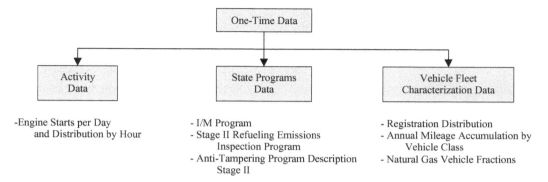

Figure A10.1   Basic data for MOBILE6.

MOBILE6 assumes that no I/M program exists. Input data include number of I/M programs that will be considered in the program run, calendar year at the start of the I/M program, calendar year at termination of the I/M program, frequency of I/M inspection (annual vs. biennial), I/M program type, and I/M inspection type.

Stage II Refueling Emissions Inspection Program: The **Effects of Stage II on Refueling Emissions** command enables the user to specify the impact of refueling emissions required by a stage II vapor recovery system. There is no default calculation of impact of a stage II program.

Stage II Antitampering Program Description: This gives the user the option to model the impact of an antitampering program using the **Anti-Tampering Programs** command. No default values are provided.

Vehicle Registration Distribution: This enables the user to supply vehicle registration distributions by vehicle age for any of the 16 composite (combined gas and diesel) vehicle types. A list of these vehicle types can be found in the main *User's Manual*.

Annual Mileage Accumulation by Vehicle Class: The **Annual Mileage Accumulation Rates** command allows the user to input the annual mileage accumulation rates by vehicle age for any of 28 individual vehicle types. Vehicle age groups are 0 to 25 and over 25 years.

Natural Gas Vehicle (NGV) Fractions: With this parameter, the user can specify the percentage of vehicles in the fleet that are certified to operate on either compressed or liquefied natural gas for each of the 28 individual classes beginning with the model year. The default fraction of NGV vehicles in the fleet is equal to zero.

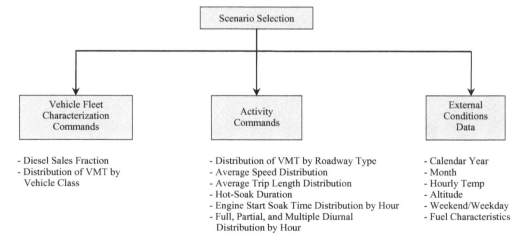

Figure A10.2   Scenario specific data for MOBILE6.

(*b*) *Scenario Selection File* This is used to assign scenario-specific values to emission variables. Various types of data needed for this file are shown in Figure A10.2.

(*c*) *Traffic-Related Data* MOBILE6 enables a relatively fine temporal distribution of traffic during the day for major traffic indicators. Hourly distributions can be input instead of 24-hour averages. Also, the fleet characterization projections of future vehicle fleet size and fraction of travel are based on a number of considerations, including vehicle age, mileage accumulation rate, and 28 vehicle classes. Data on the key traffic-related inputs (vehicle registration distribution, annual mileage accumulation rate, and the distribution of vehicle-miles traveled) are input by vehicle class and roadway type. Local data on mileage accumulation are typically more difficult to obtain because odometer readings are typically not recorded on an annual basis unless an inspection maintenance program is operational in the region under study.

A10.1.2 Sample MOBILE6 Output

Figure A10.3 below shows a sample of the output file generated by MOBILE6.

```
*************************************************************************
* MOBILE6.2.01 (31-Oct-2002)                                           *
* Input file: AFTER.IN (file 1, run 1).                                *
*************************************************************************

* # # # # # # # # # # # # # # # # # # # # # #
* Scenario Title : Master Example Input Demonstration
* File 1, Run 1, Scenario 1.
* # # # # # # # # # # # # # # # # # # # # # #

                    Calendar Year:  2006
                           Month:  Jan.
                        Altitude:  Low
             Minimum Temperature:  64.0 (F)
             Maximum Temperature:  92.0 (F)
               Absolute Humidity:  115. grains/lb
               Nominal Fuel RVP:   7.0 psi
                   Weathered RVP:  6.8 psi
             Fuel Sulfur Content:  33. ppm

              Exhaust I/M Program:  Yes
                 Evap I/M Program:  No
                     ATP Program:  Yes
                 Reformulated Gas:  No

Emissions determined from WEEKEND hourly vehicle activity fractions.

    Ether Blend Market Share: 0.500    Alcohol Blend Market Share: 0.500
    Ether Blend Oxygen Content: 0.020  Alcohol Blend Oxygen Content: 0.010
                                       Alcohol Blend RVP Waiver: No
```

Vehicle Type: GVWR:	LDGV	LDGT12 <6000	LDGT34 >6000	LDGT (All)	HDGV	LDDV	LDDT	HDDV	MC	All Veh
VMT Distribution:	0.4005	0.2435	0.0272		0.0918	0.0023	0.0014	0.2216	0.0118	1.0000
Composite Emission Factors (g/mi):										
Composite VOC :	0.763	0.932	1.684	1.007	3.207	0.929	1.724	0.882	3.00	1.108
Composite CO :	4.30	6.87	10.45	7.23	32.19	2.302	2.931	5.438	25.06	8.144
Composite NOX :	0.444	0.611	0.923	0.643	3.697	1.534	1.973	10.452	0.77	3.022
Exhaust emissions (g/mi):										
VOC Start:	0.086	0.129	0.244	0.141		0.169	0.473		0.389	
VOC Running:	0.135	0.254	0.524	0.281		0.760	1.251		2.087	
VOC Total Exhaust:	0.221	0.384	0.768	0.422	1.715	0.929	1.724	0.882	2.48	0.590
CO Start:	1.44	2.46	3.72	2.59		0.501	0.924		2.900	
CO Running:	2.86	4.41	6.73	4.64		1.802	2.006		22.161	
CO Total Exhaust:	4.30	6.87	10.45	7.23	32.19	2.302	2.931	5.438	25.06	8.144
NOx Start:	0.065	0.089	0.128	0.093		0.045	0.097		0.318	
NOx Running:	0.379	0.523	0.795	0.550		1.490	1.876		0.454	
NOx Total Exhaust:	0.444	0.611	0.923	0.643	3.697	1.534	1.973	10.452	0.77	3.022
Non-Exhaust Emissions (g/mi):										
Hot Soak Loss:	0.096	0.082	0.135	0.088	0.278	0.000	0.000	0.000	0.132	0.089
Diurnal Loss:	0.024	0.027	0.050	0.029	0.092	0.000	0.000	0.000	0.025	0.026
Resting Loss:	0.087	0.092	0.177	0.101	0.306	0.000	0.000	0.000	0.368	0.095
Running Loss:	0.318	0.310	0.490	0.328	0.645	0.000	0.000	0.000	0.000	0.275
Crankcase Loss:	0.005	0.009	0.010	0.009	0.011	0.000	0.000	0.000	0.000	0.005
Refueling Loss:	0.012	0.028	0.055	0.030	0.160	0.000	0.000	0.000	0.000	0.028
Total Non-Exhaust:	0.542	0.548	0.916	0.645	1.492	0.000	0.000	0.000	0.525	0.518

Figure A10.3   Sections of a sample MOBILE6 output file.

APPENDIX A10.2: VALUES OF THE GAUSSIAN DISTRIBUTION FUNCTION

$$G(B) = \frac{1}{\sqrt{2\pi}} \int_{-\infty}^{B} \exp\left(\frac{-B^2}{2}\right) dB$$

where $B = (x - \mu)/\sigma$.

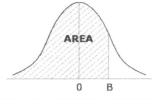

B	0	1	2	3	4	5	6	7	8	9
−3.0	0.0013	0.0010	0.0007	0.0005	0.0003	0.0002	0.0002	0.0001	0.0001	0.0000
−2.9	0.0019	0.0018	0.0017	0.0017	0.0016	0.0016	0.0015	0.0015	0.0014	0.0014
−2.8	0.0026	0.0025	0.0024	0.0023	0.0022	0.0021	0.0021	0.0021	0.0020	0.0019
−2.7	0.0035	0.0034	0.0033	0.0032	0.0031	0.0030	0.0029	0.0028	0.0027	0.0026
−2.6	0.0047	0.0045	0.0044	0.0043	0.0041	0.0040	0.0039	0.0038	0.0037	0.0036
−2.5	0.0062	0.0060	0.0059	0.0057	0.0055	0.0054	0.0052	0.0051	0.0049	0.0048
−2.4	0.0082	0.0080	0.0078	0.0075	0.0073	0.0071	0.0069	0.0068	0.0066	0.0064
−2.3	0.0107	0.0104	0.0102	0.0099	0.0096	0.0094	0.0091	0.0089	0.0087	0.0084
−2.2	0.0139	0.0136	0.0132	0.0129	0.0125	0.0122	0.0119	0.0116	0.0113	0.0110
−2.1	0.0179	0.0174	0.0170	0.0166	0.0162	0.0158	0.0154	0.0150	0.0146	0.0143
−2.0	0.0228	0.0222	0.0217	0.0212	0.0207	0.0202	0.0197	0.0192	0.0188	0.0183
−1.9	0.0287	0.0281	0.0274	0.0268	0.0262	0.0256	0.0250	0.0244	0.0239	0.0233
−1.8	0.0359	0.0351	0.0344	0.0336	0.0329	0.0322	0.0314	0.0307	0.0301	0.0294
−1.7	0.0446	0.0436	0.0427	0.0418	0.0409	0.0401	0.0392	0.0384	0.0375	0.0367
−1.6	0.0548	0.0537	0.0526	0.0516	0.0505	0.0495	0.0485	0.0475	0.0465	0.0455
−1.5	0.0668	0.0655	0.0643	0.0630	0.0618	0.0606	0.0594	0.0582	0.0571	0.0559
−1.4	0.0808	0.0793	0.0778	0.0764	0.0749	0.0735	0.0721	0.0708	0.0694	0.0681
−1.3	0.0968	0.0951	0.0934	0.0918	0.0901	0.0885	0.0869	0.0853	0.0838	0.0823
−1.2	0.1151	0.1131	0.1112	0.1093	0.1075	0.1056	0.1038	0.1020	0.1003	0.0985
−1.1	0.1357	0.1335	0.1314	0.1292	0.1271	0.1251	0.1230	0.1210	0.1190	0.1170
−1.0	0.1587	0.1562	0.1539	0.1515	0.1492	0.1469	0.1446	0.1423	0.1401	0.1379
−0.9	0.1841	0.1814	0.1788	0.1762	0.1736	0.1711	0.1685	0.1660	0.1635	0.1611
−0.8	0.2119	0.2090	0.2061	0.2033	0.2005	0.1977	0.1949	0.1922	0.1894	0.1867
−0.7	0.2420	0.2389	0.2358	0.2327	0.2296	0.2266	0.2236	0.2206	0.2177	0.2148
−0.6	0.2743	0.2709	0.2676	0.2643	0.2611	0.2578	0.2546	0.2514	0.2483	0.2451
−0.5	0.3085	0.3050	0.3015	0.2981	0.2946	0.2912	0.2877	0.2843	0.2810	0.2776
−0.4	0.3346	0.3409	0.3372	0.3336	0.3300	0.3264	0.3228	0.3192	0.3156	0.3121
−0.3	0.3821	0.3783	0.3745	0.3707	0.3669	0.3632	0.3594	0.3557	0.3520	0.3483
−0.2	0.4207	0.4168	0.4129	0.4090	0.4052	0.4013	0.3974	0.3936	0.3897	0.3859
−0.1	0.4602	0.4562	0.4522	0.4483	0.4443	0.4404	0.4364	0.4325	0.4286	0.4247

Figure A10.4   Values of the Gaussian distribution function

B	0	1	2	3	4	5	6	7	8	9
0.0	0.5000	0.4960	0.4920	0.4880	0.4840	0.4801	0.4761	0.4721	0.4681	0.4641
0.1	0.5398	0.5438	0.5478	0.5517	0.5557	0.5596	0.5636	0.5675	0.5714	0.5753
0.2	0.5793	0.5832	0.5871	0.5910	0.5948	0.5987	0.6026	0.6064	0.6103	0.6141
0.3	0.6179	0.6217	0.6255	0.6293	0.6331	0.6368	0.6406	0.6443	0.6480	0.6517
0.4	0.6554	0.6591	0.6628	0.6664	0.6700	0.6736	0.6772	0.6808	0.6844	0.6879
0.5	0.6915	0.6950	0.6985	0.7019	0.7054	0.7088	0.7123	0.7157	0.7190	0.7224
0.6	0.7257	0.7291	0.7324	0.7357	0.7389	0.7422	0.7454	0.7486	0.7517	0.7549
0.7	0.7580	0.7611	0.7642	0.7673	0.7704	0.7734	0.7764	0.7794	0.7823	0.7852
0.8	0.7881	0.7910	0.7939	0.7967	0.7995	0.8023	0.8051	0.8078	0.8106	0.8133
0.9	0.8159	0.8186	0.8212	0.8238	0.8264	0.8289	0.8315	0.8340	0.8365	0.8389
1.0	0.8413	0.8438	0.8461	0.8485	0.8508	0.8531	0.8554	0.8577	0.8599	0.8621
1.1	0.8643	0.8665	0.8686	0.8708	0.8729	0.8749	0.8770	0.8790	0.8810	0.8830
1.2	0.8849	0.8869	0.8888	0.8907	0.8925	0.8944	0.8962	0.8980	0.8997	0.9015
1.5	0.9332	0.9345	0.9357	0.9370	0.9382	0.9394	0.9406	0.9418	0.9429	0.9441
1.6	0.9452	0.9463	0.9474	0.9484	0.9495	0.9505	0.9515	0.9525	0.9535	0.9545
1.7	0.9554	0.9564	0.9573	0.9582	0.9591	0.9599	0.9608	0.9616	0.9625	0.9633
1.8	0.9641	0.9649	0.9656	0.9664	0.9671	0.9678	0.9686	0.9693	0.9699	0.9706
1.9	0.9713	0.9719	0.9726	0.9732	0.9738	0.9744	0.9750	0.9756	0.9761	0.9767
2.0	0.9772	0.9778	0.9783	0.9788	0.9793	0.9798	0.9803	0.9808	0.9812	0.9817
2.1	0.9821	0.9826	0.9830	0.9834	0.9838	0.9842	0.9846	0.9850	0.9854	0.9857
2.2	0.9861	0.9864	0.9868	0.9871	0.9875	0.9878	0.9881	0.9884	0.9887	0.9890
2.3	0.9893	0.9896	0.9898	0.9901	0.9904	0.9906	0.9909	0.9911	0.9913	0.9916
2.4	0.9918	0.9920	0.9922	0.9925	0.9927	0.9929	0.9931	0.9932	0.9934	0.9936
2.5	0.9938	0.9940	0.9941	0.9943	0.9945	0.9946	0.9948	0.9949	0.9951	0.9952
2.6	0.9953	0.9955	0.9956	0.9957	0.9959	0.9960	0.9961	0.9962	0.9963	0.9964
2.7	0.9965	0.9966	0.9967	0.9968	0.9969	0.9970	0.9971	0.9972	0.9973	0.9974
2.8	0.9974	0.9975	0.9976	0.9977	0.9977	0.9978	0.9979	0.9979	0.9980	0.9981
2.9	0.9981	0.9981	0.9982	0.9983	0.9984	0.9984	0.9985	0.9985	0.9986	0.9986
3.0	0.9987	0.9987	0.9987	0.9988	0.9988	0.9989	0.9989	0.9989	0.9990	0.9990

Figure A10.4   (*continued*)

CHAPTER 11

Noise Impacts

With silence favor me.

—*Horace (65–8 B.C.)*

INTRODUCTION

Noise, defined as unwanted or excessive sound, is one of the most widely experienced environmental externalities associated with transportation systems. Excessive noise can adversely affect real-estate value and, more importantly, can cause general nuisance and health problems. An important feature of noise pollution is that noise generated at a particular time is not affected by previous activity, nor does it affect future activities; unlike other pollutants, noise leaves no residual effects that are evidential of its unpleasantness. For this reason, there may be a tendency to overlook or to underestimate the problem of noise pollution.

In this chapter we present the fundamental concepts of sound, identify the sources of transportation noise, and discuss the environmental (weather) factors that affect highway noise propagation. Methodologies and software packages are presented for noise impact estimation. We also discuss transportation noise mitigation and noise barrier performance analysis and present legislation and regulations affecting transportation noise.

11.1 FUNDAMENTAL CONCEPTS OF SOUND

11.1.1 General Characteristics

The perception of sound, whether it is from a simple musical instrument or from complicated spectra, such as that of traffic noise, is evaluated by the human ear on the basis of four major criteria: loudness, frequency, duration, and subjectivity.

(*a*) *Loudness* Noise intensity or loudness is related to the pressure fluctuations amplitude transmitted through the air. Pressure fluctuations cause contraction of the eardrum and generate the sensation of sound. The lowest pressure fluctuation that the ear can sense (the minimum threshold of hearing) is 2×10^{-5} N/m^2 (3×10^{-9} psi), and the highest is approximately 63 N/m^2 (9.137×10^{-3} psi), considered the threshold of hearing pain. This represents a pressure change of over 10,000,000 units. Figure 11.1 shows typical sound pressure levels (SPLs).

The use of a large range of pressure fluctuations in reporting can be awkward. Also, as a protective mechanism, human auditory response is not linearly related to pressure fluctuations. In view of these two limitations of using sound pressure fluctuations as a measure of loudness, sound pressure *levels*, in terms of *decibels* (dB), are typically used. Mathematically, the transformation of sound pressure to sound pressure levels is carried out as follows (Cohn and McVoy, 1982; Wayson, 2003):

$$\text{SPL(dB)} = 10 \log_{10} \frac{p^2}{p_0^2} \qquad (11.1)$$

where SPL(dB) is the sound pressure level in decibels, p_0 the reference pressure (2×10^{-5} N/m^2 or 3×10^{-9} psi), and p the sound pressure of concern.

Example 11.1 Noise from a source located a distance D from a receptor is felt over an area A. The noise strikes the receptor with intensity I (sound force per unit area). Assume that when the receptor is moved farther away, to a distance $2D$, the noise is felt over an area $4A$. This assumption is true for noise emanating from a single point, as will be seen in a subsequent section. Prove that the acoustical effect of the doubled distance is a reduction in sound pressure level by 6 dB.

SOLUTION The sound pressure level at a receptor a distance i from the noise source is $10 \log I_i/I_r = 10 \log I_i - 10 \log I_r$, where I_r is the reference intensity.

Initial distance $= \text{SPL}_1 = 10 \log \dfrac{I_1}{I_r} = 10 \log I_1 - 10 \log I_r$

Final distance $= \text{SPL}_2 = 10 \log \dfrac{I_2}{I_r} = 10 \log I_2 - 10 \log I_r$

The difference in SPL caused by doubling the distance is

$$10 \log I_1 - 10 \log I_r - 10 \log I_2 + 10 \log I_r$$
$$= 10 \log I_1 - 10 \log I_2 = 10 \log \frac{I_1}{I_2}$$

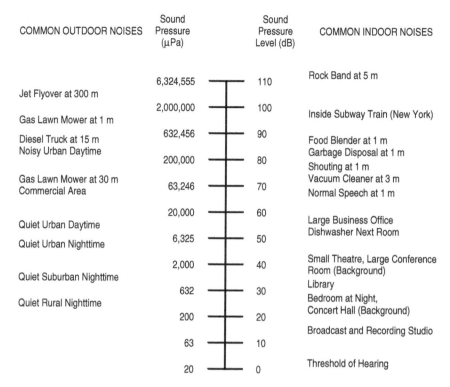

Figure 11.1 Typical noise levels. (From FHWA, 1980.)

But $I_1/I_2 = 4$ because $A_1/A_2 = \frac{1}{4}$ (I_r is the reference intensity). Therefore, the difference in SPL caused by doubling the distance $= 10 \log(4) = 6$ dB.

11.1.2 Addition of Sound Pressure Levels from Multiple Sources

Decibels add in a logarithmic rather than linear fashion. The rule of thumb for decibel summation is such that doubling the source strength (sound pressure) means an increase in the sound pressure level at the location of the recipient by only 3 dB. Similarly, halving the source strength results in a 3-dB decrease in the sound pressure level at the recipient's location. Addition of sounds from several sources can be carried out using the equation

$$SPL_{total} = 10 \log_{10} \sum_{i=1}^{n} 10^{SPL_i/10} \qquad (11.2)$$

Example 11.2 A certain jet flying overhead a residential area at a height of 300 m has a SPL of 105 dB. Find the total SPL of two of such jets flying overhead at that height.

SOLUTION   Using equation (11.2) yields

$$SPL_{total} = 10 \log_{10} \left(10^{SPL_1/10} + 10^{SPL_2/10}\right) = 108 \text{ dB}$$

In typical outdoor conditions, changes in sound levels not exceeding 3 dB are barely noticeable by receptors. Furthermore, a change of 3 dB is generally perceived to be a twofold increase in the sound level. This suggests that for a receptor to discern a change in traffic noise levels objectively, there must be a significant change in transportation patterns, such as the percentage of trucks, pavement type, or speed.

(a) Frequency   The frequency of sound, a change in the rate of pressure fluctuations in the air measured in terms of pressure changes per second or oscillations per second, has the unit hertz (Hz). A large range of frequencies can be heard by the human ear, extending from about 20 to 20,000 Hz. Differences in the rate of pressure fluctuations constitute the tonal quality of sound and help the receptor (human ear) identify the sound source (e.g., a braking sound as opposed to a horn sound). Sound frequency (f), wavelength (λ), and speed (c) are related as $f = c/\lambda$. Within the human hearing range, frequencies are not detected equally well by the human ear. Sounds at

low frequencies (below 500 Hz) and higher frequencies (above 10,000 Hz) are not heard as well as those in the intermediate range of frequencies. Therefore, mere description of a sound by its loudness is not sufficient and needs to be complemented by some description of its frequency spectra.

Sound frequency groups, typically referred to as *octave bands*, are used to describe sounds and to provide detailed descriptions of their frequency components. However, in the assessment and evaluation of transportation sounds, a broader approach is typically used. Contributions from all frequency bands are first adjusted to approximate the way the ear hears each range, then the contributions are summed to yield a single number. Three common scales have typically been used (Figure 11.2). Scale A describes the way the human ear responds to moderate sounds; scale B is the response curve for stronger sounds; and scale C describes the way the human ear responds to very loud sounds. The nonlinear response of the ear at low and high frequencies is illustrated in these graphs. Most regulations and evaluations relevant to transportation systems utilize the A scale.

(b) Duration   Noise is described more completely when combined with descriptions of loudness and frequency. A collision between two vehicles is typically loud, but it lasts only a fraction of a second. At the other extreme, noise due to continuous traffic operation may not be intense, but it is continual. Variation of traffic noise with time is considered important for assessing such noise, and some effective descriptors of the temporal variation of sound have been developed as follows: maximum sound level, $L_{max}(t)$; statistical sound levels, $L_{xx}(t)$; equivalent sound level, $L_{eq}(t)$; and day/night level, L_{dn}.

L represents the fact that each descriptor is a sound pressure level with units of decibels.

The parameter (t) indicates that each descriptor is given for a specific period of time.

L_{max} represents the maximum noise level that occurs during a definite time period and allows for a slightly more complete description of noise when combined with the loudness and frequency description. For example, 60 dB (A-weighted) $L_{max}(1h)$ defines the highest sound level, frequency, and weighting during a defined one-hour time period.

For the statistical descriptor (L_{xx}) the subscript xx indicates the percentage of time that the level listed is exceeded. For instance, a sound level of 65 dB(A-weighted) $L_{10}(1h)$ means that in a given 1-hour time period, sound pressure levels exceed 65 dB (on an A-weighted scale) 10% of the time. The numerical value can be any fraction of the time, but L_{10}, L_{50}, and L_{90} corresponding to 10%, 50% and 90% of the time, are mainly used (Figure 11.3). L_{90} is commonly used as the background level and is the sound pressure level that is exceeded 90% of the time.

L_{eq}, the equivalent sound pressure level, is the steady state (e.g., constant, nonvarying) sound level that contains the same amount of acoustic energy as a time-varying sound level in a given time period. Using L_{eq}, it is possible to add the effects of sounds simultaneously emanating from more than one source. L_{eq} has therefore become the metric of choice in the United States for highway noise analysis.

The descriptor L_{dn}, the day/night sound level (DNL), is by definition a 24-hour metric that accounts for factors others than duration, such as the time of day that the sound occurs. L_{dn} consists of hourly L_{eq} (A-weighted) values

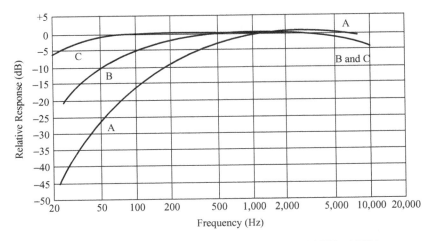

Figure 11.2   Frequency response curves. (From FHWA, 1980.)

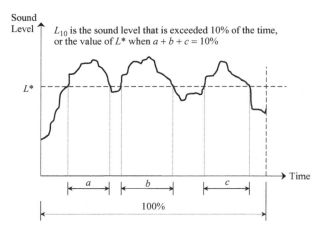

Figure 11.3 Graphical interpretation of L_{10} (not to scale).

and the energy averaged over the entire 24-hour period, with a certain decibel level penalty added to sound occurring in each hour during the night time (10 p.m. to 7 a.m.)

(c) Effects and Subjectivity Individuals have different responses to sound. A sound type that is pleasing to one person may be a nuisance (noise) to another. Also, noise annoyance levels are subjective, and criteria are usually based on attitudinal surveys. Furthermore, the degree of noise unpleasantness may be influenced by the time or place at which it occurs. For example, the flow of truck traffic through a residential area may be more offensive at night than on a weekday afternoon. A single loud noise can result in acute hearing loss. However transportation-related noise is rather chronic in nature and may result in reduced hearing ability only after long-term exposure. In the short term, noise annoyance or irritation is the issue, but in the long term, noise can lead to problems in emotional well-being and can cause discomfort by interfering in sleep or causing disruption in the daily lives of humans. Studies show that noise prevents deep sleep cycles considered necessary for complete refreshment, causes tension due to continual intrusion, affects communication, and decreases the learning abilities of students in class sessions interrupted by noise events (Wayson, 2003).

11.2   SOURCES OF TRANSPORTATION NOISE

The most common transportation noise is from highway operations (autos, trucks, buses) but there can also be significant noise from other modes (planes, trains, and water vessels). Sources of transportation noise are as follows:

- *Vehicle–air interaction.* When a vehicle is in motion, friction between the vehicle's body and the surrounding air induces a gradient in the air pressure field and thereby generates noise.
- *Tire–pavement interaction.* Tire–pavement noise generation is a direct result of the friction and small impacts that occur as the tire rolls along the highway or runway pavement surface. Such noise is generally more pronounced for concrete pavements and less for asphalt concrete pavements. In the case of rail transportation, friction between the steel wheel and the guideway often generates noise, particularly in curve areas.
- *Vehicle engines.* Vehicle engine noise levels are generally higher in areas of higher speed or with geometric designs that encourage vehicle acceleration or deceleration. Also, larger transportation vehicles (trucks, large aircraft, and ocean liners) generate more noise than do their smaller counterparts.
- *Vehicle exhaust systems.* Exhaust systems on vehicles lead to higher noise levels, especially in cases of malfunctioning noise-control devices (mufflers). Exhaust noise levels are closely related to noise from vehicle engines; higher speeds, more frequent speed changes, and larger vehicles are associated with higher levels of exhaust noise.
- *Vehicle horns and brakes.* Vehicle horns can constitute a significant and irritating source of urban traffic noise, particularly in traffic cultures where frequent horn blowing is practiced. Brakes also constitute a significant noise source, particularly for large trucks.

Noise generated from a highway traffic stream can propagate over considerable distances and has an acoustic spectrum that can typically range from 125 to 4000 Hz. This frequency range is discernible by human ears, and highway noise and thus can cause great discomfort to humans. In the next section we discuss certain factors that impede or enhance the noise propagation from transportation facilities to neighboring property.

11.3   FACTORS AFFECTING TRANSPORTATION NOISE PROPAGATION

Temperature variations between the lower and upper atmospheric belts affect noise propagation. The speed of sound is reduced when sound waves move into denser media. Sound waves therefore generally bend toward cooler temperatures. On a typical hot day (hot lower belt, cool upper belt) the temperature decreases with increasing altitude; the sound waves generated at ground level tend to bend upward, creating a *shadow zone*. In such shadow zones, sound levels are reduced by as much as 20 dBA at distances exceeding 500 ft, where an observer at a receptor site may see a sound source but fail to hear it. On a typical cold day (cool lower belt, hot upper belt), temperatures are lower at points closer to the ground than at points higher in the atmosphere (such as early morning or over calm waters); and the sound waves tend to bend toward the site surface, bounce off the site surface, and travel much farther than expected. That is why sound is said to carry well over water. This also explains why sound propagates much faster under nighttime conditions.

Also, wind direction influences sound propagation. When the wind blows against the direction of sound, the sound waves generated at ground level tend to bend upward, creating a shadow zone. When the wind blows in the direction of the sound wave, sound waves tend to bend toward the site surface, bounce off the surface, and travel much farther than it does under normal conditions.

Other important noise propagation factors are the nature of the source (line vs. point sources) and the distance of the noise receptor (affected persons) from the source. Geometric spreading is an important geometric characteristic that describes the reduction of noise with increased distance from the noise source.

11.3.1 Nature of Source, Distance, and Ground Effects

(*a*) *Point Sources*   In cases where the noise origin is a single location, the source is referred to as a *point source*, as illustrated in Figure 11.4. Examples of point sources include a boat whistle, a single truck cruising on a highway, a locomotive with an idling engine, or a single aircraft flying overhead. For a point source, sound energy spreads in a spherical surface fashion ($4\pi r^2$). That is, the propagation of noises from point sources is governed by the *spherical spreading* phenomenon, which (for monopole sources) follows the inverse square law; the intensity and the root-mean-square pressure decrease proportionally as the inverse of the square root of the distance from the source. This means that if the sound source is at an initial distance r from the receptor and the sound signal from a source strikes an initial area A, that at

a final distance $2r$ the sound strikes a final area $4A$. Thus, by doubling the distance, the intensity (sound power per unit area) is reduced by a factor of 4. The acoustical effect of this reduction in sound intensity is a 6 dB reduction in sound pressure level as shown below

$$\Delta SPL(dB) = 10\log_{10}(r_1/r_2).$$

where $\Delta SPL(dB)$ = difference in sound pressure levels
r_1, r_2 = distance of point source from points 1 and 2, respectively

The difference in SPL from point 1 to point 2 is

$$\Delta SPL(dB) = 10\log_{10}(1/2)^2 = -6 \text{ dB}$$

Therefore, the *rule for point sources* is: *For every doubling of the distance between noise source and receptor, the SPL decreases by 6 dB, and for every halving of the distance between the noise source and the receptor, the SPL increases by 6 dB, all other factors remaining the same.*

(*b*) *Line Sources*   A transportation line facility, such as a highway (with a uniform traffic flow) or a railway (along which a long train is moving), represents a linear extrusion of a point source in space, thus constituting a *line source*. Noise propagation from line sources can be described by the *cylindrical spreading* phenomenon. The line source actually consists of an infinite number of closely spaced point sources; therefore it is appropriate to consider only the spread of the noise away from the source in a single plane. As such, the spread of sound energy is proportional to the circumference of a circle. The circumference of a circle is given by $2\pi r$, and using the same mathematical procedure as for a point source, the reduction of noise from a line source can be expressed as follows:

$$\Delta SPL(dB) = 10\log_{10}(r_1/r_2) \text{ or } 10\log_{10}(d_1/d_2)$$

where r or d is the distance from the line source. When $d_2 = 2d_1$, ΔSPL (dB) $= 10\log_{10}(1/2) = 3$ dB. Therefore, the *rule for line sources* is: *For every doubling of the distance between noise source and receptor, the SPL decreases by 3 dB, and for every halving of the distance between the noise source and the receptor, the SPL increases by 3 dB, all other factors remaining the same.*

For line sources such as highways traffic noise is expected to decrease by 3 dB for each doubling of distance from the highway. However, the highway is not actually in free space but close to Earth's surface. Thus, interaction of the sound wave with features of Earth's surface causes excess attenuation above what would be expected from mere geometric spreading. Excess attenuation effects are related to the soil type, nature of the ground cover, and the surface topography. Ground effects are generally difficult to predict. However, it has been

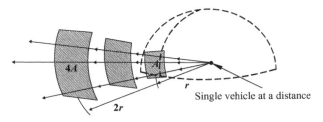

Figure 11.4   Point source propagation geometries.

determined that the value of 3 dB per doubling of distance is more typical of reflective surfaces abutting the highway (e.g., water or concrete), and a value of approximately 4.5 dB for each doubling of distance has been found to be applicable for absorptive surfaces (e.g., vegetative cover).

The spatial relationship between a transportation noise and the receptor not only determines the attenuation due to geometric spreading, but also determines the characteristics of the noise path, such as obstructions to the sound path. Spatial relationships are usually accounted for by using an x,y,z Cartesian coordinate system.

11.3.2 Effect of Noise Barriers

Obstructions in the path of a sound such as noise barriers, cause its diffraction or reflection (Figure 11.5), thus causing reduction of the sound levels. The area of decreased sound is called the *shadow zone*. Sound attenuation is maximum immediately behind the object and decreases with the distance behind the object as the sound wave reforms. Sound may also be reflected by obstructions in its path because such obstructions cause a redirection of the sound energy.

11.4   PROCEDURE FOR ESTIMATING NOISE IMPACTS FOR HIGHWAYS

Figure 11.6 presents a general procedural framework for estimating highway noise impacts and for determining whether the existing impacts are acceptable. A basic assumption of the methodology is that the roadway under investigation can be represented as one infinite linear element with constant levels and value of traffic parameters and roadway characteristics.

The first step is to approximate the highway configuration to an infinite straight line perpendicular to the line between the receptor and roadway centerline. The next step is to specify the maximum acceptable noise level (from legislation). For each vehicle class, the reference energy mean emission level (REMEL) is then established. Then the traffic and propagation parameters are used to estimate the unshielded noise experienced at the receptor site due to each noise source, assuming that there were no shield. To obtain the shielded noise due to the effect of any shielding structure(s), the roadway shielding parameters are used to estimate the reduced noise experienced by the receptor. Shielding parameters are described in Table 11.1. Figure 11.7 shows the barrier parameters for a simple barrier, a depressed roadway, and an elevated roadway. The total shielded noise level for all vehicle classes in the traffic stream is then determined by logarithmic summation. The result is then compared to the design noise level to ascertain whether noise standards are violated. The application of this framework is illustrated in the use of the

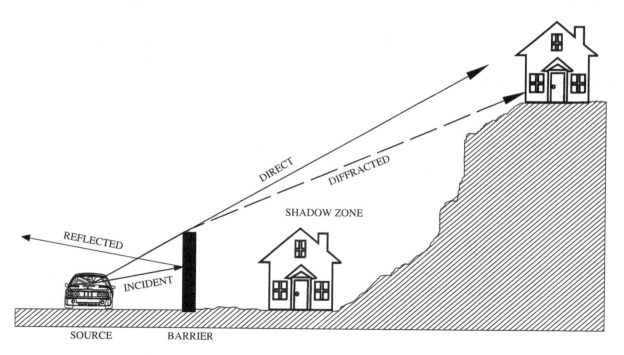

Figure 11.5 Noise barrier effects.

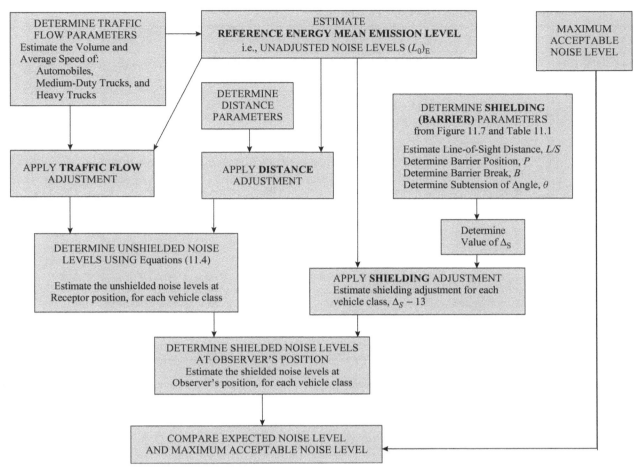

Figure 11.6   Procedure for noise impact estimation.

FHWA noise analysis methodology presented in the next section.

11.5   APPLICATION OF THE PROCEDURE USING THE FHWA MODEL EQUATIONS

Barry and Reagan (1978) provided equations for FHWA noise analysis, most of which are still in use today (FHWA, 2004). A basic *reference energy mean noise emission level* for each vehicle class is first estimated and then adjustments are made for additional acoustical effects due to flow characteristics (e.g., volume and speed), the distance between the roadway and the receptor, the length of the roadway, and the effect of noise shielding or ground effects. Attenuation due to temperature gradients, winds, and atmospheric absorption also occur, but these phenomena are not included. When the distance from the centerline of traffic to the receptor is greater than 15 m, the noise impact at any receptor

position due to a vehicle class i, is

$$L_{eq}(h)_i = \underbrace{(L_0)_{E,i}}_{\substack{\text{reference energy}\\\text{mean emission level}}} + \underbrace{10 \log_{10} \frac{N_i \pi D_0}{S_i T}}_{\substack{\text{traffic-flow}\\\text{adjustment}}} + \underbrace{10 \log_{10} \left(\frac{D_0}{D}\right)^{1+\alpha}}_{\substack{\text{distance}\\\text{adjustment}}}$$

$$+ \underbrace{10 \log \left(\frac{\psi\alpha(\phi_1, \phi_2)}{\pi}\right)}_{\substack{\text{finite roadway}\\\text{adjustment}}} + \underbrace{\Delta_s}_{\substack{\text{shielding}\\\text{adjustment}}} \qquad (11.3)$$

where $L_{eq}(h)_i$ = hourly equivalent sound level for the ith vehicle class

$(L_0)_{E,i}$ = reference energy mean emission level for vehicle class i [see Eqs. (11.4)]

N_i = number of class i vehicles passing a specified point during time T (1h)

S_i = average speed for the ith vehicle class (km/h)

Table 11.1   Definition of Shielding (Barrier) Parameters

Parameter[a]	Definition
Line-of-sight, L/S	In a straight line from receptor (observer) position to noise source. For roadway sources, L/S is drawn perpendicular to the roadway.
	At the source end, L/S must terminate at the proper source height:
	0 ft for autos and medium trucks.
	8 ft for heavy trucks.
	At the receptor end, L/S must terminate at ear height (5, 15, 25, ... ft) above the ground, depending on receptor location.
	The L/S distance is the slant length of the L/S, not the horizontal distance only.
	Figure 11.7 *a* to *c* illustrates how L/S may be determined.
Break in the line-of-sight, B	Is the perpendicular distance from the top of the barrier to the line-of-sight.
	If the line-of-sight slants, the break distance will also slant.
	Figure 11.7 *a* to *c* illustrates how B may be determined.
Barrier position, P	Is the distance from the perpendicular break point in the line-of-sight to the closer end of the line-of-sight. This is also a slant distance.
	Figure 11.7 *a* to *c* illustrates how P may be determined.
Angle subtended θ	Is measured at the receptor in the horizontal plane.
	Is the angle subtended at the ends of the barrier (see Figure 11.7*d*).
	For a barrier always parallel to the roadway, an infinite barrier would subtend 180°.
	For finite barriers, the angle may also be 180° if:
	the barrier ends bend away from the roadway.
	the receptor cannot see the roadway past the barrier ends due to obstacles such as the terrain.

Source: Kugler et al. (1976).

[a]Although the L/S distance and the barrier position distance vary slightly for high and low sources, either one may be used in practice. However, the break in the L/S distance must be measured accurately for high (heavy truck) and low (automobile and medium-duty truck) sources separately.

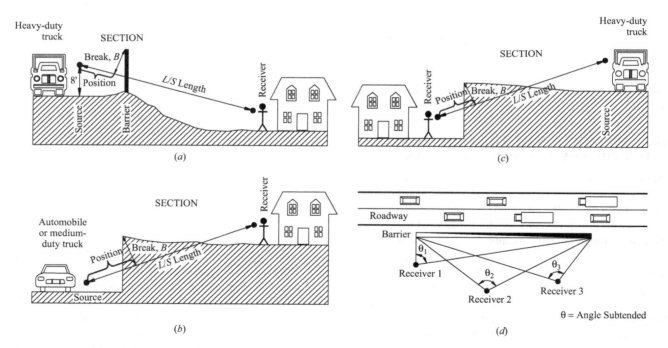

Figure 11.7   Barrier parameters: (*a*) simple barrier; (*b*) depressed roadway; (*c*) elevated roadway; (*d*) plan view.

T = time period over which L_{eq} is sought, in hours (typically, 1 h)

D = perpendicular distance traffic lane centerline to receptor

D_0 = reference distance at which the emission levels are measured; in the FHWA model, D_0 is 15 m

α = site-condition parameter, that indicates the hardness or softness of terrain surface

Ψ = adjustment for finite-length roadways

Δ_s = shielding attenuation parameter due to noise barriers, rows of houses, densely wooded area, etc. (dBA)

11.5.1 Reference Energy Mean Emission Level

For each vehicle class, the reference energy mean emission level (REMEL) can be calculated using equations (11.4).

Automobiles (A):

$$(L_0)_E = 38.1 \log_{10}(S) - 2.4 \qquad (11.4a)$$

Medium-duty trucks (MT):

$$(L_0)_E = 33.9 \log_{10}(S) + 16.4 \qquad (11.4b)$$

Heavy-duty trucks (HT):

$$(L_0)_E = 24.6 \log_{10}(S) + 38.5 \qquad (11.4c)$$

where S is the average vehicle speed in kilometers per hour of each vehicle type. Equations 11.4 are statistical models developed on the basis of a 4-state noise inventory. New REMEL equations can be developed using data obtained through FHWA-prescribed measurement procedures (FHWA, 1978).

11.5.2 Traffic Flow Adjustment

Typically, transportation noise emanates from a continuous stream of vehicles and not a single vehicle. Therefore, an adjustment is necessary to account for this effect as follows:

$$10 \log_{10} \frac{N_i \pi D_0}{S_i T}$$

For example, when $T = 1$ h, $D_0 = 0.015$ km, and S_i is in km/h, the adjustment becomes

$$10 \log_{10} \frac{N_i D_0}{S_i} - 25$$

11.5.3 Distance Adjustment

Predicting the noise level at distances greater than 15 m requires adjustment of the reference energy mean emission

levels to account for the additional distance. The distance adjustment, generally referred to as the *drop-off rate*, is expressed in terms of decibels per doubling of distance (dB/DD). Since the reference energy mean emission levels are equivalent sound levels, the distance adjustment factor can be expressed as:

$$10 \log_{10}(D_0/D)^{1+\alpha}$$

The parameter α is equal to zero under the following conditions: (1) where the source or receptor is located 3 m or more above the ground, irrespective of ground hardness; or (2) whenever the line-of-sight (a direct line between the noise source and the receptor) averages more than 3 m above the ground; or (3) whenever the top of the barrier is 3 m or more in height, irrespective of source or receptor height or ground hardness; or (4) where the height of the line-of-sight is less than 3 m but there is a clear (unobstructed) view of the highway, the ground is hard, and there are no intervening structures.

The parameter α is equal to 0.5 when the view of the roadway is interrupted by isolated buildings, clumps of bushes, or scattered trees, or when the intervening ground is soft or covered with vegetation. For $\alpha = 0$ and 0.5, doubling the distance from the noise source results in a reduction of 3 and 4.5 dbA, respectively.

11.5.4 Adjustment for Finite-Length Roadways

It is often necessary to adjust the basic sound level (REMEL) to account for the energy contribution only from the roadway section visible to the observer at the receptor site. Also, it may be necessary to separate a roadway into sections to account for changes in topography, traffic flow, shielding, and so on. To make these adjustments, the roadway must first be divided into segments of finite length. The finite-length roadway adjustment depends on (1) the orientation of these highway segments relative to the receptor and (2) ground effects.

(a) Orientation of the Highway Segment The positional relationships between the roadway segment and a receptor facing the highway segment are determined by the following procedure: ϕ_1 and ϕ_2 (degrees) are negative if measured to the left of the perpendicular, positive if measured to the right. $\Delta\phi = \phi_2 - \phi_1$. In all cases $\Delta\phi$ is positive and is numerically equal to the included angle subtended by the roadway relative to the receiver. There are three possible cases (illustrated in Figure 11.8): case A, ϕ_1 is negative, ϕ_2 is positive; case B, ϕ_1 is negative, ϕ_2 is negative; case C, ϕ_1 is positive, ϕ_2 is positive.

(b) Ground Effects The nature of the ground affects the adjustment for finite-length roadways just as it does

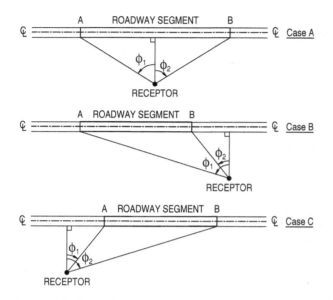

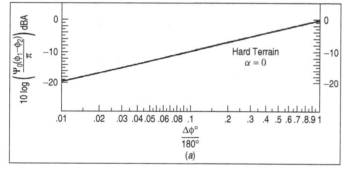

Figure 11.8 Possible positions of receptor relative to roadway segment.

for distance adjustments. The adjustment factor is $10\log_{10}[\psi\alpha(\phi_1, \phi_2)/\pi]$, where $\psi\alpha(\phi_1, \phi_2)$ is a factor related to the finite length of the roadway, ϕ_1 and ϕ_2 are the angles defined in Figure 11.8, and α is the ground hardness parameter. For a terrain with hard (perfectly reflective) ground, $\alpha = 0$ and the adjustment factor is $10\log_{10}(\Delta\phi/\pi)$. For terrains with soft (absorptive) ground, the adjustment term is in the form of a complex function that can be represented by a family of curves from which the adjustment factor values can be obtained. The function for adjustment is shown in equation (11.3) and is illustrated graphically for both the hard and soft sites in Figure 11.9.

11.5.5 Shielding Adjustment

A shielding adjustment is necessary for cases where some physical object is located between the road and the receptor, thus interfering with the propagation of the transportation sound and reducing the level of noise reaching the receptor. Such shields may be natural

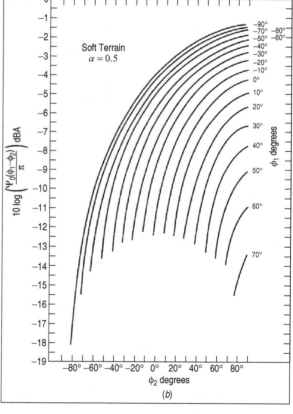

Figure 11.9 Adjustment factors for finite-length roadways and ground effects. (From Barry and Reagan, 1989.)

or human-made, intentional (e.g., earth berms, noise barriers, walls) or unintentional (e.g., large buildings, rows of houses, dense woods, hills). By interrupting sound propagation, noise shields create an acoustic shadow zone. *Field insertion loss* is the difference in the noise levels at a given location with and without a noise shield. When a shield is constructed specifically (i.e., noise barriers), the elements of design include barrier attenuation, barrier shape, and field insertion loss.

(*a*) *Barrier Attenuation and Shape*   The barrier attenuation, shape, and length provided by a freestanding wall can be expressed as a function of a parameter known as the *Fresnel number*, which is a dimensionless quantity occurring in optics, in particular in diffraction theory:

$$\Delta_{B_i} = 10 \log \left(\frac{1}{\phi_R - \phi_L} \int_{\phi_L}^{\phi_R} 10^{-\Delta_i/10} d\phi \right) \quad (11.5)$$

where Δ_{B_i} is the change in noise levels (attenuation) provided by the barrier for the ith class of vehicles, ϕ_R and ϕ_L are angles that establish the relationship (position) between the barrier and the receptor (Figure 11.10), and Δ_i is the point source attenuation for the ith class of vehicles and is given by

- $\Delta_i = 0$

  when $N_i^* \le -0.1916 - 0.0635\varepsilon$;

- $\Delta_i = 5(1 + 0.6\varepsilon) + 20 \log \dfrac{\sqrt{2\pi |N_o|_i \cos\phi}}{\tan \sqrt{2\pi |N_o|_i \cos\phi}}$

  when $-0.1916 - 0.0635\varepsilon \le N_i^* \le 0$;

- $\Delta_i = 5(1 + 0.6\varepsilon) + 20 \log \dfrac{\sqrt{2\pi (N_o)_i \cos\phi}}{\tanh \sqrt{2\pi (N_o)_i \cos\phi}}$

  when $0 \le N_i^* \le 5.03$;

- $\Delta_i = 20(1 + 0.15\varepsilon)$

  when $N_i^* \le 5.03$.   (11.6)

where ϕ is the angle subtended by the perpendicular line from the receptor to the barrier and the line connecting the receptor and the source.

$N_i^* = (N_0)_i \cos\phi$, and ε (a barrier shape parameter) is 0 for a freestanding wall, 1 for an earth berm, where N_0 is the Fresnel number determined along the perpendicular line between the source and the receptor and $(N_0)_i$ is the Fresnel number of the ith class of vehicles determined along the perpendicular line between the source and the receptor. Mathematically, the Fresnel number is defined as

$$N_0 = 2 \frac{\delta_0}{\lambda} \quad (11.7)$$

where δ_0 is the path length difference measured along the perpendicular line between the source and the receptor, λ is the wavelength of the sound radiated by the source. The path-length difference, δ_0, is the difference between a perpendicular ray traveling directly to the receptor and a ray diffracted over the top of the barrier (Figure 11.10),

$$\delta_0 = A_0 + B_0 - C_0 \quad (11.8)$$

where A_0, B_0, and C_0 are the distances shown in Figure 11.11. Note that if the height of the noise source or the observer changes, the path-length difference will also change. For barrier calculations only, the vehicle

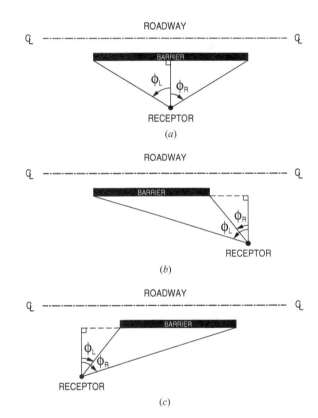

Figure 11.10   Possible positions of barrier relative to receptor: (*a*) case 1; (*b*) case 2; (*c*) case 3.

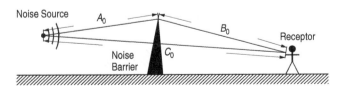

Figure 11.11   Path-length difference, $\delta_0 = A_0 + B_0 - C_0$.

noise sources are assumed to be located at the following positions above the lane centerline: automobiles, 0 m; medium-duty trucks, 0.7 m; and heavy-duty trucks, 2.44 m.

Knowing N_0, ϕ_R, and ϕ_L, the integral in equation (11.5) can be solved. Alternatively, charts provided by Barry and Reagan (1978) could be used. A sample of this is provided in Appendix A11. For infinitely long barriers (i.e., $\phi_L = -90^o$ and $\phi_R = +90^o$), the attenuation provided by the barrier can be read from Figure 11.12 for the positive and negative values of N_0, respectively.

(b) *Field Insertion Loss*   For a reflective ground between the highway and the receptor, the site condition parameter, $\alpha = 0$ (Figure 11.13a), and the drop-off rate is relatively low. If the ground is absorptive, $\alpha = \frac{1}{2}$, an additional attenuation is obtained. When a barrier exists between the highway and the receptor, the top of the barrier appears to be the noise source to the receptor (Figure 11.13b) and the drop-off rate is relatively high. The net reduction due to the barrier (often erroneously referred to as barrier attenuation) is the *field insertion loss*:

$$\text{field insertion loss} = L_{\text{before}} - L_{\text{after}} \qquad (11.9)$$

As the receptor moves away from the barrier, the ground effects obviously start occurring at some point. FHWA

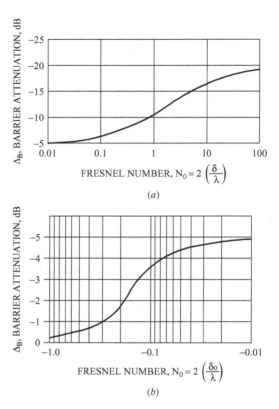

Figure 11.12   Barrier attenuation vs. Fresnel number for noise barriers of infinite length: (a) positive Fresnel number; (b) negative Fresnel number.

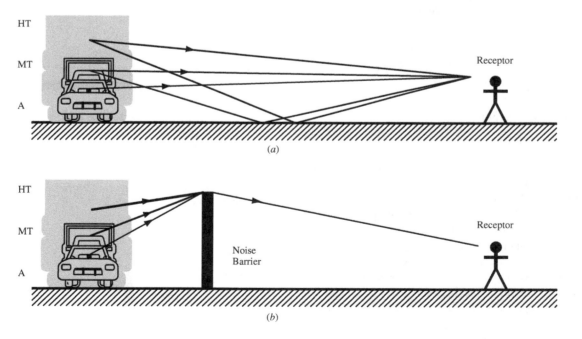

Figure 11.13   Noise paths (a) without and (b) with barrier.

(1978) recommends the use of 1.5 dB/DD as the field insertion loss value for all receptor locations.

11.5.6 Combining Noises from Various Vehicle Classes

In a typical traffic stream, there are several different vehicle classes using the same lane. There is a need, therefore, to combine noise levels from the different noise sources into an equivalent noise level using the relationship presented in equation 11.2, as shown below:

$$L_{eq}(h) = 10 \log \left(10^{L_{eq}(h)_A/10} + 10^{L_{eq}(h)_{MT}/10} \right.$$
$$\left. + 10^{L_{eq}(h)_{HT}/10} \right) \qquad (11.10)$$

The same equation can be used for finding the combined (equivalent) sound level of noises at different distances from the receptor (Barry and Reagan, 1978; Fleming et al., 1995).

Example 11.3 A freeway is proposed to bypass the central business district of a city, but will be located near areas with residences, schools, and hospitals in the city's suburbs. In preparation for a public hearing, the city's planning authority has requested an evaluation of the noise impacts with and without a noise barrier at the locations affected. The city engineer has supplied a perspective drawing (Figure E11.3.1), (which shows the depressed location of the freeway) and a data summary for your use. Use the FHWA model equation (11.3), to determine whether the L_{eq} level at the property line exceeds the maximum mean sound pressure level (dBA) recommended by FHWA (Bowlby et al., 1985, Table 8) for a "residential" type of land use. Assume that the receivers are located 250 ft from the centerline of the freeway on either side. Note that the sidewall of the freeway (1000 ft in length) can be considered as a barrier for 500 ft on each side of the receiver.

Compute the sound level at the receptor in a segment of the freeway 1000 ft (305 m) under the following conditions: (a) without a barrier (i.e., free field) and (b) with a concrete barrier 8 ft (2.44 m) high. The terrain between the freeway and the observer is paved ($\alpha = 0$). The width of the lane is 12 ft (3.66 m). The distance between the lane centerline and the barrier is 32 ft (9.75 m), and the distance between the observer and the centerline of the lane is 250 ft (76.2 m).

SOLUTION   (a) *Without a barrier*: The reference energy mean emission levels $(L_0)_{E,i}$ for each vehicle class i is determined as follows:

Automobiles: $38.1 \log_{10}(90) - 2.4 = 72.1$ dBA

Medium-sized trucks: $33.9 \log_{10}(80) + 16.4 = 80.9$ dBA

Heavy trucks: $24.6 \log_{10}(80) + 38.5 = 85.3$ dBA

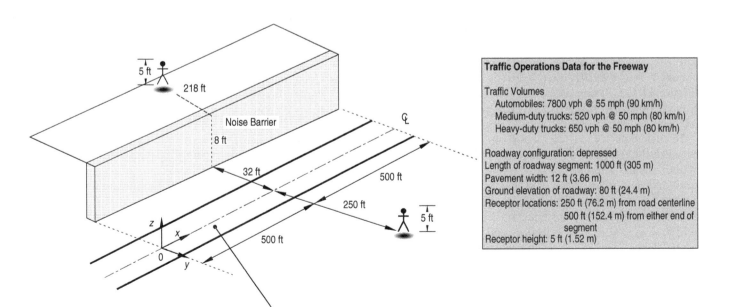

Figure E11.3.1 Freeway layout for noise analysis.

For automobiles:

Traffic flow adjustment: $10 \log(7800 \times 15/90) - 25$
$= 6.1$ dBA

Distance adjustment: $10 \log[(15/(76.2)^{1+0}]$
$= -7.1$ dBA

Adjustment for finite road length: ϕ_2 and ϕ_1 are both 64.572°, because the receptor is equidistant from the start and end of the segment and case A is applied; $\Delta\phi = \phi_2 - \phi_1 = 64.572 - (-64.572) = 129.144°$ and 129.44°/180° is 0.72. The adjustment for finite road length, then, is -2 dBA (from Figure 11.9).

Shielding adjustment: There is no shielding adjustment in this case, so $\Delta_s = 0$

Therefore, the $L_{eq}(h)$ for automobiles is $L_{eq}(h)_A = 72.1 + 6.1 - 7.1 - 2 = 69.1$ dBA. Similar calculations for medium- and heavy-duty trucks yield $L_{eq}(h)_{MT} = 66.7$ dBA and $L_{eq}(h)_{HT} = 72.1$ dBA, respectively. Therefore, the total hourly equivalent noise from all three vehicle classes is

$$L_{eq}(h)_{EB} = 10 \log(10^{6.91} + 10^{6.67} + 10^{7.21})$$
$$= 74.7 \text{ dBA}$$

(b) *With a barrier of given specifications:* height of barrier = 2.44 m, height of receptor = 1.524 m. Vehicle heights from road surface: automobile, 0; medium-duty trucks, 0.7 m; heavy-duty trucks, 2.44 m. Distance between the centerline of the lane and barrier = 9.754 m, and distance between the observer and barrier = 76.2 m − 9.754 m = 66.446 m. Using the information above, A_0, B_0, and C_0 can be calculated (Figure 11.11) and the value of δ for three types of vehicles are obtained: $\delta_{oA} = 0.291$, $\delta_{oMT} = 0.155$, and $\delta_{oHT} = 0.001$ (Figure 11.12). Using equation 11.7, the Fresnel numbers are obtained: $N_{oA} = 0.935$, $N_{oMT} = 0.5$, and $N_{oHT} = 0.003$.

To calculate $N_i^* = (N_0)_i \cos\phi$, Φ is equal to zero because the receptor is located directly in the middle of the barrier. Therefore, $N_i^* = (N_0)_i . N_0$ is greater than zero and less than 5.03, and $\varepsilon = 0$ since the barrier material is concrete. The values of Δ_{B_i} for A, MT, and HT are obtained from equation 11.5: $\Delta_{B_{AUTOMOBILES}} = -12.7$, $\Delta_{B_{MEDUIM \ TRUCKS}} = -10.0$, and $\Delta_{B_{HEAVY \ TRUCKS}} = -5.0$. Thus, $L_{eq}(h)$ for each vehicle class is calculated as follows: $L_{eq}(h)_A = 69.1 - 12.7 = 56.4$ dBA, $L_{eq}(h)_{MT} = 66.7 - 10 = 56.4$ dBA, and $L_{eq}(h)_{HT} = 72.1 - 5.0 = 67.1$ dBA. The total $L_{eq}(h)$ is then found using equation (11.10) as 67.8 dBA.

The reduction in noise level due to the barrier is $74.7 - 67.8 = 6.9$ dBA.

11.6   APPLICATION OF THE PROCEDURE USING THE TRAFFIC NOISE MODEL (TNM) SOFTWARE PACKAGE

11.6.1 The Traffic Noise Model

A number of software packages have been developed using the framework discussed in Section 11.5. In the early 1980s, the FHWA sponsored the development of STAMINA and OPTIMA, a set of models for noise analysis and cost-effectiveness evaluation of alternative noise barrier designs (Bowlby et al., 1983). STAMINA subsequently evolved into its current-day version, the traffic noise model (TNM). TNM, a state-of-the-art computer program with updated and expanded noise parameters (Menge et al., 1998), is currently used at many agencies for modeling and predicting highway noise and for designing highway noise barriers (FHWA, 2004). TNM contains the following functional components:

- Noise modeling for five standard vehicle types: automobiles, medium-duty trucks, heavy-duty trucks, buses, motorcycles, and for user-defined vehicles
- Noise modeling for constant-flow and interrupted-flow traffic using a database developed from field data
- Noise modeling of the effects of different pavement surface types
- Sound-level computations based on a one-third octave-band database and algorithms
- Graphically interactive noise barrier design and optimization
- Modeling of attenuated sound over/through rows of buildings and dense vegetation
- Multiple diffraction analysis
- Parallel barrier analysis
- Contour analysis, including sound-level contours, barrier insertion loss contours, and sound-level difference contours

These components are supported by an experimentally calibrated acoustic computation methodology and a comprehensive database. TNM includes a graphical interface for easy data and output management (USDOT, 2004). The data input is menu-driven using a digitizer, mouse, and/or keyboard. Also, users can import STAMINA 2.0 or OPTIMA files, as well as roadway design files saved in CAD and DXF formats.

The TNM model divides the road section into segments. For each segment, the software calculates the sound levels at user-defined receptor locations for each of the specified vehicle types that use the road. These sound levels are then aggregated at the receptor location to obtain the total sound impact. TNM features include handling capabilities for complex highway geometry and some atmospheric effects. The output of the model is L_{eq} and L_{10} (A-weighted) 1-hour sound levels at receptor locations specified. These values can then be compared to existing standards to ascertain compliance to noise regulations. Furthermore, TNM allows the user to design effective noise barriers by changing barrier heights, eliminating barrier segments, and establishing other design scenarios and evaluating the effects of these scenarios in terms of overall noise levels at the receptor locations specified. In addition, a cost file can be included by the user to determine the relative cost of the barrier design scenarios. That way, consideration of both effectiveness and cost in the noise barrier configuration and design is facilitated (FHWA, 2004).

Example 11.4 Compute the noise level with and without the barrier for the freeway segment in Example 11.3, using the TNM software package.

SOLUTION The TNM input data, which is based on the given data for the problem, is shown in Figure E11.4.1. The corresponding output, which is shown as Figure E11.4.2, indicates that the without-barrier sound level at the specified receptor position is 73.2 dbA. For the 8 ft high noise barrier, TNM estimates a 67.1 dBA sound level at the specified receptor position. This indicates that the sound reduction due to the barrier is 6.1 dBA. The noise levels predicted using the FHWA equations is comparable to those given by the traffic noise model.

11.7 ESTIMATING NOISE IMPACTS FOR OTHER MODES

11.7.1 Transit Noise and Vibration

Sources of transit noise include vehicle operations and increased vehicle traffic due to a transit improvement. With regard to vehicle operations, the noise is from propulsion units (such as diesel bus engines and electric traction motors in rapid transit cars), road tires, and the steel wheels of railcars. Also, guideway structures make noise when they vibrate under moving transit vehicles. Other noise sources include vehicle equipment that continues to operate even when the transit vehicle is stationary, such as fans, radiators, and air-conditioning pumps. Furthermore, noise is generated by transit maintenance operations and ventilation fans in transit stations and subway tunnels.

Depending on the transit project type and scale, the stage of project development, and the environmental setting, three analysis levels may be used in a noise impact evaluation: *screening*, to determine the need

for a more extensive noise impact analysis and to identify any noise-sensitive uses; *general assessment*, to estimate the predicted ambient noise levels at sensitive locations and the affected populations; and a *detailed noise analysis* using established models. Noise prediction techniques are described in the FTA's *Transit Noise and Vibration Impact Assessment* document, guidance manual, and spreadsheet (Harris Miller Miller & Hanson, 1995) and the FHWA highway traffic noise model for busway projects (FHWA, 2004).

11.7.2 Air Transportation

Noise pollution associated with air transportation is evaluated in a manner similar to that for highways. Pollution from aircraft occurs primarily when they are close to the ground (Figure 11.14). As with highway noise, it is required to estimate the expected increase in noise levels and to compare the new levels to the established criteria. A level of 65 dB (A-weighted) L_{dn} is the established impact criterion and is based on yearly average operations at the airport. Airport noise is usually predicted using the integrated noise model (Flythe, 1982) or the NOISEMAP computer model (Moulton, 1990). Similar to the case for highways, adjustments to reference levels are made for geometry, traffic, and environmental conditions. Air traffic noise levels at user-defined locations can be estimated with the model and expressed in the form of noise contours as shown in Figure 11.15. Contours may be plotted for a specified noise threshold level to determine areas near the airport that are affected significantly.

The Federal Aviation Administration (FAA) has developed the Airport Noise Compatibility Planning Toolkit, (FAA, 2006) which is designed to aid regional offices in assisting state and local officials and interested organizations for airport noise compatibility planning around U.S. airports. The toolkit may be used in conjunction with the noise prediction capabilities of Integrated Noise Model (INM).

11.7.3 Rail Transportation

Rail noise can be predicted on the basis of the following relationship:

$$L_{max} = 30\log(V) + C \qquad (11.11)$$

where: V is the speed of the train in miles per hour, and C is a constant. In this situation, L_{max} is at a defined location for a single passby. Adjustments for site-specific conditions (e.g., distance) must be calculated as for highway vehicles. For the newer high-speed rail, the recommended relationship is $L_{max} = 40\log V$ (Wayson and Bowlby, 1989). Noise measurements at railway yards may be used for future predictions by "scaling" the levels

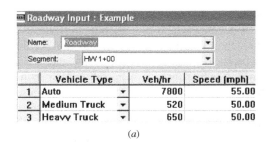

(a)

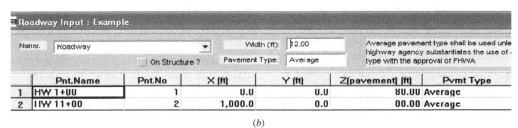

(b)

Receiver Input : Example

Default Receiver Settings

Dwelling Units: 1   Height Above Ground (ft): 80.00

	Receiver Name	Seq. #	X (ft)	Y (ft)	Z(ground) (ft)	Dwelling Units	Height (ft)
1	Receptor	2	500.00	249.87	80.00	1	5.00

(c)

Barrier Input : Example

Name: Roadway Barrier   Pert. Increment (ft): 0.00   # Pert. Up: 0   # Pert Dn: 0

Barrier Type: Wall   Height (ft): 0.00   Min. Height (ft): 0.00   Max. Height (ft): 99.99

	Pnt.Name	Pnt.No	X (ft)	Y (ft)	Z(bottom) (ft)	Height (ft)	Increment (ft)	#Up	#Dn
1	Barrier 1+00	1	0.0	32.0	80.00	8.00	2.00	3	3
2	Barrier 2+00	3	100.0	32.0	80.00	8.00	2.00	3	3
3	Barrier 3+00	4	200.0	32.0	80.00	8.00	2.00	3	3
4	Barrier 4+00	5	300.0	32.0	80.00	8.00	2.00	3	3
5	Barrier 5+00	6	400.0	32.0	80.00	8.00	2.00	3	3
6	Barrier 6+00	7	500.0	32.0	80.00	8.00	2.00	3	3
7	Barrier 7+00	8	600.0	32.0	80.00	8.00	2.00	3	3
8	Barrier 8+00	9	700.0	32.0	80.00	8.00	2.00	3	3
9	Barrier 9+00	10	800.0	32.0	80.00	8.00	2.00	3	3
10	Barrier 10+00	11	900.0	32.0	80.00	8.00	2.00	3	3
11	Barrier 11+00	2	1,000.0	32.0	80.00	8.00	2.00	3	3

(d)

Figure E11.4.1 Samples of input files for the traffic noise model: (*a*) volume and speed data by vehicle class; (*b*) highway segment boundary coordinates and pavement data; (*c*) receptor coordinates and characteristics; (*d*) barrier coordinates and characteristics.

according to expected future use. The Federal Transit Administration (FTA) developed prediction guidelines for urban rail noise (Harris Miller Miller & Hanson, 1995).

The *High-Speed Ground Transportation Noise and Vibration Impact Assessment Manual and Spreadsheet*, released by the Federal Railroad Administration, provides necessary criteria and procedures. This resource can be used for analyzing potential noise and vibration impacts resulting from various types of proposed high-speed ground transportation projects, including high-speed trains using traditional steel-wheel on steel-rail technology and magnetically levitated systems (FRA, 2005).

Sound Levels : Noise Example:													

Civil Engineering, Purdue University

M. Issa M.

28 May 2006

TNM 2.5

Calculated with TNM 2.5

RESULTS: SOUND LEVELS

PROJECT/CONTRACT:

RUN:

BARRIER DESIGN:

Example 11-3

Transportation Noise Analysis Example

INPUT HEIGHTS

Average pavement type shall be used unless a State highway agency substantiates the use of a different type with approval of FHWA.

ATMOSPHERICS: 68 deg F, 50% RH

Receiver Name	No.	#DUs	Existing LAeq1h	No Barrier LAeq1h Calculated	Crit'n	Increase over existing Calculated	Crit'n Sub'l Inc	Type Impact	With Barrier Calculated LAeq1h	Noise Reduction Calculated	Goal	Calculated minus Goal
			dBA	dBA	dBA	dB	dB		dBA	dB	dB	dB
Receptor	2	1	73.0	73.2	66	0.2	10	Snd Lvl	67.1	6.1	6	0.1

Dwelling Units	# DUs	Noise Reduction Min dB	Avg dB	Max dB
All Selected	1	6.1	6.1	6.1
All Impacted	1	6.1	6.1	6.1
All that meet NR Goal	1	6.1	6.1	6.1

Figure E11.4.2   Traffic noise model output (with and without barrier)

Figure 11.14   Noise from air traffic continues to be a nuisance at residential areas proximal to airports.

11.7.4 Marine Noise

The Marine Mammal Protection Act of 1972 was passed to regulate marine noise and its potential impacts. However, there is increasing concern regarding the possibility that human-generated noise through marine transportation and associated activities are causing destruction of marine mammals or interfering with their normal activities (NRC, 2003). Recent reports on the beaching of whales and dolphins beached due to deployment of human-generated sounds at nearby locations, have raised questions about the impact of ocean noise. Unfortunately, there seems to be inadequate data at the present time to establish any causal relationships.

11.7.5 General Guidelines for Noise Impact Evaluation of New Transportation Improvements

Analysis of noise impacts are of interest to two major stakeholders: the transportation agency or provider and the general public. Key elements of a traffic noise impact study are as follows (USDOT, 1995):

- Determination of the impact criteria and identification of noise-sensitive land-use features
- Determination of existing noise levels
- Prediction of future noise levels for each study alternative
- Impacts for each study alternative
- Identification and evaluation (cost-effectiveness) of alternative noise abatement strategies

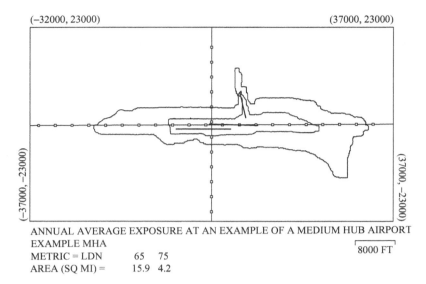

(−32000, 23000)   (37000, 23000)

(−37000, −23000)   (37000, −23000)

ANNUAL AVERAGE EXPOSURE AT AN EXAMPLE OF A MEDIUM HUB AIRPORT
EXAMPLE MHA
METRIC = LDN   65   75   8000 FT
AREA (SQ MI) =   15.9 4.2

Figure 11.15   Typical noise contour plot from INM. (From FAA, 2006.)

After the impact criteria and noise-sensitive land-use features have been identified, existing noise levels should be established by field measurements for all developed land uses (Lee and Fleming, 1996). Field measurements are considered important because background noise is typically a composite from many sources, and noise prediction models are applicable only to noise emanating from a specific source. In the case of highway transportation, for example, if the field measurements establish that existing noise levels at the locations of interest are due primarily to highway traffic, existing noise levels may be calculated using the FHWA highway traffic noise prediction model. Next, noise levels expected to occur as a result of any new construction should be estimated for each alternative. Existing and predicted noise impacts for each alternative are then compared to identify the noise levels expected to occur solely as a result of the transportation improvement. If necessary, the cost-effectiveness of various alternative noise mitigation strategies should also be evaluated.

11.8   MITIGATION OF TRANSPORTATION NOISE

Noise mitigation techniques can be classified into three groups: at-source, propagation path, and at-receptor. The practical solution of a noise problem may involve techniques from more than one group. Noise abatement techniques at the source (vehicles) are in a continual phase of development. Examples include the direct control of engine and other vehicle noise, open–graded or rubber-based asphalt pavement for tire noise control. Highway traffic noise reduction may also occur at-source by adopting certain traffic management techniques, such as speed reduction and limiting truck operation at streets in sensitive areas. Near airports, noise may be controlled at the source through the implementation of aircraft noise specifications.

Noise reduction may also be achieved by enforcing sufficient right-of-way distances. Establishment of greenbelt buffer zones contributes to reduction of noise levels at the receptor locations. This option is popular with nearby residents but is rather costly in large urban areas because of the extent of relocation required and the cost of land.

A more promising abatement method for highways and railways may be achieved by changes in the vertical or horizontal alignment so that the transportation facility avoids noise-sensitive areas. Another more cost-effective abatement measure is the diffraction of the sound wave by noise barriers, which has been used extensively in North America, Europe, Japan, and Australia. Typical traffic noise barrier designs are shown in Figures 11.16 and 11.17.

Regarding the rail noise abatement, diffraction methods, reduction of rail squeal, dampening of track noise, and increased separation distance have all been applied, and rail noise barriers have been found to be reasonably efficient because of possible placement in the vicinity of the guided path (Wayson, 2003). New designs with nonflat surfaces and crumb rubber coatings show promising improvements over traditional flat concrete or wood panels. Some barriers consist of lightweight hollow panels made of composite material filled with crumb rubber.

Figure 11.16 Highway noise barrier: post and panel. (From FHWA, 2005.)

Figure 11.17 Highway noise barrier: combination of noise berm and noise wall. (From FHWA, 2005.)

Solid noise barriers have been found more effective than fences or vegetation (Cowan, 1999). However, it should be realized that even the most effective sound barriers provide only 15 to 20 dBA of noise reduction because a significant amount of sound energy travels over and around barriers. Also, noise barriers are most effective only up to 200 to 300 ft; if the distance between a noise source and receptor exceeds 300 ft, a noise barrier is not likely to be effective.

Sometimes it is not practical or possible to lessen traffic noise at source or along the propagation medium. In these cases, noise insulation may be provided at the receptor. For instance, with improved building insulation (e.g., in walls and roof, double-paned or thicker-pane windows, acoustic vents, storm doors) a typical wood-framed residence can gain a decrease of 40 dB (Wayson, 2003). Many homes, clinics, and schools near highways and airports have been insulated similarly.

11.8.1 Noise Barrier Cost Estimates

The effectiveness (noise reduction) and costs (per unit area) of noise barriers differ by barrier shape, inclination, orientation, and most important, by their material, length, and height. As such, the effectiveness of noise barriers in reducing unwanted sound to tolerable levels must be accompanied by an analysis of their construction and maintenance costs to ascertain their overall cost-effectiveness. The preliminary costs of noise barriers include the engineering costs (feasibility studies, design, etc.) and construction costs (excavation, embankment, materials, surveying, mobilization, demobilization, and traffic maintenance), construction contingency, and administration. As an alternative to such detailed item-by-item accounting costs, approximate aggregate noise barrier cost estimates can be established on the basis of barrier material type, length, and height. From historical data (USDOT, 1995) it is seen that barrier costs range from \$0.55 to \$5.44 million per mile. More recent cost data are on the order of \$25 to \$30 per square foot in 2005 dollars. Better still, barrier-cost models can be developed using historical contract data. Using the actual construction cost and cost per mile of a type II noise barrier from 17 U.S. states, a plot of cost per mile as a function of length is generated (Figure 11.18). The figure demonstrates the existence of scale economies: Shorter noise barriers have higher unit costs than those of longer noise barriers. A noise barrier cost model (in 2002 dollars) developed from the data is shown below:

$$\text{cost(\$M)/mi} = -0.7269\ln(\text{length of barrier, in mi})$$
$$+ 4.5117 \tag{11.12}$$

Although equation (11.12) does not show the cost variation by material type, this factor can be a significant determinant of noise barrier costs (Table 11.2). The

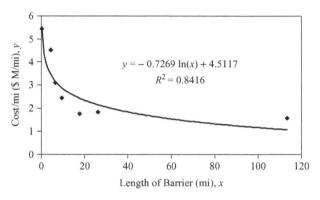

Figure 11.18 Plot of cost/mi as a function of barrier length.

Table 11.2   Noise Barrier Construction Material Average Unit Cost by Height[a]

Height (ft)	Concrete	Block	Wood	Metal	Berm	Brick	Combination	Absorptive	All Materials
≥ 30	26.26	7.16	5.97	—	1.19	—	11.94	—	15.52
27–29	23.88	—	—	—	—	—	16.71	—	21.49
24–26	19.10	13.13	22.68	22.68	—	—	21.49	16.71	19.10
21–23	29.84	—	34.62	—	3.58	—	15.52	25.07	27.46
18–20	28.65	27.46	15.52	14.33	10.74	22.68	19.10	26.26	25.07
15–17	23.88	25.07	21.49	17.91	5.97	29.84	21.49	33.43	23.88
12–14	23.88	20.29	21.49	15.52	5.97	22.68	23.88	34.62	21.49
9–11	26.26	25.07	17.91	26.26	5.97	32.23	20.29	44.17	22.68
6–8	21.49	20.29	21.49	21.49	7.16	32.23	22.68	27.46	20.29
<6	25.07	22.68	19.10	34.62	14.33	—	48.94	118.18	22.68
All	25.07	22.68	20.29	19.10	5.97	27.46	20.29	31.04	22.68

Source: Adapted from FHWA (2002).
[a] Values are in 2005 dollars per square foot.

average unit cost ranges between $13 and $23 and $ per square foot for all barrier materials, with the exception of earth berms, which have an average cost of $5/ft². The most commonly used concrete barrier has a cost of $18/ft², while a brick barrier has an average cost of $20/ft². The average costs for wood, metal, absorptive, and combination barriers are $14, $13, $23, and $15 per square foot, respectively. There is a substantial difference in average barrier construction cost across the states. These unit costs by state and year of construction are provided in FHWA (2002).

Example 11.5   In anticipation of increased traffic volumes due to planned road rehabilitation, a 1.5-mile-long noise barrier is proposed on the south-side portion of Lafayette Road, located adjacent to the Swam Farm neighborhood. Prepare a rough cost estimate for the noise barrier. Use equation (11.12) or the plot provided in Figure 11.18. Assume typical barrier construction.

SOLUTION   Using the equation: cost ($M)/mi = −0.7269ln(barrier length) + 4.5117 = −0.7269ln(1.5) + 4.5117 = $4.217/mi

Construction cost estimate = ($4.217/mi)(1.5)

$$= \$6.22 \text{ M}$$

Example 11.6   An interstate exit ramp is planned for construction near the rapidly growing city of Decatur and construction is due to start next year. An elementary school is located near the site of the proposed ramp. In order to shield the school from the ramp traffic noise, a

brick noise barrier wall system is planned. The barrier will be 600 ft in length and 13 ft in height. Determine the approximate cost of the proposed noise barrier.

SOLUTION   From Table 11.2, the average cost of a 13-ft-high brick barrier = $22.68/ft². Thus,

$$\text{cost} = (\text{area of barrier})(\text{average unit cost})$$
$$= (600 \times 13)(\$22.68/\text{ft}^2)$$
$$= (7800\text{ft}^2)(\$22.68/\text{ft}^2) = \$176,904$$

11.9   LEGISLATION AND REGULATIONS RELATED TO TRANSPORTATION NOISE

In the United States, federal legislation for noise pollution was passed in the 1960s and 1970s and is still in effect. The Housing and Urban Development Act of 1965, reinforced with the Noise Control Act of 1972, mandated the control of urban noise. Also, the Quiet Communities Act of 1978 better defined and added to the requirements of the Noise Control Act. This environmental legislation required the consideration of noise pollution for all modes of transportation and engendered the development of methodologies for noise measurement, assessment, and evaluation.

The EPA has defined a desirable neighborhood noise level as 55 dB (A-weighted) L_{dn} (USEPA, 1974).

The FHWA has established limits for the consideration of noise mitigation for new or expanded highway projects. Such limits are referred to as *noise abatement criteria* and are a function of land use and maximum 1-hour L_{eq} values (USDOT, 1995), as shown in Table 11.3. The

Table 11.3 FHWA Noise Abatement Criteria

Activity Category	$L_{eq}(h)^a$	$L_{10}(h)^a$	Description of Activity Category
A	57 (exterior)	60 (exterior)	Lands on which serenity and quiet are of extraordinary significance and serve as an important public need and where the preservation of those qualities is essential if the area is to continue to serve its intended purpose.
B	67 (exterior)	70 (exterior)	Picnic area, recreation areas, playgrounds, active sports areas, parks, residences, motels, hotels, schools, churches, libraries, and hospitals.
C	72 (exterior)	75 (exterior)	Developed lands not included in categories A and B above.
D	—	—	Undeveloped lands.
E	52 (interior)	55 (interior)	Residences, motels, hotels, public meeting rooms, schools, churches, libraries, hospitals, and auditoriums.

Source: USDOT (1995).

[a]Values are hourly A-weighted sound levels in decibels. Either $L_{eq}(h)$ or $L_{10}(h)$, but not both, may be used on a project. The sound levels indicated above are to be used only to determine impacts. Values are absolute levels where abatement must be considered.

Table 11.4 Railway Noise Standards

All Locomotives Manufactured on or before December 31, 1979

Paragraph and Section	Noise Source	Noise Standard, A-Weighted Sound Level (dB)	Noise Measure[a]	Measurement Location
201.11(a)	Stationary, idle throttle setting	73	L_{max}(slow)	30 m (100 ft)
201.11(a)	Stationary, all other throttle settings	93	L_{max}(slow)	30 m (100 ft)
201.12(a)	Moving	96	L_{max}(fast)	30 m (100 ft)

All Locomotives Manufactured after December 31, 1979

Paragraph and Section	Noise Source	Noise Standard, A-Weighted Sound Level (dB)	Noise Measure[a]	Measurement Location
201.11(b)	Stationary, idle throttle setting	70	L_{max} (slow)	30 m (100 ft)
201.11(b)	Stationary, all other throttle settings	87	L_{max} (slow)	30 m (100 ft)
201.12(b)	Moving	90	L_{max} (fast)	30 m (100 ft)
201.11(c) and 201.12(c)	Additional requirement for switcher locomotives manufactured on or before December 31, 1979 operating in yards where stationary switcher and other locomotive noise exceeds the receiving property limit of:	65	L_{90} (fast)[b]	Receiving property
201.11(c)	Stationary, idle throttle setting	70	L_{max} (slow)	30 m (100 ft)
201.11(c)	Stationary, all other throttle settings	87	L_{max} (slow)	30 m (100 ft)
201.12(c)	Moving rail cars	90	L_{max} (fast)	30 m (100 ft)
201.13(1)	Moving at speeds of 45 mph or less	88	L_{max} (fast)	30 m (100 ft)
201.13(2)	Moving at speeds greater than 45 mph	93	L_{max} (fast)	30 m (100 ft)

Source: FRA (2002).

[a] L_{max} = Maximum sound level; L_{90} = statistical sound level exceeded 90% of the time.

[b] L_{90} must be validated by determining that $L_{10}-L_{99}$ is less than or equal to 4 dBA.

requirement is that when the noise abatement criteria are approached or exceeded, noise mitigation must be considered.

The Control and Abatement of Aircraft Noise and Sonic Boom Act of 1968 mandated noise emission limits on aircraft beginning in 1970. The standard for new aircraft created stage classifications of aircraft based on their noise emission levels. Stage I (relatively noisy) aircraft have mostly been phased out in the United States. New regulations, in the form of 14CFR91 (*Transition to an All Stage III Fleet Operating in the 48 Contiguous United States and the District of Columbia*) and 14CFR161 (*Notice and Approval of Airport Noise and Access Restrictions*) call for the fast phase-in of the quieter stage III aircraft. In 1979, the Aviation Safety and Noise Abatement Act placed more responsibility on local and regional airport authorities. All stage III aircraft in current operation must meet separate standards for runway takeoffs, landings, and sidelines, which range from 89 to 106 dBA depending on the aircraft's weight and number of engines (Bearden, 2000). The Airport Noise Control and Land Use Compatibility Planning process included in federal aviation regulations allows federal funds to be allocated for noise abatement purposes. The FAA has also implemented a program that requires computer modeling for environmental analysis and documentation. Impacts are defined to occur if the L_{dn} is predicted to be above 65 dB (A-weighted).

The Federal Railroad Administration has developed noise emission standards for railway transportation (Table 11.4) to circumvent hindrances to interstate commerce caused by inconsistent local ordinances.

In addition to the administrative regulations of the USDOT, other criteria or regulations may be applicable, such as the guidelines established by the Department of Housing and Urban Development (HUD) to protect housing areas. The HUD site acceptability standards use L_{dn} (A-weighted) and noise levels are acceptable if less than 65 dB, normally unacceptable from 66 to 75 dB, and unacceptable if above 75 dB. In addition to the federal government, many state governments have also issued noise control guidelines. Local municipalities, through various zoning and administrative guidelines, may restrict the development of residential and other noise-sensitive land uses in the vicinity of noise generators, such as highways and airports. However, noise controls are not always enforced at the local level (Wayson, 2002).

SUMMARY

The human ear evaluates the perception of sound on the basis of loudness, frequency, duration, and subjectivity.

Loudness is related directly to the amplitude of the pressure fluctuations transmitting through the air and is expressed in sound pressure levels that are added logarithmically. Noise duration and the variation of traffic noise with time are considered important for noise assessment, and some effective descriptors of temporal sound variation include the maximum sound level, various statistical sound levels, equivalent sound level, and day–night level. For the various modes of transportation, the sources of noise generated from a moving flow unit (car, truck, airplane) include interaction between the flow unit and the surrounding medium (air), the tire (or undercarriage mechanism of the flow unit), the guideway (pavement, rail track, runway), the engine of the flow unit, the speed of the flow unit, the exhaust systems of the flow unit, the horns, and the brakes of the flow unit.

Transportation noise impact assessment is based on the use of reference emission levels that are established based on average values of noise levels and frequency spectra occurring from defined transportation sources, which are subsequently corrected for geometric, environmental, and traffic-related variables. After the impact criteria and noise-sensitive land-use features have been identified, existing or expected noise levels should be established. Noise mitigation techniques can be classified into three groups: at-source propagation path, and at-receptor; and the practical solution for a given noise problem may involve techniques from more than one group.

EXERCISES

11.1. Determine the impact of moving a receptor to a distance three times the original distance from a point source of transportation noise. The impacts should be stated in terms of an increase or decrease in sound pressure level at the receptor. State any assumptions made. Assume that the noise source is a line source, such as a constant hum of traffic on an urban expressway, and determine the impact of moving the receptor to a distance three times the original distance from the line source.

11.2. A residential neighborhood near an airport typically receives a noise level of 105 dB from each jet landing or taking off. It is also adjacent to a major expressway, from which it receives a noise level of 80 dB. What is the worst combined sound pressure level received by the neighborhood?

11.3. Assume an infinite highway in a residential area carrying, at an average speed of 55 mph, 3000 automobiles, 200 medium-weight trucks, and 150 heavy trucks per hour during the daytime.

The distance between the observer and the center-line of the lane is 200 ft. Determine whether the L_{eq} level at the property line exceeds the maximum mean sound pressure level (dBA) recommended by FHWA for a residential type of land use.

11.4. A 4.5-km- long barrier wall system is proposed along a section of State Road 334. Estimate the construction cost of the project. Assume typical barrier construction.

11.5. A wood noise barrier wall system is proposed along State Road 10 to attenuate the noise level upon a Jefferson neighborhood. The barrier will be 8800 ft long and 20 ft high. Prepare a rough construction cost estimate of the facility.

REFERENCES[1]

*Barry, T. M., Reagan, J. A. (1978). *FHWA Highway Traffic Noise Prediction Model*, FHWA-RD-77-108, Federal Highway Administration, U.S. Department of Transportation, Washington, DC.

Bearden, D. M. (2000). *RS20531: Noise Abatement and Control: An Overview of Federal Standards and Regulations*, CRS Rep. for U.S. Congress, Washington, DC http://www.ncseonline.org/NLE/CRSreports/Risk/. Accessed May 10, 2006.

Bowlby, W., Higgins, J., Reagan, J., Eds. (1983). Noise barrier cost reduction procedure, in *STAMINA 2.0/OPTIMA User's Manual*, FHWA-DP-58-1, Federal Highway Administration, U.S. Department of Transportation, Washington, DC.

Bowlby, W., Wayson, R. L., Stammer, R. (1985). *Predicting Stop and Go Traffic Levels*, NCHRP Rep. 311, Transportation Research Board, National Research Council, Washington, DC.

*Cohn, L. F., McVoy, G. R. (1982). *Environmental Analysis of Transportation Systems*, Wiley, New York.

Cowan, J. P. (1999). Planning to minimize highway noise impacts. *Proc. American Planning Association National Planning Conference*, Seattle, WA, http://www.asu.edu/caed/proceedings99/. Accessed Dec. 30, 2005.

*FAA (2006). Airport Noise Compatibility Planning Toolkit, Federal Aviation Administration, Washington, DC, http://www.faa.gov/about/office_org/headquarters_offices/aep/planning_toolkit/.

FHWA (1978). *Determination of Reference Energy Mean Emission Levels*, FHWA-OEP/HEV-78-1, Federal Highway Administration, Washington, DC.

———— (1980). *Fundamentals and Abatement of Highway Traffic Noise*, Federal Highway Administration, Washington, DC.

———— (2002). *Highway Traffic Noise Barrier Construction Trend*, Federal Highway Administration, U.S. Department of Transportation, Washington, DC, http://www.fhwa.dot.gov/environment/noise/barrier/trends.htm. Accessed Dec. 30, 2005.

———— (2004). *Memorandum—Highway Traffic Noise: Release and Phase-in of the FHWA Traffic Noise Model Version 2.5*, Federal Highway Administration, U.S. Department of Transportation, Washington, DC, http://www.fhwa.dot.gov/environment/noise/ tmn. Accessed May 10, 2006.

———— (2005). *Noise Barrier Type*, Federal Highway Administration, U.S. Department of Transportation Washington, DC, http://www.fhwa.dot.gov/environment/noise/4.htm. Accessed Dec. 30, 2005.

Fleming, G., Rapoza, A., Lee, C. (1995). *Development of National Reference Energy Mean Emission Levels for the FHWA Traffic Noise Model (FHWA TNM), Version 1.0*, FHWA-PD-96-008/DOT-VNTSC-FHWA-96-2, John Volpe National Transportation Systems Center, Cambridge, MA.

*Flythe, M. C. (1982). *INM, Integrated Noise Model, Version 3 User's Guide*, FAA-EE-81-17, Federal Aviation Administration, Washington, DC.

FRA (2002). Railroad noise emissions compliance regulation, *Code of Federal Regulations*, Title 49, Vol. 4, Pt. 210, Subpart A, *General Provisions*, revised as of Oct. 1, 2000, Federal Railway Administration, Washington, DC, http://www.fra.dot.gov/us/content/173.

*FRA (2005). *High Speed Ground Transportation Noise and Vibration Impact Assessment*, Tech. Rep. 293630-4 Prepared by Harris Miller Miller & Hanson, Inc, and Parsons Transportation Group for the Federal Railway Administration, Washington, DC.

*Harris Miller Miller & Hanson (1995). *Transit Noise and Vibration Impact Assessment*, Tech. Rep., U.S. Department of Transportation, Washington, DC, http://ntl.bts.gov/data/rail05/rail05.html. Accessed May 30, 2006.

*Kugler, B. A., Commins, D. E., Galloway, W. J., Beranek, B., Neuman (1976). *Highway Noise: A Design Guide for Prediction and Control*, NCHRP Rep. 174, Transportation Research Board, National Research Council, Washington, DC.

*Lee, C., Fleming, G. (1996). *Measurement of Highway-Related Noise*, Tech. Rep. FHWA-PD-96-046/DOT-VNTSC-FHWA-96-5, U.S. Department of Transportation, Washington, DC.

*Menge, C. W., Rossano, C. F., Anderson, G. S., Bajdek, C. J. (1998). *FHWA Traffic Noise Model Version 1.0, Technical Manual*, FHWA-PD-96-010 and DOT-VNTSC-FHWA-98-2, John Volpe National Transportation Systems Center, Cambridge, MA.

*Moulton, H. T. (1990). *Air Force Procedure for Predicting Aircraft Noise Around Airbases: Noise Exposure Model (NOISEMAP) User's Manual*, AAMRL-TR-90-011, Armstrong Laboratory, Wright-Patterson AFB, OH.

NRC (2003). *Ocean Noise and Marine Mammals*, Committee on Potential Impacts of Ambient Noise in the Ocean on Marine Mammals, National Research Council, Washington, DC.

*USDOT (1995). *Highway Traffic Noise Analysis and Abatement Policy and Guidance*, U.S. Department of Transportation, Washington, DC.

———— (2000). *Highway Traffic Noise Barrier Construction Trends*, Office of Natural Environment Noise Team, Federal Highway Administration, U.S. Department of Transportation, Washington, DC.

*———— (2004). *FHWA Traffic Noise Model® User's Guide (Version 2.5 Addendum)*, U.S. Department of Transportation, Washington, DC, http://www.trafficnoisemodel.org.

[1]References marked with an asterisk can also serve as useful resources for noise impact evaluation.

USEPA (1974). *Information on Levels of Environmental Noise Requisite to Protect Public Health and Welfare with an Adequate Margin of Safety*, EPA S-550/9-74-004, U.S. Environmental Protection Agency, Washington, DC.

Wayson, R. L. (2003). Environmental considerations during Transportation planning, in *The Civil Engineering Handbook*, ed. Chen, W. F., Liew, J. Y. R., CRC Press, Boca Raton, FL.

Wayson, R. L., Bowlby, W. (1989). Noise and air pollution of high speed rail systems, *ASCE J. Transp. Eng.*, Vol. 115, No. 1, pp. 20–36.

ADDITIONAL RESOURCES

Bronzaft, A. L. (1989). Public health effects of noise, Paper 89-101-3, *Proc. 82nd Annual Meeting of the Air and Waste Management Association*. Anaheim, CA.

FHWA (1979). *The Audible Landscape: A Manual for Highway Noise and Land Use*, Federal Highway Administration U.S. Department of Transportation, Washington, DC.

———— (1994). *Summary of Noise Barriers Constructed by December 31, 1992*, Federal Highway Administration U.S. Department of Transportation, Washington, DC.

———— (2000). *Highway Traffic Noise in the United States: Problem and Response*, Federal Highway Administration U.S. Department of Transportation, Washington, DC, http://www.fhwa.dot.gov/environment/probresp.htm. Accessed Dec. 30, 2005.

FlCUN (1980). *Guidelines for Considering Noise in Land Use Planning and Control*, Federal Interagency Committee on Urban Noise, Washington, DC.

Foss, R. N. (1979). Double barrier noise attenuation and a predictive algorithm, *Noise Control Eng.*, Vol. 11, No. 1, pp. 40–44.

Han, Z., Carson, D. (1999). *A Spray Based Crumb Rubber Technology in Highway Noise Reduction Application*, Rubber Pavements Association, Tempe, AZ.

Harris, R. A., Cohn, L. F. (1986). Use of vegetation for abatement of highway traffic noise, *ASCE J. Urban Plan. Dev.*, Vol. 111, No. 1, pp. 34–38.

Martin, F. N. (1991). *Introduction to Audiology*, 4th ed., Prentice Hall, Englewood Cliffs, NJ.

Saurenman, H., Nelson, J. T., Wilson, G. P. (1982). *Handbook of Urban Rail Noise and Vibration Control*, UTMA-MA-06-099-82-1, U.S. Department of Transportation, Washington, DC.

APPENDIX A11: NOISE ATTENUATION CHARTS FOR BARRIERS DEFINED BY N_0, ϕ_L, AND ϕ_R

MAXIMUM FRESNEL NUMBER, $N_0 = 0.01$

RIGHTMOST BARRIER ANGLE, ϕ_R^o

ϕ_L	−80	−70	−60	−50	−40	−30	−20	−10	0	10	20	30	40	50	60	70	80	90
−90	−5.0	−5.0	−5.0	−5.1	−5.1	−5.1	−5.1	−5.1	−5.1	−5.1	−5.1	−5.1	−5.1	−5.1	−5.1	−5.1	−5.1	−5.1
−80	—	−5.0	−5.1	−5.1	−5.1	−5.1	−5.1	−5.1	−5.1	−5.1	−5.1	−5.1	−5.1	−5.1	−5.1	−5.1	−5.1	−5.1
−70	—	—	−5.1	−5.1	−5.1	−5.1	−5.1	−5.1	−5.1	−5.1	−5.1	−5.1	−5.1	−5.1	−5.1	−5.1	−5.1	−5.1
−60	—	—	—	−5.1	−5.1	−5.1	−5.1	−5.1	−5.1	−5.1	−5.2	−5.2	−5.2	−5.2	−5.2	−5.1	−5.1	−5.1
−50	—	—	—	—	−5.1	−5.1	−5.1	−5.2	−5.2	−5.2	−5.2	−5.2	−5.2	−5.2	−5.2	−5.2	−5.1	−5.1
−40	—	—	—	—	—	−5.1	−5.2	−5.2	−5.2	−5.2	−5.2	−5.2	−5.2	−5.2	−5.2	−5.2	−5.1	−5.1
−30	—	—	—	—	—	—	−5.2	−5.2	−5.2	−5.2	−5.2	−5.2	−5.2	−5.2	−5.2	−5.1	−5.1	−5.1
−20	—	—	—	—	—	—	—	−5.2	−5.2	−5.2	−5.2	−5.2	−5.2	−5.2	−5.2	−5.1	−5.1	−5.1
−10	—	—	—	—	—	—	—	—	−5.2	−5.2	−5.2	−5.2	−5.2	−5.2	−5.2	−5.1	−5.1	−5.1
0	—	—	—	—	—	—	—	—	—	−5.2	−5.2	−5.2	−5.2	−5.2	−5.1	−5.1	−5.1	−5.1
10	—	—	—	—	—	—	—	—	—	—	−5.2	−5.2	−5.2	−5.2	−5.1	−5.1	−5.1	−5.1
20	—	—	—	—	—	—	—	—	—	—	—	−5.2	−5.2	−5.1	−5.1	−5.1	−5.1	−5.1
30	—	—	—	—	—	—	—	—	—	—	—	—	−5.1	−5.1	−5.1	−5.1	−5.1	−5.1
40	—	—	—	—	—	—	—	—	—	—	—	—	—	−5.1	−5.1	−5.1	−5.1	−5.1
50	—	—	—	—	—	—	—	—	—	—	—	—	—	—	−5.1	−5.1	−5.1	−5.1
60	—	—	—	—	—	—	—	—	—	—	—	—	—	—	—	−5.1	−5.1	−5.0
70	—	—	—	—	—	—	—	—	—	—	—	—	—	—	—	—	−5.1	−5.0
80	—	—	—	—	—	—	—	—	—	—	—	—	—	—	—	—	—	−5.0

LEFTMOST BARRIER ANGLE, ϕ_L

MAXIMUM FRESNEL NUMBER, $N_0 = 0.50$

RIGHTMOST BARRIER ANGLE, ϕ_R^o

	−80	−70	−60	−50	−40	−30	−20	−10	0	10	20	30	40	50	60	70	80	90
−90	−5.7	−8.3	−6.8	−7.2	−7.5	−7.8	−8.1	−8.3	−8.5	−8.7	−8.8	−8.9	−9.0	−9.0	−9.0	−8.9	−8.8	−8.5
−80	—	−7.0	−7.4	−7.8	−8.2	−8.4	−8.7	−8.9	−9.1	−9.2	−9.3	−9.4	−9.4	−9.4	−9.3	−9.2	−9.1	−8.8
−70	—	—	−8.0	−8.3	−8.6	−8.9	−9.1	−9.3	−9.5	−9.6	−9.6	−9.7	−9.7	−9.7	−9.6	−9.5	−9.2	−8.9
−60	—	—	—	−8.7	−9.0	−9.3	−9.5	−9.6	−9.8	−9.9	−9.9	−9.9	−9.9	−9.9	−9.8	−9.6	−9.3	−9.0
−50	—	—	—	—	−9.3	−9.6	−9.7	−9.9	−10.0	−10.1	−10.1	−10.1	−10.1	−10.0	−9.9	−9.7	−9.4	−9.0
−40	—	—	—	—	—	−9.8	−10.0	−10.1	−10.2	−10.2	−10.3	−10.2	−10.2	−10.1	−9.9	−9.7	−9.4	−9.0
−30	—	—	—	—	—	—	−10.1	−10.2	−10.3	−10.3	−10.3	−10.3	−10.2	−10.1	−9.9	−9.7	−9.4	−8.9
−20	—	—	—	—	—	—	—	−10.3	−10.4	−10.4	−10.4	−10.3	−10.3	−10.1	−9.9	−9.6	−9.3	−8.8
−10	—	—	—	—	—	—	—	—	−10.5	−10.5	−10.4	−10.3	−10.2	−10.1	−9.9	−9.6	−9.2	−8.7
0	—	—	—	—	—	—	—	—	—	−10.5	−10.4	−10.3	−10.2	−10.0	−9.8	−9.5	−9.1	−8.5
10	—	—	—	—	—	—	—	—	—	—	−10.3	−10.2	−10.1	−9.9	−9.6	−9.3	−8.9	−8.3
20	—	—	—	—	—	—	—	—	—	—	—	−10.1	−10.0	−9.7	−9.5	−9.1	−8.7	−8.1
30	—	—	—	—	—	—	—	—	—	—	—	—	−9.8	−9.6	−9.3	−8.9	−8.4	−7.8
40	—	—	—	—	—	—	—	—	—	—	—	—	—	−9.3	−9.0	−8.6	−8.2	−7.5
50	—	—	—	—	—	—	—	—	—	—	—	—	—	—	−8.7	−8.3	−7.8	−7.2
60	—	—	—	—	—	—	—	—	—	—	—	—	—	—	—	−8.0	−7.4	−6.8
70	—	—	—	—	—	—	—	—	—	—	—	—	—	—	—	—	−7.0	−6.3
80	—	—	—	—	—	—	—	—	—	—	—	—	—	—	—	—	—	−5.7

LEFTMOST BARRIER ANGLE, ϕ_L^o

Source: Barry and Reagan (1978).

CHAPTER 12

Impacts on Wetlands and Other Ecosystems

Nature does nothing uselessly.

—Aristotle (384–322 B.C.)

INTRODUCTION

Ecology is the science that deals with the physical and functional relationships between biotic elements (flora and fauna) and their abiotic environment (habitat). Transportation projects and policies can adversely affect the quantity and quality of wetlands, woodlands, and other ecological systems. It is therefore appropriate that several pieces of legislation passed over the past few decades, including SAFETEA-LU in 2005, duly recognized the importance of preserving the natural environment in the planning and evaluation of transportation systems. SAFETEA–LU emphasized environmental stewardship, provided state and local agencies a formal role in the environmental review process, and increased funding for environmental programs (Binder, 2006). In this chapter, we first present the basic concepts of ecological systems and discuss the various mechanisms by which such systems could be affected by transportation developments. We also present a set of performance measures and a procedural framework for assessing ecological impacts, focusing on wetlands. Finally, mitigation measures, related federal legislation, and available software packages for ecological impact assessment are discussed.

12.1 BASIC ECOLOGICAL CONCEPTS

12.1.1 Concept of Ecosystems

A *community* is a collection of plants and animals coexisting in a specific domain and is typically defined on the basis of predominant plant or animal species.

A *habitat*, the physical location where an organism resides for shelter and food, may be terrestrial, aquatic (freshwater/coastal), or arboreal. A *biotope* is defined as an area of relatively uniform physical conditions providing habitat(s) for a specific assemblage of plants and animals. An *ecosystem* is defined as a biotic community and its abiotic environment, and its area may range from very small to very large. Ecosystem conditions range from hot and humid rainforests to frigid, dry, and icy wastelands. In a given ecosystem, plant and animal combinations exist that are best adapted to the existing set of physical and chemical conditions. Ecosystem classification is typically based on the general biotope type, followed by the biotope's dominant species or physical feature as follows (Morris and Therivel, 2001; USEPA, 2004):

- *Terrestrial* (or *continental*) *ecosystems*, which are characterized further on the basis of dominant vegetation or habitat type (or lack thereof). They include forest ecosystems, agricultural ecosystems, meadow ecosystems (savannas, meadows, steppes), woodlands, scrubs, grasslands, heathlands, mires, marshes, rock exposures, wastelands, and so on.
- *Freshwater ecosystems*, which involve inland waters and include natural or human-made still water and flowing water (e.g., *lentic* ecosystems such as ponds and lakes or *lotic* ecosystems such as creeks, streams, and rivers), ditches, canals, springs, and geysers.
- *Coastal* (or *oceanic*) *ecosystems*, which are characterized by the meeting point between land and sea or ocean. The coastal zones consist of three subzones: *littoral* (intertidal or shore), *supralittoral* (maritime), and *sublittoral* (marine).
- *Wetlands,* which are unique ecological systems that represent the transition zone from terrestrial to aquatic habitats, linking land and water. They typically include marshes, swamps, bogs, and fens. We discuss wetland types further in Section 12.1.3.

Ecosystems may also be classified by the type of dominant organism and may therefore be termed a *plant community* or an *animal community*, or may even be named after a specific dominant species of flora or fauna. A *food chain* (Figure 12.1) is the hierarchy of feeding patterns of various organisms in an ecosystem. Transportation facility construction and operations can severely disrupt food chains in an area. As such, it is particularly relevant to consider food chain interactions in assessing the ecological impacts of transportation activities. Food webs are cyclic food chains—vegetation is consumed by herbivorous animals that serve as prey for larger predators (carnivores) such as eagles; then

313

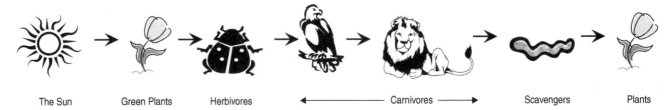

| The Sun | Green Plants | Herbivores | ← Carnivores → | Scavengers | Plants |

Figure 12.1 A typical food chain in a terrestrial habitat.

both predators and prey, through excretion or death and decomposition, provide nutrients for scavengers and ultimately to vegetation. The cyclic nature of food chains is indicative of energy flows from one organism to another: Vegetation transforms solar energy into chemical energy (through *photosynthesis*), which is then transferred to herbivores and then to carnivores. Ecological systems whose biotic populations are depleted by transportation activities suffer disruptions in their food chain interactions and breaking of their energy cycle, leading to difficulty in replacing the lost energy and consequently, destabilization.

Ecological productivity is defined as the rate at which an ecosystem utilizes abiotic elements (soil nutrients, water, sunlight, carbon dioxide, etc.) for producing living matter. By destroying or degrading the physical environment or by reducing the quantities of abiotic matter available in an area, transportation system activities can impair ecological productivity. *Ecological succession* refers to the change in the biotic or abiotic characteristics of a community over a period of time. The effects of transportation activities are often detrimental to ecological succession. Examples include clearing the right-of-way (which leads to destruction of the habitat in that locality) and planting of inappropriate vegetation in highway medians or rail/runway right-of-way (this may invite animal species that differ from those that existed before construction). The natural balance of plant and animal life in an ecosystem is achieved through various types of symbiosis among its elements. This balance can be disrupted severely if existing biotic or abiotic elements are eliminated or if new ones are introduced into the ecosystem through interventions such as transportation activities. In extreme cases, such disruptions can lead to ecological collapse and the demise of many native species. *Ecological stability or resilience* is a measure of the ability of an ecosystem to recover from such deleterious disruptions.

12.1.2 Physical Base

The *physical base* of an ecological environment consists of land, air, and water that provide support for plant and animal life. Land consists of three major components: bedrock, surficial geologic materials, and soils. *Bedrock* refers to the consolidated parent material that generally lies deep below Earth's surface. *Surficial geologic materials* are unconsolidated deposits that generally lie between bedrock and the surface. *Soil*, the surface material formed primarily from bedrock decomposition and surficial geologic materials, often includes organic material. Soil plays a vital role in the sustenance of an ecosystem because it serves as a substrate for all terrestrial vegetation and as a habitat for most terrestrial animal species.

Destruction or degradation of the physical base has a far-reaching effect on ecological systems. Transportation activities generally have relatively little impact on bedrock but may disturb geologic materials and subsurface deposits through tunneling and quarrying. Surficial soils are disturbed through removal of topsoil and vegetal cover, slope excavation, and open mining of borrow material at gravel pits and quarries for constructing road, runway, and rail track bases and embankments, and for producing concrete and asphalt mixes. By removing overburden material, these activities expose weathered and erosion-prone subsurface material to the erosive forces of wind and rain, leading to increased weathering, slope failure, erosion at certain terrestrial habitats, and pollution of aquatic habitats through sedimentation. Also, construction of new transportation facilities leads to decreased area of pervious surfaces, increased runoff, and consequently, increased soil erosion. The components of the physical base—land, air, and water—play vital roles in ecosystem sustainability, and their degradation through transportation construction and operations constitutes a significant loss to the plants and animals that depend on them.

12.1.3 Wetland Ecosystems

Wetlands are special ecosystems that receive additional consideration in environmental impact planning and legislation due to their important ecological, social, and economical functions and benefits (Table 12.1). These functions include polluted water remediation (thereby

Table 12.1 Functions and Benefits of Wetlands

Ecology	Opportunities for groundwater recharge; natural purification of water and water supply; weather-stabilizing properties; habitation and/or sanctuary for various species; shoreline protection; protection from storm/flood through storage discharge reduction during storm/flood peak flows
Human values	Recreational activities (e.g., fishing, hunting, boating, hiking, and bird-watching); forestry and agricultural harvests; opportunities for ecological education; aesthetic features; commercial fishing

Source: Adapted from Cambridge Systematics and Bernardin (2003) and USEPA (2004).

improving water quality), hazard management (flood control, drought relief, and shoreline stabilization), and ecological protection and preservation (habitats/sanctuaries for many rare, threatened, or endangered species of plants and animals). There is a wide spectrum of wetland types, which are characterized by regional and local attributes such as climate, soils, topography, hydrology, water chemistry, vegetation, and other factors, including human disturbance (USEPA, 2004). In the United States, information on wetland types and locations can be obtained from the Natural Resources Conservation Service office, local public works or planning departments, and the U.S. Fish and Wildlife Service's National Wetland Inventory Web site (www.nwi.fws.gov). As shown in Figure 12.2, wetlands can be classified as *marine* (open ocean), *estuarine* (mixture of ocean water and fresh water), *lacustrine* (lake), *riverine* (river), and *palustrine* (swamps, marshes, bogs, etc.), as identified by Erickson et al. (1980).

12.2 MECHANISMS OF ECOLOGICAL IMPACTS

12.2.1 Direct vs. Indirect Mechanisms

Transportation impacts on wetlands and other ecosystems may be direct or indirect. *Direct mechanisms* include land conversion (which in turn attracts land-use changes that further deplete or fragment the natural environment), physical destruction of habitats through earthwork and water dredging, vehicle collision with wildlife, and the release of transportation-related chemicals and solids into the environment. *Indirect mechanisms* are unintended actions that occur due to transportation activities and include habitat fragmentation, introduction of nonnative species, and reduction in the population of specific plant or animal species (USEPA, 1996). The end result is the imbalance of the ecosystem and disruption of the food chain, which may ultimately lead to depletion or extinction of certain plant and animal species in the ecosystem. Figure 12.3 summarizes the mechanisms of transportation impacts on ecology.

12.2.2 Impact Mechanism by Species Type

The mechanism of impacts depends on the species type under investigation. Species may generally be categorized as large animals, birds of prey, small game, water-based animals, special species listed in federal regulations, native vegetation, field crops, and aquatic plants (Jain et al., 2001). *Large animals* can include wild animals (deer, moose, bears, cougars, wolves, etc.) or domestic animals (sheep, cattle, goats, horses, etc.) that weigh more than 50 pounds as adults. Examples of *birds of prey* include hawks, vultures, owls, falcons, eagles, and ospreys, which are carnivores and feed on small animals. *Small game* is defined as upland birds and animals, that weigh less than 30 pounds when fully grown, and are generally hunted for leisure, such as rabbits, quail, squirrels, and pheasants. *Water-based animals* include fish (minnows, trout, salmon, tuna, etc.), shellfish (freshwater and saltwater: mussels, clams, oysters, shrimp, crabs, etc.), and waterfowl (ducks, geese, pelicans, swans, cranes, gulls, etc.). Common field crops that are grown commercially to feed human and domestic animals include wheat, corn, soybeans, tomatoes, and vegetables. *Federally listed species* include animal and plant species that are classified as threatened and endangered. Lists of such species can be obtained from the Fish and Wildlife Service of the U.S. Department of the Interior. Several commonly-recognized listed species include the southern bald eagle, timber wolf, California condor, whooping crane, and grizzly bear. A discussion of the various direct and indirect impact mechanisms, by species type, is provided in Table 12.2.

12.3 ECOLOGICAL IMPACTS OF ACTIVITIES AT VARIOUS PDP PHASES

At various stages of the transportation project development process (PDP), there are direct and indirect impacts on plants, animals, and/or the physical base of wetlands and other ecosystems. Establishing the PDP phase(s) of interest can help an analyst to identify and focus on the specific mechanisms and types of ecological impact.

Figure 12.2   Types of wetlands:

(a) *Marine* (in the vicinity of an ocean)
  Types: intertidal, subtidal
  Examples: bays, sounds, coastlines
(b) *Estuarine* (coastal wetland where
    fresh and salt water meet)
  Examples: tidal salt marsh, tidal
    freshwater marshes, mangrove
    swamps
(c) *Lacustrine* (shallow water)
  Types: littoral, limnetic
  Examples: low-lying
    areas surrounding lakes, ponds, and
    reservoirs; wet meadows

(d) *Riverine* (flowing water)
  Types: intermittent, upper and lower
    perennial, tidal
  Examples: drainage basins along freshwater
    rivers, streams, and creeks; forested
    swamps; shrub swamps
(e) *Palustrine* (standing or very slowly flowing
    water)
  Types: intertidal, subtidal
  Examples: nontidal marshes, bogs,
    peatlands.

(Image credits: (a) Courtesy of Amelia Leubscher, Creative Commons Attribution 2.0
license, (b) Courtesy of Farl, Creative Commons Attribution-NoDerivs 2.0 license,
(e) Courtesy of Tommy Wong, Creative Commons Attribution 2.0 license)

Sinha et al. (1991) provided detailed discussions of ecological impacts at each stage, as summarized below.

12.3.1 Locational Planning and Preliminary Field Surveys

At the system planning and design phase, no physical implementation is involved—as such, there are no direct ecological impacts. However, decisions at these phases can influence the degree of encroachment or loss of wetlands and other ecosystems during the construction and operational phases of the facility. The selection of transportation infrastructure locations, corridors, and alignments and the design of main and ancillary structures should be carried out to avoid ecologically sensitive areas

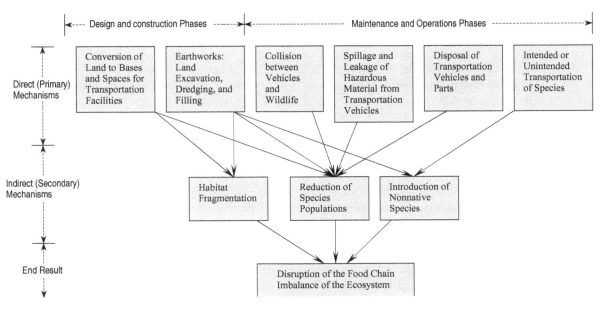

Figure 12.3 Mechanisms of transportation impacts on ecology.

Table 12.2   Direct and Indirect Impact Mechanisms by Species Type

Species	Direct (Primary) Impact Mechanism	Indirect (Secondary) Impact Mechanism
Large animals	Construction of new transportation facilities (roads, runways, rail tracks, etc.)	Reduces vegetative cover which provides food for large herbivorous animals
	Habitat segmentation due to construction of line facilities (roads, rail lines, etc.)	Limits animal movements and activities, such as restricting access to food and water locations and enclosing animals within unhealthy and undersized spaces
	Removal of vegetative cover (temporarily or permanently) during construction	Destroys vegetative cover that serves the purposes of travel, food, water, habitat, shelter, reproduction, and nurturing of young large animals
	Herbicide application during right-of-way (ROW) maintenance	Reduces habitat, shelter, and food through plant destruction and contamination of food and water supply
	Collisions between animal and vehicle; habitat encroachment by vehicular travel	Reduces the habitat suitability for food, reproduction, nurturing of young animals, etc.
Birds of prey	Physical degradation of habitat due to land-clearing activities for construction of transportation facilities	Reduces food, water, habitat, shelter, and space required for the survival of birds
	Destruction of nests or nest-bearing vegetation during ROW maintenance (herbicide application, felling, mowing)	Destroys unhatched eggs; increases fatality rate and reduces survival rate of young birds; limits areas available for reproduction/rearing of young birds
	Interference by transportation personnel and equipment at the vicinity of nesting areas	Restricts and alters movement of birds, predominantly eagles, ospreys, condors, and several types of falcons
	Continual noise from heavy equipment during construction and from vehicular travel during operation	Results in birds leaving their nests temporarily or permanently and disappearing from the areas affected

(*continued overleaf*)

Table 12.2 (*continued*)

Species	Direct (Primary) Impact Mechanism	Indirect (Secondary) Impact Mechanism
Small game	Land grading, resulting in removal of native vegetation and rearrangement of terrain/surface textures.	Reduces quantity and quality of habitat, shelter, food, and water required for the survival of small game
	Herbicide application to manage roadside vegetation during ROW maintenance	Destroys small game habitats temporarily and permanently: repeated applications result in permanent abandonment of the area affected
Water-based animals	Cutback of water quantity due to dredging, channelization, and water-related construction of transportation facilities	Reduces (through siltation) food and oxygen availability and accessibility for specific water-based animals
	Increases in accumulated concentrations of herbicide residues due to surface runoff or soil percolation of contaminated materials from land-based transportation vehicles; contamination of marine or lacustrine habitats due to chemical leakage and spillage from water vessels	Increases water acidity or alkalinity, resulting in gill damage in fish; damages, asphyxiates, or poisons (through action of chemicals) specific water-based animals; destroys habitat and food supply; reduces the reproductive capability of adult fishes and survival rate of young fishes
Field crops	Application of herbicides (during ROW maintenance) on land in the vicinity of an agricultural area	Reduces diversity of local vegetation; adversely affects people and animals that depend on field crops for sanctuary or food
	Acquisition of major agricultural lands for transportation project ROW and other purposes	Reduces land available for agricultural activities and production
	Transportation-related hydraulic features (reservoir construction/operation, runoff control structures, etc.)	Reservoirs and impounds may result in the elevation of groundwater level, flooding of root systems, and severe damage or destruction of crops
Listed species	Similar to those discussed for large animals, predatory birds, and natural land vegetation attributes	
Native vegetation	Land-clearing activities for construction; herbicide application; operation of off-road vehicular traffic; paving activities	Destroys and reduces areas covered by natural land vegetation.
Aquatic plants	Draining swamps and marshes for constructing/expanding transportation facilities alters water quantity and quality (see entries for water-based animals), erosion, etc.	Land clearing and acquisition reduces habitat and survivability for aquatic plants; Erosion increases sediment loading, reduces sunlight penetration, impairing the photosynthetic activities of aquatic plants.
		Exposure of aquatic plant roots (due to water-level changes) to the drying effects of sunshine and air or cause flooding such that periphery-dwelling species are denied air for extended periods of time.

Source: Adapted from Jain et al. (2001).

or to minimize ecological impacts at such areas when they are inevitable. Location decisions affect the magnitude of the direct impact mechanisms (e.g., how much land should be appropriated for the project; the volumes of excavation, dredging, or filling; which natural wildlife habitats will be traversed), consequently affecting the indirect factors (e.g., extent of wildlife habitat fragmentation or species depletion). During the site investigation stage of project feasibility analysis, ecological disturbances may occur, particularly during intrusive geotechnical, geophysical, and geodetic field investigations.

12.3.2 Transportation System Design

For aquatic ecosystems, transportation facility designs that lead to reductions in land retention of surface runoff or disruptions in species' migratory patterns can alter the trophic dynamics of such ecosystems. Furthermore, designs that lead to interruptions of tidal flows in coastal wetlands may result in (1) reduced productivity of marsh plants, which affects birds and other terrestrial animals that depend on the wetland area for food and sanctuary, and (2) reduced detrital loading of estuaries, which decreases the food supply for downstream and ocean-dwelling detritivores. Transportation facilities that disrupt water flow patterns can also adversely affect inland wetland ecosystems. Extensive cut-and-fill can significantly affect the level of surface water and groundwater and can consequently disrupt the health of nearby wetlands. Inadequate culvert design can lead to flooding, which in turn affects water levels. Even slight changes in the water level can seriously impair the survivability of certain wetland flora (and fauna that depend on wetland flora). Furthermore, wetland food webs typically include terrestrial food chains; therefore, inconsiderate transportation system design that affect inland wetlands adversely can ultimately impair the sustainability of inland terrestrial ecosystems.

12.3.3 Construction

When facility construction (and maintenance) is carried out without duly recognizing the natural patterns of the environment (such as migratory or reproduction behaviors of species and seasonal flow augmentation of streams by groundwater aquifers), the ecological impacts can be severe, as discussed below.

(*a*) *Toxification and Eutrophication*   By serving as nutrients, toxicants, or irritants to specific key populations in the food web, construction materials and supplies can have adverse impacts on the ecology. The leakage and spillage of fluids from construction equipment, emission of engine exhausts, and disposal of construction spoil can degrade food and water sources of biota. Furthermore, such toxic materials, through the process of *trophic magnification*, can exert lethal impacts on higher-order consumers in the food web in terrestrial or aquatic environments. For terrestrial ecosystems, such potentially toxic materials may come from exposed soils and substrates, on-site storage areas for construction materials and equipment supplies,

and on-site areas for materials processing. Using material of an organic nature or content to fill embankments and other areas located in a watershed can lead to leaching and subsequent increase in downstream concentrations of nutrient for aquatic flora, thereby enhancing *eutrophication* of downstream impoundments.

(*b*) *Changes to the Physical Base*   Construction activities such as right-of way clearing and construction in or near surface water courses can adversely affect threatened and endangered species. Felling trees in the facility right-of-way allows more sunlight to penetrate previously covered areas, thus modifying the ground microclimate, altering the diversity of flora and fauna species, facilitating the growth or demise of certain species populations, and ultimately affecting the balance of predator and prey populations. Tree felling could also lead to increased susceptibility of remaining trees to wind damage and increased rates of evapotranspiration, due to exposure of previously covered vegetation to the desiccating action of winds.

(*c*) *Movement of Soil Masses (Earthworks)*   When soil masses are excavated, moved from one place to another, filled, and compacted, ecosystems are affected by the resulting food web changes and trophic dynamics. The movement of soil masses can result in changed levels of the local water table and therefore influence plant associations that depend on the water content of soils. This, in turn, affects the animals that depend on such plants for habitat or food supply. Aquatic environments where earthworks are carried out experience increased water turbidity. In the case of estuarine wetlands, placement of soil from transportation projects covers some benthic populations, affects tidal currents, and can disturb the abiotic environment and changes in salt concentrations. Finally, soil compaction activities lead to reduced permeability of surficial soils, decreased recharge of groundwater sources, increased surface runoff, and increased erosion.

12.3.4 Operations

In the course of transportation facility use, ecological impacts occur largely through indirect mechanisms. First, operations affect the noise level, air quality, and water quality. Second, the environmental degradation in terms of these three parameters, in turn, causes a reduction in ecological quality. In a few cases, transportation operations and facility use may result in direct impacts: for example, plane and auto collisions with birds and deer, respectively.

12.3.5 Maintenance

(*a*) *Vegetation Control* During routine maintenance, facility right-of-way is typically mowed and cleared of shrubbery. Machine mowing operations can result in soil compaction in the median and right-of-way areas and thereby reduce the rate of surface water percolation and increase the likelihood of localized erosion. The noisy and physically intrusive nature of mowing operations can disrupt the nesting of birds and animals. Finally, the application of herbicides and use of paints and preservatives can not only obliterate certain target plant species on terrestrial ecosystem but also impair animal reproduction, cause stress among sensitive nontarget species, and alter the physical, chemical, and biological characteristics of the immediate habitat.

(*b*) *Pollution through Maintenance Stockpiling or Operations* Stockpiling of maintenance materials and supplies, such as salts, may result in (1) leaching of chemicals through local soils, with ensuing effects on neighboring vegetation and soil organisms; (2) wind dispersal of toxic materials and irritants, with potential effects on other terrestrial biota; and (3) alteration of physical, chemical, and biological attributes of existing habitat at the stockpile area. Spillage and leakage of petrochemicals and other volatile compounds used for facility maintenance can affect plants and animals and their habitats. Where aquatic or wetland ecosystems are involved, the respiration of subsurface aquatic organisms such as insect larvae is impaired, and floating or emergent vegetation is coated with a chemical film. In running-water courses, chemical slicks degrade habitats at downstream areas. Over a prolonged period, increased concentrations of potentially toxic materials in the mud of wetland beds could cause trophic magnification where such materials enter into detritivores and subsequently into higher-order consumers. This could lead to the depletion of species, particularly those of protected, threatened, or endangered status.

12.4   PERFORMANCE GOALS FOR ECOLOGICAL IMPACT ASSESSMENTS

Sinha et al. (1991) and Ortolano (2001) suggested the following broad performance goals from which appropriate measures can be derived: diversity of the ecosystem, state of habitat fragmentation, survival of "significant" species, diversity of species, ecosystem stability, and quality or productivity of the ecosystem. Performance measures from one or more of these goals could be used to evaluate the impact of a transportation intervention. If more than one is used, different weights may be attached to each goal to reflect the relative importance, but care should be taken to avoid double counting or overlapping effects. Each performance goal consists of performance measures, as discussed below.

12.4.1 Diversity of the Physical Base of the Ecosystem

Abiotic diversity, the variation in composition and structure of the ecosystem physical base, can affect the functional processes between the physical environment and its biological hosts. A closely related ecological parameter, *regional* (or *landscape*) *diversity*, refers to the variety in biological communities (i.e., distribution, sizes, shapes, etc.) and the relationships between one another as well as their physical base. This concept helps to explain why species survivability is related to their ability to migrate. Through land conversion and earthworks, transportation projects destroy relatively vulnerable physical bases, thus reducing landscape diversity and impairing species survivability. The determination of ecosystem diversity at a regional level is a qualitative process that is based on field sampling of similar areas of different ecosystems and assessing the diversity of each area based on the number of different species found in that area.

12.4.2 State of Habitat Fragmentation

The state of habitat fragmentation is the state of an ecosystem's physical base divided by natural barriers or by human-made structures or activities created by highway and railway construction. For example, a dense network of roads in a region yields a highly fragmented habitat, whereas a sparse road network in a similar region yields a lightly fragmented habitat (Figure 12.4). A river delta (a piece of land with multiple divisions by river paths) is a naturally fragmented habitat. Through fragmentation, a habitat is rendered into pieces of land that may be too small to sustain certain animal species, particularly larger animals. New interfaces between the ecosystem and human-made developments can cause increased exposure of the ecosystem to human activity and weather vagaries (sunlight penetration and wind erosion), restricted movements of animals, and introduction of new (and often, predatory) species.

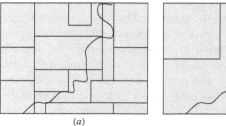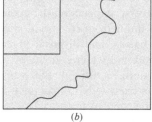

Figure 12.4 Habitat fragmentation stages: (*a*) high state of fragmentation, (*b*) low state of fragmentation.

12.4.3 Significant Species and Habitats

The selection of specific species or habitats to be included in the ecological impact assessment could be based on their rarity, their contribution to the ecosystem, or their inherent ability to provide indication of threats or opportunities in the ecosystem. *Rare species* and *habitats* contribute to genetic diversity and ecosystem stability. Also, the fact that they face extinction draws urgency to the need for their protection, and consequently, their inclusion in impact assessments. *Keystone species* are those whose removal would probably lead to significant disruptions in the structure and relationships in an ecosystem. *Indicator species* have certain unique characteristics that enable identification of any prevailing unfavorable or favorable environmental elements, such as air pollution and nutrient levels, respectively. For example, leaves of spinach plants are known to develop surface lesions when exposed to high ozone concentrations.

12.4.4 Diversity of Species

Through direct and indirect mechanisms, transportation interventions can significantly reduce the population of species that are relatively more vulnerable than others and therefore significantly disrupt the *species diversity* (the relative population distributions of different species in an ecosystem). Mathematically, ecosystem species diversity can be measured in terms of *Simpson's index* (Ortolano, 2001) as follows:

$$D = \frac{N(N-1)}{n_1(n_1-1) + n_2(n_2-1) + \cdots + n_s(n_s-1)} \quad (12.1)$$

where D is Simpson's diversity index for the ecosystem, s the number of different species in the ecosystem, n_i the number of individual organisms in species i, and N the total number of organisms in all species in the ecosystem.

Example 12.1   Consider three independent communities 1, 2, and 3 with the same total number of organisms, N. Assume that each community has a specified number of species A, B, C, D, and E. Based on the information in Table E12.1, calculate Simpson's diversity index for each community and briefly discuss the impact of absolute and relative changes in a species population on overall diversity.

SOLUTION   For community 1,

$$D_1 = \frac{1000(1000-1)}{4 \times 250 \times 249} = 4.01$$

Table E12.1   Ecosystem Species Diversity: Sample Calculations

	Community		
	1	2	3
Species A population, n_A	250	200	810
Species B Population, n_B	250	200	70
Species C population, n_C	250	200	30
Species D population, n_D	250	200	50
Species E population, n_E	0	200	40
Total population of individual organisms, N	1000	1000	1000
Simpson's diversity index, D (from eq. 12.1)	4.01	5.02	1.50

Similarly, the diversity indices for other communities are estimated as shown in Table E12.1.

Community 2, with a Simpson's index of 5.02, is more diverse than community 1. Also, community 1 is more diverse than community 3 even though community 1 has only four species types. It is therefore clear that Simpson's diversity index depends on the absolute population of species as well as their population relative to that of other species in the ecosystem. In the extreme case where there is only one species type in the ecosystem, the index takes its minimum possible value of 1.0. Also, in the other extreme case where each organism belongs to a different species, the index takes a value of infinity.

Transportation interventions, directly or indirectly, can cause reductions in species populations (in a disproportionate manner among species) and therefore can affect species diversity. In some cases, a higher diversity index is obtained even if all species suffer population reductions.

12.4.5 Ecosystem Stability

Stable ecosystems exhibit an "organized self-correcting capability to recover toward an end state that is normal and favorable for the ecosystem" (Regier, 1993). As such, the *stability* or *resilience* of an ecosystem is defined as its ability to resist disturbance and stress or to return to equilibrium after it has been subjected to stress. Using field studies and mathematical models, Shrader-Frechette and McCoy (1993) established that increased diversity is not necessarily associated with increased stability. Apart from reducing ecosystem quality, transportation actions could also degrade the ability of an ecosystem to recover from subsequent natural or human-made disruptions.

12.4.6 Ecosystem Quality or Productivity

The productivity of a terrestrial ecosystem can be represented by soil fertility because a more fertile substrate is more capable of supporting a larger number of plants and therefore, animals. The analyst can obtain agricultural maps that indicate areas of different levels of soil fertility, and can use this as a basis for assessing the effect of land conversions for purposes of transportation system construction or expansion. The impact is greatest when the most fertile lands in a region are taken for transportation purposes. Also, through construction earthworks and subsequent erosion of exposed surfaces, fertile soils may be transported from one point to another. Such redistribution of nutrients in ecosystems can be assessed by examining soil fertility maps together with maps of soil types and their erodibility. For aquatic ecosystems, population density (the number of individuals per unit area or volume) is an indicator of productivity. Ecosystem quality indicates the total living matter produced in an area and also gives an indirect indication of the existing pollution levels.

12.5   PROCEDURE FOR ECOLOGICAL IMPACT ASSESSMENT

Figure 12.5 illustrates the overall framework that could be used for ecological impact assessment. In these assessments, it may be useful to consult wildlife experts and professionals for advice on evaluation aspects such as geographical and temporal scopes, data collection and analysis, and interpretation of results.

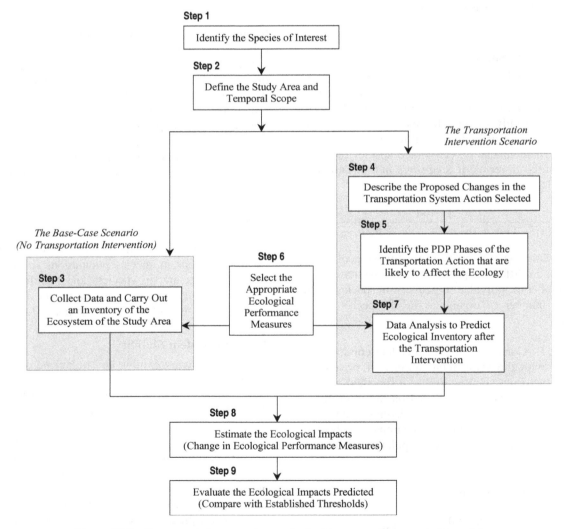

Figure 12.5 Framework for assessing ecological impacts of transportation actions.

Step 1: Identify the Species of Interest In this step, the analyst selects the species on which the ecological impact assessment will be based. The analyst should identify plants and animal species that are likely to be affected by the transportation action. In Section 12.2.2 we discuss the broad types of species to be considered. Where no such information is available, animal species that are likely to exist at the study area can be determined on the basis of existing vegetative types. Field investigations are useful to validate such lists, but their effectiveness may be limited by the seasonal migratory patterns of certain wildlife species. After establishing the lists of selected animal and plant species, the analyst must determine the food, water, habitat, and movement requirements of these species.

Step 2: Define the Study Area and Temporal Scope of the Impacts Ecological impact assessment could start as early as the system and location planning phases of the transportation development process. At that phase, the assessment could be carried out on a regional scale so that location alternatives can be established to bypass ecologically sensitive habitats or species. The analysis should continue to the project level, where the ecological impacts of specific design features are analyzed.

Step 3: Collect Ecological Data and Carry Out an Inventory of the Study Area Ecosystem

(a) *General Data Sources* Maximum use should be made of existing publications on ecological conditions at or near areas of interest. It should be noted that the quantity of animals and plants in a habitat can change significantly throughout the seasons. For example, some rare species may only be observable for a short period each year. Furthermore, in the course of their movements, migratory species use certain habitats only for very short periods. Care should thus be taken to ensure that the collected data duly recognize such naturally changing populations of species. Otherwise, the data collected would represent a static snapshot of species populations in time, when in reality they are highly dynamic. For data on the study area, the analyst should contact federal, state, provincial, and local agencies (such as forestry services, departments of agriculture, and geological survey agencies), private entities, and universities. Data on soil types, temperature, dominant vegetal cover type, animal concentrations and their seasonal movements, and so on, may exist in hardcopy or GIS forms. Also, ordnance survey maps can provide useful information on the locations of woodlands, forests, grasslands, and other types of vegetal cover in the United Kingdom (Morris and Therivel, 2001). Analysts in developing countries may seek the assistance of specialized agencies of the United Nations and other nongovernmental organizations.

(b) *Data Types* Geological and soil survey maps can help analysts identify vegetal cover type on the basis of surficial geology and soils and therefore obtain estimates of species populations. Such data can be supplemented by aggregate data from aerial photographic prints and satellite images. Information on local plant and animal species distributions is typically available at local biological data centers at various governmental agencies. Useful information may also be obtained from informal or semiformal sources such as local botanical or ornithological clubs, arboreta and conservation organizations, farmers, community leaders, and school teachers. Table 12.3 describes some techniques for collecting ecological data relating to each of several common species types.

Step 4: Describe the Proposed Changes in the Selected Transportation System Action The type and scale of the transportation action should be identified because they influence the mechanism and scope of the ecological impact to be determined. All expected structures, including main, auxiliary, and other structures related to the transportation project should be considered. For example, for culvert or bridge construction, ecological impacts at the crossing site as well as the embankments, wing walls, or entire lengths of the riprapped culvert/bridge inlet and outlets should be assessed.

Step 5: Identify the PDP Phases of the Transportation Action That Are Likely to Affect the Ecology In addition to construction, other phases, such as operations, can result in significant ecological impacts. For example, increased speed limits on a highway passing through a woodland area can increase the number of vehicle–animal collisions.

Step 6: Select the Appropriate Ecological Performance Measures To establish a set of characteristics to be used for measuring the extent and magnitude of ecological impacts, the performance measures discussed in Section 12.4 can be considered. The quantitative value of each performance measure before and after the transportation action should be assessed. Table 12.4 presents performance measures that could be used for common types of species.

Step 7: Analyze Data to Predict the Ecological Inventory after the Transportation Intervention Data from past observations may be used to develop statistical regression or simulation models for predicting the ecological impacts of planned projects. The complexity and dynamic nature of ecosystem responses to external stimuli may not always be captured easily in a mathematical model. An alternative method for ecological impact prediction, therefore, is to solicit the professional opinions of biologists and ecologists based on their prior predictive experience, familiarity with similar ecosystems, formal

Table 12.3   Data Collection Techniques by Species Type

Species	Data Collection Techniques
Large animals	Small area: Direct inspection and count in the entire area Large area: Direct inspection and count in arbitrary plots and project counts over the total area of appropriate habitation. Open topography: Estimates based on aerial photographic prints by experienced photointerpreters. Other techniques: Consult local wildlife biologists from federal/state wildlife agencies.
Birds of prey	Common species: Usable population statistics and information can be retrieved from local wildlife biologists from federal/state wildlife agencies. Less common species: Estimates and locations of nesting, breeding, and feeding areas can be obtained from biologists affiliated with the Audubon Society, Natural Heritage Program, or related wildlife protection agencies. The change in sizes of nesting and feeding areas can be estimated using GIS or planimeter based on before and after superimposition of aerial photographs, mosaics, or topographic maps.
Small game	Similar to those discussed for birds of prey. Consultation with the U.S. Fish and Wildlife Service is mandatory if the species are threatened or endangered.
Water-based animals	Fish and shellfish: Measure water quality parameters such as dissolved oxygen content, pH level, and coliform bacteria count. Waterfowl: Measure change in length of shoreline of suitable habitat and number of individual water bodies using GIS or a mechanical tool (map measurer) based on before and after superimposition of aerial photographs, project plans, or maps. Using this information, a wildlife biologist/specialist can provide an estimate of changes in the quantity of pairs of nesting waterfowl that can be supported by the habitat.
Field crops	The acreage of field crop land to be acquired can be estimated directly using GIS or a planimeter. These numbers can be also established from local offices of the Farm Services Agency (FSA) of the U.S. Department of Agriculture. The size (area) of field crop to be affected by ROW herbicidal spraying can be estimated as a function of type of application system, wind direction, and velocity.
Listed species	The procedure for counting or estimating the number of listed species is similar to that discussed for large animals, predatory birds, and natural vegetation with assistance from wildlife biologists, zoologists, botanists, and plant physiologists.
Native vegetation	Change in acreage of native vegetation can be estimated directly using GIS or planimeter based on before and after superimposition of aerial photographs or maps of vegetative cover by a skilled photointerpreter.
Aquatic plants	The change in acreage of aquatic plant habitat area and the change in total available aquatic habitat can be estimated and derived, respectively, using GIS or a planimeter based on before and after superimposition of large-scale aerial photographs by a skilled photointerpreter. Infrared photography and remote sensing are particularly helpful in such procedures.

Source: Adapted from Jain et al. (2001).

education and knowledge, and knowledge of the species under investigation. A Delphi approach could be used to refine the survey results. The outcome of data analysis should be in terms of the measures of ecological performance selected, such as destruction, injury, or relocation of plants and animals, and destruction or degradation of their physical base.

Some of the performance measures discussed in Section 12.4, such as those of a biochemical nature, are rarely used at some agencies, due to the practical difficulties in measuring transportation impacts on specific target species in terms of such measures. For specific transportation activities whose impacts may require the consideration of such measures, relatively simple models such as those that predict the chemical concentrations derived from the physical processes of transport and dilution may be used. The main concern, typically, is the ecological performance in terms of animal populations,

Table 12.4 Performance Measures on the Basis of Species Type

Species	Possible Performance Measures[a]
Large animals, birds of prey, small game, listed species	Number and diversity, amount (acres) of land available for ranging, feeding, nesting, or shelter or for all combined.
Water-based animals	Dissolved oxygen content, coliform bacteria levels, acidity levels (pH), heavy metal concentrations, and pesticide concentrations detrimental to fish and shellfish life; quantity of suitable nesting habitat for waterfowl which is associated with the length of shoreline.
Field crops, native vegetation	Size of area affected (acreage).
Aquatic plants	Change in amount of water area suitable for the growth of aquatic plants, elements of water quality that accelerate or restrict plant growth.

Source: Adapted from Jain et al. (2001)
[a]To be measured before and after the transportation activity.

particularly animals having economic and social significance, or those protected by national legislation or international treaties (Lohani et al., 1997). Most ecological assessments of transportation activities are carried out on the basis of larger, more measurable performance measures. Three methods for doing this are as follows: population dynamics modeling (which considers species recruitment and survival rates), habitat-based methods, and gap analysis.

(a) *Population Dynamics Modeling* Population dynamics models serve to predict changes in animal population over time. In assessing the impacts of a transportation activity, an analyst should first predict changes in the factors that affect species recruitment (procreation and immigration) and survival. The fundamental population dynamic model equation (Walters, 1986) is

$$N_{t+1} = s_{at}N_t + s_{jt}R_t \qquad (12.2)$$

where N_{t+1} is the total number of animals predicted at a time $t + 1$ in the annual cycle, N_t is the total number of animals at time t in the annual cycle, R_t is the recruitment to the population during the time cycle between phases t and $t + 1$, s_{at} is the survival rate of animals (N_t) during the time cycle between phases t and $t + 1$, and s_{jt} is the survival rate of new recruits (R_t) during the time cycle between phases t and $t + 1$.

After estimating the recruitment and survival parameters, equation (12.2) may be used to predict changes in population (Figure 12.6). For each species, equation 12.2 enables the projection of population changes as a function of time. The recruitment and survival rates are typically modeled as functions of other ecological parameters and outside interventions such as transportation activities (Lohani et al., 1997). For example, a recruitment model for a rabbit population could be a function of the existing population, size and quality, fertility rates, availability and quality of breeding habitat, net migration, and availability of food (suitable vegetation). Also, survival rates may be expressed as models that are functions of existing population, habitat quality, and water availability and

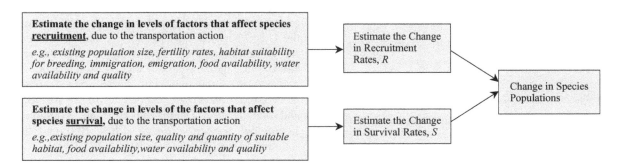

Figure 12.6   Methodology for assessing change in species population based on recruitment and survival rates.

quality. A detailed description of how such a model could be developed and validated is provided in Walters (1986).

Example 12.2 Every year, from the spring season to the summer season, 720 new deer are born in a certain woodland area with a total deer population of 2500. The average survival rates of the deer and their newborn are 0.8 and 0.68, respectively. The construction of a railway line through the area (completion slated for the spring of 2006) is expected to affect the quality of life of the existing deer and their survivability adversely by exterminating the deer directly and by contributing to a reduction in their fertility rates, degradation of habitat suitability for deer breeding, and reduction of their food and water availability and quality. On the basis of such impacts, a team of expert ecologists has estimated that the new facility will reduce deer recruitment by 10% and will cause a decrease in the survival rates of existing and newborn deer by 40% and 15%, respectively. Using the population dynamics model, estimate the change in deer population due to the railway project. Assume that no new deer enter the study area from outside.

SOLUTION   Estimated population at the end of spring 2006, $N_t = 2500$.

Without the transportation project (no-build scenario):

Deer survival rate, $s_{at} = 0.8$

Recruitment to the deer population, $R_t = 720$

Survival rate of new recruits, $s_{jt} = 0.68$

Therefore, the expected population in summer,

$$N = (2500 \times 0.8) + (720 \times 0.68) = 2490$$

With the transportation project:

Deer survival rate,

$$s_{at} = 60\% \text{ of } 0.8 = 0.48$$

Recruitment to the deer population,

$$R_t = 90\% \text{ of } 720 = 648$$

Survival rate of new recruits,

$$s_{jt} = 85\% \text{ of } 0.68 = 0.578$$

Therefore, the expected population in summer,

$$N = (2500 \times 0.48) + (648 \times 0.578) = 1574$$

The transportation project will thus cause a reduction in the deer population by $2490 - 1574 = 916$.

This represents a 37% lower population compared to the no-build scenario.

(b) *Habitat-Based Evaluation* The Habitat Evaluation Procedure (HEP) developed by the U.S. Fish and Wildlife Service (USFWS, 1980) can be used to assess the impact of changes in habitat conditions on the potential of a given area to sustain certain animal species. This method is consistent with the biological concept of habitat *carrying capacity* (the maximum number of individual organisms that can be supported by a given habitat for an indefinite period of time). The major steps involved are discussed below (Ortolano, 2001):

1. Divide the study area into clusters of homogeneous cover types. The cover types may be terrestrial communities (such as forests, woodlands, and grasslands) or aquatic zones (such as rivers and lakes). The areas are grouped such that there is minimum variation of cover type within clusters and maximum variation between clusters. The analysis is then carried out separately for each cluster.
2. Identify various species to be evaluated. As there are typically several species in a given area, evaluative species are selected on the basis of whether they are a rare or a keystone species. For example, a specific species may be selected for its critical role in the food chain or because it is in danger of extinction. Each species type selected is then considered separately in the analysis.
3. Describe the baseline habitat conditions. Habitat conditions are generally expressed in habitat units (HU), computed as follows:

$$\text{HU} = \text{AREA} \times \text{HSI}$$

where AREA is the area of the habitat under investigation and HSI is the habitat suitability index.

The habitat suitability indices are developed by various agencies for different species, corresponding to certain sets of physical and chemical characteristics based on expert opinions of ecologists and biologists. For example, the U.S. Fish and Wildlife Service has developed charts for computing habitat suitability indices for red-tailed hawks on the basis of such characteristics as the percentage of the area with herbaceous canopy cover, percentage herbaceous canopy 2 to 18 in. in height, and tree density. Details on HEP are available in the *Ecological Services Manual* (ESM 102) and the HEP training course HEP500 (Lohani et al., 1997).

Example 12.3 A survey of expert ecologists was carried out to derive the suitability of a certain habitat for species

Table E12.3   Results of Ecologists Survey

Expert	Level of Characteristic (%)	Habitat Suitability Index
1	0	0
	20	0.02
	40	0.10
	60	0.41
	80	0.84
	100	1.00
2	0	0
	20	0.12
	40	0.17
	60	0.51
	80	0.85
	100	1.00
3	0	0
	20	0.11
	40	0.32
	60	0.78
	80	0.97
	100	1.00
4	0	0
	20	0.17
	40	0.50
	60	0.71
	80	0.96
	100	1.00
5	0	0
	20	0.20
	40	0.22
	60	0.63
	80	0.95
	100	1.00

survivability under various conditions. Each condition is associated with a certain level of a given characteristic. The survey yielded results shown in Table E12.3. Develop a habitat suitability curve for the species under study.

SOLUTION   The responses from the survey can be averaged and plotted, as shown in Figure E12.3.1.

Determine the ecological impacts. For the species and region selected, the impact of the proposed project may be expressed as a relative or percentage change in habitat conditions before and after the project development.

There are two ways by which an analyst can measure the decrease (or rarely, the increase) in habitat units following a transportation project:

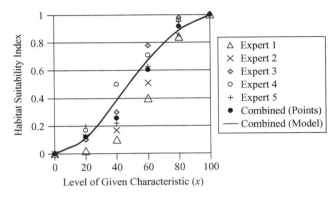

Figure E12.3.1   Species habitat suitability curve.

(i) The immediate jump in the habitat index:

$$\Delta HU = HU_B - HU_A$$
$$= (AREA_B \times HSI_B) - (AREA_A \times HSI_A)$$

where B represents the period before the project and A the period after the project conditions.

(ii) The long-term change in the habitat index, which is represented by the shaded area between the before- and after-project HU-time graphs shown in Figure 12.7

If the HU-time functions are known or can be estimated using statistical regression of field data, the ecological impacts in terms of change in area under the HU-time curve can be estimated as follows:

$$\int_{t_1}^{t_2} [HU_B(t)]\, dt - \int_{t_1}^{t_2} [HU_A(t)]\, dt$$

It may be more appropriate to use the long-term measure where the habitat suitability or area (or both) changes significantly over time, in addition to the immediate jump (decrease) in response to the transportation project.

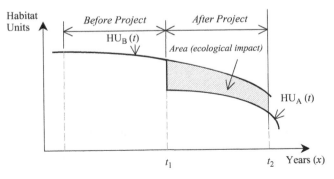

Figure 12.7   Long-term ecological impacts of transportation actions.

Example 12.4 A planned transportation project is expected to decrease the area of a nearby wetland from 3 acres to 2.31 acres. A local ecologist also predicts that the availability of clean water for the animals and plants will decrease from 90% to 65%. Determine the reduction in the habitat suitability index assuming that water availability is the ecological characteristic discussed in Example 12.3. Use the curve developed in Example 12.3. Calculate the overall short-term change in habitat units following the transportation project.

SOLUTION

The ecological characteristic is water availability.

Before the transportation project: area = 3 acres, level of ecological characteristic = 90%; thus, $HSI_B = 1.0648$ $\{1 - \exp[-(0.00028)(90)^2]\} = 0.95$ (from the curve in Example 12.3); therefore, $HU_B = (3)(0.95) = 2.85$.

After the transportation project: area = 2.31 acres, level of ecological characteristic = 65%; thus, $HSI_A = 0.74$; therefore, $HU_A = (2.31)(0.74) = 1.71$.

Immediate ecological damage (in terms of habitat units) $= (2.85 - 1.71)/2.85 = 40\%$ reduction in habitat units.

Example 12.5 Table E12.5 shows the past (years 1999–2006) and expected (years 2007–2012) habitat suitability indices and areas (acres) of a wetland ecosystem that will be traversed by a new railway project. The indices represent the average estimates and predictions from a number of ecological experts. Determine the long-term ecological damage expected to be caused by the rail project.

SOLUTION   The changes in overall habitat units over the years are shown in Figure E12.5.

Ecological damage

$$= \int_7^{13} (-0.1866t + 12.7040)\, dt$$

$$- \int_7^{13} (-0.0334t^2 + 0.5341t + 4.1590)\, dt$$

$$= \left(\frac{-0.1866t^2}{2} + 12.7040t \right)_7^{13}$$

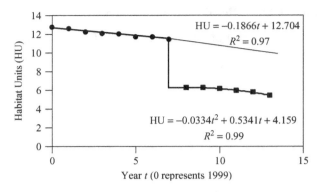

Figure E12.5   Yearly distribution of habitat units.

$$- \left(\frac{-0.0334t^3}{3} + \frac{0.5341t^2}{2} + 4.1590t \right)_7^{13}$$

$$= 65.028 - 36.359 = 28.669 \text{HU-years}$$

(c) *Gap Analysis*   Strictly speaking, gas analysis is a method of identifying (not necessarily measuring) the impact of development on ecosystems (Ortolano, 2001). However, with its map overlay feature, gap analysis is often a first step in impact evaluation. In gap analysis, vegetation maps are created using satellite imagery that distinguishes areas with different types of vegetation, grouped in the form of polygons. Maps showing the distribution of animal species are also used. A map of the proposed development, as well as the boundaries of its expected ecological impacts, is also obtained. By overlapping both maps, the analyst can identify areas that will be affected ecologically by the planned transportation project. A survey of ecological experts can be used to generate extent and severity data from which GIS maps can be developed. This method and others have often been used to identify areas that will be affected by planned projects (Farrall, 2001; Agrawal et al., 2005).

Step 8: Estimate the Ecological Impacts (Change in Level of Performance Measure)   The analyst must determine whether the estimated changes in ecological conditions are significant enough to merit a serious review of the transportation project plan. For new transportation facilities being planned, the absolute change in the level of performance measure before and after facility construction

Table E12.5   Data on Habitat Suitability Index and Wetland Area

Year	1999	2000	2001	2002	2003	2004	2005	2006	2007	2008	2009	2010	2011	2012
HSI	0.85	0.84	0.82	0.81	0.81	0.79	0.79	0.77	0.56	0.56	0.55	0.53	0.52	0.49
Area (acres)	15	15	14.9	14.9	14.8	14.8	14.8	14.8	11.2	11.2	11.2	11.2	11.1	11.2

is taken as the impact of the planned project. For existing transportation facilities that are planned for expansion or improvement, the ecological impacts are (1) the absolute change in the level of the performance measure before and after facility expansion or improvement only if there is no change in the level of use after the project, and (2) changes in the level of the performance measure per unit use (because increased use due to increased capacity may lead to increased ecological impacts but decreased ecological impacts per vehicle or per passenger). In other words, due consideration given to ecological impacts at the expansion planning stage may cause a decrease in the ecological "price" of travel.

Step 9: Evaluate the Ecological Impacts Predicted
In this task, the analyst reviews laws, regulations, and policies selected for the protection of specific habitats, special species, or the ecosystem in general compared with ecological conditions predicted (on the basis of performance measures) and determines whether any standards will be violated by the transportation project proposed. The assumption is that if environmental or chemical standards are not violated, the corresponding environmental effects are not likely to be significant. This assumption has not yet been fully tested and may not always be valid (Lohani et al., 1997). The analyst may solicit the input of active stakeholders, such as environmental groups, governmental agencies responsible for ecological protection, and the general public. Also, the analyst may solicit the expert opinions of botanists, zoologists, and ecologists for a rational and scientific perspective of the impacts.

12.6   KEY LEGISLATION

12.6.1 Endangered Species Act of 1973

The ecology-related legislation with the greatest impact on transportation project EIS is the Endangered Species Act (ESA), which was passed in 1973 to protect and recover endangered and threatened species of fish, wildlife, and plants. *Endangered species* are *"species which are in danger of extinction throughout all or a significant portion of its range,"* while *threatened species* are *"species which are likely to become an endangered species within the foreseeable future throughout all or a significant portion of its range"* (USFWS, 2006). The list of endangered or threatened species is updated periodically and published in the *Federal Register* and the Web site of the U.S. Fish and Wildlife Service. ESA requires federal agencies to consult with the responsible agency that has jurisdiction over an endangered species: terrestrial and freshwater organisms under the Interior Department's U.S. Fish and Wildlife Service (FWS) and marine species such as

salmon and whales under the Commerce Department's National Marine Fisheries Service (NMFS). This is to ensure that a proposed transportation intervention will not pose a threat to any endangered or threatened species or habitat. The ESA also requires that the federal agency, or its contractor, carry out and prepare a biological assessment report to identify any affected listed species, which is a separate document but could be incorporated in the text of (or in an appendix) to an environmental assessment (EA) or an environmental impact statement (EIS). The ESA was amended significantly in 1978, 1982, and 1988, while retaining its general framework and structure. The complete ESA document of 1993 and its history can be obtained from the U.S. Fish and Wildlife Service Web site (www.fws.gov/endangered/).

12.6.2 Laws Related to Wetlands and Other Habitats

Various ecology- and wetland-related legislation in the United States since the 1920s has had a profound impact on wetlands and other habitats. In the 1970s, wetlands were finally recognized as a significant ecological resource, and specific focus on the protection of such ecosystems was thus set in motion by (1) Executive Order 11990 (Wetlands Protection), requiring all federal agencies to "take action to minimize the destruction, loss, or degradation of wetlands and enhance the natural and beneficial values of wetlands" in the course of their duties; and (2) Executive Order 11998 (Floodplain Management), requiring similar protection for floodplains, including avoiding activity in floodplains when possible. Federal legislation and policies for protecting wetlands from further destruction have since gained momentum. These include laws geared toward land acquisition, incentives for ecological preservation, regulation of natural resource use, restoration of the ecology, and other considerations. In 1997, the Clean Water Initiative was announced. This initiative included the goal of having a net gain of 100,000 wetland acres annually, starting in 2005. The initiative included protection and improvement of wetlands through enhanced federal restoration programs and the expansion of wetland restoration incentives to landowners. In 2002, the Supreme Court limited jurisdiction of the Clean Water Act to isolated and nonnavigable intrastate wetlands, lakes, rivers, and streams. In response to the ruling, HB 5949 reasserted state authority to regulate tributaries and other inland wetlands. Table 12.5 provides the major federal regulations relevant to ecology and wetlands in the United States.

12.7   MITIGATION OF ECOLOGICAL IMPACTS

Mitigation measures for ecological impacts may be categorized as follows: preproject mitigation (avoidance, minimization) and the less-preferred postproject mitigation

Table 12.5   History of Ecology- and Wetland-Related Federal Legislation

Year	Legislation	Description
1899	Rivers & Harbors Appropriation Act	Regulated actions affecting navigation in United States waters, including wetlands.
1918	Migratory Bird Treaty Act	Protected most common wild birds.
1929	Migratory Bird Conservation Act	Approved acquisition of land and water areas for use as reservations for migratory birds.
1934	Fish and Wildlife Coordination Act	Authorized interagency cooperation to protect, rear, stock, and increase the supply of game and fur-bearing animals, and to study the effects of polluting substances on wildlife.
1940	Bald Eagle Protection Act	Protected the bald eagle and golden eagle by generally prohibiting their capture, possession and commerce.
1964	Wilderness Act	Established criteria and restrictions on activities that can be undertaken on a designated land and water in the National Wildlife Refuge System for conservation purposes.
1965	Land and Water Conservation Fund Act	Regulated admission and special recreation user fees at recreational lands and established a fund to finance state and federal acquisition of lands and waters for recreational and conservation purposes.
1965	Anadromous Fish Conservation Act	Authorized investigations, engineering and biological surveys, research, stream clearance, construction, maintenance and operations of hatcheries, devices, and structures for improving movement, feeding and spawning conditions of fishes.
1966	Department of Transportation Act	Required the conservation of the countryside, publicly owned park and recreation lands, wildlife and waterfowl refuges, and significant historic sites.
1968	Wild and Scenic Rivers Act	Established rules for identifying wild and scenic rivers and their special management.
1969	National Environmental Policy Act	Encouraged prevention of damage to the environment and biosphere, encouraged the understanding of ecological systems and vital natural resources.
1971	Wild Horses and Burros Protection Act	Encouraged the protection of wild and free-roaming horses and burros from capture, branding, harassment, or death.
1972	Marine Mammal Protection Act	Encouraged conservation of marine mammals such as the sea otter, walrus, polar bear, dugong, manatee, cetacean and pinniped.
1972	Federal Insecticide, Fungicide, and Rodenticide Act	Established control on pesticides application to protect habitat and wildlife.
1972	Marine Protection Research and Sanctuaries Act	Established limits on ocean dumping of any material that could adversely affect human health, welfare, amenities, the marine environment, ecological systems, or economic potential.
1973	Endangered Species Act	Encouraged conservation of threatened/endangered fauna and flora and their habitats.
1976	Toxic Substances Control Act	Enabled EPA to track the pathways of industrial chemicals produced or imported into the United States.
1980	Comprehensive Environmental Response, Compensation and Liability Act (CERLA)	Provided for the liability, compensation, cleanup and emergency response of sites contaminated by hazardous substances through Hazardous Substance Superfund.
1986	Emergency Wetlands Resources Act	Encouraged conversation of wetlands to sustain the human values and benefits provided by wetlands.
1986	Superfund Amendments and Reauthorization Act	Encouraged use of permanent remedies and innovative treatment technologies in cleaning up hazardous waste sites.

Table 12.5 (*continued*)

Year	Legislation	Description
1988	Great Lakes Coastal Barrier Act	Established protection for undeveloped coastal barriers and related areas by prohibiting direct or indirect federal funding of projects in such areas that might support development and minimizing damage to fish, wildlife, and other natural resources in these areas.
1989	North American Wetlands Conservation Act	Provided funding and administrative direction for implementation of the North American Waterfowl Management Plan and the Tripartite Agreement on wetlands between Canada, Mexico and the U.S.
1990	Oil Pollution Act	Enabled EPA to prevent/respond to catastrophic oil spills and leaks from vessels and oil storage facilities.
1990	Coastal Zone Management Act Reauthorization Amendments	Established controls of nonpoint source pollution for activities located in coastal zones to protect estuarine and marine habitats and species.
1990	Coastal Wetlands Planning, Protection and Restoration Act	Provided supports and funds for coastal wetlands restoration and conservation projects, especially in Louisiana.
1991	Intermodal Surface Transportation Efficiency Act (IS-TEA)	Provided for environmental conservation through highway funds to enhance the environment, such as wetland banking, mitigation of damage to wildlife habitat.
1996	Federal Insecticide, Fungicide and Rodenticide Act	Provided federal control/monitoring of pesticide distribution, sale, and use.
1996	Magnuson-Stevens Fishery Conservation and Management Act	Provided for conservation of essential fish habitat needed for spawning, breeding, feeding or growth to maturity.
1996	Federal Agriculture Improvement and Reform Act (Farm Bill)	Included programs to conserve wetlands on agricultural land.
1998	Transportation Equity Act for the 21st Century (TEA-21)	Authorized funding to conserve the environment, including water quality improvement and wetlands restoration.
2005	Safe, Accountable, Flexible, Efficient Transportation Equity Act: A Legacy for Users (SAFETEA-LU)	Addressed environmental protection in transportation through environmental stewardship and streamlining.

Source: Adapted from Canter (1995), Evink (2002), and other sources.

(rectification, preservation, and compensation) as discussed by Sinha et al. (1991). Avoiding ecologically sensitive areas is the best approach. In many cases it is not possible to avoid such areas, so appropriate measures need to be recommended to minimize adverse impacts. In April 2004, a new federal policy was declared that would protect 2 million acres of wetlands and increasing the total wetland area in the U.S. by 1 million acres over the period 2004–2009. The policy also advocated *wetland banking* (replacement of destroyed wetlands by new wetlands of at least equal quality at other locations). In recommending replacement wetland construction as a palliative to mitigate expected ecological damage by a proposed transportation project, the analyst should, to the extent possible, ensure that the replacement wetland has ecological parameters (biodiversity, resilience, productivity, ecological function, and hydrological/hydraulic connectivity to the environment) similar to those of the original natural wetland.

12.7.1 Mitigation at Various Phases of the Project Development Process

(a) *System Planning Phase* Subject to resource and other constraints, attempts should be made to select transportation modes, alignments, and facility plans whose construction and operations would directly or indirectly cause the least adverse ecological effects. Ecological preservation and sustainability should be advocated, encouraged, and pursued.

(b) *Locational Planning Phase* In locating an alignment within a corridor, planners should give due consideration to the existing locations of wetlands, woodlands, bushlands, grasslands, and forests, as well as the population and diversity of plants and animals supported by such habitats. Habitats to be avoided during corridor and alignment location are those containing regionally unique abiotic and/or biotic components, such as rare, threatened, or endangered species. Areas designated as protected areas by the national or state/provincial government should be avoided. Where there is no choice but to locate a transportation facility in an ecologically sensitive area, a restitution component to replace the affected area, such as replacement wetlands, should be included as part of the project. In some cases, land buffers can be established between the transportation alignment and sensitive areas. Legislation to control land use can also minimize the impacts of land developments associated with transportation improvements in remote areas.

(c) *Design Phase* Facility elements can be designed in a manner that helps palliate the ecological impacts. The design of structures can include features that provide erosion control (e.g., slope surface protection at embankments, riprap lining of bridge/culvert inlets). The common design of structural features related to wildlife protection include fences, culverts (drainage, stream, or dry), bridges (underpass or extended), extended existing transportation facility, viaducts, and overpasses (Evink, 2002). Less common structural and nonstructural techniques include signage, motion sensors, warning devices, road lighting, interception feeding, deer reflectors, infrared sensing devices, and PVC pipe pole installation. Furthermore, habitat reengineering techniques can be employed and typically include habitat restoration or preservation, such as land banking, habitat purchasing, tree and native grass planting, and stream undertaking. Material types, construction activity timings, and facility use schedules can be specified to minimize adverse ecological effects associated with the construction, operations, and maintenance phases of transportation facility development. These include restricting pesticide use in the right-of-way, routing of vehicles carrying hazardous substances, specifying land clearing and excavation (including blasting) operations at specific times of the year, season, or day when ecological damage is expected to be minimal, and selection of hardy plant species for vegetating the right-of-way and median. The effect of transportation operations and the area taken by transportation facilities at ecologically sensitive areas may be reduced through operational policies and designs, such as speed limits, noise-absorbing pavement materials, or noise barriers, bridging instead of filling, reducing embankment base widths (by increasing slope steepness), and geometric design features that encourage low vehicular operating speeds.

The design task need not be restricted to the facility itself but may extend to its immediate environment.

For example, the design can include stabilization of soils within and beyond the facility right-of-way through seeding, planting, or fertilizer application. Also, hydraulic structures associated with the facility (e.g., side ditches, culverts and bridges, turnouts) can be designed to augment water levels in nearby wetlands, particularly during periods of drought. Furthermore, artificial ecological features can be provided as part of the facility design, such as artificial nests and other shelters for ecosystem wildlife. Finally, new wetlands can be designed as part of the transportation facility design process.

(d) *Construction Phase* Construction supervision should be carried out to ensure compliance with environmental design specifications and staging requirements and other special design features geared toward ecosystem preservation. At this phase of the PDP, an environmental expert can monitor the project site and sensitive proximal ecological resources, prepare environmental reports, and carry out periodic sanitary surveys of construction site facilities. This phase may include construction of artificial ecological features, including wetlands.

(e) *Maintenance Phase* The extent and type of ecological impact mitigation at the maintenance phase will be influenced to some extent by ecological considerations at the design phase of the PDP. Hydraulic erosion control and ecological structures, such as culverts, side drains, turnouts, and fish ladders, need to be inspected and maintained regularly. Vegetated areas within and near the transportation facility should be maintained with a view to minimizing harm to flora and fauna and their habitat. As a part of routine maintenance, ecological monitoring could be carried out by inspecting receiving wetlands, streambeds, and other aquatic biota for any signs of diminished ecological quality.

12.8 METHODS AND SOFTWARE PACKAGES FOR ECOLOGICAL IMPACT ASSESSMENT

A list of software packages for estimating the impact of transportation activities on surrounding ecosystems, mostly wetlands, is given in Table 12.6. Some of these are discussed below.

12.8.1 Wetland Functional Analysis

Wetland Functional Analysis (WET II), a package developed for the FHWA by the U.S. Army Corps of Engineers, addresses the functional features of wetland ecosystems, such as wildlife habitat and quantity and quality attributes of the wetlands (Adamus et al., 1987, 1991). Analysts can use the package to predict the

Table 12.6   Selected Methods and Software Used in Ecological Impact Assessment

Method/Software	Purpose	Application
HEP/HSI (habitat evaluation procedure/habitat suitability indices)	Evaluates the quality and quantity of available habitat for selected wildlife species; HEP may be used in three planning activities: wildlife habitat assessments (including both baseline and future conditions), trade-off analyses, and compensation analyses.	Mainly wetland, aquatic, and terrestrial habitats.
HGM (hydrogeomorphic) approach	Evaluates wetland functions in the Section 404 Regulatory Program and other regulatory, planning, and management cases.	All types of wetlands.
Hollands–Magee method	Evaluates wetland functions in the Section 404 Regulatory Program and other regulatory, planning, and management cases.	Nontidal wetlands in the glaciated northeast and midwest regions of the United States.
IBI (index of biological integrity)	Evaluates the biological integrity of a habitat using samples of living organisms and to assess the impacts of human activities on biological systems; developed for use in managing aquatic resources (e.g., to establish use designations for water bodies, biological water quality standards, or goals for restoration).	Various types of habitats, including wetlands, streams, and lakes.
IVA (indicator value assessment)	Measures the socially important function performance within a wetland and the relative social importance and value of that wetland within a planning region or watershed; may be used to (1) evaluate potential impacts from different development actions, (2) recognize the needs for compensation within a planning region, and (3) evaluate the potential of different wetlands for enhancement.	All types of wetlands.
NEFWIBP (New England freshwater wetlands invertebrate biomonitoring protocol)	Offers a standardized and inexpensive tool for evaluating the impact of development on constantly flooded freshwater wetlands; may also be used to catalog wetlands within a watershed, to assess the effectiveness of restoration measures, to monitor the progress of created wetlands or mitigation procedures, and to guide watershed management through risk assessment.	Nontidal freshwater and constantly flooded wetlands in the New England region.
PAM HEP (Pennsylvania modified 1980 habitat evaluation procedure)	Measures the baseline habitat conditions for a species, assess the direct impacts of construction activities on these conditions, and identify a mitigation plan to alleviate these impacts.	Mainly wetland, aquatic, and terrestrial habitats in the Pennsylvania region.

Source: USACE (2001)

qualitative likelihood (high, medium, or low) of wetland performance given specific ecological performance functions before and after a transportation intervention. These functions include groundwater recharge and discharge, flood-flow alteration, sediment stabilization, sediment and toxicant retention, nutrient removal or transformation, aquatic diversity and abundance, wildlife diversity and abundance, recreation, and uniqueness and

natural heritage concerns, and species-specific fish and wildlife habitat assessments (Thiesing, 2005). WET II offers a quick screening of all transportation alternatives that interrupt any of these wetland functions.

12.8.2 Hydrogeomorphic Classification Method

The Hydrogeomorphic Classification Method (HGM) is a reference-based wetland functional assessment technique that offers the analyst a set of baseline data on a typical wetland (Brinson, 1993). The analyst compares the attributes of a wetland located in the proximity of a transportation project with those of a reference system, and thus estimates relative impacts in terms of wetland performance (Evink, 2002).

12.8.3 Habitat Evaluation Procedures Software

The principles underlying the HEP package are explained in Section 12.5, step 7(b). Given the area of available habitat and the habitat suitability index (HSI) for various types of species, HEP can assess the ecological impacts of alternative transportation projects or policies. Furthermore, the analyst can use HEP to assess the effectiveness of efforts that seek to mitigate adverse ecological impacts.

SUMMARY

Ecosystems, which are typically classified on the basis of the general biotope type, the dominant species, or a physical feature, include terrestrial, freshwater, coastal, arboreal, and wetlands. Wetlands receive special attention due to their vital ecological functions of remediation, hazard management, and ecological preservation. For any ecosystem, transportation projects and policies can disrupt the physical and functional relationships between biotic elements and their abiotic environments. Direct or indirect transportation disturbances to ecosystems, through mechanisms such as physical base degradation and direct depletion of biotic elements, can lead to food chain disruptions and consequently, energy cycle impairment. In serious cases, such disturbances can result in adverse ecological succession, loss of productivity, or even ecological destabilization, particularly in nonresilient ecosystems. The mechanism of impact depends on the species type under investigation. Ecological impacts may be categorized by the stage of the transportation development process at which they occur. In this chapter we presented a framework for ecological impact assessment that begins with a definition of the analysis area and temporal scope and the affected species, a description of the transportation intervention and relevant PDP phases, data collection, data analysis to determine the ecological impact levels, and an evaluation of the impact levels vis-à-vis established standards. Data analysis can be carried out using models such as the population dynamics model, gap analysis, or the habitat-based evaluation method. Mitigation can be carried out through avoidance and minimization or through rectification, preservation, and compensation, and can be incorporated as part of each phase of a PDP. Common computer software packages for ecological impact assessment of transportation activities or for evaluating the

efficacy of mitigation activities include WET-II, HGM, and HEC.

EXERCISES

12.1. The existing population distribution of animal species in a certain woodland ecosystem is estimated as follows: cougars, 540; white-tailed deer, 4530; opossum, 2320; caribou, 35; and skunks, 2500. Two years after construction of a railway line through this ecosystem, the percentage reductions in the animal populations are estimated as follows: cougars, 12%; white-tailed deer, 14%; opossum, 15%; caribou, 8%; and skunks, 15%. Determine the change in species diversity due to the railway project.

12.2. A new railway bridge has been proposed for construction across the Wabash River to alleviate traffic congestion on the existing bridge. A primary concern, however, is that construction of the new bridge will cause erosion of exposed soil surfaces, sedimentation, and release of oil and other construction chemicals that will ultimately lead to a significant depletion of the population of silver carp in the waters traversed by the bridge. Construction is expected to take place in the summer of 2007. It has been estimated that every year from the summer to fall season, 2000 new silver carp hatch out from their eggs and that the total silver carp population in the area is 72,000. The average survival rate of the silver carp and their new offspring are 0.7 and 0.45, respectively. It has also been estimated that the construction will reduce recruitment rate by 35% and will cause decreases in survival rates of the existing and newborn silver carp by 30% and 35%, respectively. Using the population dynamics model, estimate the change in silver carp population due to the railway bridge project. State any assumptions.

12.3. A survey of expert ecologists was carried out to derive the suitability of a certain wetland habitat for survivability of a specific endangered species under various conditions. Each condition is associated with a certain level of food availability, ranging from 0 (no food) to 100 (abundant food). The survey yielded the results shown in Table EX12.3. Develop a habitat suitability curve for the endangered species that inhabit the area under investigation.

12.4. As part of an ambitious multimodal transportation improvement program, it is planned to expand an existing seaport by constructing additional docks, constructing a multimodal transfer facility at the seaport, and widening and upgrading the existing

Table EX12.3   Expert Survey Results

	Expert 1					Expert 2					Expert 3				
Food availability	0	25	50	75	100	0	25	50	75	100	0	25	50	75	100
Habitat suitability index	0	0.21	0.52	0.78	1	0	0.33	0.45	0.68	1	0	0.19	0.63	0.91	1

Table EX12.6   Historical Data on HSI and Wetland Areas

Year	1999	2000	2001	2002	2003	2004	2005	2006	2007	2008	2009	2010	2011	2012
HSI	0.68	0.68	0.68	0.67	0.67	0.67	0.66	0.65	0.65	0.64	0.64	0.64	0.63	0.63
Area (mi^2)	3.5	3.5	3.4	3.4	3.4	3.4	3.4	2.9	2.9	2.9	2.8	2.8	2.8	2.7

highway that links the seaport to the area's industrial center. The area generally has light forest ecology with scattered palustrine wetlands. The highway crosses three major rivers and numerous small streams and creeks.

(a) Discuss the possible direct and indirect ecological impacts of the project at the various stages of the transportation development project.

(b) What steps can be taken to mitigate these adverse effects?

(c) Suggest five ways by which the effectiveness of the proposed mitigation efforts in part (b) can be measured.

12.5. As part of a planned airport runway extension, the area of a nearby 15-acre wetland is expected to suffer a 35% reduction in area, and 90% of the affected area will be replaced with a newly constructed wetland nearby. It is estimated that even with the new wetland, the overall availability of water will decrease from 100% by 30%. (a) Determine the reduction in the habitat suitability index (assuming that water availability is the sole ecological characteristic under consideration). Use the curve developed in Example 12.3. (b) Calculate the overall short-term change in habitat units following the runway project.

12.6. Table EX12.6 shows the habitat suitability indices and areas of a wetland ecosystem that will be affected by a new highway project implemented in 2006. What is the long-term ecological damage due to the project?

REFERENCES[1]

*Adamus, P. R., Clairain, E. J., Smith, R. D., Young, R. E. (1987). *Wetland Evaluation Technique (WET)*, Vol. I and II, Tech. Rep. WRP-DE-2, U.S. Army Corps of Engineers, Waterways Experiment Station, Vicksburg, MS.

*Adamus, P. R., Stockwell, L. T., Clairain, E. J., Morrow, M. E., Rozas, L. D., and Smith R.D. (1991). *Wetland Evaluation Technique (WET)*, Vol. I, *Literature Review and Evaluation Rationale*, Tech. Rep. WRP-DE-2, U.S. Army Corps of Engineers, Waterways Experiment Station, Vicksburg, MS.

Agrawal, M. L., Maitra, B., Ghose, M. K. (2005). Ecological impacts of a highway project using GIS, *GIS Dev. Geospatial Dev. Portal*, Mar., http://www.gisdevelopment.net

Binder, S. (2006). The straight scoop on SAFETEA-LU, *Public Roads*, Mar.–Apr.

Brinson, M. M. (1993). *A Hydrogeomorphic Classification for Wetlands*, Tech. Rep. TR-WRP-DE-4, Wetlands Research Program, U.S. Army Corps of Engineers, Waterways Experiment Station, Vicksburg, MS.

Cambridge Systematics; and Bernardin, Lochmueller & Associates (2003). *I-69 Evansville to Indianapolis: Final Environmental Impact Statement*, Indiana Department of Transportation, Indianapolis, IN. http://www.deis.i69indyevn.org/FEIS/. Accessed Jan. 2006.

*Canter, L. W. (1995). *Environmental Impact Assessment*, McGraw-Hill, New York.

*Erickson, P. A, Camougis, G., Miner, N. H. (1980). *Highways and Wetlands: Interim Procedural Guidelines*, Vol. 1, and *Impact Assessment, Mitigation, and Enhancement Measures*, Federal Highway Administration, U.S. Department of Transportation, Springfield, VA.

*Evink, G. L. (2002). *Interaction Between Roadways and Wildlife Ecology*, NCHRP Synth. Rep. 305, National Cooperative Highway Research Program, Transportation Research Board, National Research Council, Washington, DC.

Farrall, M. H. (2001). *Ecological Impact Assessment of Road Projects*, Road Ecology Center, University of California, Davis, CA.

*Jain, R. K., Urban L. V., Gary S., Balbach, H. E. (2001). *Environmental Assessment*. 2nd ed., McGraw-Hill, New York.

*Lohani, B., Evans, J. W., Ludwig, H., Everitt, R. R., Carpenter, R. A., Tu, S. L. (1997). *Environmental Impact Assessment for Developing Countries in Asia*, Vol. 1, *Overview*, Asian Development Bank, Manila, Philippines, www.adb.org/Documents/Books/Environment_Impact/env_impact.pdf. Accessed May 12, 2005.

*Morris, P., Therivel, R. (2001). *Methods of Environmental Impact Assessment*, 2nd ed., E. & F.N. Spon, London.

*Ortolano, L. (2001). *Environmental Regulation and Impact Assessment*, Wiley, New York.

[1]References marked with an asterisk can also serve as useful resources for ecological impact evaluation.

Regier, H. A. (1993). The notion of natural and cultural integrity, in *Ecological Integrity and the Management of Ecosystems,* ed. Woodley, S., Kay, J., Francis, G. St. Lucie Press, Delray Beach, FL, pp. 3–18.

Shrader-Frechette, K. S., McCoy, E. D. (1993). *Method in Ecology: Strategies for Conservation,* Cambridge University Press, New York.

*Sinha, K. C., Teleki, G. C., Alleman, J. E., Cohn, L. F., Radwan, E. A., Gupta, A. K. (1991). *Environmental Assessment of Land Transport Construction and Maintenance,* Prepared for the Infrastructure and Urban Development Department, World Bank, Washington, DC.

*Thiesing, M. A. (2005). *An Evaluation of Wetland Assessment Techniques and Their Applications to Decision Making,* U.S. Environmental Protection Agency, Region II, New York.

USACE (2001). *Wetland Procedure Descriptions,* U.S. Army Engineer Research and Development Center, U.S. Army Corps of Engineers, Washington, DC. http://el.erdc.usace.army.mil/emrrp/emris/. Accessed Feb. 28, 2006.

* USEPA (1996). *Indicators of the Environmental Impacts of Transportation,* EPA 230-R-96-009, Office of Policy, Planning, and Evaluation, Washington, DC.

USEPA (2004). *Wetlands Overview,* EPA 843-F-04-011a,Office of Water, U.S. Environmental Protection Agency, Washington, DC http://www.epa.gov/owow/wetlands/pdf/overview.pdf. Accessed June 2, 2006.

*USFWS (1980). *Habitat Evaluation Procedures (HEP),* Division of Ecological Services, U.S. Fish and Wildlife Service, Department of the Interior, Washington, DC.

———— (2006). *ESA Basics: 30 Years of Protecting Endangered Species,* Endangered Species Program, U.S. Fish and Wildlife Service, Department of the Interior, Arlington, VA, http://www.fws.gov/endangered/pubs/ESA%20BASICS050806.pdf. Accessed June 2, 2006.

Walters, C. (1986). *Adaptive Management of Renewable Resources,* Macmillan, New York.

ADDITIONAL RESOURCES

Bergman, J. (1999). *Environmental Impact Statements,* 2nd ed., Lewis Publishers, Boca Raton, FL.

Eccleston, C. (2001). *Effective Environmental Assessments,* Lewis Publishers, Boca Raton, FL.

Erickson, P. A., Camougis, G., Robbins, E. J. (1978). *Highways and Ecology: Impact Assessment and Mitigation—Final Report,* Washington Department of Transportation, Federal Highway Administration, Springfield, VA.

Erickson, P. A., Holcomb, B. P. (1980). *Investigation of the Relationship Between Land Use and Wildlife Abundance,* Vol. 1, *Literature Survey,* Institute for Water Resources, Springfield, VA.

Forman, R. J. J., Heanue, K., Jones, J., Swanson, F. (2002). *Road Ecology: Science* and Solutions Island Press Washington DC.

Garbisch, E. W. (1986). *Highways and Wetlands: Compensating Wetland Losses,* Offices of Research and Development, Federal Highway Administration, Springfield, VA.

Journals of the Road Ecology Center, University of California, Davis, CA, http://johnmuir.ucdavis.edu/road_ecology/.

International Association for Impact Assessment (IAIA). www.iaia.org

Ledec, G., Goodland, R. J. A. (1988). *Wildlands: Their Protection and Management in Economic Development,* World Bank, Washington, DC.

Marriott, B. (1997). *Environmental Impact Assessment: A Practical Guide,* McGraw-Hill, New York.

Shipley Group (2003). *How to Write Quality EISs and EAs,* 3rd ed., Shipley Group, Woods Cross, UT.

Sinha, K.C., Varma, A., Souba, J., Faiz., A. (1989). Environmental and ecological considerations in *land transport: A Resource Guide. Infrastructure and Urban Development Department,* Policy Planning and Research Staff, The World Bank, Washington, DC.

CHAPTER 13

Impacts on Water Resources

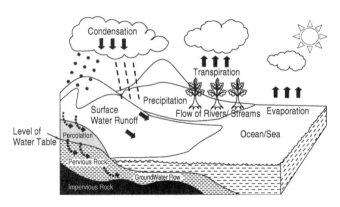

Figure 13.1 Hydrologic cycle.

Thousands have lived without love, not one without water.
—*W. H. Auden (1907–1973)*

INTRODUCTION

Water is an essential resource that is both critical for sustaining life and vital for many human activities, including agriculture, domestic and industrial use, and recreation. It is therefore desirable that the construction, maintenance, and operation of transportation facilities do not significantly deplete or degrade surface or groundwater resources or cause undue disruptions in natural water flow patterns. In this chapter we present performance measures for assessing water resources impacts, and for each transportation mode, identify the stages in the transportation project development process where significant impacts can occur. We also provide a procedural framework for evaluating hydrological or water resources impacts of transportation activities, present key legislation for water protection and conservation, and discuss the mitigation of transportation's adverse effects on water resources.

Hydrologic Cycle: The impacts of transportation on water resources can best be understood against background knowledge of the *hydrologic cycle*. This cycle describes the movement of water, at or below Earth's surface and in the atmosphere, which occurs through seven processes (McCuen, 2005) (Figure 13.1). *Evaporation* refers to the conversion of water molecules into water vapor and is released in the form of vapor into the atmosphere. *Transpiration* is the emission of water from plant leaves. As water vapor rises, it cools and eventually settles on tiny particles of dust in the air, and water particles collect and form clouds, a process known as *condensation*. Condensed water vapor droplets are propelled by air currents from one place to another, and upon cooling,

become saturated with moisture and undergo *precipitation* in the form of rain, snow, or hail. *Runoff* refers to the flow of precipitated water over Earth's land surface into creeks and ditches, and *infiltration* is the process of rainwater or snowmelt entering Earth's surface through pores of exposed permeable soil and through cracks and joints in surface bedrock. Most of this water percolates lower until it reaches groundwater, where it recharges the water table. Groundwater, which is contained in confined or unconfined subterranean pockets and lenses of varying sizes, is recharged continually through percolation and is discharged through streams, natural springs, and artificial wells or boreholes. During each of these processes, or in the course of moving from one process to another, water quality, quantity, and/or movement patterns are vulnerable to direct or indirect disruption or degradation by transportation construction, maintenance, and operations.

13.1 CATEGORIES OF HYDROLOGICAL IMPACTS

13.1.1 Source of Impacts

Transportation impacts on water resources may be due to the fixed transportation facility itself or to vehicles that use the facility. For example, physical transportation structures may disrupt natural water flow patterns, causing flooding in certain areas and water deprivation in others. Runoff from highway or runway surfaces or parking lots may cause water pollution, and the manufacture, upkeep and operation of transportation vehicles involve significant quantities of water. The production of each automobile, for instance, requires 300,000 liters of water on average (Rothengatter, 2003).

337

13.1.2 Impact Types

(a) *Impact on Water Quality (Polluting or Degrading Effects)* This refers to contamination of water that renders it unfit for use by humans or other living organisms. Such contamination may occur through physical, chemical, or biological impact mechanisms. *Physical mechanisms* include impedance or acceleration of water flow, leading to decreased or increased water availability, physical contact between pollutants and water, and subsequent spreading or mixing. Examples include spread of oil slicks and suspended solids. *Chemical mechanisms* include chemical reactions between pollutants and naturally occurring chemicals in the environment or pollutant-catalyzed reactions that would otherwise not take place. *Biological mechanisms* include the disruption of ecological patterns, which in turn affect the quantity or quality of ground or surface water bodies.

(b) *Impact on Water Course (Flow Pattern Effects)* This refers to change in the natural direction of water flow either within a hydrologic phase (e.g., change in stream flow patterns caused by marine channelization) or in moving from one hydrologic phase to another (e.g., change in the locations at which surface runoff joins stream flow due to railway or highway embankment construction).

(c) *Impact on Water Quantity (Deprivation Effects)* These effects refer to a redirection of water within or across hydrologic processes that results in reduced availability of water to humans and organisms for consumption at one location, and/or too much water (flooding) at other locations, possibly causing loss of life or property. For example, construction of highways, airports, and parking lots may lead to an increased fraction of water evaporating from the land surface but a reduction in the fraction of water percolating into the ground, thus depriving groundwater bodies of recharge.

13.1.3 Water Source Affected

(a) *Surface Water Systems* Surface water (rivers, creeks, streams, lakes, lagoons, etc.) is generally vulnerable to transportation impacts, either through reduced water quality or quantity, or disruption of natural flow patterns. Although they do not recover easily from disruptions in flow volumes and patterns, surface water bodies have an inherent natural capacity to recover relatively quickly from quality degradation through the self-purification processes of dilution, sedimentation, flocculation, biodegradation, aeration, and uptake by aquatic plants and animals.

(b) *Groundwater Systems* Compared to surface water, groundwater is relatively less vulnerable to transportation activities, but when groundwater impacts occur, they are relatively more difficult to identify and mitigate. Polluted groundwater undergoes some natural purification as it percolates through soil columns prior to reaching the water table. However, due to the relatively low flow rates associated with groundwater, pollutant removal and dilution are often retarded, and groundwater pollution therefore can remain a localized problem for significant lengths of time. As such, there could be a substantial time interval before groundwater pollution can be observed (Morris and Therivel, 2001).

13.1.4 Transportation Mode and Activity

Activities associated with each transportation mode—highways, rail, air, and marine transportation—can affect water resources adversely. These activities include construction, maintenance, and abandonment of infrastructure facilities; manufacture, maintenance, and disposal of vehicles and parts; and operation. In the next section we discuss in detail for each transportation mode the mechanisms by which transportation vehicles and facilities affect water quality, quantity, and flow patterns in various phases of facility development.

13.2 HYDROLOGICAL IMPACTS BY TRANSPORTATION MODE

13.2.1 Highway Impacts

(a) *Highway Construction* Permanent appropriation of land for highway construction or expansion seems to have the greatest impact on water resources. Other construction-related temporary use of land, such as for depots and road hauls, drilling and excavation activities, disposal of excess material, discovery of hazardous material in the right-of-way, and construction machinery use, may also have significant impacts. The construction of paved surfaces greatly increases the area of impervious surfaces and reduces infiltration of surface water into the ground. For completely natural ground cover, less than 10% of stormwater typically runs off into nearby receiving waters. An increase in paved area will result in increased volume and rate of runoff. For example, about 20% of stormwater runoff can be expected when 10 to 30% of the site is paved (US EPA, 1982; 1996).

Land clearing, blasting, ground excavation, and cut-and-fill operations associated with construction lead to emissions that affect water quality in terms of turbidity, suspended solids concentration, and color of receiving waters. Eroded sediments from exposed ground at cleared areas are deposited at various locations downstream and result in diminished capacities of natural or human-made canals and reservoirs. A long-term hydrological impact study of the H-3 highway construction in Oahu, Hawaii, Hill (1996) determined that the suspended-sediment loads are significantly higher during construction than at other times. The quality of natural water can also be degraded directly by exhaust emissions from construction machinery and haulage vehicles, and spillage during refueling. During construction, contractors working in the right-of-way often encounter unexpected hazardous materials in the form of asbestos, petroleum products, pesticides, cyanides, corrosives, and biological and radioactive wastes stored in underground tanks (NCHRP, 1993) and may accidentally release or spill such compounds, thereby leading to surface water and groundwater contamination.

(b) Street and Highway Maintenance During maintenance, water quality, rather than quantity or flow patterns, suffers the greatest impacts. The most pervasive agent of water degradation is rock salt, the most common deicing agent used in winter road maintenance throughout the United States (USEPA, 1994). Rock salt consists primarily of sodium chloride, but 10% is insoluble residues and concentrations of trace metals, including iron, nickel, lead, zinc, chromium, and cyanide (DEFRA, 2005). In the long term, rock salt promotes pavement disintegration, pavement and bridge reinforcement corrosion, and corrosion of vehicle bodies, all of which subsequently release rust and other material into surface water and groundwater, thereby altering the water chemistry of receiving lakes, rivers, and wetlands (USEPA, 1996). Another polluting source associated with maintenance is bridge paint. In the course of cleaning a bridge spanning a water body, leaded paint chips can be released into the receiving waters below (Figure 13.2). This is a major water quality concern in certain states (Dupuis, 2002).

(c) Highway Operations Pollutants associated with highway operations include heavy metals, suspended solids and particulates, liquids, oxygen-consuming compounds, nutrients, and microorganisms. In many cases, these pollutants are deposited on the roadway directly through vehicle emission or leakage. Liquid pollutants include oil, grease, antifreeze, hydraulic fluids, and cleaning agents, and solid pollutants include debris from brake linings, tires, and other products released by the wear and tear of frictional parts (USEPA, 1996). Over 95% of the solid matter on roadways are deposited by vehicle tires that picked up such pollutants from construction sites, parking lots, farms, and dirt roads (Barrett, 1993). Surface contaminants, including litter, trash, and debris from road users,

Figure 13.2 Rust, paint, deicing chemicals, and other pollutants can impair water quality.

constitute a significant pollution problem, as these solid and fluid substances are washed out during surface runoff and end up in receiving water bodies. Furthermore, transportation operations can contaminate groundwater through leakage from underground tanks bearing transportation fuels (USEPA, 1996). Rainfall simulators have been used to evaluate the impact of transportation operations on the quality of nearby surface water bodies (Barrett, 1993).

13.2.2 Railway Impacts

As shown in Table 13.1, the impacts of railway construction on water resources are similar to those of highway construction discussed in Section 13.2.1. Such impacts include interruption or contamination of natural drainage during construction activity (Carpenter, 1994). Also, hazardous material spillage during operations is a common concern for all transportation modes. During the early 1990s, an annual average of 1100 hazardous material spills occurred during railroad operations in the United States (USEPA, 1996). Other rail sources of water pollution include fuel spills during the refueling of trains, deicing operations on train tracks, herbicide application, release of chemicals during operation or from routine cleaning and maintenance operations, and discharge from toilets (DEFRA, 2005).

13.2.3 Air Transportation Impacts

As seen in Table 13.1, the effects of airport construction and expansion are generally similar to those of highways and railways. Also, airport maintenance and operations can greatly affect the surrounding water quality. In winter, over 12 million gallons of deicing products are applied annually for purposes of runway and aircraft

Table 13.1   Impacts of Transportation Activities on Water Resources

Mode	Transportation Activities	Nature of Impact
Highway	Construction and maintenance of pavements, bridges, tunnels, and parking garages and lots	Embankments and cut sections cause retraining of surface water courses, thus disrupting their natural flow patterns
		Construction and maintenance dust and sediments pollute water bodies
		Transport of deicing compounds (rock salt) into surface water bodies
		Transport of solid matter through highway runoff into surface water bodies
	Manufacture of motor vehicles and parts	Toxic releases and other emissions during manufacture
		Direct use of water in vehicle manufacture
	Highway operations (road vehicle travel)	Hazardous material spills during transport
		Tailpipe and evaporative emissions
		Fugitive dust emissions from roads
		Emissions of refrigerant agents from vehicle air conditioners
		Road surface debris from motor vehicles and road users wash off into streams
	Maintenance of motor vehicles	Contaminant releases during terminal operations: tank truck cleaning, maintenance, repair, and refueling
		Contaminant releases during vehicle cleaning, maintenance, repair, and refueling
		Leaking underground storage tanks containing petroleum products
		Use of water for vehicle washing
	Disposal of motor vehicles and parts	Scrappage of vehicles
		Improper disposal of motor oil and other vehicle fluids
		Disposal of tire, lead–acid batteries, and other consumables
Rail	Construction and maintenance of railway tracks and bridges	Emissions during construction and maintenance
	Manufacture of rail cars and parts	Toxic releases and other emissions during manufacture
	Rail transportation operations (rail travel)	Exhaust emissions
		Spillage of hazardous materials during transport incidents
	Maintenance and support operations for rail cars	Releases during terminal operations: car cleaning, maintenance, repair, and refueling
		Emissions from utilities that provide power for rail
	Disposal of rail cars and parts	Rail car and parts disposal
		Abandonment of rail tracks
Air	Construction, maintenance, or expansion of airports or runways	Emissions during construction and maintenance
		Releases of deicing compounds
		Airport runoff
	Manufacture of aircraft and parts	Toxic releases and other emissions during manufacture
	Air transportation operations (air travel)	High-altitude emissions
		Low-altitude/ground-level emissions
		Hazardous materials incidents during transport
	Airport facility operations	Emissions from ground support equipment involved in aircraft loading, cleaning, maintenance, repair, and refueling
	Disposal of aircraft and parts	Airplane and parts disposal

Table 13.1 (*continued*)

Mode	Transportation Activities	Nature of Impact
Marine	Construction and maintenance of marine navigation infrastructure	Direct deterioration of water quality from dredging or other navigation improvements
		Contamination from disposal of dredged material
	Manufacture of maritime vessels and parts	Toxic releases during manufacture
	Maritime transportation operations (vessel travel)	Nitrogen oxide and sulfur oxide emissions by vessel engines
		Hazardous materials spills during transport
		Overboard dumping of solid waste and sewage
		Release of ballast water containing alien species
	Maritime vessel maintenance and support	Antifouling chemicals to prevent biological growth on vessel hulls during terminal operations
	Disposal of maritime vessels and parts	Scrappage of old vessels and dilapidated parts

Source: USEPA (1996), Ortolano (1997), Jain et al. (2001), Rothengatter (2003), Talley (2003), Wayson (2003).

maintenance (USEPA, 1996). With regard to aircraft deicing, it is estimated that 50 to 80% of the solution applied falls to the apron, where it is subsequently washed off into surface waters or percolates into the ground (D'Itri, 1992). Glycol, a common deicing compound, is generally considered to have low toxicity and is readily biodegradable. However, it does exert a high biological oxygen demand and can therefore severely deoxygenate receiving water bodies, thus damaging biotic communities (DEFRA, 2005). Other water pollutants released during aircraft maintenance include lead paint and various solvents. From airport operations such as aircraft cleaning, maintenance, repair, fueling, baggage handling, and other cargo support services, there can be significant emissions, fuel spills, oil leakages, and so on, from ground support equipment (USEPA, 1996). During rainfall events, such contaminants are washed off from runways and taxiways, aprons, roads, and parking lots and in the absence of stormwater treatment facilities, are deposited into nearby surface waters. Maintenance of runway pavements and rights-of-way, through herbicide application for instance, can also lead to pollution of nearby waters. Furthermore, during plane loading and unloading, or when aircraft are airborne, pollutants can be released that ultimately find their way into surface waters directly or indirectly (USEPA, 1996).

13.2.4 Marine Transportation Impacts

The impact of marine transportation on water resources is related primarily to the quality, and to a lesser extent, the quantity and flow patterns. Examples include direct deterioration of water quality from navigation improvements, contamination from dredged material disposal, toxic releases from water-based cargo through leakage or dumping, and scrapping of vessel parts. Other common marine transportation sources of water pollution are deliberate or accidental releases of solid waste, sewage, and hazardous materials, leakage, and spills (USEPA, 1996). Deliberate spillage occurs in the form of operation dumping: After discharging its liquid chemical (often, oil) cargo, a vessel takes ballast water into its cargo tanks to ensure vessel stability on the return trip; then at or near the loading port, the water–chemical mix is discharged, thus polluting the receiving water body (Talley, 2003). Accidental spillage occurs in the form of leaking tanks and burned, or sunken ships. Between 1996 and 2000, there were over 8000 reported oil spills in U.S. navigable waters annually; in 2000, the total volume of oil spilled by vessels was approximately 1 million gallons (Gibson, 2001). Spills of hazardous materials from maritime vessels are one of the most publicized impacts of water-based transportation. An example is the 11-million gallon crude oil spill by the Exxon *Valdez* oil tanker in 1989, which resulted in severe water pollution and tremendous loss of marine life (Miller and Tyler, 1990).

13.3 PERFORMANCE MEASURES FOR HYDROLOGICAL IMPACT ASSESSMENT

Assessing the water resources impact of transportation is best carried out on the basis of appropriate performance measures as discussed below (Sinha et al., 1991; Jain et al., 2001).

13.3.1 Measures Related to Water Quantity and Flow Patterns

(*a*) *Aquifer Safe Yield* This performance measure, typically used to assess impacts on groundwater quantity, is defined as the difference between withdrawal and recharge rates. The aquifer safe yield, which is a critical issue in regions that depend on groundwater for water supplies, is indicative of water that can be withdrawn per unit time: for example, acre-feet per year. Aquifer safe yield can be impaired directly by the construction of transportation facilities, such as parking lots, runways, and highways, because these activities cause increased impervious surface areas, thus reducing the infiltration of surface water into the ground.

(*b*) *Flow Variations* Flow variations are of greatest significance at their extreme conditions: low and high flows. During low flows, a stream's natural assimilative capacity is greatly reduced, amplifying the negative impacts of natural and human-induced waste loads. During high flows, the major concerns are inundation, erosion, and sedimentation. Flow variation impacts on water resources can be expressed in terms of the change in velocity of flow (ft/s) and rate of discharge (ft^3/s). Transportation activities such as earthwork (cutting or filling), site clearance, paving, bridge and culvert construction, ditch and drain construction, and turnout construction often result in rearrangement of ground surface and topography, and therefore alter stream flow patterns, causing too little water availability in certain areas and too much in other areas.

13.3.2 Measures Related to Water Quality

(*a*) *Oil Contamination* Oil is a petroleum product that is discharged from transportation vehicles and vessels and reaches surface waters either directly or through runoff, where it spreads out on the water surface, thus impairing its quality and suitability for supporting aquatic life. Oil discharge affects the dissolved oxygen content, aesthetics, taste, and odor of water bodies.

(*b*) *Suspended Solids* Insoluble solid contaminants from transportation sources typically remain suspended in water, causing turbidity. Excessive turbidity reduces solar radiation intensity and penetration into surface water bodies and can therefore degrade aquatic life. Transportation activities such as dredging, construction of hydraulic structures, and gravel washing, cause the direct release of suspended solids into water bodies and cause such effects indirectly through surface paving and landscaping. These activities result in changed surface runoff patterns, and typically increase flow rates and erosion.

(*c*) *Acidity and Alkalinity* The pH value is an important indicator of environmental quality. A high pH value signifies an alkaline condition, and a low pH value represents an acidic condition, a pH value of 7 represents a neutral condition. Spills or disposal of acid and alkaline material into waters can change the pH of the receiving water, rendering it hazardous to aquatic life and lowering its capacity to assimilate organic wastes. At pH levels below 5 and above 9, fish mortality may occur.

(*d*) *Biochemical Oxygen Demand (BOD)* BOD is a direct bioassay measure of the amount of oxygen required for biological decomposition of organic matter in a water body. Pollutants from transportation activities that increase the BOD of receiving waters impair the quality of such water sources. BOD is typically measured by the quantity of oxygen consumed (mg/L) by microorganisms during a 5-day period at 20°C.

(*e*) *Dissolved Oxygen (DO)* To maintain life, all life forms need oxygen, directly or indirectly. Most warm aquatic organisms require a DO concentration above 5 mg/L. A lack of DO will generate anaerobic conditions, resulting in unfavorable odor and visual appearance. Transportation activities that affect DO include site preparation, demolition, dredging, and excavation. DO concentrations may be altered by oil discharged from routine operations and maintenance of aircraft, watercraft, automotive equipment, and railroad rolling stock, by means of an oil film that prevents oxygen from dissolving in water.

(*f*) *Dissolved Solids* A high level of dissolved solids degrades water quality by altering the physical and chemical characteristics of the water and exerting osmotic pressure on organisms living in such waters. Transportation-related activities that may lead to an increased level of dissolved solids include mining and quarrying for construction materials, winter salting operations, waste disposal into surface waters and landfills, and accidental spillage of chemicals during transportation operations.

(*g*) *Nutrients* Eutrophication, a process whereby water bodies receive nutrients (such as phosphorus and nitrogen) through human activity or natural processes (erosion of soil containing nutrients), enhances plant growth (often known as *algal blooms*), forming surface water scum and algae-littered beaches and banks. Dead and decomposing algae cells reduce dissolved oxygen in the water, causing the death of other aquatic organisms. Transportation-related activities such as mining for aggregates, tunneling, blasting, and quarrying into phosphate rocks may affect surface-water or groundwater systems by contributing to increased phosphorus from surface runoff. Dredging of waterways leads to the discharge of nutrients from in the mud bottom, subsequent enrichment of the water, and ultimately, eutrophication.

(*h*) *Toxic Compounds*   Activities that might cause the release of toxic compounds into water resources include the discharge of waste from vehicle, vessel, and aircraft manufacture, maintenance, and repair shops through electroplating, galvanizing, metal finishing, and cooling tower blowdown processes. Other activities include aggregate mining, accidental spills of chemicals, and dumping disposed transportation vehicles (or parts thereof) into landfills, which subsequently leach toxic compounds. In particular, wastes that contain heavy metals (mercury, copper,

silver, lead, nickel, cobalt, arsenic, cadmium, chromium), ammonium compounds, cyanides, sulfides, fluorides, and petrochemical wastes can cause serious water pollution effects. Other toxic substances are herbicides used in right-of-way maintenance.

13.4   PROCEDURE FOR WATER QUALITY IMPACT ASSESSMENT

The hydrological or water resource performance measures discussed in Section 13.3 provide a basis by which analysts can assess the impacts of transportation activities on water quality, quantity, and flow patterns. An overall framework that can be used for water resources impact assessment is illustrated in Figure 13.3.

Step 1: Define the Study Area and Temporal Scope of the Analysis   In assessing transportation impact on water resources, proper timing is critical because the

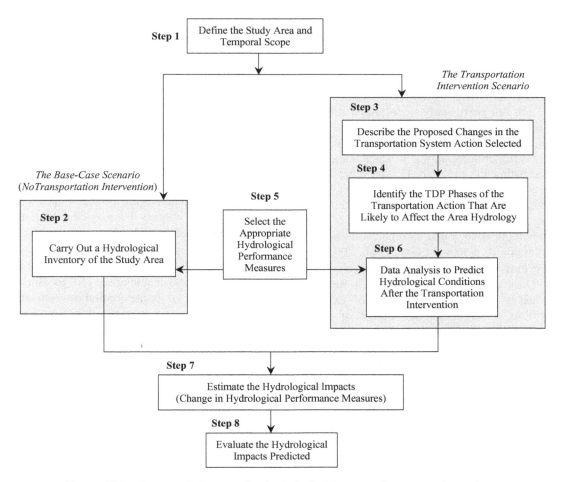

Figure 13.3   Framework for assessing hydrological impacts of transportation actions.

vulnerability of water bodies to polluting and deprivation effects may vary by time of year, and if hydrological field surveys are carried out at the wrong time, incorrect conclusions may be drawn. Also, assessments should be carried out separately for each stage of the project development process (PDP): At the system and location planning phases, a hydrological or water resources impact assessment should be carried out on a regional scale so that facility location alternatives can be established to bypass hydrologically sensitive areas. At the design phase, the impact assessment should be undertaken to evaluate the local water resource impact of specific design features of the transportation project.

Furthermore, the selection of appropriate spatial and temporal scopes for the impact assessment should be influenced by the performance measure under consideration (if such measures have already been chosen at this stage of the analysis). Impacts related to aquifer safe yield are most significant at locations that depend on groundwater for supply, have a high water table, or have significant seasonal precipitation and infiltration rates (Jain et al., 2001). Analysts may ascertain the existence of these conditions by contacting local geological survey offices and state water resource agencies. Adverse transportation impacts on water quantity involve deprivation of water at areas where it is needed, and/or redirection of water to areas that do not need it, thus causing flooding. Water deprivation impacts are felt particularly at locations that are prone to prolonged periods of drought. These periods typically occur in summer, when biological activity is high and dissolved oxygen concentrations in water bodies are low, thus exacerbating the problem. On the other hand, impacts associated with excessive water flow conditions often occur in floodprone areas. With regard to nutrient effects of transportation activities, the dependence of algal growth on temperature is such that eutrophication is generally more pronounced in summer (Jain et al., 2001).

Step 2: Carry Out a Hydrological Inventory of the Study Area   Data can be obtained through detailed field surveys and a comprehensive search of existing literature regarding the locations and nature of water sources in the study area. Such information is generally available in hard-copy or GIS map forms at federal, state, provincial, and local water resource agencies. Analysts in developing countries may seek the assistance of specialized agencies of the United Nations and other nongovernmental organizations. Other sources of data include geological maps, aerial photographs, and satellite images local conservation organizations, and farmers located near the transportation project.

In accordance with U.S. federal law the number and location of wild and scenic stream and river crossings by a proposed transportation project must be identified. A list of protected rivers can be obtained from sources such as the National Park Service's National Wild and Scenic Rivers System Web site (http://www.nps.gov/rivers/) and state agencies responsible for environmental conservation and protection (Cambridge Systematics and Bernardin, 2003). Also, the number and areas of floodplain crossings or parallel encroachments by the transportation project right-of-way should be determined.

The analyst should collect data on groundwater characteristics (such as depth, variability, and quality) where the water table lies within 5 feet of the proposed subgrade or where surface waters are recharged by groundwater due either to discrete springs or to seepage (Sinha et al., 1991). Such data are generally available at local water resource agencies or geological surveys in the form of maps or borehole and well logs.

Step 3: Describe the Changes Proposed in the Selected Transportation System Action   Establishing the type and scale of the transportation intervention is important because it helps to identify the appropriate mechanism and scope of the hydrological impact. The main, auxiliary, and related structures of the transportation project should be considered. For example, for culvert or bridge construction, hydrological impacts at both the crossing site as well as entire lengths of the rip-rapped inlets and outlets should be assessed.

Step 4: Identify the PDP Phases of the Transportation Action Likely to Affect Ground and Surface Waters   In hydrological impact assessment for new construction, analysts typically consider only the transportation project implementation phase. However, the facility operations and preservation phases are also associated with prolonged and cumulative impacts that may be at least as deleterious as the implementation phase.

Step 5: Select the Appropriate Hydrological Performance Measures   At this step, the analyst establishes a set of hydrological characteristics for measuring the extent and magnitude of transportation impacts on the water resources. This may consist of performance measures associated with any one or more of the following performance attributes or others: aquifer safe yield, flow variations, oil contamination, radioactivity, suspended solids, biological oxygen demand (BOD), pH, dissolved solids, toxic compounds, dissolved oxygen, and nutrients. The performance measures are discussed in Section 13.3.

Step 6: Analyze the Data to Predict Hydrological Conditions after the Transportation Intervention   This step involves the use of various mathematical models to estimate the changes in water quantity, quality, and flow patterns and volumes in response to the proposed transportation activity. Details of such analysis are provided in Section 13.5.

Step 7: Estimate the Hydrological Impacts (Change in Performance Measures) From step 5 (existing situation) and step 6 (after the transportation intervention), the expected change in the hydrological performance measure is determined simply as the difference between the existing and postintervention scenarios.

Step 8: Evaluate the Hydrological Impacts Predicted This step involves a comparison of the predicted impacts (postintervention hydrological conditions) obtained from step 7 against established environmental standards. In this task, the analyst reviews laws, regulations, and policies geared toward the protection of water resources compared with the predicted hydrological condition (on the basis of the performance measures) and determines whether any standards will be violated by the transportation intervention. In Section 13.7 we discuss the federal standards regarding hydrological impacts. The analyst may seek the inputs of related and concerned organizations, environmental groups, governmental agencies responsible for the protection of water bodies, and the general public. Also, the analyst may solicit opinions of water resource experts for a rational and scientific perspective of the impacts. In summing up the analysis, the analyst must determine whether the estimated hydrological impacts are significant enough to merit further review of the transportation project plan.

13.5 METHODS FOR PREDICTING IMPACTS ON WATER RESOURCES

13.5.1 Impacts on Water Quantity

Methodologies for predicting the water resource impacts of transportation activities differ by the nature of the water source affected (ground vs. surface water), impact category (water deprivation vs. water degradation vs. flow pattern disruption), and performance measure (aquifer yield, toxic substance concentration, etc.). Some of these methodologies have been automated in the form of computerized modeling software. To apply any of these packages to a specific transportation project, the analyst must gather environmental information to establish baseline values and to determine the values for the model's variables and parameters, respectively (Lohani et al., 1997). Canter (1996) and Canter and Sadler (1997) provided overviews of water impact prediction methods. A summary description of such models is provided in this section.

(a) Predicting Impacts on Surface Runoff Quantities
Fundamental Runoff Formula: Mathematical models are available for predicting surface water runoff, on the

basis of the extent of land cover, degree of surface permeability and the extent of sewerage provision. These models generally use the same basic principle, which involves a balance between hydrological inputs and outputs to surface runoff over a fixed time period (Lohani et al., 1997):

$$\text{runoff} = \text{precipitation} - \text{evapotranspiration}$$
$$- \text{infiltration} - \text{storage} \qquad (13.1)$$

Variations of the basic model above are those that exclude one or more of the foregoing variables in case where they are considered negligible (equal to zero). Input data types for the general models include rainfall, air temperature, drainage network configuration, soil types, ground cover, land use, and management measures. An application scenario for this model is where construction of paved highways and runways lead to increase in the impervious fraction of land area: This results in decreased infiltration rate and consequently, increased surface runoff volumes.

Rational Formula: The rational formula, a variation of the fundamental runoff formula presented in the preceding section, is used to compute the peak discharge flow rate (Wayson, 2003). The model is as follows:

$$Q_p = c i A \qquad (13.2)$$

where Q_p is the peak discharge (ft^3/s); c the runoff coefficient, an empirical coefficient representing the relationship between rainfall rate and runoff rate, and is a function of type of land cover, slope, and/or hydrologic soil classification (see examples in Table 13.2); i the rainfall intensity (in/h) for a storm duration equal to the *time of concentration*, typically estimated using rainfall intensity–duration–frequency curves developed at most state transportation agencies; and A the drainage area (acre) (Burke and Burke, 1994). The time of concentration is the time after which the runoff rate equals the excess rainfall rate.

The maximum runoff rate in a catchment is reached when all parts of the watershed contribute to the outflow. This happens when the time of concentration is reached. The rational method assumes that rainfall intensity and runoff coefficient remain constant over the entire drainage area and that the peak discharge and rainfall intensity predicted have equal probability of occurrence or return period (Corbitt, 1999; Wayson, 2003). Also, the entire drainage area should be considered as a single unit. The estimation of flow is for the most downstream point (ITC, 2005). The rational method is applicable to any watershed

Table 13.2   Coefficients for the Rational Method of Runoff Estimation

Ground Cover	Runoff Coefficient, c	Ground Cover	Runoff Coefficient, c
Business		Pasture	0.12–0.62
Downtown	0.70–0.95	Forest	0.05–0.25
Neighborhood	0.50–0.70	Cultivated land	0.08–0.41
Residential		Pavement	
Single-family	0.30–0.50	Asphaltic	0.70–0.95
Multiunits, detached	0.40–0.60	Concrete	0.80–0.95
Multiunits, attached	0.60–0.75	Brick	0.70–0.85
Residential (suburban)	0.50–0.70	Drives and walks	0.75–0.85
Apartment	0.50–0.70	Roofs	0.75–0.95
Industrial		Lawns, sandy soil	
Light	0.50–0.80	Flat, 2%	0.05–0.10
Heavy	0.60–0.90	Average, 2–7%	0.10–0.15
Parks, cemeteries	0.10–0.25	Steep, >7%	0.15–0.20
Playgrounds	0.20–0.35	Lawns, heavy soil	
Railroad yard	0.20–0.35	Flat, 2%	0.13–0.17
Unimproved	0.10–0.30	Average, 2–7%	0.18–0.22
Meadow	0.10–0.50	Steep, >7%	0.25–0.35

Source: Singh (1992), Corbitt (1999), and other sources.

whose time of concentration does not exceed 20 minutes (Wayson, 2003). Generally, i can be expressed as a function of return period, T (years), and time of concentration, t_c (minutes), as shown in the example below. The value of t_c can be influenced by land-use changes such as transportation development.

Example 13.1   Due to the planned construction of a large intermodal transportation transfer facility at the outskirts of a city in a midwestern region, 15% of a 30-acre wooded area will be converted to concrete parking lots and streets, lawns (average lawns with 2 to 7% heavy soil) will cover 5%, and the roofs of the facility buildings and shelters will cover 2% of the wooded area. Assuming that the entire drainage area can be considered as a single drainage unit, estimate the expected change in runoff volume due to the new facility. Assume that the times of concentration before and after the transportation project are 30 and 18 minutes, respectively, and assume a return period of 10 years. Use the midpoints of runoff coefficient ranges shown in Table 13.2. The rainfall intensity function for the region is as follows:

$$i = \frac{2.1048 T^{0.1733}}{(t_c/60 + 0.47)^{1.1289}}$$

SOLUTION   Before the project, the rainfall intensity:

$$i_{\text{before}} = \frac{2.1048 \times 10^{0.1733}}{(30/60 + 0.47)^{1.1289}} = 3.25$$

After the project, the rainfall intensity:

$$i_{\text{after}} = \frac{2.1048 \times 10^{0.1733}}{(18/60 + 0.47)^{1.1289}} = 4.21$$

Results are shown in Table E13.1. The change in runoff volume = $33.82 - 14.63 = 19.19$ ft^3/s. Thus, the project resulted in a 131% increase in surface runoff quantity.

(b) Models That Predict Changes in Water Flow Patterns Hydrodynamic models (equations that describe the movement of water) are derived from the three-dimensional Navier–Stokes equations and can be applied in the before and after transportation activity scenarios to establish any differences in water flow patterns. For relatively simple hydrodynamic scenarios, standard computer models are available, but for more complex scenarios, the help of hydrological experts may be solicited. Hydrodynamic models can help predict sediment behavior, salinity, temperature, water quality, and surface water movements. To calibrate, validate, and apply mathematical hydraulic models, data on system geometry, inflows and outflows (time

Table E13.1   Results of Runoff Computations

Description	Before Project				After Project			
	Area I: Wooded Area	Area II: Concrete Streets	Area III: Lawns	Area IV: Roofs	Area I: Wooded Area	Area II: Concrete Streets	Area III: Lawns	Area IV: Roofs
Area, A (acres)	30	0	0	0	23.4	4.5	1.5	0.6
Runoff coefficient, c	0.15	0.825	0.2	0.85	0.15	0.825	0.2	0.85
Rainfall intensity, i (in/hr)	3.25	3.25	3.25	3.25	4.21	4.21	4.21	4.21
Runoff from each area (ft³/s)	14.625	0	0	0	14.78	15.63	1.263	2.147
Total Runoff (ft³/s)	14.63				33.82			

series), initial hydrological conditions, channel bed conditions, water levels, and wind conditions are generally needed (Lohani et al., 1997).

(c) Models That Predict Changes in Groundwater Quantity

The flow of groundwater (and changes thereof) can be predicted using mathematical models. These models are based primarily on analytical or numerical solution of equations for conservation of mass combined with Darcy's law. The law states that the flow of fluid through a saturated porous medium (e.g., water through an aquifer) is dependent on the hydraulic gradient (change in piezometric head over a distance) and the hydraulic conductivity (permeability) of the medium, as follows (Lohani et al., 1997):

$$Q = KA \frac{\partial H}{\partial L} \tag{13.3}$$

where Q is the total discharge/flow (m³/day), K the permeability or hydraulic conductivity (m/day), A the cross-sectional area to flow (m²), and $\partial H / \partial L$ the hydraulic gradient (the change in the water table elevation per unit change in the horizontal direction for unconfined aquifers).

It is important to note that once surface water reaches the groundwater regime through infiltration and percolation, the factors affecting groundwater flow (i.e., permeability and cross-sectional area of the porous medium, and the hydraulic head) are hardly influenced by surface transportation projects. On the other hand, construction of subterranean transportation facilities such as road and rail tunnels and underground terminals can significantly reduce the cross-sectional area of any porous

rock media that they encounter and thereby disrupt groundwater flow, depending on the level of confinement of the porous medium.

Example 13.2   It has been decided to locate a new underground subway terminal at a site that is deemed hydrologically sensitive because it traverses an unconfined aquifer that supplies the city's water supply. From geological profiles of the area and design drawings for the facility, it is estimated that 7% of the cross-sectional area of the aquifer would be taken up by the terminal walls. Calculate the expected percentage reduction in groundwater flow due to the construction proposed. Prior tests indicate the following underground conditions: fractured rock material with a K value of 1000 m/day for the rock medium, and borehole readings indicating a hydraulic gradient of 0.05. Assume that after the construction, the permeability of the medium is reduced by 10% but that the hydraulic gradient remains unchanged.

SOLUTION   From equation (13.3): *Flow before the project*:

$$Q = KA(\partial H / \partial L) = 1000A(0.05)$$
$$= 50A \text{ m}^3/\text{day}$$

Flow after the project:

$$Q = KA(\partial H / \partial L) = (0.90)(1000)0.93A(0.05)$$
$$= 41.85A \text{ m}^3/\text{day}$$

Percentage reduction in groundwater flow

$$= \frac{50A - 41.85A}{50A} = 16.3\%$$

13.5.2 Impacts on Water Quality

(*a*) *Models That Predict Changes in Water Quality*   Most water quality models are based on the concept of *mass balance*, also known as *material balance*, which states that the mass that enters a system must either leave the system or accumulate within the system through the principle of conservation of mass, illustrated in Figure 13.4. The rate of change in total contaminant mass in a compartment over time is given by (McKay and Peterson, 1993; Lohani et al., 1997)

$$\frac{dM}{dt} = (I + D + F + J) - (X + R + T) \qquad (13.4)$$

where I is the mass inflow rate into the compartment (mass/time), D the discharge into the compartment (mass/time), F the mass formation rate due to biochemical activity in the compartment (mass/time), J the transfer from other compartments (mass/time), X the outflow from the compartment (mass/time), R the degrading reaction (mass/time), and T the transfer to other compartments (mass/time). For the long-term average rates when the system is in equilibrium, the left-hand side of equation (13.4) is set to zero.

For the purposes of the modeling, a volume space is identified as being comprised of several compartments, such as water, air, or biota. Subsequently, a simple mass balance equation is established. This equation stipulates that the change in the amount of a chemical in the compartment is equal to the difference in input and output amounts of that chemical (mass/volume).

Canter (1996) presented a mass balance formulation for dissolved oxygen to include transport exchanges with the atmosphere, with the biota due to photosynthesis and respiration, and with the water column and sediments. In environmental impact assessments for transportation projects, only those chemical discharges associated with a specific contaminant may be of concern. In such cases, a mass balance model can be used to estimate the changes in concentrations of the contaminant in the study area.

A variation of the mass balance equation is the Gupta et al. (1981) model, which provides an approximate estimate of the accumulated pollutant load as a function of initial load, accumulation rate, length of highway and duration of accumulation:

$$P = P_0 + K_1 H_L T \qquad (13.5)$$

where P is the load of pollutant after accumulation (lb), P_0 the load of pollutant before accumulation (lb), K_1 the accumulation rate (lb/mile per day) = (0.007)(average daily traffic, ADT)$^{0.89}$, H_L the length of the highway (miles), T the duration of accumulation (days).

Example 13.3   A bypass is being planned around a town. The expected traffic volume is 30,000 ADT, and the length of the bypass is 6.5 miles. The accumulation period is 6 days. Assuming the initial pollutant load is negligible, what is the expected pollutant load after a storm due to the bypass construction?

SOLUTION   Using equation (13.5), the load of surface pollutant after a storm due to the construction of the bypass is:

$$P = (0.007)(30,000)^{0.89}(6.5)(6)$$

$$= 2,636 \text{ lbs}$$

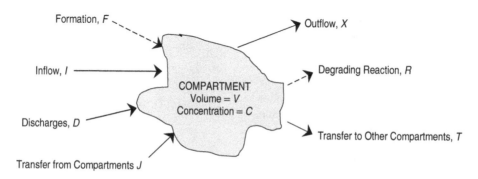

Figure 13.4   Mass balance of a pollutant.

Each event of a rain storm with accumulation period of 6 days will thus add a pollutant load of 2,636 lbs to the water bodies in the vicinity of the proposed bypass.

(b) *Models That Predict the Occurrence and Extent of Erosion and Sedimentation in Surface Waters* *Soil erosion*, defined as the detaching of soil particles from a soil surface which occurs when soil is exposed to the power of rainfall energy and flowing water, has a significant impact in the realm of transportation activities. Erosion may result in loss of land productivity, increased level of sedimentation in waterways, and flash floods in the construction area, and it provides a substrate for toxic chemicals carried into receiving waters.

Many erosion models have been designed for prediction of the average annual soil loss from a study site, but the most widely used is the *universal soil loss equation* (USLE), developed by the U. S. Department of Agriculture (USDA) for estimating annual soil erosion from agricultural fields under specific conditions. This model is also pertinent to nonagricultural conditions such as at construction sites (Wischmeier and Smith, 1958; 1978). The equation incorporates six major factors that influence the soil erosion rate, based on empirical research and statistical analysis of field studies for prediction of erosion. The USLE provides an effective and invaluable tool for planners and decision makers to use to assess the impact of transportation-related construction projects on soil erosion. The universal soil loss equation is defined as (Wischmeier and Smith, 1978):

$$A = RK(LS)CP \qquad (13.6)$$

where A is the average annual soil loss, R the rainfall-runoff erosivity factor, K the soil erodibility factor, LS the slope length factor, C the cover management factor, and P the support practice factor. Details of these parameters are given below (Renard et al., 1997).

1. A: the spatial average and temporal average soil loss, generally expressed in tons/acre per year.

2. R: the rainfall-runoff erosivity factor (hundreds of ft·tonf·inch per acre per year) is the number of the rainfall erosion index units in an average year's rain plus any significant runoff factor from snowmelt. The R value can be obtained from the CITY database file at the offices of the USDA, and the Natural Resources Conservation Service (NRCS) [formerly known as the Soil Conservation Service (SCS)]. When published values of R are not available, it can be computed as $R = \sum EI_{30}(10^{-2})$, where E is the total kinetic energy of a storm (ft·tonf per acre) and I_{30} is the maximum 30-minute rainfall intensity (in/h), utilizing the energy table provided in Wischmeier and Smith (1958).

3. K: the soil erodibility factor [ton·acre·hour (hundreds of acre·ft·tonf·inch)$^{-1}$] is the soil loss rate (tons/acre) of a specific soil type and horizon as measured on a standard plot of land caused by rainwater. The K value is an index from 0.001 (nonerodible) to 1 (erodible) based on soil structure, soil particle size distribution, permeability, organic matter content, and iron content. It is the average soil loss in tons/acre per unit area for a particular soil in cultivated, continuous fallow with an arbitrarily selected slope length of 72.6 ft and a slope steepness of 9%. Values of K for undisturbed and disturbed soils can be obtained from soil-survey information published on the Web (http://soils.usda.gov/survey/) by the NRCS and computations of soil erodibility nomograph, respectively. Sample values of K for various soil types are shown in Table 13.3.

Table 13.3 K-Factor Data (Organic Matter Content)

Textural Class	Average	Less Than 2%	More Than 2%
Clay	0.22	0.24	0.21
Clay loam	0.30	0.33	0.28
Coarse sandy loam	0.07	—	0.07
Fine sand	0.08	0.09	0.06
Fine sandy loam	0.18	0.22	0.17
Heavy clay	0.17	0.19	0.15
Loam	0.30	0.34	0.26
Loamy fine sand	0.11	0.15	0.09
Loamy sand	0.04	0.05	0.04
Loamy very fine sand	0.39	0.44	0.25
Sand	0.02	0.03	0.01
Sandy clay loam	0.20	—	0.20
Sandy loam	0.13	0.14	0.12
Silt Loam	0.38	0.41	0.37
Silty clay	0.26	0.27	0.26
Silty clay loam	0.32	0.35	0.30
Very fine sand	0.43	0.46	0.37
Very fine sandy loam	0.35	0.41	0.33

Source: Stone and Hilborn (2000).

Table 13.4   NN Values for LS Factor Calculation

S	<1	1 ≤ slope < 3	3 ≤ slope < 5	≥5
NN	0.2	0.3	0.4	0.5

4. *LS*: the slope length factor, where *S* is the slope angle (%) and *L* is the length of the slope (ft). The factor *LS* is expressed as a ratio of the erosion from that experienced on a gradient of 9% and a length of 72.6 ft. *LS* can be calculated based on the *NN* values provided in Table 13.4 using the equation (Stone and Hilborn, 2000):

$$LS = [0.065 + 0.0456(\text{slope})$$
$$+ 0.00654(\text{slope})^2] \left(\frac{\text{slope length}}{72.6}\right)^{NN} \quad (13.7)$$

5. *C*: the cover (cropping) management factor (dimensionless) is the ratio of soil loss from an area with a specified cover and management to the corresponding loss from a clean-tilled continuously fallow condition. The *C* value ranges from 0.001 for a well-managed woodland to 1.0 for no cover. *C* values can be determined on the basis of the type of land cover and land use, as shown in Table 13.5.

6. *P*: the ratio of soil loss (dimensionless) with a support practice such as contouring, stripcropping, or implementing terraces compared to up-and-down slope cultivation; ranges from 0.001 for effective contouring, terracing, and other erosion control for tilled land to 1.0 for an absence of erosion control. *P* reflects the effects of practices that will reduce the amount and rate of water runoff and thus reduce the amount of erosion. Sample values of *P* for various support practice factors are shown in Table 13.5.

Table 13.5   Crop Type and Support Practice Factor

Land-Use or Land-Cover Type	C Factor	P Factor
Fallow	1	1
Laterite cap	1	1
Agricultural crop (rice)	0.1	0.03
Settlement	0.1	1
Dry fallow	1	1
Open forest	0.8	0.8
Water bodies	0.1	1

Source: Agrawal et al. (2003).

Example 13.4   A proposed road of length 5400 ft and 3% slope is expected to affect a field of open forest. The soil is heavy clay with average organic matter content. Using the USLE method, estimate the average annual soil loss expected at the site due to the construction. Use an *R* factor of 90.

SOLUTION   Calculation of soil erosion using USLE: $A = RK(LS)CP$. From Table 13.3, the *K* factor of heavy clay with an average organic matter content = 0.17. Using equation (13.7) and an *NN* value from Table 13.4, the *LS* factor for the affected site of length 5400 ft and 3% slope is

$$LS = [0.065 + 0.0456(\text{slope}) + 0.00654(\text{slope})^2]$$
$$\times \left(\frac{\text{slope length}}{72.6}\right)^{NN}$$
$$= [0.065 + (0.0456)(3) + (0.00654)(3)^2]$$
$$\times \left(\frac{5400}{72.6}\right)^{0.4} = 1.46$$

From Table 13.5 the *C* and *P* factors for the construction site are both 0.8. Therefore,

$$A = RK(LS)CP = (90)(0.17)(1.46)(0.80)(0.80)$$
$$= 14.3 \text{ tons/acre per year}$$

In 1985, a revised USLE (RUSLE) was developed with enhanced technology for factor evaluation and introduction of new data while retaining experimental procedures similar to those of the original USLE, aided by a computer program. The improvements include an extension of the rainfall-runoff erosivity factor (*R*) database in the western United States, a time-varying soil erodibility factor (*K*), altered topographic factors for slope length and steepness (*LS*), a revised cover-management factor (*C*), and an extended support practice factor (*P*) to reflect rangelands, contouring, stripcropping, and terracing (Renard et al., 1997). In-depth information and guidance to conservation planning with the RUSLE is contained in *Agricultural Handbook No. 703* (Renard et al., 1997) and in guidelines provided by Toy et al. (1998).

Example 13.5   As part of construction of a large runway located upstream, grass cover (erodibility of 0.05) was stripped away to expose sandy silt topsoil (erodibility of 0.8). The value of *R*, a measure of rainfall intensity, is 80, the slope angle is 6%, and the length of the slope is 100 ft. Due to ongoing construction activities, there is no temporary vegetative cover or erosion control after the topsoil is exposed. Assume perfect erosion control–management practice and well-maintained vegetative cover before the project. Calculate the average soil loss (tons/acre) over the seven-month segment of the construction project during which the topsoil is exposed.

SOLUTION From equation 13.7, $LS = 0.674$. Before exposure of the topsoil, the annual average soil loss [from equation (13.6)] was

$$A = RK(LS)CP$$

$$= (80)(0.05)(0.674)(0.001)(0.001)$$

$$= 2.7 \times 10^{-6} \text{ ton/acre per year}$$

For a 10-acre area over a seven-month period, soil loss $= (2.7 \times 10^{-6}) \times 10(7/12) = 1.6 \times 10^{-5}$ ton. After exposure of the topsoil, the annual average soil loss [from equation (13.6)] was $(80) \times (0.8)(0.674)(1)(1) = 43.14$ tons/acre per year. For a 10-acre area over a seven-month period, the soil loss $= (43.14)(10)(7/12) = 251.65$ tons. Therefore, it is evident that the topsoil would lead to severe erosion and sedimentation of the adjacent stream.

(c) *Changes in Groundwater Quality* Pollution of groundwater occurs through discharge of effluents or disposal of waste onto land, leaching of contaminants into groundwater, changes in the quality of surface water, and deposition of air pollutants on land. The effects of these pollutants vary from first-order effects of leaching into soil and groundwater to second-order effects such as changes in groundwater regime and quality. Tracer experiments may be used to predict dispersion of pollutants in groundwater.

(d) *Models That Predict Fully Mixed In-Stream Pollutant Concentration* Traditionally, estimation of fully mixed in-stream pollutant concentrations has provided a means to assess whether pollutants discharged from a continuous point polluting source surpass water quality criteria for aquatic life, wildlife, and human health. Dupuis (2002) presented expressions for estimating the fully mixed in-stream pollutant concentrations for various types of receiving waters as shown below.

Streams and Rivers:

METHOD 1 The in-stream concentration of pollutants can be estimated given the average stream and runoff flows and the concentrations of pollutants in the upstream and runoff flows. This is calculated for chlorides as

$$C = \frac{Q_{rw}C_{rw} + Q_{sw}C_{sw}}{Q_{rw} + Q_{sw}} \tag{13.8}$$

where C is the in-stream chloride concentration (mg/L), Q_{rw} the average stream flow (ft^3/s), C_{rw} the average upstream chloride concentration (mg/L), Q_{sw} the average runoff flow (ft^3/s), and C_{sw} the average runoff chloride concentration (mg/L).

For chlorides, the variable C_{sw} can be calculated based on monitoring expected chloride concentrations from annual salt application data:

$$C_{sw} = \frac{S_{ar}}{nV_{sw}} \tag{13.9}$$

where S_{ar} is the annual salt application mass (kg), n the estimated number of winter storms with deicing, and V_{sw} the average storm runoff volume (ft^3).

Example 13.6 An interstate highway crosses a major river at the Green River Bridge. In the winter, sodium chloride is applied regularly to deice the bridge surface. A chloride monitoring station is established a distance downstream of the bridge. Assume that the average upstream chloride concentration is measured as 3 mg/L and the average runoff flow is 50 ft^3/s. The average stream flow is 1200 ft^3/s. Annually, 50 kg of sodium chloride is applied. In a typical year there are 14 winter storms that require deicing, and the average storm runoff volume is 400 ft^3. Determine the in-stream chloride concentration.

SOLUTION The annual salt application mass, $S_{ar} = 50$ kg. The estimated number of winter storms with deicing, $n = 14$. The average storm runoff volume, $V_{sw} = 400$ ft^3 (conversion factor: 1ft^3 = 28.317 L). Therefore, the average chloride concentration expected in the runoff due to salt use is given by

$$C_{sw} = \frac{S_{ar}}{nV_{sw}} = \frac{50,000}{(14)(400)(28.317)} = 0.315 \text{ g/L}$$

The average stream flow, $Q_{rw} = 1200$ ft^3/s, the average upstream chloride concentration, $C_{rw} = 0.003$ g/L, the average runoff flow, $Q_{sw} = 50$ ft^3/s, and the chloride concentration [from equation (13.8] is

$$C = \frac{(1200 \times 0.003) + (50 \times 0.315)}{1200 + 50} = 0.0155 \text{ g/L}$$

METHOD 2 (FIRST-ORDER DECAY MODELS) Decay models are typically used to describe the gradual reduction of pollutant concentration over time due to degradation or dilution. These models are therefore useful for monitoring the levels of deicing compounds (urea-based deicers are commonly used in highway and runway maintenance and in aircraft operations) and the die-off rates of bacteria such as fecal coliforms that are common in highway runoff. The steady-state first-order decay model for a stream or river is (Young et al., 1996; Dupuis, 2002):

$$C_x = C_0 e^{-K(x/U)} \qquad (13.10)$$

where C_x is the concentration a distance x downstream of the polluting source, C_0 the initial complete-mix concentration at the point of discharge, K the decay rate, and U the stream velocity. To model bacteria die-off, a decay rate value between 0.7/day to 1.5/day can be used. Exposure time t (days) is used to represent the term x/U (NC-DENR, 2002).

Example 13.7   Runoff containing deicing compounds is regularly discharged into a creek located near a major airport. The initial complete-mix concentration of these pollutants at the point of discharge is 550 mg/L, and the decay rate for these compounds has been estimated as 4 per day. The average stream velocity calculated at the site was 3.5 ft/s. Find the concentration at a fish spawning area located 1 mile downstream of the source of discharge.

SOLUTION   The required concentration is calculated as

$$C_1 = 550 \exp\left\{-\left[\frac{4}{(24)(3600)}\right]\left[\frac{5280}{(3.5)}\right]\right\}$$
$$= 512.9 \text{ mg/L}$$

Wetlands, Lakes, and Reservoirs:
METHOD 1   The expected pollutant concentrations for receiving water sources such as wetlands, lakes, and reservoirs after a storm event can be calculated as follows; (assuming the transportation facility in question is the only source of discharge):

$$C = \frac{Q_{sw} T C_{sw}}{V} + C_0 \qquad (13.11)$$

where C is the concentration of pollutant in the receiving water system after mixing (mg/L), V the volume of water in the receiving water system (ft^3), C_{sw} the concentration of pollutant in the stormwater (mg/L), Q_{sw} the storm event flow into the lake (ft^3/s), T runoff duration, and C_0 the initial concentration of pollutant in the receiving water system before the storm event (mg/L). For a conservative pollutant that does not change overtime due to chemical, physical, or biological reactions, the long-term equilibrium concentration in the receiving water body is:

$$C_{eq} = 0.001119 \left(\frac{W}{Q_{out}}\right) \qquad (13.12)$$

where C_{eq} is the equilibrium concentration of the receiving water system (mg/L), W the annual pollutant loading from the transportation facility (kg/yr), and Q_{out} the annual outflow from the receiving water system (ft^3/s).

Example 13.8   A highway passing near Tawpingo Lake is the only source of a certain pollutant to that water body. The estimated volume of water in the lake is 0.3 million cubic feet, the concentration of the pollutant in the stormwater is 80 mg/L, and the volume of storm event flow into the lake is 30 ft^3/s. The runoff duration is 15 minutes. If the initial concentration of pollutant in the lake before the storm event is 0.8 mg/L, determine the expected pollutant concentration in the lake after the storm event.

SOLUTION   From equation (13.11), the pollutant concentration expected in the lake after a storm event is given by:

$$C = \frac{(30)(15 \times 60)(80)}{300,000} + 0.8 = 8 \text{ mg/L}$$

METHOD 2   For a completely mixed body of water such as lake, wetland, or reservoir, the steady-state first-order decay model is (Dupuis, 2002):

$$C = \frac{W}{(Q + KV)} \left\{1 - \exp\left[-\left(\frac{Q}{V} + K\right) T\right]\right\}$$
$$+ C_0 \exp\left[-\left(\frac{Q}{V} + K\right) T\right]$$

where C is the fully mixed lake concentration (mg/L), C_0 the initial lake concentration (mg/L), W the pollutant load during the time interval (kg), Q the lake inflow/outflow (m^3/day), V the lake volume (m^3), K the decay rate constant, and T the time (in days).

Example 13.9   A large pond located near an airport receives storm runoff that contains uric deicing compounds. The initial pond concentration is 0.3 mg/L, the pollutant load during the time interval is 50 kg/day, the net pond outflow is 80 m^3/day, the pond volume is 30,000 m^3, and the decay rate constant is 0.4 per day, Find the fully mixed pond concentration after 15 days.

SOLUTION

$$C = \frac{50,000}{(80 + 0.4 \times 30,000)} \left\{1 - \exp\left[-\left(\frac{80}{30,000} + 0.4\right)(15)\right]\right\}$$
$$+ 0.3 \exp\left[-\left(\frac{80}{30,000} + 0.4\right)(15)\right]$$
$$= 4.1299 \text{ mg/L}$$

(e) *Models That Predict In-Stream Pollutant Concentrations at Zones of Initial Dilution*   In certain cases, the short-term lethal effects of pollutants are of greater interest than the fully mixed concentration. This is particularly true at localized areas, where transportation surface runoff discharges into the receiving waters, typically termed the *zone of initial dilution* (ZID). Methods for determining the size of ZID exist at state agencies and are given in several standard texts in hydrology (Dupuis, 2002).

13.6 MITIGATION OF WATER RESOURCE IMPACTS

The mitigation of hydrological impacts may be carried out from the perspective of impact criterion water source (surface vs. groundwater), and the phase of transportation project development process. Mitigation can occur on two fronts: preemptive measures (applied before the pollutants emitted are released into the receiving water bodies), and palliative measures (applied after such pollutants have been released and dispersed). The latter is relatively more tedious, expensive, and time consuming.

13.6.1 Mitigation Measures by Impact Criterion

With regard to aquifer yield, it is necessary to control any activity related to the transportation project that is likely to affect land surface runoff and infiltration and thereby decrease water availability to proximal aquifers. At the location planning stage, it is useful to investigate the groundwater hydrology at or near each alternative project location, and the final project location should be chosen to minimize such adverse effects. To minimize impacts due to water flow variations, transportation activities that are related to land-use changes and water impoundments and operations should be duly considered to minimize postproject water flow variations from mean natural flow quantities and directions. Pollutant impacts can be minimized by controlling all direct discharge into natural waters. For example, surface runoff from oil-handling areas should be treated before discharge. Lagooning of oil wastes and land disposal of oily sludges should be restricted or controlled to avoid possible contamination of groundwater. During construction, gravel-washing activities, mine tailings, and dust may be controlled by utilizing available technology. With regard to BOD impact mitigation, all transportation operation wastes containing organic material should be processed. Recommended treatment methods for dissolved solids include removal of liquid and disposal of residue by controlled landfilling to avoid leaching from the fills. All surface runoff around mines or quarries should be collected and disposed of

appropriately. Brine may be disposed of using deep-well injection or other acceptable means.

13.6.2 Mitigation Measures by Nature of Water Source

Transportation line facilities such as highways, runways, and railways typically traverse wide swaths of land over which many different surface water and groundwater conditions are encountered. Therefore, the appropriate mitigation measures will depend on the particular location as well as the project type and scope. For each type of condition encountered, mitigation measures may be recommended at the location planning stage or carried out at the phase of facility construction, operations, or maintenance. These generally involve preemptive measures that decrease the magnitude of the impacts or palliative measures that strive to reduce the severity of impacts that have already occurred.

(a) *Groundwater Impacts*   At the operations phase that involves disposal of used transportation vehicles and parts, careful selection of disposal sites can help minimize the effect of leaching and consequent groundwater pollution. At the maintenance phase, leachate generation from petrochemical and herbicide storage facilities could be checked using liners and leachate collection systems. For right-of-way (ROW) and median maintenance activities that involve the use of herbicides, the timing, rate, and extent of applications can be planned such that the risk and exposure of water resources to adverse impacts will be minimal. To compensate for possible groundwater deprivation due to a transportation project, construction of a wetland could be made a part of the project.

(b) *Surface Water Impacts*   During the construction phase of a transportation project, erosion could be minimized by using on-site sediment-retention basins, geotextiles, or by planting rapid-growing vegetation at areas of stripped topsoil that are unlikely to be covered for a significant length of time, particularly in a season where rains are expected. During the maintenance phases that involve ROW and median vegetation control with herbicides, measures similar to those discussed for groundwater could be adopted. To compensate for surface water deprivation and degradation due to transportation operations, hydraulic structures could be constructed as a permanent part of the project. These structures include sediment basins, vegetative filters, wetlands, and deep ponds. Other palliative measures include the use of sediment removal and macrophyte (weed) harvesting for restoring lakes and reservoirs that have suffered surface water quality degradation or eutrophication. In extreme cases of water pollution from transportation activities, pollutant load in stormwater runoff could be reduced using a small physical–chemical treatment plant built for this purpose, and capable of chemical flocculation, settling, and filtering. Physical elements of such plants may include detention basins and ponds, sand filter beds, wetlands, infiltration basins, and percolation basins (Morris and Therivel, 2001).

13.6.3 Mitigation Measures by PDP Phase

(a) *Location Planning*   During location planning, areas sensitive to water resource impacts could be avoided as much as possible.

(b) *Design*   To compensate for surface water deprivation and degradation due to a transportation project, appropriate hydraulic structures could be designed as part of the project.

(c) *Construction*   During construction, vegetation is stripped off the land, and soils are moved to prepare the site for the new development. New streets, utilities, and buildings or other human-made structures are then constructed or reconstructed. During this phase, care should be taken to avoid soil erosion by stormwater runoff and there should be proper disposal of construction waste, such as concrete delivery truck wash water, unused asphalt, old timber and plaster, wiring, piping, and roofing materials.

(d) *Operations*   This phase involves primarily the disposal of used transportation vehicles and their parts into landfills or open areas that may, in the long term, generate leachates that could pollute surface runoff or groundwater bodies. Also, debris from transportation vehicles using the facility can lead to pollution of nearby water bodies. These can be minimized by adopting policies that control vehicle disposal and prohibition of littering by facility users, among other measures.

(e) *Maintenance*   From a water resources standpoint, the most deleterious aspect of transportation facility maintenance is the use of herbicide to control vegetation in ROWs and other areas. This can be minimized by adopting physical or biological means of vegetation control.

13.6.4 Discussion of Mitigation

Regardless of the type of receiving water or impact criterion, the mitigation of water resource impacts of transportation projects should be a continuing process throughout the facility life cycle, involving a number of

tasks. These include the prevention of water degradation, deprivation and flow pattern disruption, reduction or prevention of contact between pollutants and precipitation and surface runoff (thus minimizing the migration of pollutants off-site) using nonstructural practices and structural facilities, source disposal and treatment of runoff to reduce pollutant load transported by stormwater downstream, and follow-up treatments such as intercepting stormwater runoff downstream of all source and on-site controls to provide final follow-up treatment. To compensate for surface water deprivation and degradation due to a transportation project, construction of appropriate mitigating hydraulic structures could be made a part of the project. Nonstructural mitigation practices may include enforcement of building and site development codes, street sweeping, leaf pickup, deicing programs, infiltration practices (such as swales and filter strips, porous or modular pavement, percolation trenches, and infiltration basins), filter basins and filter inlets, follow-up water quality detention basins (dry), follow-up water quality retention ponds, follow-up wetland treatment, and enforcement of local government rules and regulations.

13.7   WATER QUALITY STANDARDS

The USEPA publishes its national recommended water quality criteria for the protection of aquatic communities, wildlife, and human health. These criteria are developed on the basis of requirements established by Section 304(a)(1) of the Clean Water Act. In developing these criteria, the EPA considers the effects of specific pollutants on shellfish, fish, plankton, wildlife, plant life, aesthetics, and recreation in any body of water. This includes specific information on pollutant concentration and dispersal through biological, physical, and chemical processes. These criteria provide guidelines for each state or tribe for the development of general or site-specific water quality standards (USEPA, 2006). The EPA water standards can be accessed at the following Web address: http://www.epa.gov/waterscience/criteria/.

13.8   LEGISLATION RELATED TO WATER RESOURCE CONSERVATION

Since the early 1970s, a number of laws and policies have sought to ensure that transportation planners and decision makers duly consider the impact of transportation activities on natural water sources. Some of these laws are discussed below.

- *National Environmental Policy Act (1969)*. In establishing a national policy regarding environmental protection, NEPA ushered in a new period of environmental legislation. This act requires all agencies to assess the environmental impact of implementing any project requiring federal action.

- *Wild and Scenic Rivers Act (1969)*. This act establishes the Wild and Scenic River System and protects rivers designated for their pristine and scenic value from activities that may affect those values adversely.
- *Clean Water Act (1972)*. The Clean Water Act is the primary authority for water pollution control programs and is aimed at restoring and maintaining the chemical, physical, and biological integrity of natural water resources. Among other provisions, the act sets national goals to eliminate the discharge of pollutants into navigable waters and protects the quality of water for aquatic and terrestrial organisms.
- *Marine Protection, Research, and Sanctuaries Act (1972)*. This act provides a permitting process to control the ocean dumping of dredged material.
- *Coastal Zone Management Act (1972)*. This act was passed in response to the public concern for balanced preservation and development activities in coastal areas.
- *Section 404 Regulatory Program (1972)*. The Section 404 Regulatory Program states that it is unlawful to discharge dredged or fill material into rivers, lakes, streams, tidal waters, and most wetlands without the necessary authorization, permit, or exemptions.
- *Safe Drinking Water Act (1974)*. In seeking to protect the nation's sources of drinking water, this act authorized the EPA to develop regulations for protecting underground sources of drinking water and to establish maximum contaminant levels to protect public health.
- *Resource Conservation and Recovery Act (1976)*. RCRA gives the EPA authority to regulate the transportation and disposal of hazardous wastes, prohibits open dumping of wastes, and regulates underground storage tanks, among others.
- *Comprehensive Environmental Response Compensation and Liability Act (1980)*. CERCLA authorized the EPA to respond to releases of hazardous wastes, established regulations to control inactive hazardous waste sites, established liability for releases of hazardous wastes from inactive sites, provided an inventory of inactive hazardous waste sites, and established appropriate action to protect the public from possible dangers at such sites.
- *Superfund Amendments and Reauthorization Act (1986)*. SARA revised and extended CERCLA and provided for emergency planning and preparedness, community right-to-know reporting, and toxic chemical release reporting. SARA also established a special

program for restoration of contaminated lands, somewhat similar to the Superfund under CERCLA.
- *Pollution Prevention Act (1990)*. The basic objective of this act is to establish a national policy for preventing or reducing pollution at the source wherever feasible, and it directs the federal EPA to undertake certain steps in that regard.

13.9 SOFTWARE FOR WATER RESOURCES IMPACT ASSESSMENT

Most computer-based models that can be used to assess the water quality impacts of transportation activities were developed and are maintained by the U.S. Army Corps of Engineers Waterways Experiment Station (USACE WES), the U.S. Department of Agriculture-Agricultural Research Services (USDA-ARS), the U.S. EPA Center of Exposure Assessment Modeling (CEAM), and Center for Subsurface Modeling Support (CSMoS). Information and copies of these models can be obtained from their web sites at http://el.erdc.usace.army.mil/, http://www.epa.gov/ceampubl/, and http://www.epa.gov/ada/csmos.html. The Civil and Environmental Engineering Department at Old Dominion University also documented a list of computer models from different sources and these models can be accessed through their electronic Civil/Environmental Model Library (CEML), (CEE-ODU, 2006): http://www.cee.odu.edu/cee/model/.

SUMMARY

In this chapter we identify and describe the various categories of hydrological impacts, such as the polluting source (facility or vehicle), water attribute (water quality, flow pattern effects and water quantity), the water source affected (surface water or groundwater systems), nature of the impact mechanism (physical, chemical, or biological), and the transportation mode and activity (construction, maintenance, and abandonment of infrastructure facilities, manufacture, maintenance, and disposal of vehicles and parts, and facility operations). Performance measures for assessing water resource, impacts of each transportation mode are then identified. We also provide a procedural framework for evaluating the water resource impacts of transportation activities. Steps include definitions of the study area and temporal scope for the analysis, hydrological inventory, identification of PDP phases that could affect water resources, and selection of hydrological performance measures. The framework also includes data analysis to predict hydrological impacts (change in the values of performance measures). Data analysis for hydrological impacts evaluation, later singled out for more detailed discussion, includes models that help in estimating changes in surface runoff and groundwater quantities, occurrence and extent of erosion and sedimentation in surface water, changes in groundwater quality, and predicting fully mixed in-stream pollutant concentration. We then discuss the mitigation of hydrological impacts from the perspective of impact criterion and the nature of water source (surface vs. groundwater). The chapter describes how the mitigation of water resources impacts can occur on two fronts: preemptive measures (applied before the pollutants are released into the receiving water bodies) and palliative measures (after pollutants have been released and dispersed). Water quality criteria for the protection of aquatic communities, wildlife, and human health published by the U.S. EPA and the key legislation related to water resources impacts of transportation are presented. Finally, we identify a few software packages that could be used to evaluate the water resource impacts of transportation projects.

EXERCISES

13.1. A new airport is proposed at a location that has surface water and groundwater resources. Describe how you would carry out an assessment of the water resources impacts of the proposed project. List five performance measures that could be used for the impact assessment.

13.2. The construction of a large parking terminal for a proposed park-and-ride facility will result in the conversion of a 9.5-acre grassland into 4.5 acres of concrete pavement, and 4 acres into lawns. Assuming that the entire drainage area could be considered as a single drainage unit, calculate the change in runoff volume due to the facility construction. The average rainfall intensity is 4 in/h. For each land-use type, use the midpoints of runoff coefficient ranges provided in Table 13.2.

13.3. It has been estimated that 5% of the cross-sectional area of an aquifer would be taken up by an underground transportation facility. What is the expected percentage reduction in groundwater flow due to the proposed project? Assume a K value of 800 ft/day for the rock medium and a hydraulic gradient of 0.05. Assume that after the project, the permeability of the medium is reduced by 5% but the hydraulic gradient remains unchanged.

13.4. It is proposed to upgrade a 25-mile county road into a state highway. The current traffic volume is 1200 ADT, and the projected traffic volume after completion highway is 8000. Assuming an accumulation period of 5 days, make a rough estimate of the expected change in pollutant load due to the project after each storm. The pre-accumulation pollutant load in both cases is 350 lbs.

13.5. As part of construction activities for a new interchange ramp for an existing Interstate highway, a grass cover (erodibility of 0.072) was removed but no temporary vegetative cover or erosion control was subsequently provided. The exposed soil has an erodibility of 0.55. A perennial creek is located downstream of the proposed interchange and is expected to be affected by the construction. The value of R, a measure of rainfall intensity, is 75, the slope angle is 4%, and the length of the slope is 120 ft. Assume perfect erosion control and management practice and well-maintained vegetative cover before the project. The construction will last for 5 months, and the topsoil is expected to remain exposed for 65% of that period. Calculate the average soil loss (tons/acre) over the construction period. Assume a cropping management factor (C) of 0.80.

13.6. Application of deicing compounds (1000 kg of sodium chloride per year) to the deck surface of a highway bridge over a large stream leads to chloride pollution of that water body. The average stream flow is 1500 ft^3/s and the average runoff flow is 35 ft^3/s. The average storm runoff volume is 400 ft^3. From readings at chloride monitoring stations established upstream and downstream of the bridge, it is observed that the average upstream chloride concentration is 1.2 mg/L. In a typical year, there are eight winter storms that are severe enough to merit salting activities of the deck. Determine the in-stream chloride concentration.

13.7. A river located near a busy freeway section receives storm runoff containing biodegradable pollutants from that highway. The initial complete-mix concentration of these pollutants at the point of discharge is 430 mg/L, and the decay rate for the pollutant is 3.3 per day. The stream velocity is 2.8 ft/s. Determine the pollutant concentration at a point 1.5 miles downstream of point of discharge.

13.8. Surface runoff from the Beltway International Airport is the only source of pollutants received by a nearby wetland. It is estimated that there are 0.13 million cubic ft of water in the wetland. The pollutant concentration in the storm water is 75 mg/L, and the volume of storm event flow into wetland is 23 ft^3/s. The runoff duration is

11 minutes. If the initial pollutant concentration in the wetland before the storm event is 0.65 mg/L, determine the expected pollutant concentration in the wetland after a storm event.

13.9. A new runway is planned as part of expansion of an existing airport. It is expected that the reservoir situated near the location of the proposed runway would receive runway storm runoff containing uric deicing compounds. Currently, the concentration of the pollutant in the lake is 0.21 mg/L, the pollutant load is 3.5 kg per day, the net inflow/outflow is 80 m³/day, the lake volume is 13,500 m³, the decay rate constant for the pollutant is 2 per day, and the time is 12 days. Find the fully mixed concentration of the pollutant in the lake.

REFERENCES[1]

Agrawal, M. L., Dikshit, A. K., Ghose, M. K. (2003). Impact assessment on soil erosion due to highway construction using GIS, *Electron. J. Geotech. Eng.*, Vol 8C.

Barrett, M. (1993). *A Review and Evaluation of the Literature Pertaining to the Quantity and Control of Pollution from Highway Runoff and Construction*, Center for Transportation Research, University of Texas, Austin, TX.

Burke, C., Burke, T. T. (1994). *HERPICC Stormwater Drainage Manual*, 1994 Highway Extension and Research Project for Indiana Counties and Cities, West Lafayette, IN.

Cambridge Systematics; Bernardin, Lochmueller & Associates (2003). *I-69 Evansville to Indianapolis: Final Environmental Impact Statement*, Indiana Department of Transportation, Indianapolis, IN, http://www.deis.i69indyevn.org/FEIS/. Accessed Jan. 2006.

*Canter, L. W. (1996). *Environmental Impact Assessment*, McGraw-Hill, New York.

*Canter, L. W., Sadler, B. (1997). *A Tool Kit for Effective EIA Practice: Review of Methods and Perspectives on Their Application*, Suppl. Rep., International Study of the Effectiveness of Environmental Assessment, Environmental and Ground Water Institute, University of Oklahoma, Institute of Environmental Assessment UK, and the International Association for Impact Assessment.

Carpenter, T. G. (1994). *The Environmental Impact of Railways*, Wiley, Chichester, UK.

CEE-ODU (2006). *Prologue*, Civil/Environmental Computer Model Library (CEML), Old Dominion University, Norfolk, VA, http://www.cee-odu-edu/cee/model/Assessed Nov. 15, 2006.

*Corbitt, R. A. (1999). *Standard Handbook of Environmental Engineering*, McGraw-Hill, New York.

D'Itri, F. A. (1992). *Chemical Deicers and the Environment*, Lewis Publishers, London, UK.

DEFRA (2005). *Water Quality: A Diffuse Pollution Review*, Department for Environment, Food and Rural Affairs, London.

*Dupuis, T. V. (2002). *Assessing the Impacts of Bridge Deck Runoff Contaminants in Receiving Waters*, Vol. 2, *Practitioner's Handbook*, NCHRP Rep. 474, Transportation Research Board, National Research Council, Washington, DC.

Gibson, P. (2001). *Oil Spills in U.S. Navigable Waters, 1990–1999*, American Petrolium Institute, Washington, DC.

Gupta, M. K., Agnew, R. W., Gruber, D., Kreutzberger, W. (1981). *Constituents of Highway Runoff*, Vol. IV, *Characteristics of Runoff from Operating Highways*, Tech, Rep. FHWA/RD-81/045, Federal Highway Administration, U.S. Department of Transportation, Washington, DC.

Hill, B. R. (1996). *Streamflow and Suspended-Sediment Loads Before and During Highway Construction, North Halawa, Haiku, and Kamooalii Drainage Basins, Oahu, Hawaii, 1983–91*; Water-Res. Inf. Rep. 96–4259, U.S. Geological Survey, Washington, DC.

ITC (2005). *Determination of Peak Runoff: Theoretical Background*, International Training Centre for Aerial Survey, Enschede, The Netherlands, http://www.itc.nl/ilwis/applications/application11.asp. Accessed Dec. 26, 2005.

*Jain, R. K., Urban L. V., Gary S., Balbach, H. E. (2001). *Environmental Assessment*, 2nd ed., McGraw-Hill, New York.

Lohani, B., Evans, J. W., Ludwig, W., Everitt, R. R., Carpenter, R. A., Tu, S. L. (1997). *Environmental Impact Assessment for Developing Countries in Asia*, Vol. 1, *Overview*, Asian Development Bank, Manila, Philippines.

McCuen, R. H. (2005). *Hydrologic Analysis and Design*, 3rd ed., Pearson Prentice Hall, Upper Saddle River, NJ.

*McKay, D., Peterson, S. (1993). Mathematical models of transport and fate, in *Ecological Risk Assessment*, ed. Suter, G. W., II, Lewis Publishers, Ann Arbor, MI.

Miller, G., Tyler, J. (1990). *Living in the Environment: An Introduction to Environmental Science,* Wadsworth, Belmont, CA.

*Morris, P., Therivel, R. (2001). *Methods of Environmental Impact Assessment*, 2nd ed., E.&F.N. Spon, London.

NC-DENR (2002). *Total Maximum Daily Load for Fecal Coliform Bacteria to Little Troublesome Creek*, North Carolina Cape Fear River Basin, North Carolina, Department of Environmental and Natural Resources, Raleigh, NC.

NCHRP (1993). *Hazardous Wastes in Highway Rights of-Way*, National Cooperative Highway Research Program, Transportation Research Board, National Research Council, Washington, DC.

*Ortolano, L. (1997). *Environmental Regulation and Impact Assessment*, Wiley, New York.

*Renard, K. G., Foster, G. R., Weesies, G. A., McCool, D. K., Yoder, D. C. (1997). *Predicting Soil Erosion by Water: A Guide to Conservation Planning with Revised Universal Soil Loss Equation (RUSLE)*. USDA Agric. Hdbk 703, U.S. Department of Agriculture, Washington, DC.

Rothengatter, W. (2003). Environmental concepts: physical and economic, in *Handbook of Transport and the Environment*, ed. Hensher, D. A., Button, K. A., Elsevier, Amsterdam, The Netherlands.

Singh, V. P. (1992). *Elementary Hydrology*. Prentice Hall, Englewood Cliffs, NJ.

Sinha, K. C., Teleki, G. C., Alleman, J. E., Cohn, L. F., Radwan, E., Gupta, A. K. (1991). *Environmental Assessment of Land Transport System Construction and Maintenance*, Tech. Rep.,

---

[1]References marked with an asterisk can also serve as useful resources for water resources impact evaluation.

Infrastructure and Urban Development Department, World Bank, Washington, DC.

Stone, R. P., Hilborn, D. (2000). *Universal Soil Loss Equation (USLE)*, Ministry of Agricultural, Food and Rural Affairs, Ontario, Canada, http://www.omafra.gov.on.ca/english/engineer/facts/00-001.htm#tab1. Accessed Feb. 21, 2006.

Talley, W. K. (2003). Environmental impacts of shipping, in *Handbook of Transport and the Environment*, ed. Hensher, D. A., Button, K. A., Elsevier, Amsterdam, The Netherlands.

Toy, T. J., Foster, G. R., Galetovic, J. R. (1998). *Guidelines for the Use of the Revised Universal Soil Loss Equation (RUSLE) Version 1.06 on Mined Lands, Construction Sites, and Reclaimed Lands*, Technical Coordinator, Office of Technology Transfer, Office of Surface Mining and Reclamation, Western Regional Coordinating Center, Denver, CO, http://www.ott.wrcc.osmre.gov/library/hbmanual/rusle. htm#downloadhandbook. Accessed Feb. 21, 2006.

USEPA (1982). *Results of Nationwide Urban Runoff Program* U.S. Environmental Protection Agency, Washington, DC.

———— (1994). *National Water Quality Inventory: 1992 Report to Congress*, U.S. Environmental Protection Agency, Washington, DC.

*———— (1996). *Indicators of the Environmental Impacts of Transportation*, EPA 230-R-96-009, Office of Policy, Planning and Evaluation, U.S. Environmental Protection Agency, Washington, DC.

———— (2006). *Water Quality Criteria*, U.S. Environmental Protection Agency, Washington, DC, http://www.epa.gov/waterscience/criteria/. Accessed Feb. 12, 2006.

Wayson, R. L. (2003). In Environmental Considerations During Transportation Planning, *The Civil Engineering Handbook*, ed. Chen, W. F., Liew, J. Y. R., CRC Press, Boca Raton, FL.

Wischmeier, W. H., Smith, D. D. (1958). Rainfall energy and its relationship to soil loss, *Trans. Am. Geophys. Union*, Vol. 39, No. 2, pp. 285–291.

*———— (1978). *Predicting Rainfall Erosion Losses: A Guide to Conservation Planning*, U.S. Department of Agriculture, Washington, DC.

*Young, G. K., Stein, S., Cole, P., Kammer, T., Graziano, F., Bank, F. (1996). *Evaluation and Management of Highway Runoff Water Quality*, FHWA-PD-96-032, Federal Highway Administration, U.S. Department of Transportation, Washington, DC.

ADDITIONAL RESOURCES

International Association for Impact Assessment, www.iaia.org

Jones, P. H., Jeffrey, B. A., Watler, P. K., Hutchon, H. (1986). Environmental Impact of Road Salting: State of the Art, Ontario Ministry of Transport and Communications, Downsview, ON.

Sadler, B. (1997). *EIA process strengthening: perspective and priorities*, in *Report of the EIA Process Strengthening Workshop*, Canberra, *Australia* Apr. 4–7, 1995, Environmental Protection Agency, Canberra, Australia for the International Study of the Effectiveness of Environmental Assessment.

CHAPTER 14

Visual Impacts

I think that I shall never see a billboard as lovely as a tree. Perhaps, unless the billboards fall, I'll never see a tree at all.

— *Ogden Nash (1902–1971)*

INTRODUCTION

The construction and operation of transportation facilities often have profound aesthetic or visual impacts. Visual impacts are typically measured in terms of the extent to which the new facility beautifies or blends in with its surrounding environment or how it obscures an aesthetically pleasant natural or human-made feature. By adding elements of local surroundings into its design, a transportation facility is rendered more aesthetically pleasing and compatible with its environment and imparts a feeling of well-being to those who experience its view.

In the past, roadways and transit systems were built without adequate attention paid to aesthetics during design, as most emphasis was placed on speed, safety, vehicle operating costs, and economic efficiency. In recent years, however, transportation agencies increasingly give due consideration to issues of aesthetics. At the federal level, highway aesthetics was first directly addressed by legislation in the form of protective measures for scenic roads and parkway views through the Historic Preservation Act of 1966 that spawned "the view from the road" programs in the mid-1960s. Since then, a number of legislative acts have addressed the issue of visual performance of highway and transit projects. Initiatives by the FTA and FHWA (Neuman et al., 2002) in context-sensitive design (CSD) have also aided aesthetic considerations in transportation project planning and design. Using interdisciplinary and collaborative approaches that solicit stakeholder input, CSD strives to balance aesthetic, environmental, and community needs with the primary

functions of mobility, accessibility, and safety. Federal initiatives have in turn encouraged state and local agencies to place greater emphasis on the role of aesthetics in existing and proposed transportation facilities and to develop and implement guidelines for integrating aesthetics into facility design. In most developing countries, however, consideration of aesthetics in transportation facility design generally does not always receive due attention.

In this chapter we present the principles of visual performance, identify the various factors that affect the visual performance of transportation facilities, and discuss the mechanisms by which such factors could either enhance or degrade visual performance. A methodology is presented for assessing the visual impacts of transportation projects, and the chapter concludes by identifying various ways to mitigate visually deficient transportation corridors and areas.

14.1 PRINCIPLES OF VISUAL PERFORMANCE

14.1.1 General Principles

Aesthetics, represented by visual performance, refers to the quality and character of *visual experience*. Visual experience, in turn, is a compound of *visual resources* (such as the proposed project and its human-made or natural setting) and viewer response. The components of a visual environment are the visual resources and the visual characteristics (Figure 14.1). The *visual performance impact* due to a transportation project is therefore an interaction of the resulting change in total visual resource and any change in *viewer response*. Degradation of visual experience due to a new transportation facility could take the form of *visual intrusions* (failing to blend in with the existing environment) or *visual obstruction* (blocking the view of aesthetically pleasing features).

The level of facility aesthetics is important in establishing the scope of visual assessments. FHWA (1988) identifies such levels:

- *Internal aesthetics:* follows traditional visual design theory to examine a project by itself.
- *Relational aesthetics:* considers the visual relationships between a project and specific elements of its human-made or natural surroundings and can influence community acceptance of a project.
- *Environmental aesthetics:* examines the aesthetics of the overall environment affected, of which the proposed project is only a part.

On a more micro level, it may be useful to identify the specific topological elements of a transportation facility and/or its environment that affect visual patterns and

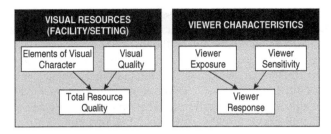

Figure 14.1 Components of the visual environment.

quality. These elements were identified by Lynch (1960) as follows: *paths* [these represent linear landscape elements along which vehicular or pedestrian travel occurs, such as guideways (railways or roadways), sidewalks, iron railings that separate the sidewalk from the carriageway, etc.]; *edges* (which represent linear elements of the landscape, but are seen as boundaries, such as walls of adjacent properties, and highways and streets that segregate distinct areas of a city or landscape); *districts* (distinctive areas of a community that have a consistent feature or underlying character, such as shopping malls, residential areas, and parks); *nodes* (points on the transportation network that link path features, such as interchanges and intersections); and *landmarks* (point locations that are typically viewed at a considerable distance from the transportation facility).

14.1.2 Performance Measures for Visual Performance Assessment

Although visual experience is subjective and varies across individuals, a set of performance criteria and measures can be developed for assessing visual performance to a fair degree of consistency. Criteria that are provided in Smardon and Hunter (1983), FHWA

(1986), and Ortolano (1997) are categorized as shown in Figure 14.2 and are discussed thereafter.

(*a*) *Visual Character* Visual character is comprised of visual pattern elements and pattern character. The *visual pattern elements* are the primary visual attributes of objects: color, form, line, and texture.

- *Color* refers to consistencies between a facility's colors, hues, values, and chrome with those of its environment.
- *Form* of an object is its virtual mass, bulk, or shape and refers to the compatibility between the facility's dimension and shape and its environment.
- *Line* pertains to the edges of the facility or parts thereof and refers to the compatibility of the facility's edges, bands, and introduced silhouette lines with its environment.
- *Texture* is the apparent surface coarseness and may refer to the compatibility between the facility's surface textural grain, density, and pattern regularity.

In some texts, the overall term *landscape compatibility* is used to indicate how well the new facility fits into the overall landscape from the perspectives of the visual pattern elements: color, form, line, and texture. Viewer appreciation of visual pattern elements can be influenced by the viewer's distance from the object.

Visual pattern character refers to the visual contrast between a transportation facility and its visual environment (setting). Two objects may have similar visual pattern elements but may exhibit very different visual characters. Visual characters are scale, dominance, diversity, continuity, and variety.

- *Scale contrast* is the extent to which the facility blends into its environment from the perspective of

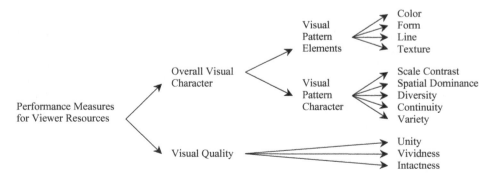

Figure 14.2 Performance measures for visual resources.

its size relative to the sizes of other features in its environment. The scale contrast of a new facility is considered excellent if the contrast is small or it introduces a small scale of activity; and it is considered very poor if it involves a major scale introduction/intrusion.

- *Spatial dominance* is similar to scale contrast but on a larger dimension and is the extent to which the project elements would be dominant in views of larger landscape and cityscape. This is also described as the dominance of the project in the setting or the landscape situation backdrop. An excellent rating (low spatial dominance) would be given to a facility that does not dominate; a poor rating (high spatial dominance) to one where the facility features too prominently in the composition of the landscape and therefore completely dominates the landform, water, or sky backdrops; and intermediate ratings where the facility is subordinate or codominant with some other natural feature.
- *Diversity* is a function of the frequency, variety, and intermixing of the visual pattern elements of the facility with its setting. Also termed as *setting contrast* (the extent to which a project's visual pattern elements contrast with or blend in with its existing natural or human-made background).
- *Continuity* is the uninterrupted flow of pattern elements in a landscape and the maintenance of visual relationships between landscape components that are immediately connected or related.
- *Variety* is the richness and diversity of physical objects and interrelationships within the landscape.

(*b*) *Visual Quality* Visual quality is simply the *excellence* of the viewing experience. Although this may be a subjective measure, there is generally consensus regarding views that have high visual quality (e.g., city skylines, waterfalls, fall leaf colorations). Visual quality may be assessed using one of several approaches: (1) whether (and possibly, the extent to which) an area is designated a site of natural history (parks, scenic rivers, etc.), (2) opinion surveys of viewers, and (3) indicators of visual quality: vividness, intactness, and unity. Assessments of all three measures should be "high" in order to conclude existence of high visual quality.

- *Unity* is the degree to which the various visual resources of the landscape join to form a coherent, harmonious visual pattern.
- *Vividness* is the "memorability" of the visual impression received from contrasting landscape elements as they combine to form a striking and distinctive visual pattern.

- *Intactness* is the integrity of visual order in the natural and human-built landscape and the extent to which the landscape is free of visual encroachment.

From the foregoing list of visual character and quality attributes, the analyst can develop a set of performance measures to evaluate a given project. Care should be taken to avoid the choice of performance measures that overlap, such as landscape compatibility and setting contrast. Figure 14.3 provides examples of various attributes of visual character and quality.

14.2 FACTORS AFFECTING VISUAL PERFORMANCE AND IMPACT MECHANISMS

14.2.1 Factors

The factors that influence the visual performance of a transportation project can be categorized as follows:

1. *Transportation facility characteristics.* Include the facility type, dimensions, shape, texture, and other features. These characteristics may render a facility appealing or repulsive, to various degrees of intensity.
2. *Stage of the project development process.* Completed facilities naturally have greater visual appeal than in their partially completed states.
3. *Extent of the exposure.* The greater the exposure of an unsightly transportation facility from public view, the higher the degree of visual degradation. Similarly, the greater the exposure of a visually appealing facility to public view or the greater the exposure of visually pleasing landscapes from the facility, the more favorable the facility's visual impacts. For example, for bridge underpasses, the undersides (which are typically not designed with aesthetic considerations) are mostly visible to the viewing public and may therefore cause the entire bridge to be perceived as having poor visual performance.
4. *Viewer sensitivity.* This reflects the level of concern exhibited by likely viewers of the proposed facility and its environs for a good aesthetic experience. Viewer sensitivity, in turn, is influenced by viewer category (pedestrian, facility user, tourist, or resident), age, gender, background, frequency and duration of viewing, and type of activity that the viewer is engaged in while experiencing the scene in question (commuting, work, recreation, etc.). Viewer sensitivity can also be influenced by the viewers' proximity to the facility and their level of aesthetic training or skills.
5. *Landscape characteristics.* Types of landscapes or their features that are generally considered desirable

Figure 14.3 Examples of visual character attributes: (*a*) Intactness and unity: Water transportation often offers a comprehensive viewshed of a city's visual environment. The above viewshed illustrates a fair degree of visual intactness and unity. (*b*) Form: Good visual form is reflected in harmony between horizontal and vertical alignment and the natural terrain. (*c*) Continuity: The visual continuity of the natural environment can be enhanced by a well designed highway. (*d*) Visual character: A transportation facility can be visually dominant when its pattern elements (form, color, line) are in significant contrast with its setting. (*e*) Scale: The visual scale of this highway is consistent with the scale of its rural setting due to the relatively gentle and grassy side slopes that merge into the setting. (*f*) Vividness (memorability): Viewed from the transportation facility, vivid or memorable landscapes (man-made or natural, such as city skyline or waterfalls, respectively) can enhance visual quality. (Image credits: (*a*) Courtesy of David Prieto, Creative Commons Attribution-ShareAlike 2.0; (*b*) Courtesy of Eric Weaver, Creative Commons Attribution-ShareAlike 2.0; (*c*) Courtesy of Alyson Hurt, Creative Commons Attribution 2.0; (*d*) Courtesy of Jehane, Creative Commons Attribution-ShareAlike 2.0; (*e*) Courtesy of Robert Chan, Creative Commons Attribution-ShareAlike 2.0; (*f*) Courtesy of Juliane von Prondzinsky, Creative Commons Attribution-NoDerivs 2.0.)

include areas known for their scenic beauty, parks, recreational areas, areas with historic and culturally preserved structures, and entry to population centers.

In assessing the visual impact of an existing or proposed facility, therefore, it is important to account for the factors noted above. These factors affect visual performance through the impact mechanisms discussed next.

14.2.2 Impact Mechanisms

The visual impacts of transportation projects may be short or long term. Short-term effects generally include construction-related visual degradation from construction stockpiles and debris, equipment, materials, signage, and staging area in the construction zone; the long-term effects are related to the facility itself. Transportation facilities affect the overall aesthetic quality of an area through the following mechanisms:

- Addition of sizable new physical elements on the visual landscape through construction of new transportation facilities or the expansion of existing facilities. These can either intrude or blend in with their surrounding environment.
- Communities adjacent to freeways, railways, or air terminals encounter views of passing transportation vehicles. Depending on viewer sensitivity, such views may either degrade or enhance viewing pleasure.
- Blocking of existing visually pleasing natural or human-made features (such as landmarks, open space, community areas of interest) or visually repulsive features (such as blighted areas) through new construction or expansion of transportation facilities.
- Removal of existing visually pleasing or repulsive structures and other features located in the right-of-way during new facility construction or expansion.
- Addition of visual clutter to the landscape due to provision of new transportation features, such as road signs, overhead traffic sign posts, and lines. In some areas, particularly in developing countries, commercial billboards and junkyards that are placed near or along major transportation facilities can be a source of visual degradation.
- Replacement of unsightly transportation infrastructure with upgraded facilities, primarily for reasons of capacity or safety enhancement but with aesthetic improvement as a secondary benefit.
- Provision of visually pleasing features as part of the transportation project, such as lighting (in urban areas) and landscaping of medians and roadsides.

- Elevated railways or freeways can affect the privacy of people in houses and other buildings located below.

The mechanisms described above apply to any transportation mode with slight variations across modes. Carpenter (1994) identified several ways in which railways can affect visual experience.

14.3 PROCEDURE FOR VISUAL IMPACT ASSESSMENT

The steps involved in assessing the impacts of transportation project on visual performance are presented in Figure 14.4. The methodology enables the visual performance impact evaluation of an existing transportation facility and its surroundings with and without a proposed improvement and can also be used to compare the visual impacts of alternative designs for a project at a given site or at multiple sites. The framework is developed on the basis of methodologies discussed by Sinha et al. (1989), Carpenter (1994), Forkenbrock and Weisbrod (2001), CTS (2003), and Florida DOT (2003). The overall assessment can be categorized into four primary tasks: inventory, simulation, evaluation, and mitigation.

Step 1: Establish the Visual Analysis Areas and Prepare the Visual Inventories This step involves establishing analysis areas (distinct viewing settings at various points along the project corridor or around the project area) and the preparation of visual inventories for the existing (or preimplementation) condition. A proposed project may be *lineal* (highway, railway, waterway, hiking trail, etc.) or *nodal* (such as an airport terminal, transit terminal, or parking garage). For nodal facilities, distinct view settings may comprise the various perspectives of the facilities, particularly the areas that are frequently seen by facility users (the most common being the facility's front view). For lineal facilities, the entire project must be divided into distinct segments (each segment and its environment constitutes a separate analysis area), and the assessment must be carried out for each analysis area.

For each perspective of a nodal facility and for each lineal facility segment, the proposed project may affect not only the immediate surroundings of the facility at that location but also a wide area well beyond the facility's physical boundaries. All areas visible from the proposed facility, or from which the transportation facility is visible, may be included in the study area. A *route inventory* is used for lineal facilities; an *area inventory* is used for nodal facilities. Topographic maps and aerial photographs typically form the basis for the preparation of visual inventories and are supplemented by field surveys, sketches,

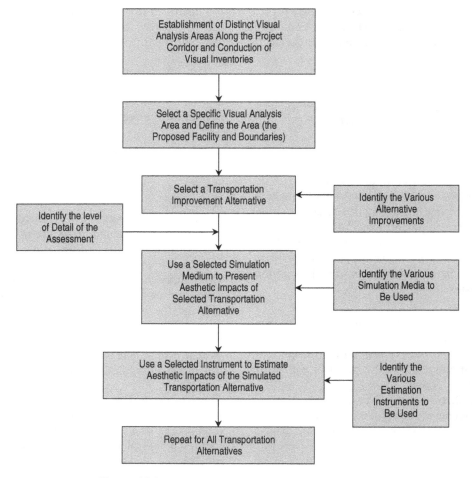

Figure 14.4 Framework for visual impact assessment.

and ground-level photos or video clips. During inventory preparation, distinct viewsheds (views seen from a particular location) are established. The components of a visual inventory are the overall visual character (patterns of natural and manmade structures and features), scenic or visual quality (attractiveness of the natural or manmade features and structures), viewing conditions (viewpoints from which the proposed project or its environment can be seen), viewing characteristics (frequency and duration of viewing), and viewer sensitivity (level of concern exhibited by the likely viewers of the project and its environment after completion), and visual policies (policies, guidelines, or standards) established by the agency providing the facility or the community affected (Sheppard, 1989). CHSRA (2004) suggests a study area size for visual resource assessment as the quarter-mile loci from corridors and stations, except in instances where there are scenic viewing points or overlooks within 1 mile of

a project. For below-surface transportation projects, the potential for visual impacts occurs at the interface with surface facilities, such as underground transit stations.

Step 2: Determine the Transportation Alternatives The next step is to consider each of the several transportation improvement alternatives under investigation. For instance, a river crossing may be achieved using a suspension bridge or a cable-stayed bridge. Also, a mountain may be traversed using a tunnel or by excavating an open section through it. It is unlikely that aesthetic considerations alone would have enough influence to sway the decision to adopt one transportation alternative over another, unless there are negligible differences in levels of other performance measures across the alternatives or for facilities where visual performance far exceeds all other performance measures in importance: for example, environments or facilities that inherently attract tourists or provide viewing pleasure due to their natural beauty or historical–cultural significance.

Step 3: Identify the Scope and Level of Detail of the Visual Performance Assessment   This step may be considered an extension of the visual inventories. To address the different types of visual issues and to identify which issues require analysis for a specific project, FHWA (1988) provided a scoping questionnaire. This tool can be used to generate data for visual performance assessments for highway and transit projects (Figure 14.5). Checklists used by the California Department of Transportation, presented in Figure 14.6, provide an example of how to

decide what level of visual assessment may be required to meet federal or other jurisdictional requirements.

Step 4: Present the Existing Scene and Simulation of the Situation Proposed   This step involves a description of the physical features or operation of the transportation system using an appropriate simulation technique: still images (photomontage, artist sketch, etc.) or dynamic images. Visual impact coverage should contain enough information about the visual characteristics of the project and the people who will frequently view the visual resources to be offered by the facility and its environment (FHWA, 1988). Documenting visual impacts can be

---

SCOPING QUESTIONNAIRE FOR VISUAL ASSESSMENT

CONSIDER EACH OF THE EIGHTEEN QUESTIONS BELOW AND SELECT THE RESPONSE THAT MOST CLOSELY APPLIES TO THE PROJECT IN QUESTION.

1. Project Characteristics
A. What are the major project design standards (capacity, access, speed, geometry)? What are the alternatives?
B. What is the typical geometric profile (roadway, roadside slopes and drainage, right-of-way, guideway)? What major structures and appurtenances will be required? What are the alternatives?
C. What other facilities (such as rest areas, maintenance yards, or stations) are part of the project? What construction areas (borrow pits, spoil areas) will be needed? What are the alternatives?
D. What secondary effects (such as development at interchanges, station, etc. or conversion of land from rural to urban uses or from residential to commercial) may result from the project?

2. Visual Environment of Project
A. What landscape components (landform, water, vegetation, and human-made development) are characteristic of the regional landscape and the immediate project area?
B. From which locations are the project likely to be seen?
C. What visually distinct landscape units can be identified within the immediate project area?

3. Significant Visual Resource Issues
A. How would the project alternative affect the landscape components present within the visual environment?
B. What is the existing visual character of the project environment (e.g., form, line, color, texture and dominance, scale, diversity, continuity), and how compatible would project alternatives be with this character?
C. What levels of visual quality exist at the current time (may be evaluated using criteria such as vividness, intactness, unity, and other indicators), and how much would project alternatives affect these?

4. Significant Viewer Response Issues
A. What major viewer groups are likely to see the project?
B. What is the viewer exposure to project alternatives for different groups (numbers, distance, duration and speed of view, etc.), and how does each alternative affect important existing views?
C. How are viewer activity and awareness likely to affect the attention that different groups pay to the project and its visual environment? (Include both viewers from the road (such as vehicle operators) and off the road (such as pedestrians).
D. Are any visual resources in the project environment are particularly important to local viewers? Are there any districts, sites, or features that are regionally or nationally recognized for their historical/cultural significance?
E. Is the project thought to threaten or support expectations for the future appearance of any areas it traverses?

5. Visual Impacts and Impact Management
A. In summary, what significant visual impacts, if any, appear likely? (Include both adverse and beneficial impacts.)
B. What alternative might avoid, minimize, or reduce any adverse visual impacts, and by how much?
C. What actions might rectify or compensate for adverse visual impacts, and by how much?

Figure 14.5   Scoping questionnaire for visual assessments for transportation projects. (Adapted from FHWA, 1988.)

VISUAL IMPACT ASSESSMENT CHECKLIST GUIDE

Consider each of the ten questions below and select the response that most closely applies to the project in question. Each response has a corresponding point value. After the checklist is completed, the total score will represent the type of VIA document suitable for the project.

It is important that this scoring system be used as a preliminary guide only and not be used as a substitute for objective analysis on the part of the user. Although the collective score may direct the user toward a certain level of analysis document, circumstances associated with any one of the ten question areas may necessitate elevating the VIA to a greater level of detail.

A. CHANGE TO THE VISUAL ENVIRONMENT

1. Will the project result in a noticeable change in the physical characteristics of the existing environment?
(Consider all project components and construction impacts, both permanent and temporary, including landform changes, structures, noise barriers, vegetation removal, railing, signage, and contractor activities)

High level of change ☐   Moderate level of change ☐   Low level of change ☐

2. Will the project complement or contrast with the visual character desired by the community?
(Evaluate the scale and extent of the project features compared to the surrounding scale of the community. Is the project likely to give an urban appearance to an existing rural or suburban community? Is the change viewed as positive or negative? Research planning documents, or talk with local planners and community representatives to get a rough idea of what type of visual environment local residents envision for their community.)

Highly incompatible ☐   Somewhat incompatible ☐   Not compatible ☐

3. What types of project features and construction impacts are proposed? Are bridge structures, large excavations, sound barriers, or median planting removal proposed?
(Certain project improvements can be of special local interest, causing a heightened level of public concern and requiring a more focused visual analysis.)

High concern ☐   Moderate concern ☐   Low concern ☐

4. Will the project changes likely be mitigated by normal means such as landscaping and architectural enhancement, or will avoidance measures be necessary to minimize adverse change?
(Consider the types of changes caused by the project, i.e., can undesirable views be screened or will desirable views be permanently obscured?)

Project alternative may be needed ☐   Extensive mitigation likely ☐   Normal mitigation ☐

5. Will this project, when seen collectively with other projects, result in an aggregate adverse change in overall visual quality or character?
(Identification of contributing projects should include any projects (both departmental and local) in the are a that have been constructed within the last couple of years and those currently envisioned or planned for future construction. The window of time and the extent of area applicable to possible cumulative impacts should be based on a reasonable anticipation of the viewing public's perception.)

Impacts likely in 0-5 years ☐   Impacts likely in 6-10 years ☐   Cumulative impacts unlikely ☐

Figure 14.6   Visual impact assessment checklist guide. (Adapted from Caltrans, 2003.)

VISUAL IMPACT ASSESSMENT CHECKLIST GUIDE (Continued)

B. VIEWER SENSITIVITY

1. What is the potential that the project proposal may be controversial within the community or opposed by any organized group?
(This can be researched initially by talking with departmental and local agency management and staff familiar with the affected community's sentiments as evidenced by past projects and/or current information. Factor in your own judgment as well.)

High potential ☐   Moderate potential ☐   Low potential ☐

2. How sensitive are potential viewer groups likely to be regarding visible changes proposed by the project?
(Consider among other factors the number of viewers within the group, probable viewer expectations, activities, viewing duration, and orientation. The expected viewer sensitivity level may be scoped by applying professional judgment and by soliciting information from other staff of the transportation agency, local agencies, and community representatives familiar with the affected community's sentiments and demonstrated concerns.)

High Sensitivity ☐   Moderate Sensitivity ☐   Low Sensitivity ☐

3. To what degree does the project appear to be consistent with applicable laws, ordinances, regulations, policies, or standards?
(Although the state may not be obligated to adhere to local planning ordinances, these documents are critical in understanding the importance the local communities place on aesthetic issues. The environmental planning branch of the transportation agency may have copies of the planning documents that pertain to the project. If not, this information can be obtained by contacting the local planning department. Relevant documents can be found online at websites of transportation agencies.)

Incompatible ☐   Moderately compatible ☐   Largely compatible ☐

4. Are any permits going to be required by outside regulatory agencies (i.e., federal, state, or local) that will necessitate a particular level of visual impact assessment?
(Anticipated permits, as well as specific permit requirements, may be determined by contacting the environmental planner, project engineer, or other relevant staff responsible for the project. The analyst should coordinate with the agency's representative responsible for obtaining the permit prior to communicating directly with any permitting agency.)

Yes ☐   Maybe ☐   No ☐

5. Will the project development team or public benefit from a more detailed visual analysis in order to help reach consensus on a course of action?
(Consider the proposed project features, possible environmental impacts, and probable mitigation recommendations.)

Yes ☐   Maybe ☐   No ☐

C. DETERMINING THE TYPE OF VISUAL IMPACT ASSESSMENT REQUIRED
The total score this preliminary visual checklist will indicate the general level of visual impact assessment that should be performed for the project. Once the level of recommended assessment is identified, the user should double-check the results by comparing each of the ten question are as to the total score in order to confirm that the level of document appears sufficient and reasonable in each case.

Total Score 25-30 – Prior to preparing a VIA, a formal visual scoping study that meets or exceeds any FHWA or FTA requirement is recommended to alert the project development team to potential highly adverse impacts and to develop new project alternatives to avoid those impacts.

Total Score 20-24 – A fully developed VIA that meets or exceeds any FHWA or FTA requirement is recommended. This technical study will probably receive extensive public review.

Total Score 15-19 – An abbreviated VIA would be appropriate in this case. The assessment would describe project features, impacts, and mitigation requirements. Visual simulations would be optional.

Total Score 10-14 – A brief visual assessment report in a simple memo form may be adequate.

Figure 14.6   (*continued*)

carried out using a graphical illustration of the still structure (and its environment) by an artist, photomontage techniques, computer modeling (Figures 14.7 and 14.8), and GIS-based simulation models (Yamada et al., 1986; Ortolano, 1997; Stamps, 1997; Burkhart et al., 1998; Singh, 1999). Forkenbrock and Weisbrod (2001) suggest that photomontage techniques are particularly useful when time or resources are limited, when the proposed change would add a new visual element to the street scene (e.g., an elevated structure), when the change will block existing views of significant landmarks or green space in the area, or when the character of the street scene will be altered (e.g., an upgraded intersection).

The analyst can use virtual computer models if there are adequate time and monetary resources to do so. Also, simulation software can be used to describe the various views encountered as one travels along the transportation roadway or guideway (for lineal facilities) or as one moves

Figure 14.7 Computer simulation of rail exchange yard, port of Vancouver. (From Lauga & Associates, 2006.)

Figure 14.8 Computer simulation of Highway 12 reconstruction, Long Lake, Minnesota. (From Minnesota DOT, 2003.)

around a nodal facility, thereby offering perspectives that are more dynamic than sketches, photos, or still computer images. In extreme cases where there is a complete lack of visual data for the project area, images of existing similar facilities at other locations with similar environments may be used to serve as a basis for the aesthetic impact assessment of a proposed transportation improvement.

The selection of appropriate simulation media depends on the project scale, the physical environment, the number of alternatives, the availability of resources, and the analyst's familiarity with the techniques. For transportation projects that cover a small area and involve only a small number of alternative designs, simple artist sketches may be appropriate. On the other hand, for a large number of transportation alternatives, where each alternative is extensive in terms of scope and coverage, and where each alternative is expected to have significant aesthetic impact, the use of computer simulations or 3D GIS-based visualisations is recommended. FHWA (2005) has developed a guide that discusses the use of commonly available software tools for providing visualizations that facilitate understanding and communicating the visual impacts of facility designs. A recent report provides best practices and experience to date within transportation agencies that are developing and incorporating visualization into the project development process (Hixon, 2006).

Step 5: Estimate the Visual Impacts   After the aesthetic impacts have been described using an appropriate simulation medium, the next step is to identify the level of desirability for each simulated exhibit of the proposed transportation system in its environment. This is typically done using a selected estimation instrument such as a simple questionnaire survey. The respondents of the survey should not only be the direct users of the transportation facility, but also persons who regularly encounter the image of the facility in their regular activities and whose visual perception of the study area can be affected by the presence of the proposed facility. These persons could include area residents, local business owners, city officials, the general public, or other interested parties along the area affected. Respondents are asked about their perspectives on the aesthetic appeal of each alternative system design. The survey instrument should preferably include the images of the transportation alternatives and may be mailed out to target respondents or published on the Internet. The use of Internet-based surveys facilitates the presentation of several alternatives to the respondent using a variety of simulation media, including dynamic images.

(a) *Scaling*   Each respondent should assess the impacts based on a scale of increasing or decreasing appeal.

Responses may be categorical (high, medium, or low appeal), or quantitative (on a scale of 0 to 10 or 0 to 100, 0 being very unappealing and 10 or 100 being very appealing). For instance, the California High-Speed Rail Authority (CHSRA, 2004) ranked the potential shadow impacts of elevated structures and the light and glare impacts of each design alternative in terms of "high," "medium," and "low" ratings.

(b) *Weighting* Ratings indicated by the responses can be adjusted by various influencing factors, such as the length of time that each respondent typically would encounter the image on a daily basis, and then all responses can be collated to generate a single rating index or value that represents the level of aesthetic appeal of the selected transportation alternative.

Example 14.1 Figure E14.1 shows a newly constructed permanent soil nail retaining wall along an urban highway near a university town in California. In a weighting survey of experts, 10 graphic artists and architects were asked to give the importance they attach to each of the following visual performance measures: landscape compatibility, 0.5; scale contrast, 0.3; and spatial dominance, 0.2. For landscape compatibility, relative weights were assigned to each subcriterion within that performance measure as follows: color, 0.4; form, 0.3; line, 0.2; and texture, 0.1. Then, in a survey of users (students, commuters, and residents of the area), the design was shown to these persons to rate them on the basis of the performance measures. Determine the overall visual performance index for the design under consideration.

SOLUTION Adopting a form from Smardon and Hunter (1983), the overall visual performance index of the retaining wall can be estimated in terms of the performance

Figure E14.1 Visual representation of proposed retaining wall. (From FHWA, 2001.)

measures on the basis of ratings obtained from the user survey. The rating scores, given in Table E14.1, indicate average values. The combined rating for the landscape compatibility performance measure for example, is given by

$$VP_{land\ compatibility}$$
$$= \frac{(0.4 \times 8) + (0.3 \times 7) + (0.2 \times 8) + (0.1 \times 6)}{0.4 + 0.3 + 0.2 + 0.1}$$
$$= 7.5$$

The overall weighted rating for visual impact assessment is found as follows:

$$VP = \frac{(0.5 \times 7.5) + (0.3 \times 8) + (0.2 \times 4)}{0.5 + 0.3 + 0.2} = 6.95$$

The visual performance rating of 6.95 out of a maximum of 10 indicates a "moderate" to "good" visual performance.

Example 14.2: Consider two hypothetical alternative designs that are proposed for a bridge crossing a large river in a rural mountainous region (Figure E14.2). Assume that the alternative designs are generally similar in operational function and safety performance but that their visual quality and patterns are significantly different. The simulated images were shown to a review panel, which then rated each alternative on the basis of each of the performance criteria considered, on a scale of 0 to 10.

The average ratings are shown in Table E14.2. Determine the best alternative on the basis of the visual performance considerations. Use the same weights given in Example 14.1.

SOLUTION Combined visual performance (sum of weighted ratings):
Alternative a:

$$VP_a = \{0.5 \times [(0.4 \times 5) + (0.3 \times 7) + (0.2 \times 5)$$
$$+ (0.1 \times 4)]\} + (0.3 \times 6) + (0.2 \times 6) = 4.25$$

Alternative b:

$$VP_b = \{0.5 \times [(0.4 \times 6) + (0.3 \times 9) + (0.2 \times 7)$$
$$+ (0.1 \times 6)]\} + (0.3 \times 7) + (0.2 \times 9) = 5.5$$

For visual performance, alternative *b* can therefore be considered the better option. It should be noted that in practice an extensive set of images from various angles is used for each design.

Step 6: Present the Results of the Analysis The entire process is repeated for each transportation improvement

Table E14.1   Ratings of Visual Performance

Visual Element (Performance Measure)	Visual Sub-elements	Indicators or Clues	Sub-element Weight	Rating Score (from Scale)	Weighted Rating Score	WRS for Performance Measure
Landscape compatibility	Color	Consistency between facility colors, hues, values, and chroma with those of its environment	0.4	8	3.2	
Rating Scale *Excellent—10* *Good—7* *Moderate—5* *Poor—3* *Very poor—0*	Form	Compatibility between facility dimension and shape and its environment	0.3	7	2.1	7.5
	Line	Compatibility of facility edges, bands, and introduced silhouette lines with its environment	0.2	8	1.6	
	Texture	Compatibility between facility surface textural grain, density, and regularity of pattern	0.1	6	0.6	

Scale contrast
Rating Scale

Excellent (small object or scale of activity)—10 *Good (significant object or scale)—8* *Moderate (one of several major scales or major objects in confined setting)—4* *Poor/very poor (major scale introduction/intrusion)—0*			1	8	8	8

Spatial dominance (landscape situation backdrop)
Rating Scale

Excellent (object does not dominate)—10 *Good (object is subordinate to some other natural feature)—8* *Moderate (object is co-dominant with a natural feature)—4* *Poor/very poor (object is very prominent in the composition of the landscape; dominates the landform, water, or sky backdrops)—0*			1	4	4	4

Source: Adapted from Smardon and Hunter (1983).

Table E14.2   Average Visual Ratings of Two Final Alternatives

Design Alternative	Landscape Compatibility				Scale Contrast	Spatial Dominance
	Color	Form	Line	Texture		
a	5	7	5	4	6	6
b	6	9	7	6	7	9

Figure E14.2   Simulated images of alternative bridge designs.

alternative, and a table of results can be generated indicating the cost and estimated level of impact of each alternative.

14.4   LEGISLATION RELATED TO VISUAL IMPACT

The Historic Preservation Act of 1966 and the 1966 Department of Transportation Act, Section 4(f) brought due recognition to transportation aesthetics and precipitated consideration of visual resource impact mitigation in the transportation planning process. This was done in a bid to minimize the potential harm that might result from transportation facilities to the natural beauty of the countryside, public parks and recreational lands, wildlife and waterfowl refuges, and historic sites. In 1969, the National Environmental Policy Act (NEPA) applied environmental awareness policies to all types of federally supported projects and all types of project settings. NEPA and the Council on Environmental Quality regulations identified aesthetics as one of the elements or factors in the human environment that must be considered in determining the effects of a highway or transit action. NEPA requires the development of techniques that appropriately weigh aesthetic values

in transportation agency decision making. In addition, Title 23, Section 752.2 of the Code of Federal Regulations states: "Highway aesthetics is a most important consideration in the federal-aid highway program. Highways must not only blend in with our natural, social, and cultural environment, but also provide the pleasure and satisfaction in their use" (FHWA, 1986, 1988). The Federal Transit Agency's environmental impact regulation (Title 23, Section 771 of the Code of Federal Regulations), issued jointly with FHWA, describes two types of mass transit projects that normally have significant effects on the environment, including visual performance: new construction or extension of fixed-rail transit facilities (e.g., rapid rail, light rail, commuter rail, automated guideway transit), and new construction or extension of a separate roadway for buses or high-occupancy vehicles not located within an existing highway.

14.5   MITIGATION OF POOR VISUAL PERFORMANCE OF EXISTING FACILITIES

If an existing transportation facility or a proposed improvement (or part thereof) is generally deemed aesthetically unappealing by viewers, appropriate mitigation measures should be undertaken. In consultation with residents and other users, various mitigation alternatives can be evaluated for their suitability in terms of the cost and enhancement of visual performance. To ensure the full realization of mitigation actions, visual assessment activities should be coordinated with the subsequent design, construction, and maintenance phases of project development. Examples of such mitigation measures are presented in Table 14.1 (Sinha et al., 1989; FHWA, 2001; Sound Transit, 2005).

14.6   VISUAL PERFORMANCE ENHANCEMENT: STATE OF PRACTICE

There are many examples of visual impact assessments conducted by various state transportation agencies. The California Department of Transportation has prepared assessments for the Interstate 15 managed lanes, Route 88

in Amador County, and the Bakersfield-to-Los Angeles high-speed train (CHSRA, 2004). In the state of Washington, the DOT carried out visual impact assessments for a number of highway projects, including the I-90 Snoqualmie Pass East Project (Washington State DOT, 2004) and the Palouse County Scenic Byway (Washington State DOT, 2002). The Massachusetts Bay Transportation Authority incorporated visual and aesthetic considerations in the Environmental Impact Study of the Silver Line Phase III project (MBTA, 2005). In New York State, the Department of Environmental Conservation has developed guidelines that distinguish between state and local concerns and establish measures geared toward elimination, reduction, or compensation for poor visual qualities (New York State DEC, 2000). A comprehensive effort in assessing and enhancing visual performance of transportation projects carried out in Minnesota (CTS, 2005) produced an Internet-accessible video tool for this purpose.

Table 14.1   Visual Impact Mitigation Measures

PDP Stage	Mitigation Measures for Poor Visual Performance
Location planning	Select and/or modify routes to avoid or reduce the need to acquire and clear a new right of way.
	Integrate facilities with area redevelopment plans, particularly nodal facilities such as stations and terminals.
	Minimize obstructions to scenic views.
	Avoid adverse impacts to scenery of high visual performance by tunneling or bypassing.
Design	Use interdisciplinary design teams to incorporate aesthetic considerations in the design of project elements.
	Minimize the elevation or height of elevated guideways where possible (without sacrificing vertical clearances) to limit their visibility. Minimize, wherever possible, the extent of parking areas associated with the project.
Construction	Minimize clearing area for construction, construction staging, stockpiling, and storage.
	Reduce temporary construction light and glare impacts by aiming and shielding light sources.
	Screen views of construction equipment and materials.
	Minimize construction-related dust.
	After project completion, restore landscapes disturbed by construction-related activities to preconstruction condition.
Maintenance and operations	Plant appropriate vegetation in and adjoining the project right of way to replace existing street trees and greenbelts and to provide screening for sensitive visual resources and viewers.
	Replant remainder parcels with grass or simple plantings; maintain them; and pursue their redevelopment for land uses that prove feasible and consistent with neighborhood plans, such as residential, commercial, or open-space uses.
	Use source shielding in exterior lighting at stations and ancillary facilities, such as maintenance bases and park-and-ride lots, to ensure that light sources (such as bulbs) are not directly visible from residential areas, streets, and highways and to limit spillover light and glare in residential areas.

14.6.1 Context-Sensitive Design Practices

Context-sensitive design (CSD) involves using inputs from technical professionals, the local community, interest groups, and the general public in the development of transportation solutions. The use of CSD principles has helped communities to address issues of safety, mobility, historic and natural resource preservation, aesthetics, and the environment in general. Examples of successful application of CSD principles that helped preserve or enhance the visual performance of the communities affected are presented below.

Paris Pike in Kentucky's bluegrass region (Figure 14.9) was rebuilt using the natural landscape patterns of the area as the framework for addressing historic, scenic, and natural resources in the area (Irving, 2003). The design team worked with residents in designing the road,

(a)

(b)

Figure 14.9 Views of Kentucky's Paris Pike. (a) Grass shoulders along Kentucky's Paris Pike reduce overall pavement width and lessen stormwater runoff. Compared to paved shoulders, grassed shoulders create visually narrower road sections that complement rural settings and are conducive to lower travel speeds. (b) Carefully constructed rock fence along the new Paris Pike, built with the same materials and methods as historic rock fences in the region. Rock fences help blend the highway into the surrounding landscape. (From Irving, 2003)

communicating directly with individual property owners, displaying three-dimensional computer models of roadway designs, and using electronic polling to gauge public opinion. Also, transportation officials hired stone masons from Scotland to teach local artisans how to build and replace dry-laid stone walls along the scenic roadway (AASHTO, 2005). For Pennsylvania's Danville–Riverside Bridge replacement project (Figure 14.10), aesthetic treatments included cut-stone architectural surface treatments at all piers and abutments, decorative masonry lighting on the bridge structure, and gateway pylons of brick and mortar at the touchdown points for the new bridge, which was consistent with the design of nearby historic buildings (ORCCSS, 2006).

14.6.2 Policies and Guidelines for Visual Performance Preservation and Enhancement

Sipes (2005) discussed policies, guidelines, and design standards for visual impact assessment and enhancement at various transportation agencies. For example, the Florida legislature directed the state's DOT to include aesthetics in the development of all highway projects and suggested that local governments and municipalities include aesthetic considerations in their planning activities. The Ohio DOT (1999), through its Design Aesthetics Initiative, encourages the use of pattern, color, texture, and landscaping to make a road noise barrier or bridge visually appealing to motorists and residents. In Minnesota, the Aesthetics Initiative Measurement System (AIMS) is used to understand and monitor how travelers perceive the attractiveness of highway corridors and to make appropriate recommendations regarding facility planning, design, construction, and maintenance. In making aesthetics a central component of its highway design, the state of Nevada adopted a master plan for aesthetics to improve the quality of life for its residents as well as the public image of

Figure 14.10 Danville–Riverside bridge project in Pennsylvania. (From ORCCSS, 2006.)

Figure 14.11 Visual considerations in transportation design: (*a*) addition of color to highlight the traffic island can enhance visual character of the scene; (*b*) concrete pavers easily introduce patterns into medians; (*c*) because of their visibility, attractive signals and signage can have a positive influence on overall design; (*d*) brick is compatible with residential areas but can be visually dominating—vegetation in front can help reduce the apparent height of architectural structures; (*e*) experience shows that appropriate themes can enhance a corridor and discourage graffiti vandalism; (*f*) design imprints on noise barriers along a highway. (From FHWA, 2001; Texas DOT, 2001.)

the facility, through artwork, landscaping, street design, signage, and other treatments along the roadside. In Maryland, aesthetics was identified as one of the considerations for incorporating environmental design as part of highway planning. The California Department of Transportation, a pioneer in the use of aesthetics in facility design,

has established a program to foster the incorporation of transportation art and aesthetics into highway structures (Caltrans, 2003).

The Texas DOT developed a *Landscape and Aesthetics Design Manual* (Texas DOT, 2001) that offers guidelines for all highway and street project development, including

Table 14.2   Examples of Visual Performance Considerations in Design

Subject	Possible Visual Performance Considerations
Main facility structure	Veneers made from local materials or taken from the natural environment of the area can be used to relate the structure to its environment and also to highlight special areas (Figure 14.11a, and b).
Geometric features and colors	Contrasting textures and colors can be used to visually mark different zones of activity, such as cross-walks, traffic islands, etc. (Figure 14.11a, and b).
Sidewalks	Modular paving units such as bricks and concrete can be used to create decorative walkways or medians (Figure 14.11b).
Road furniture	Lighting, graphics, signage, and other information devices can be incorporated into the transportation structure. For example, plates bearing street names can be placed on the structure rather than being mounted on posts (Figure 14.11c).
	Ornamental fencing can be used instead of unsightly fence structures (Figure 14.11d).
	Graffiti-resistant themes, paint, and surface finishes can be used for fencing and other structures along roadways and guideways (Figure 14.11e).
Concrete surfaces	Texture or color tints can be added to concrete surfaces to ensure better blend with their natural environment (Figure 14.11f).
Vegetation in median, right-of-way, etc.	Special decorative but hardy trees, shrubs, bushes, and wild flowers can be planted along guideways, roadways, and interchanges (Figure 14.11f).
Noise barriers	Noise barriers, retaining walls, and other fencewalls along the guideway or roadway can be (1) leveled and capped to eliminate unsightly irregularities in their levels, (2) surface textured or colored in such a way as to blend with their natural or human-made surroundings (Figure 14.11d and f), (3) decorated to reflect the nature of their environment, such as designing imprints on their surfaces (Figures 14.11f).
Adjacent properties	Appropriate structures and vegetation can be used to screen adjacent properties from view (Figure 14.11d and f), thus enhancing privacy, aesthetics, and noise control.

common structural elements and transportation features. Elements of these guidelines are presented in Table 14.2 and illustrated in Figure 14.11.

14.6.3 Cost of Visual Performance Enhancements

The challenge of maximizing the visual performance benefits of transportation projects may add to the final cost of the transportation facility. However, the additional expense is expected to be minimal compared to the overall cost of the project. For example, the Ohio DOT estimated that the cost of incorporating or improving aesthetics typically amounts to less than 1% of the overall project cost (Sipes, 2005). Funding support for aesthetic enhancement efforts by state and local agencies may come from the Federal Transportation Enhancement Program, which is geared toward effectively integrating transportation facilities into their surrounding communities

and natural environment, thereby increasing the value of a project and rendering making it more aesthetically pleasing.

The benefits of visual performance improvements may not be readily quantifiable in monetary terms. Where tourism is involved, increased revenue could be used as a proxy for benefits. In other cases, the increase in community well-being due to enhanced aesthetics may be difficult to measure in monetary terms, but an attempt could be made using the willingness-to-pay approach, where residents, road users, and pedestrians are asked how much they are willing to pay to see a specific improvement in the visual performance of the transportation facility and its immediate environment.

SUMMARY

Assessments of aesthetic impacts of new or expanded transportation projects should consider the visual characteristics of the new facility, the people who encounter the project view, and the visual resources of the project area. The level of detail required for a visual impact assessment of a proposed project typically depends on the scale of the project, the physical environment, and the availability of resources. The methodology presented in this chapter for aesthetic impact assessment consists of several steps which start by defining specific visual analysis areas within the project corridor or project area. Then the required level of detail of the assessment is identified, and a simulation medium is selected to generate an image of how the transportation facility would blend into the existing surroundings. The opinions of community residents invited to view the hypothetical image are solicited using a questionnaire survey regarding how they would rate the enhancement or degradation of visual performance brought upon by the proposed project. Through an analysis of the survey results, the transportation alternative associated with the highest increase in overall visual performance can be identified. Where there is only one proposed transportation alternative or where an existing facility is being analyzed, there may be a need to make recommendations to mitigate any visual degradation of the environment. In that case it may be necessary to screen visual quality enhancement treatment for their relative cost-effectiveness and select the best treatment. We discussed the efforts by several states that have sought to enhance the visual performance of their transportation systems by developing and implementing guidelines that integrate aesthetics into facility design. The practice of using context-sensitive design principles in highway design and construction provides an opportunity not only for mobility and safety improvements, but also for community well-being and overall environmental quality enhancement.

EXERCISES

14.1. Of the performance measures presented in Section 14.2.1, which four would you consider most appropriate for evaluating the visual impacts of (**a**) a proposed rehabilitation of an airport terminal, and (**b**) selecting a route alignment through a scenic area? For each case, what relative weights (total of 1 unit for each case) would you assign to each performance measure?

14.2. It is proposed to renovate or relocate an intermodal facility terminal that serves road and rail transportation. Four alternative designs are submitted for consideration. The facility is located near a busy urban freeway and is visible to road users as well as residents of nearby apartment complexes. Identify four alternative methods of describing to an audience the visual characteristics of the facility proposed.

What are the merits and demerits of each method? What performance measures would you establish for assessing the visual impacts of this project?

14.3. Assume that a freeway bypass is being planned near your city. What are some of the visual performance concerns that are likely to arise?

REFERENCES[1]

AASHTO (2005). *Taking the High Road: The Environmental and Social Contributions of America's Highway Programs*, Center for Environmental Excellence, American Association of State Highway and Transportation Officials, Washington, DC.

*Burkhart, R., Soehgen, B., Schulz, B. (1998). Computer imaging for transportation projects, *ITE J.*, Vol. 68, pp. 46–48.

*Caltrans (2003). Standard environmental reference (SER), Chap. 27, in *Visual and Aesthetics Review*, California Department. of Transportation, Sacramento, CA, http://www.dot.ca.gov/ser/. Accessed July 20, 2005.

Carpenter, T. G. (1994). *The Environmental Impact of Railways*, Wiley, Chichester, UK.

CHSRA (2004). *Bay Area to Merced: Aesthetics and Visual Quality Technical Evaluation,* Program EIR/EIS, California High Speed Rail Authority, U.S. Department of Transportation, Federal Railway Administration, http://www.cahighspeedrail.ca.gov/eir/report/EIR_TOC.asp. Accessed Sept. 15, 2005.

CTS (2003). *Visual Impact Assessment for Context Sensitive Design*, Center for Transportation Studies, University of Minnesota, Minneapolis, MN, http://www.cts.umn.edu/education/csd/video. Accessed Dec. 15, 2005.

*FHWA (1986). *Guidance Material on the Preparation of Visual Impact Assessments*, Federal Highway Administration, U.S. Department of Transportation, Washington, DC.

*⸻ (1988). *Visual Impact Assessment for Highway Projects*, Federal Highway Administration, U.S. Department of Transportation, Washington, DC.

*⸻ (2001). *Roadway Aesthetic Treatments 2001: Photo Album Workbook*, Federal Highway Administration, U.S. Department of Transportation, Washington, DC.

*⸻ (2005). *Design Visualization Guide*, Federal Lands Highway Division, Federal Highway Administration, U.S. Department of Transportation, Washington, DC.

*Forkenbrock, D. J., Weisbrod, G. E. (2001). *Guidebook for Assessing the Social and Economic Effects of Transportation Projects*, NCHRP Rep. 456, Transportation Research Board, National Research Council Washington, DC.

Florida DOT (2003). *Project Development and Environment Marval*, part 2, chapter 15, Florida Department of Transportation, Tallhassee, FL.

Irving, L. (2003). A new approach to road building, *Public Roads*, Vol. 67, No. 1, July–Aug.

*Hixon III, C. L. (2006). *Visualization for Project Development*, NCHRP Synthesis Report 361, Transportation Research Board, National Research Council, Washington, DC.

[1]References marked with an asterisk can also serve as useful resources for visual quality assessment.

Lauga & Associates Consulting Ltd. (2006). *Computer Simulation of Rail Exchange Yard*, Vancouver Wharves Ltd., Port of Vancouver, BC, Canada, www.port-plan.com. Accessed May, 3, 2006.

Lynch, K. (1960). *Image of the City,* MIT Press, Cambridge, MA.

MBTA (2005). *Silver Line Phase III: Supplemental Draft EIS/EIR*, Massachusetts Bay Transportation Authority, Boston, MA, http://www.allaboutsilverline.com. Accessed July 13, 2005.

Minnesota DOT (2003). Newsletter 98, Minnesota Department of Transportation, St, Paul, MN, www.newsline.dot.state.mn.us. Accessed May 3, 2006.

*Nassauer, J. L., Larson, D. (2004). Aesthetic initiative measurement system (AIMS), *Procs. 84th Annual Meeting of the Transportation Research Board*, Washington, DC.

NCTR (National Center for Transit Research) and Florida DOT. (2002). *Community Impact Assessment and Environmental Justice for Transit Agencies: A Reference Guide for Transportation* http://ntl.bts.gov/lib/_for_Transit.pdf. Accessed Sept. 15, 2005.

Neuman, T. R., Schwartz, M. Clark, L., Bednar, J., Edaw, Inc. (2002) *A Guide to Best Practices for Achieving Context Sensitive Solutions*, NCHRP Rep. 480, Transportation Research Board, National Research Council, Washington, DC.

*New York State DEC (2000). *Assessing and Mitigating Visual Impacts,* Department of Environmental Conservation, Albany, NY, http://www.dec.state.ny.us/website/dcs/policy/visual2000.pdf. Accessed July 3, 2005.

Ohio DOT (1999). *Governor Announces ODOT Design Aesthetics Initiative*, Ohio Department of Transportation, Colombus, OH, Internet news release, Dec. 10, http://www.dot.state.oh.us/news/1999/12-10-99.htm. Accessed Sept. 3, 2005.

ORCCSS (2006). *CSS Case Studies: Danville–Riverside Bridge and Approach—Pennsylvania*, Online Resource Center for Context Sensitive Solutions, Washington, DC, http://www.contextsensitivesolutions.org/content/case_studies/kentucky_penn/. Accessed May 1, 2006.

*Ortolano, L. (1997). *Environmental Regulation and Impact Assessment*, Wiley New York.

*Sheppard, S. R. J. (1989). *Visual Simulation: A Users Guide for Architects, Engineers, and Planners*, Van Nostrand Reinhold, New York.

Sinha, K. C., Varma, A., Souba, J., Faiz, A. (1989). *Environmental and Ecological Considerations in Land Transport: A Resource Guide*, Tech. Paper, INU41, World Bank, Washington, DC.

Singh, R. R. (1999). Sketching the city: a GIS-based approach, *Environ. Plan.*, Vol. 26B, pp. 455–468.

Sipes, J. L. (2005). Curb appeal, *Public Roads,* Vol. 69, No. 2, Sept.–Oct.

*Smardon R. C., and Hunter, M. (1983). Procedures and methods for wetland and coastal areas visual impact assessment, in *The Future of wetlands. Assessing Visual-Cultural Values*, Ed: Smardon R. C., Allanheld Osmun Publishers, Totawa N.J.

Sound Transit (2005). *Final Supplemental Environmental Impact Statement on the Regional Transit Long-Range Plan,* St, Seattle, WA, http://www.soundtransit.org/projects/longrange/finalseis.asp. Accessed Sept. 3, 2005.

Stamps, A. E., III (1997). Some streets of San Francisco: preference effects of trees, cars, wires, and buildings, *Environ. Plan.*, Vol. 24B, pp. 81–93.

Texas DOT (2001). *Landscape and Aesthetics Design Manual*, Texas Department of Transportation, Austin, TX.

Washington State DOT (2002). *Palouse-Scenic Byway Application*, Visual Analysis Discipline Report, Washington Department of Transportation, Tacoma, WA.

Washington State DOT (2004). *Interstate 90 - Snoqualmie Pass East*, Visual Impact Assessment Discipline Report, Washington Department of Transportation, Tacoma, WA.

Yamada, H., Shinohara, O., Amano, K., Okada, K. (1986). Visual vulnerability of streetscapes to elevated structures *Environ. Behav.*, Vol. 18, No. 6, pp. 733–754.

ADDITIONAL RESOURCES

Adobe Photoshop CS2 http://www.adobe.com/digitalimag/engineering.html is is a professional image-editing software that provides capabilities for visualization of hypothetical alternative transportation images using techniques such as photomontage.

Silicon Graphics, Inc. (www.sgi.com), Harvard Graphics (www.harvardgraphics.com), ESRI's Arcview (www.esri.com/software/arcgis), and Intergraph (http://data.geocomm.com) provide computer modeling and simulation software for describing hypothetical visual images.

Burkart, R., Soehgen, B., Schulz, B. (1998). Computer imaging for transportation projects, *ITE J.*, Vol. 68, Feb., pp. 46–48. Offers a detailed discussion of several computer-aided techniques that were used in the visual analysis component of an environmental impact statement that are typically prepared for transportation agencies.

CTS (2005). *Visual Impact Assessment for Context Sensitive Design*, Center for Transportation Studies, University of Minnesota, Minneapolis, MN, http://www.cts.umn.edu/education/csd/video. This Web site contains a video resource developed by Minnesota Department of Transportation in conjunction with the Federal Highway Administration that presents procedures for visual impact assessment.

Valuation methods for environmental impacts, http://www.czp.cuni.cz/vzdel/letni_skola/.

Smardon, R. C., Palmer, J. F., Fellman, J. P. (1986). *Foundations for Visual Project Analysis*, Wiley, New York, NY.

NCTR (National Center for Transit Research) and Florida DOT. (2002). *Community Impact Assessment and Environmental Justice for Transit Agencies: A Reference Guide for Transportation* http://ntl.bts.gov/lib/_for_Transit.pdf. Accessed Sept. 15, 2005.

Nassauer, J. L., Larson, D. (2004). Aesthetic initiative measurement system (AIMS), *Procs. 84th Annual Meeting of the Transportation Research Board*, Washington, DC.

Nevada DOT (2002). *Pattern and Palette of Place: A Landscape and Aesthetics Master Plan for the Nevada State Highway System*, Nevada Department of Transportation, Carson City NV, http://www.nevadadot.com/. Accessed July 13, 2005.

Rahman, O. M. A. (1992). Visual quality and response assessment: an experimental technique, *Environ. Plan.*, Vol. 19B, pp. 689–708.

USDOT (2005). Federal Highway Administration's Context-Sensitive Design/Thinking Beyond the Pavement Web site, U.S. Department of Transportation, Washington, DC, www.fhwa.dot.govenvironment/flex/index.htm. Accessed Sept. 3, 2005.

CHAPTER 15

Impacts on Energy Use

Energy is eternal delight.

—*William Blake (1757–1827)*

INTRODUCTION

Until the industrial revolution, energy sources for land transportation were muscles of horses and donkeys, and for sea transportation, wind power. The invention of the steam engine in the early nineteenth century ushered in rail-based land transportation that was powered mostly by coal and steam-powered boats for sea transportation. By the late nineteenth century, a new form of fuel, petroleum, had been discovered, and at the turn of that century, gasoline, a petroleum product, was being used as fuel for newly invented internal combustion engines. Industrial mass production in the early twentieth century led to reduced automobile costs and spawned automobile use. Since that period, and particularly after the 1950s, developments such as the decline in energy production costs, new highway construction, urban sprawl, and increase in automobile sizes contributed to a sharp rise in energy use, particularly gasoline. By 1970, the average mileage of an American car was only 13.5 miles per gallon, and a gallon of gasoline cost less than 25 cents. Global political developments in the 1970s led to steep upsurges in oil prices and motivated the search for efficient gasoline use and for alternative and renewable energy sources. In the mid-1980s, new suppliers entered the oil market, leading to a general lowering of oil prices. Corrected for inflation, the price of gasoline in 2006 was one-half its 1980 price.

Although oil prices are relatively low at the current time, the world economy remains vulnerable to price and supply disruptions. Since 1995, oil imports in the United States have exceeded domestic production. With the transportation and industrial sectors currently dominating energy consumption, projections suggest that the transportation sector will gradually become the sole dominant consumer of energy use in future, reaching 40% of total energy needs by 2025 (Figure 15.1a). With respect to petroleum in particular, the share of transportation sector consumption (66% as of 2003) is expected to increase further in the future (Figure 15.1b). With increased industrialization and motorization of the developing countries, particularly India and China, the future demand for petroleum throughout the world can be expected to increase rapidly beyond its current level.

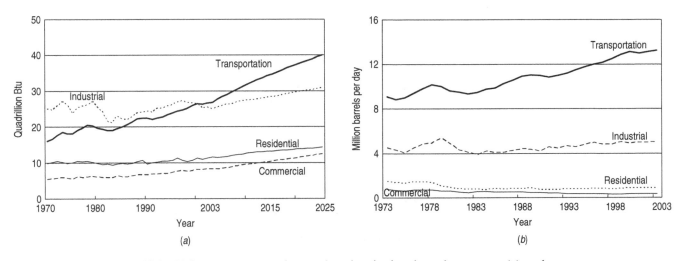

Figure 15.1 U.S. energy consumption trends and projections by end-use sector: (*a*) total energy (past and projected), 1970–2025; (*b*) petroleum, 1974–2003. (From EIA, 2005a.)

As petroleum is a non-renewable resource, it is clear that the impacts of transportation programs, policies, and projects on energy consumption must receive due consideration in the evaluation process.

Transportation Energy Use by Mode and Fuel Type:
A breakdown of U.S. transportation energy use by mode and fuel type is provided in Table 15.1. In measuring and comparing energy consumption, the British thermal unit (Btu) is widely used. One Btu represents the amount of energy required to increase the temperature of 1 pound of water (equivalent to 1 pint) by 1°F, approximately the heat produced from burning one matchstick. Results of petroleum energy computation can also be expressed as "equivalent barrels of crude oil." One barrel of crude oil contains approximately 42 gallons of gasoline or diesel. One gallon of gasoline is equivalent to approximately 125,000 Btu, and 1 gallon of diesel is equivalent to approximately 139,000 Btu.

Table 15.1   U.S. Transportation Energy Consumption in 2003,[a] by Mode and Fuel Type (10^{12} Btu)

	Gasoline	Diesel Fuel	Liquefied Petroleum Gas	Jet Fuel	Residual Fuel Oil	Natural Gas	Electricity	Total
HIGHWAY	**16,387.0**	**5,138.1**	**57.3**	**0**	**0**	**13.6**	**0.8**	**21,596.80**
Light vehicles	**15,863.8**	**364.1**	**40**	**0**	**0**	**0**	**0**	**16267.9**
Automobiles	9,203.0	51.7						9254.7
Light trucks[b]	6,637.0	312.4	40					6989.4
Motorcycles	23.8	0.0						23.8
Buses	**6.4**	**165.6**	**0.2**	**0**	**0**	**13.6**	**0.8**	**186.6**
Transit	0.1	74.3	0.2	0	0	13.6	0.8	89.0
Intercity[c]		28.3						28.3
School	6.3	63.0						69.3
Medium/heavy trucks	**516.8**	**4,608.4**	**17.1**	**0**	**0**	**0**	**0**	**5,142.3**
NONHIGHWAY	**194.1**	**852.2**	**0**	**2,186.60**	**570.6**	**685.6**	**345.9**	**4,835.0**
Air	**30.7**	**0.0**	**0**	**2,186.60**	**0**	**0**	**0**	**2,217.3**
General aviation	30.7			110.7				141.4
Domestic air carriers				1,749.40				1,749.4
International air carriers[d]	0.0			326.5				326.5
Water	**163.4**	**298.0**	**0**	**0.0**	**570.6**	**0**	**0**	**1,032.0**
Freight		257.8			570.6			828.4
Recreational	163.4	40.2						203.6
Pipeline						685.6	274.6	960.2
Rail	**0.0**	**554.2**	**0**	**0.0**	**0**	**0**	**71.3**	**625.5**
Freight (Class I)		533.9					0	533.9
Passenger		20.3					71.3	91.6
Transit		0.0					48.7	48.7
Commuter		10.0					16.3	26.3
Intercity[c]		10.3					6.3	16.6
HWY & NONHWY TOTAL	**16,581.1**	**5,990.3**	**57.3**	**2,186.6**	**570.6**	**699.2**	**346.7**	**26,431.8**
OFF-HIGHWAY	**733.8**	**1,469.6**	**0**	**0.0**	**0**	**0**	**0**	**2,203.4**
Agriculture	42.2	464.9						507.1
Industrial & commercial	216.6	248.9						465.5
Construction	34.2	741.6						775.8
Personal & recreational	440.5	5.8						446.3
Other	0.3	8.4						8.7
TOTAL	**17,314.9**	**7,459.9**	**57.3**	**2,186.6**	**570.6**	**699.2**	**346.7**	**28,635.2**

Source: Davis and Siegel (2006)

[a]Figures are for civilian consumption only. Totals may not include all possible uses of fuels for transportation (e.g., snowmobiles).

[b]Light trucks refer to two-axle, four-tire trucks.

[c]For Intercity buses, data is for year 2000.

[d]One half of fuel used by domestic carriers in international operation.

15.1 FACTORS THAT AFFECT TRANSPORTATION ENERGY CONSUMPTION

There are several factors that affect transportation energy consumption. Economic growth, for example, is a key factor and is typically reflected in an increase in vehicle trips and vehicle sales. Also, increases in fuel prices or taxes often result in reduced personal travel demand or demand shifts from one transportation mode to another, thus affecting overall energy consumption. Another factor that affects energy consumption is governmental regulation. Stringent emission or fuel consumption standards cause changes in the vehicle market in terms of vehicle mix by class and cohort distribution as well as technological specifications. Other factors that affect energy consumption are improvements in vehicle technology, facility improvements, and operational policies (for example, speed limits).

Project-level transportation system interventions cause changes in energy consumption on a relatively local scale. Such interventions typically lead to changes in vehicle-miles of travel and average fuel consumption rates. On a macroscopic scale, interventions between economic and transportation systems in a region are modeled to predict changes in energy use.

15.1.1 Fuel Prices and Taxes

Fuel prices influence the demand for travel and therefore affect the use of fuel and ultimately, energy consumption, even when consumption rates remain the same. However, under some conditions, travel demand elasticities with respect to fuel price may not be large enough to cause any significant reduction in travel demand and thus, energy consumption (Goodwin et al., 2004). Table 15.2 presents fuel consumption elasticities with respect to fuel price.

15.1.2 Fuel Economy Regulation

First enacted by the U.S. Congress in 1975, the *corporate average fuel economy* (CAFE) standards specify minimum fuel efficiency (or fuel economy) requirements for new cars and light trucks in a bid to control energy consumption. The mandated standards are as follows: passenger cars, 27.5 miles/gallon (mpg) (since 1996); and light-duty trucks, which include pickups, vans, and sport utility vehicles, 20.7 mpg (1996–2004), 21.0 mpg (2005), 21.6 mpg (2006), and 22.2 mpg (2007).

15.1.3 Vehicle Sales by Class

Over the past few decades, the vehicle class that has shown the greatest changes in sales volume is the sports utility vehicle (SUV), which generally has relatively low fuel efficiencies. This trend has led to an overall decrease in average fuel economy of the highway vehicle fleet. Trends in light-duty vehicle sales are summarized in Table 15.3.

15.1.4 Vehicle Technology

Vehicle fuel efficiency is influenced by vehicle technology and features, such as the type of internal combustion engine, engine combustion rate and burn temperature, and the vehicle type, size, and curb weight. To comply with CAFE requirements automakers have implemented several technological innovations that improve fuel efficiency. The average fuel efficiency of model year 2005 vehicles was 25.0 mpg (28.9 mpg for autos and 21.0 mpg for trucks), the highest since 1996, but 5% lower than the 1988 level. The trend in vehicle fuel efficiency from 1975 through 2005 is shown in Figure 15.2. The average fuel efficiency (FE_{avg}) can be calculated using the equation

$$FE_{avg} = \text{total sales} \Big/ \Big(\sum \text{sales}_i / FE_i \Big)$$

where sales_i is the vehicle sales of model year i and FE_i is the average fuel economy of vehicles of model year i.

Using year 2003 as the base year, projections of average fuel efficiency for selected transportation vehicle classes are shown in Figure 15.3. For all transportation modes, fuel efficiency is expected to increase, due to improvements in engine technology, use of light-weight materials, and use of alternative fuels. From 2003 to 2025, the fuel efficiencies for light-duty vehicles and freight trucks are projected to improve by 5 to 10%, and for new aircraft, fuel efficiency is projected to increase by 19%.

15.1.5 Road Geometry

As discussed in Chapter 7, vertical and horizontal curves affect fuel consumption rates. The relationships in Figure 15.4 indicate the extra energy needed to maintain speed on vertical grades. These are based on HERS Fuel consumption equations.

Table 15.2 Elasticities of Demand with Respect to Fuel Price (Dollars/Liter)

Measure of Demand	Short-Term Elasticity	Long-Term Elasticity
Fuel consumption (total)	−0.25	−0.64
Fuel consumption (per vehicle)	−0.08	−1.10
Vehicle-km (total)	−0.10	−0.29
Vehicle-km (per vehicle)	−0.10	−0.30

Source: Goodwin et al. (2004)

Table 15.3   Annual Sales and Market Shares of Light-Duty Vehicles at Selected Years

Type of Vehicle	1980	1990	2000	2003	2003 (%)
Small pickup	516,412	1,135,727	1,071,730	744,040	4.6
Large pickup	1,115,248	1,116,490	1,968,710	2,077,330	12.7
Vans: small and large	341,714	1,331,570	1,640,890	1,387,502	8.5
SUVs: small, medium-sized, and large	243,163	930,838	3,625,623	4,408,542	27.0
Subtotal: Light-duty trucks/vans/SUVs	2,216,537	4,514,625	8,306,953	8,617,414	52.8
Subcompact, compact auto	4,685,213	5,433,870	4,328,667	3,723,040	22.8
Midsized auto	3,073,103	2,511,503	3,352,198	2,624,346	16.1
Large auto	1,336,190	1,279,092	1,297,237	1,350,670	8.3
Subtotal: Automobiles	9,094,506	9,224,465	8,978,102	7,698,056	47.2
Total light-duty vehicles	11,311,043	13,739,090	17,285,055	16,315,470	100.0

Source: ORNL (2005).

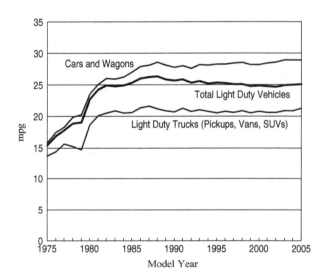

Figure 15.2   Adjusted fuel economy (three-year moving average) by model year. (From Davis and Diegel, 2006.)

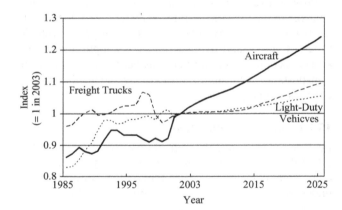

Figure 15.3   Transportation stock fuel efficiency index by mode, 1985 to 2025. (From EIA, 2005a.)

For horizontal alignments, the degree of curvature influences the rate of fuel consumption because vehicles expend extra energy to overcome centrifugal forces along curves. The relationships in Figure 15.5 show the excess fuel consumption of light-duty vehicles, in gal/1000 vehicle-miles, under various degrees of curvature and speed (Caltrans, 1983). For a given speed, the sharper the curve, the greater is the fuel consumption.

15.1.6 Transportation Intervention

The type of a transportation intervention can influence project-level energy consumption, as discussed below.

(*a*) *New Construction or System*   A HOV facility or a bus or rail rapid transit can significantly affect transportation energy consumed along a corridor, because of the higher energy efficiencies of these modes compared to single-occupancy automobiles. However, energy use by public transit depends on *load factor* values which can vary significantly according to the demand level. For example, a bus with 50 passengers consumes approximately a tenth of what an average automobile uses (per passenger-mile) (VTPI, 2005). In certain regions of the United States, however, energy consumption per passenger-mile may be slightly higher for transit systems than for autos, due to the low load factors of the former.

(*b*) *Operational Intervention*   Improvements in traffic operations can significantly improve energy use. Examples

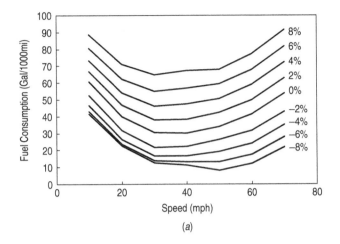

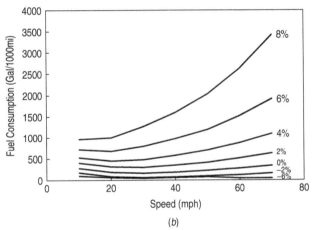

Figure 15.4 Constant speed fuel consumption (CSFC) for grade. (*a*) automobiles; (*b*) trucks. (From FHWA, 2000; 2002.)

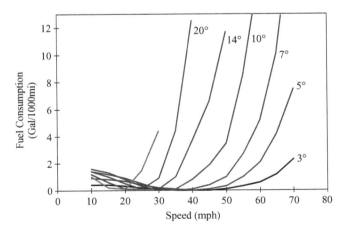

Figure 15.5 Excess fuel consumption at horizontal curves, by grade and speed. (From Caltrans, 1983.)

include changes in speed limits, introduction of Intelligent Transportation System (ITS) technologies, arterial signal coordination, and so on.

(*c*) *Preservation and Expansion of Existing Systems* Transportation interventions that help preserve existing assets or/and expand their capacities can reduce delay and therefore contribute to reduced energy use.

15.1.7 Other Factors

Other transportation-related interventions that can affect energy use include vehicle inspection and maintenance and fuel-use policies and incentives. Furthermore, policies regarding vehicle retirement can affect energy rates because older vehicles typically are less fuel efficient.

15.2 ENERGY INTENSITY

The concept of energy intensity helps enable fair comparison of different energy sources among transportation modes. Figures 15.6(*a*) and (*b*) show temporal trends of energy intensity (per passenger-mile and per freight ton-mile, respectively). While transit buses consume the most per passenger-mile, intercity buses are least energy intensive. Air transportation indicated an improvement in the rate of energy use in recent years in both passenger-mile and freight ton-mile categories. In terms of ton-miles, rail is the least energy intensive, while heavy truck consumes the most.

Example 15.1 If the target energy intensity of transit buses in a city is 4127 Btu/passenger-mile and the average fuel efficiency of the buses is 7 miles per gallon of diesel, estimate the minimum bus occupancy needed to achieve the target energy intensity.

SOLUTION From the information in Section 15.1, 1 gallon of diesel is equivalent to 139,000 Btu. If threshold occupancy is X passengers per vehicle,

$$\frac{4127 \text{ Btu}}{1 \text{ passenger-mile}} = \left(\frac{139,000 \text{ Btu}}{1 \text{ gallon}}\right)$$
$$\times \left(\frac{1 \text{ gallon}}{7 \text{ vehicle-miles}}\right)\left(\frac{1 \text{ vehicle}}{X \text{ passengers}}\right)$$

Thus,

$$X = \frac{139,000}{(7)(4127)} \approx 4.8 \text{ passengers/bus}$$

15.3 FRAMEWORK FOR ENERGY IMPACT ANALYSIS

Energy impacts of a transportation system include both direct and indirect consumption, as shown in Figure 15.7.

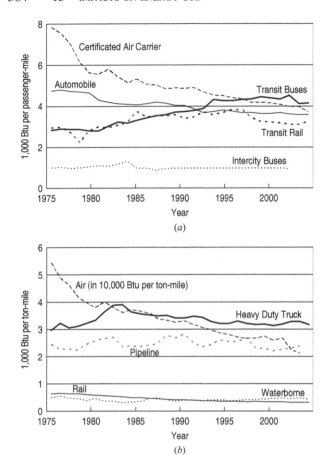

Figure 15.6 Trends in energy intensity for selected modes: (*a*) per passenger-mile. (From Davis and Diegel, 2006.); (*b*) per ton-mile. (From EERE, 2006.)

15.3.1 Direct Consumption

The use of energy for vehicle propulsion is referred to as direct consumption and such energy consumption can

be evaluated using aggregate or disaggregate approaches. Aggregate approach utilizes average energy consumption rates for various vehicle classes under different transportation modes. Intensity rates for passenger and freight transportation are provided in Tables 15.4 and 15.5, respectively. In the aggregate approach, the energy consumption rate (such as gal/VMT) is multiplied by the amount of use (VMT) to yield the total energy consumed. This is done for each mode and is carried out for with and without intervention scenarios.

The disaggregate approach, on the other hand, deals with each vehicle in a traffic stream and measures direct energy consumption on the basis of models that relate fuel consumption rates with vehicle operating characteristics (such as speed, acceleration, and delay) and facility condition (such as guideway grade and curvature). For the highway mode, disaggregate models are typically incorporated in traffic simulation models such as TRANSYT-7F, FREFLO, and VISSIM.

The user cost of energy consumption has been discussed in Chapter 7 as a part of vehicle operating cost impacts. In the present chapter, the societal cost associated with the use of non-renewable energy resources is being evaluated, and therefore we estimate the energy impact of a proposed intervention as the difference in total energy consumed with and without the intervention. Care should be exercised so that double counting of energy impacts can be avoided.

15.3.2 Indirect Consumption

Indirect consumption includes all energy used in constructing and running a transportation system, including vehicle manufacture and maintenance and facility construction, maintenance, and operation. Table 15.6 summarizes the average rate of indirect energy consumption

Table 15.4   Direct Energy Consumption of Passenger Transportation (Average Rate by Mode)

	Rate of Use		
Mode	Per Vehicle-Mile Traveled	Per Vehicle-Hour	Per 1000 Passenger-Miles
Commuter rail	12.0 kWh (41,002 Btu)	375.7 kWh (1,283,231 Btu)	382.9 kWh (1,307,748 Btu)
Heavy rail	6.4 kWh (21,790 Btu)	127.0 kWh (433,824 Btu)	277.4 kWh (947,391 Btu)
Light rail	8.1 kWh (27,663 Btu)	123.9 kWh (423,177 Btu)	345.5 kWh (1,180,051 Btu)
Bus	0.28 gal (38,800 Btu)	3.6 gal (504,118 Btu)	31.4 gal (4,371,338 Btu)
Automobiles, sport utility vehicles, and light trucks	0.048 gal (5,952 Btu)	N/A	42.0 gal (5,255,000 Btu)

Source: Based on data from Shapiro et al. (2002), FTA (2005), USEPA (2005), and Davis and Diegel (2006).

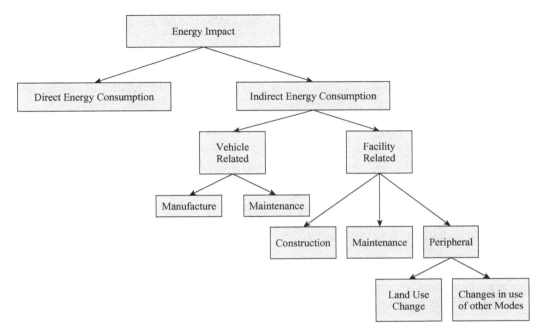

Figure 15.7 Energy impact analysis categories. (From Caltrans, 1983.)

Table 15.5 Intercity Freight Movement and Energy Use in the United States, 2003

Mode	Ton-Miles (billions)	Energy Use (trillion Btu)	Rate of Use (Btu/ ton-mile)
Trucks	1051	3653	3476
Class I railroads	1551	534	344
Waterborne commerce	606	253	417

Source: Based on data from Davis and Diegel (2006).

by activity, facility, and mode. Components of indirect energy consumption are summarized below.

(*a*) *Vehicle Manufacture* Manufacturing energy includes energy consumed in all phases of the vehicle manufacturing process. Given the amount of energy used in producing a single vehicle and the life of the vehicle in terms of miles driven, energy used per vehicle-mile can be estimated. For example, if 120 million Btu of energy is used in producing an automobile that has an average life of 80,000 miles, the average manufacturing energy for the automobile, in terms of vehicle-miles can be found as follows: 120 million/80,000 miles = 1500 Btu/vehicle-mile.

(*b*) *Vehicle Maintenance* Energy is consumed in vehicle maintenance, particularly for routine wear and replacement, guideway-related wear, and operation of repair facilities for vehicles. This energy can also be estimated in terms of vehicle-miles of travel.

(*c*) *Facility Construction* Construction energy can be estimated as the sum of itemized energy factors, including excavation, backfill, dredging, structures, surface/pavements, signs, lights, landscaping, and material transport. Table 15.6 indicates that on average, a dollar's worth (2005 dollars) of transportation facility construction consumes approximately 16,000 Btu. Assuming a 50-year facility service life, the annualized construction energy is 16,000/50 years = 320 Btu/year .

(*d*) *Facility Operation and Maintenance* The energy expended on facility maintenance is influenced by transportation facility type and cost. For highway transportation, for example, asphalt pavement maintenance energy needs can generally exceed that of its rigid counterparts, and urban sections require greater energy than rural sections (Table 15.6).

(*e*) *Peripheral Effects* Peripheral impacts of transportation interventions on energy may be due to changes in land use, changes in fuel type, changes in local energy needs,

Table 15.6   Indirect Energy Consumption in Highway Transportation: Average Rates

Activity	Mode	Factor
Construction	Automobiles and trucks (manufacturing)	1,410 Btu/vehicle-mile
	Bus (manufacturing)	3,470 Btu/vehicle-mile
	Roadway (construction)	15,778 Btu/$
	Signals, illumination, miscellaneous	3,981 Btu/$
Maintenance	Automobiles and trucks	1,400 Btu/vehicle-mile
	Bus	13,142 Btu/vehicle-mile
	Annual roadway maintenance (portland cement concrete)	163.4 million Btu/lane-mile (urban)
		66.1 million Btu/lane-mile (rural)
	Annual roadway maintenance (asphalt)	177.6 million Btu/lane-mile (urban)
		80.3 million Btu/lane-mile (rural)

Source: Caltrans (1983), updated to Year 2005 dollars.

and so on. However, these impacts occur over significant periods of time and may be difficult to quantify.

15.4   PROCEDURES FOR ESTIMATING ENERGY CONSUMPTION

A methodological framework for energy impact estimation is represented in Figure 15.8. First, the scope of analysis must be established. The first two approaches, A and B, are macroscopic analyses that use an average rate of energy consumption; whereas approach C is a microscopic analysis that involves simulation of individual vehicle operations.

15.4.1 Macroscopic Assessment: Approach A

In this approach, both direct and indirect energy consumption are considered in a comprehensive manner as applied below:

Application: Macroscopic Assessment of Direct and Indirect Energy Consumption: It is proposed to improve an existing highway transportation corridor in a bid to reduce congestion and to enhance connectivity with adjacent interstate highways and local roads. Four alternatives, including a no-build option, are being evaluated. Table 15.7 is a summary of alternative characteristics. It is assumed that the peripheral impacts are negligible.

- *Alternative I: no build.* No capital improvements are included in this alternative. It is projected that traffic conditions would worsen, ultimately deteriorating to a point where the congestion spreads to other arterials in the network.
- *Alternative II: TSM (transportation system management)/expanded bus service.* This alternative includes

various improvements, such as increasing capacity and speed by removing roadside parking, and reducing bus headways, due to the addition of approximately 50 buses during peak periods, and synchronization of traffic signals.

- *Alternative III: full build.* This alternative provides exclusive bus lanes on the corridor and construction of six connectors to adjacent highways (total 22 miles) in addition to treatments described for alternative II.
- *Alternative IV: reduced build.* This alternative includes exclusive bus lanes in the corridor and construction of three connectors to adjacent highways (total 14.6 miles) in addition to treatments described for alternative II.

Direct energy consumption is calculated as the product of the modal consumption rate and the modal VMTs. For example, in alternative I, for passenger cars, the amount of travel is 3,202,164,000 vehicle-miles (Table 15.7) and 5952 Btu (Table 15.4) is consumed per vehicle-mile. The energy consumed by all passenger cars in a year is then determined as (3,202,164) × (5952) = 19,059,281 million Btu/year. In a similar manner, the energy consumption associated with other vehicle classes and other modes can be estimated for each transportation alternative. Table 15.8 presents the direct energy consumption amounts by alternative and shows that alternative III presents positive savings in direct energy compared to the no-build alternative.

The level of indirect energy consumption can be expressed in terms of energy consumed per vehicle-mile, per dollar spent, or per lane-mile (or line-mile). For lump-sum cost items such as construction cost, energy consumption is typically proportional of the amount spent.

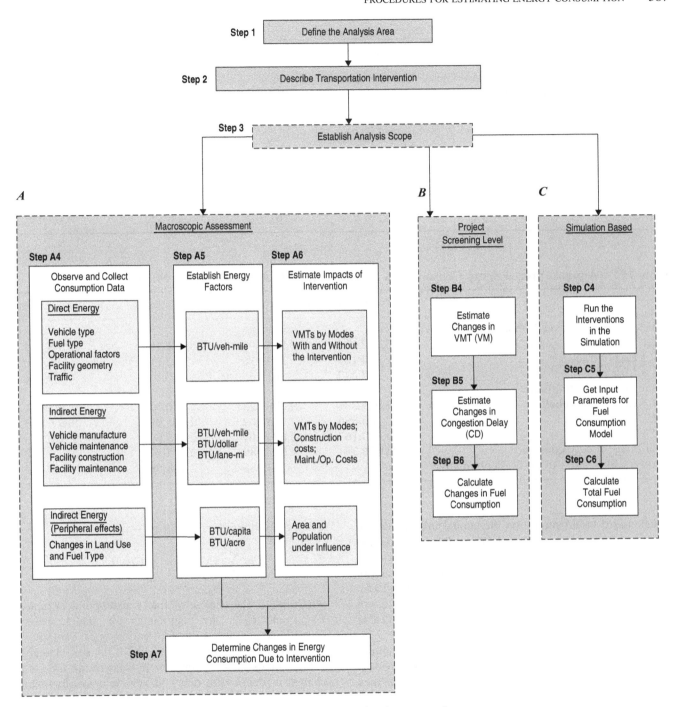

Figure 15.8 Framework for estimating energy impacts.

If it is assumed that the energy consumed during the construction period can be distributed over the facility service life (assumed to be 50 years in the example presented in Table 15.9), the equivalent uniform annual level of construction-related energy use can be determined as illustrated in Section 5.3.2(c). Annual construction costs for roadway and TSM related signal and other equipment in Table 15.9 are calculated (rows 1.3 and 1.4). Vehicle manufacturing and maintenance energy are measured on the basis of vehicle-miles traveled; this can

Table 15.7   Description of Alternatives

Description	I: No build	II: TSM/Bus	III: Full Build	IV: Reduced Build
Construction cost (millions of 2005 dollars)	—	—	340	230
TSM cost (millions 2005 dollars)	—	30	30	30
Estimated service life (years)	50	50	50	50
Pavement type	Asphalt	Asphalt	Asphalt	Asphalt
Number of lanes	4	4	6	6
Length of road segment	30	30	52	44.6
Annual vehicle-miles traveled (1000 veh-mi/yr)				
Light-duty vehicles	3,202,164	3,225,592	3,182,197	3,288,483
Heavy trucks	94,181	94,870	93,594	96,720
Buses	67,272	67,765	66,853	69,086

Table 15.8   Annual Direct Energy Consumption by Alternative (1000 Vehicle-Miles, Million Btu/Year)

	I: No Build	II: TSM/Bus	III: Full Build	IV: Reduced Build
1. Vehicle miles traveled (thousands)				
1.1 Light-duty vehicles	3,202,164	3,225,592	3,182,197	3,288,483
1.2 Heavy trucks	94,181	94,870	93,594	96,720
1.3 Buses	67,272	67,765	66,853	69,086
2. Btu consumed (millions)				
2.1 Light-duty vehicles	19,059,281	19,198,723	18,940,434	19,573,051
2.2 Heavy trucks	2,206,856	2,223,002	2,193,095	2,266,345
2.3 Buses	2,610,167	2,629,264	2,593,891	2,680,528
Total direct energy consumed	23,876,304	24,050,989	23,727,420	24,519,924
Change with respect to no-build alternative	—	−174,685	148,884	−643,620

be calculated based simply on the annual mileage driven and rate of energy consumption (rows 1.1, 1.2, 2.1, and 2.2). For annual roadway maintenance (row 2.3), energy consumption can be expressed as a function of the facility size, e.g., number of lane-miles of roadway (Table 15.6). As expected, all the alternatives have higher indirect energy consumption than the no-build alternative (Table 15.9).

Summing the two components of energy consumption, direct and indirect energy consumption levels, it is evident that the Full Build option provides a small energy savings over the other alternatives (Table 15.10).

15.4.2 Project Screening Level Model: Approach B

This approach considers direct energy only, and is incorporated in IMPACTS, FHWA developed spreadsheets for screening-level evaluation of multimodal corridor alternatives (DeCorla-Souza, 1999; FHWA, 2005). This model associates traffic congestion with energy consumption

using the following relationship:

$$FC = \sum_i C_{VM,i} VM_i + C_{CD,i} CD_i \qquad (15.1)$$

where FC is the change in fuel consumption in gallons, $C_{VM,i}$ and $C_{CD,i}$ are coefficients for vehicle class i, in gallons per veh-mile and gallons per veh-hour of congestion delay, respectively, VM_i is the change in vehicle miles traveled for vehicle class i, and CD_i is the change in congestion delay in vehicle-hours for vehicle class i.

Table 15.11 provides values of the coefficients C_{VM} and C_{CD} that are used for assessing the impact of congestion on fuel consumption. The values were found to be consistent with the data from other sources. For example, from Table 15.4, the average energy consumption rates for automobiles and buses are 5952, and 38,800 Btu per vehicle-mile, respectively. Also, average Btu per vehicle-mile for heavy trucks is found to be 23,461

Table 15.9   Annual Indirect Energy Consumption by Alternatives (Million Btu/Year)

	I: No Build	II: TSM/Bus	III: Full Build	IV: Reduced Build
1. Construction				
1.1 Automobile manufacturing	4,647,847	4,681,852	4,618,865	4,773,136
1.2 Bus manufacturing	233,435	235,143	231,979	239,728
1.3 Roadway	0	0	107,288	72,577
1.4 TSM	0	2,388	2,388	2,388
Subtotal	4,881,282	4,919,383	4,960,520	5,087,829
2. Maintenance				
2.1 Auto maintenance	4,614,884	4,648,647	4,586,107	4,739,284
2.2 Bus maintenance	884,093	890,562	878,580	907,925
2.3 Roadway	21,312	21,312	55,411	47,526
Subtotal	5,520,289	5,560,521	5,520,098	5,694,735
Total indirect energy consumed	10,401,571	10,479,904	10,480,618	10,782,565
Change vs. no build	—	−78,333	−79,048	−380,994

Table 15.10   Annual Total Energy Consumption by Alternatives (Million Btu/Year)

	I: No Build	II: TSM/Bus	III: Full Build	IV: Reduced Build
Direct energy consumed	23,876,304	24,050,989	23,727,420	24,519,924
Indirect energy consumed	10,401,571	10,479,904	10,480,618	10,782,565
Total energy consumed	34,277,875	34,530,893	34,208,038	35,302,481
Change vs. no build	—	−253,018	69,837	−1,024,614

(Davis and Diegel, 2006). By dividing these values by the Btu-gallon equivalents for gasoline and diesel, gallons per vehicle-mile are obtained as 0.05, 0.17, and 0.28 for automobiles, heavy trucks, and buses, respectively. These values confirm the values for C_{VM} in Table 15.11.

Application: Project Screening-Level Analysis for a Multimodal Corridor Improvement: To mitigate

Table 15.11   Congestion-Related Fuel Consumption Coefficients

Vehicle Class	Coefficients	
	C_{VM}	C_{CD}
Automobiles	0.04	0.42
Heavy trucks	0.16	1.87
Buses	0.25	—

Source: FHWA (2005).

congestion at a certain transportation corridor, three alternatives are considered at the project screening level: expanding the existing bus system; constructing two additional lanes, and converting the existing highway into a toll road. In this example, using equation (15.1) and coefficient information in Table 15.11, alternatives are analyzed and compared from the perspective of fuel consumption. Project information is given in Tables 15.12 and 15.13.

For each alternative, changes in travel demand (VMT) and congestion delay (CD) are estimated. Changes in travel demand are generally caused by diversions between modes and induced demand due to reduced congestion. Induced demand is estimated as shown below (FHWA, 2005):

$$\text{induced demand} = \frac{\text{decrease in passenger car equivalent (pce-VMT)}}{1 - 1/(\text{avg. speed} \times \text{marginal delay per added VMT} \times \text{elasticity})} \text{(veh-mi)}$$

(15.2)

Table 15.12   Demand Characteristics

	ADT (veh/day) (and Modal Share)	Average Trip Length (miles)	Total Daily VMT (veh-mi)	Occupancy (persons/vehicle)	Passenger Car Equivalent (pce)
Auto	131,000 (0.94)	12.0	1,572,000	1.2	1.0
Truck	6,950 (0.05)	14.0	97,300	—	3.0
Bus	1,700 (0.01)	15.0	25,500	18.0	2.0
Total	139,650 (1.00)	—	1,694,800	—	—

Table 15.13   Supplementary Information for the Projects

Number of weekdays/year	250
Demand elasticity with respect to travel time (VMT)	−0.5
Average highway speed (mph)	37.50
Marginal delay (h/1000 pce-VMT)	61.26[a]

[a]This value indicates that an additional 1000 pce-VMT induces 61.26 hours of additional delay.

The congestion delays (CDs) by modes are

$$CD(auto) = \left(\begin{array}{c} \text{total change in highway} \\ \text{travel time, pce-veh-h} \end{array} \right)$$
$$\times \text{(fraction of auto)}$$
$$- \text{(induced auto VMT)}$$
$$\times \left(\frac{1}{\text{avg. highway speed}} - \frac{1}{60} \right) \text{(veh-h)}$$
$$(15.3)$$

Equation (15.3) assumes that average highway speeds lower that 60 mph cause congestion delay for induced traffic.

$$CD(truck) = \left(\begin{array}{c} \text{total change in highway} \\ \text{travel time, pce-veh-h} \end{array} \right)$$
$$\times \frac{\text{fraction of truck, \%}}{\text{pce for trucks, veh-h}} \quad (15.4)$$

$$CD(bus) = \left(\begin{array}{c} \text{total change in highway} \\ \text{travel time, pce-veh-h} \end{array} \right)$$
$$\times \frac{\text{fraction of bus, \%}}{\text{pce for buses, veh-h}} \quad (15.5)$$

where

total change in travel time

= (marginal delay, h/pce-VMT)

× (net change, pce-VMT)

= 61.26(from Table 15.13)(h/1000 pce-VMT)

× (net change, 1,000 pce-VMT)   (15.6)

ALTERNATIVE I: EXPANDING THE BUS SYSTEM BY INTRODUCING 50 NEW BUSES   Table 15.14 provides details of Alternative I. The addition of 50 new buses is expected to increase ridership by 6,000 person-trips per day and in-vehicle auto travel time will be reduced by 5 minutes per trip.

1. *Changes in VMT* (vehicle-miles of travel). For this alternative, VMT changes consist of three sources: change in bus demand, change in auto demand due to diversion to bus, and induced auto demand due to congestion reduction after diversion. The change in bus VMT can be calculated simply as follows:

increase in bus VMT

= (bus ridership after adding new buses − before)

$$\times \frac{\text{average bus mile}}{\text{bus occupancy}}$$

$$= (36,600 - 30,600) \left(\frac{15.0}{18.0} \right) = 5000 \text{ veh-mi}$$

Changes in auto VMT consist of (a) change in auto VMT due to diversion to bus, which can be calculated

Table 15.14   Characteristics of Alternative I

I_a: Bus ridership without improvement (person-trips/day)	30,600
I_b: Bus ridership with improvement (person-trips/day)	36,600
I_c: Fraction of new bus riders who were auto drivers	0.50
I_d: Average length of diverted auto trips (miles/trip)	12.0

Table 15.15   Savings in Fuel Consumption (Alternative I)

	C_{VM}	Decrease in VM	C_{CD}	Decrease in CD	Total Daily Savings in Fuel Consumption (gal) per Day
Automobiles	0.04	19,308	0.42	729	1,079
Heavy trucks	0.16	—	1.87	10	18
Buses	0.25	−5,000	—	3	−1,250
Total	—	—	—	—	−154

knowing the change in bus ridership (I_a, I_b) and the fraction of new bus riders diverted from auto users (I_c), and (b) demand induced due to congestion reduction, which is calculated on the basis of the changes in bus and auto VMT, and equation (15.2):

decrease in auto VMT (due to diversion)

= (bus ridership after − before)

$$\times \text{ fraction } \times \frac{\text{average auto mile}}{\text{auto occupancy}}$$

$$= (36,600 - 30,600)(0.5) \left(\frac{12.0}{1.2} \right)$$

$$= 30,000 (\text{veh-mi})$$

decrease in pce-VMT

decrease in auto VMT due to modal shift

− increase in bus VMT × bus-pce

$$= 30,000 - (5000)(2) = 20,000$$

From equation (15.2),

induced auto VMT

$$= \frac{20,000}{1 - 1/[(37.50)(61.26/1000)(-0.5)]}$$

$$= 10,692 \text{ veh-mi}$$

Total decrease in auto VMT $= 30,000 - 10,692 = 19,308$ veh-mi

There is no change in truck VMT.

2. *Changes in CD* (congestion delay). From equation (15.6), the total decrease in highway travel time is calculated as shown below:

$$\text{pce-veh-h} = \left(\frac{61.26}{1000} \right)(20,000 - 10,692) = 570$$

Equations (15.3) to (15.5) give the congestion delay for each mode:

decrease in congestion delay (auto)

$$= (570)(0.94) + (30,000 - 10,692)$$

$$\times \left(\frac{1}{37.50} - \frac{1}{60} \right) = 729 \text{ veh-h}$$

decrease in congestion delay (truck)

$$= (570) \left(\frac{0.05}{3.0} \right) = 10 \text{ veh-h}$$

decrease in congestion delay (bus)

$$= (570) \left(\frac{0.01}{2.0} \right) = 3 \text{ veh-h}$$

3. *Savings in fuel consumption.* With coefficients from Table 15.4 and results 1 and 2, the change in fuel consumption is estimated as shown in Table 15.15.

ALTERNATIVE II: CONSTRUCTING TWO ADDITIONAL LANES Adding two lanes is expected to divert 5% of bus users to auto, to increase average travel speed from 37.50 mph to 52.50 mph, and to reduce vehicle-hours of travel from 45,193 to 32,282 per day (Table 15.16.)

1. *Changes in VMT* (vehicle miles traveled). Assuming no changes in bus VMT, this alternative concerns only

Table 15.16   Characteristics of Alternative II

	Before Adding Lanes	After Adding Lanes
Average weekday speed (mph)	37.50	52.50
Vehicle hours of travel (VHT)	45,195	32,282

induced auto demand due to improved speed after road expansion. Induced demand is estimated as follows:

induced auto demand

$$= \frac{\left(\begin{array}{c} \text{veh-h of auto users before} \\ -\text{veh-h of auto users after} \end{array}\right)}{\left(\begin{array}{c} \text{marginal delay to others per added VMT} \\ -\dfrac{\text{veh-h of auto users after}}{\text{total VMT} \times \text{elasticity}} \end{array}\right)}$$

$$= \frac{45,195 - 32,282}{61.26/1000 - 32,282/1,694,800/(-0.5)}$$

$$= 129,966 \text{ veh-mi}$$

2. *Changes in CD* (congestion delay). Based on equation (15.6) and considering changes in vehicle-hours after adding two lanes, the total decrease in highway travel time is calculated as

$$\text{pce-veh-h} = 45,195 - 32,282 - \left(\frac{61.26}{1000}\right)(129,966)$$

$$= 4951$$

average speed after additional delay due to induced traffic

$$= \frac{\text{VMT for all segments}}{(\text{veh-h before}) + (\text{change in veh-h})}$$

$$= \frac{1,694,800}{45,195 - 4951} = 42.11 \text{ mph}$$

Equations (15.3) to (15.5) give the congestion delay for each mode:

decrease in congestion delay (auto)

$$= (+4951)(0.94) - (129,966)\left(\frac{1}{42.11} - \frac{1}{60}\right)$$

$$= +3734 \text{ veh-h}$$

decrease in congestion delay (truck)

$$= (4951)\left(\frac{0.05}{3.0}\right) = 83 \text{ veh-h}$$

decrease in congestion delay (bus)

$$= (4951)\left(\frac{0.01}{2.0}\right) = 25 \text{ veh-h}$$

3. *Savings in fuel consumption*. From results 1 and 2, changes in fuel consumption are estimated as shown in Table 15.17.

ALTERNATIVE III: IMPOSING $1 TOLLS   Alternative III is to introduce tolls to the system. In this option it is expected that a particular portion of auto trip demand shift to bus and that overall improvement in travel speed will induce additional trips, as in previous alternatives. Information for this alternative is given in Table 15.18.

1. *Changes in VMT* (vehicle-miles traveled). VMT changes in this alternative consist of three parts: change in auto trips caused by toll, change in bus demand due to diversion from auto, and induced auto trips due to congestion reduction after diversion. The VMT changes are estimated as shown below:

decrease in auto VMT

$$= (\text{auto ridership before} - \text{after})\frac{\text{average auto mile}}{\text{auto occupancy}}$$

$$= (157,200 - 135,000)\left(\frac{12.0}{1.2}\right)$$

$$= 222,000 \text{ veh-mi}$$

increase change in bus VMT (due to diversion)

$$= (\text{auto ridership before—after}) \times \text{fraction}$$

$$\times \frac{\text{average bus mile}}{\text{bus occupancy}}$$

$$= (157,200 - 135,000)(0.4)\left(\frac{15.0}{18.0}\right)$$

$$= 7400 \text{ veh-mi}$$

Table 15.17   Changes in Fuel Consumption: Alternative II

	C_{VM}	Decrease in VM	C_{CD}	Decrease in CD	Total Daily Savings in Fuel Consumption (gal)
Automobiles	0.04	−129,966	0.42	3,734	−3,630
Heavy Trucks	0.16	—	1.87	83	154
Buses	0.25	—	—	25	—
Total	—	—	—	—	−3,476

Table 15.18   Characteristics of Alternative III

Auto trips without disincentive (person-trips/day)	157,200
Auto trips with disincentive (person-trips/day)	135,000
Fraction of the change in auto person-trips that shifted to bus	0.4

decrease in pce-VMT

= decrease in auto VMT − increase in

bus VMT × pce for bus

= 222,000 − (7400)(2)

= 207,200

From equation (15.2),

induced auto VMT

$$= \frac{207,200}{1 - 1/[(37.50)(61.26/1000)(-0.5)]}$$

= 110,766 veh-mi

Therefore

total decrease in auto VMT = 222,000 − 110,766

= 111,234 veh-mi

There is no change in truck VMT.

2. *Changes in CD* (congestion delay). From equation (15.6), the total decrease in highway travel time after the introduction of tolls is calculated as

$$\text{pce-veh-h} = \left(\frac{61.26}{1000}\right)(207,200 - 110,766) = 5908$$

Equations (15.3) to (15.5) give the congestion delay for each mode:

decrease in CD (auto) = (5908)(0.94)

+ (222,000 − 110,766)

$$\times \left(\frac{1}{37.50} - \frac{1}{60}\right)$$

= 6665 veh-h

$$\text{decrease in CD (truck)} = (5908)\left(\frac{0.05}{3.0}\right) = 98 \text{ veh-h}$$

$$\text{CD (bus)} = (5908)\left(\frac{0.01}{2.0}\right) = 30 \text{ veh-h}$$

3. *Savings in fuel consumption.* From results 1 and 2, changes in fuel consumption are estimated as shown in Table 15.19.

Table 15.20 summarizes the results for the three alternatives. The results show that alternative III, which introduces tolls to the system, is the best *from the perspective of energy conservation.*

15.4.3 Microscopic Simulation: Approach C

This approach involves a detailed energy consumption analysis by simulating the movement of individual vehicles in this transportation network. Equations for estimating fuel consumption are incorporated in the simulation and total fuel consumption of all vehicles in the network is calculated on the basis of their speed profiles. For urban areas, for example, the average-speed model for auto fuel consumption is:

$$F = a_0 + \frac{a_1}{V} \tag{15.7}$$

where $F =$ is the average fuel consumption (gallons per 1000 miles), V the average speed of the trip, and a_0 and a_1 are vehicle-specific coefficients. Coefficient a_0 reflects the amount of fuel required to overcome rolling resistance, drag, inertia, and grade resistance, and is approximately proportional to the vehicle

Table 15.19   Savings in Fuel Consumption: Alternative III

	C_{VM}	Decrease in VM	C_{CD}	Decrease in CD	Total Daily Savings in Fuel Consumption (gal)
Automobiles	0.04	111,234	0.42	6,665	7,249
Heavy trucks	0.16	—	1.87	98	184
Buses	0.25	−7,400	—	30	−1,850
Total	—	—	—	—	−5,583

Table 15.20   Summary of Project Screening Analysis

Alternative	Total Daily Savings in Fuel Consumption (gal)	Total Annual Savings in Fuel Consumption (gal)
I	−154	−38,419
II	−3,476	−869,006
III	5,583	1,395,736

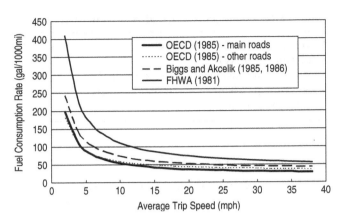

Figure 15.9   Relationships between energy consumption and speed.

weight; a_1 is an estimate of vehicle-idle fuel rate. This rather simple model has been found to give satisfactory results in cases where the average speed ranges between 6 and 37 mph (approximately 10 and 60 km/h). Several calibration results obtained from various sites, road types, and vehicle classes are summarized in Table 15.21 and the relationship between speed and fuel consumption rate based on those models is shown in Figure 15.9.

The average speed model has been expanded to include vehicle operational characteristics representing power demand in the relation with acceleration profile. Two different versions of expansion are given here. First, Biggs and Akcelik (1986) proposed a model that expressed

Table 15.21   Coefficients for Average-Speed Fuel Consumption Models[a]

Source	$F \text{(gal/1000 mi)} = a_0 + \dfrac{a_1}{f(V)}$	Model Coverage by Mode	Year and Region
OECD (1985)	$F = 18.5 + \dfrac{359.0}{V}$	Entire traffic stream, main highways	1981, in the Netherlands
	$F = 27.4 + \dfrac{308.0}{V}$	Entire traffic stream, highways other than main highways	1981, non-main highways in the Netherlands
Biggs and Akcelik (1986)	$F = 25.1 + \dfrac{338.1}{V}$	Small cars only	1985, Australia
	$F = 31.4 + \dfrac{422.7}{V}$	Entire traffic stream	1985, Australia, recommended as default model
	$F = 37.7 + \dfrac{507.2}{V}$	Large cars only	1985, Australia
FHWA (1981)	$F = 36.2 + \dfrac{746.3}{V}$	Average for passenger cars in the United States	1981, United States
	$F = \dfrac{1,000}{0.48 + 1.12\sqrt{V}}$	Single unit trucks only	1981, United States
	$F = 170 + \dfrac{2430}{V}$	Tractor-trailers only	1981, United States

Sources: FHWA (1981) Fwa and Ang (1992).
[a]V is in mph and F is in gallons/1000-miles.

changes in fuel consumption rate (per distance) as a function of acceleration and grade as follows:

$$F = \begin{cases} \dfrac{\alpha}{V} + \beta_1 R_T + \beta_2 (a_e R_{IG})_{a_e > 0} & \text{for } R_T > 0 \\ \dfrac{\alpha}{V} & \text{for } R_T \leq 0 \end{cases}$$

where F is the fuel consumption per unit distance, α the idle fuel consumption per unit time, V the average speed of the trip, and R_T the total tractive force required $= (R_D + R_I + R_G)$, R_D the total drag resistance, R_I the inertial resistance, R_G the grade resistance, $R_{IG} = R_I + R_G$, $a_e = a + (G/100)g$, a is the vehicle acceleration, G the percent grade, g the acceleration due to gravity, and β_1, β_2 are coefficients.

In a similar study, Rao and Krammes (1994) identified and calibrated a fuel consumption model for the FHWA-developed transportation simulation model, FREFLO, as follows:

$$F = 12.76 + \frac{700}{V} + 0.0023V^2 + 39.21a_e$$
$$+ 0.0033(a_e)^2 \quad \text{if } a_e > 0$$

$$F = 12.6 + \frac{700}{V} + 0.0023V^2 + 39.21a_e$$
$$\text{if } a_e < 0 \text{ and } 12.76 + 0.0023V^2 + 39.21a_e \geq 0$$

$$F = \frac{700}{V} \quad \text{if } 12.76 + 0.0023V^2 + 39.21a_e < 0$$

$$(15.8)$$

where F is the fuel consumption (gallons per 1000 miles), V the average speed on the trip, and a_e the acceleration (in ft/s^2).

There are several transportation simulation software packages (e.g., SimTraffic and CORSIM), where fuel consumption is estimated on the basis of speed and acceleration rate of individual vehicles. The unit fuel consumption rates used in SimTraffic are shown in Table 15.22. After the speed profile of each vehicle is

Table 15.22 Fuel Consumption Rates (10^{-5} gallons per second) Used in SimTraffic

Vehicle Class	Speed ft/s (mph)	Acceleration (ft/s²)						
		−9	−6	−3	0	3	6	9
Cars	0 (0.0)	13	13	13	13	21	28	35
	10 (6.8)	17	17	17	17	43	67	91
	20 (13.6)	18	18	18	20	73	113	148
	30 (20.5)	19	19	19	24	93	127	179
	40 (27.3)	18	18	18	30	113	211	223
	50 (34.1)	17	17	17	30	135	255	255
	60 (40.9)	18	18	18	36	174	244	244
	70 (47.7)	20	20	20	43	212	228	228
Trucks	0 (0.0)	17	17	17	17	17	35	40
	10 (6.8)	17	17	17	49	103	163	284
	20 (13.6)	17	17	17	62	161	320	5366
	30 (20.5)	17	17	17	72	249	5329	5328
	40 (27.3)	17	17	17	90	5258	5467	5465
	50 (34.1)	17	17	17	118	5329	5328	5327
	60 (40.9)	17	17	17	149	5406	5405	5404
	70 (47.7)	17	17	17	185	5328	5327	5327
Buses	0 (0.0)	9	9	9	9	9	135	243
	10 (6.8)	9	9	9	15	112	230	5269
	20 (13.6)	9	9	9	27	176	5276	5276
	30 (20.5)	9	9	9	45	255	5298	5297
	40 (27.3)	9	9	9	105	340	5341	5341
	50 (34.1)	9	9	9	95	5235	5235	5234
	60 (40.9)	9	9	9	123	5278	5278	5278
	70 (47.7)	9	9	9	164	5306	5306	5305

determined, it is used to calculate the amount of fuel consumed during the time increment.

Application: Microscopic Simulation for an Urban Arterial Improvement: The following alternative improvements are being considered for a section of U.S. Highway 52, an urban arterial (Figure 15.10): signal coordination only, and adding left and right turn bays in addition to the signal coordination. There are three

signal-controlled intersections along the section under study: at Klondike Road, Morehouse Road, and Cumberland Avenue. Also, there are two stop-controlled intersections: County Road N 400 W and McCormick Lane. Observed peak hour volumes by movement are presented in Figure 15.11.

Based on traffic volumes and turning movements given, SimTraffic can provide optimal values of splits, offsets and cycle lengths for traffic signal timing along the arterial, on the basis of delay, number of stops, travel time, and vehicle-miles traveled for the entire system of approaches, intersections, and the arterial. The corresponding speed profile of each individual vehicle is tracked and the total fuel consumption is calculated using input relationships such as those shown in Table 15.22. For purposes of illustration, a summary of fuel consumption values by approach for the Cumberland and U.S. 52 intersection is presented in Table 15.23.

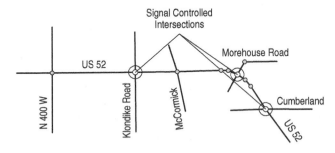

Figure 15.10   Example arterial.

BASE CASE: SIGNAL TIMING OPTIMIZATION FOR INDIVIDUAL INTERSECTIONS   In the base case, signal timing of each

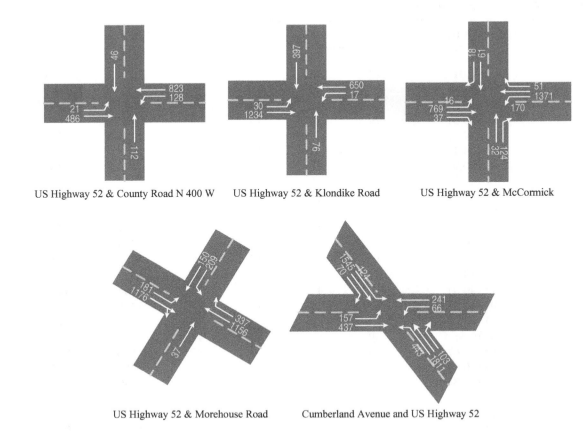

Figure 15.11   Intersection volumes (peak hours).

Table 15.23   Fuel Consumption (gals/h) at the Intersection of Cumberland Avenue and U.S. 52 by Approach

	EBL	EBT	EBR	WBL	WBT	NBL	NBT	NBR	SBL	SBT	SBR	All
Base Case	11.2	6.1	25.1	1.4	1.7	30.7	92.1	5.4	3.5	38.7	0.9	219.1
Alternative I: Signal Coordination Only	4.1	1.8	6.8	2.5	4.7	26	75.7	3.7	3.8	44.5	1.1	181.7
Alternative II: Signal Coordination and Addition of Turning Bays	2.8	1.4	5.1	1.6	2.6	17.7	59	3.1	1.9	12.9	0.4	112.3

intersection is optimized individually without considering coordination with other intersections.

IMPROVEMENT ALTERNATIVE I: SIGNAL COORDINATION This alternative involves coordination of signals along the arterial. As observed in Tables 15.23 and 15.24, signal coordination can lead to significant reduction in systemwide fuel consumption.

IMPROVEMENT ALTERNATIVE II: SIGNAL COORDINATION AND ADDING LEFT- AND RIGHT-TURN BAYS   In view of the levels of service observed for each movement, Alternative II involves additional left- and right-turn bays for the Cumberland Avenue and U.S. 52 intersection. Significant reductions in fuel consumption are observed in some approaches as summarized in Table 15.23. Table 15.24 compares the results to conclude that Alternative II (signal timing coordination and addition of turn bays) is the best alternative from the perspective of energy consumption. Given this evaluation for the peak hours and assuming 3 hours of peak hours per day and 250 workdays per year, the yearly savings of fuel consumption of alternatives are calculated as shown below.

Alternative I: (36.3 gallons/hr)(3 hrs/day)

$\times$ (250 days/year) $=$ 27,225 gallons per year

Alternative II: (101.3 gallons/hr)(3 hrs/day)

$\times$ (250 days/year) $=$ 75,975 gallons per year

15.5   THE NATIONAL ENERGY MODELING SYSTEM

The National Energy Modeling System (NEMS), developed by the Energy Information Agency of the U.S. Department of Energy, is used to assess the impacts of multimodal transportation programs and policies, on energy consumption. The diagram depicting the conceptual relationship between different sectors and submodules within each sector is shown in Figure 15.12. The components of the transportation sector model are listed below.

- The LDV (light-duty vehicle) module models energy consumption by vehicles in this category. It consists of manufacturers' technology choice based on fuel market conditions, consumer choices based on fuel efficiency and cost, LDV stock accounting, and travel demand (VMT) estimates
- The air travel module is based on estimates of air travel demand and aircraft fleet efficiency
- The freight transportation module (truck, rail, marine) forecasts energy use in the freight transportation

Table 15.24   Fuel Consumption Changes along U.S. 52 Arterial

Fuel Consumption (gals/hr)	Base Case (Do Nothing)	Signal Coordination	Signal Coordination and Adding Turn Bays
County Road N 400 W Intersection	32.7	31.11	32.9
Klondike Road Intersection	32.7	33.4	33.9
McCormick Intersection	42.2	43.9	44.5
Morehouse Road Intersection	19.5	19.9	21.4
Cumberland Avenue Intersection	219.2	181.7	112.3
Network total	346.3	310.0	245.0
Change from base (ratio in parentheses)	0 (0%)	36.3(10.4% reduction)	101.3 (29.3% reduction)

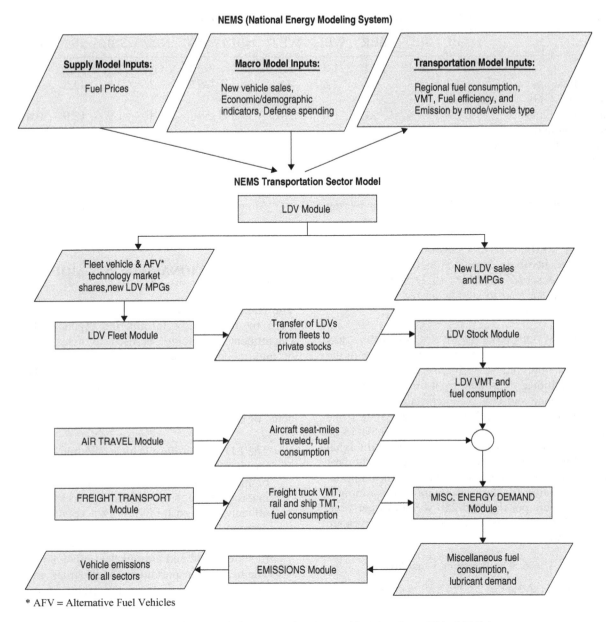

Figure 15.12   NEMS Transportation Sector Models. (From EIA, 2005b.)

sector based on industry output, production level, trade indices, and others
- The miscellaneous energy use module includes military use, mass transit, and recreational use

The model considers the following exogenous factors of energy consumption:

- Economic indices: GDP, income level, production level by industry, and so on

- Fiscal policies on fuel taxes and subsidies
- Fuel economy levels by vehicle class
- CAFE (corporate average fuel economy) levels
- Vehicle sales by technology type
- Demand for vehicle performance within vehicle classes
- Fleet vehicle sales by technology type
- Market shares of vehicles using alternative fuels

- Changes in emission- and safety-related regulations and standards
- VMT changes

The most recent application of NEMS is the forecast of energy consumption to 2025 at the national level based on a starting year of 1995 (EIA, 2005b).

15.6 APPROACHES TO ENERGY CONSUMPTION ESTIMATION–A COMPARISON

Various approaches for estimating the energy impacts of transportation projects, policies, or programs are summarized in Table 15.25. While the NEMS is useful for national or regional level evaluation of policies and programs, project level analysis can be done using less data intensive approaches. The choice of a specific approach will depend on the type of transportation intervention and the available data.

15.7 ENERGY AND TRANSPORTATION: WHAT THE FUTURE HOLDS

As Figure 15.13 indicates, transportation energy demand will continue to increase in the foreseeable future. With the growing worldwide demand for energy, particularly at rapidly industrializing countries including India and China, the continued dependence of transportation, particularly on petroleum based energy sources, poses a serious

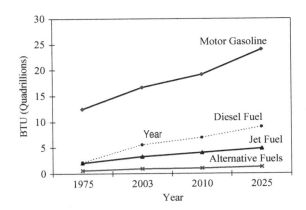

Figure 15.13 Transportation energy consumption, 1975 to 2025. (From EIA, 2005a.)

problem. Possibilities for coping with the looming energy crisis can be sought both in the demand and supply sides of the energy market (Beimborn, 2005; Greene and Decicco, 2005). Demand-side options include fuel economy standards, improved traffic operations, pricing tools, and other demand management; whereas supply-side initiatives mostly involve alternative fuels.

Changes in fuel economy standards (CAFE) can significantly affect energy consumption. Given that there have not been considerable changes in requirements since mid-1980s, NHTSA, at the current time, is proposing

Table 15.25   Modeling Approaches for Energy Consumption Estimation

	National and Regional-Level Models	Project Level Models		
		Macroscopic Assessment	Project-Screening Level Model	Microscopic Simulation Models
Perspective	Energy as a part of The national/regional economy system as a whole	Direct and indirect energy impact of transportation projects	Direct consumption only	Individual intersections, road segments
Inputs—data requirement	Economic indices: GDP, income level, etc; Regulations/standards Fuel economy Vehicle sales/demand	VMT by modes Construction cost Service life Average energy consumption rates	VMT changes Congestion delay changes Average energy consumption rates	Network and traffic information incorporated in traffic simulation
Outputs	Energy consumption at national/regional level by sector, mode, vehicle class	Energy consumed for each alternative	Energy consumed for each alternative	Energy consumed for each alternative

higher standards of fuel economy. According to this proposal, manufacturers would have a transition period (2008 to 2010), and would comply with the reformed rule beginning in 2011. Overall fuel savings would be about 8.1% when the rule is fully phased-in or about 10.7 billions gallons of fuel over the lifetime of the vehicles in those model years.

There is great interest in the country for alternative fuels. The U.S. Department of Energy (DOE) defines alternative fuels as fuels that are substantially nonpetroleum and yield energy security and environment benefits. DOE currently recognizes the following as alternative fuels: methanol and denatured ethanol (alcohol mixtures that contain no less than 70% of the alcohol fuels); natural gas (compressed or liquefied); liquefied petroleum gas; hydrogen; coal-derived liquid fuels; fuels derived from biological material; and electricity (including solar energy).

It has been projected that alternative fuels will displace 207,000 barrels of oil equivalent per day in 2010 and 280,400 barrels per day (2.2% of light-duty vehicle fuel consumption) in 2025. However, the share of gasoline in demand is expected to be sustained because of relatively low prices compared to the inflation rate and increase in popularity of low-fuel-efficiency vehicles, such as sports utility vehicles (EIA, 2005a). Figures 15.13 and 15.14 present forecasts of future demand of alternative energy sources.

At the present time, a major impediment to large-scale adoption of alternative fuels and new technologies is the high cost of the fuels, their delivery systems, and the vehicles that use them. As such, there are relatively few of such vehicles currently in use. For example, there are only approximately 50,000 vehicles in the United States that use compressed natural gas (CNG). The most promising application of alternative

fuel technologies in the near term is for fleets of buses and niche vehicles such as military vehicles, postal, and vehicles belonging to the state or federal government (Brecher, 2001). The electric utility industry is in active partnership with federal agencies in pursuing efficient charging or refueling technologies with low cost and widely available infrastructure.

On a more systemwide scale, there is great potential to reduce overall transportation energy consumption as well as energy use (per person or per person-mile) through improved traffic operations and by enhancing the performance of multimodal transportation systems. Recently, the U.S. Congress passed the Energy Policy Act of 2005, providing subsidies or tax incentives for alternative fuel production and alternative vehicle ownership.

SUMMARY

In this chapter, we examined factors that affect energy consumption, including fuel prices and taxes, regulation on fuel efficiency and emission, vehicle sales by class, vehicle technology, and transportation project-level intervention. The chapter also discussed the concept of energy intensity as a basis for comparing energy efficiencies across various transportation modes and vehicles. To evaluate the energy impact of transportation projects, we introduced different modeling frameworks, ranging from a macroscopic project screening level to simulation-based microscopic analysis. We concluded by providing observations and predictions of future changes in the energy market and associated technologies that will influence energy consumption in the transportation sector.

EXERCISES

15.1. For a congested corridor, an HOV lane is introduced for vehicles with more than one occupant. After introduction of this facility, it is observed that 10% of SOV (single-occupancy vehicles) demand (vehicle-miles) is shifted to two-person carpools. If the energy intensities of automobiles are 5500 Btu/vehicle-mile and 3500 Btu/passenger-mile before the improvement, calculate the energy intensities after the improvement. Assume no change in total demand in passenger-mile and fuel efficiency.

15.2. The Rao and Krammes' model [equation (15.8)] has different formulations depending on speed and acceleration conditions. How can this relationship be interpreted in the perspective of vehicle kinetics? What is the major difference between this model and the average speed model [equation (15.7)] in terms of the impact of speed? Assume that $a_e = 0$.

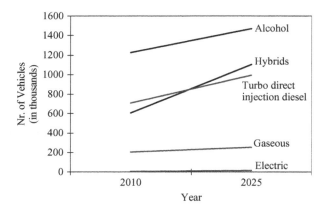

Figure 15.14 Predicted sales of advanced technology light-duty vehicles by fuel type. (From EIA, 2005a.)

Compare the model formulation with that of the average speed approach.

15.3. For the multimodal corridor described in Section 15.4.2, the option of an LRT (light-rail transit) system is considered. The LRT related data are given below:

> LRT ridership (person trips/day) is 25,960
> bus and LRT ridership after LRT (person trips/day) is 50,000
> average trip length for LRT (mi/trip) is 12.0
> average occupancy for LRT (persons per vehicle) is 25.0
> fraction of new transit ridership that were auto users is 0.5
> fraction of bus VMT eliminated is 0.5

Compare the LRT with other alternatives mentioned in Section 15.4.2 with respect to fuel consumption.

15.4. Consider the project-screening level problem in Section 15.4.2. What can be said about the impact of induced demand on the fuel consumption level?

REFERENCES[1]

Beimborn, E. (2005). *Transportation Energy: Supply, Demand and the Future*, Center for Urban Transportation Studies, University of Wisconsin, Milwaukee, WI.

Biggs, D. C., Akcelik, R. (1985). An Interpretation of the Parameters in the Simple Average Travel Speed Model of Fuel Consumption, *Australian Road Research*, Vol. 15, No. 1, pp. 46–49.

Biggs, D. C., Akcelik, R. (1986), Models for Estimation of Car Fuel Consumption in Urban Traffic, *ITE Journal*, Vol. 56, No. 7, pp. 29–32.

Brecher, A. (2001). *Transportation and Energy Issues*, U.S. Department of Transportation National Transportation Systems Center, Cambridge, MA.

Caltrans (1983). *Energy and Transportation Systems*, Tech. Rep. FHWA-CA-TL-83-08, California Department of Transportation, Sacramento, CA.

Davis, S. C., Diegel, S. W. (2006). *Transportation Energy Data Book*, 26th ed., Oak Ridge National Laboratory, U.S. Department of Energy, Washington, DC.

EIA (2005a). *Annual Energy Outlook, 2005—With Projection to 2025*, Energy Information Administration, U.S. Department of Energy, Washington, DC.

*_____ (2005b). *The Transportation Sector Model of the National Energy Modeling System: Model Documentation Report*, Energy Information Administration, U.S. Department of Energy, Washington, DC.

*_____ (1981). *A Method for Estimating Fuel Consumption and Vehicular Emissions on Urban Arterials and Networks*,

Tech. Rep. FHWA-TS-81-210, Federal Highway Administration U.S. Department of Transportation, Washington, DC.

_____ (2005). IMPACTS Spreadsheet home page, http://www.fhwa.dot.gov/steam/impacts.htm.

EERE (2006). *Indicators of Energy Intensity in the United States*, Energy Efficiency and Renewable Energy, U.S. Department of Energy, Washington, DC. http://intensityindicators.pnl.gov. Accessed Dec. 2006.

FHWA (1981). *A Method for Estimating Fuel Consumption and Vehicular Emissions on Urban Arterials and Networks*, Tech. Rep. FHWA-TS-81-210, U.S. Department of Transportation, Washington, DC.

FHWA (2002). *Highway Economic Requirements System–State Version (HERS-ST) V. 3.54*, Tech. Rep. FHWA-IF-02-060, U.S. Department of Transportation, Washington, DC.

FTA (2005). *National Transit Database, 2004*, Federal Transit Administration, U.S. Department of Transportation, Washington, DC.

*Fwa, T. F., Ang, B. W. (1992). *Estimating Automobile Fuel Consumption in Urban Traffic*, Transp. Res. Rec. 1366, Transportation Research Board, National Research Council, Washington, DC.

Goodwin, P., Dargay, J., Hanly, M. (2004). Elasticities of road traffic and fuel consumption with respect to price and income: a review, *Transp. Revi.*, Vol. 24, No. 3, pp. 275–292.

Greene, D. L., Decicco, J. M. (1999). Energy and Transportation Beyond 2000 in Transportation in the New Millennium, *TR News* Nr. 205, Transportation Research Board, National Research Council, Washington, D.C.

OECD (1985). *Energy Savings and Road Traffic Management*, Organization for Economic Cooperation and Development, Paris, France.

*Rao, K. S., Krammes, R. A. (1994). *Energy-Based Fuel Consumption Model for FREFLO*, Transp. Res. Rec. 1444, Transportation Research Board, National Research Council, Washington, DC.

Shapiro, R. J., Hassett, K. A., Arnold, F. S. (2002). *Conserving Energy and Preserving the Environment: The Role of Public Transportation*, American Public Transportation Association, Washington, DC.

USEPA (2005). *Light-Duty Automotive Technology and Fuel Economy Trends: 1975 Through 2005*, Tech. Rep. EPA 420-S-05-001, U.S. Environmental. Protection Agency, Washington, DC.

*VTPI (2005). Energy conservation and emission reduction strategies, in *TDM Encyclopedia*, Victoria Transport Policy Institute, Victoria, BC, Canada, http://www.vtpi.org/tdm/tdm59. Accessed May 2006.

ADDITIONAL RESOURCES

BTS (2005). *National Transportation Statistics 2005*. Bureau of Transportation Statistics, U.S. Department of Transportation, Washington, D.C. url.bts.gov/category.etm?cat=10.

Caltrans (2001). *SR-22 West Orange County Connection: Final Report Environmental Impact Statement*, California Department of Transportation, Sacramento, CA, http://www.dot.ca.gov/dist12/sr22FEIS_Intro.htm. Accessed May 2006.

_____ (2003). Energy, Chap. 13 in *Standard Environmental Reference*, California Department of Transportation, Sacramento, CA, http://www.dot.ca.gov/s er/vol1/sec3/physical/ch13energy/chap13.htm. Accessed Sept. 2005.

---

[1]References marked with an asterisk can also serve as useful resources for evaluating the energy impacts of transportation.

DeCorla-Souza, P. (1999). *IMPACTS: Spreadsheet Software Documentation*, Federal Highway Administration, U.S. Department of Transportation, Washington, DC.

FHWA (1980). *Energy Requirements for Transportation System*, Office of Environmental Policy, Federal Highway Administration, U.S. Department of Transportation, Washington, DC.

International Energy Agency (2002). *Transportation and Energy*, Paris, France.

Levinson, H. S., Strate, H. E., Dickson, W. (1984). Indirect transportation energy, *J. Transp. Eng.*, Vol. 110, No. 2, pp. 159–174.

CHAPTER 16

Land-Use Impacts

Nothing endures but change.
—*Heraclitus (540 BC–480 BC)*

INTRODUCTION

As early as 1826, Johann Henrich von Thünen laid down the first serious treatment of land-use economics by analyzing the relationship between the price of land and the cost of transportation. It was probably the first formal attempt to recognize that activity patterns are largely a function of a region's spatial distribution and characteristics, and thus the formation and pattern of land uses are strongly influenced by the level of accessibility provided by the existing transportation system (Figure 16.1). Does the transportation system significantly affect growth and shape of the land-use patterns, or do land-use patterns influence the form and extent of the transportation system? In either case, the answer is affirmative.

As shown schematically in Figure 16.1, the provision of transportation infrastructure affects the intensity and distribution of land-use patterns in an area by altering its level of accessibility, which is reflected in both the price and intensity of the developments in the area. The mechanism of transportation impacts on land use may be direct and/or indirect. For example, a highway project in

an area may involve direct appropriation of land and consequently may alter accessibility to that land. The project may also have indirect impacts if land development that follows the highway project also has impacts on land-use types, patterns, or distribution in outlying areas even though the highway does not pass through those areas. Changes in land use, at a given location and time can be due to a variety of factors including transportation interventions, socioeconomic changes, and implementation of new policy or control. As such, the process of isolating the effects of transportation systems on the land-use form can be complicated. For example, commercial businesses tend to cluster around interstate access points or rapid transit terminals, but it is difficult to determine the extent to which such location choice can be attributed to the high level of accessibility provided by the new facility, relative to the influence of other factors such as the local economic climate and potential market share. The type, extent, and timing of land-use changes are also influenced by the type and scope of the transportation project. For example, large-scale transportation projects, such as freeway capacity additions, are more likely to produce measurable land-use changes compared to small-scale projects such as improved signalization at arterials. In general, improvements in highway transportation tend to produce impacts that are more spatially distributed compared to those of transit improvements. This is because for the former, a greater variety of trip-makers is affected and the benefits are dispersed by the street systems connected to the highway (Parsons, 1999).

Accessibility created by new transportation facilities can have adverse impacts on land use. Highways that serve remote areas effectively open the adjacent land for new settlement and development. In ecologically sensitive areas, such development can be damaging to the land and can thus jeopardize the agricultural land-uses in such areas. The Polonoroeste project in Brazil is an example often cited by environmentalists of the dangers of uncontrolled development in rural areas (Lutzenberger, 1985).

One of the earliest pieces of legislation that required transportation agencies to analyze impacts, including land-use impacts, in advance of building transportation infrastructure projects was the National Environmental Policy Act (NEPA) of 1969. Due to this legislation, transportation agencies have sought to identify the land-use effects of their projects (ICF Consulting, 2005).

The need for land-use impact assessments of transportation decisions is stated in Sections 134(f) and 135(c) of the Intermodal Surface Transportation Efficiency Act (ISTEA) legislation, which mandates greater attention

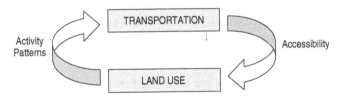

Figure 16.1 Transportation and land-use interactions.

to transportation and land-use relationships. In developing transportation planning plans and programs, each metropolitan planning organization is directed, at a minimum, to consider the following among other factors: "The likely effect of transportation policy decisions on land use and development and the consistency of transportation plans and programs with the provision of all applicable short- and long-term land use and development plans." Land use is not specifically mentioned in the Transportation Equity Act of 21st Century (TEA 21), but is implicit in Section 1204(d): "Each State shall carry out a transportation planning process that provides for consideration of projects and strategies that will protect and enhance the environment, promote energy conservation, and improve quality of life." The Safe, Accountable, Flexible and Efficient Transportation Equity Act—A Legacy for Users (SAFETEA-LU) of 2005 reaffirms land use as a factor to be considered in transportation planning of highway and transit projects.

In this chapter we first provide an overview of the relationship between transportation and land-use elements, discuss the methods and tools involved in estimating the impacts of transportation improvements on land-use, describe the analysis steps involved in the land-use impact assessment procedure and finally, offer some guidelines for mitigating adverse impacts of transportation on land use.

16.1   THE TRANSPORTATION–LAND-USE RELATIONSHIP

Investments in transportation infrastructure, such as the construction of new highways and the provision of new transit services, can alter the spatial form of land development by increasing the accessibility of land (through the introduction of new access or improvement of existing access), enhancing the mobility of land users, lowering transportation costs, and ultimately encouraging land development of various types. Accessibility can be measured as the number of travel opportunities or destinations within a particular travel radius, in terms of travel time or distance. Regions with well-developed transportation networks generally have high degrees of accessibility. Mobility is a measure of the ability to move passengers and freight efficiently between the origins and destinations in a network, and is influenced by the layout of the transportation network and its level of service. Changes in land use in turn generate activities that create a demand for travel. Then increased travel generates the need for new transportation facilities, which in turn increase accessibility and attractiveness of further development. This cycle continues until it is halted by natural limitation, policy, or system equilibrium.

For example, the construction of a new interchange in a highway corridor through undeveloped land may increase the accessibility of sites in the vicinity, attract commercial and employment activity, and become the focus for growth and development. In addition, the new interchange would likely offer existing users of the highway network time savings over their current routes and destinations, and would thereby increasing demand for new land development at these sites. Other transportation investments may produce "induced growth" in similar ways (Beimborn et al., 1999; Cervero, 2003).

In the United States, the planning and construction of transportation infrastructure is primarily a public-sector activity, while land development is mostly carried out by the private sector. Consequently, the transportation and land-use relationship is not dictated solely by activities in either sector. Furthermore, transportation is only one of many factors that influence land development decisions. There are other factors such as social, economic, and institutional considerations (Figure 16.2). There are instances in real estate development, where transit and land development investments have been jointly undertaken as part of cooperative efforts between public and private sectors (ULI, 1979; Wisconsin DOT, 1996).

16.1.1 Land-Use Impacts on Transportation

Travel demand forecasting models are built on the fundamental relationship presented in Figure 16.1. In the four-step travel demand model discussed in Chapter 3, trip generation involves forecasting productions and attractions (travel demand) as a function of population and employment. Changes in land-use patterns can be represented by changing residential and employment densities within activity centers, neighborhood classifications (contemporary vs. traditional), various measures of accessibility and land-use mix within neighborhoods and activity centers, and various other quantitative and qualitative measures. Travel characteristics can include total household vehicular travel (VMT or VHT), trip frequencies, trip lengths, and modal choice.

In an extensive review of more than 50 empirical studies, Ewing and Cervero (2001) found that:

1. *Total vehicular travel* (VMT or VHT) is primarily a function of regional accessibility. Local densities and degree of land-use mix had little impact on total travel, which implied that pockets of dense, mixed-use developments in sparsely populated areas would offer little benefits.
2. *Trip frequencies* are largely a function of socioeconomic characteristics and are largely independent of land-use variables.

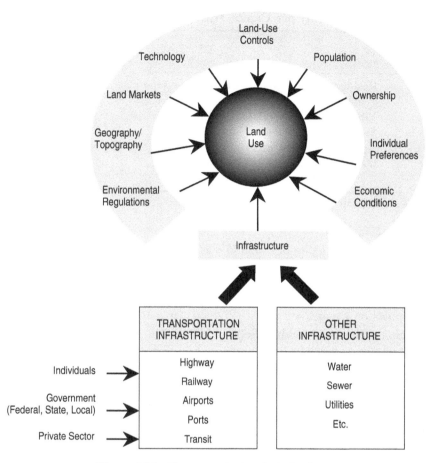

Figure 16.2 Transportation's role in land use.

3. *Trip lengths* are directly due to the built environment and indirectly due to socioeconomic characteristics. They are generally shorter in traditional urban settings characterized by concentrations of activities, diverse land-uses, and grid-like street networks.
4. *Modal choice* depends on the diversity of land-use, local population densities, and socioeconomic characteristics. However, the prevalence of walking and transit in traditional urban settings may be attributed to the self-selection nature of the sample (i.e., people who prefer to use transit or walk are likely to choose to live in traditional urban settings).

In the same study, Ewing and Cervero (2001) also pointed out that population density itself probably does not affect the way that people travel. Rather, it is what goes along with density, such as the fact that the high costs of automobile ownership at high densities due to traffic congestion and limited parking would encourage

the use of alternative modes. The authors also generalized elasticity values of different land-use variables across the studies in a meta-analysis, (Table 16.1). The measures used to represent the built environment were *density* (residents plus employees divided by land area), *diversity* (a measure of the balance between jobs and population), *design* (a combination of sidewalk completeness, route directness, and street network density), and *regional accessibility* (derived from a gravity model). Although most of the elasticity values were statistically significant, their magnitudes were rather small; for example, a 100% increase in local population density would result in only a decrease of 5% of vehicle trips. The highest elasticity value was for regional accessibility, where a 100% increase would result in a 20% decrease in vehicle-miles of travel: This supports the argument for public policy intervention and coordination of transportation planning throughout a region.

Table 16.1   Typical Elasticities of per Day per Capita Travel with Respect to Built Environment

Factor	Description[a]	Vehicle Trips	Vehicle-Miles Traveled
Local density	Residents plus employees divided by land area (acres)	−0.05	−0.05
Local diversity (mix)	Jobs divided by residential population	−0.03	−0.05
Local design	Sidewalk completeness[b] divided by route directness[c] and street network density[d]	−0.05	−0.03
Regional accessibility	Distance to other activity centers in the region (ft)	—	−0.20

Source: Ewing and Cervero (2001).
[a] Offered in USEPA (2003).
[b] Percent street frontage with sidewalks.
[c] Average ratio of walking distance from point of origin to central node, to straight-line distance.
[d] Street centerline miles per square mile.

16.1.2 Transportation Impacts on Land Use

In extending the von Thünen model, Alonso (1960) observed that different land-use types, based on their need and ability to pay for accessibility, eventually attain a spatial equilibrium where land is allocated to the use that earns the highest location rent. In recent decades, most studies on land use–transportation interaction have been based on this concept.

Tables 16.2 and 16.3 summarize the state of knowledge regarding the impacts of transportation investments and policy on land use, and vice versa. The tables include land-use elasticities and potential mitigating factors. The land-use elasticity serves as a measure of the degree to which land-use impacts are expected to occur: an action that has "high" land-use elasticity, is one whose impacts are large, not necessarily in absolute terms, but in relation to other actions. The observations reported in the tables represent general trends only and thus oversimplify the complex relationship between land-use and transportation investments.

Table 16.2 shows that major infrastructure investments such as new freeway segments or interchanges have a high potential for altering existing land-use patterns. The extent of land-use impacts depends on the extent to which capacity is increased and the overall demand for additional capacity. Large impacts are also generally expected for policies that affect the user cost of automobile travel. The greater the change in user cost, the greater the impact on land use. The magnitude of land-use impacts also depends on the characteristics of the transportation action and existing land use. For example, impacts of the largest magnitudes are generally expected when a new transportation facility is constructed in a developing, fast-growing area.

Table 16.3 suggests that transit investments of minor scale are unlikely to yield significant, measurable land-use impacts. The accessibility benefits of added capacity transit investments are accrued by a far smaller share of the travel market and therefore cannot be expected to have the same range of impacts as those of added highway capacity projects, although localized changes can be significant. Furthermore, because highways are often multimodal, their impacts on land use tend to be more diffused; and thus may be harder to measure than transit impacts which are likely to be localized (as transit only serves passengers) and thus more definable.

16.1.3 Land-Use Impacts in terms of Monetary Costs

Land-use impacts can be difficult to monetize, in part because it is difficult to predict changes in land-use patterns resulting from a particular transportation policy or planning decision, and also because the various related impacts (economic, social, and environmental) are mostly difficult to monetize. As such, there are few existing monetary estimates of the costs associated with changes in land-use patterns.

Bein (1997) calculated the environmental costs of policies and projects that change land-use patterns. Table 16.4 shows the monetary values assigned to the ecological benefits lost when land is paved or developed, or gained when green space is preserved. For each hectare of land converted from its current use (left column) to another use (top row), the dollar amount (in the intersection cell) indicates the change in external environmental benefits. Indirect land-use impacts (measured within 500 m of a road) can be considered to account for half of these costs. For example, converting wetlands to pavement represents an

Table 16.2   Summary of Land-Use Impacts of Highway Investment and Policies

Action	Land-Use Elasticity	Land-Use Impact	Mitigating Factors
New facilities (i.e., highway corridors, interchanges)	High	Redistribution of metropolitan growth to highway corridors. Decentralization of population and employment. Increased land values and concentration of development around interchanges.	Local and regional economic conditions. Degree of impact on regional accessibility. Congestion levels. Local land use policies.
Added lanes, intersections	High	Same as above, but to a lesser degree.	Same as above.
System management	Low	None likely.	Levels of congestion and latent demand.
Congestion pricing	High	Unknown. Possible shift of population and jobs toward more accessible locations. Possible shift of population and employment to exurban areas.	Local and regional economic conditions. Magnitude and spatial extent of pricing policy. Degree of congestion. Availability of alternative modes and routes.
Parking pricing, management	High	Unknown. Possible increased development of major employment centers. Likely increased development density.	Local and regional economic conditions. Magnitude and spatial extent of pricing policy. Long-run incidence of parking fees. Availability of alternative modes.
Vehicle and/or fuel tax	Moderate	More compact development if cost of driving is high enough to encourage use of other modes.	Magnitude of tax. Availability of alternative modes.
Transportation demand management	Low	None likely.	NA[a]
Safety improvements	Low	None likely.	The extent to which the improvement changes capacity or accessibility.

Source: Adapted from Parsons (1999).

[a]NA, not applicable.

environmental cost of $55,000 per hectare (cost of direct impact) plus $27,500 per hectare (cost of indirect impact).

Table 16.5 shows the estimated land-use costs associated with motor vehicle use. Litman (2002) developed these values to estimate the impacts of policy and planning decisions on land use. However, the author states that these values are crude estimates, and it may be inappropriate to apply them to the evaluation of specific projects without due caution.

16.2   TOOLS FOR ANALYZING LAND-USE CHANGES

There is a wide range of analysis tools and strategies for investigating land use–transportation system interactions. Some of these methods are straightforward and can be achieved through common survey techniques, comparisons, or basic quantitative analysis. Others are relatively complex and require specialized software and training.

Table 16.3   Summary of Land-Use Impacts of Transit Investments and Policies

Action	Land-Use Elasticity	Land-Use Impact	Mitigating Factors
New rail facilities	Moderate	Increased land values and development density. Redistribution of development to downtown, station areas. Decentralization of population.	Local land-use policies. Degree of impact on accessibility. Local economic conditions. Station access and local circulation patterns. Corridor congestion levels.
Rail extensions, rail stations	Moderate	Same as above, but to a lesser degree.	Same as above.
New high-capacity arterial bus lines, bus rapid transit stations	Moderate	Possible redistribution of development to major bus transit corridors.	Local economic conditions.
Changes in local service	Low	Possible redistribution of development to transit corridors.	NA[a]
Fare policy changes	Low	None expected.	NA
Safety improvements	Low	None expected.	Whether the improvement changes perceptions about passenger safety.

Source: Adapted from: Parsons (1999).
[a]NA, not applicable.

Table 16.4   Land Conversion Costs per Hectare per Year[a]

Land-Use Categories	Wetlands	Pristine Wildland/ Urban Greenspace	Second Growth Forest	Pasture/ Farmland	Settlement/ Buffer	Pavement
Wetlands	0	−11,000	−22,000	−33,000	−44,000	−55,000
Pristine wildland/urban greenspace	11,000	0	−11,000	−22,000	−33,000	−44,000
Second growth forest	22,000	11,000	0	−11,000	−22,000	−33,000
Pasture/farmland	33,000	22,000	11,000	0	−11,000	−22,000
Settlement/buffer	44,000	33,000	22,000	11,000	0	−11,000
Pavement	55,000	44,000	33,000	22,000	11,000	0

Source: Adapted from Bein (1997).
[a]Original costs expressed in 1994 dollars, converted to 2005 dollars using the National Association of Real Estate Investment Trusts (REITs) Historical Price Index (http://www.nareit.com/library/global/daily.cfm).

Selecting the most appropriate tools and strategies should be guided by the quality and availability of data and other resources, and level of desired sophistication of the analysis. NCHRP Report 423A (Parsons, 1999) offers guidelines on the use of each tool and strategy and discusses example cases of past applications.

16.2.1 Qualitative Tools

Qualitative methods can be used alone or with quantitative methods to evaluate transportation–land use interactions. These methods are useful for evaluating situations where conflicting societal values exist and can help identify and clarify the underlying issues. Also, they are useful in cases

Table 16.5   Land-Use Impact Cost Estimates (Dollars per Vehicle-Mile)

Cost Category	Estimate (cents/VMT)
Environmental costs of paving land for roadways	2.5
Aesthetic degradation and loss of cultural sites	0.5
Social costs	2.5
Municipal (or public service costs)	2.3
Transportation (both user and external costs)	6.2
Total sprawl cost	14.0
Automobile sprawl costs (total reduced 50% for other contributing factors to urban sprawl)	7.0

Source: Adapted from Litman (2002).

where data for quantitative methods are not available. Types of qualitative methods include expert panels and Delphi process, interviews and surveys, and comparative case studies.

(a) Expert Panels and the Delphi Process   The use of expert panels aggregates the experience of land planners, urban designers, and other local professionals to predict the most likely range of land-use outcomes from specific transportation policies or investments. Expert judgments can be obtained through several rounds of questionnaires either to arrive at consensus or to clarify differences of opinion. Panel responses can be obtained either through face-to-face group meetings or separately from each panel member through a Delphi process. The Delphi method was used in a study in San Jose, California (Cavalli-Sforza et al., 1982) to predict the land-use impacts of three transportation alternatives: highways, bus and HOV lanes, and light rail. The study recruited 12 panelists with expertise in transportation and land use. These experts worked as economists, engineers, planners, or public administrators or were community activists involved in transportation and land-use issues. The Delphi process was also used in Wisconsin to assess potential land-use impacts from State Trunk Highway 26, where a panel of experts was convened from county and municipal planners and engineers, university staff, local economic development professionals, and representatives of the farming, real estate, and environmental communities (Wisconsin DOT, 2005). The Maryland DOT also used an expert panel to evaluate the land-use impacts of various

transportation options, including a highway option and a light-rail option for the Highway 301 Corridor Study. The panel consisted of 6 individuals with expertise in land economics, transportation, land use, and real estate (Parsons, 1999).

(b) Interviews and Surveys   In many cases, transportation–land use interactions are evaluated by an individual or a team of experts who conduct interviews, collect and analyze data, and conduct field visits. The economic and land-use impacts of various corridor projects for example, can be determined by analyzing data on the local economy and land uses, interviewing business people and planners, comparing the situation with that of previous similar studies, and utilizing knowledge acquired or lessons learned from past experiences regarding the transportation action impacts on land development.

(c) Case Studies   Another qualitative approach is to conduct case studies of locations where similar types of transportation projects or policies have been implemented. A case study may involve interviews, site visits, and data compilation. Analysts look for patterns among comparable cases and for reasons that some cases deviate from these patterns. For a comparison to be valid, cases should be similar in size, project type, location, demographic statistics, and population and economic growth rates. Although comparisons may be imperfect, having a real example of a comparable situation can lend credibility and tangibility to an assessment of land-use impacts. For example, for proposed interchange projects, the impacts of similar interchange developments at other locations in the past can be studied. Similarity can be in terms of traffic volume, demographics, and existing land-use controls.

16.2.2 Quantitative Tools

A number of quantitative tools can be used for assessing the impact of transportation projects on land use. These tools range from simple decision rules and gravity models to sophisticated analytical methods that are often automated. Also, GIS can serve as a useful tool for such analyses by providing data on the type of the land use, demographics, and employment data that are also used in computer models. These methods are discussed below.

(a) Allocation Rules   There are several types of allocation rules. These differ by the amount of data and analysis, and the inherent assumptions. Allocation rules work best for typical or average areas and for widespread activities such as retailing and residential development. Activities

that have a limited number of locations, such as industrial development, are better assigned using other methods. The simplest rule, *constant share*, assumes that all zones share in the growth of population or jobs in proportion to the amount of vacant land zoned for that purpose. For example, if a zone has 7% of all vacant industrial-zoned land in the region, it is assigned 7% of new industrial jobs. This method requires relatively little data and is relatively simple. However it may not capture the actual development process where various parts of the region develop at different rates due to differences in attractions. Two other methods, *share of growth* and *shift share*, are based on the patterns of recent growth. Both methods are trend extrapolation techniques, as they use evidence from the recent past to determine the allocations for the future. Although these methods are more realistic than assigning constant shares, they may be applicable only for short-time-period estimations for regions with consistent growth because they do not capture the shifts of development to new areas or the slowing growth rate for areas with mature transportation systems or one-time events, such as the opening of a large industrial plant or shopping center. Finally, the *gravity model* (Lowry, 1988) considers specific factors that influence the attractiveness of areas, such as distance to population and employment centers and other types of attractions, such as health facilities, educational institutions, airports, and cultural events. Other factors, such as income levels and tax rates, that also affect attractiveness can be incorporated into gravity models provided that the relative weight of each variable can be estimated. A gravity model approach was used to estimate the potential impacts of a highway extension on residential development in the Route 531 corridor near Rochester, New York (Hirschman and Henderson, 1990). An example of the use of the gravity model, adapted from the aforementioned study, is given below.

Application of the Gravity Model: A proposed rapid transit system is expected to provide faster travel to a downtown employment center from the suburbs of a city. Table 16.6 shows the number of employees in each suburb that commute to the employment center and the reduction in average travel time due to the improvement. Table 16.7 shows the accessibility indices associated with the downtown employment center from each of the residential locations, as well as a composite index of accessibility to the downtown employment center. The composite accessibility index is the weighted average of the accessibility from each residential location, weighted by the number of employees commuting to the downtown area from each suburb. These indices can be computed using the gravity model, as follows:

$$A_j = \sum_i \frac{E_i}{t_{ij}^2} \tag{16.1}$$

where A_j is the accessibility of the downtown employment center j, E_i the number of employees commuting to downtown from location i, and t_{ij} the generalized time or cost of travel from i to j.

The results suggest that the proposed rapid transit system will increase access to the downtown employment center by 26%. The gravity model formulation presented by Hirschman and Henderson (1990) can be applied to expand the previous analysis to estimate the potential

Table 16.6 Gravity Model Input Data on Accessibility to a Major Downtown Employment Center

Residential Location (*i*)	Total Attraction to Downtown Employment Center (*j*)[a] E_i	Average Travel Time to Downtown Employment Center (*j*), t_{ij}	
		Existing Transportation Network (Base Case)	New Rapid Transit System (Proposed)
Downtown	10,000	6	6
Northern suburb	12,000	30	20
Southern suburb	10,000	25	15
Eastern suburb	14,000	35	25
Western suburb	16,000	40	25

[a]Number of trips per day from origin *i* to destination *j*, assuming to be same as the number of employees residing in location *i*.

Table 16.7   Gravity Model Calculations of Accessibility to a Major Downtown Employment Center

Residential Location (i)	Gravity Model Accessibility Index, E_i/t_{ij}^2		
	Existing Transportation Network (Base Case)	New Rapid Transit System (Proposed)	Percent Change
Downtown	277.8	277.8	0
Northern suburb	13.3	30.0	125
Southern suburb	16.0	44.4	178
Eastern suburb	11.4	22.4	96
Western suburb	10.0	25.6	156
Composite accessibility index (A_j)	55.1	69.4	+26

impacts of the downtown transportation access improvement on residential development. If the total population growth projected for the region in 10 years is 18,000, the distribution of the additional growth can be estimated using the gravity model, as shown below. Standard land-use densities for housing can then be applied to convert the estimated population changes to acreages, as shown in Table 16.8.

$$G_i = G_t \frac{L_i A_i}{\sum_i L_i A_i} \qquad (16.2)$$

where G_i is the population growth increment allocated to residential location i, G_t the total growth projected for the region, L_i the (availability of) developable land in residential location i, and A_i the accessibility index to the downtown employment center j from location i.

The past few decades have seen the development of a number of land-use models to represent land-use transportation interaction. These models, many of which are based on the two-way constrained gravity model conceived by Lowry (1964), use an iterative process to estimate the distribution of population and service employment within a region or an urban area at a given point in time. The Lowry model has constraints that ensure that the estimated distribution of jobs (i.e., attractions) and residents (i.e., productions) in each analysis zone within a region is the same as that given in initial estimates. Two basic assumptions inherent in the Lowry model, which are related to factors governing the location of activities in a region, are as follows: (i) the location of an individual's work place strongly influences the individual's choice of residential location, (ii) economic activities are classified

Table 16.8   Gravity Model Calculations of Residential Development

Residential Location (i)	Developable Land, (acres) L_i	$L_i A_i$	$L_i A_i / \sum L_i A_i$	Population Growth Allocated to Residential Location i G_i	Land-Use Change (vs. Base Case) (acres)
Downtown[a]	100	27,778	0.345	6,216	317
Northern suburb[b]	500	15,000	0.186	3,357	883
Southern suburb[b]	300	13,333	0.166	2,984	785
Eastern suburb[b]	400	8,960	0.111	2,005	528
Western suburb[b]	600	15,360	0.191	3,437	905
Total		Sum = 80,431		18,000	3,418

[a]Residential land-use density of 19.6 persons/acre was assumed.
[b]Residential land-use density of 3.8 persons/acre was assumed.

as basic or service activities. It is assumed that the location of basic activities is determined by external factors, while service activities are assumed to be largely a function of the distribution of the local population. An example that illustrates the general principles underlying the Lowry model is provided in Masser (1972). In the Lowry model, the first iteration begins by using the distribution of basic employment as the input to the attraction-constrained gravity model and then predicts the distribution of resident workers. Then the estimated number of resident workers is multiplied by the *activity rate* to yield estimates of residential populations. An activity rate is a measure of the relationship between the labor force and the working age population, and is expressed as the proportion of the population under consideration that is in employment. The production-constrained model then determines the destination of the service trips made by the trip-makers from residential locations. The number in the service employment is then estimated by multiplying the trips to service locations by the ratio of service jobs to the total population. The second iteration begins by using the distribution of service employment (instead of the distribution of basic employment) as the input to the attraction-constrained gravity model and generates further distributions of residential population and service employment. This iteration process is repeated until an equilibrium estimation distribution is reached (typically, after four or five iterations). The impact of changes in transportation systems is modeled through changes in travel time and/or cost affecting zonal attractiveness.

(*b*) *Decision Rules* Typically, simple decision rules are utilized to specify the relationship between transportation and land use. Worksheets and guidebooks have been prepared for transferring measures developed from empirical studies (Parsons, 1999). These decision rules save considerable time by minimizing the collection and analysis of new data. However, the context in which such rules were developed must be comparable to the context in which they are to be applied. A weakness of all decision rules is that they are static; they assume that the relationships that were valid in the past will continue in the future. An example of decision rules is the set of estimates of employment and household density needed to support various types of transit service by Pushkarev and Zupan (1977). The rule of thumb suggests that a minimum density of seven dwelling units per acre is needed to support intracity bus service. A study in the Research Triangle region of North Carolina used rules of thumb to assess the land-use changes that would be needed to support a proposed rail transit system (Barton-Aschman and Hammer, 1990).

(*c*) *Statistical Methods* Statistical methods vary from basic descriptive tools to complex models that investigate the effects of multiple variables. A review of available literature shows that regression models and discrete choice models have been used primarily in land-use and development impact studies. Both types of models are discussed below.

An example of the use of *regression models* is the national study conducted by Hartgen and Kim (1998) to estimate the extent of commercial development at rural and small town Interstate exits. Models were estimated for five development types (gas stations, convenience stores, fast-food restaurants, sit-down restaurants, and motels) using classification and regression techniques. Table 16.9 lists the significant variables in the models estimated by development type (defined within 1 mile of an exit). Elasticity values are shown in parentheses.

Other examples of past application of this tool include evaluating whether rail transit influences property values and determining the size of central business districts needed to support commuter rail ridership. Table 16.10 summarizes the results of studies that applied regression techniques to evaluate the impact of transportation system changes on residential and commercial property values. A detailed discussion of property value analysis is provided in NCHRP Report 456 (Forkenbrock and Weisbrod, 2001). To illustrate the use of Table 16.10, consider the construction of a new rail station in Atlanta, Georgia. The total commercial property value impact can be $75/m^2$ (1994 dollars) or $138/m^2$ (2005 dollars), decrease for each meter away from the stations on average.

Discrete choice models can be applied to predict the location decisions of households and firms as a result of a change in an area's transportation system. Such models can predict the probability that a given household would move to a different residential location, on the basis of household characteristics such as age of household head, presence of children, number of workers, housing tenure, and ratio of housing costs to income, and the physical and social characteristics of a community. Similarly, employment relocation choice can be predicted using a discrete choice framework, as a function of business characteristics (such as industry size), characteristics of potential zones for land-use development (such as accessibility, density, and employment levels), and characteristics of vacant land (quantity and cost). Levine (1998) employed discrete choice modeling to assess the relationship between residences and jobs. A limitation of this approach is that individual-level data that are required can generally be obtained only through surveys.

Table 16.9   Estimated Regression Models by Development Type

Dependent Variable	Significant Independent Variables	R^2
Average number of gas stations	Total number of development units at exit (0.39); number of convenience store services within 1 mile at exit (0.26)	0.68
	Exit population share of market within 3 miles/town population (−0.85)	0.48
Average number of convenience stores	Traffic counts at interstate highway: west and north direction (4.62); traffic counts at interstate highway: east and south direction (−3.94)	0.74
	Exit distance weighted competition share (0.58); number of discount store services within 1 mile at exit (0.11); number of gas store services within 1 mile at exit (0.97)	0.64
Average number of fast-food restaurants	Total number of development units at exit (1.71); Number of establishments at exit 2 (0.14); number of sit-down restaurants within 1 mile at exit (−0.64)	0.66
Average number of sit-down restaurants	Exit competition share (1.23); total number of development units at nearby exits (0.49); average median household income within 1 mile at exit (−0.71); cross-street traffic counts toward town (0.35)	0.70
Average number of motels	Total number of development units at nearby exits (0.14); average distance from exit to a city (0.66)	0.60

Source: Hartgen and Kim (1998).

(*d*) *Economic Models*   Economic models simulate an area's economy and estimate the impact of major economic changes such as population growth, industrial expansion or recession, and lower transportation costs on various sectors of the economy. These models can provide information on the regional economy that can be used with land-use allocation models to estimate the land-use impacts of transportation investments. The major types of regional economic models are input–output models, econometric models, and combinations of the two. In Chapter 9 we describe the use of these models as well as some of the most widely used, commercially available regional economic models.

(*e*) *Land-Use Models*   Several land-use models are in operational use in the United States and abroad. The major types of them are stand-alone land allocation processes or integrated land use–transportation models. Integrated land use–transportation models link land-use allocation processes (e.g., DRAM/EMPAL) with travel demand models. Such models are typically developed for an entire metropolitan region. Through an iterative process, these integrated models predict an equilibrium land use–traffic pattern for a future year. Using regionwide forecasts of population and employment, these models allocate housing and business development to small zones based on transportation accessibility, land prices, and land availability. Using historical data on transportation accessibility

and land development and prices, the models are calibrated to represent the decision-making characteristics of a given metropolitan area. Many of the integrated models are used for large regions and are highly data- and labor-intensive. Some commonly used land-use models are DRAM/EMPAL MEPLAN, TRANUS, TELUS, METROSIM, LUTRIM, HLFM II+, UrbanSim, and the California Urban Futures (CUF) model. Most of these models not only allow integration with GIS and/or travel demand models, but also can be used to evaluate land-use impacts of a wide range of projects and policies. For example, the city of Sacramento used two transportation–land use models, MEPLAN and TRANUS, to evaluate the land-use impacts of several policies, such as HOV and HOT lanes, various transit investments, transit-oriented development, and roadway pricing (Johnston et al., 2000). A detailed discussion of these models, in terms of requirements for use, operational features, applications, strengths, and weaknesses is provided in the NCHRP Report 423A (Parsons, 1999).

16.3   PROCEDURE FOR LAND-USE IMPACT ASSESSMENT

General steps to assess land-use changes are applicable for a single transportation project or for a set of multiple projects in a given region. However, multiple projects can have competitive or synergistic effects on land use. For example, highway added-capacity projects could improve

Table 16.10   Summary of Regression Studies of Property Value Effects

Study	Transportation Factor/Mode	Effect Observed
Residential Property Values (effects observed after project completion)		
Traffic Volume		
Grand Rapids, MI (Bagby, 1980)	Change in traffic volume in a residential neighborhood	Property values decreased roughly 2% per additional 100 vehicles per day on residential streets.
Baton Rouge, LA (Hughes and Sirmans, 1992)	Difference in traffic volume on a street	On high-traffic streets, each additional 1000 vehicles per day reduced property values by 1% in urban areas and 0.5% in suburban areas.
Highway		
Washington, DC (Langley, 1981)	Distance from the Capital Beltway	Property values increased with the distance from the highway out to a distance of 1125 feet and then decreased by $3000—$3500 per house beyond that distance (1977 dollars).
Washington State (Palmquist, 1982)	Distance from a newly constructed highway	Property values increased 15–17% where there was highway access, but properties located nearby decrease 0.2–1.2% per dBA of traffic noise.
Orange County, CA (Boarnet and Chalermpong, 2001)	Distance to nearest ramp of a new toll highway	Property values reduced by $1 to $4 for each foot of distance away from a highway (1982 dollars).
Heavy rail		
Southern New Jersey (Boyce et al., 1972)	Travel-time savings	Positive increase of $149 in the price of a home for each dollar of value in time savings (1971 dollars).
Toronto, Canada (Bajic, 1983)	Distance from heavy-rail station	$2,237 premium for homes close to a station (1971 dollars).
Washington, DC (Benjamin and Sirmin, 1996)	Distance from heavy-rail station	Apartment rents decreased by 2.4 to 2.6% for each one-tenth mile distance from a metro station.
San Francisco Bay area, CA (Cervero, 1996)	Distance from heavy-rail station	+10–15% in rent for rental units within $\frac{1}{4}$ mile of a metro station
San Francisco Bay area, CA (Sedway Group, 1999)	Distance from heavy-rail station	Apartment rents near stations higher by 15 to 26% than apartments distant from stations.
Greater Toronto Area, Canada (Haider and Miller, 2000)	Distance from heavy-rail station	Prices of houses within 1.5 km distance of a subway line higher by 4000 Canadian dollars (1995), on average.
Commuter rail		
Montgomery County, PA (Voith, 1993)	Distance from train station	$7279 to $9605 premium per house (1990 dollars) associated with CBD-oriented train service.
Boston, MA (Armstrong, 1994)	Availability of commuter rail service	Increase in single-family residential property values of approximately 6.7% in a community that has a commuter rail station.
Light rail (LRT)		
Portland, OR (Al-Mosaind et al., 1993)	Distance from light-rail station	+10.6% premium for homes within 500 m of LRT stations
Portland, OR (Chen et al., 1998)	Distance from light-rail station	+10.5% single-family home price differential for homes near LRT stations.
Santa Clara County, CA (Cervero, 2002)	Distance from light-rail station	Premiums for large apartments within $\frac{1}{4}$ mile of LRT stations as high as 45%.

Table 16.10   *(continued)*

Study	Transportation Factor/Mode	Effect Observed
Buffalo, NY (Hess and Alemeida, 2006)	Distance from light-rail station	Premiums for houses within $\frac{1}{4}$ mile of LRT stations between 4–11% of the median assessed home value.

Commercial/Office Rents (effects observed after project completion)

Highway

Austin, TX (Kockelman and ten Siethoff, 2002)	Distance to frontage road network	Assessed property and land values decreased by $510,000/ acre per square mile away from the road network (real dollars in years 1982–1999).

Heavy rail

Washington, DC (Rybeck, 1981)	Distance from heavy-rail station	9–14% premium for sites close to a station.
San Francisco, CA (Landis and Loutzenheiser, 1995)	Distance from heavy-rail station	No effect in San Francisco or Oakland; elsewhere, rents increased 16% for sites up to $\frac{3}{8}$ mile from a station.
Atlanta, GA (Bollinger et al., 1996)	Distance from heavy-rail station	Rents increase 4% for sites close to a station.
Atlanta, GA (Nelson, 1999)	Distance from heavy-rail station	Commercial prices per square meter fell by $75 for each meter away from heavy rail stations (1994 dollars).

Commuter rail

Santa Clara, CA (Cervero and Duncan, 2002)	Distance from commuter rail station	More than 120% premium for commercial land in a business district within $\frac{1}{4}$ mile of a commuter rail station.

Light rail

Santa Clara, CA (Weinberger, 2000)	Distance from light-rail station	Rent values increased 3–6% for sites within a mile of a light-rail station.
Santa Clara, CA (Weinberger, 2001)	Distance from light-rail station	Almost 15% higher commercial rents for properties within $\frac{1}{2}$ mile of light-rail stations.

Source: Diaz (1999), Forkenbrock and Weisbrod (2001), Forkenbrock and Sheeley (2004), Smith and Gihring (2006).

accessibility in multiple areas, creating competition among those areas for new development. Alternatively, multiple transportation projects could collectively support the same patterns of land use (i.e., a mix of transit improvements). Various steps involved in land-use impact assessment are summarized below based on the information provided by Parsons (1999), Louis Berger Group (2004), D'Ignazio and Hunkins (2005), Wisconsin DOT (1996), and Stanley (2006).

Step 1: Define the Project Area The project study area should be large enough to contain all areas whose land development is expected to be affected by the transportation intervention. For example, for interchange construction projects, as a rule of thumb, all land within a $\frac{1}{2}$-mile radius of the interchange, at a minimum, should

be included in the project study area. However, it will probably be unnecessary to conduct the same level of detailed analysis for all subareas of the study area. For example, areas closer to the project are more likely to show a wider range of effects than are areas farther away. The Wisconsin DOT (1996) identified three options for defining the project study area, as follows:

(a) *Option 1: TrafficShed (Radial Routes to and from an Attraction)   The project study area is defined as the entire area served by the transportation project to reach a major destination. A trafficshed for a transportation facility is analogous to the watershed of a river. First, a most important destination, such as a city center, is identified. All origins that plausibly connect to the main destination via the transportation facility are included*

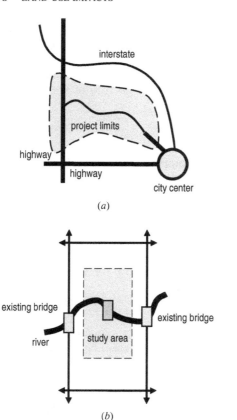

(a)

(b)

Figure 16.3 TrafficShed: (*a*) Example 1; (*b*) example 2. (Adapted from Wisconsin DOT, 1996.)

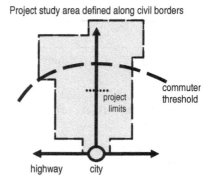

Figure 16.4 Commuter shed example. (Adapted from Wisconsin DOT, 1996.)

in the facility's traffic shed, as shown in Figure 16.3*a*. Figure 16.3*b* illustrates a case where a traffic shed may not include the entire area (larger rectangle) because of the presence of a barrier to travel, in this case a river. A drawback of Option 1 is that it may define large study areas, and for much of the study area, the relationship between land-use changes and the transportation project may not be readily identifiable.

(b) *Option 2: CommuterShed (Radial Routes to and from an Attraction) The project study area is defined as the area served by the transportation project for commuting to a major destination.* In this method, the trafficshed is first defined. The project study area is limited to areas within a preset commuting range of or threshold to the major destination (Figure 16.4). The premise is that a project is expected to affect only residential development and related service development to the level that persuades commuting to a major destination. The commutershed method is useful for those projects crossing from a rural or urbanizing area into an urban area. There are several possible ways to define the commuting range.

The range is based on travel time or existing sources of data on commuter activity, such as census data or origin–destination surveys. For metropolitan areas, 30 minutes from the specified destination may be a good rule-of-thumb definition for the outer limits of a commutershed based on travel time. The outer limit of a commutershed based on existing commuting patterns is determined by first selecting a commuting threshold. The rule-of-thumb threshold is 25% or more of the employed inhabitants commuting to the central developed area.

(c) *Option 3: 20-Year Growth Boundary The project study area is defined as the area expected to develop in the next 20 years.* At most communities, twenty-year growth boundaries have already been defined as part of their water or sewer service area plans, metropolitan planning organization plans, or local land-use plans (Figure 16.5). This method is likely to be useful for bypass, beltline, or other urban transportation projects. Option 3 may be preferable if all project alternatives are scoped within the 20-year growth boundary; problems may arise if some of the alternatives extend beyond the 20-year growth boundary. Another drawback of this method is that it does not address the question of whether a project will inherently cause a change in the community's 20-year growth boundary.

(d) *Option 4: Interview The project study area is defined by asking "experts" what land area may be affected by the project.* Unlike others, this method will produce a project study area. Information on this technique is presented in Section 16.2.1.

Step 2: Analyze the Existing and Future Patterns and Trends for Land Use and Development The purpose of this step is to produce a general description of the overall character of the project study area, the significant land-use trends, the current demand for development

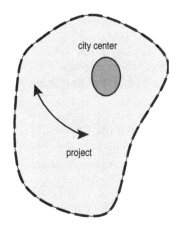

Figure 16.5 Growth boundary example. (Adapted from Wisconsin DOT, 1996.)

within the study area and the relative attractiveness of the area to development, as well as the future for the project study area if the land-use trends continue, assuming that the transportation corridor under study remains unchanged. The analyst will need to distinguish between developed and agricultural/rural areas and to identify areas of natural resource interest. For most projects, the analyst will need to collect data on existing population, employment, housing, and public policies as well as existing travel conditions and accessibility levels of the study area. Existing databases, surveys, statistical trend analysis, and GIS are useful at this stage of analysis (ICF Consulting, 2005). For projects affecting small study areas, manual methods of tabulating data and mapping may suffice.

Step 3: Develop the Transportation Alternatives or Policy Assumptions The Wisconsin DOT (1996) identified the following six general categories of project design characteristics that can impact land use to varying degrees:

1. *Location.* One of the most significant design characteristic affecting land-use changes is the location of a transportation facility. New alignments of existing roadways, new roadways, and bypasses as well as new transit facilities can significantly influence present and future commercial, industrial, residential, and central business district land development patterns. Also, profound changes in adjacent land-use characteristics can be expected from the location of new or significantly improved intersections or interchanges or transit facilities and rapid transit terminals.

2. *Access management.* The degree of highway access control can affect the siting and design of adjacent area

development even if it does not affect the total level of development. In general, the more stringent the access control, the more likely the transportation facility will influence the location of land uses only at the access points allowed.

3. *Capacity.* Increased capacity of a transportation facility can affect the dispersal of residential, commercial, and industrial development, particularly if it decreases zone-to-zone travel time.

4. *Travel patterns.* Different project alternatives can have different effects on redistribution of traffic. For example, if an alternative improves traffic capacity in an area, attracting traffic from slower or more crowded streets, increases in traffic volumes can affect the desirability of adjacent land for residential use. On the other hand, if traffic volume or flow at a facility is reduced, leading to reduced congestion, the adjacent area may be rendered capable of supporting additional higher-intensity development.

5. *Traffic control.* When traffic control devices facilitate access to a parcel of land, the land is rendered more desirable for development.

6. *Other.* Improvements in design characteristics, such as on-street parking, shoulders, noise barriers, landscaping, traveler accommodations/amenities, and drainage features, can all have an influence on preserving or altering land uses in adjacent areas.

It should be noted that there may be some overlap of design features in the six categories, and a feature may have characteristics from more than one category. The overlap of design features may also occur between the categories of access management and traffic control; capacity and traffic control; or capacity and travel patterns.

Step 4: Estimate the Potential Changes in Travel Patterns and Accessibility The purpose of this step is to produce forecasts about the direction and volume of travel behaviors, accessibility, and the impact of the transportation project on travel cost. The project may affect the movement of people (e.g., a transit project), goods (an intermodal freight facility), or both (highway projects). This has implications for the size of the impact area. Transit projects, for example, tend to have localized impacts, whereas highway projects tend to have more diffused impacts due to the number and nature of travelers who use the facilities. Travel demand models and freight models are likely to be required for this step of analysis. A discussion of travel demand models is provided in Chapter 3.

Step 5: Estimate the Regional Population and Employment Growth Resulting from Changes in Travel Patterns and Accessibility This step uses local population and employment trends, broader state and national economic industry trends, and economic forecasting models to establish future population and employment trends for various scenarios, and to analyze shifts in population and jobs. The magnitude of future growth depends to a large extent on the study area. If the transportation project is to be implemented in a growing area, it will have the potential of causing significant changes in land uses. In contrast, if the study area is expected to have a low growth rate, there is much less potential for land-use change. Qualitative methods can be used for estimating total population and job growth for any size of geographic area. For larger areas, statistical methods, and regional economic and demographic models are the key tools.

Step 6: Estimate the Potential Changes in Trends or Patterns of Land Use and Development   This step uses information on land availability, cost of development, and attractiveness of various areas to forecast the types, quantities, and location of new development in the study area with and without the project. The analysis conducted in the previous steps provides the base on which the influence of the proposed alternatives are analyzed. In this stage, expert interviews and panels, statistical trend analysis, and/or integrated transportation and land-use models can be used to make forecasts of whether the growth patterns will change due to the project and where development will be located within the study area. For example, the New Hampshire Department of Transportation used the Delphi expert panel approach to assess land-use impacts of the I-93 widening project where the study area could not be well defined, whereas an economic model (REMI) was used for the new Spaulding Turnpike improvement project, where the study area was well delineated (Vanasse Hangen Brustlin and ESNR, 2000).

Step 7: Mitigation Strategies for Adverse Impacts of Transportation Alternatives on Land Use and Development   Mitigation strategies can be applied for avoiding, minimizing, rectifying, reducing, and/or compensating with a substitute any undesirable or environmentally damaging land-use changes from transportation improvements. They can also be applied as part of a broader strategy to ensure that land-use and transportation interactions occur in ways that support economic, social, and environmental goals. Some of the ways to address undesirable land-use or related impacts of transportation investments include:

1. Right-of-way design measures and alignment adjustments to mitigate the adverse proximity effects that result

in dramatic land value changes which eventually lead to undesirable developments.

2. Measures such as noise barriers, tunneling, and elevated roadways where appropriate, help preserve the original features of land, and thus minimize the potential for adverse effects.

3. Access management programs and implementation of statutory access controls (i.e., county highway access controls, driveway permits).

4. Corridor planning activities to address land-use and transportation issues along a highway or transit corridor and near existing, redesigned, or proposed access points, such as interchanges and transit stations.

5. Land purchase or banking. Through this mechanism, a transportation agency budgets for the purchase and protection of important habitat in an area that is not yet experiencing development pressures and where there are no current plans for transportation investment, for habitat protection and other environmental benefits.

6. Land-use policies or controls, including zoning regulations, growth management regulations, transfer of development rights, development fees and exactions, subvision/land division ordinances, comprehensive plans, official maps, and so on. These tools can also be useful to implement land-use growth and development management visions and goals, as well as regional strategies to manage growth. A comprehensive discussion on land-use regulations and controls is provided by Ortolano (1997).

7. Principles of context-sensitive design can provide guidance to best practices for achieving desirable solutions to specific land-use and related problems (CH2M Hill, 2002).

8. Community input and involvement can be very helpful, as they provide the opportunity to identify, discuss, and minimize possible adverse land-use impacts of a transportation project.

Step 8: Present the Results of the Analysis   The entire process is repeated for each transportation alternative. The results can be generated to show the following information:

- Description of each alternative
- Estimated level of impacts of each alternative on land use and development
- Possible mitigation measures for each alternative
- Rankings of the various alternatives against their potential effects

Some examples of land-use impact assessments that have been conducted by various state transportation

Table 16.11 Results for Analysis of the Five Alternatives

Direct Land-Use Impacts (acres)	Alternative				
	I	II	III	IV	V
Farmland	1675	4380	4650	5460	4120
Forests	150	1000	1290	820	1280
Developed lands	200	115	185	120	395
Other (including open water, quarries, bare rock, urban grasses, and shrubland)	100	95	15	20	35
Total	2125	5590	6140	6420	5830

Source: Cambridge Systematics and Bernardin (2003).

agencies as part of an environmental impact statement (EIS) include the state trunk highway 26 EIS prepared for the Wisconsin DOT (2005), the I-69 Evansville-to-Indianapolis corridor study prepared for the Indiana DOT (Cambridge Systematics and Bernardin, 2003), the western bypass major investment study in the Portland metropolitan area prepared for the Oregon DOT (1995), and the southeast corridor light-rail transit EIS in Dallas County, Texas (FTA and DART, 2003).

16.4 CASE STUDIES: LAND-USE IMPACT ASSESSMENT

To illustrate how the analytical framework presented in Section 16.3 could be used in practice, we consider two different projects, a highway corridor construction project involving multiple counties and a light-rail transit project along a highway corridor. Potential land-use impacts of these projects are assessed to fulfill the requirements of an environmental impact statement (EIS).

16.4.1 Evansville-Indianapolis I-69 Highway Project

1. *Review of land-use impacts.* (a) Land-use plans adopted by counties in the study area are to be reviewed and (b) the alternatives are evaluated to determine consistency with the plans and to quantify the direct and indirect impacts of each alternative on different land-use types, specifically forest, farmland, wetlands, and developed areas. For this type of project, the direct impacts are due to the right-of-way needs of the various highway alternatives. Indirect impacts may include impacts related to induced changes in the pattern of land use, population density, or growth rate. A review of the comprehensive plans for the counties within the I-69 corridor (Evansville to Indianapolis) study (Cambridge Systematics and Bernardin, 2003) identified industrial and

commercial growth along transportation corridor in each county including potential interchange locations that could stimulate and enhance these growth patterns; land near interchanges to be used for high-quality nonresidential mixed-use development; and the need for a local service (frontage access) road system to provide access to future commercial and industrial land uses.

2. *Estimation of direct land-use impacts.* Right-of-way needs for the working alignment were estimated, including potential interchanges and rest areas in the total acreage affected. GIS tools were used to identify the land uses such as natural forest areas, farmlands, wetlands, residential, commercial and industrial developed areas, and so on that are likely to be affected by the project. Table 16.11 presents the results of this analysis step for the five alternatives considered in the I-69 corridor (Evansville to Indianapolis) study.

3. *Estimation of indirect land-use impacts.* County-level data on population and employment growth patterns were obtained. These forecasts were assigned to the subcounty areas based on existing development patterns and subsequently, these forecasts provided input to a travel demand model to estimate the changes in travel patterns and accessibility due to each alternative. A regional economic model (REMI) was used to assess the relative economic impacts of each alternative in terms of regional net changes in population and employment, resulting from changes in travel patterns and accessibility. The impacts of the new population and employment "induced" by the improved highway corridor on the transportation network were determined with the use of a transportation planning model. A land-use model was then used to make forecasts of additional land-use changes that would result from this increment of population and employment given the estimates of added population and employment. Finally, models such as those developed by

Table 16.12   Results for the Analysis of the Five Alternatives Considered in I-69

Indirect Land-Use Impacts (acres)	Alternative				
	I	II	III	IV	V
Farmland	455	615	800	770	765
Forests	105	200	345	255	400
Wetlands	15	15	20	20	30
Total	575	830	1165	1045	1195

Source: Cambridge Systematics and Bernardin (2003).

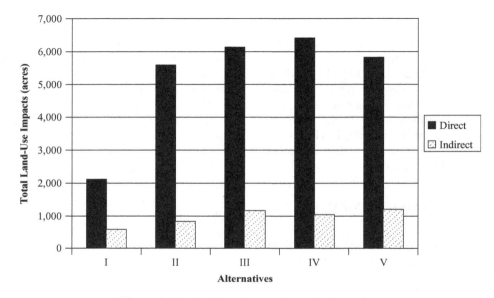

Figure 16.6   Land-use impacts of each alternative.

Hartgen and Kim (1998) [presented in Section 16.2.2(c)] were used to forecast commercial development at potential interchanges. The estimated changes were converted into acreages using standard land-use densities for housing and employment (e.g., residents or employees per hectare). Table 16.12 presents the results of this analysis step for the five alternatives considered in the I-69 corridor (Evansville to Indianapolis) study.

4. *Generation of tables and/or graphs.* The impacts (in acres) of the right-of-way needs (direct impacts) for each alternative as well as the acreages that are estimated to be converted to residential, commercial, and industrial land use as a result of the highway (indirect impacts) are shown in Figures 16.6 and 16.7. The alternatives were ranked by their potential land-use impacts, a rank of number "1"

for the alternative with the least area of land taken (see Table 16.13).

16.4.2 Light-Rail Transit Project

To illustrate the analytical steps involved in the land-use impact assessment for a transit project, the case associated with the construction and operation of a light-rail transit (LRT) project to improve transit service in the southeastern corridor of the Dallas Area Rapid Transit (DART) service area is considered. An analysis of a no-build alternative is done to provide a baseline comparison for the LRT alternative. The no-build alternative includes the highway and transit facilities that already exist in the southeastern corridor. The LRT alternative consists of an approximate 10.2-mile extension of LRT service, connecting downtown Dallas with five other communities.

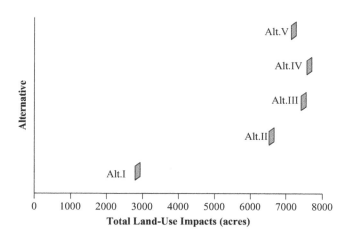

Figure 16.7 Total land-use impacts of each alternative.

The capital cost for the build alternative (LRT) is estimated to be approximately $450 million (2002 dollars). The steps followed in the assessment (FTA and DART, 2003) of land-use impacts of the LRT vs. no-build alternative are discussed below.

1. *Define the project area.* The area within 1 mile of the build alternative (LRT) was defined as the study corridor.

2. *Determine the consistency with land-use plans.* This analysis step determines the consistency of each alternative with the local land-use plans and policies such as the *growth policy plan* implemented by the city of Dallas. This long-range plan includes development policies such as density bonuses necessary to support higher levels of development. The no-build alternative would not be consistent with this plan because it would not support the recommended increased development potential of the corridor. On the other hand, the LRT alternative would be consistent because it will utilize the development potential that would be stimulated by LRT stations.

3. *Estimate the regional land-use and development impacts.* Rapid population and employment growth and a disproportionate growth in VMT have led the region, through the North Central Texas Council of Governments (NCTCOG), to adopt policies supporting sustainable development. Whether both alternatives support the policies for sustainable development as outlined by NCTCOG, was examined. It was found that the no-build alternative would have no effect on regional land use and development. The LRT alternative may shift some types of new development and redevelopment from outlying areas to transit station areas. As indicated in Table 16.10, property values around LRT stations would rise as a result of the expansion of the light rail system. It can therefore be expected that the LRT project would improve the mobility and quality of life for residents, and also would increase the region's attractiveness to businesses considering locating within the region.

4. *Estimate the corridor-level land-use and development impacts.* This analysis step examines whether current land-use trends in the study area would continue with the LRT. A great variety of types of land uses were identified in the study corridor: residential, office, and commercial development. No changes in current land uses are anticipated with the no-build alternative. On the other hand, the presence of a major and highly accessible transit service such as LRT can have long-term impacts on the distribution and density of land uses in the area. The land-use effects of LRT include introduction of fixed LRT station facilities and services, positive impacts on land uses and property values, and attraction of new development, employment and residents in the corridor.

5. *Estimate the station vicinity impacts on land use.* This step assesses the direct and indirect land-use impacts near LRT stations in the study corridor as a result of the alternatives. Direct impacts on land use are readily identified with the station location. Indirect impacts on land use generally can be identified through assumptions

Table 16.13 Results of Direct Impacts of Each Alternative

Land-Use Impacts (acres)	Alternative				
	I	II	III	IV	V
Direct land-use impacts	2125	5590	6140	6420	5830
Indirect land-use impacts	575	830	1165	1045	1195
Total	2700	6420	7305	7465	7025
Rank by total land taken	1	2	4	5	3

Source: Cambridge Systematics and Bernandin (2003).

Table 16.14   Direct Effects in Relation to Potential Acquisitions

Location	Current Property Use	Approximate Acreage/ % of Parcel Affected/ E or A[a]
A	Commercial	0.02/4%/E
B	Parking	0.01/17%/E, A
C	Multi-family residential	0.4/100%/A
D	Single-family residence	0.01/4%/A
E	Vacant	0.2/100%/A
F	Latino cultural center	0.1/16%/E, A

Source: FTA and DART (2003).
[a] *E* easement portion of the parcel will be acquired; *A* portion or complete parcel will be acquired.

about the capacity for change; in this case, these effects were assumed to occur within 1500 ft of the station. With the no-build alternative, most of the land uses would not change as a result of the transit centers. However, both direct and indirect effects will occur with implementation of the LRT alternative. Direct effects will occur in relation to acquisitions and considered resulting from the construction of LRT stations and related access facilities (i.e., bus bays, park-and-ride lots), as shown in Table 16.14. Indirect effects will occur as land development or redevelopment actions take place in response to the presence and availability of LRT service.

6. *Develop mitigation measures.* Measures proposed for mitigating the impacts of potential LRT right-of-way acquisitions are considered in this step. Property owners will be paid fair market value for property acquired. In cases where relocation will be necessary for right-of-way acquisitions for stations, comparable facilities for relocation existing in the area will be reviewed with each business owner.

SUMMARY

Land use and transportation are inextricably linked. Land use generates activities that create a demand for travel, and travel generates the need for new facilities, which in turn increases accessibility and attracts further development. Transportation improvements make land more accessible and thus increase the likelihood that it will be developed or redeveloped, and transportation agencies are increasingly being asked to assess the land-use impacts development impacts of their projects, and to mitigate any adverse impacts. Land-use impact analysis generally takes the form of comparing future land use with and

without the transportation project in question. A wide range of analysis tools and strategies can be used, and these range from common survey techniques, comparative case studies and basic quantitative analysis, to more complex models requiring specialized software and training. Different tools and strategies are applied at different stages in the analysis process depending on the quality and availability of resources, and level of desired sophistication of the analysis. When a transportation project leads to undesirable land-use effects, mitigation strategies are employed to address such impacts. Also, these strategies are applied as part of a broader strategy to ensure that land use–transportation interactions occur in ways that support economic, social, and environmental community goals.

EXERCISES

16.1. A proposed transportation system improvement would provide faster travel to a downtown commercial district from the northern suburbs of the city. Apply the gravity model described in Section 16.2.2 (a) to measure the accessibility improvement to the downtown commercial area (in terms of change in travel times) given the data on households and travel times shown in Table EX16.1.

16.2. Owners of properties that are located close to a proposed rail transit station are concerned about the project's effect on residential and commercial property values. Table 16.7 shows past findings as to the impact of transportation system changes on residential and commercial property values. Assuming that similar effects would occur near the project in question, conduct a comparative

Table EX16.1 Data on Accessibility to the Commercial Area

Market Area: Place of Residence	Total Households	Average Travel Time to Commercial Area (min.)	
		Base Case	With Transit System Proposed
Downtown	15,000	25	20
Northern suburb	9,000	55	45
Southern suburb	4,000	33	33
Eastern suburb	6,000	35	32
Western suburb	11,000	20	20

analysis to estimate the property value effects of the new rail transit system in the area. State any other assumptions that you make. Are the property owners' concerns justified?

16.3. Consider the case of a new interchange in a rural area. You are hired to forecast the potential commercial development (i.e., gas stations, convenience stores, fast-food restaurants, sit-down restaurants, and motels) that would be generated due to the project. Use the models developed by Hartgen et al. (1998) [presented in Section 16.2.2(c)]. Conduct a sensitivity analysis applying the elasticity values presented in Table 16.9. How different is the effect of the total number of development units at the exit on the number of gas stations and fast-food restaurants? How different is the effect of the total number of development units at nearby exits on the number of motels and sit-down restaurants? In each case, compare and discuss the results. State any assumptions that you make.

REFERENCES[1]

*Al-Mosaind, M.A., Dueker, K. J., Strathman J. G. (1993). Light-rail transit stations and property values: A hedonic price approach, *Transp. Res. Rec. 1400*, Transportation Research Board, National Research Council, Washington, DC.

Alonso, W. (1960). A theory of the urban land market, Papers Proc. *Reg. Sci. Assoc.*, Vol. 6, www4.trb.org/trb/crp.nsf/All+Projects/NCHRP+25-25#3. Accessed Oct. 2005.

Armstrong, R. J. (1994). Impacts of commuter rail service as reflected in single-family residential property values, *Transp. Res. Rec. 1466*, Transportation Research Board, National Research Council, Washington, DC.

Bagby, G. (1980). Effects of traffic flow on residential property values. *Journal of American Planning Association*, Vol. 46, No. 1, Chicago, IL.

Bajic, V. (1983). The effects of a new subway line on housing prices in metropolitan Toronto. *Urban Studies*, Vol. 2., Glasgow, Scotland.

*Barton-Aschman Associates, Inc., Hammer, Siler, George Associates (1990). *Research Triangle Regional Transit/Land-Use Study*, Public Transportation Division, North Carolina Department of Transportation, Raleigh, NC.

*Beimborn, E., Horowitz, A., Vijayan, S., Bordewin, M. (1999). *An Overview: Land Use and Economic Development in Statewide Transportation Planning*, Federal Highway Administration, U.S. Department of Transportation, Washington, DC.

Bein, P. (1997). *Monetization of Environmental Impacts of Roads*, Ministry of Transportation and Highways, Victoria, BC, Canada, www.th.gov.bc.ca/bchighways. Accessed Oct. 2005.

Benjamin, J. D., Sirmin G. S. (1996). Mass transportation, apartment rent and property values, *J. Real Estate Res.* Vol. 12, No. 1.

Boarnet, M. G., Chalermpong S. (2001). New highways, house prices, and urban development: a case study of toll roads in Orange County, CA. *Housing Policy Debate*, Vol. 12, No. 3, Fanny Mae Foundation, Washington, DC.

Bollinger, C., Ihlanfeldt, K., Bowes, D. (1996). *Spatial variation in office rents within the Atlanta region*. Policy Research Center, Georgia State University, Atlanta, GA.

*Boyce, D., Allen, B., Mudge, R., Slater, P., Isserman, A. (1972). Impact of rapid transit on suburban residential property values and land development. Final Report to U.S. Department of Transportation, Washington, DC.

*Cambridge Systematics, Bernardin, Lochmueller & Associates (2003). *I-69 Evansville to Indianapolis, Indiana Final Environmental Impact Statement*. Indiana Department of Transportation and Federal Highway Administration, http://www.deis.i69indyevn.org/FEIS/. Accessed Jan. 15, 2006.

Cavalli-Sforza, V., Ortolano, L. Dajani, J. S. Russo, M. V. (1982). *Transit Facilities and Land Use: An Application of the Delphi Method*. Program in Infrastructure Planning and Management, Stanford University, Stanford, CA.

Cervero R. (1996). Transit-based housing in the San Francisco bay area: market profiles and rent premiums, *Transp. Quart.* Vol. 50, No. 3, Eno Foundation Transportation, Inc., Washington, DC.

Cervero, R. (2002). Benefits of proximity to rail on housing markets: experiences in Santa Clara county, *J. of Pub. Transp.*, Vol. 5, No. 1, National Center for Transit Research, Tampa, FL.

*Cervero, R., Duncan, M. (2002). Transit's value added: Effects of light commercial rail services on commercial land values, Presented at 81st Annual Meeting of the Transportation Research Board, National Research Council, Washington, DC.

*Cervero, R. (2003). Road expansion, urban growth, and induced travel: A path analysis, *J. Am. Plan. Assoc.*, Vol. 69, No. 2.

*CH2M Hill (2002). *A Guide to Best Practices for Achieving Context Sensitive Solutions*, NCHRP Report 480. Transportation Research Board, National Research Council, Washington, DC.

[1]References marked with an asterisk can serve as useful references for assessing land-use impacts of transportation projects.

*Chen, H., Rufolo, A., Dueker, K. (1998). Measuring the impact of light rail systems on single family home values: an hedonic approach with GIS Application, *Transp. Res. Rec. 1617*, Transportation Research Board, National Research Council, Washington, DC.

*D'Ignazio J., Hunkins, J. (2005). *Integrating Planning and NEPA: Linking Transportation and Land Use Planning to Indirect and Cumulative Impacts*, prepared by the North Carolina State University for the North Carolina Department of Transportation Raleigh, NC, presented at the 85th Transportation Research Board Annual Meeting, Washington, DC.

*Diaz, R.B. (1999). Impacts of rail transit on property values. American Public Transportation Association, Proceedings of the Conference on Commuter Rail/Rapid Transit, Toronto, Canada.

Ewing, R., Cervero, R. (2001). Travel and the built environment: a synthesis. *Transp. Res. Rec. 1780*. National Academy Press. Washington DC.

Forkenbrock, D., Sheeley J. (2004). *Effective Methods for Environmental Justice Assessment*. NCHRP Report 532. Transportation Research Board, National Research Council, Washington, DC.

*Forkenbrock, D.J., Weisbrod G. (2001). *Guidebook for Assessing the Social and Economic Effects of Transportation Projects*, NCHRP Report 456, Transportation Research Board, National Research Council, Washington, DC.

FTA and DART (2003). *Southeast Corridor Light Rail Transit in Dallas County, Texas: Final Environmental Impact Study and Section 4(f) Statement*, Federal Transit Administration and Dallas Area Rapid Transit Dallas, TX.

*Haider, M., Miller, E.J. (2000). Effects of transportation infrastructure and locational elements on residential real estate values—Application of spatial autoregressive technique. *Transp. Res. Rec. 1722*. Transportation Research Board, National Research Council, Washington, DC.

Hartgen D.T., Kim J.Y. (1998). Commercial development at rural and small town interstate exits. *Transp. Res. Rec. 1649*, National Academy Press, Washington, DC.

Hess, D. B., Almeida T. M. (2006). Impact of proximity to light rail rapid transit on station-area property values in Buffalo. Transportation Research Board 85th Annual Meeting, National Research Council, Washington, DC.

*Hirschman, I., Henderson M. (1990). Methodology for Assessing Local Land Use Impacts of Highways, *Transp. Res. Rec. 1274*, Transportation Research Board, National Research Council, Washington, DC.

Hughes, W., Sirmans, C.F. (1992). Traffic externalities and single-family house prices. *J. Reg. Sci.*, Vol. 32, No. 4 (November).

*ICF Consulting (2005). *Handbook on Integrating Land Use Considerations into Transportation Projects to Address Induced Growth*, prepared as part of *NCHRP Project 25-25, Task 3:* Analysis of Assessment and Mitigation Strategies for Land Development Impacts of Transportation Improvements, National Cooperative Highway Research Program, Transportation Research Board, National Research Council, Washington, DC.

Johnston, R. A., Rodier, C. J. Choy, M., Abraham, J. (2000). *Air Quality Impacts of Regional Land Use Policies: Final Report for the Environmental Protection Agency*. Department of Environmental Science and Policy, University of California–Davis, Davis, CA.

*Kockelman, K., Siethoff B. (2002). Property values and highway expansions: an investigation of timing, size, location, and use effects. *Transp. Res. Rec. 1812*. Transportation Research Board, National Research Council, Washington, DC.

Landis, J., Loutzenheiser, D. (1995). *BART @ 20: BART access and office building performance*. Working Paper 648. Berkeley, CA: Institute of Urban and Regional Development, University of California, Berkeley, CA.

Langley, J. C. (1981). Highways and property values: The Washington beltway revisited. *Transp. Res. Rec. 812*. Transportation Research Board, National Research Council, Washington, DC.

Levine, J. (1998). Rethinking accessibility and jobs-housing balance. *J. Am. Plan. Assoc.*, Vol. 64, No. 2 (Spring).

Litman, T. (2002). Land-use impacts in Sect. 5.14, *Transportation Cost and Benefit Analysis: Techniques, Estimates and Implications*. Victoria Transport Policy Institute, Victoria, BC, Canada, www.vtpi.org.

*Louis Berger Group, Inc. (2004). *Indirect and Cumulative Impact Assessment Guidance: Integrated NEPA/SEPA/401 Eight-Step ICI Assessment Process*. North Carolina Department of Transportation and North Carolina Department of Environment and Natural Resources. www.ncdot.org/doh/preconstruct/pe/NEPA401Guidance.doc. Accessed Jan. 2006.

Lowry, I. (1964). *A Model of Metropolis*, RM-4035-RC, Rand Corporation, Santa Monica, CA.

Lowry, I. (1988). *Planning for Urban Sprawl*. Spec. Rep. 220, Transportation Research Board, National Research Council, Washington, DC.

Lutzenberger, J. (1985). The World Bank's Polonoroeste Project: a Social and Environmental Catastrophe. *Ecologist*, Vol. 15.

Masser, I. (1972). *Analytical Models for Urban and Regional Planning*, David and Charles, Newton Abbot, UK.

Nelson, A. C. (1999). Transit stations and commercial property values: A case study with policy and land-use implications. *J. Pub. Transp.*, Vol. 2, No. 3. Tampa, FL.

Oregon DOT (1995). *Western Bypass Study Alternatives Analysis*. Oregon Department of Transportation and Washington County, Portland, OR.

Ortolano, L. (1997). *Environmental Regulation and Impact Assessment*. Wiley, New York.

Palmquist, R.B. (1982). Impact of highway improvements on property values in Washington State. *Transp. Res. Rec. 887*. Transportation Research Board, National Research Council, Washington, DC.

*Parsons Brickerhoff Quade and Douglas (1999) *Land Use Impacts of Transportation: A Guidebook*, NCHRP Report 423A, Transportation Research Board, National Research Council, Washington, DC.

Pushkarev, B., Zupan, J. (1977). *Public Transportation and Land Use Policy*, Indiana University Press, Bloomington, IN.

Rybeck, W. (1981). Transit-induced land values. *Economic Development Commentary*, Council for Urban Economic Development, Washington, DC.

*Sedway Group (1999). Regional impact study. Commissioned by Bay Area Rapid Transit District (BART), Rail Transit and Property Values, Transit Resource Guide, American Public Transportation Association, www.apta.com/research/info/briefings/briefing_1.cfm, Accessed Dec 2006.

Smith J.J., T. A. Gihring (2006). Financing Transit Systems through Value Capture: An Annotated Bibliography. Victoria Transport Policy Institute (www.vtpi.org), BC, Canada.

*Stanley M. (2006). *NCHRP 25-25 Task 11 Indirect and Cumulative Impacts Analysis*. Requested by American Association of State Highway and Transportation Officials (AASHTO)—Standing Committee on the Environment. National Cooperative Highway Research Program, Transportation Research Board, National Research Council, Washington, DC. Available at: http://www4.trb.org/trb/crp.nsf/reference/boilerplate/Attachments/$file/25-25(11)_FR.pdf

ULI (1979). *Joint Development: Making the Real Estate—Transit Connection*. Urban Land Institute, Washington. DC.

USEPA (2003). *EPA's Smart Growth INDEX in 20 Pilot Communities: Using GIS Sketch Modeling to Advance Smart Growth*. 231-R-03-001. Office of Policy, Economics, and Innovation (1808T), Development, Community, and Environment Division, U.S. Environmental Protection Agency, Washington, DC. www.epa.gov/dced/pdf/Final_screen.pdf.

Vanasse Hangen Brustlin, ESNR (2000). *Interstate 93 Improvements Salem to Manchester, New Hampshire*. New Hampshire Department of Transportation, Federal Highway Administration, U.S. Department of Transportation, Washington, DC.

Voith, R. (1993). Changing capitalization of CBD-oriented transportation systems: evidence from Philadelphia, 1970–1988. *J. Urb. Econ.*, Vol. 33, No. 3. Elsevier, Amsterdam, The Netherlands.

Weinberger, R. R. (2000). Commercial property values and proximity to light rail: calculating benefits with hedonic price model. Presented at the 79th Annual Meeting of the Transportation Research Board, National Research Council, Washington, DC.

Weinberger, R. R. (2001). Commercial rents and transportation improvements: Case of Santa Clara County's light rail. WP00RW2, Lincoln Institute of Land Policy, Cambridge, MA.

*Wisconsin DOT (1996). *Indirect and Cumulative Effects Analysis for Project Induced Land Development*. Technical Reference Guidance Document. Wisconsin Department of Transportation, Madison, WI.

Wisconsin DOT (2005). *Wisconsin State Trunk Highway 26 Environmental Impact Statement*, Wisconsin Department of Transportation, Madison, WI, and Federal Highway Administration, Washington, DC.

CHAPTER 17

Social and Cultural Impacts

Things do not change; we change.
—*Henry David Thoreau (1817–1862)*

INTRODUCTION

Compared to most other types of transportation system impacts, social and cultural impact assessment is a relatively inexact science because social environments differ from place to place and the impacts depend on the manner of social change interpretation, the level of anticipation, and the resilience of the affected population. FHWA (1982) defines *social impacts* as the destruction or disruption of human-made resources, social values, community cohesion, and availability of public facilities and services; displacement of people, businesses, and farms; and disruption of desirable community and regional growth. Another definition by IOCGP (2003) is "the consequences to human populations of any public or private actions that alter the ways in which people live, work, play, relate to one another, organize to meet their needs and generally cope as members of society." A Federal Transit Administration document describes social effects as the changes in physical layouts, demographics, and sense of neighborhood in local communities (FTA, 2005). According to the Section 106 Compliance Plan of the National Historic Preservation of 1966, a transportation project is considered to have adverse effects on cultural environment if "it alters, directly or indirectly, any characteristics of a historic property in a manner that would diminish the integrity of the property's location, design, setting, materials, workmanship, feeling, or association." The emphasis on sociocultural consideration of transportation system impact in evaluation and decision making was provided by legislative action such as the 1970 Federal Highway Act and the 1970 National Environmental Policy Act (NEPA) and was fostered by a number of executive orders in the 1990s. In the context of the developing world, multilateral lending agencies such as the World Bank require borrower countries to undertake social impact assessments to ensure that funded projects will yield significant favorable impacts on the lives of people in those countries in terms of sociocultural, institutional, historical, and political effects (World Bank, 2003). Starting in 1968, when the World Bank stressed the issue of poverty alleviation, social analysis has gained a prominent role in the agenda of international lending agencies and development organizations, including the United Nations, the Asian Development Bank, and the European Economic Commission.

Distributive effects, analysis of which is an important aspect of sociocultural impact assessment, can refer to the variation in impact severity of a transportation project as one moves away from the project area but is more often taken to mean the variation in the impact severity across community groups, population groups, or/and ethnic groups in the overall area where the transportation project is located (Chatterjee and Sinha, 1976). Distributive effects can also include how such distance- and community group–based variations change over time.

The analysis of distributive effects of transportation projects, especially with respect to sociocultural impacts, is particularly critical when the project (1) requires unusually large amounts of right-of-way in an urban area; (2) would involve the displacement of a large number of households, businesses, community amenities, historic districts, and landmarks; (3) conflicts with local transportation or land-use plans; (4) would cause a significant change in traffic characteristics (volume, speed, percentage trucks, etc.); and (5) would unduly and unfairly reduce the welfare of vulnerable segments of the population.

The World Bank (2003) argues that social impact assessments should be a continuous process occurring throughout the cycle of project development, from appraisal (or planning) stages to implementation. The bank identifies five dimensions of inquiry, or *entry points*, for social impact assessments: social diversity and gender, institutions, rules and behavior, stakeholders, participation and social risk, and states that the relative scope of each dimension depends on the circumstances and context of a particular project. Many studies have been carried out to examine the relationships between the spatial distribution of the environmental disbenefits of transportation and sociodemographic attributes of affected communities, particularly those that are disadvantaged and marginalized.

17.1 MECHANISMS OF TRANSPORTATION IMPACTS ON THE SOCIAL AND CULTURAL ENVIRONMENTS

The impact of transportation projects and policies may cause desirable or undesirable impacts on the social or cultural capital of an area in three major ways: direct, indirect, or/and cumulative.

17.1.1 Direct Impacts

(a) *Relocation Effects*   The acquisition of rights-of-way for new or expanding facilities requires additional relocation of houses, businesses, and community facilities. The loss of a family home and real estate, leaving a familiar neighborhood, or the physical, emotional, and financial stresses of moving can be overwhelming for families or individuals. Also, relocations dismantle the social fabric by removing the formal and informal social networks established by residents for physical or psychological support. Businesses that typically suffer the effects of relocation include grocery shops, banks, and shopping centers, and community facilities that include schools, churches, and recreation areas.

If a project requires relocation of a disproportionate number of businesses and community facilities, residents will be forced to seek services or even jobs outside their communities, resulting in increased commutes to access the services offered by the relocated facilities. As the households are relocated, community facilities suffer reduced demand or enrollment, decreased operational cost-effectiveness, and ultimately, possible closure.

Population segments that are most sensitive and susceptible to relocation impacts are the elderly, low-income families, long-time residents and homeowners, handicapped persons, and minority and ethnic group members. For residents who move frequently, relocation is relatively less harmful than it is for residents who are more stable and established in the community. Also, relocation is more disruptive for residents with school-aged children, especially if they have to transfer to new schools.

Often, a severe problem is the lack of available and suitable housing for dislocated persons. With regard to business relocations, a major issue is whether there is adequate land available to which firms can relocate and remain economically viable (Caltrans, 1997). Even where vacant land with proper zoning may be available, the new location may not meet the specific needs of the business in question. Large-scale transportation expansion projects typically lead to displacement of businesses that rely on highway traffic for patronage, such as gas stations, motels, and restaurants. Very often, these businesses fail to find other suitable locations along a busy roadway

and are thus unable to attract adequate customers to stay in business. In most instances, however, the business clientele is quickly absorbed by similar businesses in the immediate area. Generally, if businesses relocate to other areas in the community and do not suffer loss of viability, the unemployment impacts are only temporary. However, relocation of businesses to areas outside of the community can lead to unemployment and the concomitant loss of multiplier effects. Also, even when businesses relocate to another area within the community, some workers may not be willing to travel or relocate to the new location, thus the business would lose employees. Also, businesses that have established loyal clientele over a period of time may need time to reestablish their customer base when they relocate to new areas. For national chains, such time for reestablishing clientele is often minimal compared to the time needed by small businesses.

With regard to cultural resources, alignment of the transportation facility at sensitive areas necessitates relocation of such resources. This can cause physical destruction of all or part of the property, alteration of the character of a cultural resource, removal of a cultural resource from its original location, or negligent handling of a cultural property that causes its deterioration or destruction.

(b) *Barriers*   New, widened, or extended line facilities (roads and rail tracks) affect the structure, function, and social pattern of the surrounding neighborhoods because they cause separation of households, businesses, and community facilities, or reduce access between such entities. For example, after a project is implemented, it may be more difficult or impossible to access social facilities by foot or cycle, and vehicle trips to such facilities may take more time. The transportation facility can constitute a physical and psychological barrier that is difficult to cross, particularly for the elderly, young children, and other residents who travel on foot or by bicycle. The barrier effect often leads to isolation of community facilities, services, and institutions.

(c) *Integrative Features*   Certain transportation improvements involve, include, or result in provision of increased pedestrian walkways, bikeways, and related facilities. These projects serve to integrate the community and therefore have beneficial impacts on the sociocultural environment of an area.

17.1.2 Indirect Impacts

Indirect impacts arise not from the physical presence of the transportation facility but from its increased usage due to travel generated or induced. Increased traffic can lead to psychological encumbrances that reduce the extent and quality of social interaction in the community. Other indirect effects of a transportation project on a community include increased noise, dust, and debris, and reduced safety of pedestrians, particularly children. Furthermore, transportation projects can reduce the number or locations of parking spaces temporarily (during construction when they are used up by construction personnel) or permanently (when parking spaces are taken for the new transportation facility). Loss of parking for customers and delivery trucks can disrupt the operation of community facilities and services such as schools, hospitals, and businesses including restaurants and small retailers that depend on adjacent on-street parking for delivery trucks. In rural areas, transportation projects may lead to significant social impacts as they open up these areas to new settlements. The new settlers may face difficulties in social adjustment in the new area. On the other hand, an influx of new settlers can dramatically change the demographics of small rural communities served by the new facility, which can lead to loss of the community identity and erosion of traditional value systems and lifestyles. Experience of the World Bank in developing countries suggests that tribal societies, particularly those fully or partially isolated from outside influences, are sensitive primarily to the influx of new settlers and other external intrusions brought about by transportation projects (World Bank, 2003). Also, improved access fostered by a transportation project can lead to the opening up of rural areas, affecting their social character. For example, with access to urban areas, independent farmers and hunters may opt to migrate there to become wage earners or petty retailers (Sinha et al., 1989).

17.1.3 Cumulative Impacts

A third type of impact comprises the combined effects produced as seemingly minor project impacts assume greater significance when they are considered together with the effects of other past, present, or reasonably foreseeable future actions (Florida DOT, 2000). Also, there can be a counterbalancing effect of certain beneficial and adverse impacts that may be direct or indirect.

17.2 TARGET FACILITIES AND GROUPS, AND PERFORMANCE MEASURES

17.2.1 Target Facilities and Groups

In assessing the sociocultural impact of transportation projects, the analyst should first identify the target facilities and populations that would be affected. The facilities typically considered include schools; religious institutions; playgrounds, parks, and recreational areas; hospitals, clinics, and other medical facilities; residential and social facilities for the elderly; social service agencies; and

libraries. Generally, all persons within the impact area are considered in the analysis. When environmental justice is an issue, focus should be placed on certain specific population segments, such as the elderly, disabled, nondrivers and transit-dependent persons, minority groups, and low-income or poverty-stricken individuals and households. It is useful to note that poverty extends beyond income deprivation to include deprivation of basic capabilities (Sen, 2000). Other target groups include those that are vulnerable to conflict, violence, or economic shocks.

17.2.2 Performance Measures

Performance measures for social and cultural impacts may differ in scale, severity, or intensity depending on the community resources available, the nature of the community, and so on. Specifically, performance measures may differ in spatial extent or temporal duration. For example, on the basis of certain performance measures, communities may "return to normalcy" in a relatively short time after the project implementation, while for other performance measures, return periods may be longer. For a given performance measure, the desirability (beneficial or adverse) and the intensity of impacts may vary among different communities and population groups, depending on their resilience, diversity, level of sociocultural capital, and so on. For example, a transportation improvement may produce generally positive sociocultural effects for some groups or communities but may have adverse impacts for others. In many cases, low-income and other groups with relatively little effective electoral representation find themselves at the short end of such situations. These groups are often affected disproportionately by strategic national and regional plans and decisions that typically culminate in outcomes such as relocations to make way for transportation projects. Selected performance measures that could be useful in sociocultural impact assessments are provided in Table 17.1.

Community cohesion describes the social network and actions that provide satisfaction, security, camaraderie, and identity to members of a community or neighborhood. For many people, community cohesion is vital to the success of family life and contributes to feelings of satisfaction and fulfillment in community life (Forkenbrock and Weisbrod, 2001). In a bid to facilitate objective evaluation of transportation alternatives, the analyst may be tempted to establish a mathematical index or rating to describe the level of performance measures for social and cultural impacts. However, a great deal of circumspection is recommended in such efforts. Caltrans (1997) reports that for a quarter of a century, several transportation agencies countrywide have used a *stability index* to measure levels of community cohesion. Such indices are

Table 17.1 Performance Measures for Social and Cultural Impacts

Performance Category	Performance Measure
Social	
Population change	Population size density and change
	Ethnic racial composition and distribution
	Relocating people
	Influx and outflows of temporaries
	Percentage of seasonal residents
Community and institutional structures	Voluntary associations
	Interest-group activity
	Size and structure of local government
	Historical experience with change
	Employment and income characteristics
	Employment equity of disadvantaged groups
	Local and regional and national linkages
	Industrial and commercial diversity
	Presence of planning and zoning
Political and social resources	Distribution of power and authority
	Conflict newcomers and old-timers
	Identification of stakeholders
	Interested and affected parties
	Leadership capability and characteristics
	Interorganizational cooperation
Community and family changes	Perceptions of risk, health, and safety
	Displacement and relocation concerns
	Trust in political and social institutions
	Residential stability
	Density of acquaintanceships
	Attitudes toward proposed action
	Family and friendship networks
	Concerns about social well-being
Community resources	Change in community infrastructure
	Indigenous populations
	Change in land-use patterns
Cultural	
Community cultural resources	Historical buildings and districts
	Sacred and religious buildings and sites
	Archaeological sites and treasures

Source: IOCGP (2003).

based on the length of time that residents have lived in a community; the longer the length of time in the community, the greater the stability index. The stability index computation may be biased because it may exclude renters, who often are low-income persons, and minorities, or frequent movers, but nevertheless are a part of a cohesive community. A number of past studies have utilized cultural performance measures. For example, in the final environmental impact statement for the Interstate 69 Evansville-to-Indianapolis highway (Cambridge Systematics and Bernardin, 2003), the criteria for assessing the cultural impacts of each alignment included the possibility and extent of encroachment of archeological sites, areas of historic schools, Amish communities, Mennonite communities, and establishments registered (or eligible to register) with the National Register of Historic Places.

In the developing world, the sociocultural impacts of transportation and other major projects can be expressed in terms of the change in social and economic assets and the capabilities of people, particularly the low-income and vulnerable, and the extent to which the project helps to reduce social tensions, conflict, and political unrest. That is not to say that transportation projects can prevent armed conflicts. However, transportation projects can help address issues of poverty, inequality, and lack of cross-ethnic interactions that are among the root causes of ethnic tensions and unrest.

17.2.3 The Issue of Poverty Alleviation in Developing Countries

A country's ability to fully exploit its potential to achieve economic development and to improve the welfare of its residents, particularly those with low incomes, is closely linked to the state of its transportation system (World Bank, 2002). Transportation plays a critical role in developing the economy and strengthening the sociocultural fabric and is also critical for day-to-day subsistence: Poor households depend heavily on transportation facilities to move their water, fuel, farm produce, and fertilizer efficiently and also to have access to markets, jobs, and health clinics. The pervasive and complex influence of transportation makes it difficult to track and measure the exact and ultimate impacts of transportation interventions on the welfare of low-income households and communities. However, some general patterns have clearly emerged. The World Bank, for instance, has found that children in poor households, particularly daughters, are more likely to attend primary and secondary schools when the region has good and affordable transportation systems and services. Also, improved transportation systems facilitate the participation of low-income residents in social, cultural, and political processes and thereby generally help such people to accumulate adequate human, physical, financial, and social assets to get out of poverty. King and Alderman (2001) reported that investments that reduce distance or time to school contribute to increased female enrollment rates by reducing the opportunity cost of schooling for girls; in Ghana, India, Malaysia, Pakistan, Peru, and the Philippines, for example, the distance to school was found to be a greater deterrent to a girl's schooling than that of a boy. Similarly, increasing access to local health care facilities reduces the time that women and girls need to spend on in-home care for sick or aging family members. The World Bank's *A Sourcebook for Poverty Reduction Strategies* (World Bank, 2002) provides transportation decision makers with guidelines for using poverty reduction as a performance measure in evaluating transportation projects in developing countries.

17.3 EQUITY AND ENVIRONMENTAL JUSTICE CONCERNS

In a perfect world, all persons, irrespective of their social, economic, or cultural background, would incur similar proportions of benefits accrued and costs incurred arising from a transportation project. In the real world, however, transportation projects result in very different distributions of adverse and beneficial impacts that are dispersed spatially as well as across various communities. For example, the adverse impact of most air pollutants is most intense at the relatively small area that is immediately proximal to the facility. Also, some communities, by virtue of their proximity to the project or because of their unique sociocultural practices or income status, may suffer more adverse consequences relative to the benefits they accrue from the project. From these issues arise the concept of environmental justice which seeks to promote basic human values of fairness and human rights. Environmental justice can be formally defined as the fair treatment and meaningful involvement of all people, regardless of race, color, national origin, or income with respect to the development, implementation, and enforcement of environmental laws in general (Bass, 1998; Quan, 2002). In the context of transportation, environmental justice refers to the distribution of benefits and costs arising from transportation projects, programs, and policies (Forkenbrock and Sheeley, 2004). Also, TRB (2002) refers to environmental justice as the equitable distribution of both negative and positive ecological, economic, and social impacts across racial, ethnic, and income groups.

Environmental justice appears to be rooted in two elements of *Rawls' theory of justice* (Rawls, 1999; Khisty, 1996; Alsnih and Stopher, 2003): (1) all social primary goods, such as liberty, opportunity, income, and wealth, are to be distributed equally; and (2) if such goods are not distributed equally, they are to be distributed to favor the disadvantaged. FHWA has consistently stressed the importance of environmental justice considerations in local planning and gave general guidelines to MPOs to ensure such considerations in planning. Consistent with U.S. DOT principles, there are three core principles of environmental justice: (1) avoid, minimize, or mitigate disproportionately high and adverse human health or environmental, social, and economic effects on any population segment; (2) ensure full and fair participation by communities potentially affected by the transportation decision-making process; and (3) prevent the denial of, reduction of, or significant delay in the receipt of benefits by any segment of the population.

Environmental justice principles and issues may differ from country to country. In the United Kingdom, for instance, environmental justice issues arise because problems of the environment are a component of social exclusion and are therefore a component of social justice issues (Agyeman, 2001). In China, models in environmental justice simply relate to occupational and peasantry status (Alsnih and Stopher, 2003), as the society is not as fragmented by race, ethnicity, and probably income as it is in many Western countries. However, with ongoing rapid changes in social hierarchy and as the citizens become increasingly aware of their environment and rights, the issue of environmental justice is expected to become increasingly prominent in China (Quan, 2002). On the global scene, the changing economic conditions are expected to yield greater gaps in income levels and will probably lead to sociospatial segregation (Wessel, 2000). Coupled with the increasing awareness of personal and community rights, such trends are likely to result in increased agitation and prominence of environmental justice issues worldwide. In a study in Oslo, Rietveld (2003) identified two roles of equity in transportation policies: (1) as a side effect (unintended) of policies and projects to address transportation problems and (2) as the primary motivation for a transportation project whose explicit aim is to improve transportation infrastructure in underdeveloped regions or communities. In considering environmental justice issues in sociocultural impact analysis, the unintended effects of the transportation project should be identified and given due attention (Goodwin, 2003). Multilateral lending agencies such as the World Bank undertake social impact assessments for transportation projects to ensure that funded projects yield significant poverty reduction impacts, equitable economic opportunity, and widely shared benefits. This involves leveling the playing field so that the population segments affected can express their opinions and participate in the development opportunities established or fostered by the project (World Bank, 2003). In this context, the World Bank describes social sustainability as the provision of equitable economic opportunity for the diverse social groups residing in the project area if its social benefits are widely shared among those population segments, and if its design is compatible with the culture and institutions of the local population affected. The term "compatibility" does not imply that all forms of the existing culture are inherently good.

Similar to the case for social impacts, the impetus for environmental justice considerations in transportation decision making include Title VI of the Civil Rights Act of 1964 and Executive Order 12898 by President Clinton in 1994. In 1997, the U.S. DOT issued an Order on Environmental Justice requiring state DOTs to implement Executive Order 12898 by incorporating principles of environmental justice in all programs, policies, and activities carried out by that agency. Environmental justice strives to ensure that the perspectives of affected residents, particularly the less powerful in society, are given due attention (Fritz, 1999). The attainment of environmental justice is a key equity-related performance measure. A concept that is closely related to environmental justice is distributive effects analysis. Distributive effects are measurable adverse and beneficial outcomes of a transportation plan, program, or project that do not affect all members of a population equally (Chatterjee and Sinha, 1976). Analysis of such effects helps to identify and address the issue of environmental justice. As Table 17.2 indicates, the impacts related to environmental justice cover a broad range including community cohesion, air quality, visual quality, and so on.

The disruptive effects of transportation projects on the surrounding sociocultural environment affects all types of societies but may be even more pronounced for low-income communities and certain ethnic groups that have (1) community services and facilities that cater specifically to their tastes, culture, and value systems; (2) a more intricate fabric of social interaction and dependence; and (3) certain unique values and practices. In such societies, individuals and families tend to be more interdependent on each other for services such as ride sharing and child care through informal barter systems. Families with less personal wealth and resources can be more affected by the disruptive effects of transportation projects because they lose part or all of their support system and may have to pay for services that were bartered before the transportation project implementation. For such communities, therefore, the relocation of households or community facilities and the direct and indirect effects of physical or psychological barriers can have more severe consequences than they have for traditional communities.

The adverse impacts of transportation on certain segments of the population are not restricted to community cohesion alone. The mobility of people can also be affected. A transportation project that traverses a low-income area may increase the mobility of high-income and other people passing through the area but may also inhibit the mobility of low-income residents.

As members of minority and low-income groups typically rely more on other modes besides auto travel, they often do not reap benefits of reduced travel times and vehicle operating costs that highway projects offer.

Table 17.2   Impacts Related to Environmental Justice

Impacts of Transportation Projects	Environmental Justice Concerns with Respect to Low-Income and/or Minority Communities
Changes in traveler costs	Some community residents may be faced with no option but to undertake longer, costlier, and more difficult commutes after the transportation project due to relocations of businesses and houses.
Transportation choice	Low income and minority populations tend to use nonmotorized and transit modes more heavily than do other communities.
	Patronage of local businesses may depend heavily on pedestrian and transit access.
Accessibility	Communities with lower-income households tend to be less mobile; as a result, their options for employment and other activities are constrained.
Community cohesion	Long residential histories, strong community ties (e.g., where residents exchange child care or other services), and fewer housing choices deepen the effects of transportation disruptions and relocations in these communities.
	Locally owned businesses tend to suffer from disruptions of community cohesion because they are dependent on local clientele.
	These populations may have unique value systems and community preferences significantly different from these outsiders would predict.
Air quality	Some communities, particularly inner-city low-income areas, may be exposed to higher concentrations of pollutants emitted from transportation vehicles, due to their close proximity to such facilities.
	Suburban areas, where higher-income groups usually reside, are typically exposed to lower concentrations of transportation pollutants because such areas are located relatively far from freeways, city streets, and other major roadways.
Traffic noise	Baseline noise levels in these communities may already be higher than in other communities (due to proximity to existing highways or industrial areas).
	Housing characteristics such as poor-quality construction, less insulations, and open windows in the summer may allow more traffic noise into the indoor environment.
Visual quality	Cultural influences may help form unique community visual quality standards that may be significantly different from those outsiders would predict.

Source: Forkenbrock and Weisbrod (2001) and others.

Cairns et al. (2003) identified key principles for ensuring outcome equity (equitable distributions of adverse and beneficial effects): (1) equality [that everyone receives an equal share of net benefits (benefits minus costs)], (2) ability to pay (that persons are entitled to receive all the benefits they can pay for, assuming that they compensate for any costs incurred by other persons), (3) maximum benefit (that most persons obtain the greatest possible benefit), and (4) priority to the disadvantaged and vulnerable populations (ensuring that existing inequalities can be remedied by focusing on the needs of such persons). Examples of socio-economic case studies involving environmental justice and distributive effects are presented by Bullard and Johnson (1997) and Kennedy (2000).

In further arguing for due consideration of both costs and benefits in environmental justice analysis, Forkenbrock and Weisbrod (2001) present a diagram (Figure 17.1) that illustrates distribution of costs and benefits in a region

- *Universal set:* the overall population in the region
- *Set 1:* all persons who benefit from the transportation project (reduced travel time, costs of vehicle operation, safety, etc.)
- *Set 2:* all persons who suffer increased costs due to the project
- *Set 3:* minority or low-income persons

Figure 17.1 demonstrates the possible situation where many people who benefit from the project do not incur the

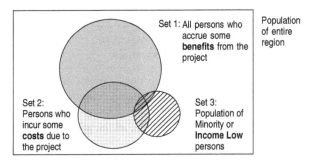

Figure 17.1   Distribution of costs and benefits of transportation projects.

cost and where some persons who incur the project costs do not benefit from the project. The overlap of sets 1 and 2 are those who incur both costs and benefits. The overlaps of set 3 with sets 2 and 1 represent persons

of minority and low-income status who are made worse off by the project and who benefit from the project, respectively. The overlap of all three sets represents persons of minority and low-income status that reap some benefits but also incur some costs of the project; for such persons, the project may yield higher benefits compared to costs (progressive) or lower benefits compared to costs (regressive). A mismatch between cost contributions and benefit sharing can only be revealed after detailed data collection regarding the extent of impact for each group.

17.3.1 An Example of the Distribution of Project Costs and Benefits

One of the early studies of the distributive effects of transportation projects examined the expected benefits and costs accruing to the residents of selected zones in Atlanta, Georgia due to the construction of the rail rapid transit system (Dajani and Egan, 1974). The rapid transit benefits

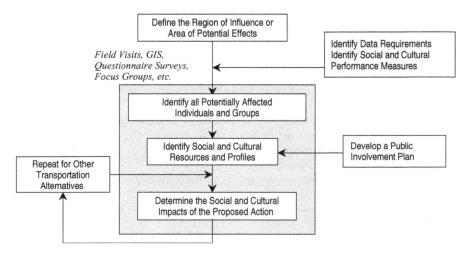

Figure 17.2   General methodology for social and cultural impact assessment.

Table 17.3   Net Zonal Annual Benefits of Atlanta Mass Transit System in 1983 ($1973) by Income Group

Zone	Average Income	Net Annual Benefits ($)	Net Annual Benefits per Household	Net Annual Benefits per Annual Trip Maker	Distance to Transit Station
258	5,396	759,925	733	104	0.5
167	6,353	627,295	630	108	0.9
147	8,711	805,630	1,044	122	2.7
342	8,838	87,123	61	7	1.4
316	10,308	366,123	344	35	0.9
80	11,683	32,908	137	15	2.7
308	18,173	900,429	224	21	1.4
185	18,595	255,597	562	54	0.8

Source: Dajani and Egan (1975).

were estimated in comparison with the existing highway and bus systems and included savings in travel time, savings in automobile capital, operating, and insurance costs, and savings in parking costs and transit fare. The user costs included the contribution due to the proposed sales tax for financing the transit system and the transit fare. Table 17.3 presents the net annual benefits projected for 1983 for 8 out of 399 zones in the study area. In the example, the medium-income zones are estimated to receive less net benefits than lower- and higher-income zones. The scope or details of the information on the distribution of benefits and costs can vary depending on the travel analysis technique used (Chatterjee and Sinha, 1976).

17.4 PROCEDURE FOR SOCIAL AND CULTURAL IMPACT ASSESSMENT

Assessing the sociocultural impacts of proposed transportation projects can be carried out using the framework shown in Figure 17.2. Sociocultural assessments are inherently inexact; therefore, the outcome can be influenced by the analyst's experience and perspectives. As such, a flexible approach that duly incorporates public involvement is necessary, and it is imperative that the analyst spends considerable time in the study area to gain an intimate knowledge of the sociocultural patterns and value systems, and to adequately recognize the potential direct and indirect impacts of a transportation project (Forkenbrock and Weisbrod, 2001). The methodology provided below is a general guideline, and the steps to be followed in a specific case will depend on the nature of the project, the area affected, and the experience of the analyst.

Step 1: Define the Project Impact Area Sociocultural effects may be far-reaching in area and long-lasting in time. However, for purposes of analysis, spatial and temporal boundaries for the assessment should be established. In demarcating the study area, step 1 identifies all communities likely to be affected by the transportation project. In some literature, the study area is referred to as the *region of influence* or *area of potential effects*. The study area can be influenced by the available level of aggregation of socioeconomic data. For example, where data are available only at the county level, the study area can be defined in terms of county boundaries. The smaller the level of aggregation, the more reliable is the analysis. The study area should include communities within and immediately surrounding the project area and may also include communities located a considerable distance from the project site whose sociocultural characteristics may be affected by the project. For example, if the

project involves large-scale construction efforts where construction workers may commute from long distances, the study area could include such outlying areas. The study area can also be influenced by the type and scale of a transportation project. For example, for a line facility (rail or highway), the width of the study area may be large for some segments and narrow at others, depending on the social and cultural capital that exists along the segment. Within the study area, neighborhood boundaries can be defined using physical barriers, land-use patterns, political or jurisdictional divisions, selected demographic characteristics, and/or resident perceptions (Florida DOT, 2000). The extent of the study area may be finalized in consultation with established organizations, such as state or local historic preservation offices.

Step 2: Identify All Potentially Affected Public Groups Groups that may be affected by proposed transportation actions may include nearby residents and businesses, those who are forced to relocate or alter their land-use plans to make way for a project, and those who may have an interest in the project even though they are not located in its proximity. Public response to the transportation action proposed can be obtained with the use of public involvement techniques such as public hearings, interviews, and surveys. Using census data on ethnic composition, income levels, car- and home-ownership rates, and the like, the analyst can use statistical tools such as cluster or discriminant analysis to identify distinct communities that exist in the study area. Some expert judgment is necessary to complement the results of such statistical analyses. Noting that accurate depiction of community and neighborhood boundaries can be a challenging task due to its inherently arbitrary nature, Forkenbrock and Weisbrod (2001) recommended that analysts should work closely with local government staff and neighborhood organizations to identify the groups potentially affected, particularly the vulnerable segments of the population.

Step 3: Describe the Community Profiles and Inventory the Sociocultural Resources For each affected group within the study area, an inventory of the sociocultural resources should be established. This task involves describing the existing conditions and trends of the social and cultural environment. Tools and methods for carrying out this task are described in Section 17.5. Published reports that yield useful sociocultural information are often released by the U.S. Census Bureau and metropolitan planning organizations; and interviews and surveys can provide unique perspectives that cannot be obtained from traditional data collection methods. Site visits by an analyst can offer firsthand views and experiences of the sociocultural interactions and relationships in a community. Maps can provide descriptions of the physical homogeneity (or otherwise) of the affected communities in the study area. Other useful information sources include past field studies or surveys in the study area or at similar study areas.

Social resources to be identified include households, community facilities (e.g., day care centers, parks, schools, clinics, social rehabilitation centers, community centers), and businesses, particularly those owned by individuals or families in the study area (e.g., bookstores, barbershops, hair salons, groceries). Comprehensive and impartial field observations and data collection can enable deeper understanding and appreciation of the values and practices of low-income and minority populations. Cultural resources include buildings and other structures of architectural value, sites of historical significance, and archaeological sites such as burial grounds and other areas that show evidence of prehistoric or historic human presence or activity. For the cultural resources inventory, the analyst should identify properties that are listed or eligible for listing in the *National Register of Historic Places*. This should be complemented with information from ethnic organizations, local academic institutions and museums, historical and archaeological societies, and state or local archaeological and historic resource surveys and inventories. Documentation of social and cultural resources is typically presented in the form of visual maps, tables, graphs, and narrative texts (ACHP, 2005).

Step 4: Apply Analysis Tools for Predicting Sociocultural Impacts The most critical stage of the sociocultural impact evaluation involves prediction of the social and cultural impacts of a project (i.e., determining the expected levels of social and cultural capital after the project implementation and assessing these levels vis-à-vis the preproject conditions). This can be done on the basis of the performance measures listed in Table 17.1 for each identified sociocultural group or community in the study area. Consistent with the principles of environmental justice, instances where certain groups or communities are seen to be affected to a significantly greater degree than others, should be identified. Performance measures could include changes in population, community cohesion and interaction, isolation, displacement, environmental justice, social values, and quality of life. For cultural impact assessment, performance measures may include physical destruction or damage to all or part of a property; alteration of a property, including restoration, rehabilitation, repair, and maintenance; removal of a property from its historic location; neglect of a property which causes its deterioration or destruction; and transfer, lease, or sale of a property. While they are more difficult to estimate precisely, secondary or indirect impacts (Section 17.1.2)

as well as cumulative impacts (Section 17.1.3), are very important to address in the sociocultural impact assessment. Particular attention should be given to the timing of the changes expected in the social environment (i.e., temporary changes due to construction-phase disruptions, or permanent changes due to relocation of families as a result of land purchases), as well as to the interconnections between community impacts. In Section 17.5, we provide details on the use of each tool and strategy and examples of past applications.

(a) *Evaluate the Levels of Social and Cultural Impact Predicted* This step involves determining the significance of the changes identified in the social environment. After the direct and secondary project-induced impacts are predicted, an overall value needs to be established that is based primarily on judgments made either by experts (using, for example, Delphi techniques) or by the public affected (using, for example, comparable cases and interviews). According to Canter (1996), evaluating social impacts should include the following:

- *Application of screening criteria* to assess the nature of the impact (i.e., if the impact is likely to occur, who will be affected, where, and how); scale, severity, and extent of the impact (i.e., if the local community is sensitive to the impact and its absolute magnitude); and the potential for mitigation, including the duration of the impact over time and its reversibility (i.e., if the pre-implementation levels of the impact type can be reverted to in the short- or long-term), the associated economic costs, and any institutional barriers.
- *Consideration of relevant standards and criteria* from professional groups and government institutions. Caution should be exercised to apply those standards for comparable cases and in conjunction with other tools.
- *Comparison with spatial (e.g., regional or national) and temporal averages (e.g., historic growth rates).* An impact is judged significant if it causes the deviation of a predetermined indicator from the corresponding regional average. The assumption is that for the without-intervention (no build) alternative, the value of the performance indicator is close to the regional average. These thresholds, however, may be inappropriate for an analysis of social impacts at a local level or for communities experiencing temporally unstable growth rates.

(b) *Mitigate Any Adverse Social Impacts That Are Anticipated* For the transportation alternative in question, the levels of social and cultural performance criteria, individually or combined (weighted), are compared with established performance threshold values (if any) or are evaluated by a knowledgeable group of experts in case no thresholds are available. Even where the most desirable alternative has significantly adverse impacts on the social and cultural resources of the area affected, mitigation measures should be recommended. Mitigation principles are discussed in Section 17.6.

Example 17.1   A transit system improvement is planned for a large city. Two alternatives being considered are metro-rail (subway) and bus rapid transit (BRT). Expert consultations (including state historic preservation officials), field inspections, and aerial photos have been used to assess the potential impacts as follows: Both alternatives improve the access of local residents to local community facilities and parks, thus improving social interaction (BRT would provide greater access than metro-rail would); changes in land use that would then probably impair community cohesion [the magnitude of the impact would be lower for rail than for BRT since changes would occur only in isolated areas (such as rail stations)]; and displacement of residential housing and business establishments due to right-of-way needs (these impacts appear to be greater for rail than for BRT, due to increased right-of-way acquisition). Neither project is anticipated to cause any disproportionate impacts to minority or low-income communities. However, the rail alternative is expected to increase inconvenience for commuters with

disabilities, who may experience multiple transfers to reach their destinations. In addition, during construction, pedestrians, transit commuters, and cyclists are expected to experience inconvenience and increased safety risks associated with street crossings. Some archaeological sites will be affected by the rail project, and the BRT alternative is expected to have adverse effects on a historic cathedral located nearby, due to increased traffic. Finally, the BRT alternative necessitates changes in bicycle travel patterns by providing alternative routes for cyclists. You are asked to assist the city's decision makers in assessing the potential social and cultural impacts associated with each alternative based on the predefined impact-rating criteria presented in Table E17.1.1. The relative importance of various criteria, as obtained through a consensus of decision makers, is represented by weights.

SOLUTION   A decision matrix displaying the weights and the ratings of each alternative for all criteria is developed. The final step involves multiplying the weights by the ratings to obtain a composite evaluation score for each alternative. Table E17.1.2 summarizes the results of this analysis and can be presented to decision makers to assist in the selection process.

On the basis of the composite evaluation of the data presented, it is seen that the combined impact of both alternatives on the social and cultural environment is expected to be on a limited-to-moderate scale (between

Table E17.1.1   Ratings of Various Levels of Sociocultural Impact

Criterion	Rating	
	Beneficial Impacts	Adverse Impacts
No impact	0	0
Minimal impact: very low probability of occurrence of impact; impact of minimal severity and extent; minimal mitigation necessary (in case of negative impact).	1	−1
Limited impact: low probability of occurrence of impact; impact of limited severity and extent; limited mitigation necessary (in case of negative impact).	2	−2
Moderate impact: moderate probability of occurrence of impact; impact of moderate severity and extent; moderate mitigation necessary (in case of negative impact).	3	−3
Significant impact: significant probability of occurrence of impact; impact of significant severity and extent; significant mitigation necessary (in case of negative impact).	4	−4
Major impact: high probability of occurrence of impact; impact of major severity and extent; major mitigation necessary (in case of negative impact).	5	−5

Table E17.1.2   Sociocultural Impacts of Rail and BRT

Social and Cultural Criteria	Weight	Metro Rail	BRT
		Impact Rating	
1. Impacts on social interaction	1	2	3
2. Impacts on community cohesion	7	−3	−4
3. Impacts on pedestrian and bicycle safety	5	−3	−5
4. Displacement or relocation impacts	6	−5	−2
5. Environmental justice	4	−1	−1
6. Impacts on historic properties	2	−2	−4
7. Impacts on archaeological resources	3	−4	−1
Composite evaluation rating		−3.0	−2.75

Table E17.2   Assessment of Community Cohesion

Indicators of Community Cohesion	Arlington Village	Summerville
	Predominant Existing Social Conditions	
Interaction among neighbors	Frequent and intense	Frequent
Use of community facilities	Regular and high reliance	Regular and low reliance
Long-serving community leadership	Present	Present
Participation in local organizations	Active	Active
Identification with the community	Established neighborhood name and boundary	Similar to that for Arlington
Desire to stay in the community	Strong	Moderate
Satisfaction with the community	Highly satisfied	Satisfied
Homogeneity (income, ethnicity, age, etc.)	Homogeneous in terms of income and ethnicity	Homogeneous in terms of income and age
Family- vs. singles-oriented communities	Family-oriented	Singles-oriented
Length of residency compared with other variables (e.g., satisfaction with community)	Long-term, voluntary residence	Short-term

2 and 3). The BRT alternative has a slightly lower unfavorable overall composite rating and is therefore more desirable from a sociocultural standpoint. However, because the difference in overall rating is so small, further evaluation should be carried out by considering specific impact items before selecting the superior transportation alternative.

Example 17.2   A new highway link is proposed purposely to serve a planned bus rapid transit system to link a city's central business district to the suburbs. GIS analysis shows that two neighborhoods, Arlington Village and Summerville, will be most affected by the project through land appropriations and other direct and indirect impact mechanisms. Using the data shown in Table E17.2, assess the relative impacts of the planned project on the cohesion of these communities.

SOLUTION   Compared to Summerville, Arlington Village has residents that interact more, are more involved in community issues, have an established neighborhood identity and boundary, have a strong desire to stay in their neighborhood, are highly satisfied with their community, have a high proportion of family residences, and have a long history of voluntary residence in the community. As such, from the social perspective, Arlington is expected to be affected to a greater extent than Summerville.

Example 17.3 Two alternative transportation projects are being considered for implementation in an urban area. Sociocultural experts have prepared a social and cultural impact checklist and have rated the extent and permanency of such impacts (Table E17.3). Assuming that all of the criteria have equal importance, comment on the desirability of the alternatives in terms of their sociocultural impacts.

SOLUTION   From the sociocultural standpoint, it is seen that alternative 1 yields ten adverse impacts with five of them being permanent, while alternative 2 has nine adverse impacts, four of which are permanent. Alternative 2 can therefore be considered somewhat more desirable than alternative 1.

Example 17.4 As part of a planned airport runway extension, it is proposed to acquire, by eminent domain, additional nearby land from an old residential neighborhood. This development will necessitate the relocation of a significant number of households, businesses, and other social services. Identify the social groups that would possibly be affected by the relocation, and discuss the nature of their difficulties (adapted from CUTR, 2006).

SOLUTION

Using the census data, the groups identified are as follows:

(a) *Groups:* elderly, physically and mentally disabled, low-income, households with school-aged children, non–English speaking, ethnic and racial minority, and long-term residents.

Difficulties: Possible difficulties to be faced on these groups include the lack or shortage of affordable, safe, and clean housing; financial, social, and emotional impacts; a sense of loss when compelled to relocate; broken community support and social networks; disruptive school transfers; anxiety, alienation, and difficulty in forming new friendships at new locations; increase in length and time for work and other trips; and hardships due to the loss of businesses that cater to vulnerable segments.

(b) *Businesses:* The types of affected businesses are small businesses that cater to local clientele, typically family-owned, ethnic, or minority-owned.

Difficulties: Include time, effort, and cost to find and obtain a suitable replacement site; cost of building or redesigning new sites; moving expenses; loss of customers; advertisement costs to attract new business and old clientele; time and cost to replace employees not able to continue working at the new location; and increased employee commutes to reach the new location.

Example 17.5 As seen in Example 17.4, relocation is one of the more serious adverse impacts of new transportation projects, particularly when an extensive amount of land needs to be appropriated for the facility right-of-way. However, there are some desirable sociocultural effects of relocation. Identify some of the beneficial effects of relocation.

SOLUTION   Possible beneficial impacts of such projects may include the following (CUTR, 2006):

• Increase in property values due to the new transportation development and the removal of blighted areas
• More desirable new residential and business units for displaced persons and businesses
• More desirable new sites for relocated businesses.
• Removal of hazardous or non-conforming structures and features in the transportation corridor.

17.5 TOOLS FOR SOCIOCULTURAL IMPACT ASSESSMENT

The choice of tools and methods for a specific social assessment will depend on several factors such as the project area and the quality of the existing social development information specific to the project and the study area. Resource constraints and the time frame for the social impact assessment will also affect the choice of assessment tools. Unlike quantitative tools, qualitative tools are typically used in cases of complex and poorly understood social phenomena and are particularly useful for describing multidimensional interpersonal interactions and the nonincome dimensions of poverty that are more difficult to capture in quantitative terms.

17.5.1 Qualitative Tools

Qualitative tools for sociocultural impact assessment include those described below (Canter, 1996; Apogee and Parsons Brinckerhoff, 1996; Caltrans, 1997; Forkenbrock and Weisbrod, 2001; World Bank, 2003).

(a) *Expert Consultation* Professionals within and outside the fields of social and cultural studies can serve as excellent sources by providing their perspectives on the inventory of existing sociocultural capital as well as the expected nature, extent, and severity of sociocultural impacts. Tools for consultation include roundtable and brainstorming sessions, focus groups, and Delphi techniques for consensus building. In soliciting expert opinion

Table E17.3   Checklist of Sociocultural Impacts

Social Item	Alternative 1		Alternative 2	
	Impacts (Y/N)	Permanent/ Temporary	Impacts (Y/N)	Permanent/ Temporary
Project creates a barrier that divides the neighborhood or limits access to all or part of the neighborhood	N	—	Y	Permanent
Project effects on special groups (e.g. elderly, persons with disabilities, racial/ethnic/religious groups) within the neighborhood	Y	Permanent	Y	Permanent
Project reduces the social interaction that occurs within the neighborhood	N	—	Y	Permanent
Displacement of residents negatively affects the perceived quality of life in the neighborhood	N	—	Y	Permanent
Project affects access to, or results in the removal of, neighborhood facilities or services that are needed and valued by the neighborhood	Y	Permanent	N	—
Facilities and services subject to removal or relocation are able to remain in or within proximity of the neighborhood	N	—	N	—
Project results in an increase in noise, vibration, odor, or pollution that reduces social interaction in the neighborhood	Y	Temporary	Y	Temporary
Communal areas (e.g., parks, playgrounds) used by residents are negatively affected by construction of the project	Y	Temporary	Y	Temporary
Availability and convenience of transit services reduced as a result of the project	Y	Permanent	N	—
Project negatively affects pedestrian and nonmotorized mobility within the neighborhood	Y	Permanent	N	—
Vehicular mobility within the neighborhood is affected negatively by this project	Y	Temporary	Y	Temporary
Vehicular traffic increases as a result of the project	Y	Permanent	Y	Temporary
Vehicular traffic increase creates unsafe conditions for nonmotorized transportation within the neighborhood	Y	Temporary	N	—
"Blind or isolated" areas be created that are difficult to monitor for criminal activity as a result of the project	Y	Temporary	N	—
Emergency response routes are affected negatively as a result of the project	N	—	Y	Temporary

Source: Adapted from CUTR (2006)

through surveys, it is useful to include descriptive checklists. Also, relevant reports and publications by experts can be reviewed as part of the consultative process.

(b) Field Solicitations (Neighborhood Surveys, Interviews, and Questionnaires) Field solicitation is a flexible tool that provides the opportunity to collect firsthand information on the social and cultural resources in the study area and to ascertain whether community members perceive the transportation project as a threat or as an opportunity. Using this tool, the analyst can identify certain sociocultural impacts, such as the types, destinations, and durations of trips that are important to community residents, and their pre- and post-project capabilities (or inhibitions) to make desired trips. Although field solicitations can be time-consuming and labor-intensive, they generally provide useful and revealing insights that are vital for assessments of this type. Colony (1972) focused on a displaced population before and after a relocation exercise in Cleveland, Ohio and conducted interviews of relocated households to assess the social, psychological, and economic impacts. More recently, field interviews in Boston revealed that residents of the city's north end were averse to a proposed replacement of an elevated highway structure by a tunnel because they perceived the existing structure as a desirable barrier that prevents gentrification and thereby helps preserve the ethnic character of their neighborhood (Forkenbrock and Weisbrod, 2001).

Field solicitations must be preceded by (1) identification of the target respondents, which may include community and neighborhood leaders and the general public, and (2) careful design of a survey questionnaire. The survey instrument could include subjects such as locations of community facilities (businesses, social centers, recreational areas, places of worship, etc.), and pedestrian or cycling routes. The survey instrument should be easy to complete, avoid strong language, reserve sensitive questions for the last stages of the survey, and avoid long questions that could cause respondents to lose interest in the survey. Detailed guidelines for designing the survey instrument are provided by Babbie (1990) and Forkenbrock and Weisbrod (2001). The results of the survey can be used to create a database that can be queried for any specific item.

Other mechanisms for soliciting public opinion in the affected communities, depending on the available time, expense, organization, and resources, include focus groups, fishbowls, charrettes, deliberative polling, and nominal group workshops. Details of these mechanisms and the conditions under which they are appropriate are provided by Forkenbrock and Sheeley (2004).

(c) Field Solicitations for Organized Involvement of Public Bodies Public bodies include citizen advisory groups, public meetings, community events, and participants at special workshops organized for the purpose of impact evaluation. In some cases, if the analyst confers with organized groups rather than individuals, additional information beyond traditional one-on-one methods can yield deeper insights. The participation of organized public bodies can provide insights about the community profile, community issues and attitudes, and any foreseeable impacts of the proposed transportation action on community facilities.

(d) Field Solicitation (On-Site-Analysis) The value system and intricate social and cultural fabric of a community cannot really be assessed completely without personal visits and tours of the study area. Several field trips should be undertaken to observe the performance criteria associated with community cohesion, social interactions, and cultural resources. Unless there are security concerns, such visits should not be from a vehicle but preferably through walking the common routes, recreational areas, and places of social gathering, such as malls, recreational facilities, barbershops, and so on, where residents can easily be encountered, approached, and interviewed. The analyst should seek evidence of social interdependence and interaction, such as the existence of community committees, neighborhood watch, level of pedestrian activity, children at play in or out of playgrounds, condition of houses and lawns, shared parking facilities, local newspaper articles, and columns and letters where residents comment on community issues, Also, during on-site tours, the analyst should, with the help of an appropriate map, identify all structures and take note of the characteristics of currently operating businesses (i.e., number of employees, location of employee residences, location of intended move, etc.). The analyst should establish a contact through which interviewees can provide additional information at a later time. The social vulnerability of households should be noted. Households are vulnerable if a shock (such as relocation) is likely to push them below or farther below a predefined welfare threshold (such as the poverty line).

(e) Comparative Analysis Information on the sociocultural impacts of transportation actions at other comparatively similar locations can be an inexpensive approach. Examples such analyses solving environmental justice and distributive effects are Kennedy (2000) and Bullar and Johnson (1997).

17.5.2 Quantitative Tools
Quantitative tools can provide supportive data in some cases, and they are discussed below.

(*a*) *Visual Tools for Image or Data Analysis*   Maps depicting physical characteristics, demographics, and project alternatives, as well as social and cultural resources, can be plotted and superimposed to create a composite image that enables a more precise assessment of the sociocultural impacts of a transportation project. Using aerial photographs or GIS tools, an analyst can provide a visual picture of how a proposed transportation project would affect households, businesses, community facilities, activity centers, and cultural resources. These tools can also help identify the extent to which the transportation project implementation may cause sociocultural resources to become isolated from the population (or parts thereof) they are meant to serve. Using GIS, the analyst may carry out *overlay analysis* to integrate different data layers and to permit visual presentations of various scenarios of transportation project locations and designs and their respective impacts on sociocultural resources. Bahadur et al. (1998) and Werner (1998) used GIS to assess the demographic effects of bus route changes and other sociocultural impacts, respectively, on the basis of environmental justice. In the final environmental impact statement for the Interstate 69 Evansville-to-Indianapolis highway, GIS tools were used to analyze the impact of the project on historic sites for 12 alternative project alignments (Cambridge Systematics and Bernadin, 2003). GIS is useful in social and cultural impact assessments in general and distributive effects and environmental justice in particular because it enables spatial analysis and display capabilities. The scale of GIS features should be duly noted by the analyst because some precision could be lost through aggregation of spatial attributes in GIS maps.

(*b*) *Statistical Analysis (Demographic Impact Prediction Analysis)*   Curve-fitting and regression-based techniques can be used to predict the number, distribution, and characteristics of people expected to move into or away from the study area (or each of its constituent communities). This tool has been used in past studies, such as assessment of changes in pedestrian safety due to changes in the physical environment (Timmermans et al., 1992).

(*c*) *Computer Modeling*   This can be used to simulate and predict social and cultural impacts, such as changes in access for low-income and minority communities to economic or sociocultural destinations, due to a transportation project. Using computer modeling, Almanza and Alvarez (1999) evaluated the impacts of a proposed light-rail system on low-income and minority communities in the metropolitan area of Austin, Texas and found that the construction and operations of the new system would significantly limit access to transportation facilities and community resources,

increase noise and air pollution, reduce property values, and cause relocation of some residents.

17.6   MITIGATION OF ADVERSE SOCIOCULTURAL IMPACTS

Mitigation to avoid or minimize potential adverse impacts is an important aspect of sociocultural impact assessment. Mitigation strategies to be selected depend on the nature of the transportation project, scale of the project, and the type, distribution, and sizes of the population groups affected. If an undertaking results in adverse impacts, actions that reduce or compensate for the damage to cultural resources are necessary.

Typical mitigation measures fall within the following categories (Canter, 1996; Caltrans, 1997; Florida DOT, 2000):

- *Preemptive:* altering the project design or alignment at the planning or design phases of the PDP so that the anticipated adverse impact does not occur.
- *Minimization:* modifying the project alignment or design through redesign, reorientation to reduce the extent or severity of the adverse impact, such as shifting the alignment; or depressing or elevating the facility.
- *Mitigation:* alleviating or offsetting an existing or inevitable adverse impact. This generally includes the repair, rehabilitation, or restoration of the affected resource directed toward retaining the qualities that made the resource valuable from a community or historic standpoint, or replacement of an appropriated resource. An example is reconstruction of a demolished school or placing a historic resource at a new location, or partial recovery or salvage of a historic property (archaeological, architectural, etc.) when the property cannot be relocated and must be demolished.
- *Enhancement:* adding a desirable or attractive feature through preservation and maintenance operations or other activity. Includes provision of trees and other landscaping, scenic and rest areas, adding artwork to structures, phasing the project implementation to minimize community disruption, providing temporary or permanent access to residents as and where needed, providing bicycle and pedestrian paths and crossings, providing lighting and street signs, and fair compensation of properties that are taken from the community.

17.6.1 Sociocultural Impact Mitigation: State of Practice

In its guidelines on community, culture, and the environment, USEPA (2002) provides impact assessment methods and tools and presents community case studies across the nation. The Washington State DOT (2003) has established recommended best practices, including context-sensitive solutions. In addition, the Minnesota DOT (1999) and Howard/Stein-Hudson and Parsons (1996) offer guidelines for effective public involvement. The Colorado DOT (2003) provides tools for enhancing Colorado's statewide and regional transportation planning process, including environmental justice considerations such as measuring the distribution of benefits from transportation plans and transportation investments and enhancing public involvement. Examples of studies where social and cultural impact assessments were conducted as part of an environmental impact statement by a state transportation agency include the I-69 Evansville-to-Indianapolis corridor study prepared for the Indiana DOT (Cambridge Systematics and Bernardin, 2003); and the I-405 congestion relief and bus rapid transit program prepared for the Washington state DOT (2001). The World Bank has established generic terms of reference to guide analysts in conducting social impact assessment for transportation and other sectoral developments (World Bank, 2003).

In urban areas, measures to mitigate sociocultural impacts include provision of noise barriers, pleasing landscape designs, and pedestrian traffic crossings. Since the appropriation of land bearing private or public facilities is often unavoidable, measures to ensure proper relocation and replacement of such facilities can greatly reduce the magnitude of the social impact. This will require background knowledge and input that may best be obtained through community interaction at the planning stage. Consequently, it is essential that the planning and timing of the proposed project be communicated clearly and accurately to the groups affected. There are several techniques that may greatly facilitate open communication, free exchange of information, and hopefully, improvements in the planning process. Finally, the specific timing of the project's stages, such as acquisition and demolition, may be adjusted to lessen the community burden and facilitate relocation. Providing displaced residents with assistance in finding appropriate and affordable new housing can largely reduce the stress levels and some social impacts. As a last resort, when existing housing supplies are inadequate, consideration must be given to creating additional housing through major rehabilitation or new construction projects. It is therefore obvious that sociocultural impacts and land-use impacts of transportation projects (Chapter 16) are often related.

For rural transportation projects, several measures are available to reduce the potentially adverse social impacts of rural highways. Adjustments in alignment may be made to avoid sensitive areas, such as the known territories of tribal populations. Also, consideration should be given to any known wildlife migrations such as herd movements. Such avoidance measures are the best means of protecting the interest of tribal people wanting to maintain their traditional cultures. A possible regulatory measure may be to establish land-use controls so that culturally sensitive areas and other lands can be preserved. Other regulatory and enforcement measures may be used to control the development practices along the transportation route. Such measures could greatly reduce the adverse effects caused by spontaneous and uncontrolled development by squatters along the transport routes. The World Bank (2003) has issued guidelines for considering appropriate measures to protect tribal peoples affected by development projects.

17.7   LEGISLATION RELATED TO SOCIOCULTURAL IMPACTS

Consideration of impacts on the social environment was initiated by the National Environmental Policy Act (NEPA) in 1970 and the issuance of CEQ guidelines, which arose out of concerns about the secondary impacts of development activities on existing community facilities and activities. The Federal Highway Act of 1970, Section 23 USC 109(h), lists the types of adverse social impacts that require investigation and documentation; these include the destruction or disruption of human-made resources, social values, community cohesion, and the availability of public facilities and services; displacement of people, businesses, and farms; and disruption of desirable community and regional growth (FHWA, 1982). The Federal Transit Administration (FTA) Office of Planning recognizes that "transit projects affect the social environment in several ways and may change the physical layout, demographics, and sense of neighborhood in local communities" (FTA, 2005). Other legislation related to assessment of social and community impacts include the Intermodal Surface Transportation Efficiency Act (ISTEA) of 1991. In incorporating Sections 109(h) and 128 of Title 23 of the U.S. Code on Highways, ISTEA required that the social and economic impacts of proposed federal-aid projects be determined, evaluated, and eliminated or minimized as part of environmental documentation for project development, and such impacts should include community cohesion, availability of public services and facilities, adverse employment effects, injurious displacement of people, businesses, and farms, and disruption of desirable community and regional growth.

Cultural resources include historic and archeological resources and are undoubtedly an important part of national values and character. Therefore, their preservation needs to be considered in the decision-making process for transportation projects. Legislative and executive mandates on the need to preserve and enhance cultural resources have been expressed in the Department of Transportation Act of 1966, the Federal-Aid Highway Act of 1968, NEPA of 1969, the National Historic Preservation Act of 1966, Executive Order 11593 of 1971, the Archeological and Historic Preservation Act of 1974, the American Indian Religious Freedom Act of 1978, the Archeological Resource Protection Act of 1979, the Surface Transportation and Uniform Relocation Assistance Act of 1987, and ISTEA of 1991. In particular, NEPA of 1970 addressed potential impacts on the cultural environment by indicating the responsibility of the federal government to "preserve important historic, cultural, and natural aspects of our national heritage." Furthermore, Section 106 of the National Historic Preservation Act of 1966 requires that federal agencies take into account the impact of federal undertakings on historic properties included in or eligible for listing in the National Register of Historic Places. Finally, regulations by the Federal Council on Environmental Quality (40 CFR, Part 1500–1508.14) and the Advisory Council on Historic Preservation (36 CFR, Part 800) were promulgated to ensure that in the development of federal undertakings, the effects on historic and archeological resources are duly considered.

With regard to environmental justice as it relates to the impacts on sociocultural resources and other impact types, the 1964 Civil Rights Act (Title IV) probably served as the first legislation intended, at least implicitly, to ensure that the rights of minority segments of the population were duly considered in the planning and execution of federal projects. The act and its related statutes required that there should be no discrimination in federally assisted programs on the basis of race, color, national origin, age, gender, or disability. Information needed to assess possible discrimination (and thus to address Title IV issues) can be obtained during the evaluation stage of transportation project development. The Americans with Disabilities Act of 1990 (which extended the Civil Rights Act to the disabled), prohibits discrimination in public transportation and other services and stipulated involvement of the disabled in the development of such projects. For example, in planning for improvements to urban roadways or pedestrian facilities, the input of the disabled is vital. Also, the Uniform Relocation Assistance and Real Property Acquisition Policy Act of 1970 (Title 49 CFR Part 24), as amended in 1987, required equal treatment of persons displaced from their homes, businesses, farms, and so on, by federal and federally

assisted programs, and established uniform and equitable land acquisition policies. In 1994, President Clinton's Executive Order 12898 (Federal Actions to Address Environmental Justice in Minority Populations and Low-Income Populations) required each federal agency to develop an agencywide environmental justice strategy to ensure that low-income and minority populations are not subject to disproportionately high and adverse environmental effects (USDOT, 1997).

SUMMARY

Transportation investments cause significant impacts on communities and society as a whole. Social impact analysis involves assessment and evaluation of the potential direct and indirect benefits and disbenefits that can be expressed in terms of acquisitions and displacements, neighborhoods, community cohesion and social interaction, environmental justice, community facilities, social values, and quality of life. Also, cultural impacts can be assessed in terms of degradation to historical assets (buildings, structures, sites, objects, and districts) and archaeological resources. There are federal, state, and local laws and regulations to encourage consideration for the protection of the social and cultural capital in transportation systems evaluation. The assessment of social and cultural impacts of transportation action typically coincides with environmental reviews and provides the public as well as the decision maker with information on how various project alternatives will affect community and cultural resources. To accomplish the intent of legislation and regulations, it is necessary that project area(s) of potential effect be established and that certain levels of investigation of a community's profile and historic and archeological resources are accomplished during the transportation project development process. The investigation and discussion should be commensurate with the importance of the severity, as well as the magnitude, of the project's impacts. Groups that may be affected adversely by a transportation action need to be identified at an early stage of the process. As with the evaluation of adverse effects, avoidance, minimization, and mitigation measures need to be considered at an appropriate level of detail whenever possible. Proactive involvement of community residents is vital in sociocultural impact assessments because it ensures that transportation decisions are responsive to community concerns and goals and thereby enhance community acceptance of the proposed project. Furthermore, environmental justice is better served when transportation decision makers identify and address possible disproportionate adverse impacts on specific segments of the population. Social impact assessment facilitates systemic participation of relevant stakeholders in project planning, design, and implementation. Such assessments increase the likelihood that the intended social benefits of the project are realized, such as increased and/or equitable access to development opportunities.

EXERCISES

17.1. List the methods by which a transportation agency can reduce the adverse impacts of social and cultural impacts for (**a**) planned projects and (**b**) existing facilities.

17.2. It is proposed to extend a commuter rail system to connect more outlying suburbs to the downtown of a metropolitan city. Describe how you would carry out a sociocultural impact assessment of the project.

17.3. Two alternative transit projects have similar costs and user benefits but are expected to have markedly different sociocultural impacts because they are designed to serve different areas. What typical performance measures could be used to evaluate these alternative transit projects?

17.4. Discuss the role of public involvement in social and cultural impact assessment of transportation projects.

17.5. (**a**) Define a *community profile*, and discuss how it could be developed for a given community. (**b**) For a social impact assessment of a proposed railway extension that passes through an existing community, discuss the data types needed in developing a community profile.

17.6. For a rapidly growing city and its suburbs, two alternative transportation projects to help ease peak-hour congestion are being considered. However,

Table EX17.6   Relocation Impact Checklist

Consideration Item	Weight	Alternative 1		Alternative 2	
		Impact (Y/N)	Comments	Impact (Y/N)	Comments
Long-time neighborhood residents (5+ years tenure)	1/15	N	15	Y	10
Elderly residents (65+ years old)	1/15	Y	5	Y	3
Disabled residents	1/15	N	—	Y	1
Low-income residents (generally, poverty level)	1/15	N	—	Y	30
Ethnic or racial minority residents	1/15	Y	35	N	—
Non-English-speaking residents	1/15	N	—	N	—
Households with school-aged children	1/15	Y	3	Y	
Adequate, comparable replacement housing or building sites available for relocatees in or near their current neighborhood. How close?	1/15	Y	2 miles	Y	1.3 miles
Impairment of access to employment due to relocation.	1/15	Y	Increase in travel time and distance by 10-min and 2-mi, respectively	N	—
Impairment of access to schools, medical care, child care or other essential goods and services due to relocation.	1/15	Y	Increase in travel time and distance by 5-min and 2-mi, respectively	N	—
Relocation of a community facility such that the purpose for the facility is reduced or otherwise impaired.	1/15	N	—	Y	0.8 mile
Relocation of a business that depends upon its specific location for business.	1/15	Y	5	N	—
Project requires the acquisition of right-of-way from public lands.	1/15	N	—	N	—

Adapted from CUTR (2006)

it has been established that each alternative will involve some degree of relocation of existing households, social facilities (churches, schools, and a museum), and businesses to make way for the facility. Based on the relocation impact checklist in Table EX17.6, carry out an evaluation to identify the superior alternative from the sociocultural viewpoint.

REFERENCES[1]

ACHP (2005). *National Register Criteria*, Advisory Council on Historic Preservation, Washington, DC, http://www.achp.gov/nrcriteria.html. Accessed Nov. 29, 2005.

Agyeman, J. (2001). Ethnic minorities in Britain: short change, systematic indifference and sustainable development, *J. Environ. Policy Plan.*, Vol. 3, pp. 15–30.

Almanza, S., Alvarez, R. (1999). *The Impacts of Siting Transportation Facilities in Low-Income Communities and Communities of Color*, U.S. Department of Transportation, Washington, DC, www.fta.dot.gov/fta/library/policy/envir-just/backcf.htm. Accessed Nov. 29, 2005.

Alsnih, R. Stopher, P. R. (2003). Environmental justice applications in transport, in *Handbook of Transportation and the Environment*, ed. Hensher, D. A., Button, K. J., Elsevier, Amsterdam, The Netherlands.

*Apogee, Parsons Brinckerhoff Quade & Douglas (1996). *Community Impact Assessment: A Quick Reference for Transportation*, Tech. Rep. FHWA-PD-96-036, Federal Highway Administration, U.S. Department of Transportation, Washington, DC.

*Babbie, E. (1990). *Survey Research Methods*, 2nd Ed., Wadsworth, Belmont, CA.

Bahadur, R., Samuels, W. B., Williams, J. W. (1998). Application of geographic information systems in studies of environmental justice, *Proc. 1998 ESRI International Users Conference*, Environmental Systems Research Institute, Redlands, CA, http://gis.esri.com/library/userconf/proc98/PROCEED/TO150/PAP128/P128.HTM. Accessed Nov. 29, 2005.

Bass, R. (1998). Evaluating environmental justice under the National Environmental Policy Act, *Environ. Impact Assess. Rev.*, Vol. 18, No. 1, pp. 83–92.

Bullard, R., Johnson, G. (1997). *Just Transportation*, New Society Publishers, Gabriola Island, BC, Canada.

*Cairns, S., Greig, J., Wachs M. (2003). *Environmental Justice and Transportation: A Citizen's Handbook*, UCB-ITS-M-2003-1, Berkeley Institute of Transportation Studies, University of Southern California, Berkeley, CA.

*Caltrans (1997). *Caltrans Environmental Handbook*, Vol. 4, *Community Impact Assessment*, California Department of Transportation, Sacramento, CA.

Cambridge Systematics, Bernardin, Lochmueller & Associates (2003). *I-69 Evansville to Indianapolis: Final Environmental Impact Statement*. Indiana Department of Transportation, Indianapolis, IN, http://www.deis.i69indyevn.org/FEIS/. Accessed Jan. 14, 2006.

*Canter, L.W. (1996). *Environmental Impact Assessment*, 2nd ed. McGraw-Hill, New York.

Chatterjee, A., Sinha, K. C. (1976). Distribution of the benefits of Public Transit Projects, *ASCE Journal of Transp. Eng.*, No. 3, Vol. 101.

Colony, D. (1972). *Study of the Impacts on Households of Relocation from a Highway Right-of-Way*, Hwy. Res. Rec. 399, Transportation Research Board, National Research Council, Washington, DC, pp. 12–26.

Colorado DOT (2003). *Colorado Environmental Justice Research Project Report*, Tech. Rep. CDOT-DTD-R-2003-12, prepared by DMJM+HARRIS for the Colorado Department of Transportation, Denver, CO, http://www.dot.state.co.us/publications/EnvironmentalJustice/EJ.htm. Accessed Jan. 17, 2006.

CUTR (2006). Community Impact Assessment, Chapter 5: Social Impacts, Center for Urban Transportation Research, University of South Florida, Tampa, FL. www.cutr.usf.edu/pubs.CIA

Dajani, J. S., Egan, M. M. (1974). Income distribution effects of the Atlanta mass transit system, presented at the 53rd Annual Meeting of the Highway Research Board, Washington, DC.

*FHWA (1982). *Social Impact Assessment: A Sourcebook for Highway Planners*, Vol. III, *Inventory of Highway Related Social Impacts*, Federal Highway Administration, U.S. Department of Transportation, Washington, DC.

*Florida DOT (2000). *Community Impact Assessment: A Handbook for Transportation Professionals*, Florida Department. of Transportation, Tallahassee, FL.

*Forkenbrock, D. J., Sheeley, J. (2004). *Effective Methods for Environmental Justice Assessment*, NCHRP Rep. 532, National Academy Press, Washington, DC.

Forkenbrock, D. J., Weisbrod, G. E. (2001). *Guidebook for Assessing the Social and Economic Effects of Transportation Projects*, NCHRP Rep. 456, National Academy Press, Washington, DC.

Fritz, J. M. (1999). Searching for environmental justice: national stories, global possibilities, *Social Justice*, Vol. 26, pp. 174–189.

FTA (2005). *Social and Economic Impacts*. Office of Planning, Federal Transit Administration, U.S. Department of Transportation, Washington, DC, http://www.fta.dot.gov/office/planning/ep/subjarea/se.html. Accessed Nov. 29, 2005.

Goodwin, P. (2003). Unintended effects of policies, in *Handbook of Transportation and the Environment*, eds. Hensher, D. A., Button, K. J., Elsevier, Amsterdam, The Netherlands.

Howard/Stein-Hudson Associates, Parsons Brinckerhoff Quade & Douglas (1996). *Public Involvement Techniques for Transportation Decision-Making*, FHWA-PD-96-031, Federal Highway Administration and Federal Transit Administration, Washington, DC, http://www.fhwa.dot.gov/reports/pittd/cover.htm. Accessed Jan. 17, 2006.

*IOCGP (2003). Principles and guidelines for social impact assessment in the USA. *Impact Assess. Project Appraisal*, Vol. 21, No. 3, pp. 231–250, Interorganizational Committee on Principles and Guidelines for Social Impact Assessment, Beech Tree Publishing, Surrey, UK.

Kennedy, L. (2000). Environmental justice and where it should be addressed in the 21st century concerning the transportation industry: historical perspective and summary, *Proc. Conference on Refocusing Transportation Planning for the 21st Century*. Transportation Research Board, National Research Council, Washington, DC.

Khisty, C. J. (1996). Operationalizing Concepts of Equity for Public Project Investments, *Transp. Res. Rec. 1559*,

---

[1]References marked with on asterisk can also serve as useful resources for social and cultural impact assessment.

Transportation Research Board, National Research Council, Washington, DC, pp. 94–99.

King, E. M., Alderman, H. (2001). Education, Focus 06–Brief 06, *in 2020 Vision for Food, Agriculture, and the Environment*, International Food Policy Research Institute, Washington, DC.

Minnesota DOT (1999). *A Guide to Public Involvement at Mn/DOT*, Minnesota Department of Transportation, St. Paul, MN, http://www.dot.state.mn.us/pubinvolve/pdf/sep10hev. pdf. Accessed Jan. 17, 2006.

Quan, R. (2002). Establishing China's environmental justice study needs, *Georgetown Environ. Law Rev.*, Vol. 14, pp. 461–487.

Rawls, J. (1999). *A Theory of Justice*. Oxford University Press, Oxford, UK.

Rietveld, P. (2003). Efficiency, equity, and compensation, in *Handbook of Transportation and the Environment*, ed. Hensher, D. A., Button, K. J., Elsevier, Amsterdam The Netherlands.

National Historic Preservation Act (1966) Section 106 regulations, 36 CFR Part 800, Protection of Historic Properties, of the National Historic Preservation Act. http://www.achp.gov/regs-rev04.pdf. Accessed December 10, 2005.

Sen, A. (2000). *Development as Freedom*, Anchor Publishing, Virginia Beach, VA.

Sinha, K. C., Varma, A., Souba, J., Faiz, A. (1989). *Environmental and Ecological Considerations in Land Transport: A Resource Guide*. Infrastructure and Urban Development Department, Policy Planning and Research Staff, The World Bank, Washington, DC.

Timmermans, H., van der Hagen, X., Borgers, A. (1992). Transportation systems, retail environments and pedestrian trip chaining behavior: modeling issues and applications, *Transp. Res.*, Vol. 26B, No. 1.

TRB (2002). *Environmental and Social Justice Surface, Transportation Environmental Research: A Long-Term Strategy*, Spec. Rep. 268, Transportation Research Board, National Research Council, Washington, DC http://www.its.berkeley. edu/publications/ejhandbook/ejhandbook.html. Accessed Jan. 17, 2006.

USDOT (1997). *Department of Transportation order to address environmental justice in minority populations and low-income populations*, Docket OST–95–141 (50125), *Fed. Reg.*, Vol. 62, No. 72 (Apri. 15), pp. 18377–18381, U.S. Department of Transportation, Washington, DC.

US EPA (2002). *Community Culture and the Environment: A Guide to Understanding a Sense of Place*. EPA 842-B-01-003, Office of Water, U.S. Environmental Protection Agency, Washington, DC, http://www.epa.gov/ecocommunity/pdf/ ccecomplete.pdf. Accessed Jan. 17, 2006.

Washington State DOT (2001). *I-405 Congestion Relief and Bus Rapid Transit Program: Final Environmental Impact Statement*. Washington Department of Transportation, Olympia,

WA, http://wsdot.wa.gov/projects/i-405/feis/. Accessed Jan. 17, 2006.

_____ (2003). *Building Projects That Build Communities*, developed by the Community Partnership Forum for the Washington Department of Transportation, Olympia, WA, http://www.wsdot.wa.gov/biz/csd/BPBC_Final/.Accessed Jan. 17, 2006.

Werner, R. J. (1998). Equity in public transit: a test statistic for buffer problems, *Proc. 1998 ESRI International User's Conference*, Environmental Systems Research Institute, Redlands, CA, http://gis.esri.com/library/userconf/proc98/. Accessed Nov. 29, 2005.

Wessel, T. (2000). Social polarization and socio-economic segregation in a welfare state: the case of Oslo, *Urban Stud.*, No. 37.

*World Bank (2002). *A Sourcebook for Poverty Reduction Strategies*, International Bank for Reconstruction and Development, Washington, DC.

*_____ (2003). *Social Analysis Sourcebook: Incorporating Social Dimensions into Bank-Supported Projects*, Social Development Department, World Bank, Washington, DC.

ADDITIONAL RESOURCES

Caltrans (2003). *Environmental Justice in Transportation Planning and Investments: Desk Reference*, prepared by ICF Consulting and Myra L. Frank & Associates for the California Department. of Transportation, Sacramento, CA, http://www.dot.ca.gov/hq/tpp/offices/opar/EJDeskGuide-Jan03.pdf. Accessed Jan. 17, 2006.

Cambridge Systematics (2002). *Technical Methods to Support Analysis of Environmental Justice Issues*, Tech. Rep. NCHRP 8–36(11), Transportation Research Board, National Research Council, Washington, DC.

FHWA (1988). *Guidance on the Consideration of Historic and Archaeological Resources in the Highway Project Development Process*, Office of Environmental Policy, Federal Administration, U.S. Department of Transportation, Washington, DC.

Forkenbrock, D. J., Schweitzer, L. A. (1999). Environmental justice and transportation planning, *J. Am. Plan. Assoc.*, Vol. 65, No. 1.

LBG, Inc. (2002). *Desk Reference for Estimating the Indirect Effects of Transportation Projects*, National Academy Press, Washington, DC.

USDOI (1998). *The Secretary of the Interior's Standards and Guidelines for Federal Agency Historic Preservation Programs*, pursuant to the National Historic Preservation Act Published in Final Federal Register, U.S. Department of the Interior, Washington, DC.

USDOT (2003). *Environmental Justice Effective Practices*, U.S Department of Transportation, Washington, DC, http://www. fhwa.dot.environment/ejustice/effect.

CHAPTER 18

Evaluation of Transportation Projects and Programs Using Multiple Criteria

Say not, 'I have found the truth,' but rather, 'I have found a truth.'

—*Khalil Gibran (1883–1931)*

INTRODUCTION

Up to this point, we have discussed the principles of evaluating transportation systems using a limited set of performance criteria at a time. In Chapters 5, 6 and 7, for example, we presented evaluations that are based separately on travel time, safety, and vehicle operating cost, and Chapter 8 combined these three criteria along with agency cost to evaluate projects on the basis of economic efficiency where only monetized impacts can be considered. However, it is often sought to make transportation decisions on the basis of a wider range of performance criteria that reflect the concerns of all key stakeholders (i.e., agency goals, perspectives of facility users, concerns of the society as a whole). In this chapter we discuss various techniques that can be used to make decisions when there are multiple criteria with different dimensions, both monetary and nonmonetary.

The first task in multiple criteria evaluation is to assess how decision makers attach relative levels of importance (or *weights*) to these criteria. The next task in multicriteria evaluation is *scaling* where each criterion is converted from its original dimension to one that is uniform and commensurate across all performance criteria.

After all performance criteria have been weighted and scaled, the challenge remains to *combine* the impacts for each transportation alternative. In this *amalgamation* step, an appropriate operation is used to yield a combined level of "desirability" for each alternative so that the best

choice can be identified. Several tools and techniques are employed for amalgamation, such as mathematical value or utility functions, rating and ranking, and cost-effectiveness.

Is some cases, there is no truly dominant alternative (one that is superior to all others in terms of each and every criterion). For example, one cannot absolutely maximize service levels and at the same time absolutely minimize agency costs. In multicriteria evaluation, therefore, it is often useful to establish formulations that reflect institutional or policy constraints that are typically expressed in terms of the individual performance criteria such as budgets (an agency cost constraint) and minimum average thresholds (a facility condition/performance constraint). Specifically, a decision-making mechanism based on multiple criteria can (1) help structure an agency's decision-making process in a clear, rational, well-defined, documentable, comprehensive, and defensible manner; and (2) help the agency to carry out "what-if" analyses and to investigate trade-offs between performance criteria.

In this chapter we discuss the general steps involved in the multicriteria evaluation and decision-making process (Figure 18.1). Certain approaches for multicriteria decision making, such as cost-effectiveness analysis, may not explicitly include all elements of this process. The initial steps of the process, defining the alternative transportation actions and establishing the appropriate performance criteria, are addressed in Chapters 1 and 2, respectively.

18.1 ESTABLISHING WEIGHTS OF PERFORMANCE CRITERIA

In multicriteria decision making, a key step is the explicit or implicit assignment of relative weights to each performance criterion to reflect its importance compared to other criteria; for example, to what extent is safety improvement more important than travel-time reduction, increase in facility condition, vehicle operating cost decrease, increased economic development, improved aesthetics, and so on? The following methods can be used to establish the weights: (1) equal weighting, (2) direct weighting, (3) regression-based observer-derived weighting, (4) the Delphi approach, (5) the gamble method, (6) pairwise comparison, and (7) value swinging.

18.1.1 Equal Weighting

The equal weighting approach, which assigns the same weight to all performance criteria, is simple and easy to implement. For example, it has been common practice at many agencies to simply sum up agency and user costs to obtain a single cost value upon which a decision is made, an approach that assumes implicitly that agency cost and

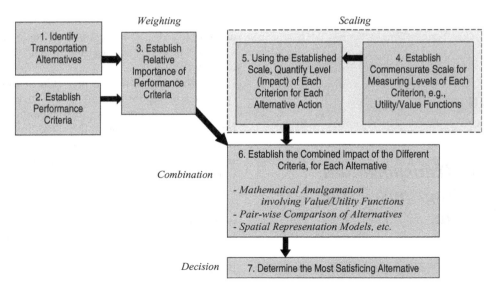

Figure 18.1   Typical steps in multicriteria decision making.

user costs have the same weight. The equal weighting approach may yield flawed results because it does not incorporate the relative preferences that may exist among criteria. For example, some agencies attach a higher level of importance to agency cost than to user cost because unlike the latter, the former is driven by agency budget and involves agency spending of actual funds.

18.1.2 Direct Weighting

In the direct weighting method, decision makers assign numerical weight values directly to performance criteria (Dodgson et al., 2001). Two approaches for direct weighting are:

1. *Point allocation.* A number of points (say, 100) are allocated among the performance criteria in proportion to their importance. Point allocation can be *global* (where the decision makers assign specific weights within a given range whose ends represent the lowest and highest levels of importance), or *local* (where the performance criteria are first placed into categories and then weights are assigned to each category, and then to criteria within each category).
2. *Ranking.* This involves a simple ordering of performance criteria by decreasing importance as perceived by the decision makers.

Of these two methods of direct weighting, point allocation is typically preferred because unlike ranking, it yields a cardinal rather than an ordinal scale of importance. Cardinality is a useful property because it

gives better meaning to the relative importance of the criteria. The local method of the point allocation approach is particularly useful when there are a large number of criteria. Direct weighting methods are generally easy to implement and are therefore useful for initial estimation of relative weights. There are other more comprehensive methods that can be used to capture more precisely the decision makers' relative preferences, as discussed below.

18.1.3 Regression-Based Observer-Derived Weighting

Regression-based observer-derived weighting is based on unaided subjective evaluations of alternative actions and their overall impact (in terms of inherently combined performance criteria), followed by analysis of the results using statistical regression (Hobbs and Meier, 2000) to identify the implicit relative weights. For each transportation alternative, survey respondents, such as agency decision-making personnel or facility users, are requested to assign scores of overall "benefit" or "desirability" for a given combination of performance criteria that is accrued by a given transportation alternative. Using statistical regression, a functional relationship is then established on the basis of each respondent's overall desirability (indicated total score) for each alternative (the response variable) and the scores assigned to individual criteria (the explanatory variables). Mathematically, this is stated:

$$\text{Minimize} \sum \varepsilon_i^2$$

subject to

$$\text{TV}_i = \sum_{j=1}^{J}(w_j V_{ji}) + \varepsilon_i \qquad \text{for } i = 1, 2, \ldots, n$$

transportation alternatives

(18.1)

where TV_i is the outcome of the transportation action i in terms of its overall benefit or desirability, w_j the weight of criterion j, V_{ji} the value of criterion j attained by undertaking transportation alternative i, J the total number of performance criteria, and ε_i the predictive error. Equation (18.1) seeks to obtain the line of best fit through the observed points (responses) that minimizes the sum of squared deviations from the line. Values of TV for each combination of performance criteria (due to a given transportation action) are provided by each respondent. The values of w that yield the line of best fit are the calibrated coefficients of the model and represent the relative weights of the performance criteria.

Example 18.1 Three alternative improvement projects are being considered by a transit agency. The impacts have been estimated as shown in Table E18.1.1. Seven members of the agency's management committee were asked to indicate their overall preference of the levels of these performance criteria on a scale of 1 (worst) to 10 (best), and to indicate the desirability of each improvement project in terms of an overall combined performance score on the same scale. The results were as shown in Table 18.1.2.

SOLUTION   A regression equation can be developed of the form

$$\text{TV}_i = w_{\text{COST}} V(\text{COST}_i) + w_{\text{TIME}} V(\text{TIME}_i)$$

Using MINITAB, SAS, SPSS or other appropriate statistical software and the survey data, the regression equation can be determined as follows:

$$\text{TV} = 0.209 V(\text{COST}) + 0.746 V(\text{TIME}) \quad R^2 = 0.988$$

Table E18.1.1   Project Costs and Time Savings

Alternative	Cost (thousands of dollars)	Decrease in Travel Time (min)
A	800	30
B	1000	25
C	500	15

Table E18.1.2

Alternative i	Impacts in Terms of Performance Criteria Scores		Total Score Assigned by Each Respondent, (TV_i)						
	$V(\text{COST}_i)$	$V(\text{TIME}_i)$	1	2	3	4	5	6	7
A	5	10	9	8	8	7	9	9	8
B	4	9	8	8	7	7	8	9	8
C	8	5	6	5	5	6	7	5	4

The coefficient of the variables are the relative weights. Therefore $w_{\text{COST}} = 0.209$, and $w_{\text{time}} = 0.746$

18.1.4 Delphi Technique

In Sections 18.1.1 to 18.1.3 we have described relative weighting of performance criteria on the basis of inputs from individual survey respondents and the use of aggregation techniques that synthesize the individual priorities (weights) to yield collective priorities. Aggregation helps to address inconsistencies in assigned weights. These methods fail, however, to address possible concerns of the survey respondents, who wish to know how their individual weight assessments stand vis-à-vis the assessments of other experts, and may thus wish to incorporate the others' assessments into theirs to yield modified weights. The Delphi technique (Dalkey and Helmer, 1963) addresses this issue. Delphi is a widely used group decision-making tool that aggregates the perspectives from individual experts for consensus building and ultimately for a holistic final assessment. In this technique, the results from the first set of questionnaire surveys are analyzed and summarized, and the summary statistics are presented to the respondents. The respondents review their original individual responses relative to the summary statistics (average and standard deviation) and make any needed adjustments to the weights they assigned originally. This cycle of iterations continues until there is no change in scores. The final scores are then averaged to yield the relative weights. In most cases, a consensus emerges after two iterations.

Example 18.2 In a bid to evaluate alternative designs and locations for a proposed parking garage in a certain city, the following performance criteria were considered: garage capacity, cost-effectiveness, and appropriateness of location.

SOLUTION   The relative weights were assessed using the direct method (described in Section 18.1.2) by a group of planners. The weights were averaged and the results (with 1 standard deviation from the mean) are shown in Figure E18.2. It can be observed that the standard deviations (vertical bars) are generally smaller after the second round, which indicates an increased level of agreement among the planners after they have revised their original responses.

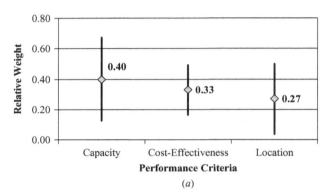

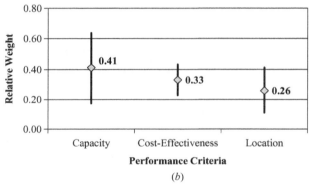

Figure E18.2 (*a*) First round; (*b*) second round.

18.1.5 Gamble Method

The gamble method (Anderson et al., 2002) assigns a weight for one performance criterion at a time by asking survey respondents to compare their preference for a guaranteed outcome (a "sure thing") against an outcome that is not guaranteed (a "gamble"). This method involves the following steps:

1. Carry out an initial (and tentative) ranking of performance criteria in order of decreasing importance. Set the first criterion at its most desirable level and all other criteria at their least desirable levels.
2. Compare between the following two outcomes:
   (a) *Sure thing*. The outcome is that the criterion in question is at its most desirable level while all other criteria are at their least desirable levels,

   (b) *Gamble*. In this outcome, all criteria attain their most desirable levels *p*% of the time, and attain their least desirable levels (1 − *p*)% of the time,

At a certain level of *p*, the two situations (sure thing and gamble) are equally desirable to the decision maker. At that level, the value of *p* is noted and taken to represent the weight for the performance criterion in question.

Step 2 is repeated for all other criteria until the weights have been determined for all criteria.

The gamble method is particularly useful for determining the relative weights of performance criteria in the outcome risk scenario (where exact outcomes of transportation actions are unknown but their outcome probabilities are known). A disadvantage of this method is that it may be difficult to comprehend or administer.

Example 18.3 Townsville City's transportation planners are assessing alternative bus routes on the basis of the following performance criteria: bus headway (varies from 5 to 15 minutes) and population served (varies from 5000 to 10,000).

SOLUTION   To establish the relative weight for the headway criterion, the following situations are considered:

1. *Sure thing*. The outcome is such that the criterion in question (headway) is guaranteed to be at its most desirable level (5 minutes) and the other criterion (population served) is guaranteed to be at its worst level (5000).
2. *Gamble*. There are two outcomes with probabilities as follows:

(a)   A *p*% chance of an outcome that has both criteria at their most desirable levels (5 minutes headway; population served 10,000).

(b)   A *q*% chance of an outcome that has both criteria at their least desirable levels (15 minutes headway; population served 5000), *q* = 1 − *p*.

Then the probability *p* is varied gradually until the threshold probability (at which the planner is indifferent between the two situations) is determined. The value of *p* that characterizes the indifference of the decision maker toward the two scenarios is the relative weight. Suppose, for illustrative purposes, the planners are found to be indifferent between the following outcomes:

1. A guaranteed outcome where the bus headway is 5 minutes and the population served is 5000.
2. Two possible outcomes with the following probabilities: a 60% chance of a 5-minute headway and a

population served of 10,000 ($p = 0.6$) and a 40% chance of a 15-minute headway and a population served of 5000 ($q = 0.4$).

Given this response, the relative weight for "headway" performance criterion is determined to be 0.6. The steps can be repeated to determine the relative weight of the population served criterion.

18.1.6 Pairwise Comparison of the Performance Criteria

Weighting can be carried out using pairwise comparison of performance criteria, and a common tool for doing this is the analytical hierarchy process (AHP). AHP establishes the weights of performance criteria by allowing the survey respondent (decision maker) to consider objective and subjective factors in assessing the relative importance of each criterion (Saaty, 1977). Using AHP, decision makers can develop weights that reflect their experience and knowledge in a natural and intuitive manner. In AHP, complex structures representing performance criteria are organized in hierarchical clusters facilitate pairwise comparisons between the criteria at each hierarchical level to estimate their relative weights. Pairwise comparisons between two performance criteria i and j can be represented using the following reciprocal matrix:

$$A = \begin{bmatrix} 1 & a_{12} & \cdots & a_{1n} \\ 1/a_{12} & 1 & \cdots & a_{2n} \\ \vdots & \vdots & \ddots & \vdots \\ 1/a_{1n} & 1/a_{2n} & \cdots & 1 \end{bmatrix}$$

where each entry a_{ij} is the decision-maker's quantified judgment of the relative importance of two criteria i and j on the basis of a scale of 1 to 9 (Table 18.1). The elements on the diagonal have a value of unity because the value 1 represents the comparison of a criterion to itself. The elements in the lower triangular matrix are reciprocals of the corresponding elements in the upper triangular matrix. A typical element of the matrix (for performance criteria i and j) can be defined as follows:

$$w_i/w_j = a_{ij}, \qquad \text{for } i, j = 1, 2, \ldots, n \qquad (18.2)$$

n is the total number of performance criteria. w_i and w_j are the relative weights of the pair of performance criteria.

After establishing the pairwise comparison matrix, the next step is to derive the relative weights. For this, Saaty (1994) proposed a procedure using matrix theory. The procedure, known as the *eigenvector approach*, establishes

Table 18.1   Ratios for Pairwise Comparison Matrix

Comparison	X/Y Ratio
Criterion X is extremely more important than criterion Y	9
Criterion X is strongly more important than criterion Y	7
Criterion X is moderately more important than criterion Y	5
Criterion X is slightly more important than criterion Y	3
Criterion X is equally important to criterion Y	1
Criterion X is slightly less important than criterion Y	1/3
Criterion X is moderately less important than criterion Y	1/5
Criterion X is strongly less important than criterion Y	1/7
Criterion X is extremely less important than criterion Y	1/9

the relative weights on the basis of recorded judgments using a reciprocal matrix. If $\mathbf{w}$ is the vector that represents the weights, the matrix relationship

$$A\mathbf{w} = n\mathbf{w} \qquad (18.3)$$

stipulates that $\mathbf{w}$ is an eigenvector of the matrix A with eigenvalue n. In reality, however, the eigenvalue of the matrix A may not be equal to n, due to the inconsistency in the decision-makers' responses. Inconsistency exists if the following triangular relationship does not hold:

$$a_{ij} = a_{ik} \times a_{kj}, \qquad \text{for } i, j = 1, 2, \ldots, n \qquad (18.4)$$

Such situations, which may result in the development of an inconsistent reciprocal matrix, generally arise when the entries a_{ij} are based on subjective judgments of the survey respondents (decision makers) rather than on exact measurements. As a result, the relationship in equation (18.3) is modified as follows:

$$A\mathbf{w} = \lambda\mathbf{w} \qquad (18.5)$$

where λ is a set of eigenvalues of the matrix A such that $\sum_{i=1}^{n} \lambda_i = n$.

When the relationship in equation (18.3) holds and the reciprocal matrix is perfectly consistent, all the eigenvalues (λ_i) are zero with the exception of one which is equal to n (the largest eigenvalue of the matrix in this scenario). On the other hand, when the reciprocal matrix is inconsistent (due to subjectivity of judgments), the largest eigenvalue, λ_{max}, is close to n, and the remaining eigenvalues are close to zero. Under these circumstances, to determine a unique set of relative weights, a vector $\mathbf{w}$ has to be found that satisfies the following relationship:

$$A\mathbf{w} = \lambda_{max}\mathbf{w} \qquad (18.6)$$

The vector $\mathbf{w}$ is the eigenvector that corresponds to the maximum eigenvalue, λ_{max}. The elements of $\mathbf{w}$ represent the weights for the performance criteria. To obtain a normalized solution, $\mathbf{w}$ is replaced by $\overline{\mathbf{w}}$, where

$$\overline{\mathbf{w}} = \frac{1}{\alpha}\mathbf{w} \qquad \text{where} \quad \alpha = \sum_{i=1}^{n} w_i \qquad (18.7)$$

The relative weights are computed as the components of the normalized eigenvector associated with the largest eigenvalue of their comparison matrix. The eigenvector of the reciprocal matrix can be computed using vector algebra, numerical methods, or available software (e.g., MATLAB) to yield the relative weights.

Consistency checks, an important step in AHP, assess the degree of randomness in the judgments used to develop the reciprocal matrix. The deviation of λ_{max} from n is used as a measure of the consistency with the reciprocal matrix developed. The logical consistency of the pairwise comparisons can be measured using the *consistency index* (CI), which is defined as:

$$\text{CI} = \frac{\lambda_{max} - n}{n - 1} \qquad (18.8)$$

The consistency index is then compared with the average consistency index of randomly generated reciprocal matrices (referred to as the *random index*, RI) to determine the level of inconsistency in the survey responses. Table 18.2 shows the random indices for matrices of order 1 through 10 (Saaty, 1994).

The overall consistency of AHP judgments can be determined using the *consistency ratio* (CR), which is computed as follows:

$$\text{CR} = \frac{\text{CI}}{\text{RI}} = \frac{\lambda_{max} - n}{(n-1)(\text{RI})} \qquad (18.9)$$

A consistency ratio of 0.1 or lower is considered acceptable (Saaty, 1994). If the ratio exceeds 0.1, then

Table 18.2 Relationship Between Matrix Order and Average Random Index

Order of Matrix (n)	Average Random Index
1	0.00
2	0.00
3	0.58
4	0.90
5	1.12
6	1.24
7	1.32
8	1.41
9	1.45
10	1.49

the judgments are considered random and the reciprocal matrix should be recomputed.

In summary, the AHP process for weighting involves the following steps:

1. Constructing a pairwise comparison matrix
2. Estimating the value of the eigenvector that reflects the relative weights
3. Checking for consistency

AHP can also be used to synthesize judgments and estimate priorities for alternatives.

Example 18.4 It is sought to use AHP to assign relative weights to three key bridge performance criteria: deck condition, superstructure and substructure condition.

SOLUTION The pairwise comparison matrix shown in Table E18.4.1 was obtained from a survey of 11 bridge engineers (BE). To illustrate the interpretation of the matrix, the italicized values in the table are explained as follows:

Table E18.4.1 Pairwise Comparison Matrix

	Deck Condition	Superstructure Condition	Substructure Condition
Deck condition	1	*7*	*5*
Superstructure condition	—	1	*1/3*
Substructure condition	—	—	1

- 7   Deck condition is considered to be *strongly more important* than superstructure condition.
- 5   Deck condition is considered *moderately more important* than substructure condition.
- 1/3   Superstructure condition is considered *slightly less important* than substructure condition.

These values are then processed numerically to arrive at the relative weights as follows:

1. The reciprocal matrix (A) is determined, and the complete matrix is established.
2. Column entries corresponding to each criterion (deck condition, superstructure condition, and substructure condition) are summed up (Table E18.4.2).
3. Each column entry is divided by the respective column sum to yield a new matrix, A_{norm}.
4. The rows of A_{norm} are summed up. Then the normalized relative weights are obtained corresponding to each performance criterion. For example, the normalized weight for deck condition is given by $2.17/(2.17 + 0.25 + 0.58) = 0.72$ (Table E18.4.3). The eigenvector ($\mathbf{w}$) corresponding to the eigenvalue λ_{max} in the matrix equation $A\mathbf{w} = \lambda_{max}\mathbf{w}$ is therefore given as $\overline{\mathbf{w}}^T = [0.72 \quad 0.08 \quad 0.19]$. Then compute $A\mathbf{w}^T$. In the next step, this column is used for calculating the consistency ratios.

5. The consistency ratio is then determined: $\lambda_{max} = (3.097 + 3.000 + 3.000)/3 = 3.0323$ (Table E18.4.4); from Table 18.2, the random index, RI = 0.58. Therefore, the consistency ratio is given as

$$CR = \frac{CI}{RI} = \frac{\lambda_{max} - n}{(n-1)(RI)} = \frac{3.023 - 3}{(3-1)(0.58)}$$
$$= 0.028 < 0.1$$

The consistency ratio does not exceed 0.1. This result indicates that the determined weights are not random and therefore are acceptable.

18.1.7 Value Swinging Method

The value swinging method involves the following steps (Goicoechea et al. 1982):

1. Consider a hypothetical situation where performance criteria are all at their worst values.
2. Determine the criterion for which it is most preferred to "swing" from its worst value to its best value, all other criteria remaining at the worst values.
3. Repeat steps 1 and 2 for all criteria.
4. Assign to the most important criterion, the highest weight in a selected weighting range (e.g., for a range of 1 to 100, assign a weight of 100), and then assign weights to the remaining criteria in proportion to their rank of importance.

Comments on Weighting: It is desirable that relative weights be stable so that decisions can be robust. In reality, however, the relative weights of performance criteria may change from time to time, across locations, project or facility type to among different stakeholders, and from one agency to another, to reflect different circumstances. As such, for a given evaluation problem, sensitivity analysis is often useful for investigating the stability of the final decision with respect to changes in the relative weights of the performance criteria.

Table E18.4.2   Complete Matrix (from Step 2)

	Deck Condition	Superstructure Condition	Substructure Condition
Deck condition	1.00	7.00	5.00
Superstructure condition	0.14	1.00	0.33
Substructure condition	0.20	3.00	1.00
Total	1.34	11.00	6.33

Table E18.4.3   Estimation of Normalized Weights

	Deck Condition	Super-structure Condition	Sub-structure Condition	Column Entries Divided by Corresponding Column Sums			Row Sums of Last 3 Columns	Normalized Weights
Deck condition	1.00	7.00	5.00	0.74	0.64	0.79	2.17	0.72
Superstructure condition	0.14	1.00	0.33	0.11	0.09	0.05	0.25	0.08
Substructure condition	0.20	3.00	1.00	0.15	0.27	0.16	0.58	0.19

Table E18.4.4   Computations for Consistency Ratios

	Weights (Priorities)	$A\mathbf{w}^T$	Weighted Row Sum
Deck condition	0.72	2.23	$2.23/0.72 = 3.097$
Superstructure condition	0.08	0.24	$0.24/0.08 = 3.000$
Substructure condition	0.19	0.57	$0.57/0.19 = 3.000$

18.2 SCALING OF PERFORMANCE CRITERIA

Another key aspect of multiple criteria evaluation is the establishment of a common unit or scale of measurement so that all performance criteria can be expressed in commensurate units to enable comparison or combination of the performance criteria. This step is referred to as *scaling* or *metricization*. In economic efficiency evaluation, for instance (Chapter 8), the required common metric of measurement is monetary (dollars), and therefore performance criteria such as travel time, safety, and vehicle operating cost are expressed in monetary value. On the other hand, certain criteria, such as economic development, land use, air quality, noise, and others, are not so easily monetized. Thus, there is a need in most multicriteria evaluation problems to provide a common metric for all monetizeable or nonmonetizeable performance criteria. For any given performance criterion, scaling involves the establishment of a dimensionless unit of desirability (e.g., utility, value) that decision makers can assign to each level of the criterion. The *value function approach* is adopted when the decision making is carried out under the *certainty scenario*, whereas the *utility function approach* is used when the decision making is being carried out under the *risk scenario*. These scenarios and their various approaches to scaling, presented in Figure 18.2, are discussed below.

18.2.1 Scaling Where Decision Making Is under Certainty

The techniques used for this scenario are based on *value theory* (Keeney and Raiffa, 1976). A value function is a scalar index of decision makers' preferences representing the respective values they attach to each level of a performance criterion, under conditions of certainty. If a scale of 0 to 100, for example, is used, the values 0 and 100 correspond to the worst and best levels, respectively, of the criterion, and the values attached to intermediate levels of the criterion are decided by the decision

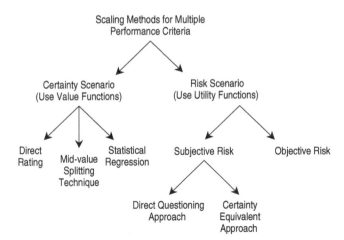

Figure 18.2   Scaling methods.

makers. A value function is therefore a mathematical representation of the decision makers' preference structure and can be linear or nonlinear. Methods for developing value functions for a performance criterion include the direct rating method, midvalue splitting technique, and regression analysis.

(*a*) *Direct Rating Method*   Direct rating is a simple method that often involves surveys to generate value functions by asking respondents (decision makers) to assign directly the values they attach to each level of a given performance criterion (Hobbs and Meier, 2000). This method is particularly useful for developing value functions for criteria that have relatively few, discrete levels.

(*b*) *Midvalue Splitting Technique*   This method solicits information from survey respondents (decision makers) regarding their "indifference" between changes in levels of a performance criterion (Keeney and Raiffa, 1979). This technique is particularly appropriate for criteria that have a large domain of possible levels.

Example 18.5   The city of Metropolis is planning to improve service along a rail transit route. Several alternative actions that involve increased numbers of stops and/or service frequencies, are being considered. The key performance criterion is the expected increase in average daily transit trips (ADTT) in thousands. An ADTT of 35,000 is considered most desirable and is assigned a value of 100 units. Develop a value function for this criterion.

SOLUTION   The midvalue splitting approach is considered appropriate for this scaling problem because the performance criterion has a wide domain, that is, it has several possible levels: from 0 to 35 (in thousands). The steps of the method are presented below.

1. Set $V(\text{ADTT} = 0) = 0$ units and $V(\text{ADTT} = 35) = 100$ units.
2. $V(\text{ADTT} = 0)$, for instance, represents the value the survey respondents (decision makers) attach to an ADTT of 0. Find X_{50} for which $V(X_{50}) = 50$ units. In this step the survey respondents are asked to indicate individually the value X_{50} such that they are equally satisfied between a change of ADTT from 0 to X_{50} and a change of ADTT from X_{50} to 35 units. Assume for the sake of illustration that the survey respondents indicate that on average, they are as satisfied with an ADTT increase from 0 to 20 as they would for an increase from 20 to 35 ADTT. Then $X_{50} = 20$.
3. Find X_{25} for which $V(X_{25}) = 25$. In this step the survey respondents are asked to indicate individually the value X_{25} such that they are equally satisfied between a change of ADTT from 0 to X_{25} and a change of ADTT from X_{25} to X_{50}. Assume for the sake of illustration that the average value of $X_{25} = 12$.
4. Find X_{75} for which $V(X_{75}) = 75$. In this step the survey respondents are asked to indicate the value X_{75} such that they are equally satisfied between a change in ADTT from X_{50} to X_{75} and a change of ADTT from X_{75} to 35. Assume for the sake of illustration that the average value of $X_{75} = 25$.
5. *Consistency check*: Determine whether there is equal satisfaction with an ADTT change from X_{25} to X_{50} as with a change from X_{50} to X_{75}. If the answer is affirmative, the results of the scaling process can be considered *consistent*. Otherwise, the survey respondents are asked to revise their responses (i.e., steps 2 to 4 are repeated). Using the values in the example, the value function of the performance criterion in question (increase in ridership, ADTT) can be constructed as shown in Figure E18.5.

(*c*) *Using Regression Analysis to Develop Value Functions*   In cases where there is a large number of respondents, regression analysis can be used to arrive at a global function that best represents their combined scaling

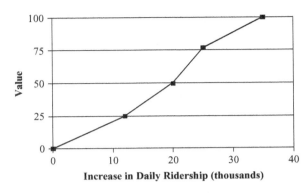

Figure E18.5   Developed value function.

preference orders. This is done by collecting all data points from individual decision makers and regressing the responses to obtain the function that exhibits the least deviation from the responses. The functional form of the value function selected can be linear, convex, concave, S-shaped, and so on.

Example 18.6   An agency conducted a survey of transportation planners and decision makers to generate data representing the value (on a scale of 0 to 100) assigned by the respondents to the additional ridership expected. The data were generated using direct rating separately for each respondent. Midvalue splitting technique could also be used. The individual data points are plotted in Figure E18.6.

SOLUTION   A polynomial functional form was used to obtain the value function. The additional ridership, ADTT, is the independent variable, and the value assigned by the respondents to respective ridership levels, V_{ADTT}, is the

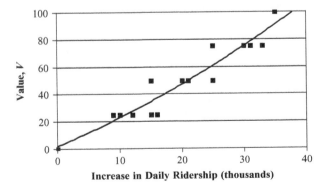

Figure E18.6   Plot of survey results.

dependent variable. The regression equation that results is

$$V_{ADTT} = 0.0293 ADTT^2 + 1.7323 ADTT$$
$$+ 0.9075 \qquad R^2 = 0.96$$

The curve representing the regression equation is shown in Figure E18.6.

18.2.2 Scaling Where Decision Making Is under Risk

In scaling, the intent is to establish a suitable scale so that levels of the performance criterion accrued by a transportation action can be ascertained. However, the inability to predict exact levels of the outcomes of transportation actions is a reality faced by agencies. This justifies the need to incorporate risk and uncertainty concepts in scaling the performance criteria. In the risk scenario (unlike the uncertainty scenario), the range and distribution of possible outcomes are known. Risk can be *subjective* or *objective*. Subjective risk is based on personal perceptions; objective risk is based on theory, experiment, or observation.

Utility functions are typically used to account for the subjective risk element in scaling the performance criteria. If the decision makers specify a certain level of "desirability" (or "utility") that they attach to each of several possible outcomes of an action, the overall utility expected for each alternative action can be calculated and the best course of action (that with the highest expected utility) can be identified for selection (Keeney and Raiffa, 1976). For a given performance criterion, a utility function provides a scale showing the decision maker preferences for different levels of that criterion. In this respect, the utility function is a generalized form of the value function because it captures the risk preferences of the decision maker for each performance criterion. The risk behavior of the decision maker can be ascertained from the utility function shape and parameter values. Possible shapes include linear, concave, convex, and S-shaped as well as others. It can be shown mathematically that a *risk-taking* decision maker has a strictly convex utility function, a *risk-averse* decision maker has a strictly concave utility function, and a *risk-neutral* decision maker has a linear utility function (Figure 18.3). The risk-taking behaviors of decision makers, in turn, reflect their *risk premium* (Winston, 1993).

In this section we first discuss techniques for developing utility functions in cases of subjective risk. These techniques generally involve a survey of decision makers, using their responses to establish the utility functions. We then describe how probability distributions can be used to develop utility functions for cases of objective risk.

The incorporation of risk and uncertainty in the scaling process does not obviate the need to consider these concepts in other aspects of the evaluation. For example, the input factors of transportation evaluation can vary widely. The impact of such variations on the final decision can be addressed using probability distributions, Monte Carlo simulation, and other tools discussed later in the chapter.

(*a*) *Developing Subjective-Risk Utility Functions: Direct Questioning Using the Gamble Approach*   To develop the utility function for a performance criterion X, the best level of the criterion (X_B) and its worst level (X_W) are assigned the following utilities: $U(X_W) = 0$ and $U(X_B) = 100$. Then the following two situations are compared:

1. Guaranteed prospect of an outcome of $X = 0.5 \times (X_B - X_W)$.
2. Risky prospect of obtaining an outcome of X_W with probability p and an outcome of X_B with probability $(1 - p)$.

The comparison is repeated by varying p until reaching a threshold point, say p^*, where the survey respondents indicate that they are indifferent between situations (1) and (2). The process is repeated for all other levels l of the criterion such as $0.25(X_B - X_W)$ and $0.75(X_B - X_W)$, and p^* is obtained for each level. A plot of p^* vs. l yields the utility function for that performance criterion.

(*b*) *Developing Subjective-Risk Utility Functions: Certainty Equivalency Approach*   The steps in this approach are as follows:

1. Select an appropriate utility scale, say, 0 to 100. Assume that the worst (least favorable) level of the criterion has zero utility [i.e., $U(X_0)$ or $U(X_{WORST}) = 0$] and that the best (most favorable) level of the criterion has maximum utility [i.e., $U(X_{100})$ or $U(X_{BEST}) = 100$].
2. Determine the level of the criterion for which the utility is halfway between the worst and the best. That is, find X_{50} or X_{MID} such that utility $U(X_{50})$ or $U(X_{MID}) = 50$. To determine X_{50}, the survey respondents (decision makers) are asked to indicate the level of the performance criterion (say, X_{50}) at which they would be indifferent between the following guaranteed situation and alternative risky situation:

(a) *Sure thing* X_{50}(probability = 1.0).

(b) *Gamble*   50% chance that the outcome is the worst level of the criterion, X_0; 50% chance that the outcome is the best level of the criterion, X_{100}. The expected utility of this gamble is $0.5U(X_0) + 0.5U(X_{100}) = 50$.

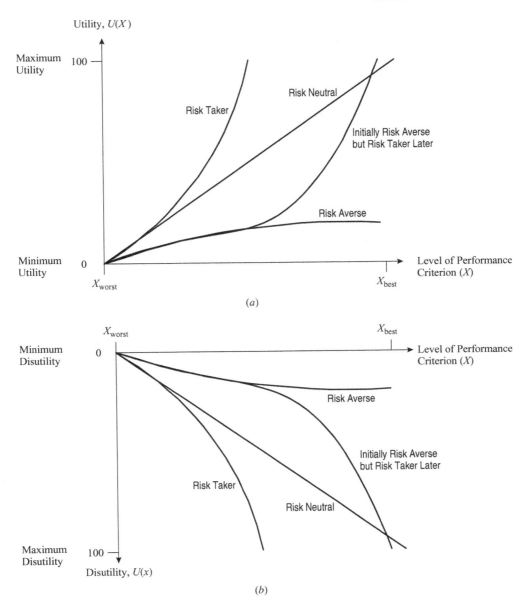

Figure 18.3   Relation between risk attitude and utility function.

The gamble is compared with the guaranteed situation (sure thing) where a specific level of the criterion X_{50} is achieved with certainty. The level at which the decision maker is indifferent between the guaranteed situation and the gamble (X_{50}) is called the *certainty equivalent* of the gamble.

3. Determine the certainty equivalent corresponding to criteria levels X_{25} and X_{75} in a similar fashion.

4. Plot the utility function corresponding to the performance criterion in question. Repeat the entire procedure for all other performance criteria under consideration.

Example 18.7   An agency seeks to develop a commensurate scale for the three performance criteria: agency cost, extent of ecological destruction, and extent of vulnerable population served (the disabled, low-income, and disadvantaged). It is sought to use the certainty equivalent approach to develop utility functions for the three criteria.

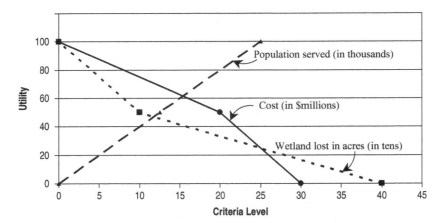

Figure E18.7   Utility functions for performance measures.

SOLUTION   (1) *Single-criterion utility function for cost:* $U_{\text{cost}}(\$30 \text{ million}) = 0$ (worst), $U_{\text{cost}}(\$0) = 1$(best).

Sure thing: The outcome is that agency cost is guaranteed to be $20 million.
Gamble: There is a 50% chance that cost = $0, and a 50% chance that cost = $30 million.

$X_{50} = \$20$ million is the certainty equivalent because the expected utility of the gamble is 0.5. Therefore, $U_{\text{cost}}(\$20 \text{ million}) = 0.5$.

(2) *Single-criterion utility function for ecological destruction* $U_{\text{ecol}}(400 \text{ acres}) = 0$ (worst), $U_{\text{ecol}}(0 \text{ acres}) = 1$(best).

Sure thing: The outcome is guaranteed that ecological destruction = 100 acres.
Gamble: There is a 50% chance that the extent of destroyed ecology = 0 acres, and a 50% chance that the extent of destroyed ecology = 400 acres.

$X_{50} = 100$ acres is the certainty equivalent because the expected utility of the gamble is 0.5. Therefore, $U_{\text{ecol}}(100 \text{ acres}) = 0.5$.

(3) *Single-criterion utility function for vulnerable population served* (in thousands): $U_{\text{pop}}(0) = 0$ (worst), $U_{\text{pop}}(25,000) = 1$ (best).

Sure Thing: There is a guaranteed outcome that the vulnerable population served = 12,000.
Gamble: There is a 50% chance that population served = 0; and a 50% chance that population served = 25,000.

$X_{50} = 12,500$ is the certainty equivalent because the expected utility of the gamble is 0.5. Therefore, $U_{\text{pop}}(12,500) = 0.5$.

The utility function for each performance criterion is shown in Figure E18.7. These plots can be made smoother by similarly computing the certainty equivalents for their intermediate points (such as X_{25} and X_{75}).

Identifying Risk-Taking Behavior Using the Certainty Equivalent Approach: The relation between the decision makers' single-criterion utility function and their risk attitude can be ascertained for the subjective risk situation. The risk premium is the difference between the value expected for the gamble and the certainty equivalent. Decision makers are described as *risk averse* if their risk premium exceeds zero, *risk neutral* if their risk premium is zero, and *risk takers* if their risk premium is less than zero.

Example 18.8 Using the information presented in Example 18.7, determine the expected value of the gamble and risk premium. Also determine the risk attitude of the decision makers with respect to all three performance criteria.

SOLUTION

Criterion: agency cost
    Certainty equivalent = $20 million.
    Expected value of gamble = $0.5 \times 0 + 0.5 \times 30 = \$15$ million.
    Risk premium $= -(15 - 20) = \$5$ million, which exceeds zero. Therefore, the decision makers have a risk-averse tendency with respect to the agency cost criterion; this is evidenced by the concavity of the utility function for this performance criterion in Figure E18.7.

Criterion: ecological destruction (in acres)

Certainty equivalent = 100 acres.

Expected value of

gamble = $(0.5 \times 400) + (0.5 \times 0)$
= 200 acres.

Risk premium = $-(200-100) = -100$ acres,
which is less than zero. With respect to the
ecological destruction criterion, therefore, the
decision makers have a risk-taking tendency;
this is reflected in the convex shape of the utility
function for this performance criterion in Figure
E18.7.

Criterion: size of vulnerable population served

Certainty equivalent = 12,500 persons.

Expected value of gamble = (0.5×0)
$+(0.5 \times 25,000) = 12,500$ persons.

Risk premium = $12,500 - 12,500 = 0$. Therefore,
the decision makers have a neutral tendency
toward risk with respect to this criterion. This is
reflected by the linear utility function curve in
Figure E18.7.

(c) *Developing Objective-Risk Utility Functions Using
Probability Distributions* In the situation where the
decision makers seek to account for, in an objective
manner, the uncertainties of the impact of a transportation
alternative in terms of a given performance criterion, the
expected utility value can be used. This parameter can be
calculated by assuming a probability distribution function
(p.d.f.) of the possible outcomes. Given the probability
of occurrence of a specific outcome in terms of a given
performance criterion, the expected utility value can then
be determined as follows:

$$E[u(X)] = \begin{cases} \sum_{x=\min}^{\max} u(x)P(X=x) \\ \qquad \text{when the p.d.f. is discrete} \quad (18.10) \\ \int_{\min}^{\max} u(x)f(x|\min < x < x|\max)\,dx \\ \qquad \text{when the p.d.f. is continuous} \quad (18.11) \end{cases}$$

Here $u(X)$ represents the utility function and x represents the outcome level corresponding to the performance
criterion X. For discrete probability distributions, the variation in x is discrete between a minimum and a maximum
value, whereas for continuous probability distributions, x
varies continuously.

**Selection of Probability Distributions for Performance
Criteria:** The distribution that best fits the possible
outcomes of a performance criterion is determined and the
parameters of the distribution function are calibrated. The

distribution could be uniform or nonuniform, depending
on the performance criterion in question. For example, the
binomial distribution can be considered for the discrete
performance criteria that have a small range of outcomes
such as bridge deck condition, traffic level of service, and
crash severity. The binomial distribution is given as

$$b(n, p, x) = \binom{n}{x} p^x (1-p)^{n-x} \quad (18.12)$$

where n is the number of possible outcomes, p the
probability of occurrence of a given outcome, and x the
number of occurrences over all possible outcomes (i.e.,
exactly x possible outcomes occur over n Bernoulli trials).
The mean and variance are:

$$\mu = np \quad \text{and} \quad \sigma^2 = np(1-p) \quad (18.13)$$

However, if the range of possible outcomes (overall
utility) is large (greater than 30), the *Poisson distribution*
can be used as an approximation to the binomial
distribution:

$$P(X=x) = b(n, p, x) = \frac{(np)^x}{x!} e^{-np} \quad (18.14)$$

The mean and variance of a Poisson distribution are
$\mu = \sigma^2 = np$.

For performance criteria that involve continuous variables such as congested lane-miles, travel-time savings,
emissions, and project costs, the outcomes are spread out
over a given range in a continuous fashion. The distribution of the possible outcomes, in terms of utility, can
be symmetric or skewed and it can be modeled as a *beta
distribution* to account for the degree of skewness and
kurtosis (Li and Sinha, 2004). The parameters in a beta
distribution include a lower limit (L), an upper limit (H),
and two shape parameters, α and β, and the density function is then given by

$$f(x|\alpha, \beta, L, H) = \frac{\tau(\alpha+\beta)(x-L)^{\alpha-1}(H-x)^{\beta-1}}{\tau(\alpha)\tau(\beta)(H-L)^{\alpha+\beta-1}}$$
$$\text{for } L \le x \le H \quad (18.15)$$

where the τ function factors serve to normalize the
distribution so that the area under the p.d.f. from L to
H is exactly equal to 1.

The mean and variance for the beta distribution are
given by

$$\mu = \frac{\alpha}{\alpha+\beta} \quad \text{and} \quad \sigma^2 = \frac{\alpha\beta}{(\alpha+\beta)^2(\alpha+\beta-1)}$$

The distribution in equation (18.15) is substituted in
equation (18.11) to determine the expected utility for

performance criteria that are described by the beta distribution. Approximate values of the shape parameters for beta distributions can be developed under various skewness and variance combinations.

18.3   COMBINATION OF PERFORMANCE CRITERIA

This step involves the amalgamation or combination (through addition, pairwise comparison, etc.) of all scaled and weighted performance criteria to determine the overall outcome of a transportation alternative and therefore to choose the best alternative. In multiple-criteria evaluation, the existence of conflicts between performance criteria leads to the issue of *Pareto optimality* and influences the manner in which the different criteria are combined to make a choice of the best transportation alternative. It is rare to find a solution (best transportation alterative) that yields the highest desired values of all benefit criteria and lowest desired values of all cost criteria. As such, a solution is Pareto-optimal if by reallocation (choosing a different transportation alternative) the decision maker cannot obtain a more desirable level of a performance criterion without obtaining a less desired level of some other criterion. A transportation alternative is not Pareto-optimal, then, if one can obtain a more desired value of a criterion without yielding a less desired value of any other criterion. After the set of all Pareto-optimal transportation alternatives (also referred to as *nondominated solutions*) has been established, the analyst seeks the alternative that best achieves a compromise between all competing objectives (i.e., the most *satisficing* solution).

The common tools for combining performance criteria and using the combined measures to choose the best alternative are (1) mathematical functions of value, utility, or cost-effectiveness; (2) ranking and rating; (3) maxmin approach; (4) impact index method; (5) pairwise comparisons; (6) mathematical programming; and (7) outranking method. In some of these methods, the penultimate tasks of scaling or weighting of performance criteria (or both) could be only implicit or even excluded altogether. For example, a cost-effectiveness analysis may proceed without explicitly weighting or scaling the criteria.

18.3.1 Combined Mathematical Functions of Value, Utility, or Cost-Effectiveness

This technique combines the individual criterion utility functions or values into a single combined function or value. The combination generally takes one of two forms:

1. A *difference* between all benefits and all costs
2. A *ratio* of all benefits to all costs

All benefits refers to the combined values of utilities of the benefit performance criteria. Similarly, *all costs* refers to the combined values of the cost criteria. The combined benefits may simply be the value of a single benefit performance criterion or may be the combined level of several scaled and weighted benefit performance criteria, such as travel-time reduction, safety enhancement, and condition improvement. Similarly, the total level of costs may be the value of a single cost performance criterion or may be the combined level of several scaled and weighted cost performance criteria, such as initial cost, recurring costs, and other related costs. Where a ratio form is used, the denominator (cost) typically comprises only agency costs, while changes in nonagency costs are incorporated in the numerator.

(a) Difference Approach   In the difference approach,

$$\text{total value or utility} = \text{total value or utility of benefits} \\ - \text{total value or utility of costs}$$

Irrespective of whether the difference or ratio approach is used, the combined value of benefits may be the summed benefits from individual benefit performance criteria (additive function) or may be the multiplied product of benefits. As explained in the discussion below, for either benefits or costs, the manner of combination is influenced by the mathematical assumptions about decision makers' preference structure. Common examples of the difference approach include the computation of net present value (NPV) and equivalent uniform annual return (EUAR) in economic efficiency analysis (Chapter 8).

$$\text{NPV} = \text{present value of all benefits} \\ - \text{present value of all costs}$$

$$\text{EUAR} = \text{equivalent uniform annual benefits} \\ - \text{equivalent uniform annual costs}$$

NPV and EUAR computations are specific forms of the combined (amalgamated) function where all the benefit performance criteria (such as travel time and other user cost savings) are assigned positive values and all the cost performance criteria (such as agency initial or life-cycle costs) are assigned negative values.

(b) Ratio Approach   In the ratio approach,

$$\text{ratio} = \frac{\text{total value or utility of benefits}}{\text{total value or utility of costs}}$$

Here, the value of all benefits appears in the numerator either as a single value from one benefit performance criterion, or as a scaled, weighted, and amalgamated value from multiple benefit criteria. The values of all costs (appearing in the denominator) are calculated similarly. An example of the combination of benefit and cost performance criteria using the additive functional form of the ratio approach is the benefit–cost ratio (BCR) computation in economic efficiency analysis (Chapter 8):

$$BCR = \frac{\text{value of all benefits}}{\text{value of all costs}}$$

Another variation of the ratio approach is the incremental benefit–cost ratio (IBCR), where the combined impact of each transportation alternative is the ratio of the incremental benefit relative to that of some base alternative, to the incremental cost relative to that of the base alternative.

(c) Cost-Effectiveness Another example of the amalgamation of benefit and cost performance criteria using a ratio approach is the cost-effectiveness method. Here, unlike the BCR approach, the benefits and costs are not necessarily expressed in the same metrics; costs are typically measured in dollars whereas benefits are typically expressed in specific units unique to the other criterion. Cost-effectiveness analysis facilitates examination of trade-offs between conflicting performance criteria. Also, this type of evaluation is typically used when the evaluation involves higher level performance attributes such as societal values and agency goals and objectives. In cost-effectiveness analysis, the best alternative is selected through compromise (Thomas and Schofer, 1970). The process often involves determining, for each alternative, what level of benefits is achieved at a given cost (typically, a budgetary ceiling or a threshold of a combined cost performance criterion), and what monetary cost or combined (monetary and nonmonetary) cost is associated with a specified level of combined benefits of a single criterion or of multiple criteria.

Discussion for Approaches (a), (b), and (c): How to Select the Most Appropriate Functional Form Irrespective of which of the above three approaches is used to combine the benefit functions and cost functions, an important issue is the selection of an appropriate mathematical form for "combining" the various benefit performance measures, or the various cost performance measures, or both. The selection will depend on whether the benefit and cost functions are of utility or value function types.

The case of value functions: A theorem in value theory states that given the performance criteria $Z_1, Z_2, \ldots, Z_p$, the additive value function shown below exists if and only if the criteria are *mutually preferentially independent* (Keeney and Raiffa, 1976):

$$v(z_1, z_2, \ldots, z_p) = \sum_{i=1}^{p} v_i(z_i)$$

where v_i is a single-criterion value function over the criterion Z_i.

For two mutually exclusive and collectively exhaustive subsets of the set $Z \equiv \{Z_1, Z_2, \ldots, Z_p\}$: X and Y, the set of criteria X is *preferentially independent* of the complementary set Y if and only if the conditional preference structure in the **x** space given **y**′ does not depend on **y**′. In other words, the preference structure among the criteria in set X does not depend on the levels of the criteria in Y. This is known as the *concept of preferential independence*. Symbolically, if $(\mathbf{x}_1, \mathbf{y}_0)$ is preferred to $(\mathbf{x}_2, \mathbf{y}_0)$, then $(\mathbf{x}_1, \mathbf{y})$ is preferred to $(\mathbf{x}_2, \mathbf{y})$ for all **y**. The set of criteria Z is mutually preferentially independent if every subset X of these criteria is preferentially independent of its complementary set of criteria.

The case of utility functions: A theorem in utility theory states that given the criteria $Z_1, Z_2, \ldots, Z_p$, the multiplicative utility function shown below exists if and only if the criteria are *mutually utility independent* (Keeney and Raiffa, 1976). k and k_i are scaling constants:

$$ku(z_1, z_2, \ldots, z_p) + 1 = \prod_{i=1}^{p} [kk_i u_i(z_i) + 1]$$

where u_i is a single-criterion utility function over the criterion Z_i.

For two mutually exclusive and collectively exhaustive subsets of the set $Z \equiv \{Z_1, Z_2, \ldots, Z_p\}$: X and Y, the set of criteria X is *utility independent* of set Y if and only if the conditional preference order for *lotteries* involving only changes in the levels of attributes in X does not depend on the levels at which the attributes in Y are held fixed. This is known as the *concept of mutual utility independence*. Symbolically, if $\langle \mathbf{x}_1, \mathbf{y}_0 \rangle$ is preferred to $\langle \mathbf{x}_2, \mathbf{y}_0 \rangle$ then $\langle \mathbf{x}_1, \mathbf{y} \rangle$ is preferred to $\langle \mathbf{x}_2, \mathbf{y} \rangle$ for all **y**. The symbol "$\langle \ \rangle$" represents a lottery: i.e., it captures the risk preference of decision maker in the presence of uncertainty. The set of criteria Z are mutually utility independent if every subset X of these criteria is utility independent of its complementary set of criteria.

Another important theorem states the existence of additive utility function: Given the criteria $Z_1, Z_2, \ldots, Z_p$, the following additive utility function exists if and only if the additive independence condition holds among the criteria (Keeney and Raiffa, 1976):

$$u(z_1, z_2, \ldots, z_p) = \sum_{i=1}^{p} k_i u_i(z_i)$$

where u_i is a single-criterion utility function over the criterion Z_i.

This means that preferences over lotteries on Z_1, Z_2, $\ldots$, Z_p depend only on their marginal probability distributions and not on their joint probability distribution.

For a given evaluation problem, the appropriateness of the multiplicative and additive functional forms for the multi-attribute value or utility function can be checked by eliciting information on decision makers' preference structures and ascertaining how well the survey data support the underlying assumptions stated in the above theorems.

(d) Indifference Curves   A concept that is closely related to the mathematical forms of the combinations of performance criteria is that of indifference curves. They can be graphically represented and they are particularly useful for examining trade-offs between criteria. To illustrate this concept, consider the situation where a transit agency is investigating five alternatives geared at increasing travel-time reliability as well as passenger comfort and convenience. These two performance criteria can be considered as being in conflict with each other because efforts to ensure on-time arrivals may lead to reduced comfort and convenience of passengers. Figure 18.4 shows the degree to which the alternative

actions separately achieve the performance criteria. It should be noted that the graph shown in the figure may differ for scaled and unscaled values of the performance measures.

When the mathematical form (showing how the scaled and weighted performance criteria are combined) is known, it is possible to ascertain the marginal rates of substitution between performance criteria. For example, assume that the only two performance criteria are travel-time reliability and passenger comfort and convenience, that the former is twice as important as the latter, and that the composite function is linear and additive: TV = 2.0 TTR + 1.0 PCC.

With this mathematical form of performance criteria amalgamation, it is possible to generate a set of *trade-off lines* (also referred to as *indifference curves*). Each indifference curve represents all the various ways by which the individual performance criteria can be combined to yield the total value (*TV*) for a given alternative (Figure 18.5).

Indifference curves not only help to investigate trade-offs between conflicting criteria but can also provide an indication of the superior alternative, the alternative that lies near the highest combined value line. Thus, in the example shown in Figure 18.5, alternative A_2 lies closest to the indifference curve of greatest total value. As such, if travel-time reliability and passenger comfort and convenience are the only performance criteria to be considered, A_2 is the best alternative.

18.3.2 Ranking and Rating Method

There are several possible approaches to ranking and rating transportation alternatives for purposes of selection.

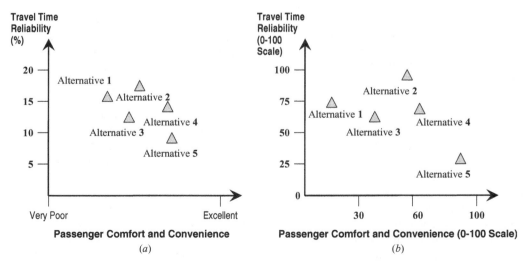

Figure 18.4 Levels of performance for alternatives: (*a*) unscaled; (*b*) scaled.

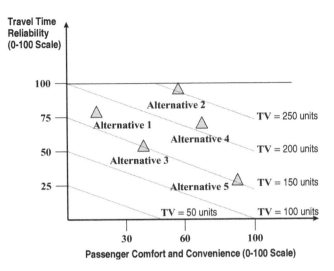

Figure 18.5 Indifference curves using mathematical forms of utility/value function for combined performance measures.

One of the early applications of this approach is the *rank-based value method* (Schlager, 1968). It is assumed that all performance criteria are measured in the same units or scale (otherwise, the analyst should carry out the necessary scaling procedures before proceeding with this method). The steps for ranking and rating are as follows:

1. Establish the weights of each performance criterion j, w_j. This is referred to as *weighting* (or *ranking* of the performance criteria)
2. Establishing a pre-established scale, the degree to which performance criterion j is achieved by implementing alternative i, O_{ij}. This is referred to as *scaling* (or rating of the alternative under criterion j).
3. Determine the score associated with performance criterion j and alternative i, $w_j O_{ij}$. This is referred to as *scoring*. For each alternative action i, sum up

scores for all performance criteria to yield a *total score* for that alternative, $\sum_{j=1}^{J} w_j O_{ij}$.
4. Choose the alternative with the highest final score.
5. In certain evaluation problems, alternatives have different probabilities of implementation. In that case, the computed final score for each alternative should be adjusted by multiplying it by its implementation probability as follows: $P_i \sum_{j=1}^{J} w_j O_{ij}$.

Example 18.9 Three alternative transportation plans are being considered for a region. Table E18.9 presents the weights, performance criteria, probability of implementation, and the extent to which each criterion would be attained upon implementing each alternative plan. Which alternative should be selected?

SOLUTION   Sample calculation: For alternative A, total score $= 0.75[(5 \times 3) + (4 \times 1) + (1 \times 2) + (3 \times 3) + (2 \times 1)] = 24$. On the basis of the total score, alternative B is most preferred.

18.3.3 Maxmin Approach

If the ratings (outcomes) are not fixed but rather, have a range or probability distribution, the classical maxmin approach can be used. Here, the minimum (least desired outcomes) are determined for all the alternatives, and the alternative that has the maxmin (maximum of these minimums) is selected. This method is particularly appropriate when the decision makers are risk averse. In another variation of this problem, there exist thresholds for each rating type (performance criterion outcome), and the decision makers seek the alternative that provides the maxmin outcome without violating any threshold. These are referred to as *satisficing noncompensatory models*. Where the ratings are associated with complete uncertainty, Bell (2006) describes classical methods (the Bellman–Zadeh fuzzy decision principle) and a new *possibility theory* that could be used to ascertain the best alternative.

Table E18.9   Examples of Rank-Based Value Method

Transportation alternative:		A	B	C
Probability of implementation:		0.75	0.97	0.82
Performance criteria weights and ratings	Economic efficiency	5	3	2	1
	Economic development	4	1	2	3
	Minimization of community disruption	1	2	1	3
	Accessibility	3	3	1	2
	Environmental impact minimization	2	1	3	2
Total score			24	27.16	24.6

18.3.4 Impact Index Method

The impact index method is a variation of the ranking and rating method. It was first used for the evaluation of alternative alignments for Interstate 75 near Marietta, Georgia (Zieman et al., 1971). The procedure considers possible errors associated with impact measurement. The impact index for an alternative i is calculated as follows:

$$I_i = \sum_{j=1}^{J} R_j S_j X_{ij} + e_j R_j S_j X_{ij} \qquad (18.16)$$

where R_j = relative weight of performance criterion j, $= w_j / \sum_{j=1}^{J} |w_j|$, in this approach, weights can be positive or negative, and the R value is obtained by dividing the weight by the sum of absolute values of all weights,

S_j = scaling factor for measurement X of the performance criterion j, $= 1/\max(X_{1j}, X_{2j}, \ldots, X_N j)$

X_{ij} = extent to which an alternative i achieves a performance criterion j,

j = $1, 2, \ldots, J$ (the total number of performance criteria under consideration),

N = number of alternatives,

w_j = unscaled weight of the performance criterion j,

e_j = random number drawn from probability distribution in the absence of a known distribution, a uniform rectangular distribution is considered; assuming $\pm 50\%$ error, the random value of e ranges as follows: $(-0.5 \le e \le +0.5)$.

For each alternative action, the deterministic value of the impact index can be obtained by setting $e = 0$. In the probabilistic approach, a uniformly distributed random number generator (assuming a rectangular error distribution) is used to compute impact index values for each alternative through Monte Carlo sampling. After generating 30 or more index values, the average impact, standard deviation, and confidence intervals can be plotted to make judgments about selecting a desirable alternative. The confidence interval for the combined impact of alternative i is as follows: $\overline{I}_i \pm (\text{STDEV}_i / \sqrt{m}) t_{1-\alpha/2, m-1}$, where $\overline{I}_i$ is the average total impact value of alternative i, STDEV_i is the standard deviation of the impact value of alternative i. α the level of significance, m the number of random impact values, and t the t-distribution value. The decision makers can then select the desirable best

alternative on the basis of these plots. The more favorable the average impact index value and the smaller the confidence interval, the more likely an alternative is chosen.

Example 18.10 A metropolitan planning organization is considering five alternative projects along a corridor. Each alternative involves a mix of activities, such as bypass road construction, LRT services, and various traffic and transit operational improvements. It is sought to select the best alternative on the basis of a number of performance criteria, as shown in Table E18.10.1. Weights are assigned to represent the relative importance of the criteria on an absolute value scale of 1 to 10; negative and positive values indicate undesirable and desirable impacts, respectively.

SOLUTION   The unscaled, scaled, and weighted impacts of each performance criterion for each transportation alternative are presented in Table E18.10.2. The absolute weight and relative weight (in parentheses) are given in column 3. The cell entries in columns 4 to 8 are as follows:

- The first entry is the unscaled impact, X, of the performance criterion in its specific units.
- The second entry (italicized) is the scaled impact $(S \times X)$ of the criterion in a common unit.
- The third entry is the weighted and scaled impact $(R \times S \times X)$ of the criterion.
- The fourth entry is the probabilistic weighted and scaled impact $[(R \times S \times X) + (e \times R \times S \times X)]$ of the criterion.

The sum of probabilistic impact indices shown in Table E18.10.2 is only for a single run. Using the given probability distribution and parameters, a Monte Carlo simulation is run several times (at each run, e takes a value between -0.5 and $+0.5$ of the weighted and scaled impact of each performance criterion and transportation alternative). In this example, the simulation was run 30 times, and the statistics (mean, standard deviation, upper and lower limits of 95% confidence interval) of the relative impact were determined and plotted (Figure E18.10). The results indicate that alternatives 1, 3, and 4 have better performance than alternatives 2 and 5. In the first group, alternative 3 has the highest mean value but a larger standard deviation. Also there is an overlap of the confidence intervals of alternatives 3 and 1. Consequently, the consideration of other criteria is required before a clear and unequivocal choice can be made among alternatives 1, 3, and 4.

Table E18.10.1 Performance of Alternatives

Performance Criterion	Unit	Weight	Expected Impact on Performance Criterion due to Alternative A_i				
			A_1	A_2	A_3	A_4	A_5
Land and wetland taken for project ROW	mi²	−5	1.1	0.8	0.4	0.68	0.72
Number of water bodies affected	number	−3	21	5	17	3	14
Agency cost (present worth)	$M	−10	308	430	505	380	415
User cost (present worth)	$M	−5	1130	2150	1320	1580	1830
Air pollution added	tons	−2	130	190	210	160	175
Noise pollution added	dB	−1	5.1	6.5	4.4	4.3	4.8
Economic development	jobs	+8	250	167	340	180	190
Community disruption	index (1–5)	−1	3	5	1	6	4
Visual quality	index (1–10)	+2	7.5	8.0	7.2	6.5	9.1

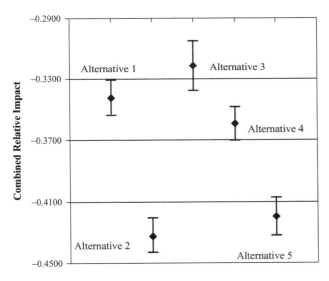

Figure E18.10 Plot of confidence intervals.

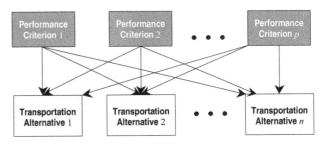

Figure 18.6 Schematic structure for evaluation using AHP pairwise comparison.

18.3.5 Pairwise Comparison of Transportation Alternatives Using AHP

Pairwise comparisons across sets of transportation alternatives and their consequences can be carried out to establish priorities among the alternatives. The analytical hierarchy process, a tool for weighting and scaling steps, can also be used for such pairwise comparisons. The structure of the decision problem can be illustrated in a hierarchical form as shown in Figure 18.6. The priorities for

alternatives are estimated on the basis of two sets of pairwise comparisons: (1) pairwise comparisons of all criteria, and (2) pairwise comparisons of all alternatives for each performance criterion—this is feasible primarily when the set of alternatives is small. The process yields a cardinal ranking of the alternatives. The following example shows how pairwise comparisons are used to evaluate and rank alternative projects on the basis of scaled and weighted multiple performance criteria.

Example 18.11 The city of Metropolis plans to establish a rapid transit service to connect the suburban areas to the downtown. Three alternative routes are being evaluated along with the do-nothing option on the basis of the following criteria: construction cost, area of lost wetland, and population within a specified distance of the transit service, all relative to the do-nothing alternative. Evaluate

Table E18.10.2   Computation of Impact Index Values

Performance Criterion, j	Unit	Weight, w_j (R_j in parenthesis)	Expected Impact on Performance Criterion due to Alternative A_i					Criterion Scaling Factor, S_j
			A_1	A_2	A_3	A_4	A_5	
Land and wetland taken for project ROW	mi^2	−5 (−0.1351)	1.1 mi *1* −0.1351 −0.1346	0.8 *0.73* −0.0983 −0.0979	0.4 *0.36* −0.0491 −0.0490	0.68 *0.62* −0.0835 −0.0832	0.72 *0.66* −0.0885 −0.0881	1/1.1 = 0.9091
Number of water bodies affected	number	−3 (−0.0811)	21 *1* −0.0811 −0.0804	5 *0.24* −0.0193 −0.0191	17 *0.81* −0.0656 −0.0651	3 *0.14* −0.0116 −0.0115	14 *0.67* −0.0541 −0.0536	1/21 = 0.0476
Agency cost of construction, operation, and maintenance	$M	−10 (−0.2703)	308 *0.61* −0.1648 −0.1605	430 *0.85* −0.2301 −0.2241	505 *1* −0.2703 −0.2632	380 *0.75* −0.2034 −0.1980	415 *0.82* −0.2221 −0.2163	1/505 = 0.0020
User cost of vehicle operation, safety, and travel time (O_4)	$M	−5 (−0.0351)	1130 *0.53* −0.0710 −0.0716	2150 *1* −0.1351 −0.1361	1320 *0.61* −0.0830 −0.0836	1580 *0.74* −0.0993 −0.1000	1830 *0.85* −0.1150 −0.1159	1/2150 = 0.00047
Air pollution	tons	−2 (−0.0541)	130 *0.62* −0.0335 −0.0325	190 *0.91* −0.0489 −0.0475	210 *1* −0.0541 −0.0525	160 *0.76* −0.0412 −0.0400	175 *0.83* −0.0450 −0.0438	1/210 = 0.0048
Noise pollution	dB	−1 (−0.0270)	5.1 *0.76* −0.0212 −0.0219	6.5 *1* −0.0270 −0.0279	4.4 *0.68* −0.0183 −0.0189	4.3 *0.66* −0.0179 −0.0184	4.8 *0.74* −0.0200 −0.0206	1/6.5 = 0.0154
Economic development	jobs	+8 (0.2162)	250 *0.74* 0.1590 0.1634	167 *0.50* 0.1062 0.1091	340 *1* 0.2162 0.2222	180 *0.53* 0.1145 0.1176	190 *0.56* 0.1208 0.1242	1/340 = 0.0029
Community disruption	index (1–5)	−1 (−0.0270)	3 *0.50* −0.0135 −0.0132	5 *0.83* −0.0225 −0.0220	1 *0.17* −0.0045 −0.0044	6 *1* −0.0270 −0.0264	4 *0.67* −0.0180 −0.0176	1/6 = 0.1667
Visual quality	index (1–10)	+2 (0.0541)	7.5 *0.82* 0.0446 0.0449	8.0 *0.88* 0.0475 0.0479	7.2 *0.80* 0.0428 0.0431	6.5 *0.71* 0.0386 0.0389	9.1 *1* 0.0541 0.0545	1/9.1 = 0.1099
Sum of probabilistic weighted and scaled impacts			**−0.3063**	**−0.4176**	**−0.2712**	**−0.3210**	**−0.3771**	

the alternatives using the impacts and pairwise weights (local priorities) provided in Table E18.11.1

SOLUTION (1) Determine the pairwise judgments between criteria (Table E18.11.2).

(2) Determine the pairwise judgments between alternatives A, B, C, and D with respect to each criterion (Table E18.11.3).

(3) Determine final scores (weighted and scaled performance levels) of alternatives A, B, C, and D

Table E18.11.1   Data for the Alternatives

Alternative	Cost (millions of dollars)	Wetland Lost (acres)	Population Served
	Criteria, i		
A	0	0	0
B	20	100	12,500
C	30	400	25,000
D	25	120	15,000

(Table E18.11.4). *Sample calculation*: For alternative A, total weighted score = $(0.6 \times 2/9) + (0.6 \times 1/9) + (0.06 \times 2/3) = 0.24$ On the basis of final scores alternative C is the best choice.

(4) Check consistency ratios for the performance criteria.

Compute $A\mathbf{w}^T$ using the normalized local weights ($\mathbf{w}$) and the reciprocal matrix (A) corresponding to each criterion (cost, wetland lost, and population served) (Table E18.11.5). Then compute the consistency ratios using equation (18.9) as follows:

Consistency ratio for cost:

$$\lambda_{max} = \frac{4.08 + 4.06 + 4.07 + 4.02}{4} = 4.0575$$

$$CR = \frac{CI}{RI} = \frac{4.0575 - 4}{(4-1)(0.90)} = 0.0213 < 0.1$$

Consistency ratio for wetland lost:

$$\lambda_{max} = \frac{4.18 + 4.23 + 4.15 + 4.12}{4} = 4.17$$

$$CR = \frac{CI}{RI} = \frac{4.17 - 4}{(4-1)(0.90)} = 0.063 < 0.1$$

Consistency ratio for population served:

$$\lambda_{max} = \frac{(4.00 + 4.01 + 4.01 + 4.01)}{4} = 4.0075$$

$$CR = \frac{CI}{RI} = \frac{4.0075 - 4}{(4-1)(0.90)} = 0.0028 < 0.1$$

Therefore, the evaluation results are consistent for all three performance criteria.

18.3.6 Mathematical Programming

Mathematical programming can be used to solve multicriteria decision-making problems by finding the optimal combination of performance criteria that maximizes an objective function against a given set of constraints. These often involve an objective Function that is comprised of different scaled and weighted performance measure. Instead of using a weighted sum or product of the multiple objectives, the weighted Tchebycheff method (Stever,

Table E18.11.2   Pairwise and Normalized Criteria Weights

(*a*) Pairwise Weights

Criteria Weights	Cost	Wetland Lost	Population Served
Cost	1	2	1/3
Wetland Lost	1/2	1	1/6
Population Served	3	6	1
Column sum	**4.5**	**9**	**1.5**

(*b*) Normalization of Weights

Criteria Weights	Column Entries Divided by Corresponding Column Sums from Table E18.11.2(a)			Row Sums for 3 Previous Columns	Normalized Weights
Cost	2/9	2/9	2/9	2/3	2/9
Wetland lost	1/9	1/9	1/9	1/3	1/9
Population served	2/3	2/3	2/3	2	2/3

Table E18.11.3   Normalized Scores for the Alternatives under Each Performance Criterion

	Alt. A	Alt. B	Alt. C	Alt. D	Column Entries Divided by Corresponding Column Sums				Row Sum of Last Four Columns	Normalized Scores
Cost										
Alt. A	1	3	7	5	0.6	0.55	0.64	0.63	2.4	0.60
Alt. B	0.33	1	2	1	0.2	0.18	0.18	0.13	0.69	0.17
Alt. C	0.14	0.5	1	1	0.08	0.09	0.09	0.13	0.39	0.10
Alt. D	0.20	1	1	1	0.12	0.18	0.09	0.13	0.52	0.13
Column sum	1.67	5.5	11	8						
Wetland Lost										
Alt. A	1	3	9	5	0.61	0.57	0.60	0.63	2.4	0.60
Alt. B	0.33	1	4	1	0.2	0.19	0.27	0.13	0.79	0.20
Alt. C	0.11	0.25	1	1	0.07	0.05	0.07	0.13	0.32	0.08
Alt. D	0.20	1	1	1	0.12	0.19	0.07	0.13	0.51	0.13
Column sum	1.64	5.25	15	8						
Population Served										
Alt. A	1	0.25	0.13	0.2	0.06	0.06	0.06	0.05	0.22	0.06
Alt. B	4	1	0.5	1	0.22	0.24	0.24	0.24	0.93	0.23
Alt. C	8	2	1	2	0.44	0.47	0.47	0.48	1.86	0.47
Alt. D	5	1	0.5	1	0.28	0.24	0.24	0.24	0.99	0.25
Column sum	18	4.25	2.13	4.2						

Table E18.11.4   Final Scores for the Alternatives

Alt.	Normalized Scores			Criterion	Weights	Alt.	Final Score
	Cost	Wetland Lost	Population Served				
A	0.6	0.6	0.06	Cost	2/9	A	0.24
B	0.17	0.2	0.23	Wetland Lost	1/9	B	0.22
C	0.10	0.08	0.47	Population Served	2/3	C	0.34
D	0.13	0.13	0.25			D	0.21

Table E18.11.5   Computations for Consistency Ratios

	Cost			Wetland Lost			Population Served		
Alt.	Normalized Scores	$A\mathbf{w}^T$	Weighted Cost	Normalized Scores	$A\mathbf{w}^T$	Weighted Value	Normalized Scores	$A\mathbf{w}^T$	Weighted Value
A	0.6	2.45	2.45/0.60 = 4.08	0.6	2.51	2.51/0.60 = 4.18	0.06	0.22	0.22/0.06 = 4.00
B	0.17	0.7	0.70/0.17 = 4.06	0.2	0.83	0.83/0.20 = 4.23	0.23	0.93	0.93/0.23 = 4.01
C	0.1	0.4	0.40/0.10 = 4.07	0.08	0.32	0.32/0.08 = 4.15	0.47	1.87	1.87/0.47 = 4.01
D	0.13	0.52	0.52/0.13 = 4.02	0.13	0.52	0.52/0.13 = 4.12	0.25	0.99	0.99/0.25 = 4.01

1989) uses distance metrics for the amalgamation process, thereby providing a theoretical basis for *goal programming*. The goal programming concept requires the decision maker to provide relative weights and targets of the performance goals. Alternatives are then ranked on the basis of the closeness of their attained performance criteria to respective established or threshold values (also referred to as *goals*). The best alternative is one whose attained performance levels are at the least distance from the goals. Figure 18.7 illustrates the goal programming concept. The relative weights of the goals and the target level for each goal are specified as inputs. The technique selects the alternative i that is associated with the minimal value of the following distance-measuring goal programming function:

$$Z = \left\{ \sum_{j=1}^{J} [w_j^* | G_j - V_j(A_{ij})|]^p \right\}^{1/p} \qquad (18.17)$$

where, Z represents the sum of deviations from the goal, $V_j(A_{ij})$ is the value function of alternative i corresponding to goal j, G_j represents the target or goal j, and w_j^* is the relative weight of goal j. By minimizing Z, the weighted deviation from the goals is minimized.

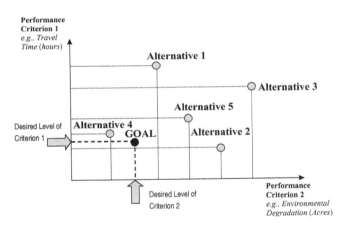

Figure 18.7   Concept of goal programming.

Various norm metrics can be used in the minimization of the goal programming function and the parameter p is varied to determine the type of distance metric being measured. The three most commonly considered metric norms in goal programming are as follows:

If $p = 1$, a *city block distance* is used as the measure of closeness to the goals; if $p = 2$ and ∞, a *Euclidean distance* and *minmax distance* (or the *infinity norm*), respectively, are used.

The infinity norm minimizes the maximum distance between the goal and the alternatives. The infinity norm of the vector Z is defined as $\max |w_j^*| G_j - V_j(A_{ij})||$. The minmax function is important because it provides a way to consider the impact of the worst deviation. However, the city block distance approach has been found to be more robust in some statistical environments because compared to the other high-power error metrics, it is less vulnerable outliers.

The following example illustrates use of the goal programming technique in combining the multiple performance criteria.

Example 18.12   In Example 18.11 we considered four alternatives for a rapid transit service connecting suburban areas to downtown Metropolis. Assume that the agency has set the following goals: a maximum project cost of \$5 million, at least 20,000 people served, and total wetland loss not exceeding 50 acres. Determine the best project alternative using the goal programming method. Compare the result with that obtained using AHP in Example 18.11.

SOLUTION   Assuming linearity, the value functions are established as shown in Table E18.12.1. The agency sets goals of \$5 million for cost, 50 acres for wetland affected, and 20,000 for population served. The linear value function assumption then results in $G_{cost} = V_{cost}(5) = 0.833$, $G_{wetland} = V_{wetland}(50) = 0.875$, and $G_{pop} = V_{pop}(20,000) = 0.8$. Assuming that normalized weights are $w_{cost} = 0.26$, $w_{wetland} = 0.26$, and $w_{pop} = 0.48$ for each alternative, the distances from goals are as follows:

Table E18.12.1   Estimation of Value Functions

	Alternative A (Do Nothing)	Alternative B	Alternative C	Alternative D
Cost	$V_{cost}(0) = 1$	$V_{cost}(20) = 1/3$	$V_{cost}(30) = 0$	$V_{cost}(25) = 1/6$
Wetland	$V_{wetland}(0) = 1$	$V_{wetland}(100) = 0.75$	$V_{wetland}(400) = 0$	$V_{wetland}(120) = 0.7$
Population served	$V_{pop}(0) = 0$	$V_{pop}(12,500) = 0.5$	$V_{pop}(25,000) = 1$	$V_{pop}(15,000) = 0.6$

- *Alternative A:* $D_A = \{(0.26)|0.833 - 1|]^p + (0.26)|0.875 - 1|)^p + (0.48)|0.8 - 0|]^p\}^{1/p}$
- *Alternative B:* $D_B = \{(0.26)|0.833 - 1/3|]^p + (0.26)|0.875 - 0.75|)^p + (0.480)|0.8 - 0.5|]^p\}^{1/p}$
- *Alternative C:* $D_C = \{(0.26)|0.833 - 0|]^p + (0.26)|0.875 - 0|)^p + (0.48)|0.8 - 1|]^p\}^{1/p}$
- *Alternative D:* $D_D = \{(0.26)|0.833 - 1/6|]^p + (0.26)|0.875 - 0.7|)^p + (0.48)|0.8 - 0.6|]^p\}^{1/p}$

Substituting for $p = 1$, 2, and ∞ yields the city block, Euclidean, and minmax distances, respectively (Table E18.12.2).

Table E18.12.2   Distance Metrics

| Transportation | Distance Metric | | |
Alternative	$p = 1$	$p = 2$	$p = \infty$
A	0.460	0.387	0.384
B	0.307	0.197	0.144
C	0.540	0.329	0.228
D	0.315	0.203	0.173

For $p = \infty$, the minmax distances from goals, D, are computed as follows:

- *Alternative A:* $D_A = \max\{0.043, 0.033, 0.384\}$
- *Alternative B:* $D_B = \max\{0.130, 0.033, 0.144\}$
- *Alternative C:* $D_C = max\{0.217, 0.228, 0.096\}$
- *Alternative D:* $D_D = \max\{0.173, 0.046, 0.096\}$

Irrespective of distance metric, alternative B has an outcome that has the least distance to the established goals and is therefore recommended for selection. Using the pairwise comparison approach (Example 18.11), the selected alternative was C.

18.3.7 Pairwise Comparison of Alternatives Using the Outranking Method

The outranking method belongs to a class of decision-making techniques that yield an ordinal ranking of the alternatives. The method can be used to compare two alternatives at a time to determine if one is sufficiently superior to the other. An alternative A_1 outranks alternative A_2 only if the following two conditions hold:

1. The *concordance index* $CI(A_1, A_2)$ exceeds a threshold value, m (referred to as a *concordance parameter*). $CI(A_1, A_2)$ is defined as the number of performance criteria for which alternative A_1 is superior to A_2.
2. The *discordance index*, $DI(A_1, A_2)$ is 0. $DI(A_1, A_2)$ is the number of criteria for which the performance of A_2 is superior to that of A_1 by an amount that exceeds an established amount q. Each performance criterion has its specific value of q.

Example 18.13   Use the outranking method to determine the transit alternative for the city of Metropolis in Example 18.11. Assume a concordance parameter m of 0.501; discordance parameters for construction cost, wetland lost, and population served are as follows: $q_{cost} = 25$, $q_{wetland} = 400$, and $q_{pop} = 8000$. Also assume the following weights: $w_{cost} = 50$, $w_{wetland} = 25$, and $w_{pop} = 90$.

SOLUTION   Determine the normalized weights of the performance criteria as follows: $w_{cost} = 50/(50 + 25 + 90) = 0.303$, $w_{wetland} = 0.152$, and $w_{pop} = 0.545$. Alternative A is superior to alternatives B, C and D with respect to construction cost and wetland lost. (This is expected, because alternative A is the do-nothing alternative.) The sum of weights of these two criteria is 0.455 (Table E18.13). Hence, the matrix elements in the first row and second, third, and fourth columns (a_{12}, a_{13}, a_{14}) are 0.455. However, alternatives B, C, and D are superior to alternative A in terms of the population served criterion. Hence, the elements a_{21}, a_{31}, and a_{41} are equal to 0.545. Other elements of the concordance matrix are determined similarly.

Table E18.13   Use of the Outranking Method

| | Concordance Matrix | | | | | Discordance Matrix | | | |
| | Alternative | | | | | Alternative | | | |
Alternative	A	B	C	D	Alternative	A	B	C	D
A	—	0.455	0.455	0.455	A	—	1	1	1
B	0.545	—	0.455	0.455	B	0	—	1	0
C	0.545	0.545	—	0.545	C	1	0	—	0
D	0.545	0.545	0.455	—	D	0	0	1	—

Discordance matrix: In the example given, it is seen that alternatives B, C, and D are superior to alternative A with respect to the population served by a value that exceeds the tolerable threshold of 8000, and hence the elements a_{12}, a_{13}, and a_{14} of the discordance matrix are equal to 1. Although alternative A is superior to alternative B with respect to the construction cost and the wetland performance criteria, the cost and wetland thresholds of $25 million and 400 acres, respectively, are not exceeded. Therefore, the element a_{21} of the discordance matrix is zero. Other elements of the discordance matrix are determined similarly.

Outranking result: alternative 1 outranks alternative 2 only if CI $(1, 2) > m = 0.501$ and DI$(1, 2) = 0$. The following outranking relations are established: B > A; D > A; C > B; D > B; and C > D. Hence, C > D > B > A; alternative C appears to be the best choice. This is consistent with the evaluation results in Example 18.11 where AHP was used for pairwise comparison.

18.4 CASE STUDY: EVALUATING ALTERNATIVE PROJECTS FOR A TRANSPORTATION CORRIDOR USING MULTIPLE CRITERIA

A study conducted by a state transportation department to analyze the traffic conditions at a certain interstate freeway corridor determined that congestion on the interstate and other arterial routes in its vicinity will increase substantially by 2020. A number of alternative projects were recommended. These included highway improvements and light-rail transit. It is sought to select the best project or strategy on the basis of multiple criteria. Input data for the evaluation are provided at various steps of the study.

1. *Identify the alternatives.* The alternatives are as follows:

- *Alternative 1:* Do nothing during the analysis period (ending year 2020). This is referred to as the baseline alternative.
- *Alternative 2:* Construct a new arterial road.
- *Alternative 3:* Construct light-rail transit (LRT).

2. *Determine the performance criteria and their relative weights.* The goals and performance criteria for evaluating the projects and in the development of alternative priorities are presented in Table 18.3. The weights indicated were obtained from the decision makers using the direct weighting method.

3. *Establish the scale.* The first step in scaling is to establish a common metric for quantifying the performance criteria. Levels of all the performance criteria were measured in comparison with the baseline conditions. For example, congestion was measured as the percentage increase in the congested lane-miles compared to the baseline condition. The data are shown in Table 18.4. The second step in scaling is to convert the levels of the performance criteria into dimensionless entities using utility functions. The utility functions were developed to account for subjective risk by incorporating the risk-taking attitude of the decision makers. The objective risk associated with the outcomes corresponding to each performance criteria is assumed to be negligible. The utility functions shown in Figure 18.8 were calibrated using results from a survey of panelists using the certainty equivalency approach. A discrete utility function was developed for scaling the socioeconomic impact into dimensionless units. Using the utility functions developed, the change in performance levels corresponding to each alternative can be converted to dimensionless values. The dimensionless utility values for each performance criterion $u_i(x_{ij})$ and for various alternative are shown in the fourth through sixth columns of Table 18.5.

Table 18.3 Performance Criteria and Weights

Performance Criteria (Goal)	Description	Weight (Scale: 0–100)
Congestion on I-5 and I-205 (C1)	Lane-miles congested on I-5 and I-205 during the evening peak period	100
System delay (C2)	Delay for vehicles on all study area roadways during the evening peak period	70
Environmental impact (C3)	PM10: particulates emission	80
Sociocultural impact (C4)	Residences and businesses displaced	60
Project costs (C5)	Highway or transit capital and maintenance cost (millions of 2001 dollars)	60

Table 18.4   Level of Criteria in Comparison with Baseline

| Performance Criterion | Performance Measure Compared to Baseline | | | Change in Performance Measure Compared to Baseline | | |
	Baseline: Alt. 1	New Arterial: Alt. 2	LRT: Alt. 3	Baseline: Alt. 1	New Arterial: Alt. 2	LRT: Alt. 3
C1	30.40%	25.20%	13.00%	0	−17.0%	−57.0%
C2	21,450 h.	17,200 h.	13,100 h.	0	−20.0%	−39.0%
C3	0%	14%	1%	0	14.00%	1.00%
C4	—	Minor change	Moderate change	No impact	Minor change	Moderate change
C5	$291M	$947M	$3087M	$0	$656M	$2796M

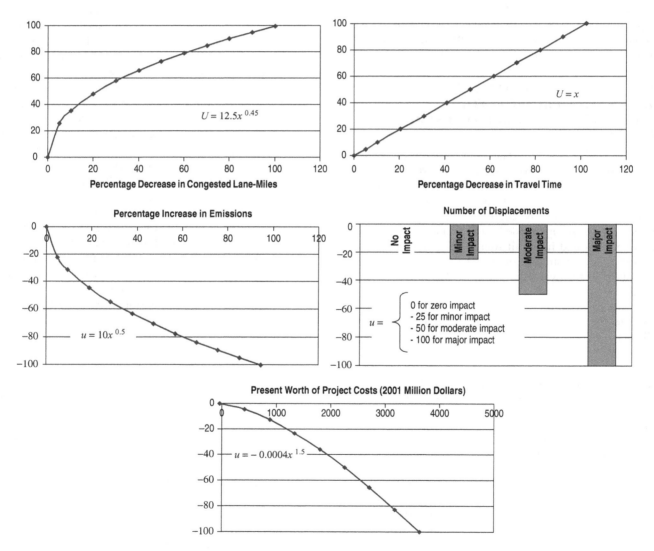

Figure 18.8   Utility functions for the performance criteria.

Table 18.5   Weights and Utility Values of the Performance Criteria

Criterion	Weight	Relative Weight	Unweighted Utilities $u_i\ (x_{ij})$ A1: Do Nothing	A2: New Arterial	A3: LRT	Weighted Utilities A1: Do Nothing	A2: New Arterial	A3: LRT
C1	100	0.27	0	44.7	77.1	0	12.07	20.82
C2	70	0.19	0	20	39	0	3.80	7.41
C3	80	0.22	0	−37.4	−10	0	−8.23	−2.20
C4	60	0.16	0	−25	−50	0	−4.00	−8.00
C5	60	0.16	0	−6.7	−59.1	0	−1.07	−9.46
TOTAL	**370**					**0**	**2.57**	**8.57**

4. *Calculate the combined impact of all performance criteria.*

For each project i, the combined utility in terms of all five criteria can be calculated using an additive utility function:

$$U_i = \sum_{j=1}^{5} w_j u_i(x_{ij})$$

where w_j is the relative weight of performance criterion j; x_{ij} the expected level of the criterion j due to project i, and $u_i\ (x_{ij})$ the utility associated with level x_{ij} (columns 7, 8, and 9 in Table 18.5). Therefore, on the basis of the total weighted utility values, LRT is the best alternative.

18.5   GENERAL CONSIDERATIONS OF RISK AND UNCERTAINTY IN EVALUATION

It is important to consider the effect of the variability of input factors and uncertainty in the outputs. In the deterministic approach, sensitivity analysis is used for this purpose, whereas the probabilistic approach does it by using subjective risk, objective risk and uncertainty techniques. With deterministic analysis, only a single value of each input factor is used in the analysis, and the output is also a single number. The probabilistic approach, on the other hand, uses a range of values for each input factor and yields a range of values for the output. In the objective risk approach, in particular, the range of values for each input factor depends on the probability distribution (and associated parameters of the distribution) for that factor. The probability distribution and parameters can be derived using analysis of historical data, expert opinion, or both.

Risk and uncertainty are concepts that are applicable to evaluation irrespective of the number of criteria involved, but their consideration becomes particularly relevant in cases of multiple criteria, due to the increasing variability of inputs and outcomes that are encountered when the evaluation involves a large number of performance criteria. For instance, variations in unit costs, travel demand, technology, and economic indicators (e.g., interest rate) can result in different outcomes in terms of the levels of performance criteria. Section 4.2.3 in Chapter 4 discusses risk as an element of agency cost, and explains that variations in such costs are quite common in practice and are attributable to uncertainties in estimation, natural disasters and other factors. The methodologies presented in Chapters 5 to 17 for evaluating transportation projects are generally deterministic in nature, but it is fairly straightforward to incorporate stochastic elements in these analyses. In Section 18.2.2 we discuss the different types of risk: subjective and objective. Generally, the process of incorporating risk is referred to as *risk analysis* (which is aptly named because there is some "risk" that the true outcome may deviate from that predicted using the deterministic approach).

18.5.1 The Case of Certainty: Using Sensitivity Analysis

In sensitivity analysis, the input factor under investigation is varied incrementally while all other factors are held constant. For each input factor, the analysis is carried out for the entire range of possible values. The best, worst, and most likely cases are then identified. A graphical presentation of the results (output vs. the input factor) is useful in assessing the impact of any possible variability of the input factor. Drawbacks to the use of sensitivity analyses include the difficulty of capturing effects of variability simultaneously among several factors (because combinations of discrete input changes require a very large number of separate analyses (Walls and Smith, 1998).

18.5.2 The Case of Objective Risk: Using Probability Distributions and Simulation

One way of incorporating risk in evaluation is to carry out a probabilistic analysis that treats inputs as ranges of values and assigns a likelihood of occurrence to those values and allows for simultaneous variability among inputs. The outputs of probabilistic analysis are also ranges of values with calculated likelihoods of occurrence. This is done using simulation, a mathematical technique that captures the effect of natural variability of model inputs on results. Values are randomly sampled from an input probability distribution, and each randomly selected input is used to determine a single outcome iteration.

Risk analysis combines probabilistic descriptions of uncertain input parameters with computer simulation to characterize the risk associated with possible outcomes (Harnett, 1975; Walls and Smith, 1998). In effect, risk analysis allows the probability of a specific outcome to be predicted. The outcome of each input parameter is modeled using a probability distribution that best fits the data observed for that variable. *Monte Carlo simulation* is a risk analysis method where random sampling procedures are used for treating deterministic mathematical situations, thus permitting incorporation into the evaluation process of the inherent risk associated with each input parameter. The general procedure for a Monte Carlo simulation is shown in Figure 18.9.

A random trial process is then initiated to establish a probability distribution function for the deterministic situation being modeled. During each iteration of the process, a value for each parameter is selected randomly from the probability distribution defining that parameter. The random values are entered into the calculation and an output value is obtained, and the process is repeated. The appropriate number of iterations for an analysis is a function of the number of input parameters, the complexity of the situation modeled, and the precision desired for the output (Herbold, 2000).

Thus, the expected outcome (e.g., crash reduction, congestion reduction, facility condition enhancement, net present value, etc.) that corresponds to a given set of input variables can be computed. The final result of a Monte Carlo simulation is a probability distribution describing the output parameter (Figure 18.10). Monte Carlo simulation can be used irrespective of criteria combination and evaluation method used. Example 18.10 shows how Monte Carlo simulation is used in the impact index method.

How does an analyst establish a probability distribution for input variables? In the absence of existing reliable probability distributions for the variables, historical data can be collected and statistical analyses of the data must be

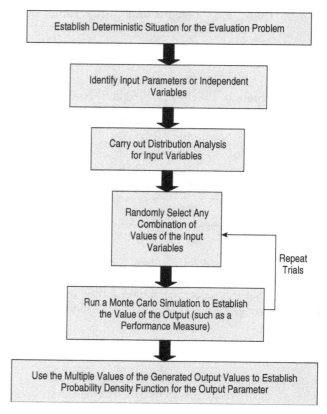

Figure 18.9 General Monte Carlo simulation approach.

carried out. Statistical analysis includes the development of frequency and cumulative frequency tables and curves, probability and cumulative probability tables and curves, and calculation of the measures of central tendency and variability. From the resulting shape of the curves, the analyst can identify the most appropriate distribution for the evaluation input factor under consideration. Common probability distributions that are encountered in engineering analysis include the normal, triangular, beta, and uniform distributions. In Section 18.2.2, details of a number of probability distributions are discussed. After having identified the appropriate probability distribution and statistical parameters for all key input factors, the analyst should then carry out simulation modeling using such information. Simulation uses randomly selected sets of values from probability distributions of input variables and calculates discrete results arrayed in the form of a distribution covering all possible outputs.

After simulation has been carried out, a variety of tools can be used to investigate the resulting impacts of input factor variations on the evaluation output. *Correlation analysis* is used to explain the relationship between the

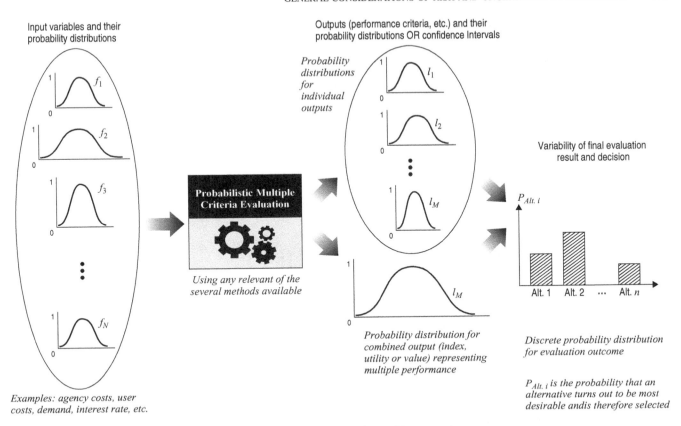

Figure 18.10   Probability distributions of inputs and outputs.

input and output variables: A positive correlation between input and output suggests a direct relationship, whereas a negative correlation suggests an inverse relationship. A *tornado graph* can be used to determine the relationship of each input variable to the output and displays the factors in order of the strength and direction of their correlation with the simulation output. *Extreme tail analysis* is used to identify the input factors that "drive" the tails of the distribution of the evaluation output by consistently producing worst- and best-case scenarios.

18.5.3 The Case of Uncertainty

For situations where project attributes are known either with certainty or under risk (where input variables have known probability distributions or can be ascertained subjectively), the procedures for choosing one project from a set of alternative projects have been discussed in preceding sections of the chapter. On the other hand, for handling uncertainty (where input values are not known with certainty and their probability distributions are also unknown), available tools are not so straightforward. Averbakh (2001) stated that for making decisions under

uncertainty, a risk-averse decision maker makes decisions based on the worst cases: A set of possible scenarios is determined deterministically (a scenario is some specific realization of the problem parameters), and the objective is to find a solution that performs reasonably well for all scenarios. In the *minmax regret* (or *robust*) version of the worst-case approach, the decision maker does not know which scenario will occur but seeks the transportation alternative that minimizes the worst-case loss in the objective function value (Bell, 2006).

Another useful model for dealing with uncertainty in multicriteria evaluation is Shackle's model, one of the original approaches to decision making under uncertainty (Shackle, 1949). It is particularly relevant in cases where the probability of project implementation depends on both the project outcome and the measure of uncertainty. Unlike the expected utility approaches (which use weighted averages of outcomes and their probabilities), Shackle's model uses three concepts: (1) the *degree of surprise concept* (instead of probability) as a measure of uncertainty, (2) the *priority index* as a mechanism to evaluate different outcomes and the corresponding degrees

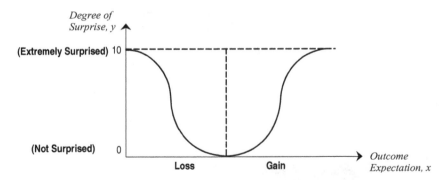

Figure 18.11   Degree of surprise function.

of surprise, and (3) the *standardized focus gain-over-loss ratio* as the basis of comparing different projects. The procedure for applying Shackle's model is discussed below.

(*a*) *Degree of Surprise Function*   This concept represents the decision maker's degree of uncertainty regarding the hypothetical outcomes of a project implementation with respect to benefits (gains) and costs (losses). Suppose that there is a range of outcomes T_k from an action ($k = 1, 2, \ldots, K$). Assign a degree of surprise y to represent the extent of the decision maker's expectation that the action would yield a certain outcome ranging from 0 (no surprise) to 10 (extremely surprised). Then the degree of surprise function $y = y(x)$ can be as shown in Figure 18.11.

(*b*) *Priority Function and Focus Values*   The priority function $\varphi = \varphi(x, y)$ represents the weighting index φ (0 for lowest priority and 10 for highest priority) that the decision maker assigns to any given outcome. Each outcome is represented by the degree of surprise coordinate (x, y). The priority function possesses properties of $(\delta\varphi/\delta x) > 0$; $(\delta\varphi/\delta y) > 0$. In other words, higher priority is given to increases in "gain" outcomes (e.g., transportation benefits) and to decreases in the degree of surprise. There also exists a *priority indifference curve* $\varphi[x, y(x)]$ that traces out different combinations of the outcome [degree of surprise pair (x, y)] with the weight φ kept constant (Figure 18.12). The priority function is defined by points at which the degree of surprise function $y = y(x)$ intersects the priority indifference curves $\varphi[x, y(x)] \equiv 0, 1, 2, \ldots, 10$. In a bid to maximize the priority function, the decision maker is faced with two maximum values, termed the *focus gain* and *focus loss* from expectation.

(*c*) *Standardized Focus Gain-over-Loss Ratio and Function*   The focus gain and focus loss values are associated with a

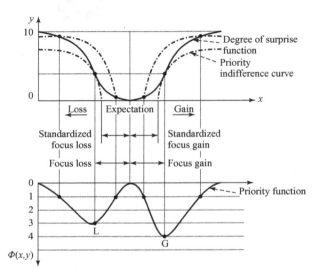

Figure 18.12   Priority function and indifference curve.

certain degree of surprise, [i.e., $y(x) \neq 0$]. It is therefore useful for the decision maker to explore the respective gain and loss values on the same priority indifference curves that are attached with a zero degree of surprise: namely, to find x for which $\varphi(x, y)$ is maximized with $y = 0$. The gain and loss values with zero degree of surprise are called *standardized* gain and loss values. As shown in Figure 18.13, different standardized focus gain-over-loss ratios are associated with different ranges of deviations in the possible outcomes. After the standardized gain-over-loss ratio function is determined, the ranking of any transportation alternative over another is carried out on the basis of the ratios of standardized focus gain-over-loss pairs. The alternative with the largest ratio is the selected. An example application of Shackle's model to transportation project selection is found in Li and Sinha (2004).

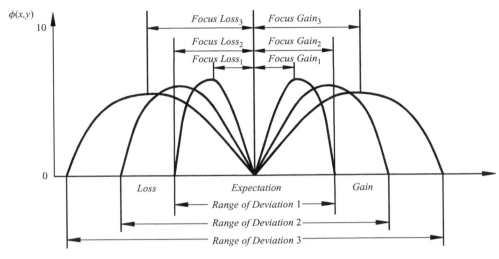

Figure 18.13 Standardized gain and loss values for different ranges of standard deviation from expectation.

SUMMARY

Due to the multiplicity of agency goals, facility user perspectives, stakeholder concerns, and in-house management systems and programs, a preliminary step in multiple criteria evaluation is to establish a list of performance criteria to be used in the evaluation. Then to account for the various levels of importance that decision makers have for different performance criteria, any of several weighting methods may be used. Also, given the differences in units used to measure the impacts of the transportation action for each performance criterion, there is a need to establish a common, often dimensionless scale so that the impacts of the action in terms of the criteria may be combined or compared with each other. After all performance criteria have been weighted and scaled either directly or implicitly, the impact of each transportation action, expressed as a desirability, utility, or value in terms of all the criteria, can then be determined. There are a variety of approaches that could be used to combine performance criteria and hence to select the best alternative on the basis of the combined impacts. Often, there is no truly dominant alternative (one that is superior to all others in terms of every criterion), certain transportation alternatives may exhibit superiority in terms of some criteria, while other alternatives may be superior in terms of other criteria. It is therefore often useful for the decision maker to carry out trade-off analysis to ascertain how much of a criterion can be "exchanged" for a given level of another. Furthermore, there is often a need to incorporate the elements of risk and uncertainty to incorporate the variability in input factors and the concomitant variation in the evaluation output and final decision.

The use of multiple criteria in evaluation can help an agency to structure its decision-making process in a clear, rational, well-defined, documentable, comprehensive, and defensible manner, and can facilitate "what-if" and trade-offs analyses. In this chapter we examined the concepts and theories involving the process of multiple-criteria decision making and presented examples and a case study to illustrate the process.

EXERCISES

18.1. Transportation planners at the city of Townsville seek to establish the relative weights of the performance criteria needed for selecting the best strategy for mitigating congestion on the I-777 interstate freeway that passes through the city. The planners intend to use the following criteria: travel time, crash rate, and air pollutant emissions. Using AHP, determine the relative weights of these criteria. The comparison matrix based on decision makers' judgments is given in Table EX18.1. Investigate the consistency of the judgments.

Table EX18.1   Comparison Matrix

	Travel Time	Crash Rate	Air Pollutant Emissions
Travel time	1	1/8	1/3
Crash rate	—	1	3
Emissions	—	—	1

18.2. For the problem in Exercise 18.1, the level of each criterion corresponding to the alternatives is given in Table EX18.2. Develop utility functions for each criterion using the certainty equivalency approach assuming appropriate responses from the decision makers. It is given that the decision maker is risk averse with respect to the crash rate and air pollutant emission criteria and has a neutral tendency toward risk with respect to the travel-time savings criterion. Select the best alternative using the additive utility function approach for amalgamation. Use the relative weights that were obtained for the three criteria in Exercise 18.1.

Table EX18.2   Expected Performance for Alternatives

Alternative	Travel-Time Savings (min)	Annual Crash Rate (crashes/ 10^6 VMT)	Air Pollutant Emissions (kg $\times 10^3$)
A	10	8	30
B	7	13	12
C	15	23	15
D	12	15	20

18.3. Use the outranking method to determine the best congestion mitigation alternative for the city of Townsville in Exercise 18.2. Assume that concordance parameter m is 0.5 (threshold value) and the discordance parameters for travel-time savings, crash rate, and pollutant emission represent the tolerable thresholds $q_{time} = 10$, $q_{crash} = 10$, and $q_{emission} = 11,000$. Use the relative weights obtained in Exercise 18.1.

18.4. The transportation planners at a metropolitan planning organization used the midvalue splitting technique to determine X_0, X_{25}, X_{50}, X_{75}, and X_{100} corresponding to a criterion X that represents transit ridership (in thousands). The values assigned by five decision makers are as shown in Table EX18.4. If

Table EX18.4   Decision Makers' Responses

Value	DM-1	DM-2	DM-3	DM-4	DM-5
0	5	5	5	5	5
25	9	10	13	8	11
50	16	20	21	14	18
75	25	25	27	20	23
100	30	30	30	30	30

it is known that a minimum of 5000 riders and a maximum of 30,000 riders will use the transit service, develop a value function for transit ridership using regression.

18.5. The construction of a new runway is being considered by the planning authorities at the Urbanville International Airport. The planners have identified three criteria that will be used in the decision-making process for the selection of the best alternative: ecological impact, economic development impact, and operating cost savings for the aircraft operators. Three alternatives were identified for runway construction. The scores assigned by planners to each alternative with respect to each criterion on a scale of 0 to 100 are given in Table EX18.5. Determine the relative weights of the criteria using regression.

Table EX18.5   Scores for Alternatives

Alternative	Individual Scores with respect to:			
	Ecological Impact	Economic Development Impact	Operating Cost Savings	Total Value
A	80	60	95	90
B	65	35	80	75
C	35	75	60	60

18.6. A new light-rail transit system is proposed for the city of Burgsville to link the suburbs and the downtown area. Three criteria have been identified by decision makers based on which the best alternative strategy for implementation of transit will be selected: forecast transit ridership, socioeconomic impact, and number of jobs created (economic development). The relative weights for ridership, socioeconomic impact, and jobs created are given as 0.25, 0.40, and 0.35, respectively. The ridership and the jobs created are measured in thousands. The socioeconomic impact is measured in terms of number of residences and businesses displaced. The following goals have been set by the city planners: 12,000 for transit ridership, 23 for households and businesses displaced, and 15,000 for jobs created. Given the following value functions shown in Table EX18.6, use goal programming to determine the best alternative.

Table EX18.6   Value Functions for Goal Programming

	Alternative A (Do Nothing)	Alternative B	Alternative C	Alternative D
Ridership	$V_{rider}(0) = 0$	$V_{rider}(11) = 0.4$	$V_{rider}(15) = 0.65$	$V_{rider}(20) = 1$
Socio-ecocomic	$V_{soc}(0) = 1$	$V_{soc}(32) = 0.75$	$V_{soc}(40) = 0$	$V_{soc}(25) = 0.5$
Jobs	$V_{jobs}(0) = 0$	$V_{jobs}(12) = 0.5$	$V_{jobs}(20) = 1$	$V_{jobs}(17) = 0.75$

REFERENCES[1]

Anderson, R. M., Hobbs, B. F., Bell, M. L (2002). Multi-objective decision-making in negotiation and conflict resolution, in *Encyclopedia of Life Support Systems*. EOLSS Publishers Co. Ltd., Oxford, UK.

*Averbakh, I. (2001). On the complexity of a class of combinatorial optimization problems with uncertainty, *Math. Program.*, Vol. 90, pp. 263–272.

*Bell, M. G. H. (2006). Rational decision-making under uncertainty in transport: minimizing maximum regret, *Proc., 5th International Conference on Traffic and Transportation Studies*, Xi'an, China, pp. 13–21.

Dalkey, N., Helmer, O. (1963). An experimental application of the Delphi method to the use of experts., *Manage. Sci.*, Vol. 9.

*Dodgson, J., Spackman, M., Pearman, A., Phillips, L. (2001). *Multicriteria Analysis: A Manual*. Department for Environment, Food and Rural Affairs, London.

Harnett, D. L. (1975). *Introduction to Statistical Methods, 2nd ed.*, Addison-Wesley, Reading, MA.

Herbold, K. (2000). Using Monte Carlo simulation for pavement cost analysis, *Public Roads*, Nov.–Dec.

*Hobbs, B., Meier, P. (2000). *Energy Decision and the Environment: A Guide to the User of Multicriteria Methods*, Kluwer Academic, Norwood, MA.

*Keeney R. L., Raiffa, K. (1976). *Decisions with Multiple Objectives: Preferences and Value Tradeoffs*, Cambridge University Press, Cambridge, UK.

Li, Z., Sinha, K. C. (2004). *Multi-criteria Highway Programming Incorporating Risk and Uncertainty: A Methodology for Highway Asset Management System*, Tech. Rep. FHWA/IN/JRTP-2003-21, Joint Transportation Research Program, Purdue University, West Lafayette, IN.

Saaty, T. L. (1977). A scaling method for priorities in hierarchical structures, *J. Math. Psych.*, Vol. 15.

————— (1994). *Fundamentals of Decision Making and Priority Theory with the Analytic Hierarchy Process*, RWS Publications, Pittsburgh, PA.

Schlager, K. (1968). *The Rank-Based Expected Value Method of Plan Evaluation*, Hwy. Res. Rec. 238, National Research Council, Washington, DC pp. 153–158.

Shackle, G.L.S. (1949). *Expectation in Economics*, 2nd Edition, Cambridge University Press, Cambridge, UK.

Stever, R.E. (1989). *Multiple Criteria Optimization: Theory, Computation, and Application*. Krieger, Melbourne, FL.

Thomas, E.N., Schofer, J.L. (1970). *Strategies for the Evaluation of Alternative Transportation Plans*, NCHRP Rep. 96, Highway Research Board, National Research Council, Washington, DC.

Walls, J., III, Smith, M. R. (1998). *Life-Cycle Cost Analysis in Pavement Design: Interim Technical Bulletin*, Tech. Rep. FHWA-SA-98-079, Federal Highway Administration, U.S. Department of Transportation, Washington, DC.

Winston L. W. (1993). *Operations Research: Applications and Algorithms*, Duxbury Press, Belmont, CA.

Zieman, J. C., Shugart, H. H., Bramlet, G. A., Ike, A., Champlin, J. R., Odum, E. P. (1971). *Optimum Pathway Analysis Approach to the Environmental Decision-Making Process*, Institute of Ecology, University of Georgia, Athens, GA.

ADDITIONAL RESOURCES

Belton, V., Stewart, T. J. (2003). *Multiple Criteria Decision Analysis*. Kluwer Academic, Norwood, MA.

Chankong, V., Haimes, Y. Y. (1983). *Multi-objective Decision-Making: Theory and Methodology*, North-Holland, New York.

Figueira, J., Greco, S., Ehrgott, M., Eds. (2004). *Multiple Criteria Decision Analysis, State of the Art Surveys*, Springer, New York, NY.

Goicoechea, A., Hansen, D. H., Duckstein, L. (1982). *Multi-objective Decision Analysis with Engineering and Business Applications*, Wiley, New York.

Journal of Multicriteria Decision Analysis, Wiley, New York, NY.

Keeney, R. L., Gregory, R. S. (2005). Selecting attributes to measure the achievement of objectives, *Oper. Res.*, Vol. 53, No. 1.

Lootsma, F. A. (1999). *Multi-criteria Decision Analysis via Ratio and Difference Judgement*, Kluwer Academic, Norwood, MA.

Speicher, D., Schwartz, M., Mar, T. (2000). *Prioritizing Major Transportation Improvement Projects: Comparison of Evaluation Criteria*, Transp. Res. Rec. 1706, Transportation Research Board, National Research Council, Washington, DC.

Steuer, R. E. (1989). *Multiple Criteria Optimization: Theory, Computation, and Application*. Krieger, Melbourne, FL.

[1]References marked with an asterisk can also serve as useful resources for multiple-objective decisionmaking.

Triantaphyllou, E. (2004). *Multi-criteria Decision Making Methods: A Comparative Study*, Kluwer, Norwood, MA.

Yoon, K.P., Hwang, C.-L. (1995). *Multiple Attribute Decision making: An Introduction*. Sage University paper series on Quantitative Applications in the Social Sciences, 07-104. Thousand Oaks, CA: Sage.

Younger, K. (1994). *Multimodal Project Evaluation: A Common Framework, Different Methods*, Transp. Res. Rec. 1429, Transportation Research Board, Washington, DC.

Zeleny, M. (1982). *Multiple Criteria Decision-Making*, McGraw-Hill, New York.

CHAPTER 19

Use of Geographical and Other Information Systems

The most successful person is the one with the best information.

—*Benjamin Disraeli (1804–1881)*

INTRODUCTION

Information management, which generally refers to the handling of information in a manner that ensures efficient access to intended end users, consists of data collection, data storage, data retrieval, and data manipulation. In recent years, the discipline has been extended to include knowledge management, which seeks to transform data into "wisdom" that helps an agency to retain the valuable experience and knowledge of departing experts. In the context of transportation systems evaluation and decision making, the task of information management is critical because, as implied in earlier chapters, incorrect investment decisions can be made if there is a lack of accurate or timely information on the costs and benefits of alternative system interventions. Therefore, at all phases of the project development process and at all stages of an evaluation process, information management is a key ingredient. In this chapter we present a framework for overall information management for transportation system evaluation with particular focus on the use of geographical information system (GIS) tools. The framework of information management can be divided broadly into three components: data, hardware, and software. The data comprise a collection of facts about the physical features, operating policies and characteristics, agency and user costs, and other attributes of the transportation system. The data can be in the form of text, numbers, still images, videos and so

on. The hardware is a physical system that is used to collect and manage the data, such as hard copy files and folders or a computer (for electronic data). For electronic data, an algorithm or a set of instructions are developed for the computer for analyzing the data and for generating information based on this analysis. This set of instructions is referred to as a *computer program* or *software*. The hardware and software components expedite the information-generation process and help in efficient management of information and the underlying database. The characteristic features of these tools are discussed in this chapter.

19.1 HARDWARE FOR INFORMATION MANAGEMENT

Modern information systems require significant hardware for data collection, manipulation, and storage. Transportation data collection equipment can include instrumented vans, handheld computers equipped with GPS, video cameras, weigh-in-motion (WIM) scales, automatic traffic recorders (ATR), and vehicle detectors (infrared, microwave, laser, radar, etc). Data storage and processing equipment can include fixed hardware, such as PCs and high performance servers, and consumables, such as magnetic media and flash drives. A small-scale information system mostly requires a personal computer, while larger systems require servers or a network of servers supporting multiple users. Other equipment can include GPS receivers, imaging, communication and navigation satellites, and other imaging system hardware. At the current time, transportation agencies are turning increasingly to the use of equipment such as video cameras, loop detectors, and microloop, infrared, radar, and other nonintrusive detection systems for data collection.

19.2 SOFTWARE AND OTHER TOOLS FOR INFORMATION MANAGEMENT

For basic information systems whose only electronic component is a database, the management of transportation data is a manual operation. However, the addition of suitable software components elevates a database to an *information management system*. Most information management systems include relational databases and can be categorized as GIS-based or GIS-compatible systems and those without a GIS component or capability.

19.2.1 Non-GIS Relational Database Management Systems

In a relational database management system (RDBMS), data are stored in several files and tables, and relationships are created and maintained across the tables. To retrieve

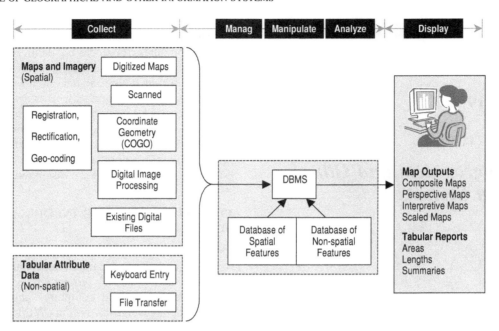

Figure 19.1   Functions and components of a GIS.

information from the database, queries are performed using tools such as Structured Query Language (SQL), which makes use of the relationships between the constituent tables in the database. For example, a typical safety management database consists of several inter-linked tables that seperately contain data on the roadway, weather, crash histories, crash costs, and so on. Due to linkages between tables, information from any one table can be cross-referenced as well as complemented with information from another. Unlike nonrelational databases, a RDBMS greatly facilitates data retrieval from one or more of the several constituent databases and therefore facilitates the generation of a *report*, a new table or database consisting of the data retrieved from multiple tables or databases.

19.2.2 Geographical Information Systems

GIS, the present-day state-of-the-art of cartography, has evolved over many years. This tool enables increasingly precise and comprehensive representation and analysis of natural and human-made land features and activities. GIS allows different physical and operational aspects of a transportation system to be modeled as layers that can be edited and manipulated using specialized software. Map outputs and displays are then generated by switching the appropriate layers on or off and assigning to each layer a predefined cartographic representation (Thurgood and Bethel, 2003). While other RDBMSs store information

in tables and provide links between the tables, GIS goes a step further by providing a geographic component for visualization and analysis purposes, an attribute that is particularly relevant and consistent with the spatial nature of transportation system inventories, operations, and impacts. Figure 19.1 illustrates the various functional and physical components of GIS.

GIS can have spatial or attribute (aspatial) data. For each of these two data types, the mechanisms by which GIS manages the data are explained below.

(*a*) *Spatial Data*   Spatial data can be assigned an explicit geographic location in the form of coordinates from a well-defined reference system. Such data can be obtained from scanned or digitized paper maps and drawings, digital files imported from computer-aided design (CAD) or other graphics systems, coordinate data recorded using a GPS receiver, data captured from satellite imagery or aerial photography, and so on. Spatial information can be presented in one of two ways: (1) as vector data in the form of points, lines, and areas (polygons); (2) as a raster in the form of uniform, systematically organized cells for example, using satellite images called *orthophotos*. These mechanisms are explained below.

Vector Data: Vector data are represented by the geometry of points, lines, or areas and associated topology. *Points*, which are considered the most fundamental and the simplest representation of geographical objects, are

Figure 19.2   Orthophoto image of I-70 and I-465 interchange, Marion County, Indiana.

dimensionless because they have no extension. Points may represent starting, break, or ending points on a line. Examples in transportation include nodal or nonlinear facilities occupying relatively small areas (bridge locations) and operational characteristics (specific crash locations). *Lines*, on the other hand, consist of points linked together with segments. A line has two points as a boundary; a starting and an ending point. In transportation, examples of line facilities are rail tracks, highways, and transit routes. An *area* is represented by a set of lines that enclose a space, thus forming a closed polygon, such as wetland areas or areas of poor air quality.

Raster Data: In a raster data set, transportation features are represented as a matrix of cells in continuous space. Each layer represents one attribute (although other attributes can be attached to a cell). Analysis is carried out by combining the layers to create new layers with new cell values. The cell is generally based on the original map scale and the minimum mapping unit. Another example of a raster data is imagery such as orthoimagery. An orthoimage is a geo-referenced image or picture of Earth that is produced from aerial photographs or other remote sensing sources, and corrected for terrain relief. Orthoimages can be used to generate vector data sets, such as types of landuse, vegetal cover, surface water resources, and transportation networks. Also, such images can serve as a basis for analyzing, updating, or referencing other data items, such as roads, intersections, or buildings, on the basis of the visibility of these features. Figure 19.2 shows an orthophoto of the I-70 and I-465 interchange in Marion County, Indiana (Indiana DOT, 2005).

Discussion: The spatial information stored in most GISs typically is a combination of vector data and raster

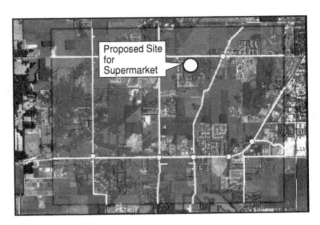

Figure 19.3   ArcMap-generated map for purposes of traffic impact evaluation.

data. The vector and raster data are presented as sets of linkable thematic layers. For example, the map shown in Figure 19.3 shows a traffic impact analysis study area in West Lafayette, Indiana for a proposed supermarket in the city prepared using the GIS software ArcMap. The information is presented through three layers: a first layer, the orthoimage, which presents the background and is used to generate vector data relating to the transportation network; a second layer, the line layer, which has vector data representing the road transportation network; and a third layer, the point features layer, which has vector data representing the intersections in the road network.

(*b*) *Attribute Data*   Unlike spatial data, attribute data are descriptive data sets that contain physical or operational information relevant to a particular point, line, or area feature, such as dimensions, traffic flow, congestion, and crash frequency. Attribute data can be linked to such locations using appropriate identifiers, such as a linear reference number or global coordinates. Such data may be stored in the form of *dataviews* or *spreadsheets* which present the data in tabular format. Dataviews can be used to create and edit data, print reports, and customize the way that data are displayed. GIS software, such as TransCAD, can store the attribute data in the form of matrices which are used to display the transportation attribute values, their spatial adjacency, cross-tabulation results, and other data. The matrices created can be edited, manipulated, and combined with other matrices to support various analytical applications. In addition to matrices and dataviews, facility attribute information can also be presented in the form of two- or three-dimensional figures, charts, graphs such as prism and three-dimensional maps, pie and bar charts, line charts, and scatterplots. Maps, dataviews, and figures can be customized as desired to display the required information.

19.2.3 Internet GIS

Internet GIS refers to network-centric GIS tools that use the Internet as a major means to access and transmit data and to enhance the visualization and integration of spatial data. This tool can be used by transportation agencies to publish spatial data on the network for public access and can also facilitate spatial data sharing within a transportation agency and between the agency and other private or public agencies.

Internet GIS offers interactive rather than static map images on the Web. Electronic maps on the Internet are often more convenient to use than the traditional paper maps and may have greater flexibility by providing search and browse functions. Users can search for a specific facility and are provided with a map centered on the search address. The map can be zoomed in or out or panned around to browse the neighboring areas. For example, MapQuest, Yahoo, and Google provide interactive mapping services on the Internet by allowing address and route queries. The level of map detail depends on the map scale and the information requested by the user.

Internet GIS can incorporate and display up-to-date real-time information. For example, TrafficWise, a real-time traffic information management system (Indiana DOT, 2005), provides for Internet users a GIS-enabled real-time traffic map (Figure 19.4) of northwestern Indiana for purposes of traffic monitoring and evaluation of traffic mitigation interventions. Furthermore, Internet GIS offers an ideal medium for transportation agencies to share data within and across agencies and also with the general public. For example, transit operators publish information on transit routes and schedules on the Web, which can be used by commuters to obtain transit route information and to evaluate transit system performance. The publishing of land-use and zoning maps on the Internet by land-use planning agencies and maps of environmentally sensitive areas by environmental agencies are resources useful to analysts involved in evaluating transportation systems. In fact, the open and transparent data sharing afforded by Internet GIS greatly reduces the institutional and physical barriers that impede the flow of data critical for transportation system planning, design, monitoring, and evaluation.

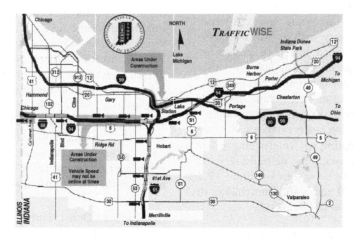

Figure 19.4   GIS-enabled Internet-available real-time traffic map in northwest Indiana.

19.2.4 Video Log Information Management Systems

A *video log* is a dynamic image-based record of transportation physical assets and operations viewed from the perspective of facility users (i.e., drivers, pedestrians, etc.) (NASA–USDOT, 2001; Florida DOT, 2005). Transportation agencies in several states, such as Pennsylvania, Ohio, Florida, Oregon, and Washington, are currently using video log systems for the purpose of monitoring and evaluation. Video log images are collected using a van equipped with a GPS unit, distance-measuring instrumentation, pavement and road condition sensors, cameras, and associated computer components to record digital images of the guideway. The spatial reference information from GPS or a linear measurement instrument is used, and coupled with time information, these data are employed to develop a query-based system for retrieval of video information at any site of interest. Linear referencing information typically includes the milepost, latitude, and longitude.

Video logs can be used to make measurements of facility characteristics such as guideway width, shoulder width, guideway condition, and other geometric or operational features. Video logs can therefore be used to generate data vital for the evaluation of transportation projects. Figure 19.5 shows the digital image control window from commercially available software that can provide still and sequential images by running them as a video file. Video log photos were used to examine crash locations and determine sight distances, signage, crosswalks, speed transition zone, presence of pedestrian signals, and so on, to study fatal run-off-the road crashes in Florida (Singh, 2005).

19.3 GIS APPLICATIONS IN TRANSPORTATION SYSTEMS EVALUATION

GIS has provided the flexibility to capture the dynamic nature of data that vary with space and time. It can provide the following information on geographic elements or features: location, characteristics, logical and geometric relationships with other features, and spatial interdependencies. This is accomplished using GIS components such as drafting, polygon processing, network analysis, spatial querying, and application development tools such as programming libraries (Thurgood and Bethel, 2003). Estimating, visualizing, and monitoring all impact types including both the physical and social assessment of projects can be greatly facilitated using GIS tools. For example, GIS is currently being used in transportation demand predictions; identifying the extent and severity of air, water, noise, and ecological impact violations; land-use and economic development impacts; spatial review of facility condition improvements corresponding to various levels of preservation funding; transit service improvements at rejuvenated

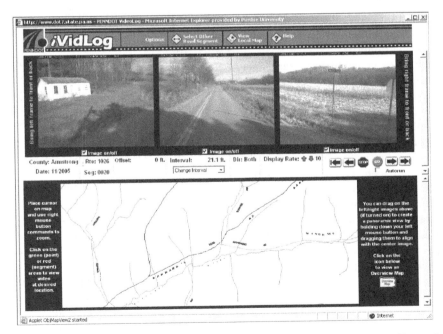

Figure 19.5 Digital image control window from a video-log system. (From Pennsylvania DOT, 2006.)

urban centers; and location planning of highways at scenic or recreational areas. Some of the major categories of GIS application in transportation systems evaluation and related areas are discussed in subsequent sections.

19.3.1 Query, Display, and Visualization of Initial Data

GIS enables quick visualization of the spatial distribution of transportation system physical assets and operational features and can therefore help analysts view the spread of impacts of transportation actions. The initial data (such as transportation demand) and the end result of the evaluation (such as air pollution impacts) are best visualized as a map or graph. Maps present an efficient mechanism for storing and communicating geographic information, and map displays can be integrated with reports, three-dimensional views, photographic images, and other multimedia outputs. For example, presentation of the spatially referenced transportation data using GIS can help highlight the transportation system's reach, coverage, modal relationships, key corridors, and relationship to the regional economic activity and environment. In the United States, the Geographic Information Services office at the Bureau of Transportation Statistics (BTS) is a national resource for transportation spatial data and analysis. The "mapping center" on the BTS Web site provides mapping and data download applications using queries, that can be used to analyze data geographically and retrieve them from the TranStats data library. The mapping center provides access to all transportation geospatial data collected and maintained by the U.S. Department of Transportation through a map-based download interface.

The use of GIS mapping to generate necessary data for project evaluation can be illustrated by considering the proposed construction of a supermarket in Tippecanoe County (shown in Figure 19.6). To assess the traffic impact, the expected volume of shopping trips from neighboring counties should be estimated. Figure 19.6 shows a GIS map indicating the traffic impact analysis area with reference to the adjacent counties. Using the distance measuring tool of GIS, the distances of the proposed supermarket site to the county boundaries, other shopping areas and major transportation facilities can be determined. Figure 19.6 also illustrates how GIS mapping can be used to generate the initial spatial and attribute data and to visualize the location of the study area before conducting traffic impact analysis.

19.3.2 Buffer Analysis

A *buffer* is an area defined by a specified distance from a map feature for purposes of identifying and analyzing

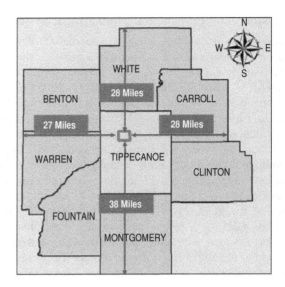

Figure 19.6   Spatial attribute visualization using GIS.

features and areas that are related to the map feature. The creation of a buffer and development of spatial relationships between the associated geographic features is referred to as *buffer analysis* or *proximity analysis* (Forkenbrock and Weisbrod, 2001). The map feature of interest could be a *point feature*, *line feature*, or area feature depending on the objective of the buffer analysis. For example, in a study of noise impacts in areas proximal to a nodal transportation facility, a buffer is created around the point feature that represents the facility. On the other hand, if it is desired to investigate the noise impact due to traffic operations on a highway route, the buffer is created along the highway route which is represented by a line feature. In buffer analysis, the analyst simply identifies the map feature of interest and then specifies the buffer width. The development and analysis of spatial relationships between the various geographic features involves the following tasks: identification of features that lie inside or outside the buffer; determination of the impact of the transportation facility or intervention at the proximal areas in terms of performance measures such as travel time, congestion, noise impact, emissions, ecological, cultural, and socioeconomic impact; determination of the existence of a transportation facility within a certain distance from zone centroids, and so on. The objective of establishing such spatial relationships is to determine the extent of the spread and/or severity of the effects in the analysis area before proceeding with a more in-depth analysis. The results of the analysis can be used by transportation agencies to ascertain issues of land ownership and cost estimation for compensation purposes.

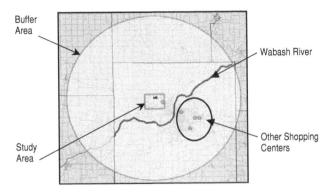

Figure 19.7   Analysis of 15-mile buffer area for traffic impact analysis.

For example, Figure 19.7 illustrates a 15-mile circular buffer area around the proposed supermarket in Tippecanoe County (Figure 19.6). In this example, a 15-mile radius was used to estimate the number of potential customers and the impact on highway vehicular traffic in the study area on the basis of acceptable shopping travel time.

A more detailed analysis of the circular region shown in Figure 19.7 can be conducted by taking into consideration other shopping options for the population in this region. The GIS tool identified the location of the neighboring shopping places within the 15-mile buffered region. With the four other shopping centers located east of the river, further analyses can be carried out to determine if any significant number of shopping trips will be made across the river to the new supermarket. Such an analysis can be done using the gravity model approach discussed in Chapters 3 and 9.

19.3.3 Overlay Analysis

In overlay analysis, information from two or more distinct data layers is combined to derive information. *Overlay* or *spatial join* operations in GIS can be used to integrate different types of data layers irrespective of the data feature type: a point, line, or polygon. For example, a layer containing information on highway routes could be integrated with another containing information on traffic analysis zones and their attributes such as population. Such spatial joins can be used to determine the need for a specific transportation infrastructure in the study area. Tasks that would generally take several months of field work can be reduced significantly using overlay operations. Furthermore, overlay analysis and terrain modeling can be utilized to determine the impact of transportation infrastructure improvements on the overall transportation network and on the natural and human-made environment. For example, the impact of a runway expansion project on

neighborhood water resources, the impact of improvement in travel time on a link on its parent transportation network, or the socioeconomic impact of a rail bridge project in an urban area can be assessed with relative ease by spatially combining data from two or more layers that pertain to these features. Another example of overlay analysis application is evaluation of the noise impacts of transportation operations: specifically, the spatial distribution and number of persons to be affected by noise pollution resulting from a proposed transportation project and the spatial distribution of the problem severity. For this situation, existing data layers can represent the distribution of population density and noise loudness (represented by noise contours). In most GIS packages, the analyst can select an appropriate tool for the overlay analysis and then proceed to select the appropriate layers of interest.

For state or regional travel demand analysis, it is necessary to examine socioeconomic characteristics vis-à-vis transportation facilities or traffic analysis zones. Figure 19.8 illustrates the spatial superimposition of the layers for urban population, interstate routes, and county boundaries in Indiana. Information on the number of urban residents in each county (made available through the overlay analysis) is a key data input for estimating the number of interstate trips attracted and produced by each county.

19.3.4 Analysis of Transportation Operations

GIS can be used in various transportation analyses such as identifying the shortest paths between points on a network, a capability that is useful in the four-step process of transportation demand estimation. The GIS software TransCAD, for example, allows the planner to carry out traffic assignment. GIS data structures can be extended to model one-way directions of streets, turn restrictions, rush-hour flow patterns, and other spatial features of the transportation system inventory and operations. Also, GIS can help identify the relationships between the operational features of a transportation network, and can be used to investigate for example, the influence of bridge failure at a link on overall system traffic conditions, the flow of evacuees from a disaster area under various levels of transportation system capacity, and so on.

19.3.5 Public Input in Transportation System Evaluation

In soliciting public perspectives on proposed projects through the Internet, town hall meetings, and other media, GIS provides a convenient mechanism for presenting alternative routes, locations, and designs that can easily be presented to and understood by the stakeholders.

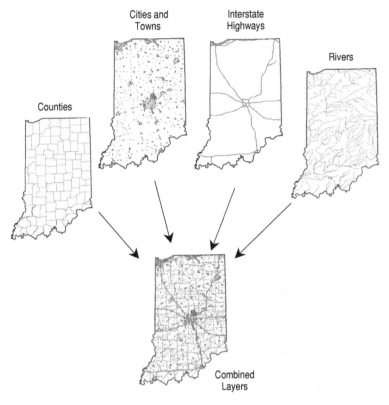

Figure 19.8   Overlaying urban centers over interstate routes, rivers, and county boundaries. (From IGIC, 2006.)

This way, public input regarding transportation system impacts can be analyzed efficiently and fed back to the transportation agency.

19.3.6 Multicriteria Decision Making

The impacts of transportation systems are typically expressed in several performance measures each of which have a temporal and spatial scope. In that respect, GIS can be useful tool in evaluation because it can (1) enable visualization of the spatial and temporal effects of transportation system changes for each performance measure and for any combination of the performance measures; (2) help identify the spatial and temporal impacts of a given performance measure over the analysis zone or period (e.g., ecological damage due to deicing operations at an airport runway—how widespread and how intense, and how the spatial extent and severity varies with the time of year); and (3) facilitate recognition of spatial and temporal patterns of the relationships between performance measures (e.g., relationships between deficient route alignments and crash frequency).

19.4   EXISTING DATABASES AND INFORMATION SYSTEMS

Figure 19.9 shows the sources of data for information systems (including GIS) that are used to support transportation systems evaluation and decision making. Data sources and type include databases, vector and alphanumeric data, satellite images, scanned maps, documents, photographs, field survey results, coordinate geometry, photogrammetry, online digitizing, and so on. The most common data sources are databases, and these are discussed further in this section. Several information management systems have been developed for transit, rail, highway, and air transportation networks. Most of these systems are RDBMS- or GIS-based systems. In this section we present some of the information management systems and their role in transportation system evaluation.

19.4.1 Information Systems and Data Items Available by Transportation Mode

For all modes of transportation, there are databases that serve as the basic building block for developing

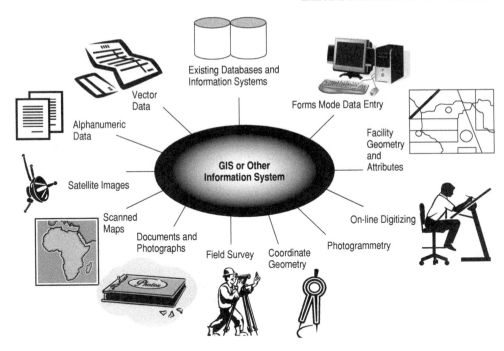

Figure 19.9 Data sources for GIS and other information systems.

information management systems, which are in turn used for data extraction and data analysis. Transportation data can be categorized as follows: inventory data, operations data (usage, safety, and congestion), and cost data (agency costs of construction, preservation, and operations and user costs items). These data are indispensable for evaluating alternative transportation facility locations and designs, construction delivery practices, preservation strategies, and operational strategies (safety, congestion, and intermodalism). While these are general data needs, each transportation mode has specific data needs, as explained below.

(a) Highway Transportation Information Systems Highway data typically include information on referencing, flow (AADT, hourly traffic, vehicle composition), inventory (shoulder width, grade, and other geometric features), and cost (agency costs of highway construction, maintenance, and operations, etc.). Specific highway management systems have their own data needs as discussed below.

Databases and Information Systems for Highway Safety Management: Safety evaluation of highway projects typically requires data on the following: crash frequency, severity, patterns, locations, time, environmental conditions, crash circumstances, vehicles involved, driver information and number of occupants, pedestrians and cyclists involved, the violation charged, and other relevant data. Commercial motor vehicle crash databases contain information on vehicle configuration, cargo body type, hazardous materials, motor carrier identification number, and past crash history. Other safety-related databases are the emergency medical services (EMS) files, which contain data on the emergency care provided to victims and ambulance responses to crashes, and the citation/conviction files, which identify the type of citation and the time, date, and location of the violation; the violator, vehicle, and enforcement agency; and adjudication action and results, including the courts of jurisdiction.

The Highway Safety Information System (HSIS) is a multistate GIS-compatible safety database that can be used to evaluate the effectiveness of safety countermeasures. HSIS contains data on 5 million crashes, traffic volume, and 165,000 miles of inventory data for highways in eight states and video photo logs for selected states (USDOT, 2006b). All the information in the database can be linked into analysis files for safety studies. HSIS is operated by the University of North Carolina Highway Safety Research Center (HSRC) and LENDIS Corporation under a contract with FHWA. The crash file contains basic information on accidents, vehicles, and occupancies on a case-by-case basis. The data include the type of accident, type of vehicle, gender and age of occupants, crash severity, and weather conditions. The roadway inventory file contains information on the roadway cross section and the road type. The traffic volume file contains information on hourly traffic data, AADT, and truck percentage. The roadway geometrics file contains information on horizontal curve and vertical grade. The intersection file contains information on highway intersections, such as the traffic control type, intersection type, signal phasing, and turn lanes. The interchange file has data on the types of interchanges and ramp characteristics, while the guardrail/barrier file contains an inventory of guardrails. Using GIS tools, crashes at a specific spot or length of roadway can be identified and analyzed (Figure 19.10). Also, crashes that are clustered around a roadway feature, such as a bridge, a signalized intersection, or a railroad crossing, can be identified and analyzed.

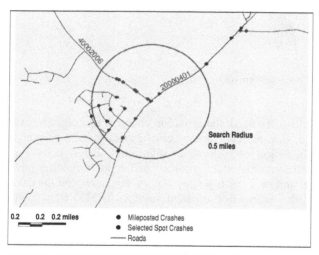

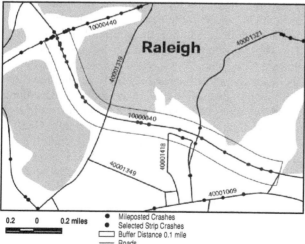

Figure 19.10 GIS-based processing of HSIS data. (From USDOT, 1999.)

Databases and Information Systems for Pavements: Data that are typically contained in pavement databases include location; type; age; contract dates and cost; material type; traffic information; performance indicators such as ride quality, cracking, rutting, and friction; and the year of testing. Such databases help agencies investigate the performance of their pavements or effectiveness of pavement treatments; in determining pavement service lives; and in evaluating the impact of factors such as traffic and weather on pavement longevity. Data can also include material characteristics, quality control and quality assurance data, asphalt and binder test data, coarse aggregate test data, fine aggregate angularity test data, and other laboratory mix test data. The Highway Performance Monitoring System (HPMS) is a national highway database that has data relevant to several management systems, such as the extent, condition, performance, use (travel), and pavement and operating characteristics of highways. The database has information useful for analyzing safety impacts, air quality impacts, and so on. In addition to the data, the HPMS framework consists of simulation models that can be used to simulate future investments and their costs, for purposes of investigating the consequences of alternative investment levels and strategies.

Databases and Information Systems for Bridges: The National Bridge Inventory (NBI) is an FHWA database that has information on bridges at crossings involving water courses, highways, railroads, pedestrian–bicycle facilities, and overpasses (FHWA, 2004). The data include the construction year, bridge dimensions, traffic, inspection dates, condition, and other information. Data are available for bridge elements and features such as deck structure types, scour, and other performance measures. The NBI bridge database can be used effectively by state and local agencies by incorporating appropriate GIS capabilities.

(b) Transit Information Management Systems The Federal Transit Administration's (FTA) National Transit Database (NTD) has information on vehicle fleet size, performance, operations, transit availability and accessibility, transit accidents, safety, causality, finances. and crime statistics for more than 400 urban areas (FTA, 2006). The database contains information on approximately 85,000 transit vehicles, 7000 miles of rail track, 2000 rail stations, and 1000 maintenance facilities. The database can be used to generate information useful for evaluating project-level transit investments. An Integrated National Transit Database Analysis System (INTDAS) is available to facilitate visualization, retrieval, and analysis of data from the National Transit Database (FTIS, 2006). An

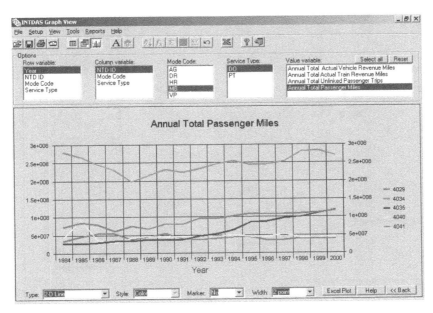

Figure 19.11   Retrieved data from INTDAS displayed in graphical format.

SQL query editor can be used to carry out tasks such as identifying transit systems that meet a certain performance threshold. INTDAS can generate reports with tables (Figure 19.11), graphs, maps, and Excel plots for individual transit systems and can therefore help identify areas needing improvements.

The FTA-sponsored GIS-based transit information management system can be used to evaluate and analyze specific transit systems and to extract information such as the route layout, service frequency, schedules, stops, wheelchair accessibility, and related information. For example, Figure 19.12 shows a map of the Washington Metro Transit routes and stations (with the city surface streets in the background). This map provides information about which surface streets are close to transit stations. Management of such information with the help of underlying databases and its presentation with the help of maps, charts, figures, and tables can play an important role in the evaluation of transit system performance.

(c) Databases and Information Systems for Air Transportation   Air transportation data include the number of enplanements; size, routing, and scheduling of aircraft; delays and on-time arrival and departure of aircraft; passengers, freight, and mail demand patterns at each airport; and aircraft crashes. These data are stored in the form of databases, which are maintained by various organizations, such as the Federal Aviation Administration (FAA), the Bureau of Transportation Statistics (BTS), and the National Aviation Safety Data Analysis Center.

Data also include capacity statistics from individual air carriers, planned and actual arrival and departure time of flights; origin, destination, and other itinerary details; and financial data, including balance sheet, income statement, cash flow, aircraft inventory, and aircraft operating expenses (BTS, 2006). The Aircraft Registry database stores records for over 320,000 registered civil aircraft in the United States (FAA, 2006c).

Databases containing aviation safety and accident data include the Aviation Safety Reporting System (ASRS), the Accident/Incident Data System (AIDS), the Near Midair Collision System, and the World Aircraft Accident Summary (WAAS) databases (FAA, 2006c). ASRS has data on unsafe occurrences and hazardous situations that are reported by pilots and air traffic controllers, and the AIDS database has data on aircraft incidents since 1978. The WAAS database provides information on major operational crashes involving air carriers, operating jets, helicopters, and turboprop aircraft. Also, the National Transportation Safety Board aviation accident and incident database contains data on civil aircraft accidents and incidents in the United States and its territories, and in international waters.

The Air Traffic Activity Data System provides information on historical air traffic operations, which includes daily, monthly, and annual counts either by facility, state, or region, or nationally (FAA, 2006b). The Aviation System Performance Metrics is an integrated database that provides information on air traffic operations, airline schedules, arrival and departure rates, operations and

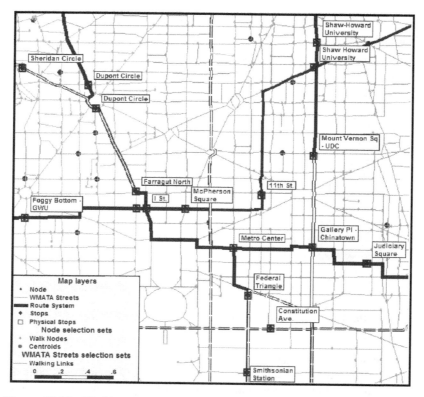

Figure 19.12   Washington metropolitan transit routes. (From www.wmata.com.).

delays, runway configuration data, weather information, runway information, and related statistics (FAA, 2006a). The database is used in analysis of the operating performance of the National Airspace System on the basis of the percentage of aircraft arriving on time, the average daily capacity for selected airports, the average daily capacity for selected metropolitan areas, the system airport efficiency rate, and the terminal arrival efficiency rate.

(d) Databases and Information Systems for Rail Transportation   The Federal Railroad Administration (FRA) sponsors, monitors, or is involved in the development of most rail databases. The FRA maintains records of railroads' monthly operations, including the train-miles run and labor-hours worked, the highway–rail crossing inventory, and railroad accidents. Accident and incident data are available for approximately 300 railroad systems in the United States (FRA, 2006). The FRA also maintains a railroad crossing database containing information on approximately 300,000 crossings. The database contains details such as referencing (railroad milepost), crossing-street name, warning device type, train speed, number of traffic lanes, average daily traffic, and number of accidents. GIS maps can be generated to

show, for example, the spatial distribution of grade-crossing accidents and for answering queries such as the total number of railroad crossings that are located within a certain distance of a football stadium. The information in the database can be used to determine the level of service on the highway network in the vicinity of the railroad crossings, especially during special events. The FRA accident database can be used to generate vital performance data for safety evaluation of rail improvements.

19.4.2 General Databases Useful for Transportation Systems Evaluation

(a) The Census Transportation Planning Package (CTPP) 2000:   Census 2000 (USDOT, 2006a) provides information on the travel patterns of millions of household units and individuals across the United States. The data include housing, population, and social and economic characteristics by blocks, block groups, census tracts, counties, and metropolitan areas. Also there are data on key transportation planning variables, such as household size, household income, vehicles per household, age and gender of workers, occupation of workers, worker earnings, usual mode to work, commuting time, work trip departure time, and work location. Maps, boundary files, and other geographic

products are available to assist users locate and identify geographic areas. The package provides transportation planners with comprehensive demographic data useful for basic planning tasks (developing or updating travel demand models, analyzing demographic and travel trends, forecasting travel, transit planning, corridor planning, air quality modeling, trend analysis, and other system analysis operations). The package thus assists in evaluating existing conditions and predicting the impacts of future transportation policies and projects. CTPP data can be used by GIS packages such as TransCAD for transportation planning and evaluation purposes.

(*b*) *National Transportation Atlas Database (NTAD):* NTAD is an assortment of geospatial databases for managing information on transportation facilities, networks, and services of national significance (BTS, 2006b). NTAD has information on various transportation modes, intermodal connectors and terminals, and key transportation structures such as bridges and tunnels. The database includes public-use airports, runways associated with public-use airports, major highway–rail intermodal freight facilities, Amtrak passenger stations, waterway and marine terminals, and highway–rail transfer facilities. For these transportation facilities, data available include capacity, traffic, and inventory. The transportation network databases include topologically connected lines that can be used to determine the locations and alignments of nationally significant roads, railroads, waterways, airways, and transitways. The nodes in the network are used to represent the terminals, interchange facilities, junctions, or intersections. The network database contains information on the highway, railway, waterway, transit, and commercial air network. The highway network database contains data for approximately 400,000 miles of federal-aid roads in the United States. The railway network database contains information on all railway mainlines and railroad yards. The waterway network database stores information on all navigable inland and intracoastal waterways; the Gulf, Great Lakes, and coastal sea lanes; and major sea lanes between the continental United States, Alaska, Hawaii, and Puerto Rico. The transit network database includes data on all guided transit networks in the continental United States, and the air network database includes information about all direct air routes between commercial airports in the country. The NTAD database has been used in several studies for evaluating various transportation systems. For example, Krishnan and Hancock (1998) used the database to develop a GIS-based methodology for evaluating alternative plans for freight distribution and assignment in Massachusetts. Also, a nationwide GIS-based analysis of congested airports was conducted using the database

(CTRE, 2006a). Another study used NTAD data to evaluate the accessibility and levels of service of alternative transit routes in Chicago, Illinois (CTRE, 2006b).

19.5 GIS-BASED SOFTWARE PACKAGES FOR INFORMATION MANAGEMENT

The common GIS software packages include TransCAD, ArcInfo, ArcGIS, and ArcIMS. TransCAD combines GIS methods for solving problems in transportation planning, management, and operations (Caliper, 2006). It is the first GIS software developed specifically for transportation professionals to store, display, manage, and analyze transportation data. The software integrates GIS and transportation modeling capabilities in a single platform. TransCAD can be used for any mode of transportation at the neighborhood, city, state, national, or worldwide level. The software provides a built-in relational database manager with methods for linking all forms of data and an assortment of tools for analyzing, interpreting, and making effective transportation graphics and presentation using maps. It can be used for transportation planning, vehicle routing, and distribution logistics.

ArcGIS, a full-featured GIS software, produced by ESRI, for visualizing, managing, creating, and analyzing geographic data, is the most widely used desktop GIS software. ArcGIS can be used to quickly build new spatial datasets, to and manage visually model the spatial database, tables, files, and other data resources from a single application; and to perform calculations with special data.

ArcInfo includes all the functionality of ArcGIS and adds advanced geoprocessing and data conversion capabilities. The software can be used for all aspects of data building, modeling, analysis, and map display for screen viewing and output. It can be used to store, edit, display, and to plot traffic simulation networks. Before a simulation is run, ArcInfo can enable easier maintenance of simulation networks by using a base network and tracking projects that will change that network over time. ArcInfo also provides a range of network editing tools. It is possible to conduct further analysis in ArcInfo using the Network and GRID modules. ArcIMS is a server-based GIS application which is used for many types of centrally hosted GIS computing. A centralized GIS application is set up at the server to provide GIS capabilities to a large number of users over the network. ArcIMS can be used to deliver and share dynamic maps and GIS data and services over the Internet. This package therefore provides a scalable framework for GIS Web publishing that can meet the needs of agency intranet and internet systems in an agency.

SUMMARY

Data management plays an important role in transportation systems analysis and in the evaluation of existing systems and proposed projects. Several information management systems have been developed by various agencies by incorporating large databases and providing various features for data management. This chapter presented an overview of GIS and how it can be used in transportation systems evaluation in conjuction with various data management systems.

EXERCISES

19.1 In a bid to cater for increased transit demand, it is proposed to extend existing transit lines further outwards into the outlying, fast growing areas of a large city. As part of a multiple criteria evaluation of the impacts of the proposed project, you are asked to develop an information warehouse that will provide the necessary data for the evaluation. Performance criteria to be considered include agency and user cost, economic development and land use, damage of ecological resources including wetlands, noise and air pollution, and sociocultural impacts. Identify and briefly discuss the types of data needed and how such data could be collected. Identify existing national databases that could serve as secondary sources of information.

19.2 For the proposed evaluation in Exercise 19.1, identify appropriate software packages that could be used to manage the data, and explain the various ways by which GIS tools and Internet resources can enhance the data analysis and visualization.

REFERENCES

BTS (2006). North American Transportation Atlas Data (NORTAD), Bureau of Transportation Statistics, U.S. Department of Transportation Washington, D.C., www.bts.gov/publications. Accessed May 2006.

BTS (2006). TranStats, *The Intermodal Transportation Database*, Bureau of Transportation Statistics, Washington, DC, www.transtats.bts.gov/DataIndex.asp. Accessed May 2006.

Caliper (2006). *Overview of the GIS Software*, TransCAD, http://www.caliper.com/tcovu.htm. Accessed May 2006.

CTRE (2006a). *Nationwide Analysis of Congested Airports*, Centre for Transportation Research and Education Iowa State University, Ames, IA, www.ctre.iastate.edu/Research/bts_wb/. Accessed May 2006.

———— (2006b). *Chicago Transit Routes vs. Vehicle Density*, Centre for Transportation Research and Education, http://www.ctre.iastate.edu/Research/. Accessed May 2006.

FAA (2006a). *Aviation System Performance Metrics*, Federal Aviation Administration, Washington, DC, www.apo.data.faa.gov/aspm/. Accessed May 2006.

———— (2006b). *FAA Operations and Performance Data*, Federal Aviation Administration, Washington, DC, www.apo.data.faa.gov/main/. Accessed May 2006.

———— (2006c). *The National Aviation Safety Data Analysis Center Databases*, Federal Aviation Administration, Washington, DC, www.nasdac.faa.gov/portal. Accessed May 2006.

FHWA (2004). *National Bridge Inventory Database*, Federal Highway Administration, U.S. Department of Transportation, Washington, DC, www.fhwa.dot.gov/bridge/nbi.htm.

Florida DOT (2005). *County Section Number Key Sheets and Video Logs*, Chap. 6, Florida Department of Transportation Tallahassel, FL, www.dot.state.fl.us/planning/statistics/rci/officehandbook/ch6.pdf. Accessed May 2006.

Forkenbrock, D. J., Weisbrod, G. E. (2001). *Guidebook for Assessing the Social and Economic Effects of Transportation Projects*, NCHRP Rep. 456, National Academy Press, Washington, DC.

FRA (2006). *Railroad Safety Information Database*, Federal Railroad Administration, Washington, DC, http://safetydata.fra.dot.gov/officeofsafety. Accessed May 2006.

FTA (2006). *National Transit Database*, Federal Transit Administration, Washington, DC, www.ntdprogram.com. Accessed May 2006.

FTIS (2006). *Integrated National Transit Database Analysis System*, Florida Transit Information System, Tallahassee, Fl., http://lctr.eng.fiu.edu/Ftis/Intdas.htm Accessed May 2006.

IGIC (2006). Map Creation Tool, Indiana Geographic Information Council. www.in.gov/igic/map

Indiana DOT (2005). Indiana Department of Transportation, Indianapolis, IN, http://www.in.gov/dot/div/trafficwise/. Accessed Dec 10, 2005

Krishnan, V., Hancock, K. (1998). Highway freight flow assignment in Massachusetts using geographic information systems, *Proc. 91st Annual Meeting of the Transportation Research Board*, Washington, DC.

Minnesota DOT (2004). *Video-Log User Manual*, Tech. Rep. MnDOT/OM-PM-2004-02, Office of Materials, Pavement Management Unit, Minnesota Department of Transportation, St. Paul, MN, www.mrr.dot.state.mn.us/pavement/PvmtMgmt/Videolog_User_Manual.pdf. Accessed May 2006.

NASA–USDOT (2001). *Remote Sensing and Spatial Information Technologies in Transportation Synthesis Report*, National Consortia on Remote Sensing in Transportation, NASA U.S. Department of Transportation, Washington, DC, www.ncgia.ucsb.edu/ncrst/synthesis/SynthRep2001/SynthesisReport2001.pdf. Accessed May 2006.

Singh, P. (2005). *A Study of Fatal Run-off Road Crashes in the State of Florida*, Florida State University, Tallahassee, Fl, etd.lib.fsu.edu/theses/. Accessed May 2006.

Thurgood, J. D., Bethel, J. S. (2003). Geographic information systems, in *The Civil Engineering Handbook*, 2nd ed., Ed. Chen, W. F., Liew, J. Y. R., CRC Press, Boca Raton, FL.

USDOT (1999). *GIS-Based Crash Referencing and Analysis System*, FHWA-RD-99-081, U.S. Department of Transportation Washington, DC, www.hsisinfo.org/pdf/99-081.pdf. Accessed May 2006.

———— (2006a). *Census Transportation Planning Package, 2000*, U.S. Department of Transportation, Washington, DC, www.fhwa.dot.gov/ctpp/. Accessed May 2006.

———— (2006b). *Highway Safety Information System*, U.S. Department of Transportation, Washington, DC, www.tfhrc.gov/about/hsis.htm. Accessed May 2006.

CHAPTER 20

Transportation Programming

First weigh the considerations, then take the risks.
—*Helmuth von Moltke (1800–1891)*

INTRODUCTION

Programming can be described as the process of selecting and scheduling facility and/or rolling stock preservation, improvement, and replacement projects for a transportation network over a period of time. A key element of such a process is the matching of needed projects to available funds, to accomplish the strategic goals and objectives set by a transportation agency. An effective programming framework is expected to provide a mechanism for selecting cost-effective projects reflecting community needs and to develop a multiyear investment strategy within budgetary constraints over a planning horizon. The framework should assist both technical and policy decision making by presenting options and the trade-offs in terms of benefits and costs.

Formal techniques for transportation programming were discussed as early as the late sixties (Stearns and Hodgens, 1969). Since the passage of the Intermodal Surface Transportation Efficiency Act (ISTEA) of 1991 and the Transportation Equity Act for the 21st Century (TEA-21) in 1998, the need for performance-based planning and programming and accountability in transportation investment decision making has intensified at all levels of government. The Safe, Accountable, Flexible, Efficient Transportation Equity Act: A Legacy for Users (SAFETEA-LU) of 2005 has made several changes in the statewide and metropolitan planning processes. Some of these changes add flexibility and efficiency; others add new requirements. For example, the Legislation required that the statewide planning process should be coordinated into metropolitan and trade and economic development

planning activities, and the metropolitan planning process should consider environmental mitigation, improved performance, multimodal capacity, and activities that enhance the environmental, cultural, and aesthetic aspects of the transportation system. Furthermore, SAFETEA-LU required that tribal, bicycle, pedestrian, and disabled persons' interests be provided an opportunity to participate in the transportation decision-making process (Binder, 2006). These and other developments, such as increasing emphasis on multimodal trade-offs and funding flexibility to consider a range of transportation options, increased focus on facility preservation and system management, and the changing roles of state, regional, and local agencies in making program and project decisions, have led to increasing complexity of the transportation programming process.

Transportation programming involves determination of the work to be performed in a specified period of time to accomplish the objectives set for that period, with due regard given to the relative urgency of work (TRB, 1978). In the current era there are new tools and techniques to deal with the technical aspects of the transportation programming process. Due consideration is given to influence from a wide range of policy, political, and qualitative factors. Emerging public–private partnership opportunities have also broadened the scope of programming decisions in terms of financing and time of project delivery.

The programming process at any level of government will depend on the specific policies and requirements of the agencies involved, including statutory requirements, federal, state, regional, and local funding programs and their eligibility requirements, agency roles and coordination mechanisms, existing formal and informal statements of policy, and established long- and short-range planning processes (Neumann et al., 1993). Although particular approaches to investment decisions often vary from state to state, region to region, and mode to mode, there is a basic framework that is common to most transportation programming processes, as shown in subsequent sections.

The framework and principles for transportation planning and programming in metropolitan areas are provided by the Federal Metropolitan Planning Regulations. Section 450.318 of the FHWA/FTA Final Rule on Statewide and Metropolitan Planning issued in the *Federal Register* in 1993 has its roots in the 1991 ISTEA and the 1990 CAA amendments (Benz, 1999).

20.1 ROLES OF PROGRAMMING

The programming process can serve various roles:

1. To make optimal investments to achieve strategic policy goals

2. To evaluate trade-offs among investment options
3. To assist in the budgeting process
4. To facilitate efficient program and project delivery
5. To provide a mechanism to assess agency performance
6. To guide business processes and give direction to agency operations

Altogether, these objectives represent an ideal programming process. However, reasons for carrying out programming differ from agency to agency. The common thread is the need for a schedule for timing various transportation interventions so that budgetary and other resources can be allocated effectively. Although quantitative tools are available, this determination, at the current time, is often made through a subjective process involving priority setting based on prior commitments, project delivery capability of the agency, and funding availability.

20.1.1 Optimal Investment Decisions

Not all agencies make explicit use of optimization tools for resource allocation decisions but an effort is always made to arrive at decisions that are rational, defensible and duly cognizant of an array of established performance criteria.

An effective allocation of resources is the primary purpose of a programming process. To fulfill this objective, a clear set of agency goals and objectives must be predetermined. Most agencies have a set of strategic goals and associated performance measures, and a programming process should be sufficiently responsive to the strategic goals and objectives. The relative importance of goals or objectives may change periodically to reflect the prevailing political and economic environment. In Chapter 2 we discussed agency goals and objectives along with performance measures typically used in the decision-making process at transportation agencies.

20.1.2 Trade-off Considerations

It is important for decision makers to comprehend the trade-offs associated with specific investment decisions. By choosing a particular mix of projects, some objectives will be satisfied at the expense of others. A programming process should be capable of ascertaining the trade-offs and present them clearly to decision makers. For example, using appropriate "if–then" analysis, an analyst can assess the consequences of a given level of budget on facility longevity, vulnerability and other performance measures. There are techniques to quantitatively assess the trade-offs, as discussed in Chapter 18. Although trade-off considerations start with quantitative factors,

programming choices often end up being influenced by qualitative considerations, such as business rules requiring geographic distribution or corridor completion.

20.1.3 Linkage to Budgeting

Programming goes side by side with budgeting. An agency must decide on the use of available internal resources to match funds from other sources, such as grants from higher levels of government or for many developing countries, loans and grants from multinational financial institutions and donor agencies. As most external sources fund only capital projects, it is necessary to determine the extent of local resources that can be committed to attract matching funds from external sources and the expense of entirely internally funded projects and operations and maintenance budgets. With increasing trends toward nontraditional financing of transportation projects using bonding and private-sector leasing, fiscal analyses are becoming an important part of programming so that revenues and cash flows can be tied to commitments.

20.1.4 Efficiency in Program and Project Delivery

It is not sufficient that programming satisfy fiscal constraints and comply with statutory requirements and stated goals and objectives. An effective programming process must also be realistic in terms of the agency's capacity to implement the program selected. Scheduling of projects within the programming period is therefore an important consideration. Also, limitations in the number and quality of available personnel can pose a formidable barrier to implementation of adopted programs, a situation that is currently exemplified by the retirement of the baby boomer generation and the consequent loss of their accumulated knowledge base. Consequently, some agencies outsource both program delivery and management functions as a way to deal with their internal manpower constraints.

20.1.5 Monitoring and Feedback

An ideal programming process should have a regular and systematic procedures for monitoring its effectiveness through user surveys and physical measurements of transportation system performance parameters. Although performance measures such as facility condition, level of service, safety, and other factors are often regularly monitored, these measurements are rarely used to perform a postimplementation evaluation of agencies' programs. To establish accountability and to assess how well the agency's strategic objectives are being achieved, a performance monitoring mechanism should be implemented within the programming process. Such a mechanism will not only establish the effectiveness of the agency program but will also provide useful data for improving the programming process itself. With the passage of SAFETEA-LU, greater emphasis is now being placed on tracking project cost and time as well as on monitoring facility levels of service in terms of safety, facility condition, and other performance measures. Also, scope changes and time delays during implementation offer require adjustments to multiyear programs.

20.2 PROCEDURE FOR PROGRAMMING TRANSPORTATION PROJECTS

Table 20.1 presents the basic elements of an overall programming (program development) process, while Figure 20.1 presents a framework showing not only the sequence of these elements but also how they are related to other critical agency tasks of planning and

needs assessment, and revenue and financial analysis. The ISTEA of 1991 established a critical need for effective and consistent links between planning, programming, and finance. The TEA-21 of 1998 and SAFETEA-LU of 2005 subsequently strengthened this requirement.

At any specific agency, the exact steps used for programming and their relationship to other agency functions may vary from those presented in the figure and may depend on the specific institutional arrangements, funding sources, and practices in the agency. Other factors affecting the variations in programming include history, geography, level of urbanization, and political culture (Stout, 1996).

Step 1: Define the Policy Guidance This step involves a definition of the program goals and objectives, which are

Table 20.1 Elements of the Program Development Process

Element	Activity
Setting program policies, goals and objectives	Establish clear and measurable statements of what the transportation agency wants to accomplish to meet its stated policies
Identification of program performance measures	Set criteria for selecting projects to measure effectiveness of program implementation and to evaluate the results in terms of system performance, costs, and benefits
Systemwide needs assessment	Identify and measure deficiencies, problems, and needs
	Identify alternatives to address these needs
	Develop candidate projects
Systemwide projects and programs selection	Evaluate proposed projects and programs according to consistent criteria
Identification of available funds	Assess expected funding levels from all sources for maintenance, preservation, and improvement
Gap analysis	Identify projects that cannot be implemented and place them as backlog for the next cycle
Priority setting	Organize the agency's work into program areas reflecting the objectives of constituent geographical units and/or types of work
	Identify priorities for each program area consistent with agency goals and objectives
	Set priorities for projects within (or across) each program area using criteria which reflect agency goals and objectives
	Develop fiscally constrained candidate programs reflecting realistic project budgets and schedules
Program trade-offs	Evaluate what the proposed program will achieve
	Evaluate trade-offs for shifting resources among program areas or project types (e.g., bridge rehabilitation vs. capacity expansion)
	Determine levels of resource allocation across program areas based on agency priorities including the results of needs analysis
Monitoring and feedback	Monitor performance after implementation
	Make adjustments to incorporate changes in scope and time delays

Source: Adapted from NCHRP 243 (Neumann, 1997).

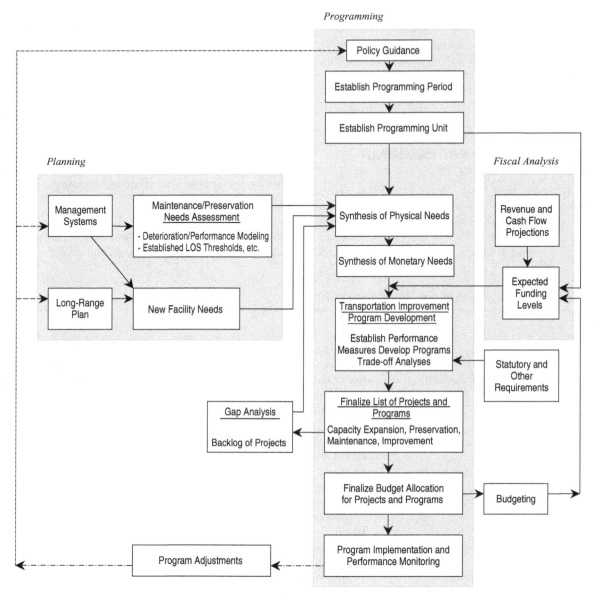

Figure 20.1 Transportation programming process and its relationship with other functions.

clear and measurable statements of what the transportation agency wants to accomplish to meet its policy goals.

Step 2: Establish the Programming Period and Unit
A *programming period* is the amount of time over which specific decisions are made for facilities. At many transportation agencies, a programming period is three years but may be as many as five to seven years for certain facility types. For example, for a new facility that requires six years for design, right-of-way acquisition, and environmental clearances, the program period needs to be seven years. A long-range *planning horizon* can be

fifteen to twenty years and span over several programming periods.

In developing programs, agencies grapple with the problem of balancing funds across various areas of need, such as geographic balance vs. transportation needs, rural vs. urban needs, capital expansion vs. asset preservation, technical needs vs. political realities, and competing transportation modes. Therefore, many agencies find it convenient to establish programming units and then develop a program for disbursing funds in each unit. A *programming unit* can therefore be defined by the geographical jurisdiction and the functional area. For example, for purposes of political or transportation administration, most countries are divided into entities such as states, provinces, prefectures, regions, and districts. Typically, each year of a programming period, these administrative entities are allocated a certain amount of funds. These funds are typically placed into several categories representing program areas. In the United States for example, states receive federal highway funds in the following categories: surface transportation program fund, hazard elimination/safety fund, transportation enhancement fund, minimum guarantee fund, and the bridge program fund. States share these funds with their local public agencies. Therefore, for a given state, a program can be developed for transportation facilities in each of its programming units: the state/local entity on one dimension and the funding category on the other dimension. In certain states or provinces, programming units are further broken down by urban–rural status, local area size, facility condition categories, and so on. Examples of programming units representing jurisdictions and functional areas include local road safety, state highway safety, interstate congestion, state highways—all functions, and metropolitan street and transit operations. The programming unit may also be parochially defined as a set of facilities targeted by a particular funding source for intervention; for example, for disbursing a special fund for historic bridges, programming may be carried out only for those bridges. Recent legislation have relaxed the constraint of programming units that tended to inhibit development of realistic programs at the state or local level. For example, ISTEA of 1991 provided flexibility to metropolitan planning organizations (MPOs) in programming federal transportation funds for multimodal projects (Younger, 1994). Subsequent legislation such as TEA-21 and SAFETEA-LU has added other areas of flexibility both at the state and MPO levels.

Step 3: Perform a Needs (Physical/Monetary) Assessment Using Planning Functions   Most transportation agencies have established procedures for identifying systemwide deficiencies, needs, and candidate projects. This activity falls within the planning function and provides vital information with which a program is developed. For a state agency, projects can be identified from various sources: the central office, districts, and MPOs. Applied systemwide or districtwide, various *management systems* can generate projects for pavements, bridges, safety, and congestion mitigation. Also, new facilities and capacity improvement projects are generated though *long-range plans*. The planners and system managers utilize key information, such as inventory listing, deficiency identification, deterioration modeling, and performance thresholds (and/or facility service lives), to establish the

physical needs (type of work needed for each facility) in the network. Performance thresholds and facility service lives can be established through questionnaire surveys of experts, information in agency design guides, or using analytical tools (Lamptey et al., 2004). Physical needs identified for facility improvement, preservation, and replacement, reconstruction, and rehabilitation are used together with established preservation cost models to estimate the monetary needs. For new facilities, a long-range plan is used to determine the physical amount of new facility work, and using these estimates and construction cost models, the monetary needs associated with the long-range plan are estimated.

For the evaluation and selection of projects, the analyst needs a set of criteria that are consistent with the performance measures used in needs assessment. The analyst must also decide whether decisions are to be made on the basis of existing performance levels before proposed interventions or the expected change in performance levels upon intervention. For example, for project need identification, the threshold value of flexible pavement condition for freeways can be 140 IRI, indicating that if any pavement section fails to meet this requirement, it should be upgraded. However, for the evaluation of project alternatives, the one that would yield the most cost-effective improvement in IRI will be selected. Projects are identified, along with their costs and timings, so that monetary needs can be matched with expected funds.

Sufficient funds are typically not available to cover all needed projects, so it is necessary to identify candidate projects selectively so that the best mix can be implemented. A variety of tools are available to accomplish this task, including ranking-based, cost–benefit, and cost-effectiveness analyses. Optimization methods can also be used for selecting a mix of projects to maximize the overall cost-effectiveness subject to a given budget.

Step 4: Perform a Fiscal Analysis   As part of the long-range plan required of all state transportation agencies and MPOs, it is necessary that agencies develop a financial plan that compares the annual revenue from existing and proposed funding sources that are dedicated to transportation uses to the annual costs of constructing, maintaining, and operating the transportation facilities. The funding available depends on the programming unit in question. For example, if the programming unit is local road safety, the funding expected could be available from the federal hazard elimination fund.

Funding levels expected during the program period should be established by program category. Projects requiring matching funds from internal resources should be identified, along with those not eligible for matching funds from higher levels of government or other external sources. For example, an agency is responsible for operating and maintenance expenditures and these items are generally not included in the programming process. However, it should be noted that there exist trade-offs between maintenance and preservation of physical facilities. Preventive maintenance can extend the service life of facilities, resulting in less frequent or less intensive preservation or renewal. In addition, there are certain operations-oriented expenditure (for example, installation of ITS equipment, etc) that should be considered in programming because they often can serve as alternatives to facility expansions or additions. An effective programming process should consider the maintenance program while allocating resources in the preservation and improvement categories. This task includes revenue projection, where the expected levels of revenue from various sources are estimated.

Step 5: Develop a Transportation Improvement Program   This task involves the establishment of performance measures, selection of projects under each program, analysis of trade-offs between various performance measures, and fund allocation to various program categories.

(a) *Program Performance Measures*   In Chapter 2 we presented a general set of performance goals and measures that can be considered in transportation systems evaluation in general. The particular set of measures to be adopted for any given situation depends on the agency, the purposes of the evaluation; and in the context of programming, the stage of the programming process (at the management system level or at a higher administrative level). An

agency selects its own set of performance measures, which should reflect current transportation issues (Speicher et al., 2000) such as stakeholder input and broad community objectives. These measures should be decided through an open process involving all stakeholders and should be updated periodically.

Table 20.2 lists system performance measures commonly used by transportation agencies in the U.S. for highway and transit systems. Measures based on facility inspection data, such as pavement performance, bridge condition ratings, and rolling stock condition, are used frequently. Also often used are traffic levels of service measures, such as volume–capacity ratios and safety measures.

Good infrastructure condition, efficient traffic flow, and healthy economic activity are all valid goals, but "quality of life" in the community is an equally important goal. Explicit inclusion of measures that are related to non-motorized travel such as pedestrian access including provision of sidewalks and quality pedestrian crossings, walkability, cycling level of service, for example, are becoming accepted as important performance measures.

The purpose of establishing program performance measures is to enable agency managers to assess the degree to which the investment program selected has been successful in terms of improved system performance, cost, and benefits. In addition, measures are established to evaluate the effectiveness of the agency's program delivery process from planning to design and construction. The specific performance measures selected will be

Table 20.2   Examples of Quantifiable Program Performance Measures

Goal	Performance Measures
Facility condition	Pavement condition ratings, bridge condition and safety ratings, sufficiency/deficiency ratings, maintenance levels of service, rolling stock condition ratings, terminals and other facility conditions
Capacity expansion	Mobility and accessibility current and projected average daily traffic, average daily transit ridership, volume–capacity or traffic level of service, peak-hour congestion, average daily truck traffic
Safety	Crash frequency, crash rate, crash density
Security	Vulnerability to human-made/natural disasters
Environmental	Air quality conformity, noise, land use, water resources, ecological impacts, including wetlands; aesthetics
Strategic issues	Agency strategic planning goals, legislative mandates, community goals, private-sector participation
Economic efficiency	Initial cost, life-cycle cost, life-cycle cost and benefits (benefit/cost, net present value, equivalent uniform annual returns)
Economic vitality	Jobs created/retained, growth in personal income, gross regional product
Quality of life	Facilitation of non-motorized travel, pedestrian access, walkability, cycling level of service.

unique to the circumstances of each agency, including that agency's infrastructure condition, resource base, and policy focus. Irrespective of which performance measures are selected, the process of setting clear standards for performance and using the results of this evaluation to inform future investment choices and management decisions is essential to ensure that an agency's investment of resources is producing the intended outcomes.

System performance measures reflect mobility, accessibility, user cost, infrastructure conditions, environmental quality, safety, and other factors. Program delivery performance measures include the duration and cost of project phases, the number of design change or construction change orders, a comparison of total cost and schedule to the program and budget targets, and productivity measures related to the volume and unit cost of the work accomplished. A reasonable balance should be established between quantitative and qualitative performance measures. The Capital District Transportation Committee (CDTC), in Albany, New York, for example, develops its programs by screening projects for minimum requirements and then by evaluating the merits of screened projects using benefit–cost as well as qualitative variables (Younger, 1994). Also, Sinha and Jukins (1980) established a set of quantitative and qualitative criteria that could be used by transportation agencies for programming their projects.

(b) *Program Development* An agency work program is developed through an iterative process producing a mix of projects that optimizes the agency's strategic goals and objectives subject to a given budget level. By exercising if–then scenarios, various alternative programs can be developed for different budgetary levels. A work program can be structured to prioritize and allocate resources in the categories of maintenance, preservation, and improvement or renewal. A well-structured program would not only provide a focus for policy and strategic direction for the agency but would generate information that could be communicated to legislative bodies and the general public regarding the investment choices and associated system impacts and trade-offs.

(c) *Trade-off Analysis* For resource allocation trade-offs and final funding decisions, it is necessary to incorporate an evaluation process for the program as a whole so that its expected performance can be assessed in terms of stated goals and objectives. Program-level evaluation can facilitate explicit trade-off analyses between categories within a mode, between modes, or between jurisdictional levels. Lambert et al. (2005) developed a methodology for prioritizing investments across several transportation modes. At the present time, most state transportation agencies deal primarily with highway projects and programs, whereas multimodal programming is more common at metropolitan planning organizations.

For explicit evaluation of programs and associated trade-offs, a number of approaches can be used. Engineering economic analysis is applicable if the evaluation criteria are priceable. Otherwise, a multicriteria summary of program impacts in terms of both monetary and nonmonetary measures can be developed and used. A quantitative approach to multicriteria and multimodal evaluation will require relative weights to be assigned to various performance criteria and a procedure to render the criteria into commensurable units to enable the use of a common scale, as discussed in Chapter 18.

Trade-off analyses help decision makers to determine how much of one performance measure can be "bought" for a given level of another: for example, "expending" funds for facility preservation and "earning", in return, an enhanced facility condition (and increased facility life), or increased mobility and accessibility. This trade off indicates how many units of user cost can be obtained for each unit of agency cost. It is important to recognize that planning, programming and design involve tradeoffs; for example, in certain situations it may be necessary to accept some congestion in order to provide a high quality pedestrian environment.

(d) *Fund Allocation* On the basis of the results of trade-off analyses, funds are allocated to various programming units across categories, modes, and jurisdictional boundaries. There are two quantitative approaches that can be used to accomplish this task. First, a simple prioritization can be employed for yearly allocation of funds, where programs and projects are selected in descending order by their priority rankings until all the given levels of funds are exhausted. Such an approach usually does not produce the most effective allocation of funds. A better approach is to use an optimization technique that can maximize a combined measure of program performance subject to given budget constraints or minimize costs subject to a set of performance constraints. Programming tools are discussed in Section 20.3. It must be noted that while quantitative approaches are available, many transportation agencies do not use them at the current time and therefore the final selection of programs and projects is often largely subjective or is based on nontechnical criteria. Irrespective of which approach is used for programming, some portion of the funds are often allocated to specific program categories, modes, or geographic regions, to satisfy prior commitments, statutory requirements, or legislative mandates.

Final allocation of funds is then used to prepare the budget. Budgeting is an iterative process linked to programming. Although programs can be for multiyear periods, the first-year program is used as the annual budget. Multiyear programs should be updated each year on the basis of new data.

Step 6: Finalize the List of Projects, Programs, and Budget Allocation   As funds are invariably limited, it is necessary to set priorities for selecting projects within a given program or among programs of various levels. For example, consider a specific program category of pavement rehabilitation. An agency can have a number of projects within this category and the task can be to select a subset of projects from this set. This type of situation arises when the funding is non transferable across programs, that is, whole or part of the funds available for one program cannot be transferred to another. If, on the other hand, the funds available can be used for any of several project types related to a particular element, such as pavements, the task can be to select projects across various types of pavement work. For example, for pavements, project types include pavement rehabilitation, pavement resurfacing, or replacement and widening. Maintenance and operating expenditures are often excluded from capital programming. The next level of programming complexity arises when the *candidate projects* (a list of all possible projects that can be applied to all facilities under consideration) include various project subtypes, different elements, and multiple modes.

Step 7: Implement the Program and Monitor Performance   A well-designed monitoring plan should be instituted to track the progress of program implementation in terms of system performance, costs, and benefits. As mentioned earlier, this element of programming has not always been pursued. When implemented, program monitoring provides an essential feedback loop into policy directions and technical assumptions made in the programming process. A monitoring plan can indicate the effectiveness of the programming process and at the same time establish accountability and enhance credibility. To accomplish its objectives, a monitoring program must be rooted in the stated goals, objectives, and policy criteria defined explicitly though a transparent process of public participation.

An important function of monitoring is to track project costs and execution times, because significant variations in these items can disrupt a multiyear program. If even one project experiences a large cost overrun or is delayed by even one year, the resulting disruption in the transportation program can be significant as it will require shifting another project slated to be implemented at that time to a later implementation time in order to stay within the yearly budget and schedule. Only through careful monitoring of project costs and implementation periods can appropriate adjustments be made in multiyear programs in the course of their implementation.

20.3   PROGRAMMING TOOLS

When there is no shortage of funds, all needed projects can be implemented. However, since the funds are invariability limited, only a few projects can be implemented each year under each budget category. Therefore, after the impacts of each planned project have been identified in terms of overall utility (a combination of benefit and cast performance functions), the problem is to select the subset of projects that will yield the highest desirability in terms of the combined utility without exceeding the budget for any year. Quantitative tools available to accomplish this task are listed below.

20.3.1 Priority Setting

The most common approach to selecting projects for a program is by setting project priorities. After the projects have been ordered in terms of priorities, they can be selected starting from the top in descending order of priorities until the funding limit is reached. Priorities can be set using methods such as: (1) economic analysis method, (2) cost-effectiveness method, and (3) utility–cost method.

(*a*) *Economic Analysis (Benefit–Cost Ratio/Net Present Value Method)*   Economic analysis requires efficiency considerations. This type of analysis may include benefit–cost ratio, net present worth value, rate of return, or other variations. Of these, the method often recommended is the net present value approach. The benefits are calculated by monetizing the various impacts associated with the project, as discussed in Chapter 8.

(*b*) *Cost-Effectiveness Analysis*   In this approach, all costs and consequences are identified explicitly and the trade-offs of various impacts are assessed. The effectiveness of each objective, whether it is travel-time reduction, air quality improvement, reduction in noise level, or reduction in number of crashes, is determined explicitly in their respective scales—no monetization is needed. Hence, unlike the economic analysis method, the nonmonetary impacts can also be considered in this approach. The following steps are involved in setting project priorities on the basis of their cost effectiveness.

1. For each performance measure, calculate the effectiveness of each project. For example, if the objective is mobility improvement, determine the travel-time savings; if the goal is air quality improvement,

determine the reduction in pollutant emissions; if the goal is noise quality improvement, determine the reduction in noise level.

2. Plot the measure of effectiveness on the y-axis and the project cost on the x-axis for each of the projects being considered in the program. For example, if there are seven projects in the transportation improvement program, the costs (in thousands of dollars) and effectiveness (say, travel-time savings in minutes) are plotted as shown in Figure 20.2.

3. Calculate the cost-effectiveness ratio of all the projects in the program:

$$CE \ ratio_I = \frac{cost \ of \ project}{effectiveness \ of \ project}$$

$$= \left(\begin{array}{c} slope \ of \ the \ line \ joining \ the \ origin \\ to \ the \ point \ representing \ a \ project \end{array} \right)^{-1}$$

$$= \frac{1}{Slope_{OP}}$$

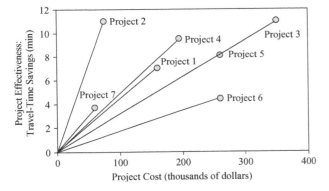

Figure 20.2 Setting project priorities using cost-effectiveness analysis.

where OP = Slope of the straight line between the origin (O) and the point representing the project (P).

The project with the least cost-effectiveness ratio has the highest priority. Some analysts may prefer to calculate cost-effectiveness ratio as follows:

$$CE \ ratio_{II} = \frac{effectiveness \ of \ project}{cost \ of \ project} = Slope_{OP}$$

In this case, the project with the highest cost effectiveness ratio has the highest priority.

4. Determine the priorities for all available projects. For example, the project priorities with respect to travel-time savings, based on the cost-effectiveness plot in Figure 20.2, are in the following order: Project P2 has the first priority, followed by P7, P4, P1, P5, P3, and P6. Projects P5 and P3 have the same priority with respect to savings in travel time.

5. Determine the project priorities with respect to all other program objectives by repeating steps 1 to 4.

The cost-effectiveness analysis approach is relatively straightforward and simple. However, it can only be used to determine cost-effectiveness with respect to each performance criterion individually. Unlike the economic analysis method, this approach cannot be used to determine an aggregated priority representing two or more performance criteria, unless effectiveness is expressed in terms of an overall utility that is comprised of the multiple performance criteria. This is discussed in the next section.

(c) Utility–Cost Method The concept of utility was introduced in Chapter 18. Using value or utility functions for various performance criteria and combining them through appropriate weights, the difficulty associated with the cost-effectiveness approach can be avoided. In the present chapter we focus on the use of multiattribute utility values to set priorities for selecting a mix of projects. A simple example of this approach is illustrated

Table 20.3 Utility–Cost Ratios for Seven Projects

Project, x_j	Multiattribute Utility, u_j (Scale 0–100)	Capital Outlay c_j (100,000's of dollars)	$p_j = u_j / c_j$
1	44	1.60	27.50
2	48	0.65	73.85
3	65	3.60	18.00
4	40	2.10	19.00
5	63	2.50	25.20
6	55	2.60	21.15
7	60	0.65	92.30

Table 20.4   Setting Project Priorities Using the Utility–Cost Ratio Method

Project, x_j	Multiattribute Utility, u_j (Scale 0–100)	Capital Outlay c_j (100,000's of dollars)	$\sum_j c_j$ (100,000's of dollars)	$p_j = u_j/c_j$
7	60	0.65	0.65	92.30
2	48	0.65	1.30	73.85
1	44	1.60	2.90	27.50
5	63	2.50	5.40	25.20
6	55	2.60	8.00	21.15
4	40	2.10	10.10	19.00
3	65	3.60	13.70	18.00

in Table 20.3, where the multiattribute utility values, capital costs, and the utility–cost ratios of seven projects are shown. The projects can be prioritized in descending order of utility–cost ratios, as shown in Table 20.4.

Using the conventional priority setting process, if the budget for the period is $800,000, the top five projects (7, 2, 1, 5, and 6) are selected for the program. However, if the budget is $1,000,000, the conventional method will choose the top five projects, leaving $200,000 of the budget unused. Such a situation provides a suboptimal solution. The process can be improved by using a heuristic search technique, as discussed in the next section.

20.3.2 Heuristic Optimization

The rule for the process is to search for a solution that can give a higher total return even though the priority ranking value (u_j/c_j) of a project selected is lower in the list, so that the entire budget is used. With a budget of $1,000,000, the conventional approach selects projects 7, 2, 1, 5, and 6, with the total return of 270 utility units (the sum of the utility values of all the projects selected). However, $200,000 of the allocated budget is left unused. On the other hand, if projects 7, 2, 5, 6, and 3 are selected, the return is 291 utility units and the entire budget of $1,000,000 is used. For a programming problem with thousands of projects, the procedure as herein described may not be practical using a manual search, and the use of a simple computer program would be necessary.

20.3.3 Mathematical Programming

A limitation of the priority-setting approaches is that they can be used for single-period programs whereas mathematical programming approaches can be used for either a single period or for multiple periods. Mathematical programming can be used to select an optimal set of projects that maximizes total benefits subject to budget and other

constraints (Sinha and Muthusubramanyam, 1981). There are several mathematical programming techniques that could be used to solve this problem, and the appropriate technique to be selected depends on the nature of the optimization problem. This is discussed below.

(a) *Linear Programming*   The linear programming technique can be used to define a class of problems where the decision variables are nonnegative real numbers, the objective function is linear, and the constraints to the objective function are linear equations or inequalities. In addition, the objective function and constraints must satisfy the proportionality assumption. Transportation programming problems that are best solved using linear programming technique include *multi-period problems*. These types of problems are those that involve scheduling of projects whose individual implementation periods exceed 1 year so that optimal yearly funding levels can be determined if it is assumed that the projects are divisible, meaning that some fraction of the project can be funded and implemented. For example, if one-third of project j is funded, the utility would be $\frac{1}{3}u_j$, and a linear programming formulation will be:

$$\text{Maximize}\quad U = \sum_{j=1}^{n} u_j x_j$$

subject to

$$\sum_j c_{jk} x_j \leq B_k, \qquad k = 1, 2, \ldots, m \qquad (20.1)$$

$$0 \leq x_j \leq 1, \qquad j = 1, 2, \ldots, n$$

where x_j is the fraction of the jth project to be funded, u_j the utility or net benefit obtained by funding j, c_{jk} the cost of the jth project in year k, and B_k the budget available in year k.

Example 20.1   Assume that there are four projects and their associated utility or benefit values are aggregated as shown in Table E20.1. The table also shows the capital outlays required for each project and yearly budgets. It is sought to develop a program for these projects so that the total utility is maximized.

SOLUTION   If all projects are funded, the total funds required are 65 and 22 units in the first and second years, respectively. A linear programming formulation of the problem is

$$\text{Maximize} \quad U = 15x_1 + 18x_2 + 20x_3 + 14x_4$$

subject to

$$13x_1 + 15x_2 + 20x_3 + 17x_4 \leq 36$$
$$4x_1 + 8x_2 + 7x_3 + 3x_4 \leq 12$$
$$0 \leq x_j \leq 1, \qquad j = 1, 2, 3, 4$$

This problem can be solved manually using the simplex method. However when there are many variables and constraints, any of the programming software available, such as Excel solver, GAMS, LINDO, and CPLEX, can be used. The optimal solution for Example 20.1 is $x_1 = 1.00$, $x_2 = 0.74$, $x_3 = 0.00$, $x_4 = 0.70$, and the total utility corresponding to the optimal solution is 38.1 units.

As stated earlier in this section, linear programming formulations are most appropriate for transportation programming when the decision variables are continuous. This means that it is applicable when the transportation agency seeks to determine the fraction of projects or programs to be funded, for example, the percentage of rail tracks to be rehabilitated. However, for most transportation problems the choice is to select or reject a project entirely. In such cases, linear programming is not

Table E20.1   Data on Project Cost and Utility

Project j	Multiattribute Utility, u_j	Capital Outlays Required	
		First Year, $c_{j,k=1}$	Second Year, $c_{j,k=2}$
1	15	13	4
2	18	15	8
3	20	20	7
4	14	17	3
Budget available		36	12

appropriate, and integer programming technique should be used.

(b) *Integer Programming*   Integer programming is a mathematical programming technique that is used when all the decision variables are integer values. Integer programming formulations are appropriate in transportation programming problems where projects are either entirely selected or entirely rejected in a programming period. In other words, there is no partial implementation of projects. For example, if the linear programming problem in Section 20.3.3 is assumed to involve binary decision variables then it can be solved using integer programming, as follows:

$$\text{Maximize} \quad U = \sum_{j=1}^{n} u_j x_j$$

subject to

$$\sum_j c_{jk} x_j \leq B_k \qquad k = 1, 2, \ldots, m$$

(total expenditure cannot exceed the yearly budget)

(20.2)

$$x_j = 0 \text{ or } 1 \quad \left(\begin{array}{c} \text{either a project is implemented} \\ \text{or it is not implemented} \end{array} \right)$$

Symbols and subscripts have their usual meanings.

Example 20.2   Solve the mathematical programming problem in Example 20.1 assuming that the decision variables are binary (integer values that can only take one of two values such as 0 and 1).

SOLUTION   The objective function and the first two constraints remain the same as expressed in the solution to Example 20.1. However, the decision variables are redefined as $x_j = 0$, 1 for all j. Solving the integer programming problem yields the following optimal solution: projects 1 and 3 should be implemented because they result in the highest utility of 35 units.

The class of transportation investment optimization problems associated with integer programming are typically termed *knapsack problems*, where the goal is to seek the best possible program (set of candidate projects to be implemented at all constituent facilities in a network and in each year to maximize networkwide utility or net benefit) in terms of a single or multiple performance measure(s) and subject to budget and other constraint(s). Considering that the single-criterion problem is merely a simplified case of the multiple-criteria problem, we proceed to provide various problem formulations for the former.

No Constraints on Annual Spending Amounts; Constraint for the Entire Programming Period: This represents the situation where a given budget is specified for the entire period and there are no constraints on the amount that can be spent each year. Although such a case is rare, it provides a globally optimal solution for a given period. The optimal transportation program (funding allocations) can be obtained by solving the following integer programming problem:

$$\text{Maximize} \quad U = \sum_{i=1}^{h} \sum_{j=1}^{n} \sum_{k}^{m} x_{ijk} u_{ijk} \qquad (20.3)$$

subject to

$$\sum_{i=1}^{h} \sum_{j=1}^{n} \sum_{k=1}^{m} x_{ijk} c_{ijk} \leq B \qquad (20.4)$$

$$\sum_{j=1}^{n} x_{ijk} = 1 \qquad \text{for all } i \text{ and all } k \qquad (20.5)$$

$$x_{ijk} = 0 \quad \text{if } k < y_i \qquad (20.6)$$

$$x_{ijk} = 0 \text{ or } 1 \qquad (20.7)$$

where c_{ijk} = cost of project j for facility i at analysis year k

B = total budget for the programming period

h = total number of facilities or candidate projects in the network selected

n = number of alternative intervention projects for facility i

m = number of years in the programming period

k = analysis year = $1, 2, \ldots, m$

y_i = year when facility i is expected to have a need for improvement (critical year)

u_{ijk} = utility value of project j for facility i at analysis year k

$$x_{ijk} = \begin{cases} 1 & \text{if project } j \text{ is implemented at} \\ & \text{facility } i \text{ in analysis year } k \\ 0 & \text{otherwise} \end{cases}$$

Other symbols have their usual meanings.

Equation (20.3) represents the objective function of the integer program. The constraints are represented by equations (20.4) to (20.7). The first constraint equation, (20.4), is the budget constraint, which states that the total expenditure (capital cost) is less than the budget available over the programming period. This is the *size* constraint of the knapsack problem. Equation (20.5) requires that only one project (including do-nothing) be selected for each facility in each year of the programming period. Equation (20.6) requires that no project should be implemented at a facility before the critical year. These are called the *choice constraints* of the knapsack problem.

Constraints on Yearly Budgets, Possible Carryover of Unspent Budget to the Following Year: Multiyear budgeting with carryover of unspent budget represents the situation where an annual budget is specified for each year of the analysis period. However, any unspent budget can be transferred to the next year. The optimal funding allocation of the funding can be obtained as follows.

$$\text{Maximize} \quad U = \sum_{i=1}^{h} \sum_{j=1}^{n} \sum_{k}^{m} x_{ijk} u_{ijk} \qquad (20.8)$$

subject to

$$\sum_{i=1}^{h} \sum_{j=1}^{n} \sum_{k=1}^{m} x_{ijk} c_{ijk} \leq \sum_{k=1}^{m} B_k \qquad \text{for all } k \qquad (20.9)$$

Symbols are as defined earlier.

Equation (20.8) is the objective function; equation (20.9) indicates that the annual allocation (capital cost) must be equal to or less than the annual budget limit plus any excess funds carried over from the previous year. The remaining constraints are similar to those shown in equations (20.5) to (20.7).

Annual Budget Constraint and One or More Nonbudget Constraints: This is a multichoice multidimensional knapsack problem (MCMDKP) that can be used to determine the best possible projects to be implemented to maximize networkwide reward (utility) and subject to yearly budget constraint and one or more nonbudgetary constraints. Nonbudgetary constraints may include condition targets, transit ridership, commuter accessibility, vulnerability, and so on. The problem can be formulated as

$$\text{Maximize} \quad U = \sum_{i}^{h} \sum_{j}^{n} \sum_{k}^{m} x_{ijk} u_{ijk} \qquad (20.10)$$

subject to

$$\sum_{i=1}^{h} \sum_{j=1}^{n} x_{ijk} c_{ijk} \leq B_k \qquad \text{for all } k \qquad (20.11)$$

$$\frac{1}{h} \sum_{i=1}^{h} \sum_{j=1}^{n} \sum_{k=1}^{m} x_{ijk} H_{ijk} \geq H_{min} \qquad \text{for all } k \quad (20.12)$$

$$\frac{1}{h} \sum_{i=1}^{h} \sum_{j=1}^{n} \sum_{k=1}^{m} x_{ijk} V_{ijk} \leq V_{max} \qquad \text{for all } k \quad (20.13)$$

Symbols are as defined earlier.

Equation (20.10) is the objective function, equation (20.11) is the budget constraint without carryover, and equations (20.12) and (20.13) are nonbudget constraints for this knapsack problem formulation. The choice constraints for this knapsack problem formulation are similar to those shown in equations (20.5) to (20.7). H_{min} are the performance targets (floor) for the network, V_{max} the performance targets (ceiling) for the network, and H_{ijk} and V_{ijk} the performance levels achieved in categories H and V.

Example 20.3 A five-year transit capital program is being developed for a large transit system consisting of subway, commuter rail, and bus services. Ten projects have been selected through an initial screening process. Each of these projects will be implemented over five years. For each project, the values of overall utility to be derived from the project completion and the capital cost of construction are shown in Table E20.3.1. Over the five year period the total cost of all projects is $963.707 million and the total budget is $600 million.

The available annual funds for the coming five years have been ascertained. Using an integer programming approach determine the optimal set of projects that should be included in the five-year program. Determine projects to be selected when:

(i) there is a yearly budget constraint, and any unused funds left over from a previous year are not available in the subsequent year,
(ii) there are the same constraints as described in (a), but there is an additional constraint that projects 1 and 7 cannot be implemented simultaneously due to logistical reasons,
(iii) there is a yearly budget constraint, and unused funds leftover from the previous year are available for use in the subsequent year.

SOLUTION (i) For the scenario without carryover provision, the integer programming problem can be

Table E20.3.1 Project Capital Requirements and Utilities

		Capital Outlay ($ millions)					
	Project	Year 1	Year 2	Year 3	Year 4	Year 5	Utility
1.	Rail station terminal improvements	25.357	33.354	26.370	24.000	16.420	70
2.	Subway downtown extension	12.410	7.620	19.800	7.410	8.526	62
3.	Rail fleet overhaul	4.200	2.300	0	6.500	4.500	45
4.	Rail rolling stock procurement	22.483	22.444	40.414	68.393	67.126	85
5.	Security improvements	2.590	2.590	2.590	2.590	2.610	40
6.	Signals and communications/electric traction systems	6.869	8.050	11.250	11.250	25.340	50
7.	Bus passenger facilities/park-and-ride	16.024	17.391	10.254	9.613	5.345	65
8.	Americans with Disabilities Act Project—platforms/stations	7.856	6.500	5.780	6.500	5.458	40
9.	Bridge and tunnel rehabilitation	21.408	20.5600	24.800	26.680	24.587	80
10	Bus acquisition program	55.100	55.103	47.576	47.576	49.240	90
	Total annual cost	174.297	175.912	193.834	210.512	209.152	—
	Annual budget	100	100	130	140	130	—

Over the five-year period: total cost of all projects = $963.707 million; total budget = $600 million.

formulated as follows:

Maximize $U = \sum_{j=1}^{10}(x_j u_j)$

subject to $\sum_{j=1}^{10}(x_j c_{jk}) \leq B_k$ for $k = 1, 2, 3, 4, 5$

$x_j \in \{0, 1\}$ for all j

Using Excel Solver, CPLEX, GAMS, and LINDO, the solution to the above integer programming problem can be obtained, as shown in Table E20.3.2.

Table E20.3.3 shows the optimal distribution of funds during the five-year programming period. The optimal value of the objective function is 452 units.

(ii) The following constraint is added to specify that projects 1 and 7 cannot be implemented simultaneously,

$$x_1 + x_7 \leq 1$$

The solution to the modified integer programming problem is as given in Table E20.3.4.

Table E20.3.2   Selected Projects for Scenario (i)

Projects	Project Status (1 means Selected; 0 means Not Selected)
1 Rail station terminal improvements	1
2 Subway downtown extension	1
3 Rail fleet overhaul	1
4 Rail rolling stock procurement	0
5 Security improvements	1
6 Signals and communications/electric traction systems	1
7 Bus passenger facilities/park-and-ride	1
8 Americans with Disabilities Act project—platforms/stations	1
9 Bridge and tunnel rehabilitation	1
10 Bus acquisition program	0

Table E20.3.3   Amount of Funding Available, Used and Unused, per Year (in $millions), Scenario (i)

	Year 1	Year 2	Year 3	Year 4	Year 5
Total Funds Available	100	100	130	140	130
Funds Used	96.714	98.365	105.844	94.543	92.786
Unused Funds	3.286	1.635	24.156	45.457	37.214

Table E20.3.4   Selected Projects for Scenario (ii)

Projects	Project Status (1 means Selected; 0 means Not Selected)
1 Rail station terminal improvements	1
2 Subway downtown extension	1
3 Rail fleet overhaul	1
4 Rail rolling stock procurement	0
5 Security improvements	1
6 Signals and communications/electric traction systems	1
7 Bus passenger facilities/park-and-ride	0
8 Americans with Disabilities Act project—platforms/stations	1
9 Bridge and tunnel rehabilitation	1
10 Bus acquisition program	0

The optimal distribution of funds is shown Table E20.3.5 and the value of the objective function is 387 units.

Table E20.3.5   Amount of Funding Available, Used and Unused, per Year (in $millions), Scenario (b)

	Year 1	Year 2	Year 3	Year 4	Year 5
Available $	100	100	130	140	130
Used $	80.690	80.974	95.590	84.930	87.441
Unused $	19.310	19.026	34.410	55.070	42.559

(iii) For the scenario, when the unused budget in the previous year is available in the next year, the integer programming problem can be formulated as follows:

Maximize $U = \sum_{j=1}^{10}(x_j u_j)$

Subject to $\displaystyle\sum_{k=1}^{5}\sum_{j=1}^{10}(x_j c_{jk})$

$$\leq \sum_{k=1}^{5} B_k \quad \text{for } k = 1, 2, 3, 4, 5$$

$$x_j \in \{0, 1\} \quad \text{for all } j$$

The solution is shown in Table E20.3.6.

Table E20.3.6   Selected Projects for Scenario (iii)

	Projects	Project Status (1 means Selected; 0 means Not Selected)
1	Rail station terminal improvements	0
2	Subway downtown extension	1
3	Rail fleet overhaul	1
4	Rail rolling stock procurement	1
5	Security improvements	1
6	Signals and communications/ electric traction systems	1
7	Bus passenger facilities/ park-and-ride	1
8	Americans with Disabilities Act project—platforms/stations	1
9	Bridge and tunnel rehabilitation	1
10	Bus acquisition program	0

Table E20.3.7 shows the unused budget during the five year programming period under the carry-over budget scenario.

Table E20.3.7   Amount of Funding Available, Used and Unused, per Year (in $millions), Scenario (iii)

	Year 1	Year 2	Year 3	Year 4	Year 5
Available $	100	200	330	470	600
Used $	93.840	181.295	301.183	440.119	583.611
Unused $	6.160	18.705	28.817	29.881	16.389

The optimal value of the objective function under scenario iii is 467 units. Hence, the system utility increases when the unused funds can be carried over from one year to the next. Also, the amounts of unused funds at the end of each year are generally lower for this scenario compared to the previous scenarios.

(c) *Goal Programming*   In some cases, a transportation agency may prefer to utilize multiple goals (objectives) separately without combining them into an aggregated utility value. In such cases, the goal programming technique can be used. With this approach the agency seeks to achieve all the performance goals to the fullest possible extent (that is, the sum of deviations of performance measure levels from their respective stated goals is minimized). As discussed in Section 18.3.6, the goal programming framework consists of the following elements: decision variables, system constraints, goal constraints, and objective function. The decision variables and system constraints are the same as those specified in linear programming. The system constraints represent absolute restrictions (such as budget constraints) that have to be satisfied before goal constraints are considered. The goal constraints are the target values of the performance measures. The objective function minimizes the weighted sum of the deviations. The weights are assigned in accordance with the importance of the goal.

Consider two highway segments. Segment 1, if improved, can provide a safety benefit of 2 units per mile; segment 2 can contribute a safety benefit of 1 unit per mile. The costs are $3 million and $2 million per mile for segments 1 and 2, respectively. The total fund available is $12 million. The agency has set a target of improving at least 8 miles of the highway system. Let x_1 and x_2 miles be the decision variables that represent the lengths that could be improved in segments 1 and 2, respectively.

The problem can be viewed as having two objectives: (1) to enhance the overall safety and try to achieve a safety target say, T, as closely as possible, and (2) to improve an 8-mile segment of highway. The system constraint is that the available budget cannot be exceeded. The goal constraints can be defined as follows:

$$2x_1 + x_2 + d_1^- - d_1^+ = T$$

safety constraint (objective 1)   (20.14)

$$x_1 + x_2 + d_2^- - d_2^+ = 8$$

goal constraint (objective 2)   (20.15)

where, d_1^- = under achievement in objective 1, d_1^+ = over achievement in objective 1, d_2^- = under achievement in objective 2, and d_2^+ = over achievement in objective 2.

The objective function is to minimize the adverse deviations. Therefore the objective function can be written as

Minimize $Z = d_1^- + d_2^-$

which represents a scenario in which the two objectives are equally important. If objective 1 is more important than objective 2, the relative importance of the two objectives can be represented by the weights w_1 and w_2, respectively, and the overall goal programming formulation can be expressed as follows:

$$\text{Minimize} \quad Z = w_1 d_1^- + w_2 d_2^-$$

subject to

$$2x_1 + x_2 + d_1^- - d_1^+ = T$$
$$x_1 + x_2 + d_2^- - d_2^+ = 8 \qquad (20.16)$$
$$x_1, x_2, d_1^-, d_1^+, d_2^-, d_2^+ \geq 0$$

This formulation reduces the goal programming problem to one of linear programming and can be solved using any of the available linear programming software packages. Sinha and Muthusubramanyam (1981) developed a highway programming algorithm using a goal programming approach. A drawback of the goal programming approach is its difficulty of handling discrete decision variables.

(d) *Dynamic Programming* Dynamic programming offers a convenient approach to solve transportation programming problems that contain such a large number of *interrelated* decision variables that the problem is considered to consist of a sequence of problems, each sequence requiring a separate solution to yield values of only a few decision variables (Jiang and Sinha, 1990). Hence, an *n*-variable problem is substituted for by *n* single-variable problems. This procedure is typically used when a number of decisions have to be made in a sequential order and when an earlier decision affects what the subsequent decisions will be (Nemhauser, 1966). Unlike other forms of mathematical programming, there is no single algorithm that can solve all dynamic programming problems.

The key elements of dynamic programming are *stages, states, decision*, and *return* or *benefit* (Cooper and Cooper, 1981). In the context of multiyear transportation programming, each year is viewed as a stage. At each stage, the transportation system is described by states such as the budget available and levels of performance measures (e.g., bridge condition, congestion level, crash frequency). Given the state of the system at the beginning of a stage, the return or benefit is maximized by selecting an optimal mix of projects, the system is transformed to the end state of the stage, and the process is repeated through the next stage. The objective of the transportation agency is to maximize the total return or benefit over all stages.

The steps involved in the dynamic programming process are as follows:

1. Decompose the problem into smaller problems and characterize the structure of an optimal solution.
2. Express the solution of the original problem in terms of optimal solutions for smaller problems.
3. Compute the objective function value of an optimal solution in a bottom-up fashion by recursive use of the solutions of smaller problems.
4. Construct an optimal solution from the information computed.

To illustrate the working of the dynamic programming technique, consider the following example: It is desired to determine the optimal spending policy which maximizes the effectiveness of the entire system over a program period. The dynamic programming technique first divides each year's available funds into several possible spending portions. Suppose that the program period, T, is two years and that the possible spending for year 1 is 50, 60, 70, 80, 90, and 100 million, and the possible spending for year 2 is 150, 140, 130, 120, 110, and 100 million, respectively. Any combination of spending for the individual years can be considered. The task of dynamic programming is to determine the optimal policy from among the possible spending combinations, such as (50, 150), (60,140), (70,130), (80,120), (90, 110), and (100, 100) and to determine the corresponding optimal mix of projects. The procedure remains the same for scenarios when T is larger than 2, say 10. The model then computes the optimal policy from year 1 to year 10 and gives the corresponding mix of projects.

Dynamic programming ensures not only optimal solution of the problem at hand but also optimal solution of the subproblems. For example, if projects have to be selected for a 10-year program period, dynamic programming gives the optimal mix of projects for the entire 10-year period as well as the optimal solution for any period less than 10 years. A major disadvantage of dynamic programming is that the presence of a large number of state variables can result in computational problems relating to information storage and computational time.

20.4 CASE STUDIES: TRANSPORTATION PROGRAMMING

Two case studies are presented in this section, one from a midwestern state transportation agency and the other from a northeastern metropolitan planning organization. In the United States, each state is required to prepare a statewide transportation improvement program (STIP) in order to receive federal funds, and within a state, each MPO is required to develop a Transportation Improvement Program (TIP) for its jurisdiction. Also, transit agencies such as Chicago Transit Authority and Washington Area Metropolitan Transit Authority are required to develop capital programs for the replacement and rehabilitation of rolling stock, terminals, workshops, and other fixed facilities.

The MPO is the designated forum for selecting transportation projects to be supported using federal highway and transit funds within its jurisdiction. In each successive federal transportation authorization starting from ISTEA to TEA-21 to SAFETEA-LU, MPO responsibilities have been expanded. Under federal law, MPO actions must now consider items ranging from security to land use plans; follow adopted public participation policies, incorporate visualization techniques, and welcome new technologies (Poorman, 2006). MPO plans must also be fiscally constrained.

Statewide planning requirements in federal law are less demanding and statewide plans are not required to include air quality conformity details, identify project level prioritization, or include fiscal analysis. STIPs are required to include MPO TIPs in their entirety. Most state transportation agencies, however, adopt and select cost-effective projects that enhance statewide objectives.

20.4.1 Programming Process at a State Transportation Agency

The state of Indiana has over 11,000 miles of state-controlled highways, including 1100 miles of interstates. It has a population of approximately 6.24 million and an area of 36,420 square miles. Its largest city is Indianapolis and it has 13 urban areas with metropolitan planning organizations (MPOs). There are 32 local bus transit systems and a regional commuter rail system along the southern shore of Lake Michigan. The annual program development process at the Indiana Department of Transportation (Indiana DOT, 2005a) is summarized in the following sections.

1. *Initiate the process.* The annual state transportation improvement program (STIP) development process of the Indiana DOT follows a time schedule and starts with a call for proposals for new projects to the Indiana DOT central office, all districts, and all MPOs. The needs of rural local areas are coordinated by the Indiana DOT districts. All project proposals include the type of work, terminals, length, design concepts, scope, and location. Proposals should also include a needs assessment of what problem this project solves, the level of support from the public,

environmental justice issues, and any planning documents relevant to the proposal.

2. *Coordinate the local areas.* Each district holds a primary public meeting to discuss the existing and proposed program of projects, focusing on consultation and cooperation with MPOs, rural planning organizations (RPOs), local elected officials, and representatives from relevant Indiana DOT central offices. The purpose of these meetings is to reach an agreed-to list of existing and proposed new projects. All district lists are then submitted into the scheduling and project management system (SPMS).

3. *Perform a statewide review and update the program.* Two sets of groups conduct the review: The Indiana Planning and Oversight Committee (IPOC) is responsible for major new capacity projects involving over $5 million, and a number of program management groups (PMGs) organized by project type such as bridge or pavement rehabilitation (Indiana DOT, 2005b). The IPOC consists mainly of the Indiana DOT executive staff and is chaired by the Indiana DOT commissioner. The IPOC holds up to six public hearings annually. PMGs include planning and programming personnel in the Indiana DOT. The primary purpose of the PMGs is to score, rank, and prioritize all projects other than major new added capacity projects. Each program category is given a budget target for a year and projects are selected according to established criteria for the category. Examples of program categories are pavement, bridge, safety, and intersection. Each PMG is based on the same principle as IPOC. All projects, including those selected by the IPOC, are then placed in a combined statewide program in a priority list based on recommendations from individual PMGs, making sure that the budget can support the current and projected improvement needs. The programming policies followed by IPOC in selecting major new projects are given in Table 20.5.

4. *Report on the program update and confirm the budget.* A draft program update report summarizes Indiana DOT's schedule of programmed projects, illustrates the effects of the new projects on the budget, and sets accepted levels of over programming. The report also includes a list of projects to be deleted from the schedule or placed on hold. The draft program is then reviewed by the Indiana DOT executive staff for approval; and after any amendments, it is sent to the districts and MPOs for comments.

At this point, the MPOs may seek public perspectives of the project through established procedures. After addressing the district and MPO comments, the final program update report is produced, indicating new projects in the SPMS and changes in the existing program.

Table 20.5   Major Project Programming Policies at a State DOT

1. Open, fair, criteria-driven process.
2. Long-range statewide planning with local approval.
3. Preservation first—greatest weight in allocating funds among state projects.
4. Transportation and economic development—transportation efficiency and effectiveness factors represent 70% of the total potential score in the selection process; economic development factors represent 30%.
5. Transportation efficiency and effectiveness criteria—include a variety of factors, such as benefit–cost ratio, project's state network connectivity, road classification, mobility enhancement, project's interstate connectivity and intermodal benefits. Preference is given to projects that expand or improve connections to water ports, airports, rail or transit facilities.
6. Safety criterion—based on the number of current and future crashes.
7. Economic development criteria—include job creation, job retention, levels of investment, and economic distress of the surrounding county. Points are allocated only if the Indiana Economic Development Corporation and Indiana DOT are assured that the economic development potential has been identified explicitly.
8. Retail and tourism—retail development is not considered. Effect on increased tourism is incorporated in job creation.
9. Customer impact—this category allows stakeholders to have a direct impact on the ranking of a project.
10. Non-Indiana DOT participation
    A. External funding points are assigned to projects based on the amount of private funding, local assistance, or funds contributed through a project-specific federal process.
    B. Interchange participation—at lease 50% of the cost of new interchanges on existing routes must come from either private, local, or other non-Indiana DOT funds.
    C. Fixed transit line evaluation—while the selection process focuses primarily on highway projects, efforts are made to evaluate transit projects using parallel criteria. For example, the expansion or construction of a new commuter rail line can be compared to a parallel highway expansion. However, projects such as transit station construction cannot be considered using parallel criteria and are evaluated on a case-by-case basis. All other criteria, such as economic development, remain the same.
11. Nontraditional projects—Indiana DOT may consider nontraditional projects such as HOV lanes, shared ride facilities, modal hubs, and other projects if they improve the operation of one of the state's major transportation corridors.
12. Bypass projects—scored using different transportation efficiency criteria, including projects expected average daily traffic. The percentage of diversion from the current facility to the bypass, v/c ratio on the current facility, number of recurrent congestion points avoided by the bypass, size of community/communities being bypassed, and major corridor completion. All other criteria remain the same as the other projects.
13. Urban revitalization—additional points are assigned to projects that contribute explicitly to the revitalization of an urban core.
14. Intelligent transportation systems (ITS)—ITS projects on the state and federal transportation network are eligible for major new capacity program funding. These projects focus primarily on incident management, including traffic management/control through ramp metering, lane control, and freeway-to-freeway diversion via dynamic message signs.

Source: Indiana DOT (2005b).

Updated budget estimates of projected federal and state funding for the following decade by fiscal year are used to establish fiscal constraint limits for all state projects in the subsequent Indiana statewide transportation improvement program (INSTIP).

5. Develop the INSTIP and coordinate with MPO TIPS. The INSTIP includes a fiscally constrained list of projects for the subsequent four-year period. This list consists not only of projects seeking federal aid, but all regionally significant projects for which a federal action is required, whether the projects are funded by federal, state, or local resources.

The content of the draft INSTIP is coordinated with the draft transportation improvement programs (TIPs) from the MPOs. The product of coordination activities is a draft fiscally constrained program of state projects in each MPO TIP covering a period of at least four, but no more than five fiscal years. The development of a TIP includes public review and comment organized by each MPO. For areas designated as maintenance or nonattainment for air quality, the MPO sends a draft TIP to the appropriate reviewing agencies for conformity consultation purposes. Each draft MPO TIP is reviewed for conformance with public involvement, air quality, and state and local area long-range plans and other requirements, and its compatibility with INSTIP. All transit projects in approved MPO TIPs are also included in the INSTIP. A TIP in an air quality maintenance or nonconforming MPO cannot be approved without a conformity finding by FHWA and FTA (Indiana DOT, 2005c).

6. *Provide for public review and prepare the Final INSTIP.* The draft INSTIP document for the subsequent three fiscal years, using the fiscally constrained agreed-to list of transportation projects, is presented for public review and comment. Indiana DOT conducts an annual public meeting at each of its districts for this purpose. All public comments are summarized in the final INSTIP document, along with a response to comments. At this stage the document is submitted to FHWA and FTA for review and comment. If approved, projects in the first three fiscal years of the INSTIP become designated as *committed projects.* The INSTIP may be amended in any manner that is agreeable to Indiana DOT, the MPOs, FHWA, and FTA.

7. *Use of quantitative approach.* The project selection follows a rating system for assigning relative weights to program goals by allocating points. Each project is then ranked within each goal through the same point system. Table 20.6 provides the point system used for ranking projects. For example, 70% of the weight is for transportation-related factors, while economic development–related factors receive 30%. If a project is earmarked through public/private, or local participating funds, it can receive an additional maximum 100 points within each of the two broad goals of transportation and economic development. These are individual criteria that can receive scores up to certain values, as shown in the table.

All projects are rated according to an established scoring procedure that is based on performance measurements. For example, the mobility criterion has four subcriteria: truck and auto AADT, volume-capacity (v/c) ratio, LOS,

and intergovernmental agreement. Each subcriterion has threshold values for specific points. If the truck AADT over the length of the project is projected to be between 10,801 and 12,000, the project receives 5 points; for an auto AADT value exceeding 72,000, 5 points are assigned. If the v/c ratio is 1.35 or more, the project receives the maximum score of 9, and a v/c ratio of 0.55 or lower gets 1. If the project improves the level of service to A, it receives the maximum score of 5. If the resulting level of service ratio of a project is F, it receives a score of zero. If a project spans the state line and there is a bistate agreement, the project gets a maximum 3 points under the criterion of intergovernmental agreement. If a project has a local government agreement, it receives 2 points under this criterion.

There is also a set of rules that affect priority setting. These rules include those that ensure balanced geographic distribution and corridor completion. For example, only one multi-year corridor can be assigned per district at any given time, and after a corridor project starts, its funding will continue until completion. Also an attempt is made to keep project size within $50–75 million range. These rules can significantly change the ranking of projects and thus can render the final ranked list different from that obtained solely on the basis of project scores.

20.4.2 Programming Process at a Metropolitan Area Level

The Capital District Transportation Committee (CDTC) is the designated metropolitan planning organization (MPO) in Albany, Rensselaer, Saratoga, and Schenectady counties in New York. With a population of approximately 780,000 people, the CDTC has three major cities: Albany, Schenectady, and Troy, and several smaller cities. Suburban trip orientation, growth in non-work travel, and the broad availability of vehicles in most house hold have led to a decline in the role of transit and walking throughout the region. While none of these trends are unique to this area, the choice of how to respond to them leaves room for pursuing options tailored to local circumstances. CDTC predicts in all measures of transportation system performance (ranging from congestion to transit access, from energy consumption to transportation/land-use compatibility), the region would be in worse shape in 2015 than in the current time. Like all MPOs, the CDTC is responsible for developing new multiyear comprehensive programs of federal-aided highway and transit projects within its jurisdiction every other year. This section of the chapter describes the transportation programming process at that organization. Further details of the process are found in CDTC (2005).

Table 20.6 Indiana DOT Major Selection Criteria for New Projects

Goal	Factors	Maximum Score
Transportation efficiency	Cost-effectiveness index: measure of the benefit–cost ratio and net present value of the investment	20
	Corridor completion: measure of a project's ability to complete statewide connectivity targets	3
	Road classification: measure of a highway's importance	5
	Congestion relief (mobility): measure of the truck and vehicle AADT volume-to-capacity ratio and change in LOS from the improvement	20
	Adjacent state or relinquishment agreement: measure of interstate connectivity	3
Safety	Measure of the crash rate, crash severity, crash frequency/density, and the change in crash rate due to the improvement	20
	Transportation points account for at least 70% of a project's base score:	*70*
Economic development	Truck volume indicator: measure of the economic impact demonstrated by trucks	5
	Jobs created: level of nonretail jobs that the project creates	10
	Job retention: evidence that the project will retain existing jobs	5
	Economic distress: points based on the severity of the unemployment rate of the county	10
	Maximum economic development score:	*20*
Customer input	Districts: measure of the priorities established at the district level	4
	MPO: measure of the priority of the local MPO or RPO	3
	Other: measure of the input of citizens either through their legislative representative or via direct documented comments to the agency.	3
	Economic development points account for up to 30% of a project's base score:	*30*
	Total transportation and economic development	100
Earmarks	Public/private/or local participating funds — bonus points (up to)	100
	Total possible points including transportation, economic development, and earmarks:	*200*

Source: (Indiana DOT 2005b).

(a) *CDTC TIP Process*   The CDTC transportation improvement program (TIP) is the product of close coordination between New York State DOT, CDTC local jurisdictions, environmental groups, and other stakeholders. The two-year TIP provides all projects within the region to be included in the New York State transportation improvement program (NYSTIP), which covers the first three years of that period. Projects must be on the NYSTIP to have access to federal funds. The TIP development involves three-staged screening, merit evaluation, and programming. The merit evaluation process combines objective and subjective evaluation of projects across a broad range of issues (impacts on travel cost, mode choice, safety, pavement condition, air quality, etc.). Programming is performed by using "fact sheets" for each candidate project and consciously attempts to select the "best" projects while balancing funding commitments by geographic area, mode, project type, etc.

The TIP process starts with the CDTC's solicitation of new projects from its constituent local jurisdictions and agencies and proceeds with a series of interagency and public meetings. The TIP is shaped by extensive public comments on the initial draft. Typically, the CDTC TIP involves an average of about $95 million per year for federal-aid highways and an average of approximately $12 million for federal transit funds, including local match. Whenever a review of the costs and schedules of existing projects in the recent TIP indicates that funding availability for area projects is limited, a solicitation is made limited to projects that benefit the highest-function highways [national highway system (NHS)], address bridges listed on rehabilitation and replacement priority list [highway bridge rehabilitation and replacement (HBRR)], or contribute to air quality improvements through transit, bicycle, pedestrian, or traffic flow actions [congestion mitigation/air quality, (CMAQ)]. In such cases, several potential projects are not considered for implementation, particularly when there is inadequate funding in the most flexible of the federal fund sources, the surface transportation plan (STP) fund, a situation that often arises due to the funding needs of prior commitments.

(b) *Screening Criteria for New Projects* Each project is required to meet a set of screening criteria (minimum requirements) before further consideration. The project must be consistent with federal, CDTC, and local plans. The following seven planning issues are stipulated in TEA-21, and all urban projects are required to address at least one of them:

1. Economic vitality of the country, the states, and metropolitan areas, especially by enabling global competitiveness, productivity, and efficiency
2. Safety and security of the transportation system for motorized and nonmotorized users
3. Accessibility and mobility options available to people and freight
4. Enhanced environment, energy conservation, and improvement of quality of life
5. Integration and connectivity of the transportation system, across and between modes, for people and freight
6. Efficiency in system management and operation
7. Preservation of the existing transportation system

The coordination involves consistency with the regional transportation plan, which includes a set of 25 planning and investment principles to guide CDTC's capital programming and congestion management system. The additional items include boundary compatibility, land-use linkage, public and sponsor support, provision of local matching funds, defined scope and timing of a project, meeting and identifying needs, and eligibility for federal aid. The needs assessment criteria are defined below.

Bridge projects are required to meet New York State Department of Transportation (NYSDOT) criteria for a deficient bridge, including condition rating (e.g., the federal sufficiency rating must be less than 50) and approach work (e.g., the approach work cost must not be more than 25% of the structure cost).

Pavement projects are to be of a scope that is consistent with the implementation of federal-aid funds. Mobility projects must address a level of service of E or worse. All candidate projects must be eligible for either the STP or the CMAQ program. Projects may include (in addition to pavements and bridges) transit projects eligible for FTA funding, carpools, park-and-ride facilities, bicycle and pedestrian facilities, traffic monitoring and control,

enhancement projects, transportation control measures, wetlands mitigation, and various planning and management efforts.

Enhancement activities must relate to surface transportation and may include:

1. Acquisition of scenic easements and scenic and historic sites
2. Scenic or historic highway programs (including provision of tourist and welcome center facilities)
3. Landscaping and other scenic beautification
4. Historic preservation, rehabilitation and operation of historic transportation buildings, structures, and facilities (such as canals)
5. Preservation of abandoned railway corridors, including the provision of bicycle and walking trails)
6. Control and removal of outdoor advertising signs
7. Archeological planning and research
8. Environmental mitigation of water pollution due to highway runoff
9. Projects to reduce vehicle caused wildlife mortality and maintaining habitat connectivity
10. Bicycle and pedestrian safety programs
11. Establishment of transportation museums

Projects eligible for CMAQ funds are those that achieve measurable emission reductions and do not involve construction of new capacity for single-occupancy vehicles. These may include vehicle inspection and maintenance programs, intermodal freight, use of alternative fuels, telecommuting, and other projects and programs with air quality benefits.

(c) *Environmental Justice Program* The issue of environmental justice is discussed in Chapter 17. The CDTC routinely performs a review of environmental justice (EJ) issues and it has implemented a standard procedure for EJ in the planning process. In February 1994, Executive Order 12898, Federal Actions to Address Environmental Justice in Minority Populations and Low-Income Populations, established this measure of equity to ensure that federally funded transportation-related programs, policies, and activities do not cause undue adverse effects on minority and low-income groups. EJ is a public policy objective that has the potential to improve the quality of life for those segments of the population whose welfare has traditionally been overlooked. The CDTC incorporates EJ issues in its programming process by focusing on specific geographical areas that may require special consideration for environmental justice and civil rights. Such special concern areas are identified by analyzing the census data and using the following criteria: high percentage of area residents in poverty level and high percentage of minority. Often, the greatest need of such an area may not be pavement or bridge or even transit improvement, but economic revitalization or relief from noise exposure.

(d) Merit Evaluation Procedure   There are three rounds of programming for new projects. Round 1 programming uses a filtering process in terms of cost-effectiveness of projects in important locations. Specifically, the following "merit items" are used:

1. *Benefit–cost (B/C) ratio*, where projects whose B/C ratios are in the top half of B/C values of a given project category pass this filter (for bicycle and pedestrian projects, a weighted score is used instead of B/C ratios).
2. *Functional classification*, where projects are assigned a passing status if the proposed work is on an NHS or a principal arterial.
3. *Priority network score*, where every project is assigned a priority network score within a category. Projects in the upper half of these scores within a given category pass this filter. The primary networks include the following types: bicycle and pedestrian, arterial (or access) management, goods movement, transit, and intelligent transportation system (ITS). For example, the goods movement primary network consists of the national highway system, including intermodal connectors and state routes that carry more than 10% trucks.

Round 2 programming ensures an opportunity for projects from any category whose benefits cannot be well quantified. After public review, in round 3, the remainder of the total funds is allocated to projects in response to public comment.

(e) Use of Quantitative Approach   Benefit–cost ratios are included in the *project fact sheet*. All benefits and costs are expressed as thousands of current dollars per year. In cases where benefits cannot be monetized, they are included as qualitative project benefits. At least five measures of project benefits are calculated, including safety, travel time, energy and user cost savings, and other benefits. Life-cycle cost savings are applied primarily to infrastructure improvements.

Bicycle and pedestrian projects are evaluated against other projects within the same category to ensure fair comparison. The items considered are potential demand for bicycle and pedestrian travel, cost-effectiveness calculated as person-miles of travel per $1000 of annualized project cost, and potential safety benefit, defined as the

reduction in the number of car–bicycle or car–pedestrian crashes.

Nonquantifiable project benefits include congestion relief; air quality; regional system linkage; land-use compatibility; contribution to community or economic development; environmental issues such as intrusion on sensitive lands (wetlands, woodlands, parklands, aquifers, and historical property); business or housing dislocations; facilitation of nonmotorized travel, goods movement, transit use, and intermodal transfers; and other considerations.

Enhancement projects are evaluated using criteria set forth in consultation with NYSDOT region 1, as listed below. Maximum points that can be assigned to the factors are shown in parentheses.

- Environmental benefits (10)
- Economic benefits (10)
- Access and patronage benefits (10)
- Transportation system enhancement (10)
- Local benefit and community cohesion (15)
- Safety benefits (10)
- Relationship to support for other plans and projects (10)
- Size of matching share (5)
- Level of community and regional support (5)
- Innovation and creativity (5)
- Mix of eligible enhancements (5)

The CDTC staff has prepared procedures on how to assign scores to projects under each criterion. It is important to note that the quantitative evaluation and ranking process is used to inform decision makers and not make decisions. After the TIP is finalized, it becomes a part of the state transportation improvement program.

20.5   KEYS TO SUCCESSFUL PROGRAMMING AND IMPLEMENTATION

20.5.1   Link between Planning and Programming

The possibility of lack of coordination between planning and programming has long been a cause for concern (TRB, 1974, 1975). Although there is a need to establish a strong link between planning and programming, making such linkages effective has often not been straightforward. For example, agencies have attempted to establish a link by requiring that the outcome of transportation programming be consistent with plans. However, this alone may not be sufficient to provide this linkage. Generally, the impediments to strengthening such link include (TRB, 1993):

- *Difference in time frames.* For planning, the emphasis has been long term, whereas programs have focused on the short term.
- *Update cycles.* At many agencies, the state of practice has been characterized by irregular updating of plans eventhough programs have typically been constantly adjusted and updated typically on the same cycle as the budget (e.g., every fiscal year or every other year).
- *Policy issues and evaluation criteria.* There is typically little or no consistency between issues addressed and the evaluation criteria used in planning and those used in programming.
- *Funding constraints.* Plans are typically not constrained by actual funding levels, whereas programs typically take cognizance of budgetary constraints (at least in the near term). However, metropolitan plans are now required to be fiscally constrained.
- *Organizational responsibility.* At many agencies, planning and programming functions are typically carried out separately by different organizational units in the agency, and there is little or no interface between them.

For agencies to overcome these barriers, TRB (1978, 1993) and Neumann et al. (1993) recommend establishing consistent criteria for various agency functions, requiring the use of management systems to generate candidate projects, updating plans and programs on a consistent cycle, establishing phased implementation strategies as part of the long-range planning process, the use of consistent financial constraints, and ensuring that planners monitor the key elements of programming constantly as they relate to projects, duly modifying system planning policies for changing times, and translating the systems planning into an updated long-range transportation plan.

20.5.2 Uncertainties Affecting Transportation Programming

Transportation programming is the result of several needs assessment and costing procedures that are inherently vulnerable to changes on the transportation environment (Neumann, 1997). A successful program requires due cognizance of factors that change with time, such as project costs, facility deterioration rates, funding availability, weather and natural disasters, and human-made attacks. Meyer and Miller (2001) identify possible approaches for addressing uncertainties in the evaluation and programming process, such as:

- Making the assumption that the useful life of a project is less than its economic life, thus ensuring that the initial capital outlay is recouped over a reduced period of time and that project benefits must be greater to justify the project.
- Adding a "risk premium" to the discount rate used in economic evaluation. Doing so would reduce the expected value of net benefits, thus requiring larger expected future benefits to justify the project.
- Staging projects over time so that after completion of a preceding stage, agencies can reevaluate the feasibility of future stages and make needed recommendations.
- Using scenarios to identify alternative future characteristics and the effect of these alternative futures on facility or system design.
- Performing sensitivity analyses for key evaluation variables to ascertain the relationship between levels of these parameters and the evaluation or programming outcome.
- Incorporating uncertainty analysis into the evaluation or programming process (see Chapter 18).

20.5.3 Intergovernmental Relationships

Typically, transportation agencies seek funding appropriations by presenting one or more alternative programs that show total needs, goals, and performance measures, and how each alternative program addresses each performance measure. Through such interactions, legislators acquire some control of the programming process to ensure that their constituents are receiving their share. The success of transportation programs can be influenced by the extent to which legislature is involved in project identification and prioritization process, but a possible danger is that legislators may select projects for reasons that are not consistent with established performance measures or selection procedure. Good communication is critical in assuring legislators that resources are used efficiently and that projects will be completed as planned. The consequences of deferred maintenance should also be communicated to legislators.

20.5.4 Equity Issues in Programming

Transportation programming implicitly involves a distribution of funds not only temporally (across programming periods) but also spatially (across various modes, functional areas, jurisdictions, regions, etc.). With regard to regional apportionment of transportation funding in particular, the issue of equity arises. Representatives of some regions and jurisdictions may feel that apportionments should be on the basis of engineering need, whereas others may feel that apportionments should be on the basis of some geographic formula. Unlike the latter, the former helps achieve cost-effectiveness of the investments but may not be equitable. It is possible to include equity concerns in the performance measures that constitute "effectiveness" so that both issues can be addressed simultaneously.

SUMMARY

Transportation agencies carry out programming in order to select and schedule facility and/or rolling stock preservation, improvement, and replacement projects on the basis of relative urgency of work. A key element of such processes is the matching of needed projects with available funds to accomplish the strategic goals and objectives set by a transportation agency for a given period. An effective programming framework is expected to provide a mechanism for selecting cost-effective projects reflecting community needs and to develop a multiyear investment strategy within budgetary constraints over a planning horizon. The framework should assist both technical and policy decisionmaking by presenting options and the trade-offs in terms of benefits and costs. Transportation agencies program their projects for several reasons, such as ensuring optimal investments to achieve strategic policy goals, evaluating trade-offs among investment options, and to facilitate budgeting and program and project delivery. Decision making could be for purposes of programming (a time-based schedule showing what to do in each of several clustered periods over an overall analysis horizon) or merely what to do in a single programming period. The general methodology for programming includes establishment of agency policies, specification of the programming period and programming unit, needs assessment using planning functions, analysis of revenue inflows, and establishment of a program development plan. The program development includes establishment of program performance measures, development of project- and network-level programs, trade-off analysis, fund allocation, program implementation, and performance monitoring. In this chapter we formulated mathematical problem structures for various problem types that vary by the nature and type of constraints and the discrete or continuous nature of the decision variables. We presented examples of how transportation programming is carried out at the state and local levels. Finally, we identified keys to successful programming and implementation.

The current transportation environment is characterized by trends toward the use of a diverse, yet often conflicting set of policy goals and objectives such as mobility, economic development, and the environment; operational accountability of agency resources; new funding flexibilities that remove barriers to considering a wide range of program choices and trade-offs;

and increased emphasis on multijurisdictional and multimodal coordination. Thus, transportation agencies are urged to strengthen the relationships between planning and programming, to include a wide range of program options and trade-offs, to include multimodal analyses, to broaden the concept of need and evaluation criteria used in the planning and programming process, and to improve accountability for program decisions by establishing a program and system performance monitoring function.

EXERCISES

20.1. An MPO developed a three-year Transportation Improvement Program (TIP) for projects in three categories: transit; sidewalk, pedestrian, and bicycle; and parking, signs and signals. Determine the optimal set of projects that can be undertaken in each year given the yearly budget constraint for each category (Tables EX20.1.1 to EX20.1.3). Solve the integer programming problem for the following scenarios: (**a**) unused budget cannot be transferred from one year to another; (**b**) unused budget in previous year can be carried over into the subsequent year.

20.2. To reduce capacity constraints, minimize operating costs of airlines, and improve air space efficiency, an airport development program is being considered at a domestic airport. The program comprises of several projects, such as runway construction, addition of new gates, renovation of the terminal building, and construction of taxiways (Table EX20.2). The entire development program must be completed in five years. C_{jk} is the cost of project j undertaken in year k, E_{jk} is the effectiveness (accounting for reduced operating costs, improved capacity, and air space efficiency) of project j implemented in year k, and B_k is the annual budget for year k. Formulate an integer programming problem and clearly define and describe the decision variables, constraints, and objective function. Assume the following:

- Once a project is selected in a year, it will be completed during that year.
- The unused budget in a year is available for expenditure in the next year.
- The only constraint in the problem is the budget constraint.
- The projects have to be completed over the five-year period.

Table EX20.1.1   Candidate Transit Projects and Annual Budgets

Project #	Transit Projects	Capital Outlay (dollars)			Total (dollars)	Utility (0–100)
		2006–2007	2007–2008	2008–2009		
1	Preventive maintenance	1,966,000	2,025,000	1,669,000	5,660,000	75
2	Tire leasing	168,000	173,000	178,000	519,000	34
3	Passenger waiting facilities and hub development	100,000	3,275,000	600,000	3,975,000	56
4	Replace service enhancement vehicles	300,000	400,000	200,000	900,000	39
5	Bus stop signs and poles	20,000	20,000	20,000	60,000	12
6	Bus storage area expansion	1,500,000	250,000	250,000	2,000,000	50
7	Maintenance facility	125,000	100,000	1,350,000	1,575,000	45
	Annual budget for project category	3,000,000	4,000,000	3,000,000	10,200,000	

Amounts shown are in dollars.

Table EX20.1.2   Candidate Projects for Sidewalks, Pedestrian, and Bicycle Facilities and Annual Budgets

Project #	Sidewalk/Pedestrian/ Bicycle Facility Projects	Capital Outlay (dollars)			Total (dollars)	Utility (0–100)
		2006–2007	2007–2008	2008–2009		
1	Accessible sidewalk program	500,000	500,000	500,000	1,500,000	43
2	City-owned sidewalk program	300,000	300,000	300,000	900,000	26
3	Onondaga Creek Walk	1,490,000	711,000	5,700,000	7,901,000	82
4	Tipperary Hill streetscape improvements	45,000	50,000	35,000	130,000	15
5	Various neighborhood improvements	300,000	300,000	300,000	900,000	50
	Annual budget for project category	2,000,000	1,000,000	4,000,000	8,000,000	

Amounts shown are in dollars.

20.3. State how the formulation of Exercise 20.2 would change in the following scenarios:

    (a) There is no annual budget constraint. The total available budget for the project over the five-year period is B.

    (b) The cost-effectiveness of each project should exceed a minimum threshold H_j.

Develop an independent formulation for each case.

20.4. It is sought to develop a congestion improvement program for a certain county. The program consists of n congestion mitigation projects at various intersections and highways in the county. Suppose that the decision variables (x_j) are continuous variables that represent the fraction of a project j that is completed. The goals of the program are as follows:

- *Improved safety*. The total number of crashes per year in the region should not exceed S.

- *Reduced congestion*. The vehicle hours saved in the system should exceed V.

- *Air quality improvement*. The amount of CO (in kilograms) emitted in the system should not exceed E.

- *Economic development*. The total number of jobs created due to the implementation of the selected projects should be more than M.

- *Displacements*. The total number of household and commercial displacements due to the implementation of projects should not exceed D.

Formulate the problem using goal programming. Clearly state the decision variables, system constraints, goal constraints, and the objective function.

Table EX20.1.3   Candidate Projects for Parking and Traffic-Related Projects and Annual Budgets

Project #	Parking/Sign/Signal/ Intersection Projects	Capital Outlay (dollars)			Total (dollars)	Utility (0–100)
		2006–2007	2007–2008	2008–2009		
1	Parking garage rehabilitation	500,000	650,000	600,000	1,750,000	45
2	Traffic signal/intersection improvements	450,000	570,000	1,080,000	2,100,000	62
3	Traffic signal interconnection	460,000	1,320,000	1,320,000	3,100,000	67
4	L.E.D traffic signal conversion	250,000	250,000	250,000	750,000	31
5	Single indication traffic signal improvements	15,000	15,000	10,000	40,000	10
6	City-wide traffic signal rehabilitation	50,000	50,000	50,000	150,000	30
7	City-wide parking meter replacement/multi bay meter system	500,000	500,000	500,000	1,500,000	25
8	Lakefront, inner harbor and downtown signage	100,000	100,000	100,000	300,000	27
	Annual budget for project category	1,500,000	2,000,000	2,500,000	7,500,000	

Capital outlay amounts shown are in dollars.

Table EX20.2   Airport Improvement Projects

Project, j	Project Description
1	Runway 12 construction
2	Taxiway 20 construction
3	Runway 13 construction
4	Addition of two new gates on the east side of the airport
5	Addition of one new gate on the south side of the airport
6	Renovation of the east side terminal building
7	Renovation of the south side terminal building
8	Taxiway 25 construction
	Annual budget

REFERENCES[1]

Benz, G. (1999). Financial and economic considerations, Chap. 9 in *Transportation Planning Handbook*, 2nd ed., Edwards, J.D. (Editor), Prentice Hall for ITE, Washington, DC.

Binder, S. J. (2006). The straight scoop on SAFETEA-LU, *Public Roads*, Mar.–Apr.

[1]References marked with an asterisk can also serve as useful resources for transportation programming.

*Cooper L., Cooper, M. W. (1981). Introduction to Dynamic Programming, Pergamon Press, New York, NY.

CDTC (2005). *Transportation Improvement Program*, Draft 2005-10, Capital District Transportation Committee, Albany, NY.

Indiana DOT (2005a). *Annual Program Development Process for INDOT State Projects (PDP-S), Version 9-01*, Indiana Department of Transportation, Indianapolis, IN.

———— (2005b). *Policies and Procedures*, Indiana Planning Oversight Committee, Indiana Department of Transportation, Indianapolis, IN.

Indiana DOT (2005c). *Program Development Process: Summary of the INDOT Federal Aid Program to Local Communities*, Programming Section, Division of Program Development Indiana Department of Transportation, Indianapolis, IN.

Jiang, Y., Sinha, K. C. (1990). *An Approach to Combine Ranking and Optimization Techniques in Highway Project Selection*, Transp. Res. Rec. 1262, Transportation Research Board, National Research Council, Washington, DC.

*Lambert, J. H., Peterson, K. D., Wadie, S. M., Farrington, M. W. (2005). *Development of a Methodology to Coordinate and Prioritize Multimodal Investment Networks*, Tech. Rep. FHWA/VTRC 05-CR14, Virginia Transportation Research Council, Charlottesville, VA.

Lamptey, G., Labi, S., Sinha, K. C. (2004). Development of alternative rehabilitation and maintenance strategies for pavement management, *Proc. 83rd Annual Meeting of the Transportation Research Board*, Washington, DC.

Meyer, M., Miller, E. (2001). *Urban Transportation Planning*, 2nd ed., McGraw-Hill, New York, NY.

Nemhauser, G. L. (1966). *Introduction to Dynamic Programming*, Academic Press, New York, NY.

Neumann, L. A., Harrison, F., Sinha, K. C. (1993). *The Changing Context for Transportation Programming*, Transp.

Res. Circ. 406, Transportation Research Board, National Research Council, Washington, DC.

*Neumann, L. A. (1997). *Methods for Capital Programming and Project Selection*, Synth. Hwy. Pract. 243, National Academy Press, Washington, DC.

Poorman, J. (2006). MPO Programming Responsibility: Correcting a Misinterpretation of Federal Law, Working Paper. Capital Transportation Committee, Albany, NY.

Sinha, K. C., Jukins, D. P. (1980). *Transportation Project Evaluation and Priority Programming: Techniques and Criteria*, Transp. Res. Circ. 213, Transportation Research Board, National Research Council, Washington, DC.

*Sinha, K. C., Muthusubramanyam, M. (1981). *Optimization Approach in Highway Programming and System Analysis*, Transp. Res. Rec. 867, Transportation Research Board, National Research Council, Washington, DC.

Speicher, D., Schwartz, M., Mar, T. (2000). *Prioritizing Major Transportation Improvement Projects: Comparison of Evaluation Criteria*, Transp. Res. Rec. 1706, Transportation Research Board, National Research Council, Washington, DC.

Stearns, P. N., Hodgens, D. A. (1969). *Programming and Scheduling Highway Improvements*, Hwy. Plan. Tech. Rep. 4, Bureau of Public Roads, Federal Highway Administration, Washington, DC.

Stout, M. L. (1996). Negotiated capital programming in New Jersey, presented at the Conference on Transportation Programming Methods and Issues, Transp. Res. Circ. 465, Transportation Research Board, National Research Council, Washington, DC.

TRB (1974). *Issues in Statewide Transportation Programming*. Spec. Rep. 146, Transportation Research Board, National Research Council, Washington, DC.

————— (1975). *Transportation Programming Process*, Spec. Rep. 157, Transportation Research Board, National Research Council, Washington, DC.

————— (1978). *Priority Programming and Project Selection*. NCHRP Synth. Hwy. Pract. 78, Transportation Research Board, National Research Council, Washington, DC.

*————— (1993). *Transportation Planning, Programming, and Finance*, Transp. Res. Circ. 406, Transportation Research Board, National Research Council, Washington, DC.

Younger, K. (1994). *Multimodal Project Evaluation: A Common Framework, Different Methods*, Transp. Res. Rec. 1429, Transportation Research Board, National Research Council, Washington, DC.

ADDITIONAL RESOURCES

Turban, E., Meredith, J. R. (2000). *Fundamentals of Management Science*, 6th ed., Irwin, Burr Ridge, IL.

Murray, D. (1995). *Multi-modal Project Application for Surface Transportation Program Funds*, Transportation Research Circular 465, Transportation Research Board, National Research Council, Washington, DC.

Song-lin, M., Zhong-jun, X., Xiang-shen, H. (2006). Capital Optimization Investment Model for Highway Network Reconstruction in China's West, Presented at the 85th Annual Meeting of the Transportation Research Board, Washington, DC.

GENERAL APPENDIX 1

Cost Indices

where C_{AY} is the cost of an activity in the year of analysis, C_{BY} the cost of the activity in the reference year, I_{AY} an index corresponding to the year of analysis, and I_{BY} an index corresponding to the reference year. For the federal aid highway construction index for example, Figure GA.1 presents past CPI values (shown as diamonds) and projected CPI values (shown as circles). The plot (or its corresponding table) can be used to estimate the values of I_{AY} and I_{BY}. Similar plots can be developed for each of the several cost indices. For example, transit costs can be updated using cost adjustment factors established by the Federal Transit Administration (FTA); user costs can be updated using consumer price indices for transportation; and user costs related to freight can be adjusted on the basis of the wholesale price index.

The standard equation for calculating the value at the year of analysis is as follows:

$$C_{AY} = C_{BY} \frac{I_{AY}}{I_{BY}}$$

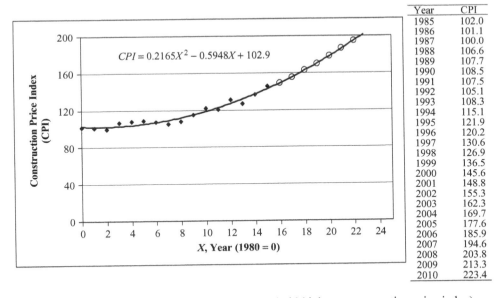

Year	CPI
1985	102.0
1986	101.1
1987	100.0
1988	106.6
1989	107.7
1990	108.5
1991	107.5
1992	105.1
1993	108.3
1994	115.1
1995	121.9
1996	120.2
1997	130.6
1998	126.9
1999	136.5
2000	145.6
2001	148.8
2002	155.3
2003	162.3
2004	169.7
2005	177.6
2006	185.9
2007	194.6
2008	203.8
2009	213.3
2010	223.4

$CPI = 0.2165X^2 - 0.5948X + 102.9$

Figure GA.1 Construction price trends (federal aid highway construction price index).

GENERAL APPENDIX 2

Performance Measures

Tables GA.1 to GA.12 present possible performance measures for various transportation program goals and objectives. Sources of this information include Sinha and Jukins (1978), Poister (1997), and Cambridge Systematics (2000).

Table GA.1 Highway Performance Measures: Operational Efficiency

Goal or Objective	Category	Performance Measure
Passenger specific	Roadway	Vehicle/passenger miles of travel (VMT/PMT) Cost per vehicle for parking Average vehicle occupancy
	Multimodal, modal comparisons	Origin–destination travel times by mode Cost of intermodal trip as a percentage of cost of auto use Change in VMT or PMT per telecommuting occasion
Freight specific	Financial	Revenue per ton-mile by mode Cost per ton-mile by mode
	Time, speed	Line-haul speed Tons transported per hour Truck delay per VMT Truck delay per ton-mile Customs and administrative processing time
	Operational	Productivity and utility by mode Mode split by ton-mile Facility use by mode
	Perception	Quality of highway service in terms of travel time, speed, delay, and scheduling convenience

Table GA.2   Performance Measures: Accessibility

Goal or Objective		Category	Performance Measure
Accessibility	Passenger or freight	Travel time, distance	Average travel time from origin to destination by mode Average trip length Accessibility index
		Roadway condition, capacity	Truck system lane-miles Number of rest areas Truck system lane-miles in acceptable condition Number of miles with ITS[a] service
		Modal choice	Overall mode split Mode split by facility or mode Percentage of change in mode splits
		Customer perception	Perceived deficiencies User identification of access issues
	Passenger specific	Population access to destinations	Percentage of population within a specified distance from their employment locations Percentage of population that can reach specified services mode
		Transportation challenged	Percentage of region's persons who have mobility impairments Existence and quality of access for persons with disabilities to all areas
		Connections, transfers	Transfer distance or time at passenger facility Connectivity deficiency
		Transit/roadway	Percent of population within a specified distance from transit routes or highways of a specific class
	Freight specific	Business access to freight service	Percent of wholesale and retail sales in the significant economic centers served by market artery routes Percent of manufacturing industries within a specified distance from interstate highways
		Quality and quantity of freight service	Number of package express carriers Capacity of package express carriers Availability of real-time cargo information
		Roadway	Average circuitry for truck trips of selected origin–destination pattern Percentage of bridges that are structurally deficient or functionally obsolete
		Intermodal service	Number of intermodal facilities Capacity of intermodal facilities

[a]ITS: Intelligent Transportation System

Table GA.3 Performance Measures: Mobility and Economic Development

Goal or Objective		Category	Performance Measure
Mobility	Passenger or freight	Travel time, speed	Origin–destination travel time by mode Average speed
		Delay, congestion	VMT by congestion level Number and percentage of lane-miles congested Delay per VMT by mode Level of service Peak volume–capacity ratio/Load factor
		Amount of travel	Average Annual Daily Traffic (AADT) Vehicle Miles Travel (VMT) per capita Vehicle Hours of Travel (VHT) per capita
		Reliability, variability	Variation in trip time
		Connections, transfers	Transfer time between modes Number of users of intermodal facilities
		Facility access	Time to access intermodal facilities Average time from facility to major highway network v/c on facility access roads
		Customer perception	Customer perception of time it takes to travel Perceived deficiencies
	Passenger specific	Multimodal	Average commuting time Proportion of persons delayed Passenger miles of travel (PMT) per capita Cost of an intermodal trip as a percent of auto use Mobility index
		Transit/roadway	Percent of lane-miles or number routes operating below acceptable LOS[a]
	Freight specific	Roadway	Line-haul speed Tonnage moved by mode
		Intermodal facilities	Average travel time/delays Average/transfer time/delays Delay of trucks at facility per ton-mile
		Other	Average cost (or speed) for a sample of shipments
Economic	Passenger or freight		Indirect jobs supported or created
Development	Passenger specific		Economic indicator for passenger movement Percentage of employers that cite difficulty in accessing desired labor supply due to transportation Employee-related percent of employers who have relocated for transportation reasons Percent of region's unemployed or poor who cite transportation access as a principal barrier to seeking employment
	Freight specific		Economic indicator for goods movement Percent of wholesale and retail sales in the significant economic centers served by market routes Percent of manufacturers/shippers relocated for transportation purposes Tonnage moved on various transportation components by mode Percent increase in intermodal facility use Business volume by commodity group

[a]LOS: Level of Service

Table GA.4   Performance Measures: Quality of Life

Goal or Objective	Category	Performance Measure
Quality of Life	Accessibility or mobility related	Percentage of region's unemployed or poor who cite transportation access as a principal barrier to seeking employment Average number of hours spent traveling Customer perception of satisfaction with commute time
	Land use related	Difference between change in urban household density and suburban household density Availability and quality of pedestrian and bicycle facilities.
	Safety related	Vehicle collisions per million VMT Customer perception of safety while in travel system Percentage of population that perceives that response time by police, fire, or emergency services has become better or worse and whether that is due to transportation factors
	Air quality related	Tons of pollutants generated Number of days that the pollution standard index or other measure of air pollution is in the unhealthful range Customer perception of satisfaction with air quality
	Noise related	Number of residences exposed to noise in excess of established thresholds Number of noise receptor sites above threshold
	Other environmental related	Customer perception of satisfaction with transportation decisions that affect the environment
	Project delivery related	Customer perception of satisfaction with involvement in preproject planning Customer perception of satisfaction with projects completed Customer perception of promises kept on project completion
	Employment practices related	Compliance with affirmative action goals
	Social and cultural related	Impact on community and institutional structures Number of area of residences/businesses/cultural areas to be relocated Impact on ethnic/racial composition and distribution Quality of sociocultural resources associated with the transportation facility
	Visual and aesthetics related	Visual character of transportation facility in its surroundings Visual quality of overall viewshed containing the facility

Table GA.5   Performance Measures: Safety and Environmental Considerations

Goal or Objective	Category	Performance Measure
Safety	Number and cost of incidents	Collision rate (fatality, injury, PDO) per million VMT
	Roadway condition related	Percent of vehicle collisions on highway system where roadway-related conditions are listed as a contributing factor Number of highway miles driven at high-collision locations Roadway segments not meeting safety standards Vehicle collisions related to bridge characteristics Customer satisfaction with snow/ice removal
	Operator behavior related	Number of collisions in which speed or traffic violation is a factor Percent of operators driving under the influence of alcohol or drugs Percent of operators complying with safety policy
	Construction related	Construction fatalities/construction costs Workzone collisions
	Incident response	Average response time for emergency services Percent of emergency road calls that get through to the agency
Environmental and Resource Conservation	Alternative modes, fuels	Overall mode split Mode split by facility or route Percent of change in mode splits Percent of vehicles using alternative fuels
	Air pollution	Tons of pollutants generated Air quality rating, emission levels, and concentration levels by pollutant type Number of days that pollution standard index is in unhealthful range Customer perception of air quality
	Fuel usage	Fuel consumption per VMT Average mileage per gallon
	Land use	Sprawl: difference between change in urban and suburban household densities Percent of developed region
	Salt usage	Amount of salt used per VMT, per mile, or per AADT
	Government actions	Customer perception of satisfaction with transportation decisions which impact the environment Number of transportation control measures planned and completed
	Miscellaneous	Collisions involving hazardous waste Number and miles of designated scenic routes
	Wetlands and other ecosystems	Number of area of wetlands or other habitals affected Level of habitat fragmentation Species population and diversity Ecosystem stability, quality, and productivity
	Water resources	Number or area of water bodies affected Physical, chemical, or biological degradation of water bodies Extent of disruption of water flow patterns

Table GA.6   Performance Measures for Intermodal Facilities

Goal or Objective	Performance Measure
Intermodal, transfer	Transfer time between modes Number of users of intermodal facilities Tons transferred per hour Delay of trucks at facility per VMT Delay of trucks at facility per ton-mile Customs and administrative processing time

Table GA.7   Measures for Highway Program Performance: System Condition and Program Delivery

Goal or Objective		Performance Measure
System condition	Pavement	Pavement quality index, condition index, roughness, etc. Remaining service life Percentage of highway pavements rated below or above a certain threshold
	Bridge	Bridge wearing surface condition rating Bridge structural condition rating Percentage of highway bridges rated below or above a certain threshold
Program delivery	Time related	Percentage of contracts planned for letting that were actually let Number of lane-miles let to contract for system preservation Number of lane-miles let to contract for capacity improvements
	Cost related	Net present value of future equipment and facility capital, operation and maintenance costs Percentage of budget allocated to system preservation Percentage of budget allocated to capacity improvements Average maintenance and preservation costs per lane-mile

Table GA.8   Measures for Postimplementation Transportation Program Performance

	Goal or Objective	Performance Measure
Financial	General	Public cost for transportation system Private cost for transportation system Percentage of variances between actual and predicted agency revenues
	Infrastructure construction, engineering administration	Benefit-cost of existing facility versus new construction Average cost per lane-mile constructed Construction productivity index Administrative cost as a percentage of total program
	Infrastructure operation and maintenance	Infrastructure maintenance expense Operational cost per toll transaction (for toll roads)
	Vehicle, traveler operations	Average cost per mile Average cost per trip Vehicle operating cost reductions
Time, speed	Infrastructure construction, operation and maintenance	Percentage of increase in number of days to complete construction contracts Units of work completed per hour worked
	Vehicle, traveler operations	Average travel time Average speed
Operational	Infrastructure construction, operation, and maintenance	Percentage of projects rated above or below a certain threshold Percentage of projects requiring few or no variance orders
	Vehicle, traveler operations	AADT or daily ridership Volume–capacity ratio or load factor Average fuel consumption per trip
Perception	Infrastructure construction, operation, and maintenance	Management/employee satisfaction with progress Management/employee satisfaction with delivery efforts
	Vehicle, traveler operations	Customer perception of satisfaction with completed projects

Table GA.9   Performance Measures for Highway Pavement Management Decisions

Goal or Objective		Performance Measure
Pavement Preservation	Project level	Pavement condition level (in terms of International Roughness Index, pavement condition rating, present serviceability index, agency-specific pavement condition indices, skid number), design ESALs or load Rate of pavement deterioration Remaining service life, in terms of time (years) or accumulated loading Jump in pavement condition (or new pavement condition) associated with each standard preservation treatment
	Network level	Percentage of highway pavement sections that are above a certain specified minimum condition level Percentage of highway pavement sections that have a remaining service life that exceeds a certain minimum level Average pavement condition for the entire network Average rate of pavement deterioration or improvement Average remaining service life Percentage of system receiving preservation treatments
Financial	General	Cost of pavement construction, rehabilitation, and maintenance Variance between engineer's estimate and winning contract bid Tort liability likelihood and cost associated with surface defects
	Agency construction and maintenance costs	Average cost per lane-mile, or per lane-mile-inch of pavement constructed Pavement construction productivity index Administrative cost as a percent of total pavement construction and rehabilitation/maintenance costs Average annual maintenance costs per lane-mile Average annual preventive maintenance costs per lane-mile Average annual corrective maintenance costs per lane-mile Average annual in-house maintenance costs per lane-mile Average rehabilitation costs per lane-mile per year or per aggregate funding period
	User costs	User costs during normal highway operations User costs during construction/rehabilitation work zones (delay, safety) Effectiveness of various preservation actions in reducing user costs during normal highway operations (e.g., VOC[a] reduction upon resurfacing)
	Life cycle cost efficiency (agency and/or user)	Life-cycle economic efficiency (benefit-cost ratio, net present value) of intervention vis-à-vis "do nothing" to existing pavement

[a]VOC: Vehicle Operating Cost.

Table GA.10   Performance Measures for Highway Bridge Management Decisions

Goal or Objective		Performance Measure
Bridge preservation	Project level	Bridge condition level (in terms of Sufficiency Rating, Health Index, National Bridge Inventory (NBI) condition rating, etc.) Rate of bridge deterioration (overall, or per bridge element) Remaining service life, in terms of time (years). Jump in bridge condition (or posttreatment level of bridge condition) associated with each standard rehabilitation activity
	Network level	Percent of highway bridges that are above a certain specified minimum condition level Percent of highway bridges that have a remaining service life exceeding a certain minimum level Average bridge condition for the entire network Average rate of bridge deterioration or improvement Average remaining service life of bridges Percentage of system receiving rehabilitation, per annum or per specified aggregate period
Financial	General	Cost of bridge construction, rehabilitation, and maintenance Variance between engineer's estimate and winning bid of bridge contracts Tort liability likelihood and costs associated with bridge inadequacies
	Agency construction and maintenance costs	Average cost per square-foot constructed Bridge construction productivity index Administrative cost as a percent of total costs of bridge construction, rehabilitation, and maintenance Average annual maintenance costs per square foot of deck area Average rehabilitation costs per square-foot of deck area
	User costs	User costs during normal bridge operations User costs during bridge work zones and detours (delay, safety, travel time)
	Life-cycle cost efficiency (agency and/or user)	Life-cycle economic efficiency (benefit-cost ratio, net present value) of intervention vis-à-vis "do nothing" to existing bridge
Safety		Geometric rating or level of functional obsolescence Inventory rating, operating rating
Protection from extreme events		Collision vulnerability rating Overload vulnerability rating Scour vulnerability rating Fatigue/fracture criticality rating Earthquake vulnerability rating Man-made disaster vulnerability rating

Table GA.11   Performance Measures for Highway Safety Management Decisions

Goal or Objective	Performance Measure
Number and rate of crashes by *crash severity* (for fatal, injury, PDO, or total) Number and rate of crashes by *crash pattern* (for rollover, head-on, sideswipe, etc.)	Crashes per AADT Crashes per mile Crashes per VMT
Roadway condition related	Percentage of vehicle collisions on highway system where roadway-related conditions are listed as a contributing factor Number of highway miles driven at high-collision locations Roadway segments not meeting safety standards Vehicle collisions related to bridge characteristics Customer satisfaction with snow and ice removal
Motorist behavior-related	Number of collisions in which speed or traffic violation is a factor Percentage of motorists driving under the influence of alcohol or drugs Percentage of drivers complying with safety policies such as seat belt laws
Construction-related	Construction fatalities and construction costs Work zone collisions and delay
Incident response	Average response time for emergency services Percentage of emergency road calls that get through to the agency
Efficiency	Crash reduction per dollar of standard safety investments Benefit-cost ratio of standard safety investments
Perception	Perception of safety and security by facility users

Table GA.12   Performance Measures for Highway Congestion Management Decisions

Goal or Objective	Performance Measure
Travel time, distance	Average travel time from facility to destination by mode Average trip length Accessibility index
Roadway condition, capacity	Number of rest areas planned versus completed System lane-miles with acceptable condition and LOS[a] Number of miles with ITS[b] service
Modal choice	Overall mode split Mode split by facility or mode Percentage change in mode splits
Customer perception	Perceived deficiencies User identification of access issues

[a]LOS: Level of Service
[b]ITS: Intelligent Transportation System

INDEX